# Estuaries of the World

**Series Editors**

Paul Montagna ⓘ, Harte Research Institute, Texas A&M University-Corpus Christi, Corpus Christi, TX, USA

Jean-Paul Ducrotoy, The University of Hull, Hull, UK

Estuaries are amongst the most endangered areas in the world. Pollution, eutrophication, urbanization, land reclamation; over fishing and exploitation continuously threaten their future. The major challenge that humans face today is managing their use, so that future generations can also enjoy the fantastic visual, cultural and edible products that they provide. Such an approach presupposes that all users of the environment share views and are able to communicate wisely on the basis of robust science. The need for robust science is pressing. Over the last decade there have been numerous advances in both understanding and approach to estuaries and more and more multidisciplinary studies are now available. The available scientific information has come from a multiplicity of case studies and projects local and national levels. Regional and global programs have been developed; some are being implemented and some are in evolution. However, despite the rapidly increasing knowledge about estuarine ecosystems, crucial questions on the causes of variability and the effects of global change are still poorly understood. Although the perception of politicians and managers of coasts is slowly shifting from a mainly short-term economic approach towards a long-term economic – ecological perspective, there is a need to make existing scientific information much more manageable by non-specialists, without compromising the quality of the information. The book series includes volumes of selected invited papers and is intended for researchers, practitioners, undergraduate and graduate students in all disciplines who are dealing with complex problems and looking for cutting-edge research as well as methodological tools to set up truly transversal science and technology projects, such as the restoration of damaged habitats.

Paul A. Montagna · Audrey R. Douglas
Editors

# Freshwater Inflows to Texas Bays and Estuaries

A Regional-Scale Review, Synthesis, and Recommendations

 Springer

*Editors*
Paul A. Montagna 🆔
Harte Research Institute, Unit 5869
Texas A&M University–Corpus Christi
Corpus Christi, TX, USA

Audrey R. Douglas 🆔
Center for Water Supply Studies
Texas A&M University–Corpus Christi
Corpus Christi, TX, USA

ISSN 2214-1553          ISSN 2214-1561   (electronic)
Estuaries of the World
ISBN 978-3-031-70881-7        ISBN 978-3-031-70882-4   (eBook)
https://doi.org/10.1007/978-3-031-70882-4

Texas General Land Office, Coastal Management Program

This Springer imprint is published by the registered company Springer Nature Switzerland AG
The registered company address is: Gewerbestrasse 11, 6330 Cham, Switzerland

If disposing of this product, please recycle the paper.

In 2014, as he was finishing his dissertation, Evan Lee Turner stopped me as I was getting a cup of coffee and waved a copy of the book *Freshwater Inflow to Texas Bays and Estuaries* (Fig. 1) at me and said something like: "you need to redo this, this is the 'bible' for every one of your graduate students, but it is 20 years old and so much has changed since it was published, it really needs updating." That brief comment was certainly the inspiration for this current volume. It was also eight years ago. While I was keen to start an update of that classic volume edited by William Longley (1994),[1] I also knew that I could never do it alone and that I was going to need a lot of support. It took about six years to coordinate all the contributors and funders so that this project could be completed.

**Fig. 1**  Cover of Longley 1994

[1]Longley WL (ed) (1994) Freshwater inflows to Texas bays and estuaries: ecological relationships and methods for determination of needs. Texas Water Development Board and Texas Parks and Wildlife Department, Austin, TX. 386 pp. https://repositories.lib.utexas.edu/bitstream/handle/2152/6728/FreshwaterInflows-+Ecological+Relationships+and+Methods+for+Determination+of+Needs+-+1994.pdf?sequence=2

More important is that the data on which the Longley book was based ends in 1988, so the data is out of date by more than 34 years. The book describes the methodology used by the State of Texas to set inflow standards for estuaries. Inflow is an environmental flow from a river to a bay, estuary, or coastal receiving water. The book had an enormous impact on the management of environmental flows to estuaries worldwide, particularly in countries where environmental water needs are recognized in statute or regulation. The book also had a significant impact on the development of a new field in science called "hydroecology" or "ecohydrology" by extending instream studies to inflow studies.

Much has changed in freshwater inflow science since 1994, but the biggest change in Texas is the result of changing legislation due to the passing of Senate Bill 3 (SB 3) in 2007, which amended provisions in House Bill 2 (HB 2) passed in 1985. The two key changes between HB 2 and SB 3 are the change from a single inflow number per year to the concept of an inflow regime and the change from a focus on protecting sport and commercial fisheries to an ecosystem-based management approach to protect persistence of habitats. These changes in management goals require an updated volume for Texas alone, but the growth of data and new scientific methods to identify inflow effects makes a new synthesis generically useful for the advancement of science and management of estuaries worldwide.

The original book was organized into nine chapters that could be viewed as five topics: (1) the legal and historical perspective; (2) the State methodology; (3) the ecology of bays and estuaries, the data available, and a summary of physical processes, nutrient cycling and primary producers, and secondary consumers; (4) an application of the State methodology to the Guadalupe Estuary; and (5) policy recommendations.

Today, there is sufficient data and analysis to provide a more detailed description of all ecological processes in all of the major Texas bay systems because of the SB 3 Bay and Basin Expert Science Team (BBEST) reports and the adaptive management activities funded by the Texas Legislature and administered by the Texas Water Development Board.

The current volume builds on the first one by updating where necessary and adding new information where it exists. The approach is to focus on ecological processes and compare differences and responses across Texas estuaries. This comparative approach is necessary because of the concept of the "estuary signature," which posits that all estuaries are unique because of the variability in the combination of physical attributes such as climate, inflow, adjacent watersheds, geomorphology, and tidal exchange. It is easy to understand how each of these five physical drivers can vary in ways that a unique combination is created for each estuary.

Corpus Christi, TX, USA                                                       Paul A. Montagna
December 2023

# Acknowledgments

Ironically, Evan Turner ended up working at the Texas Water Development Board (TWDB), which was where Longley spent his career. Many other people at the TWDB were inspirational and supportive to me for many years, including Carla Guthrie, Caimee Schoenbaechler, and Melissa Lupher. Starting with the late Gary Powell, managers at the TWDB supported my primary work on freshwater inflow to estuaries for 36 years. I also acknowledge members of the Texas Environmental Flow Science Advisory Committee (Robert J. Brandes, Franklin Heitmuller, Robert Huston, Paul Jensen, Mary Kelly, Fred Manhart, Edmund Oborny, George Ward, James Wiersema) for teaching me how much I didn't know about environmental flows, especially instream flows.

Many people provided reviews and comments for chapters, which greatly improved the content of this book, and this includes Kristina Alexander, Sandra Arismendez, William Asquith, David Bradsby, John "Chip" Breier, Ed Buskey, Jessica Chappell, Timothy Dellapena, Kevin de Santiago, John Ellis, Deana Erdner, Hudson DeYoe, Nelun Fernando, Mark Fisher, John Nielsen-Gammon, Myron Hess, Xinping Hu, Rick Kalke, Ken Kramer, Michael Lee, William Longley, Jessica Magolan, Joe Malito, Quinn McColly, Larry McKinney, Ram Neupane, Bhanu Paudel, Warren Pulich, Antonietta Quigg, Kirk Rodgers, Dan Roelke, Melissa Rohal-Lupher, Bridget Scanlon, Caimee Schoenbaechler, Thad Scott, James Tolan, Evan Turner, and George Ward.

While the current synthesis project was supported by many sources, the principal source was through Contract No. 21-155-007-C879 from the Texas General Land Office (GLO) with Gulf of Mexico Energy Security Act of 2006 funding made available to the State of Texas and awarded under the Texas Coastal Management Program. The views contained herein are those of the authors and should not be interpreted as representing the views of the GLO or the State of Texas.

Additional support for this publication was made possible by the National Oceanic and Atmospheric Administration, Office of Education Educational Partnership Program award (NA16SEC4810009 and NA21SEC4810004). Its contents are solely the responsibility of the award recipient and do not necessarily represent the official views of the US Department of Commerce, National Oceanic and Atmospheric Administration.

Finally, partial support for this project was also provided by the Harte Research Institute for Gulf of Mexico Studies and the Jacob & Terese Hershey Foundation.

# Contents

**Introduction: History of Inflow Studies in Texas** . . . . . . . . . . . . . . . . . . . . . . . . . . . . . 1
Paul A. Montagna, William L. Longley, Elizabeth Gomaa, and Jen Corrinne Brown

**Historical Perspective and Context of Freshwater Inflow Policy
and Law in Texas** . . . . . . . . . . . . . . . . . . . . . . . . . . . . . . . . . . . . . . . . . . . . . . . . . . 31
Myron J. Hess

**Climate Effects on Inflows** . . . . . . . . . . . . . . . . . . . . . . . . . . . . . . . . . . . . . . . . . . . . 55
John W. Nielsen-Gammon and Alison A. Tarter

**Hydrology, Circulation, and Salinity** . . . . . . . . . . . . . . . . . . . . . . . . . . . . . . . . . . . . . 85
Daniel Opdyke, Josef Hoffmann, Paul A. Montagna, and Joseph F. Trungale

**Groundwater-Surface Water Interactions in the Coastal Zone** . . . . . . . . . . . . . . . . 143
Audrey R. Douglas and Dorina Murgulet

**Influence of Inflows on Estuary Sediments** . . . . . . . . . . . . . . . . . . . . . . . . . . . . . . . 173
Audrey R. Douglas, Paul A. Montagna, and Timothy Dellapenna

**Nutrient-Phytoplankton Dynamics in Texas Estuaries** . . . . . . . . . . . . . . . . . . . . . . 191
Michael S. Wetz, Laura Beecraft, Molly McBride, Jamie L. Steichen,
and Antonietta Quigg

**Physical and Biogeochemical Conditions and Trends in Texas Estuaries** . . . . . . . . . 205
Xinping Hu and Hang Yin

**Coastal Wetland Habitats in Texas** . . . . . . . . . . . . . . . . . . . . . . . . . . . . . . . . . . . . . 221
James C. Gibeaut, Jessica Magolan, Pu Huang, and Paul A. Montagna

**Submerged Aquatic Vegetation, Marshes, and Mangroves** . . . . . . . . . . . . . . . . . . . 231
Kyle A. Capistrant-Fossa, Berit E. Batterton, and Kenneth H. Dunton

**Effect of Freshwater Inflow on Benthic Infauna** . . . . . . . . . . . . . . . . . . . . . . . . . . . 259
Paul A. Montagna, Richard D. Kalke, and Larry J. Hyde

**Effects of Climate-Driven Salinity Regimes on Disease Dynamics of the Eastern
Oyster, a Key Estuarine Resource and Bioindicator** . . . . . . . . . . . . . . . . . . . . . . . . 295
Kelley B. Savage, Terence A. Palmer, Paul A. Montagna,
and Jennifer Beseres Pollack

**Plankton Dynamics in Texas Estuaries** . . . . . . . . . . . . . . . . . . . . . . . . . . . . . . . . . . 309
Antonietta Quigg, Jamie L. Steichen, Laura Beecraft, and Michael S. Wetz

**Freshwater Inflow and Salinity Shape Nekton Diversity and Community
Structure Within Texas Estuaries** . . . . . . . . . . . . . . . . . . . . . . . . . . . . . . . . . . . . . . 335
Daniel M. Coffey, Gregory W. Stunz, and Paul A. Montagna

**Nitrogen and Phosphorus Budgets for Texas Estuaries**. . . . . . . . . . . . . . . . . . . . . . . 363
Danielle A. Marshall and Paul A. Montagna

**Social and Economic Values of Environmental Flows to the Coast** . . . . . . . . . . . . . . 389
David W. Yoskowitz

**Summary of Recommendations for the Future** . . . . . . . . . . . . . . . . . . . . . . . . . . . . 395
Paul A. Montagna and Audrey R. Douglas

**Correction to: Introduction: History of Inflow Studies in Texas** . . . . . . . . . . . . . . . C1
Paul A. Montagna, William L. Longley, Elizabeth Gomaa, and Jen Corrinne Brown

**Archived Datasets** . . . . . . . . . . . . . . . . . . . . . . . . . . . . . . . . . . . . . . . . . . . . . . . . 403

**Archived Documents** . . . . . . . . . . . . . . . . . . . . . . . . . . . . . . . . . . . . . . . . . . . . . . 407

**Archived Oral Histories** . . . . . . . . . . . . . . . . . . . . . . . . . . . . . . . . . . . . . . . . . . . 411

# Introduction: History of Inflow Studies in Texas

Paul A. Montagna ⓘ, William L. Longley,
Elizabeth Gomaa ⓘ, and Jen Corrinne Brown ⓘ

## Abstract

Interest in developing freshwater inflow needs or standards for Texas estuaries occurred in three eras over a 50-year period. In 1967, the concept that water is needed for preservation of bays and estuaries was first proposed. In 1975, the first law was enacted to study effects of freshwater inflow on estuaries. In the 1980s, the first detailed freshwater inflow needs studies were completed for all bays based on hydrology, salinity, and fisheries harvest. In 1994, three Texas state agencies cooperated to synthesize all the studies performed in the 1980s into one volume describing the state methodology and recommending new standards. Following major legislative changes in 2007, science and stakeholder teams were formed to recommend inflow regimes for each of the seven bay systems, and by 2014, inflow standards were adopted for all bay systems. Over time, the laws and standards evolved from a species management approach to an ecosystem-based management approach.

The original version of the chapter has been revised. A correction to this chapter can be found at https://doi.org/10.1007/978-3-031-70882-4_18

**Supplementary Information** The online version contains supplementary material available at https://doi.org/10.1007/978-3-031-70882-4_1.

P. A. Montagna (✉)
Harte Research Institute for Gulf of Mexico Studies, Texas A&M University-Corpus Christi, Corpus Christi, TX, USA
e-mail: Paul.Montagna@tamucc.edu

W. L. Longley
Texas Water Development Board (Retired), Austin, TX, USA

E. Gomaa
Department of Homeland Security, Federal Emergency Management Agency, Canton, MO, USA
e-mail: elizabeth.gomaa@fema.dhs.gov

J. C. Brown
Department of Humanities, Texas A&M University-Corpus Christi, Corpus Christi, TX, USA
e-mail: Jennifer.Brown@tamucc.edu

## Keywords

Environmental flow · Estuary · Inflow needs · Texas bays · Texas water plan

## Abbreviations

| | |
|---|---|
| ac-ft: | acre-feet |
| BBASC: | Basin and Bay Area Stakeholder Committee |
| BBEST: | Basin and Bay Expert Science Team |
| EFAG: | Environmental Flows Advisory Group |
| GIS: | Geographical Information System |
| Gulf: | Gulf of Mexico |
| MaxC: | maximum catch, same as maximum harvest |
| MaxH: | maximum harvest |
| SAC: | Science Advisory Committee |
| TLWP: | Texas Living Waters Project |
| TPWD: | Texas Parks and Wildlife Department |
| TWC: | Texas Water Commission |
| TWDB: | Texas Water Development Board |
| TWQB: | Texas Water Quality Board |
| TWRC: | Texas Water Rights Commission |
| TxEMP: | Texas Estuarine Mathematical Programming |

## Estuaries and Freshwater Inflow

Estuaries are the transition zones between land and sea where freshwater mixes with saltwater. There are many ways to define estuaries (Perillo 1995; Elliott and McLuskey 2002; McLuskey and Elliott 2007; Tagliapietra et al. 2009), and one early definition is given by Pritchard (1967) as a semi-enclosed coastal body of water, which has a free connection with the open sea and within which, sea water is measurably diluted with fresh water from land drainage. This river, stream, or creek drainage from the land to a coastal water body is "freshwater inflow." Those two end-member connections: (1) tidal flows from an ocean or sea and (2) freshwater inflow are fundamental to an estuary (Montagna et al. 2013).

Freshwater inflow is one of three types of environmental flows: instream flow, inflow, and outflow. Instream flows occur within rivers and streams. Inflows occur when rivers flow into the coastal zone, which creates brackish estuary habitat. Outflows occur when rivers or estuaries flow out onto continental shelves or nearshore coastal ocean zones. Recently, the terms "hydroecology" and "ecohydrology" have come into use to describe studies that integrate physical aspects of flow and environmental or biological responses to flow (Wood et al. 2007).

## Texas Estuaries

Most estuaries have a series of geomorphic components, such as, a river (or smaller fresh water) source, marshes (or mangroves depending on latitude), bays or lagoons, and a tidal connection (i.e., an inlet or pass). Globally, estuaries are quite different because the landscape of each component can vary, combinations and connections of these components can vary, and some components may be missing. In contrast, the components of estuaries along the Texas coast are similar because the northwestern Gulf of Mexico coast is a broad, uniform, coastal plain formed over the last 5000 to 15,000 years.

The word "estuary" does not appear on maps issued by U.S. Federal Agencies. Water bodies on the Texas coast are identified as rivers, creeks, bays, and lagoons. In fact, the Texas coast resembles one large lagoon system with seven big bays (Fig. 1). The State of Texas defined the bay systems as estuaries in a series of reports issued in the 1980s (TDWR 1980a, b, 1981a, b, c, 1982, 1983). The estuaries were named for major rivers that drained into the bay systems.

There are seven major estuarine systems along 600 km of coastline (Table 1). But it is better to define eight systems because upper and lower Laguna Madre are separated by a large mudflat named the Land Cut, and they have different freshwater inflow sources. Texas estuaries have similar geomorphic structure and physiography. Barrier islands are parallel to the mainland along the coast. Between the islands and the mainland there are lagoons that form or connect to the primary bays. The lagoons are connected to drowned river valleys that form the secondary bays that drain rivers. There are Gulf inlets through the barrier islands, which connect the sea with the lagoon behind the island. There is usually a constriction between the primary bay and a smaller secondary bay. Most secondary bays are fed by just one or two rivers draining watersheds. Rivers or creeks flow into the secondary bay and thus secondary bays have greater freshwater influence. Primary bays connected with the Gulf of Mexico and thus have greater marine influence. By convention, estuaries are named by the river that drains to the Texas coast (Longley 1994), but some authorities name the systems for the primary bays (Orlando et al. 1993).

Even though Texas estuaries are geomorphologically similar, they are hydrologically diverse due to a climatic gradient, which influences freshwater inflow to estuaries (Longley 1994). The gradient of decreasing rainfall, and concomitant freshwater inflow, from northeast to southwest, is the most distinctive feature of the coastline. Along this gradient, rainfall decreases by a factor of two, but inflow balance decreases by almost three orders of magnitude (Montagna et al. 2013). The net effect is a gradient with estuaries with similar physical characteristics but a declining salinity gradient across estuaries from north to south and within estuaries from west to east (Fig. 1). These unique geomorphological and hydrological features on a large regional scale makes the Texas coast a near-perfect "natural experiment" to study and understand the effects of freshwater inflow on estuarine biogeochemical and ecological dynamics.

**Box 1 The Drought of the Fifties**

There was no water … at the end of the day at school when we had to go to our buses, and the buses would be parked a few hundred feet away from the school in a big lot. There were times when the sandstorms were so intense, and the visibility was so limited that we would have to assemble in the classrooms and all of us would hold hands and the teacher would take us to the bus because they were concerned we could get lost and not find the bus and be lost in the sandstorm. That was how tough it was.

—Larry McKinney, former Executive Director of the Harte Research Institute, on growing up on a West Texas cotton farm during the drought of the fifties

## Beginning of Inflow Studies Before 1980

Starting in the early 1950s, the state of Texas suffered an 8-year drought so severe that rainfall dropped by as much as 40%, resulting in 244 of its 254 counties being declared disaster areas (Box 1. Note: for this box and others, the quotations come from oral history interviews, see online supplementary and archived materials for more information). Many rivers stopped running, depriving the Gulf Coast ecosystem of freshwater input and disrupting the fragile balance of its estuaries as the hyper-saline environment led to the near-disappearance of native fish, blue crabs and white shrimp, and an invasion of non-native (stenohaline) plants and animals. The one positive outcome of this natural disaster was state legislature's passage of the landmark Texas Water Planning Act (1957) that empowered the State Board of Water Engineers, established in 1913 to implement a water rights system, to create a Water Planning Division to inventory water and related data and ultimately prepare a report of water resources of the state. In 1961, Governor Price Daniels directed them to prepare a report which was titled "A Plan for Meeting the 1980 Water Requirements of Texas." The report

**Fig. 1** Long-term average salinity from 1977 to 2017 based on Texas Parks and Wildlife Department monitoring

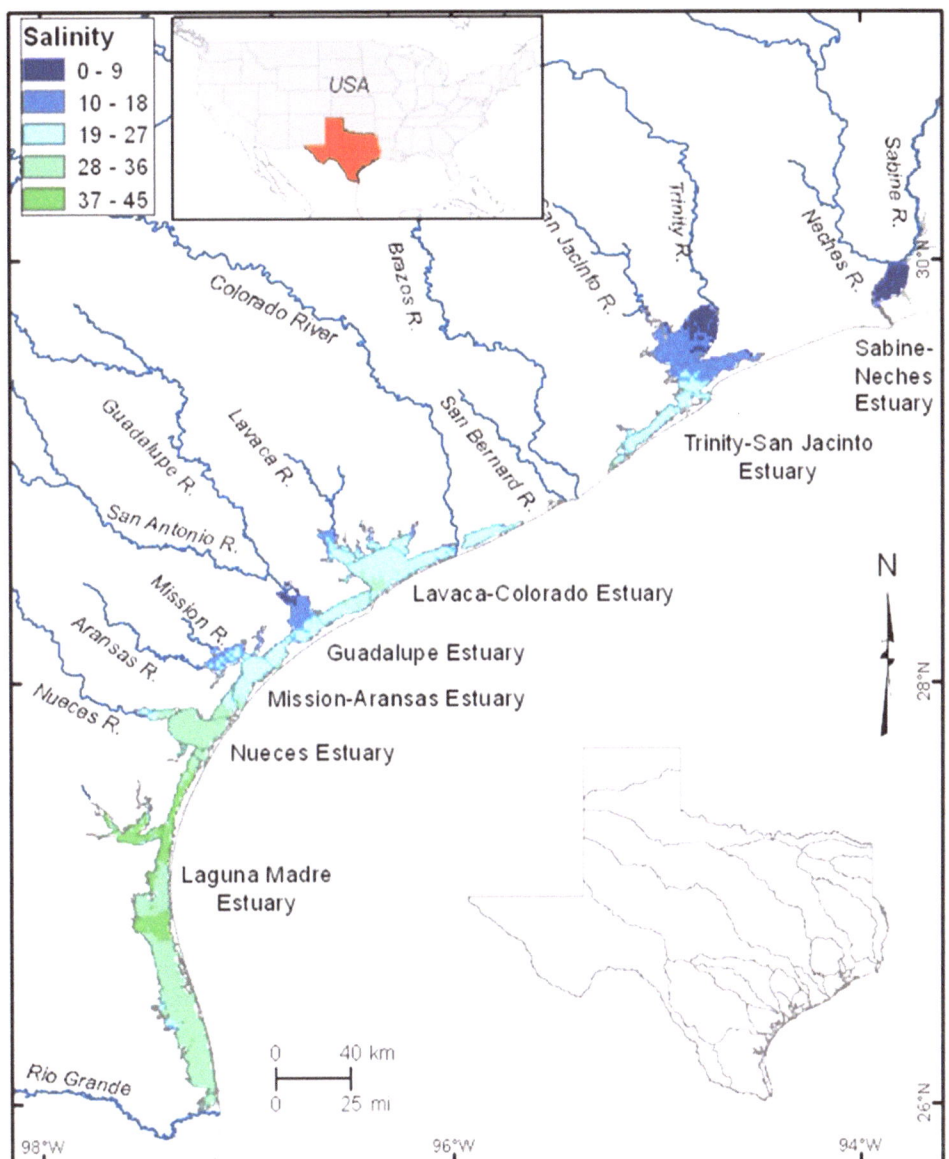

did not address freshwater inflows for the estuaries but provided a framework for future water resource data collection.

In 1957, the Texas Water Development Board (TWDB) was established by state constitutional amendment to forecast water supply needs and provide funding for water supply and conservation projects throughout the 268,581 square-mile region, which covers approximately 7% of the total water and land area within the United States. At this time, the TWDB was largely a financial institution with no real planning function except to assure that the money lent was repaid. In 1965, the TWDB was charged with responsibility for statewide water planning in addition to its financial and development tasks. The State Board of Water Engineers was renamed the Texas Water Commission (TWC) with responsibility for water rights appropriation and dispute resolution. The TWC name was later changed to the Texas Water Rights Commission (TWRC).

## First Estimate of Freshwater Inflow Needs

Among the first indications of a relationship between freshwater input and estuarine productivity were studies by Hildebrand and Gunter (1953) and Gunter and Hildebrand (1954). Their studies showed a relationship between rainfall (and by inference, inflow) and catch of white shrimp. Copeland (1966) noted the change of relative species abundance as the result of decreased river flow.

In 1964, Governor John Connally directed that a more comprehensive, long-range water plan be prepared for the state. As part of that plan, Lockwood, Andrews & Newman (1967) prepared a report for the TWDB entitled, "A New Concept-Water For Preservation of Bays and Estuaries." Their report cited legislative authority for the TWDB to include freshwater inflows as part of the plan: "Consideration shall also be given in the plan to the effect of upstream devel-

**Table 1** Texas estuary names and geomorphological components

| Estuary Name | Bay System Name | River | Secondary Bay | Primary Bay | Gulf Inlet |
|---|---|---|---|---|---|
| Sabine-Neches | Sabine Lake | Sabine, Neches | | Sabine Lake | Sabine Pass |
| Trinity-San Jacinto | Galveston Bay | Trinity, San Jacinto | Trinity | Galveston | Galveston Ship Channel |
| Lavaca-Colorado | Matagorda Bay | Lavaca, Colorado, Tres Palacios | Lavaca, Tres Palacios | Matagorda | Matagorda Ship Channel, Pass Cavallo |
| Guadalupe | San Antonio Bay | Guadalupe, San Antonio | Upper San Antonio | Lower San Antonio, Espiritu Santo, Mesquite Bay | Cedar Bayou (intermittent) |
| Mission-Aransas | Aransas Bay | Mission, Aransas | Mission, Copano | Aransas | Aransas Pass |
| Nueces | Corpus Christi Bay | Nueces | Nueces | Corpus Christi | Packery Channel, Aransas Pass |
| Upper Laguna Madre | Laguna Madre | Intermittent creeks (San Fernando, Santa Gertrudis, Los Olmos) | Baffin | Laguna Madre | Yarborough Pass (closed) |
| Lower Laguna Madre | Laguna Madre | Arroyo Colorado | | Laguna Madre | Mansfield Pass, Brazos Santiago Pass, Brownsville Ship Channel |

opment upon the bays, estuaries, and arms of the Gulf of Mexico" (Article 8280-9, V.A.C.S, Section 3b). This report related salinity levels in seven bays to freshwater inflow under five climatic conditions: wettest year (1941), high average inflows (1940—1946), average annual inflow, low average inflow (1950—1956), and driest year (1956). Salinity regimes for Texas estuarine species, based on where they were found (not physiological data), were gleaned from reports and journals. With this information a series of goals for the study were discussed (quotes are from Lockwood, Andrews and Newman 1967, p. 20):

- "Provide salinity control with reasonable limits for oyster, shrimps, speckled trout, and redfish."
- "Keep bays in humid climates good for oysters with maximum salinity of 30 psu (practical salinity units, parts per thousand used in report) (Galveston Bay); limit hypersalinities to a maximum of 40 or 45 psu in semi-arid and dry sub-humid climate (Baffin Bay, Upper and Lower Laguna Madre, Corpus Christi Bay, and Aransas Bay)."
- "In moist sub-humid climate (Matagorda Bay and San Antonio Bay) the question remains as to whether or not these estuaries will be kept good, or restored, for oysters."
- "Full advantage is taken of the runoff from coastal watersheds and in maximum use of Gulf water to minimize freshwater needs."

These goals were linked to the climatic conditions and salinity regimes where species were found, and the inflows associated with those climate conditions became the basis for defining the freshwater inflow needs (Table 2). The inflow

**Table 2** Recommended freshwater inflow supplement (million ac-ft/year) (Lockwood, Andrews & Newman 1967, p. 33). 1 ac-ft = 1233 m$^3$

| Estuary System (Bay System) | Inflow |
|---|---|
| Trinity-San Jacinto (Galveston Bay) | 1.5 |
| East Matagorda Bay | 0 |
| Lavaca-Colorado (Matagorda Bay) | 0.3 |
| Guadalupe (San Antonio Bay) | 0.3 |
| Mission-Aransas (Aransas Bay) | 0.15 |
| Nueces (Corpus Christi Bay) | 0.2 |
| Upper Laguna Madre (Baffin Bay System) | 0 |
| Total | 2.45 |

amount for each bay system was to be taken from a coastal canal under proposed climatic conditions in 2020. The coastal canal proposed under two earlier Federal studies was to provide irrigation water in the middle and lower coast but using it to supplement fresh water to the estuaries was a new proposal. These amounts of water were to be in addition to local runoff (modeled flow), developed river spills (releases from upstream reservoirs), and direct precipitation on the bay, but not return flows from municipal, industrial, and agricultural uses.

The Lockwood, Andrews & Newman (1967) report noted the need for better connection to the Gulf for some bays where 35 psu Gulf waters might moderate hypersaline conditions. In addition, the report stated that upstream uses diminishing freshwater inflow may change the character of San Antonio and Matagorda bays away from oyster production to finfish production and this would be a major and basic water planning decision (Lockwood, Andrews & Newman 1967, p. 37).

## 1968 Texas Water Plan

In November 1968, the final Water Plan was published (TWDB 1968). The plan contained a series of reservoirs, canals, and conveyances called the Texas Water System to move freshwater from areas of abundance to areas of need. A coastal canal was included in the plan (the Coastal Division) that by the year 2020 would divert 6,294,300 acre-feet per year from the Sabine River and deliver it to various sites down the coast all the way to an area near Raymondville. Bays and estuaries would receive up to 2.45 million acre-feet per year of this conveyed water, which is the amount calculated in the report by Lockwood, Andrews & Newman (1967) (Table 3).

The Total Historical Average Contribution includes river input, ungaged inflow, direct precipitation on the bay, and return flows. The Projected 2020 Average Contribution includes river input, ungaged inflow, and direct precipitation, but does not include return flow, and generally reflects reductions in inflow from upstream diversions. The Proposed Canal releases were from Lockwood, Andrews & Newman (1967), and the Projected 2020 Average Total Inflow was the sum of the previous two columns. Projected 2020 inflows would change in several systems: Trinity-San Jacinto Estuary-23%; Lavaca-Colorado Estuary-4% (but total inflow might increase due to the proposed diversion of the Colorado River mouth into the eastern arm of Matagorda Bay); Guadalupe Estuary-42% (due to projected heavy diversions from the San Antonio and Guadalupe rivers); and Nueces Estuary-20%. Inflow to East Matagorda Bay would not change as it is largely cutoff from the Gulf of Mexico and fed water from precipitation and the Intracoastal Waterway. Mission-Aransas Estuary could receive a 10% increase. No releases were anticipated for Upper Laguna Madre. Inflow needs in the plan amounted to the total amount needed and did not address finer scale details such as seasonal or monthly requirements.

Although the plan had major support from the Texas Governor, Lieutenant Governor, and Speaker of the House, there were many critics. Opponents thought the Trans-Texas Canal part of the plan to deliver water to West Texas for agricultural needs was not feasible and the delivered price of irrigation water would be too high for irrigation to be profitable. Plan proponents were wary of federal involvement and control, but they were depending on federal money to pay for moving water from the Mississippi River to the Texas border and with federal help in financing other parts of the plan. Many of the critics of the preliminary plan also opposed the 1968 plan, and environmental interests opposed the plan regarding the amount of inflow was too low to support estuarine resources. A $3.5 billion dollar bond issue for water development would have funded a number of the projects in the 1968 plan and was scheduled for voter approval less than 9 months after the plan was published. Many thought the vote on financing was to approve or disapprove the plan and the proposal was defeated 315,139 to 309,409 (Graves 1971). The overall cost of the plan was calculated to be $8.996 billion (TWDB1968, p. I-28) but some thought it would be much higher.

In 1967, before the plan was published, the TWDB initiated a cooperative program with other state agencies to collect physical, chemical, and biological data to support further efforts in water planning.

## Senate Bill 137 Comprehensive Studies

In 1975, the 64th Legislature enacted Senate Bill 137 that mandated the TWDB carry out studies of the effects of freshwater inflows on estuaries in cooperation with the Texas Parks and Wildlife Department (TPWD), TWRC, Texas Water Quality Board (TWQB), General Land Office, and Texas Coastal and Marine Council. These studies and those initiated earlier formed a basis for later analyses. Specifically, the bill stated that "the board [TDWR] shall carry out comprehensive studies of the effects of freshwater inflows upon the bays and estuaries of Texas, which studies shall include the development of methods of providing and maintaining the ecological environment thereof suitable to their living marine resources."

## Formation of the Texas Department of Water Resources

The 65th Legislature in 1977 joined the TWDB, TWRC, and TWQB into a single agency, the Texas Department of Water Resources (TDWR). The reorganization retained a board and commission structure from the three agencies: a part-time

**Table 3** Fresh water contribution to bays and estuaries (million acre-ft/year) in the Texas Water Plan 1968 (TWDB 1968, Table III-5, p III-18)

| Estuary System (Bay System) | Total Historical Ave. Contribution | Projected 2020 Ave. Contribution | Proposed 2020 Canal Releases | Projected 2020 Ave. Total Inflow |
|---|---|---|---|---|
| Trinity-San Jacinto (Galveston) | 11.6 | 7.4 | 1.5 | 8.9 |
| East Matagorda | 0.3 | 0.3 | 0 | 0.3 |
| Lavaca-Colorado (Matagorda) | 1.8 | 1.43 | 0.3 | 1.73 |
| Guadalupe (San Antonio) | 1.9 | 0.8 | 0.3 | 1.1 |
| Mission-Aransas (Aransas) | 0.59 | 0.5 | 0.15 | 0.65 |
| Nueces (Corpus Christi) | 1 | 0.6 | 0.2 | 0.8 |

citizen board for the TWDB's financial and research responsibilities, and a full-time commission for the TWRC's regulatory functions. The responsibilities of the TWQB were folded into the agency and divided between the citizen Board and commission. The administration of the agency served both the Board and commission.

The legislature may have been interested in a "one-stop-shop" arrangement for water resource management, increasing efficiency by reducing redundancy. But it introduced a problem by combining the organizations. There was concern that combining planning, financing, adjudication of water rights and permits, inspection, and regulation by the TDWR constituted a conflict of interest. At the staff level among some federal and state agencies, there was great concern that combining the agencies provided too smooth a trajectory for water rights and permits to be approved without consideration of those agencies' natural resource management responsibilities.

## 1980s Detailed Inflow Studies

A series of detailed inflow studies were published by the TDWR over the period 1980–1983 (TDWR 1980a, b, 1981a, b, c, 1983). A major problem in determining freshwater inflow needs for any estuary is how to combine many different types of information together into a whole, and how to define the objectives of the analysis. Simulation models are one possibility but require very thorough understanding and specific equations defining the relationships among the components of the model. That kind of precise quantitative detail was not available. However, there were years of fishery harvest and inflow data, salinity and nutrient measurements from several agencies, and historical data on marsh inundation that scientists knew was important in contributing organic material to the estuary and providing cover for larval species. The assumption was that if production for species of interest was maintained (with harvest acting as a surrogate for production), then the components of the estuarine system on which those species depend were being satisfied by the inflow amounts.

Martin (1987) originated the concept of using linear programming optimization models to approach the problem. The Estuarine Linear Programming Model (ELPM) was used for all estuaries except Sabine-Neches where a nonlinear programming optimization model had been developed by Lasdon (1980). The analytical routine required only a few quantitative relationships but allowed constraints such as minimum and maximum values of various components to be used in the analysis, including historical medians or means and seasonal patterns of inflow, specific inflow values for nutrient input and inundation of wetlands, and upper and lower salinity bounds to keep computed salinities within known physiological ranges in multiple regions of the estuary, and defined harvest–inflow relationships for various target species. It also allowed specification of particular objectives for the model to

achieve such as maximizing harvest, minimizing inflow, selection of target species, and others. Constraints for the model included upper and lower bounds on monthly salinities at sites where salinity measurements were taken for regression equations; upper and lower bounds of monthly inflows from historical data; upper and lower bounds for seasonal inflows used for harvest equations; and flows necessary for marsh inundation and duration.

A second analytical tool, a hydrodynamic model incorporating salinity equations, was prepared to act as a check to show that the salinity regime and gradient in the estuary resulting from the optimization model fulfilled the salinity requirements for the target species. The models used for the Lavaca-Colorado, Guadalupe, Mission-Aransas, Nueces, and Laguna Madre estuaries were developed by Masch (1971). The models for Trinity-San Jacinto Estuary were developed by Tracor (1974) with an adaptation of the conservative transport model by Masch (1971). The model for Sabine-Neches Estuary was developed by Water Resources Engineers (1975).

The mechanics of the analysis are complicated. For example, the relationship between biological production and inflow is based on commercial harvest data from the U.S. Department of the Interior, TPWD, and U.S. Department of Commerce. While commercial harvest is not a direct measure of biological productivity, it can be an indicator of secondary production. But harvest data has its own problems, for example, the method of computing the relationship between inflow and harvest was dependent on the life cycle of the species. For shrimp, with a short life span, harvest was a function of inflow the same year as the harvest; for crabs and oyster, harvest was a function of inflow 1-year antecedent to harvest. For finfish that typically live multiple years, harvest was a function of 3-year antecedent inflows for the year of harvest. Other problems with harvest involved measures of effort and whether the catch came from a particular estuary.

A difference between this study and the inflow estimates in the 1968 Water Plan involves the objective of the inflow estimate. The 1980s studies defined three alternatives that were made possible by the optimization model: Alternative I, subsistence—maintain marsh inundation, salinity control, and nutrient inflow. Alternative II, maintenance—fulfill subsistence requirements and maintain average annual harvest. Alternative III, max harvest—maximize harvest while constraining inflow not to exceed the annual historical inflow. Salinity bounds for organisms, marsh inundation frequency, and target species constraints were termed "key indicators" (TDWR 1982) and their presence were noted for the maintenance alternative for estuaries in Table 4. For the Sabine-Neches Estuary the computed inflows were outside the range of the inflows used to derive the harvest equations for the three species indicated so the computed inflow requirements were for the subsistence objective.

The 1968 Water Plan annual inflow values seemed to be minimum estimates aimed at maintaining productivity but in

**Table 4** Constraints and equations used in the analytical model for the Alternative II, maintenance; the asterisk (*) indicates a constraint for salinity limits and marsh inundation, or a statistically significant regression equation relating harvest and inflow that was used in the model. For the Sabine-Neches Estuary harvest equations for the three species indicated could not be used in the analysis

| Indicator response variables | Estuary system | | | | | | |
| --- | --- | --- | --- | --- | --- | --- | --- |
| | Sabine-Neches (TDWR 1981c) | Trinity-San Jacinto (TDWR 1981b) | Lavaca-Colorado (TDWR 1980a) | Guadalupe (TDWR 1980b) | Mission-Aransas (TDWR 1981a) | Nueces (TDWR 1981a) | Laguna Madre (TDWR 1983) |
| Salinity Limits | * | * | * | * | * | * | * |
| Marsh Inundation | * | * | * | * | * | * | |
| Monthly Inflow Limits | * | * | * | * | * | * | * |
| Seasonal Inflow Limits | * | * | * | * | * | * | * |
| Red Drum | * | * | * | * | * | * | * |
| Seatrout | * | * | * | * | * | * | * |
| All Shrimp | * | * | * | * | | | |
| All Shellfish | | * | * | * | | | |
| White Shrimp | | | | | * | * | * |
| Blue Crab | | * | | | * | * | |
| Bay Oyster | | * | | | | | * |
| Black Drum | | | | | | | * |
| Brown & Pink Shrimp | | | | | | | * |

**Table 5** Direct precipitation to bays and estuaries (million acre-ft/year) (TWDB 1968, Table III-5, page III-18)

| Estuary system | 1968 Water Plan | Direct precipitation | 1968 Water Plan w/o Precipitation |
| --- | --- | --- | --- |
| Trinity-San Jacinto | 8.9 | 1.4 | 7.5 |
| Lavaca-Colorado | 1.73 | 0.76 | 0.97 |
| Guadalupe | 1.1 | 0.4 | 0.7 |
| Mission-Aransas | 0.65 | 0.37 | 0.28 |
| Nueces | 0.8 | 0.33 | 0.47 |

**Table 6** Comparison of initial freshwater inflow needs to bay and estuaries from 1968 to 1980s (million acre-ft/year) (TWDB 1968; TDWR 1980a, b, 1981a, b)

| Bay system | 1968 Water Plan w/o Precipitation | Alternative I 1980s studies | Alternative II 1980s studies | Alternative III 1980s studies |
| --- | --- | --- | --- | --- |
| Trinity-San Jacinto | 7.5 | 6.8523 | 7.1889 | 7.02 |
| Lavaca-Colorado | 0.97 | 2.0971 | 2.8075 | 2.8109 |
| Guadalupe | 0.7 | 1.5742 | 2.0199 | 2.2622 |
| Mission-Aransas | 0.28 | 0.2552 | 0.2688 | 0.368 |
| Nueces | 0.47 | 0.434 | 0.4777 | 0.6414 |

some cases not at historical levels. A comparison can be made between the 1968 Water Plan inflow values and the 1980s studies, but the 1968 Water Plan values have to be reduced because they included direct precipitation (Table 5) on the bay while the 1980s studies do not.

Using the adjusted values, the annual inflow estimated can be compared between the 1968 Water Plan and the 1980s studies (Table 6). The results of the three alternatives were from TDWR 1980a, tables 9-6, 9-7, and 9-8 for Lavaca-Colorado Estuary; TDWR 1980b), tables 9-6, 9-7, 9-8 for Guadalupe Estuary; TDWR 1981a, tables 9-6, 9-7, and 9-8 for Mission-Aransas and Nueces Estuaries; and TDWR 1981b, tables 9-6, 9-7, 9-8 for Trinity-San Jacinto Estuary.

For Trinity-San Jacinto Estuary, all three alternatives have slightly lower annual inflow requirements than the 1968 Water Plan amount, probably due to the normally high inflow amount (Table 3). For Lavaca-Colorado and Guadalupe Estuaries, the requirements were double or triple the 1968 Water Plan's initial estimates, which mirrors the concern expressed by Lockwood, Andrews & Newman (1967) in their report about changes in these bays. For Mission-Aransas and

Nueces Estuaries, Alternatives I and II are close to the 1968 Water Plan's estimate while Alternative III requires about 34%–37% higher inflows. One problem in comparing analyses was that some target species change among alternatives, and some target species differ from one bay to another. This may be due to the suitability of the equations in the analysis or other decisions about which species were used.

Childress et al. (1975) concluded an annual gaged flow of 1.6 to 2.4 million acre-ft was associated with high shellfish production in the Guadalupe Estuary. Adding the long-term (1941-1976) average ungaged inflow (0.46 million acre-feet, TDWR 1980b, Table 4-2) to the Childress et al. (1975) estimate results in a combined inflow of 2.06 to 2.86 million acre-ft, which compares well with Alternative II and III from the TDWR (1980b) inflow needs report.

The harvest data was generally available between 1962 and 1976, a 15-year period of record. Harvest equations tended to explain 60% to 80% of the variance but it would be preferable to have a longer period of record and tighter correlations between harvest and inflow. The landings data themselves posed problems since they were recorded as catch data but with variable effort. Where possible, effort was used in the equations. At the time of the analysis, there really was no other useful productivity information source, but some questioned the data accuracy and whether it represented bay production. Inflow-salinity equations were based on monthly averages, but as with the harvest data, researchers understood at the time that more salinity data and longer period of record could improve those equations.

> **Box 2 House Bill 2**
> Lieutenant Governor [Bill] Hobby was really sort of the driving force … he basically encouraged people to come together with a consensus of water policy for the state that would include … not just building new reservoirs or new water supplies, but more actively managing our existing water supplies for conservation and for environmental protection.

—Ken Kramer, former longtime Director of the Lone Star Chapter of the Sierra Club

## Split of TDWR into Two Agencies

In 1985, Senate Bill 249 split the TDWR into two agencies, the TWDB and the Texas Water Commission (TWC) (TCEQ 2022). The legislation also directed these agencies along with TPWD "to establish and maintain a continuous data collection and analytical study program to determine bay conditions necessary to support a sound ecological environment" (Longley 1994). In 1991, the legislature combined several environmental regulatory agencies and programs into the TWC and renamed the consolidated agency the Texas Natural Resource Conservation Commission (TNRCC), but with the same cooperative water data collection and analysis effort as before.

## 1994 Freshwater Inflows to Texas Bays and Estuaries Report

Ten years after the detailed 1980s bay studies, TWDB, TPWD, and TNRCC scientists began to reexamine the earlier studies with the view of improving them and working cooperatively together. The operating statutes that drove the scientific approach were House Bill 2 (HB2), passed in 1985, following an earlier stakeholder process on water issues initiated by then Lieutenant Governor Bill Hobby (Box 2) and Senate Bill 68 (SB68) passed in 1987. HB2 came in the wake of voter rejection of proposed water development bonds and constitutional amendments as well as from the push to diversify the state's

economy during oil bust of the 1980s. Language in the TEXAS WATER CODE 11.147(b) directed the TNRCC to "… include in the permit, to the extent practicable when considering all public interests, those conditions considered necessary to maintain beneficial inflows to any bay or estuary system." Beneficial inflows was defined in TEXAS WATER CODE 11.147(a) to be "… a salinity, nutrient, and sediment loading regime adequate to maintain an ecologically sound environment in the receiving bay and estuary system that is necessary for the maintenance of productivity of economically important and ecologically characteristic sport and commercial fish and shellfish and estuarine life upon which such fish and shellfish are dependent." Researchers understood this to mean they should continue with an emphasis on maintaining production of target species on which data was being collected and on which future measures could be used to judge the adequacy of water management decisions.

## Improvements to the Methodology

An outside panel of estuarine scientists reviewed the previous studies and proposals the state agency scientists had presented for improvements. The panel generally thought the analytical method was sound but did advise using larger datasets to improve confidence levels and to update analytical techniques where possible. This led to the publication of "Freshwater Inflow to Texas Bays and Estuaries" (Longley 1994). The report followed the general methodology presented in the 1980s detailed bay studies but with various enhancements. Previously, salinity data was available only from occasional monitoring efforts by various agencies. The TWDB acquired datasondes that measured several variables, especially temperature and salinity, and worked with the TNRCC monitoring program to have several continuous recording instruments in each bay. Additional data from TNRCC Monitoring Network, Texas Department of Health Shellfish Sanitation Program, and TPWD Fishery Independent studies provided an expanded database for salinity equations needed by the optimization program.

Eleven more years of data on gaged flows, diversions, and return flows were available, and an improved model for ungaged runoff was created and used to update the inflow values. Harvest data for 11 more years of fishery harvest data were available. While many of the same problems existed with the harvest data as in the 1980s studies, it did provide a longer database. Data from TPWD's Fishery Independent monitoring program, providing unbiased estimates of abundance would probably allow better fishery equations in future, but there were not enough years of data collected over a wide range of inflows to allow use of that information for this analysis. Several different approaches were taken for fishery equations, some with high explained variances but others with less confidence.

Additional nutrient data were available from the TWDB Coastal Data System, TNRCC Statewide Monitoring network, and contracted studies to allow a thorough analysis for setting nutrient constraints. Nitrogen, phosphorus, and organic carbon budgets were prepared for a very wet and a very dry year, and movement of these nutrients into and out of the estuary and loss to sediment were calculated. It appeared nitrogen was the most critical nutrient to evaluate. The analysis allowed finding a freshwater input amount that would not result in nitrogen depletion, to maintain primary production.

In the 1980s detailed bay studies, sediment requirements for maintaining wetlands were satisfied by specifying a minimum inflow volume for a period of months to assure that wetlands and deltas were occasionally inundated. For Guadalupe Estuary, the delta was undergoing change with the lower portion undergoing an inevitable period of decay while the upper delta appeared stable. Mission Lake, almost completely surrounded by delta wetlands, was a major pathway for organisms using wetlands during their life cycle to reach the upper delta and portions of lower delta wetlands. While the lake has been shallowing due to the input of sediment through Traylor Cut, it also has been subsiding due to eustatic sea-level rise and compaction of sediments. Finding a balance point between the processes increasing and decreasing the depth of Mission Lake defined a constraint for the analysis.

## TxBlend Hydrodynamic Model

The hydrodynamic and conservative transport models used as a check on salinity conditions underwent a major change. The BLEND model was developed by William Gray of Notre Dame University to which additional input routines for tides, river inflows, winds, evaporation, and salinity concentrations were added along with other utility routines to facilitate simulation runs specific to TWDB's needs (Gray 1987). This eventually evolved into the model named TxBLEND (Longley 1994). The models used in the earlier TWDB studies were a finite-difference model that used square or triangular computational cells of uniform size to calculate the net flows and salinity transport resulting from inflow, Gulf tides, and meteorological condition. The dimensions of each cell on the grid were large, often representing about one-half mile on each side. The new model was a finite element model. The grid cells were triangular and could be of different sizes which allowed fine-scale representation of channels and passes. Guadalupe Estuary was unusual because the tidal connections were mostly through the Intracoastal Waterway of the adjacent bay systems, the Lavaca-Colorado and Mission-Aransas Estuaries that connected to Gulf passes. Calibrating the model was a lengthy process and depended on tide gauge, wind, inflow, and several field studies involving multiple teams collecting round-the-clock flow and salinity data at particular input and choke points.

## TxEMP Optimization Model

The optimization method was updated with a new analytical program (Bao et al. 1989) that was nonlinear, stochastic, and multi-objective, named the Texas Estuarine Mathematical Programming (TxEMP) model (Matsumoto et al. 1994; Powell et al. 2002). The system could better deal with nonlinear relationships and used chance constraints in the analysis, the latter to achieve a result at or above a particular degree of probability, or to not violate a constraint with less than a particular degree of probability. Chance constraints could be applied to the salinity and harvest equations since these equations were statistical representations. Chance constraints provided more flexibility in stating the problem to be solved and in computing solutions. The multi-objective aspect enabled the program to optimize more than one objective simultaneously, even when some objectives are in conflict. The program allowed computation of performance curves where all objectives and constraints were satisfied and, for a given set of conditions, there was no better solution than represented by a point on the curve. Typically, a performance curve had harvest on the vertical axis and inflow on the horizontal axis. The end of the curve toward the low end of the inflow axis defined harvest at the minimum acceptable inflow (MinQ); the end of the curve toward the high end of the inflow axis defined harvest at the maximum acceptable inflow (MaxQ); and MaxH was the highest point on the curve and was usually located between the MinQ and MaxQ regions on the inflow axis.

Several types of constraints were available in the TxEMP model. A harvest target could be set so that the harvest of each species was no less than a percentage of the historical mean. Monthly upper and lower salinity bounds could be specified for different areas in the estuary for which there were salinity equations. Constraints could be established for monthly inflows, based on historical data, so the computed inflows would not have extreme values. Seasonal (often 2-month) inflow bounds were established so the optimization model would not compute harvests outside the inflow range used to define the harvest equations (most harvest equations were based on 2-month inflow values). Sediment and nutrient constraints based on minimum annual inflow values could be included. And different weightings could be used to emphasize harvest of one or more species or even weight them on the basis of economic value. Later analysis showed that weightings alone could result in one or more species dominating the analysis so a new feature was added, harvest or biomass ratios, which would assure that the computed relative abundances of different species remained within limits that would be realistic for the estuary.

The updated optimization program allowed the analyst to have a more complete understanding of the effects of the constraints, as well as salinity and harvest equations. But it made setting the constraints more challenging and required careful interpretation of the results so decision makers could make informed judgments. The models were used in a series of analyses of all major estuaries.

## TWDB Guadalupe Estuary Inflow Analysis

The TWDB applied the models to the Guadalupe Estuary as an example analysis (Longley 1994). Several combinations of species and constraints were presented but the example of equal weighting of species and 50% chance constraint is illustrative. The species for which equations were available were eastern oyster, brown shrimp, white shrimp, blue crab, red drum, spotted seatrout, and black drum. Salinity equations for three sites were prepared and monthly salinity constraints included: 1-20 psu in upper San Antonio Bay, 5-25 psu in lower San Antonio Bay, and 10-40 psu for Espiritu Santo Bay. Other constraints are listed in Table 7.

The TxEMP model calculated a MinQ of 898.7 thousand acre-ft, a MaxQ of 1536.8 thousand acre-ft and a MaxH around 978.8 thousand acre-ft. The equation for black drum caused problems in the analysis possibly due to low quality of the data and low historical harvest, and was not used in the TxEMP optimization analysis. The computed inflows fell within the bounds of the constraints of the TxEMP model and the TxBLEND modeling showed only three instances of simulated salinities in the three areas of the estuary were higher than the desired constraints.

## TPWD Guadalupe Estuary Inflow Recommendation

Pulich et al. (1998) prepared a TPWD recommendation of inflows for the Guadalupe Estuary using the TxEMP and TxBLEND methodologies. The equations relating salinity and fishery harvest to inflow were revised and were not the same as in the TWDB example Guadalupe analysis. The target species were blue crab, eastern oyster, red drum, black drum, spotted seatrout, brown shrimp, and white shrimp. Salinity constraints for upper San Antonio Bay were 5-20 psu except for the period May through August when they were reduced to 1-15 psu. For lower San Antonio Bay, the bounds were 5-25 psu and 10-40 psu for Espiritu Santo Bay. Other constraints are listed in Table 7.

The TxEMP model calculated a MinQ of 1.03 million acre-ft per year, a MaxH of 1.15 million acre-ft per year, and a MaxQ of 1.3 million acre-ft per year. MaxQ resulted in a lower computed fishery harvest than any of the other inflow solutions. Computed harvests were 2.54 million pounds, 2.93 million pounds, and 2.28 million pounds for flows at MinQ, MaxH, and MaxQ, respectively. All of these harvests were greater than the mean historical harvest of all target species. Researchers did an extensive job of verifying the adequacy of the computed inflows for resource management. Pulich et al. (1998) developed a "toolbox" of analyses to test the output

from the analysis. Using salinity maps from the TxBLEND model, geographic information systems (GIS), additional georeferenced salinity data, and extensive databases of spatial distribution of wetlands, seagrasses, and motile species, they were able to determine where the modeled salinity levels approached critical limits of wetland species and benthic fauna under the MinQ flow regime than under MaxH, especially during critical summer months. In addition, peak density areas associated with specific salinity zones were known from the TPWD Fisheries Resource Monitoring database and were compared with the salinity zones computed by TxBLEND. Both MinQ and MaxH flow regimes resulted in lower peak density salinity zones but the peak zone reduction was greater for the MinQ regime than the MaxH regime.

In the review of inflow recommendations below, the historical inflow record was based on inflow values prepared by the TWDB over the period 1941–2020. A longer historical period of record may be more indicative of the natural range and frequency of inflows, but all the 1980s and later analyses were based on shorter periods of record from which their upper bound choices of median or mean inflows were taken. Table 8 shows the historical mean and median monthly combined inflows (gaged + ungaged + returns – diversions) to the Guadalupe Estuary and monthly inflows from the TDWR (1980b) analysis, the Longley (1994) example analysis, and the Pulich et al. (1998) recommendation. Historical median inflows for each month are much lower than the historical mean inflows, which indicates the influence of very large inflow events. The median indicates that half of the inflows in any month are above the median and half are below. Thus, the median is a better indicator of how frequently the calculated inflow needs can be achieved than the mean value. The monthly computed inflow needs in TDWR (1980b) were greater than the historical monthly median inflows in all but one month [exceedances in **bold type**] and were larger than two of the historical mean inflows [exceedances underlined] (Table 8). Upper inflow bounds were not obvious from the TDWR (1980b) study. One computed monthly inflow in Longley (1994) was greater than the historical median but equaled the upper bound inflow constraint that was set in the analysis. The Pulich et al. (1998) calculations for MinQ exceed historical median inflows in 2 months, and for MaxH exceed the historical median inflows in 5 months but were all constrained to be equal to or less than the monthly median inflow values known at the time of the analysis. Monthly and annual inflows from both the Longley (1994) and Pulich et al. (1998) studies are substantially lower than TDWR (1980b) study. The period of record for all the TDWR studies (1980a, b, 1981a, b, c, 1982) was 1941-1976. The period of record for Longley (1994) and Pulich et al. (1998) was 1941-1987, so differences in period of record, inflow upper inflow bound constraints, and equations in TDWR (1980b) may explain the exceedances. In addition, TDWR (1980b) contained a constraint for wetland inundation that was not included in the TxEMP model. This may explain the high September com-

**Table 7** Constraints for the TxEMP optimization analysis

| Study | Number of species equations | Number of salinity equations | Monthly salinity bounds | Harvest-productivity target | Monthly inflow bounds | Seasonal inflow bounds | Nutrient lower bound (annual) | Sediment lower bound (annual) | Harvest-biomass ratio | Chance constraints |
|---|---|---|---|---|---|---|---|---|---|---|
| Guadalupe Longley (1994) | 7 | 3 | Yes, for each site | 80% of Mean | 10th percentile & monthly median | Monthly sums | 286,000 ac-ft | 355,235 ac-ft | No | Yes |
| Guadalupe Pulich et al. (1998) | 7 | 3 | Yes, for each site | 80% of Mean | 10th percentile & monthly median | Modification of monthly sums | 860,000 ac-ft | 860,000 ac-ft | Yes | Yes |
| Lavaca-Colo Martin et al. (1997) | 9 | 2 | Yes, for each site | Not specified | 10th percentile & monthly average | Complicated combination | 1,710,000 ac-ft | No | Yes | Yes |
| Trinity-San Jac Lee et al. (2001) | 8 | 3 | Yes, for each site | 80% of Mean | 10th percentile & monthly median | Modification of monthly sums | 4,270,000 ac-ft | No | Yes | Yes |
| Nueces Pulich et al. (2002) | 7 | 3 | Yes, for each site | 70% of Mean | 10th percentile & monthly median | Modification of monthly sums | 117,000 ac-ft | No | Yes | Yes |
| Upper Laguna Tolan et al. (2004) | 8 | 3 | Yes, for each site | 75% of Mean | 10th percentile & monthly median | Modification of monthly sums | 20,000 ac-ft | No | Yes | Yes |
| Lower Laguna Tolan et al. (2004) | 8 | 3 | Yes, for each site | 75% of Mean | 10th percentile & monthly median | Modification of monthly sums | 67,200 ac-ft | No | Yes | Yes |
| Sabine-Neches Kuhn and Chen (2005) | 8 | 3 | Yes, for each site | 75% of Mean | 10th, 25th percentile mean & median | Modification of monthly sums | 10,281,600 ac-ft not used | No | Yes | Yes |
| Mission-Aran Chen (2010) | 8 | 3 | Yes, for each site | 70% of Mean | 25th percentile & median | Modification of monthly sums | 135,000 ac-ft not used | No | Yes | Yes |

**Table 8** Combined inflows (thousand acre-ft, 1 ac-ft = 1233.48 m$^3$) to Guadalupe Estuary (1941–2020) with computed inflow needs from three different inflow assessments

| Month | Historical Mean Inflow | Historical Median Inflow | TDWR (1980b) Alt II | Longley (1994) | Pulich et al. (1998) MinQ | Pulich et al. (1998) MaxH |
|---|---|---|---|---|---|---|
| January | 164.7 | 118.3 | **143.5** | 70.3 | 111.2 | 111.2 |
| February | 180.2 | 115.6 | **169.7** | 70.3 | **124.2** | **124.2** |
| March | 172.9 | 128.2 | **136.9** | 85.6 | 52.4 | 52.4 |
| April | 202.2 | 134.4 | **199.4** | 85.6 | 52.4 | 52.4 |
| May | 278.2 | 183.5 | **311.5** | 90.1 | **186.0** | **222.6** |
| June | 274.5 | 145.9 | **238.3** | 105.6 | 136.0 | **162.7** |
| July | 193.4 | 100.0 | 76.4 | 95.2 | 60.8 | 88.6 |
| August | 124.6 | 80.2 | **104.5** | **94.9** | 60.8 | **88.3** |
| September | 219.7 | 115.0 | **261.1** | 90.3 | 52.4 | 52.4 |
| October | 230.4 | 114.2 | **135.0** | 70.3 | 52.4 | 52.4 |
| November | 186.5 | 98.3 | **126.9** | 70.3 | 73.8 | 73.8 |
| December | 165.3 | 102.6 | **116.7** | 70.3 | 66.2 | 66.2 |
| Annual | 2392.7 | 1437.3 | 2019.9 | 998.8 | 1028.6 | 1147.2 |

puted inflow under Alternative I of the TDWR (1980b) analysis. The researchers recommended MaxH as the lowest target inflow to provide the most adequate salinity conditions for delta wetlands and peak density areas. Based on historical hydrology, computed inflow needs might be achieved at least half the time in 8 months of the year.

## Lavaca-Colorado Estuary Inflow Recommendation

Martin et al. (1997) of the Lower Colorado River Authority (LCRA) prepared an assessment of freshwater inflow needs for the Lavaca-Colorado Estuary using the TxEMP and TxBLEND methodology. Fishery equations were developed for 9 species: blue crab, brown shrimp, white shrimp, eastern oyster, Gulf menhaden, striped mullet, black drum, red drum, and southern flounder. Equations for eastern oyster and southern flounder were created using commercial harvest data. Equations for the other seven species were created using data from the TPWD's Coastal Fisheries Monitoring Program. For the fishery information to be comparable between the two data sources, the harvest data and monitoring data were converted to density (individuals per hectare). The data from the monitoring program avoided many of the problems associated with the harvest data such as effort, questions of where the actual landings were from, and completeness. Salinity equations were based on two areas in the estuary, upper Lavaca Bay and the eastern end of Matagorda Bay. Lower salinity bounds for both sites were set to 5 psu for all months except May–Aug when it was set to 1 psu. Under normal flow conditions, the upper salinity bound was set to 20 psu except for May–July when it was lowered to 15 psu. Other constraints are briefly noted in Table 7.

From the analysis, MinQ was 1.893 million acre-ft per year, MaxH was three million acre-ft per year, and MaxQ was 3.4 million acre-ft per year. The curve of inflow versus biomass density had a very flat top. The target need defined by LCRA was two million acre-ft per year, which resulted in calculated biomass density 98% of the density at the MaxH inflow regime, but required one million acre-ft less inflow. The calculated biomass density at the target inflow need of two million acre-ft per year would produce a nearly 60% higher density than the historical average density and resulted in salinities well within the salinity bounds for upper Lavaca Bay but were very close or at the upper bound salinities for eastern Matagorda Bay in the winter and summer months.

The historical mean and median inflows for the Lavaca-Colorado Estuary were used for two inflow assessments using optimization models (Table 9). The historical median inflows for each month were much lower than the historical average which shows the influence on the means of very large inflow events. For the TDWR (1980a) analysis, the inflow needs for 10 months exceed the historical median inflows [exceedances in **bold type**]; for three of those months the calculated inflow need was higher than the historical mean inflow [exceedances underlined] (Table 9). The TDWR (1980a) study included inflows to provide Lavaca Delta marsh inundation but do not account for the large spring calculated inflows; the TxEMP model did not specifically include flows for marsh inundation. While the annual inflow amount in the Martin et al. (1997) study was 800,000 acre-ft per year less than the TDWR (1980a) quantity, the calculated need was also greater than the historical median inflow in 7 months. The monthly inflow upper bound constraints in TDWR (1980a) were not entirely clear. Martin et al. (1997) used monthly mean flows as the upper inflow bounds. The hydrology data used in the Martin et al. (1997) study was extended to 1991, but since the computed inflow needs were greater than the historical median, they would probably only be achievable half the time in 5 months of the year. Differences in the hydrology period of record in the TDWR (1980a) and Martin et al. (1997) studies and choice of monthly inflow upper bounds may have had some effect on the exceedance of historical median inflows. The study recommended a target inflow of two million acre-ft per year.

**Table 9** Combined inflows (thousand acre-ft) to Lavaca-Colorado Estuary (1941–2020) with computed inflow needs from two different inflow assessments

| Month | Historical Mean Inflow | Historical Median Inflow | TDWR (1980a) Alt II | Martin et al. (1997) |
|---|---|---|---|---|
| January | 257.1 | 168.6 | 144.7 | 94.3 |
| February | 264.8 | 154.7 | **160.7** | 100.1 |
| March | 273.7 | 156.7 | 125.4 | **195.9** |
| April | 289.0 | 196.3 | **250.6** | **252.1** |
| May | 417.3 | 207.3 | **379.8** | **298.6** |
| June | 439.6 | 145.8 | **341.5** | **289.5** |
| July | 271.2 | 93.7 | **110.9** | **194.6** |
| August | 143.2 | 102.7 | **119.0** | 100.2 |
| September | 308.1 | 126.3 | **<u>336.2</u>** | **151.3** |
| October | 338.0 | 127.8 | **280.2** | **141.6** |
| November | 278.7 | 134.8 | **<u>282.8</u>** | 92.6 |
| December | 227.2 | 140.2 | **<u>275.7</u>** | 89.1 |
| Annual | 3508.0 | 1755.1 | 2807.5 | 1999.9 |

## Trinity-San Jacinto Estuary Inflow Recommendation

Lee et al. (2001) prepared a TPWD freshwater inflow recommendation for the Trinity-San Jacinto Estuary using the TxEMP and TxBLEND methodologies. Harvest data was used in the analysis and target species were blue crab, eastern oyster, red drum, black drum, spotted seatrout, brown shrimp, white shrimp, and southern flounder. Salinity equations were prepared for three areas. Bounds for the Trinity Bay site were 5-20 psu (Jan-Feb), 5-15 psu (Mar-Apr), 1-15 psu (May–Jul), 5-15 psu (Aug), 5-20 psu (Sep–Dec); bounds for Red Bluff-Clear Lake area were 5-20 psu (Jan–May), 2-15 psu (Jun), 5-15 psu (Jul), 5-20 psu (Aug), 5-25 psu (Sep–Oct), 5-20 psu (Nov–Dec); bounds for Dollar Point site were 10-30 psu (Jan–Feb), 10-25 psu (Mar–Apr), 10-20 psu (May–Jun), 10-25 psu (Jul), 10-30 psu (Aug–Dec). Other constraints are noted in Table 7.

The analysis resulted in a MinQ inflow of 4.160 million acre-ft per year and a MaxH of 5.220 acre-feet per year. The calculated MinQ and MaxQ harvests were 20.5 to 31.5% higher than the mean harvest of 8.87 million pounds per year. Annual calculated inflows were between the 25th and 50th percentile of historical inflows to the estuary. Using results from TxBLEND, GIS, and extensive TPWD databases in the manner described in Pulich et al. (1998), peak densities of species differed by less than 2% in the estuary between MinQ and MaxH regimes, but more species were favored by MaxH than by MinQ. Seasonal salinity distributions were similar most months of the year but differed in lower Galveston Bay and West Bay by 4-5 psu from April to May. Salinities in the mid-bay region in the oyster reef area did not exceed 15 psu often, which would help control the oyster pathogen *Perkinsus marinus*.

Historical mean monthly inflows were substantially greater than historical median monthly inflows except for the month of February (Table 10). Comparing computed inflows with historical inflow data, 3 monthly TDWR (1981b) inflow estimates were greater than the long-term historical median inflows [exceedances in **bold type**] and one month was greater than the historical mean inflow [exceedance underlined] (Table 10). It is not clear what upper bounds inflow were used in the TDWR (1981b) analysis but from Fig. 9-5 of the report, it appears that in April and October the calculated inflows were greater than the mean inflows from the 1941-1976 period. Inflows in the TDWR (1981b) study for April, May, June, and October appear to be strongly influenced by the marsh inundation constraint for the Trinity Delta. The calculated inflow for the Lee et al. (2001) study exceeds the long-term historical median inflow in only one month (May) for the MaxH analysis, but all inflows in that study were at or below the upper bound median inflow constraint based on data from 1941–1996; there was no specific marsh inundation constraint in the Lee et al. (2001) study. Exceedance of the historical median was the result of a longer period of inflow record and no specific Trinity Delta inundation constraint. The exceedance in the TDWR (1981b) study may be the result of a longer inflow record and higher inflow constraints including marsh inundation. The recommended inflow regime of Lee et al. (2001) was between MinQ and MaxH. Since the recommended inflows were less than the historical median inflow (except for May MaxH inflows), it seems likely that the recommended inflows could be achieved half the time, and possibly more.

## Nueces Estuary Inflow Recommendation

Pulich et al. (2002) prepared a TPWD inflow recommendation for the Nueces Estuary using TxEMP and TxBLEND methodologies. Harvest equations were prepared for seven target species: blue crab, brown shrimp, white shrimp, red

**Table 10** Combined historical mean and median inflows (thousand acre-ft) to Trinity-San Jacinto (1941–2020) with computed inflow needs from two different inflow assessments

| Month | Historical Mean Inflow | Historical Median Inflow | TDWR (1981b) Alt II | Lee et al. (2001) MinQ | Lee et al. (2001) MaxH |
|---|---|---|---|---|---|
| January | 1141.1 | 819.3 | 488.3 | 150.5 | 150.5 |
| February | 1031.2 | 1030.8 | 480.0 | 216.7 | 155.2 |
| March | 1161.5 | 826.6 | 365.0 | 363.9 | 652.8 |
| April | 1152.1 | 671.2 | **1076.0** | 352.6 | 632.5 |
| May | 1500.0 | 1165.5 | 1162.0 | 679.7 | **1273.7** |
| June | 1304.7 | 879.6 | 760.2 | 448.1 | 839.7 |
| July | 712.8 | 471.5 | 237.3 | 232.7 | 211.5 |
| August | 549.8 | 271.1 | 255.5 | 153.9 | 139.9 |
| September | 658.6 | 419.1 | 332.7 | 330.3 | 102.9 |
| October | 717.8 | 317.9 | **938.9** | 251.9 | 78.5 |
| November | 837.6 | 409.0 | **538.5** | 351.5 | 351.5 |
| December | 937.4 | 690.9 | 554.5 | 626.8 | 626.8 |
| Annual | 11704.7 | 7972.5 | 7188.9 | 4158.6 | 5215.5 |

drum, spotted seatrout, black drum, and southern flounder. Salinity equations were prepared for three sites. Monthly upper and lower salinity bounds for the sites were: Head of Nueces Estuary, 5-36 psu (Jan–Mar), 5-32 psu (Apr), 1-23 psu (May), 1-20 psu (Jun), 2-25 psu (Jul–Aug), 5-25 (Sep), and 5-30 (Oct–Dec); Mid-Nueces Bay, 5-35 psu (Jan–Apr), 5-30 psu (May), 5-25 (Jun), 5-35 (Jul–Aug), 5-30 (Sep), 5-35 (Oct–Dec); Mid-Corpus Christi Bay 20-37 psu (Jan–Dec). Other constraints are briefly listed in Table 7.

The TxEMP model calculated a MinQ of 115,640 acre-ft per year, MaxH of 138,490 acre-ft per year, and MaxQ of 167,000 acre-ft per year. MinQ and MaxH predicted harvests were close to the historical mean harvest but higher than the 70% lower bound for harvest. Extensive confirmation analyses showed that over the entire estuary the difference in inflows between MinQ and MaxH was small. The critical question of inflow here was whether the flows reach the Nueces delta which serves as nursery ground, critical habitat, and supplier of organic matter for estuarine species. Upstream water developments and diversion of water for human uses have reduced inflow through time so attention must be paid to lowering barriers to flooding of the delta and supplying pulses of flow to inundate the area. In addition to flooding the delta wetlands, MaxH flows could provide a gradient in the Nueces Bay portion of the estuary.

Historical mean monthly inflows are substantially larger than historical median monthly inflows due to occasional high inflow events such as hurricanes and tropical storms (Table 11). All computed inflows from the TDWR (1981a) report were greater than historical medians [exceedances in **bold type**], and the April and September computed inflows were greater than the historical mean inflow for that month [exceedances underlined]. It is not clear what inflow upper bounds were used in this study; the hydrology period of record (1941-1976) was shorter than the historical period. The TDWR (1981a) study had a concern for Nueces marsh inundation but determined that flooding would require peak river flows of more than 5000 cubic feet per second. Using daily hydrographs of spring and fall floods, they estimated that spring and fall floods with monthly river inflows of 79.0 and 139.0 thousand acre-ft would provide peak flows high enough to flood the marsh. These values plus ungaged and return flows were the basis of the May and September requirements, 84.00 and 148.10 acre-ft. But the historical inflow values, based on 80 years of data have shown much lower historical median inflows, putting the marsh flooding analysis in doubt. For the Pulich et al. (2002) recommendation, no calculated MinQ amounts and only one MaxH inflow were greater than the historical median inflows. All calculated inflows in that study were less than or equal to the median upper constraints in the hydrology period of record (1941–1994) at the time of the analysis. However, no specific marsh inundation constraint was included in the TxEMP analysis. The Pulich et al. (2002) recommendation was to supply the calculated MaxH flows for the period April–July (89,200 acre-ft) delivered in one or two pulsed events in any of those 4 months, and for the other 8 months to supply the MaxH flows as calculated. And if the high spring flows do not occur, then cumulative MinQ flows for September–November (27,500 ac-ft) should be used to maintain conditions in upper Nueces Bay and tidal Nueces River. On the basis of the historical median inflows, it seems likely that the MinQ inflows could be achieved half the time. The one exceedance of the MaxH result was so close to the historical median that is possible the MaxH inflows could be achieved to almost the same degree as the MinQ inflows.

## Laguna Madre Estuary Inflow Recommendation

Tolan et al. (2004) prepared a TPWD recommendation for the Laguna Madre. Because the estuary is bisected by the Land Cut, and only interconnected through the Intracoastal

**Table 11** Combined historical mean and median inflows to Nueces Estuary (1941–2020) with computed inflow needs from two different inflow assessments (thousand acre-ft)

| Month | Historical mean inflow | Historical median inflow | TDWR (1981a) Alt II | Pulich et al. (2002) MinQ | Pulich et al. (2002) MaxH |
|---|---|---|---|---|---|
| January | 23.34 | 7.00 | **9.40** | 2.23 | 2.23 |
| February | 21.93 | 6.19 | **11.10** | 2.78 | 2.78 |
| March | 22.83 | 8.61 | **10.10** | 4.41 | 4.92 |
| April | 32.04 | 7.00 | **<u>26.60</u>** | 5.18 | 5.18 |
| May | 78.29 | 37.27 | **84.00** | 32.14 | **37.77** |
| June | 78.90 | 40.12 | 38.00 | 19.99 | 36.43 |
| July | 74.40 | 26.50 | **41.90** | 6.98 | 9.82 |
| August | 39.62 | 13.01 | **25.90** | 9.75 | 9.75 |
| September | 107.39 | 31.66 | **<u>148.10</u>** | 11.04 | 9.60 |
| October | 88.78 | 22.81 | **39.80** | 8.69 | 7.56 |
| November | 36.25 | 12.51 | **31.10** | 7.78 | 7.78 |
| December | 18.68 | 6.84 | **11.70** | 4.67 | 4.67 |
| Annual | 622.44 | 219.50 | 477.70 | 115.64 | 138.49 |

Waterway, inflow needs were treated as though they were two separate estuaries. Instead of using fishery harvest data, researchers were able to use bag-seine catch data from the TPWD fishery-independent sampling program, which gave a much better estimate of productivity, not biased by different in effort and other problems. For the Upper Madre, equations relating inflow and catch were prepared for blue crab, brown shrimp, white shrimp, red drum, spotted seatrout, black drum, southern flounder, and grass shrimp. Three zones were selected for inflow-salinity equations and upper and lower salinity bounds were defined for each month: Upper Baffin Bay, 30-45 psu (Jan–Feb), 25-45 psu (Mar–May), 20-45 psu (Jun), 20-53 psu (July), 30-53 psu (Aug–Sep), 35-53 psu (Oct), and 30-45 psu (Nov–Dec); Mouth of Baffin Bay (30-40 psu (Jan–Feb), 25-40 psu (May–Jun), 32-45 psu (Jul–Oct), 28-45 psu (Nov–Dec); South Bird Island, 30-37 psu (Jan–Feb), 30-40 psu (Mar–Apr), 35-42 psu (May–Jun), 35-45 psu (Jul-Oct), 35-41 (Nov–Dec).

For Lower Laguna Madre equations relating inflow to catch were prepared for blue crab, brown shrimp, white shrimp, red drum, spotted seatrout, black drum, southern flounder, and pink shrimp. Three zones for salinity equations were selected and upper and lower salinity bounds for each month specified: South Land Cut, 30-45 psu (Jan), 30-40 psu (Feb–Apr), 25-36 psu (May), 30-40 psu (Jun–Sep), 30-45 psu (Oct–Dec); Arroyo Colorado, 20-32 psu (Jan–Apr), 20-35 psu (May), 25-35 psu (Jun), 25-26 psu (Jul–Aug), 18-36 psu (Sep), 18-32 psu (Oct), 22-32 psu (Nov–Dec); Stover Point, 25-35 psu (Jan–Mar), 30-35 psu (Apr–May), 30-40 psu (Jun), 34-42 psu (Jul–Sep), 30-42 psu (Oct), 25-35 psu (Nov–Dec). Other constraints for both areas are briefly noted in Table 7.

For Upper Laguna Madre, MinQ was 21,530 acre-ft per year, MaxQ was 31,020 acre-ft per year, and MaxC (renamed from MaxH since it was based on catch, not harvest) was 22,770 acre-ft per year. The very low calculated inflow values reflect the very low natural inflow into the Upper Laguna. For the Lower Laguna, the MinQ was 214,950 acre-ft per year, MaxQ was 248,900 acre-ft per year, and MaxC was 228,300 acre-ft per year, inflow needs about 10 times those of the Upper Laguna. Though the areas of the upper and lower Laguna Madre are large, the inflow quantities are low and, except for areas near outflows of creeks and the Arroyo Colorado, do not cause a clear salinity gradient over the entire system as in other Texas estuaries. Evaporation tends to overwhelm the effects of any inflows except during tropical storms or hurricanes. Direct precipitation on the estuary was its main source of freshwater.

Because the inflow values for Upper Laguna Madre were so small, they were combined with the Lower Laguna values to compare with historical inflows (Table 12). Monthly inflows for the TDWR (1983) report exceed the monthly historical median inflows in 4 months [exceedances in **bold type**] and exceed monthly historical mean inflows in September and October [exceedances underlined]. The dominant aquatic vegetation is submerged seagrass, so no part of the analysis was directed to marsh inundation in either study. None of the MinQ or MaxC inflows of Tolan et al. (2004) exceed the historical monthly median inflows, which were based on 1941–1996 hydrology. Therefore, the MinQ and MaxC flows can probably be achieved at least half the time. However, Kuhn and Chen note that the overall salinity structure of the Upper and Lower Laguna Madre was very similar under both MinQ and MaxH, and in a comparison of wet and dry years, there was little difference in spatial distribution of target species. For these reasons no recommendation for a specific target inflow was made.

**Table 12** Combined historical mean and median inflows to Laguna Madre Estuary (1941–2020) with computed inflow needs from two different inflow assessments (thousand acre-ft)

| Month | Historical Mean Inflow | Historical Median Inflow | TDWR (1983) Alt II | Tolan et al. (2004) MinQ | Tolan et al. (2004) MaxC |
|---|---|---|---|---|---|
| January | 38.04 | 25.78 | 18.21 | 18.31 | 18.31 |
| February | 36.83 | 24.00 | 18.68 | 17.57 | 17.57 |
| March | 37.50 | 25.34 | 21.42 | 17.95 | 21.08 |
| April | 46.57 | 30.51 | 30.49 | 20.37 | 23.94 |
| May | 71.73 | 50.77 | 38.43 | 27.77 | 29.57 |
| June | 77.92 | 49.24 | 47.19 | 24.10 | 25.30 |
| July | 89.94 | 30.67 | **37.99** | 19.85 | 19.85 |
| August | 69.39 | 31.00 | **40.13** | 16.76 | 16.85 |
| September | 135.80 | 65.98 | **152.61** | 17.39 | 18.21 |
| October | 105.37 | 51.82 | **120.92** | 20.65 | 21.53 |
| November | 47.34 | 31.27 | 27.14 | 18.63 | 20.47 |
| December | 35.40 | 24.42 | 24.80 | 17.14 | 18.42 |
| Annual | 791.83 | 440.79 | 578.01 | 236.49 | 251.10 |

## Sabine-Neches Estuary Inflow Recommendation

Kuhn and Chen (2005) prepared a TPWD inflow recommendation for the Sabine-Neches Estuary using the TxEMP and TxBLEND methodologies. In the analysis, bag seine catch was used as a measure of productivity instead of harvest, thereby avoiding many of the problems of commercial harvest data. Equations were prepared for blue crab, brown shrimp, Atlantic croaker, menhaden, red drum, spotted seatrout, spot, and white shrimp. Three salinity sampling stations were used to prepare inflow-salinity equations. Salinity bounds for the three sites were: Upper Sabine Lake, 0-10 psu (Jan–Jul), 2-10 psu (Aug), 2-15 psu (Sep–Nov); 2-10 psu (Dec); Mid Sabine Lake, 1-10 psu (Jan–May), 1-15 psu (Jun–Jul), 2-15 psu (Aug–Dec); Lower Sabine Lake, 2-20 psu (Jan–Jul), 5-20 psu (Aug–Nov), 2-20 psu (Dec). Other constraints are briefly listed in Table 7.

While the Sabine-Neches Estuary has a high level of inflow [on average around 1.13 million acre-ft per month, Longley (1994)], hydrodynamic modeling in the TDWR (1981c) report noted that the largest net flows each month occur through the Sabine-Neches and Port Arthur canals that lead to the Gulf. Consequently, only a fraction of the computed inflows actually enter Sabine Lake to support the target species and the wetlands on which they depend, though some flow through the canals may enter Keith Lake or the Intracoastal Waterway to provide fresh water to the wetlands west of the Port Arthur Canal. The MinQ flow was 7.114 million acre-ft per year, MaxC was 9.5966 million acre-ft per year, and Max Q was 11.6193 million acre-ft per year. Several analyses of the salinity patterns from TxBLEND as well as data from TPWD systematic sampling programs showed that three important species preferred salinity zones associated with the MinQ inflows but wetlands that support all species were better supported by inflows associated with the MaxC regime.

The historical mean and median inflows and the computed inflow needs are from two reports (Table 13). Unlike a number of the other studies, the annual calculated inflow from the older TDWR (1981c) study sits between the MinQ and MaxC computed inflows of Kuhn and Chen (2005). Some monthly computed inflows were higher in one study than the other and vice-versa. The TDWR (1981c) study was unable to calculate a maintenance or enhanced productivity alternative so the inflows were for the subsistence choice, which tries to fulfill salinity limits and marsh inundation needs. The May and October monthly calculated inflows from the TDWR (1981c) study were higher than both the historical median [exceedances in **bold type**] and historical mean inflows [exceedances underlined] (Table 13). Examining daily hydrographs for peak flows, TDWR (1981c) researchers determined which levels of monthly combined inflows for May and October typically had median daily peak flows high enough to flood the Sabine and Neches delta marshes and used those values as the inflow requirements. But with the 80-year historical inflow record, it is clear that historical median inflows are substantially lower than the May and October quantities used by TDWR (1981c), putting the inundation calculations in doubt. The Kuhn and Chen (2005) study had to increase the lower bounds during the first 6 months and raise the upper bound in two fall months to allow feasible solutions for the optimization model. This may be the result of extreme high inflows and the fact that only a fraction of the inflows move through Sabine Lake itself. Five calculated monthly inflows for MaxC and two of the calculated inflows for MinQ are greater than the historical median inflows and one value is greater than the historical mean. But all seven of these exceedances were at the inflow upper bound set for the analyses using hydrology for the 1941–1999 period; the difference in hydrology periods of record may account for the reduction in median values from those used in the analysis. The Kuhn and Chen (2005) recommendation was that the MaxC inflow regime was

**Table 13** Combined historical mean and median inflows to Sabine-Neches Estuary (1941–2020) with computed inflow needs from two different inflow assessments (thousand acre-ft)

| Month | Historical mean inflow | Historical median inflow | TDWR (1981c) Alt I | Kuhn & Chen (2005) MinQ | Kuhn & Chen (2005) MaxC |
|---|---|---|---|---|---|
| January | 1607.5 | 1238.6 | 570.9 | 624.0 | **1246.4** |
| February | 1506.9 | 1395.5 | 592.6 | 832.5 | **1539.2** |
| March | 1701.5 | 1598.3 | 551.0 | 998.0 | 1565.8 |
| April | 1521.0 | 1193.4 | 837.2 | 778.6 | 1136.6 |
| May | 1594.7 | 1072.4 | **1706.7** | 691.9 | 691.9 |
| June | 1228.9 | 797.8 | 768.7 | 478.7 | 478.7 |
| July | 840.6 | 573.7 | 459.7 | 424.4 | 547.3 |
| August | 578.2 | 459.1 | 403.0 | 361.7 | **466.5** |
| September | 711.9 | 506.2 | 345.7 | **574.6** | **574.6** |
| October | 595.2 | 353.1 | **1634.7** | **537.9** | **537.9** |
| November | 689.1 | 395.5 | 353.2 | 237.6 | 237.6 |
| December | 1080.3 | 714.4 | 560.4 | 574.1 | 574.1 |
| Annual | 13,655.8 | 10,298.0 | 8783.8 | 7114.0 | 9596.6 |

suited to provide the best salinity and nutrient conditions for the estuary. The January, August, and September MaxC inflows are only slightly above the historical medians so the inflow recommendation might be achievable more than half of the time for seven and possibly more months.

## Mission-Aransas Estuary Inflow Recommendation

Chen (2010) prepared a TPWD recommendation for inflows to the Mission-Aransas Estuary using the TxEMP and TxBLEND methodologies. Harvest data was used because at the time of the analysis, there was a longer record of harvest data than fisheries independent data. Equations were prepared for blue crab, brown shrimp, white shrimp, eastern oyster, red drum, spotted seatrout, black drum, and flounder. Three sites were selected for salinity regressions, and monthly salinity bounds were defined: Southwest Copano Bay, 5-21 psu (Jan), 5-20 psu (Feb–Jun), 5-22 psu (Jul–Dec); Copano Causeway, 10-23.5 psu (Jan–Apr), 8-23.5 psu (May–Jun), 6.5-27 psu (Jul), 10-27 psu (Aug–Sep), 10-26 psu (October), 6.5-23.5 psu (Nov), 10-23.5 psu (Dec); Mid-Aransas Bay, 14-27 psu (Jan–Apr), 14-30 psu (May–Jun), 16-34 psu (Jul), 20-35.5 psu (Aug–Sep), 17-32 psu (Oct), 14-28 psu (Nov–Dec). Other constraints are briefly listed in Table 7.

The analysis resulted in a MinQ flow of 58,750 acre-ft per year, a MaxH of 86,170 acre-ft per year, and a MaxQ of 95,190 acre-ft per year. The Mission-Aransas Estuary is a pulsed system; there were occasional very large inflows indicated by large differences between mean and median inflows. The influence of MinQ and MaxH was largely on the upper end of Copano Bay. Using validation tests developed by TPWD, the salinity patterns in the estuary were nearly identical between MinQ and MaxH. Comparing both salinity and species distribution data under wet and dry conditions, it appeared that both MinQ and MaxH inflow regimes should be able to maintain historical conditions. But under high inflows in the Guadalupe estuary, salinities in Aransas and Copano bays were strongly influenced by inflow from the neighboring estuary.

Historical inflows show substantial differences between mean and median inflows (Table 14). Computed inflows from the TDWR (1981a) were higher than the historical median inflows in 11 of 12 months [exceedances in **bold type**] ((Table 14)). Only MinQ and MaxH in February from the Chen (2010) study inflows were higher than the historical median inflow, but both were set to the median inflow upper bound using the hydrology from 1941–1996. Overall, the computed inflows from Chen (2010) are much lower than those from TDWR (1981a); the upper inflow bounds in the TDWR (1981a) study were not clearly stated in the report. The Chen (2010) recommendation was a target inflow within the range and monthly distribution patterns of MinQ and MaxH. Since all but one month of the MinQ and MaxH computed inflows are less than the historical median inflow, it is possible that these inflows could be achieved at least half the time.

## Comments on the Optimization Methodology and 1980s Recommendations

The computed annual inflows from the 1980s studies using the ELPM were higher than those using the TxEMP methodology, the only exception being in Sabine-Neches Estuary where only a fraction of the inflow reaches Sabine Lake, the majority travels down the canals leading to the Gulf. About 85% of the calculated monthly inflows from the 1980s studies were greater than the computed inflows from the later studies. Differences in harvest/catch and salinity equations,

**Table 14** Combined historical mean and median inflows (thousand acre-ft) to Mission-Aransas (1941–2020) with computed inflow needs from two different inflow assessments

| Month | Historical mean inflow | Historical median inflow | TDWR (1981a) Alt II | Chen (2010) MinQ | Chen (2010) MaxH |
|---|---|---|---|---|---|
| January | 17.1 | 5.2 | **5.9** | 2.94 | 2.94 |
| February | 24.3 | 4.2 | **14.4** | **5.01** | **5.01** |
| March | 22.8 | 5.2 | 4.4 | 3.05 | 3.05 |
| April | 28.8 | 4.3 | **16.5** | 2.43 | 2.43 |
| May | 61.7 | 21.4 | **39.9** | 12.86 | 19.12 |
| June | 54.8 | 19.2 | **28.4** | 10.66 | 15.83 |
| July | 55.8 | 8.9 | **25.5** | 1.41 | 1.41 |
| August | 32.8 | 8.4 | **19.5** | 2.20 | 1.88 |
| September | 96.6 | 23.3 | **55.9** | 7.36 | 17.65 |
| October | 69.0 | 15.3 | **35.6** | 4.29 | 10.31 |
| November | 31.3 | 5.5 | **7.4** | 3.76 | 3.76 |
| December | 17.9 | 3.9 | **15.5** | 2.78 | 2.78 |
| Annual | 512.9 | 124.7 | 268.8 | 58.75 | 86.17 |

constraints (especially inflow upper bounds), marsh inundation, hydrology, and possibly in the optimization models themselves may account for much of the variation in results.

There are some advantages to optimization types of analyses. They focus attention onto questions resource managers understand, e.g., quantities of water and pounds or densities of commercial species. They address the direction from the legislature: to maintain productivity of sport and commercial species and the conditions on which they depend. Additional managerial constraints can be put on the analyses as information is developed, although that can be challenging as evidenced by the nutrient and sediment discussion above. By their very nature, the inflow-salinity and inflow-harvest/catch relationships are statistical in nature; the TxEMP model uses that to advantage in the analyses. The multi-objective aspect of the method allows an analysis even when target species have competing requirements. By using multiple salinity sites in the analyses, the models allow establishment of salinity gradients known to be important in the production of desired species. Monthly calculated inflows allow modeling of circulation and salinity patterns. TPWD has developed a comprehensive toolbox of methods using these results and other data from their GIS and sampling programs to address the suitability of future modifications of inflow and circulation patterns.

There are also disadvantages to optimization analyses. The harvest data has various problems (e.g., effort, location where the data is from, and completeness) so the use of fishery-independent data is an improvement, especially if longer records of catch or biomass and hydrology are available (but in the analyses above, the ranges of percent variances explained in regression equations using harvest and fishery independent data were not substantially different). In general, the fishery equations that do not depend on multiple antecedent years of inflow make the inflow relationship appear more direct. Setting constraints can be a problem. For example, in the case of Sabine-Neches Estuary, the majority of water entering the estuary travels down the various canals to the Gulf or other areas, so the calculated constraint for nutrients was higher than any of the calculated annual inflows from the 1981 and 2005 studies. Sediment constraints were often unable to be used because of a lack of data or the presence of nearby upstream reservoirs that entrap much of the natural sediment flow. Because some constraints as well as inflow-salinity and inflow-harvest/catch equations were based on statistical measures of past inflows, additional years of hydrology, production, and salinity data may improve them.

This optimization methodology may work best for estuaries where there is enough inflow to create an estuary-wide salinity gradient and for which there are target species that have varying salinity needs as evidenced by differences in their distributions from monitoring data and their inflow-

harvest/catch equations. But for some estuaries e.g., Laguna Madre, Nueces, and Mission-Aransas, with little normal inflow into them, the usefulness of the optimization method may be limited since salinity in much of the estuary is not sensitive to small inflows, and is more affected by storms, hurricanes, wind, and evaporation. Sabine-Neches Estuary may also be an anomaly because a substantial portion of the inflow bypasses Sabine Lake.

The optimization models currently depend on measures of time in months. The inflow-harvest/catch equations, inflow-salinity equations, upper and lower monthly inflow and bimonthly inflow bounds, and upper and lower salinity bounds are all based on 1- or 2-month periods. But episodic events occur that last mere days yet have strong influence on maintaining estuarine production. The best example is the need for inundation of wetlands in the upper reaches of the estuary near sources of freshwater inflow. For these areas to be the source of organic production, biogeochemical cycling, export of detritus, and protected habitat for larval and juvenile forms, they must occasionally receive flooding inflows. Some of the TDWR 1980s studies included a constraint to allow wetland inundation although it was a very indirect way to addressing the issue. They specified one or more monthly inflows for which there was a statistical chance, there would be a peak flow great enough to inundate the marshes. In nearly all cases, however, this resulted in a monthly inflow estimate that was greater than the historical median inflow and, in some cases, greater than the historical mean inflow. The TxEMP model did not have a wetland inundation element to it, so some analysts used the results to determine an inflow regime that largely satisfied salinity requirements, but then recommended the MaxH/MaxC solution or a value between MinQ and MaxH/MaxC that would have the best chance of providing the wetland inundation needs. Both optimization methods need a way to better deal with short-term pulses to satisfy wetland inundation needs.

## 2007 Senate Bill 3 Studies

While freshwater inflow standards existed, they were not required statewide. A big change occurred with the passage of Senate Bill 3 (SB3) in 2007 by the 80th Texas Legislature (Texas State Legislature 2020; Delrosario-Gomaa 2022). For the first time, state agencies were given the authority to make available water "set-asides" for beneficial inflows to affected bays and estuaries, and a new adaptive management system was implemented. The bills required the Texas Commission on Environmental Quality (TCEQ) to adopt rules related to environmental flows. The Texas Water Code states that an environmental flow is an amount of water that should remain in a stream or river for the benefit of the environment of the river, bay, and estuary, while balancing human needs. SB3 directed the use of an environmental flow regime in developing flow standards and defined an environmental flow regime as a schedule of flow quantities that reflects seasonal and yearly fluctuations that typically would vary geographically, by specific location in a watershed, and that are shown to be adequate to support a sound ecological environment and to maintain the productivity, extent, and persistence of key aquatic habitats (SB3 2007).

**Box 3 Stakeholder Involvement**
I think the big breakthrough was finding a way to create a process that was inclusive.

—Andrew Sansom, former Executive Director of the Meadows Center for Water and the Environment

The SB3 process to create inflow regimes was complex. The process was overseen by Environmental Flows Advisory Group (EFAG), composed of policy makers: three Texas House members, three Texas Senate members, and heads of three natural resource agencies. A Science Advisory Committee (SAC) was created to advise the EFAG and recommend technical guidance to provide consistent statewide approach to create inflow regimes. Technical studies were performed for each of seven basin/bay ecosystems. For each system, the EFAG appointed a Basin and Bay Area Stakeholder Committee (BBASC), which in turn appointed a Basin and Bay Expert Science Team (BBEST) (Box 3). Each committee devised their own strategy for determining the dynamics of recommended inflow regimes. The BBEST used SAC guidance (SAC 2009a) and available data to develop a recommended environmental flow regime adequate to support a sound ecological environment. Each BBASC worked on a consensus basis, considered their BBEST's recommended environmental flow regime, and developed their own set of recommendations about flow protection standards that included economic and human and other competing water needs. Environmental flow recommendation reports were by the BBASC and BBEST were submitted to the TCEQ for rulemaking and to adopt formal environmental flow standards (TLWP 2017).

Senate Bill 3 changed the operating statute for environmental flow studies, representing a shift away from those that came after the water legislation (HB2) of 1985. There were two changes in SB3 legislation that required a new scientific and technical approach: 1) changing the environmental flow need goal from a harvestable species management approach to an ecosystem-based management approach, and 2) changing from an annual system-wide inflow value to a spatially and temporally variable inflow regime (Table 15).

One important guidance provided by the SAC (2009a) was to interpret and define a "sound ecological environment." The SAC expanded on the statue language and clarified that a "sound" environment is one that:

- sustains the full complement of native species in perpetuity,
- sustains key habitat features required by these species,
- retains key features of the natural flow regime required by these species to complete their life cycles, and,
- sustains key ecosystem processes and services, such as elemental cycling and the productivity of important plant and animal populations.

The second important guidance by the SAC was to change the conceptual model and scientific approach from earlier studies. The TxEMP modeling approach in the State Methodology described in Longley (1994) included a direct connection between flow and fisheries harvest. The SAC proposed an indirect effect between flow and biological response (Fig. 2). This approach, later dubbed the domino theory (Montagna et al. 2013), posits that flow drives abiotic resource (i.e., physical, chemical, and geological) conditions in estuaries and these create estuary habitat, and biotic resources respond to estuary habitat conditions. However, no specific ecological model with these ecosystem functions was proposed. Instead, it was suggested to use salinity–bioindicator relationships.

One new hydrology tool was created as a joint effort between SAC and the TPWD. The Hydrology-Based Environmental Flow Regime (HEFR) Methodology was the basis for the SAC guidance (SAC 2009b) on how to perform statistical analyses of a hydrologic dataset to create an environmental flow regime matrix.

**Table 15** Definitions of environmental flow needs and management goals before and after 2007

| HB 2 – 1985 Texas Water Code 11.147(a) | SB 3 – 2007 Texas Water Code 11.002(16) |
|---|---|
| In this section, "beneficial inflows" means a salinity, nutrient, and sediment loading regime adequate to maintain an ecologically sound environment in the receiving bay and estuary system that is necessary for the maintenance of productivity of economically important and ecologically characteristic sport or commercial fish and shellfish species and estuarine life, upon which such fish and shellfish are dependent. | "Environmental flow regime" means a schedule of flow quantities that reflects seasonal and yearly fluctuations that typically would vary geographically, by specific location in a watershed, and that are shown to be adequate to support a sound ecological environment and to maintain the productivity, extent, and persistence of key aquatic habitats in and along the affected water bodies. |

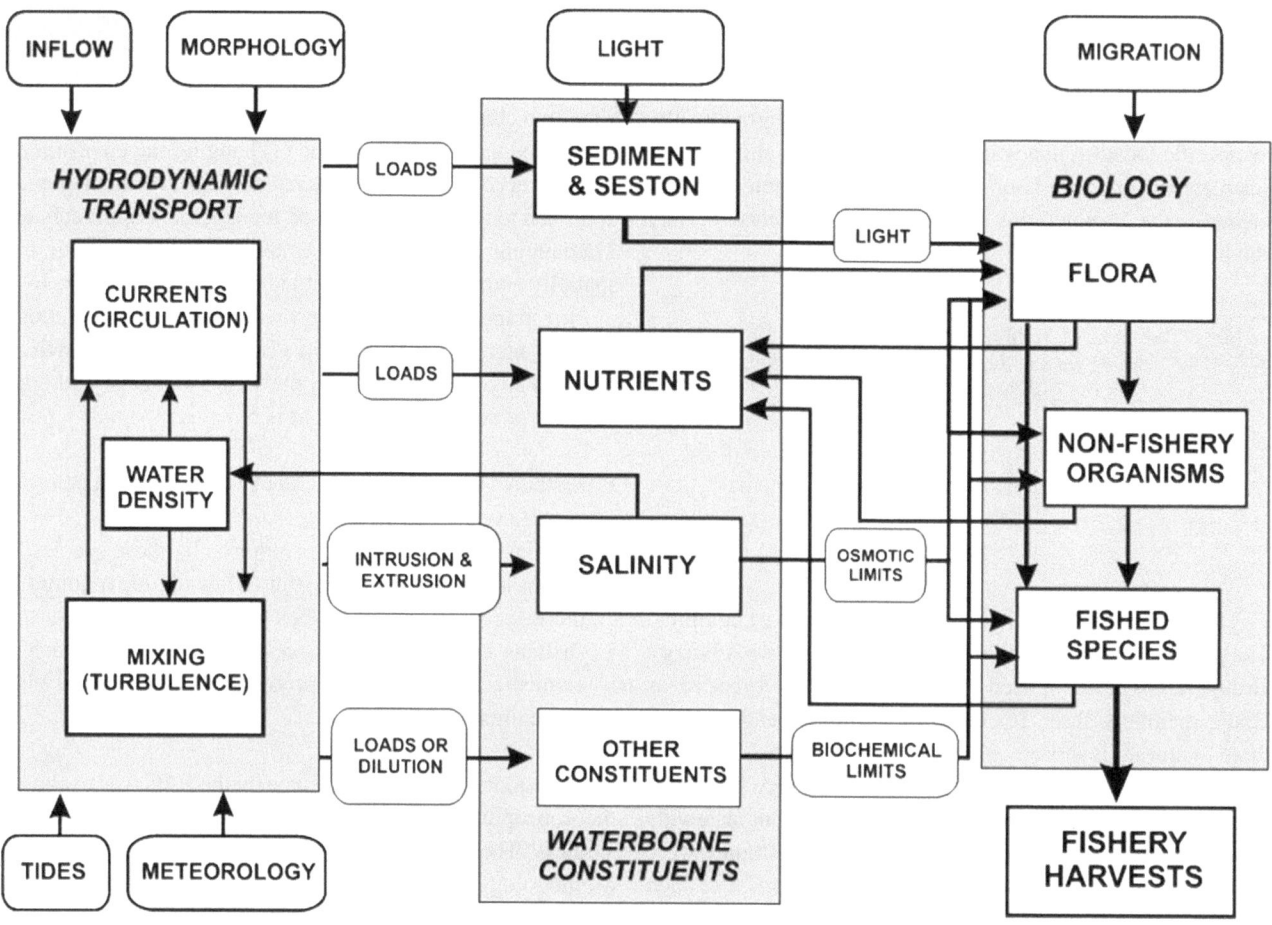

**Fig. 2** Estuary components and the role of freshwater inflow as a driver in controlling estuary conditions, and biological responses (SAC 2009a, Fig. 2.1-1)

Using SAC guidance documents available data, and best professional judgement, each BBEST and BBASC created inflow recommendations and TCEQ created inflow standards. The Basin and Bay Stakeholder Committees' reports, Expert Science Teams' reports, and the environmental flow standards for each of the seven basin/bays are listed in Table 16.

---

**Box 4 Oysters as Bioindicators**

Oysters always ended up being one of the species that were important in all the bays … we had fairly well documented ranges of salinity preference and dissolved oxygen preference, temperature preference, as well as thresholds, tolerances like if salinity gets above some level, the oysters don't thrive or if temperature gets below a level, they don't do well, and oysters were a favorite. We ended up kind of thinking of them as the canary in the coal mine for estuarine systems because they're not mobile.

Cindy Loeffler, retired Water Resources Branch Chief, TPWD

---

One of the first tasks each BBEST had to complete was to identify indicators or bioindicators of inflow effects. In nearly all cases, organisms were designated as bioindicators because they exhibit sensitivity or dependence on low salinity conditions. There was a great deal of similarity among the indicators chosen by the BBESTs (Table 17). A total of 23 indicators were identified by the BBESTs. The most common indicator that five of the seven BBESTs chose was the Eastern oyster (*Crassostrea virginica*), and four of the five chose the Atlantic Rangia clam (*Rangia cuneata*), which are both filter-feeding bivalves (Box 4). Four chose the blue crab (*Callinectes sapidus*), which is a shellfish. Abiotic indicators were chosen 12% of the time, fish were chosen 14.6% of the time, benthic plants were chosen 14.6% of the time, and benthic invertebrates were chosen 58.5% of the time. Benthic plants and animals were chosen 73.2% of the time because benthos are fixed in place and must tolerate overlying water conditions or perish. This conclusion is the same as reported in a recent review of the SB3 adaptive management studies, which recommended "focusing on non-motile benthos or nutrients, (or metrics) may provide better opportunities for

**Table 16** Summary of the Basin and Bay Stakeholder Committees' and Expert Science Teams' reports and the environmental flows rules for each of the seven basin/bays

| River Basin/Bay Area | Source | Document Title |
|---|---|---|
| Sabine and Neches Rivers, and Sabine Lake Bay | TCEQ 2012a | Environmental Flow Standards for Surface Water: Subchapter C |
| | Sabine BBEST 2009 | Environmental Flows Recommendation Report |
| | Sabine BBASC 2009 | Recommendation Report |
| Trinity and San Jacinto Rivers, and Galveston Bay | TCEQ 2011b | Environmental Flow Standards for Surface Water: Subchapter B |
| | Trinity BBEST 2009 | Environmental Flows Recommendation Report |
| | Trinity BBASC 2010a | Conditional Approach Report |
| | Trinity BBASC 2010b | Regime Approach Report |
| Colorado and Lavaca Rivers, and Matagorda and Lavaca Bays | TCEQ 2012a | Environmental Flow Standards for Surface Water: Subchapter D |
| | Colorado BBEST 2011 | Environmental Flows Regime Recommendation Report |
| | Colorado BBASC 2011 | Environmental Flows Recommendation Report |
| Guadalupe, San Antonio, Mission, and Aransas Rivers, and Mission, Copano, Aransas, and San Antonio Bays | TCEQ 2012c | Environmental Flow Standards for Surface Water: Subchapter E |
| | Guadalupe BBEST 2011 | Environmental Flows Recommendations Report |
| | Guadalupe BBASC 2011 | Recommendation Report |
| Nueces River and Corpus Christi and Baffin Bays | TCEQ 2014a | Environmental Flow Standards for Surface Water: Subchapter F |
| | Nueces BBEST 2011 | Environmental Flows Recommendations Report |
| | Nueces BBASC 2012 | Environmental Flow Standards and Strategies Recommendations Report |

(continued)

**Table 16** (continued)

| River Basin/Bay Area | Source | Document Title |
|---|---|---|
| Brazos River and its Associated Bay and Estuary System | TCEQ 2014b | Environmental Flow Standards for Surface Water: Subchapter G |
| | Brazos BBEST 2012 | Environmental Flow Regime Recommendations Report |
| | Brazos BBASC 2012 | Environmental Flow Standards and Strategies Recommendations Report |
| Rio Grande, Rio Grande Estuary, and Lower Laguna Madre | TCEQ 2014c | Environmental Flow Standards for Surface Water: Subchapter H |
| | Rio Grande BBEST 2012 | Environmental Flows Recommendations Report |
| | Rio Grande BBEST 2012 | Environmental Flows Recommendations Report |

**Table 17** Indicators of freshwater inflow effects chosen by the seven BBEST committees

| Bay System Report | Indicator Species |
|---|---|
| Sabine Lake BBEST (2009) | Eastern oyster, Atlantic *Rangia*, blue crab juveniles, Olney bulrush, Intermediate marsh, Brackish marsh |
| Galveston Bay BBEST (2009) | Eastern oyster, Atlantic Rangia, Dermo, Oyster drill, Wild celery, Gulf menhaden, blue catfish, Mantis shrimp, Pinfish |
| Brazos River BBEST (2012) | Salinity, nutrients, sediment supply |
| Lavaca and Matagorda Bays BBEST (2011) | Eastern oyster, Dermo, Oyster drill, brown shrimp, white shrimp, blue crab, Gulf menhaden and Atlantic croaker, Benthic infauna |
| Mission, Copano, Aransas, and San Antonio Bays BBEST (2011) | Eastern oyster, Atlantic Rangia, brown Rangia, white shrimp, blue crab |
| Nueces, Corpus Christi, and Baffin Bays BBEST (2011) | Eastern oyster, Atlantic Rangia, Smooth cordgrass, benthic infauna, blue crab, Atlantic croaker, nutrient cycling, sediment loading |
| Lower Laguna Madre BBEST (2012) | Seagrasses |

targeted monitoring of freshwater inflow responses" (Hardy et al. 2021).

> **Box 5 Adding Complexity to Flow Studies**
>
> One of the big benefits of the process, was in addition to collecting and assembling all this data, was to really move this state from a thinking about rivers that a single minimum flow is what you need to think about and thinking about how rivers are much more dynamic systems that you need to think about the high flows and the low flows and the seasonality and the interannual variability … I think Senate Bill 3 was a big part of moving beyond that.

—Joe Trungale, hydrologist, The Nature Conservancy

For the most part, the TCEQ adopted recommendations for standards that were made by the BBEST and BBASC groups. However, there were differences in the way the different groups made recommendations, and TCEQ did not standardize the language. This was primarily as it relates to defining seasonal periods and temporal aspects of inflow regimes (Box 5).

Seasons varied by basin and even within the basin for different measurement points (Table 18). The Colorado and Nueces basins defined seasons differently for different measurement points within their basins. All basins defined January and February as winter, and April and May as Spring. March was defined as Spring for Trinity, Colorado, Rio Grande, and Brazos; and defined as Winter for Sabine, Guadalupe, and Nueces. June was defined as Summer for Trinity, and Spring for all the rest of the basins. July and August were defined as Summer for all basins except for the Rio Grande, in which it was defined as Fall. September was defined as Summer for Sabine, Guadalupe, Nueces (some measurement points), and Brazos; and defined as Fall for Trinity, Colorado, Nueces (some measurement points), and Rio Grande. October was defined as Fall for all basins except Brazos, in which it was defined as Summer. November was defined as Fall for Trinity, Sabine, Colorado (some measurement points), Guadalupe, and Nueces (some measurement points); and Winter for Colorado (some measurement points), Nueces (some measurement points), Rio Grande, and Brazos. December was defined as Fall for Sabine and Guadalupe, and Winter for the rest of the basins. Rio Grande and Brazos, which were river basins, only defined three seasons; but Rio Grande defined July through October as Fall and Brazos defined it as Summer.

The temporal aspects of flow regimes were complex and composed of three dimensions: over components or climatic periods (such as wet and dry years), over seasons, and spatially within rivers, streams, and bay systems. The terminology for the component climatic periods for inflow regimes varied, and some were called "wet, average, dry, and subsistence," while others were labeled "levels," and some were labeled for season names (i.e., spring, summer, summer, and fall). The environmental flow standards were based on statistical evaluations of historical occurrence frequencies of different flow level hydrological categories, which also vary (Opdyke et al. 2014; Anchor QE 2021).

## Sabine and Neches Rivers, and Sabine Lake Bay

The environmental flow standards adopted by TCEQ constitute a schedule of flow quantities made up of subsistence flow, base flow, and one level of high flow pulses for ten measurement points (USGS gauged sites). There were no explicit environmental flow standards established for the estuary, but Sabine River near Ruliff is used for Sabine Lake (Table 19). Four seasons were established. The BBEST recommendation report established flow regimes from the HEFR output made up of subsistence flow, three levels of base flow (wet, average, dry), two level of high flow pulses, and overbank flows. TCEQ used the BBEST recommendations for subsistence flow for the measurement points. For base flows, TCEQ selected one value between the BBEST recommend dry and average. For the high flow pulses, TCEQ selected the lower level of the BBEST recommend flows reducing the frequency of 2 per season to 1 per season for winter and summer. The BBASC did not recommend any values of flow, but it was found by the BBASC that the HEFR analysis recommendations would substantially reduce water supply and that the flow standards would severely impact

**Table 18** Definitions of seasons for each basin as defined in the TCEQ Environmental Flow standards (TCEQ §298.). Color code: Winter = blue, Spring = green, Summer = red, and Fall = orange

| Basin | Month | | | | | | | | | | | |
|---|---|---|---|---|---|---|---|---|---|---|---|---|
| Sabine | Jan | Feb | Mar | Apr | May | Jun | Jul | Aug | Sep | Oct | Nov | Dec |
| Trinity | Jan | Feb | Mar | Apr | May | Jun | Jul | Aug | Sep | Oct | Nov | Dec |
| Colorado | Jan | Feb | Mar | Apr | May | Jun | Jul | Aug | Sep | Oct | Nov | Dec |
| | Jan | Feb | Mar | Apr | May | Jun | Jul | Aug | Sep | Oct | Nov | Dec |
| Guadalupe | Jan | Feb | Mar | Apr | May | Jun | Jul | Aug | Sep | Oct | Nov | Dec |
| Nueces | Jan | Feb | Mar | Apr | May | Jun | Jul | Aug | Sep | Oct | Nov | Dec |
| | Jan | Feb | Mar | Apr | May | Jun | Jul | Aug | Sep | Oct | Nov | Dec |
| Rio Grande | Jan | Feb | Mar | Apr | May | Jun | Jul | Aug | Sep | Oct | Nov | Dec |
| Brazos | Jan | Feb | Mar | Apr | May | Jun | Jul | Aug | Sep | Oct | Nov | Dec |

**Table 19** Environmental flow requirements used for Sabine Lake with BBEST recommendations and TCEQ standards (TCEQ 2011a; TCEQ 2012a §298.280; Sabine BBEST 2009). All flow values are for the USGS 08030500 Sabine River near Ruliff, Texas

| Regime | | Jan, Feb, Mar | | Apr, May, Jun | | July, Aug, Sep | | Oct, Nov, Dec | |
| --- | --- | --- | --- | --- | --- | --- | --- | --- | --- |
| | | Winter | | Spring | | Summer | | Fall | |
| | | BBEST | TCEQ | BBEST | TCEQ | BBEST | TCEQ | BBEST | TCEQ |
| Subsistence Flow (cfs) | | 949 | 949 | 436 | 436 | 396 | 396 | 396 | 396 |
| Base Flow (cfs) | Wet | 5063 | 1672 | 3035 | 1329 | 1430 | 737 | 1400 | 809 |
| | Average | 2565 | | 1795 | | 870 | | 970 | |
| | Dry | 1520 | | 1208 | | 670 | | 735 | |
| Large Seasonal Pulse | Trigger (cfs) | 1600 | 1600 | 3250 | 3250 | 3380 | 3380 | 2020 | 2020 |
| | Frequency (# per season) | 2 | 1 | 2 | 2 | 2 | 1 | 2 | 2 |
| | Volume (af) | 10,202 | 10,202 | 42,883 | 42,883 | 54,321 | 54,321 | 17,662 | 17,662 |
| | Duration (days) | 3 | 3 | 8 | 8 | 11 | 11 | 5 | 5 |

lake levels for reservoirs requiring new and amended permits (Sabine BBASC 2009). The BBASC recommended taking into consideration water supply, economic value of reservoir recreation, reservoir fishery resources, and other factors (such as the Sabine River Compact, etc.) before the environmental flow standards or set-asides are established (Sabine BBASC 2009).

## Trinity and San Jacinto Rivers, and Galveston Bay

The environmental flow standards adopted by TCEQ constitute a schedule of flow quantities made up of subsistence flow, base flow, and one level of high flow pulses for six measurement points (USGS gauged sites). There were explicit environmental flow standards established for the estuary with Galveston Bay for the Trinity River and the San Jacinto River recommended by the BBASC, which differs from the TCEQ standards (Table 20). The BBEST recommended inflows for the whole system and not for each of the two rivers. Four seasons were established. The BBEST recommendation report established flow regimes from the HEFR output including subsistence flow, three levels of base flow (low, medium, high), four levels of pulse flows, and overbank flows for the measurement points. The BBEST and BBASC both submitted recommendations. The TCEQ standards varied slightly from the BBEST and BBASC recommendations for subsistence flow. For base flows, the TCEQ selected one value between the BBEST and BBASC recommended low and medium values. For the high flow pulses, TCEQ selected the lower level of the BBASC recommended flows. The BBASC concluded that the BBEST (using HEFR output based on historical flows) did not identify flows adequate to maintain the health of the bay and river ecology (Trinity BBASC 2010a).

## Colorado and Lavaca Rivers, and Matagorda and Lavaca Bays

The environmental flow standards adopted by TCEQ constitute a schedule of flow quantities made up of subsistence flow, base flow, and two level of high flow pulses (small and large) for twenty-one measurement points (USGS gauged sites). There was an explicit environmental flow standards established for the estuary, Matagorda Bay and Lavaca Bay (Tables 21 and 22). Four seasons were established, varying by one month for some points. The BBEST recommendation report established flow regimes from the HEFR output made up of subsistence flow, three levels of base flow (high, medium, low), four levels of high flow pulses, overbank flows, channel maintenance flow, and long-term engagement frequencies for the measurement points. TCEQ subsistence flow varied and was greater than the BBEST recommended flows for some points, other points TCEQ agreed with the BBEST recommended flows. For base flows, TCEQ selected one value agreeing with the BBEST recommended low flow level for most points. For the high flow pulses, TCEQ agreed with the BBEST recommended flows for small seasonal pulse (2 per season), large seasonal pulse (1 per season), and an annual pulse (1 per year). The BBASC agreed with the lower flow recommendations of the BBEST with a few provisions for some points. BBASC used WAM to provide recommendations for the gauge sites and for the bays. TCEQ accepted the BBASC recommendations for the bays.

## Guadalupe, San Antonio, Mission, and Aransas Rivers, and Mission, Copano, Aransas, and San Antonio Bays

The environmental flow standards adopted by TCEQ constitute a schedule of flow quantities made up of subsistence flow, base flow, and two level of high flow pulses (small and

**Table 20** Galveston Bay environmental flow BBASC recommendations with TCEQ standards (TCEQ 2011a, b §298.225; Trinity BBASC 2010a)

| Basin | BBASC Winter Inflow Quantity (af) | BBASC Winter Target Frequency | TCEQ Winter Inflow Quantity (af) | TCEQ Winter Target Frequency | BBASC Spring Inflow Quantity (af) | BBASC Spring Target Frequency | TCEQ Spring Inflow Quantity (af) | TCEQ Spring Target Frequency | TCEQ Annual Inflow Quantity | TCEQ Annual Target Frequency |
|---|---|---|---|---|---|---|---|---|---|---|
| Trinity | 1,686,000 | 33% | 500,000 | 40% | 1,775,000 | 29% | 1,300,000 | 40% | 2,816,532 | 50% |
|  | 629,000 | 62% | 250,000 | 50% | 994,000 | 54% | 750,000 | 50% | 2,245,644 | 60% |
|  | 275,310 | 95% | 160,000 | 60% | 397,860 | 95% | 500,000 | 60% | 1,357,133 | 75% |
| San Jacinto | 896,000 | 33% | 450,000 | 40% | 534,000 | 20% | 500,000 | 40% | 1,460,424 | 50% |
|  | 349,000 | 56% | 278,000 | 50% | 289,000 | 38% | 290,000 | 50% | 1,164,408 | 60% |
|  | 1,228,302 | 95% | 123,000 | 60% | 155,390 | 95% | 155,000 | 60% | 703,699 | 75% |

| Basin | BBASC Summer Inflow Quantity (af) | BBASC Summer Target Frequency | TCEQ SummerInflow Quantity (af) | TCEQ Summer Target Frequency | BBASC Fall Inflow Quantity (af) | BBASC Fall Target Frequency | TCEQ Fall Inflow Quantity (af) | TCEQ Fall Target Frequency |
|---|---|---|---|---|---|---|---|---|
| Trinity | 713,000 | 46% | 245,000 | 40% | 919,000 | 29% | N/A | N/A |
|  | 509,000 | 50% | 180,000 | 50% | 362,000 | 46% | N/A | N/A |
|  | 211,820 | 95% | 75,000 | 60% | 110,700 | 95% | N/A | N/A |
| San Jacinto | 611,000 | 32% | 220,000 | 40% | 548,000 | 29% | 200,000 | 40% |
|  | 234,000 | 61% | 100,000 | 50% | 201,000 | 48% | 150,000 | 50% |
|  | 134,460 | 95% | 75,000 | 60% | 92,470 | 95% | 90,000 | 60% |

**Table 21** Environmental flow standards for Matagorda Bay (TCEQ 2012b §298.330; Colorado BBASC 2011)

| Inflow Regime | BBASC & TCEQ Monthly Minimum Quantity (af) | BBASC & TCEQ Spring Season Quantity (af) | BBASC & TCEQ Fall Season Quantity (af) | BBASC & TCEQ Intervening Season Quantity (af) | BBASC & TCEQ Long-Term Annual Strategy Quantity (af) | BBASC & TCEQ Annual Strategy Frequency |
|---|---|---|---|---|---|---|
| Monthly Threshold Inflow | 15,000 | N/A | N/A | N/A | N/A | 100% |
| Level 1 | N/A | 114,000 | 81,000 | 105,000 | N/A | 90% |
| Level 2 | N/A | 168,700 | 119,900 | 155,400 | N/A | 75% |
| Level 3 | N/A | 246,200 | 175,000 | 226,800 | N/A | 60% |
| Level 4 | N/A | 433,200 | 307,800 | 399,000 | N/A | 35% |
| Annual Average | N/A | N/A | N/A | N/A | 1400,000 | N/A |

**Table 22** Environmental flow standards for Lavaca Bay (TCEQ 2012b; TCEQ §298.330; Colorado BBASC 2011)

| Inflow Regime | BBASC & TCEQ Spring Inflow Quantity (af) | BBASC & TCEQ Fall Inflow Quantity (af) | BBASC & TCEQ Intervening Inflow Quantity (af) | BBASC & TCEQ Annual Strategy Frequency |
|---|---|---|---|---|
| Subsistence | 13,500 | 9600 | 6900 | 96% |
| Base Dry | 55,080 | 39,168 | 28,152 | 82% |
| Base Average | 127,980 | 91,080 | 65,412 | 46% |
| Base Wet | 223,650 | 158,976 | 114,264 | 28% |

large) for sixteen measurement points (USGS gauged sites). There was an explicit environmental flow standards established for the estuary, San Antonio Bay system and Aransas Bays, the BBEST and BBASC agreed on flow values and TCEQ accepted the recommendations for San Antonio Bay (Table 23). For Aransas Bay, TCEQ accepted the February flow and accept the quantity but not the timing of the other summer flow (Table 24). Four seasons were established, two seasons for the estuary. The BBEST and BBASC recommendation report established flow regimes from the HEFR output made up of subsistence flow, three levels of base flow (high, medium, low), three level of high flow pulses, and two

**Table 23** Environmental flow requirements for San Antonio Bay with BBEST recommendations and TCEQ standards (TCEQ 2012c §298.380; Guadalupe BBEST 2011)

| Inflow regime | BBEST & TCEQ inflow quantity (February) (af) | BBEST & TCEQ inflow quantity (March-May) (af) | Strategy target frequency |
|---|---|---|---|
| Spring 1 | N/A | 550,000-925,000 | At least 12% of the years |
| Spring 2 | N/A | 375,000-550,000 | At least 12% of the years |
| Spring 3 | N/A | 275,000-375,000 | N/A |
| Spring 4 | Greater than 75,000 | 150,000-275,000 | N/A |
| Spring 5 | Less than 75,000 | 150,000-275,000 | N/A |
| Spring 6 | N/A | 0-150,000 | No more than 9% of the years |
| Spring 2 and Spring 3 combined | N/A | N/A | At least 17% of the years |
| Spring 4 and Spring 5 combined | N/A | N/A | Less than 67% of the total |
| **Inflow regime** | **BBEST & TCEQ inflow quantity (June) (af)** | **BBEST & TCEQ inflow quantity (July–September) (af)** | **Strategy target frequency** |
| Summer 1 | N/A | 450,000-800,000 | At least 12% of the years |
| Summer 2 | N/A | 275,000-450,000 | at least 17% of the years |
| Summer 3 | N/A | 170,000-275,000 | N/A |
| Summer 4 | Greater than 40,000 | 75,000-170,000 | N/A |
| Summer 5 | Less than 40,000 | 75,000-170,000 | N/A |
| Summer 6 | N/A | 50,000-75,000 | N/A |
| Summer 7 | N/A | 0-50,000 | No more than 6% of the years |
| Summer 2 and Summer 3 combined | N/A | N/A | At least 30% of the years |
| Summer 4 and Summer 5 combined | N/A | N/A | Summer 5 no more than total 17% of the total |
| Summer 6 and Summer 7 combined | N/A | N/A | No more than 9% of the years |

levels of overbank flows for the measurement points. The BBASC submitted recommendations, subsistence flow varied and was greater than the BBEST recommended flows for some points, other points BBASC agreed with the BBEST recommended flows. TCEQ accepted the BBASC recommended flows. For base flows, TCEQ selected one value agreeing with the BBEST and BBASC recommended high flow level. For the high flow pulses, TCEQ selected the middle level of the BBEST recommended flows reducing the frequency of 2 per season to 1 per season as recommended by the BBASC.

## Nueces River and Corpus Christi and Baffin Bays

The environmental flow standards adopted by TCEQ constitute a schedule of flow quantities made up of subsistence flow, base flow, a high flow pulse, and two annual pulses for nineteen measurement points (USGS gauged sites). There was an explicit environmental flow standard established for the estuary, Nueces Bay and Nueces Delta, the BBEST and BBASC agreed and TCEQ accepted the recommendations (Table 25). Four seasons were established, varying by one month for some points. The BBEST recommendation report established flow regimes from the HEFR output made up of subsistence flow, base flow (one level), and varying levels of high flow pulses for the measurement points. The BBASC

submitted recommendations based on the HEFR output from historical flows made up of subsistence flow, three levels of base flow (high, medium, low), varying levels of high flow pulses, and overbank flows for the measurement points. The BBASC recommendations agree with that of the BBEST, with the medium level of base flows matching that of the BBEST. TCEQ subsistence flow agreed with the BBEST and BBASC recommended flows. For base flows, TCEQ agreed with the BBEST recommendation. For the high flow pulses, TCEQ accepted the BBEST and BBASC recommended flows for the lowest level for some points and the second lowest level for other points.

## Rio Grande, Rio Grande Estuary and Lower Laguna Madre

The BBEST produced two recommendation reports, one for the upper Rio Grande (Upper Rio Grande BBEST 2012) and one for the Lower Rio Grande and the estuary (Rio Grande BBEST 2012) (Table 26). The BBASC did not submit a recommendation report.

The BBEST determined that the Lower Laguna Madre (LLM) "has not developed with a substantial reliance on freshwater inflow to maintain a sound environment. The Lower Rio Grande BBEST determined that freshwater flows negatively impact the LLM under two scenarios: (a) under wet conditions, high freshwater pulses create low salinities that stress seagrass communities; and (b) under dry conditions, freshwater inflows, which now exceed 'natural' inflows are dominated by municipal and agricultural returns with resulting high nutrient loading that creates phytoplankton blooms, excessive growths of seagrass epiphytes and drifting macroalgae, all of which can reduce light availability to sea grass."

The BBEST did "not support development of environmental flow standards that would provide more water to the LLM." The TCEQ did not adopt an explicit environmental flow standards for the Rio Grande Estuary or the Lower Laguna Madre estuary. However, rules were established for the Rio Grande River (TCEQ 2014c §298.530).

Table 24 Environmental flow requirements for Aransas Bays with BBEST recommendations and TCEQ standards (TCEQ 2012c §298.380; Guadalupe BBEST 2011)

| Inflow Regime | BBEST & TCEQ Inflow Quantity (February) (af) | TCEQ Inflow Quantity (March–May) (af) | BBEST Inflow Quantity (June) (af) | BBEST Inflow Quantity (July–Sept) (af) | Strategy Target Frequency |
|---|---|---|---|---|---|
| Summer 1 | N/A | 500,000-1,000,000 | N/A | 500,000-1,000,000 | At least 2% of the years |

Table 25 Environmental flow requirements for Corpus Christi Bay with BBEST recommendations and TCEQ standards (TCEQ 2014a §298.430; Nueces BBEST 2011)

| Inflow Regime | BBEST & TCEQ Target Volume November–February (Target Frequency) | BBEST & TCEQ Target Volume March–June (Target Frequency) | BBEST & TCEQ Target Volume July–October (Target Frequency) | BBEST & TCEQ Target Volume Annual Inflow Target (Target Frequency) |
|---|---|---|---|---|
| Level 1 | 125,000 af(11%) | 250,000 af(11%) | 375,000 af(12%) | 750,000 af(16%) |
| Level 2 | 22,000 af(23%) | 88,000 af(30%) | 56,000 af(40%) | 166,000 af(47%) |
| Level 3 | 5000 af(69%) | 10,000 af(88%) | 15,000 af(74%) | 30,000 af(95%) |

**Table 26** Environmental flow recommendations for Lower Laguna Madre (Rio Grande BBEST 2012)

| Inflow Regime | Flow (ac-ft/month) | Duration (mo) | Not Exceed | Duration (d) |
|---|---|---|---|---|
| Dry Season (November–April) | 3613 - 12.901 | 3 | 12,901 | 45 |
| Wet Season (May–October) | 7888 - 38,152 | 3 | 38,152 | 45 |

## Brazos River and Its Associated Bay and Estuary System

There were only two river estuaries that were required to create environmental flow standards: the Rio Grande and Brazos River. Both rivers drain directly into the Gulf of Mexico and not into a secondary bay. The Brazos BBEST (2012) based their recommendation on the assumption that inflow into the estuary would equate to the BBEST environmental flow recommendations (EFR) for the Brazos River at Richmond. The Brazos BBASC (2012) pointed out that high-magnitude pulses are not specifically prescribed in the BBASC recommendation for the Richmond gauge, and recommended long-term monitoring of salinity, nutrient transport, and sediment transport and how it is related to estuarine health. TCEQ did not adopt explicit environmental flow standards established for the estuary portion of the lower river (TCEQ 2014b).

## Summary of SB3 Inflow Standards

The BBEST recommendation reports were presented to the BBASC to review. For the most part, the BBASCs agreed with the BBESTs inflow regime recommendations. The main concern raised by the BBASCs was that the BBESTs used the HEFR methodology to produce the recommendations. The BBASC was tasked with recommending flows that considered "human uses" and recommended lower flows in some cases, such as lower pulse flows, in fear of letting too much water through to the estuaries that might be needed for human use allocations. However, recommending the lower pulse flows and reducing the frequency of flows discounted the fact that HEFR accounted for the historical flow patterns to produce recommendations. The HEFR recommendations were to be representative of natural flow patterns and recommended pulse flows to mimic that of the natural historical flows from the inputted long-term dataset of the USGS stream gauges. The BBEST was tasked with creating recommendations for the ecosystem to preserve flows and tried to restore natural flow regimes and changing from those recommendations no longer maintains this aspect of the natural flow regimes.

For all the bays, except Galveston Bay and Aransas Bay, the TCEQ used the recommendations provided by the BBEST and BBASC committees. For Galveston Bay, TCEQ selected different values of flow and different target frequencies. For Aransas Bay BBEST and TCEQ agreed on February flows and the quantity of flow but differed on the timing of flows. It is unclear what the criteria used by TCEQ that lead to standards different from those in the recommendation reports. Standards were devised for the bays except for those where no standards were adopted such as Corpus Christi Bay, Baffin Bay, or Laguna Madre. This appears to be appropriate, because most of the time these bays are hypersaline environments, receiving little freshwater to create a transition zone as in the other bays.

One issue with the TCEQ standards arises in the inconsistency of language used. Terms varied by basin and in some cases within a basin at different USGS measurement points, such as seasons. Different terms are used to describe the inflow regimes themselves. The standards were defined in terms of inflow regimes and by season and then by other terms, with some consisting of target frequencies and some not, and some listing annual values and others not. This leads to confusion regarding the actual meaning of the standards. The legislation does not a set definition of a "*sound ecological environment*," which was the guiding principle of the flow regimes. This meant that each basin had to devise their own definition.

## Conclusion

While water quality laws and standards are common, the State of Texas is one of the few jurisdictions that also has water quantity laws and standards. There is a long history of concern over environmental flows to estuaries going back to the extended drought of the 1950s. There were three periods over which laws, rules, studies, and recommendations evolved. The initial studies in the 1960s summarized inflow by bay, statewide, and recommended single annual inflow numbers. During the middle period in the 1980s, detailed studies on salinity, nutrients, and sediments were performed and culminated in publication of the book "Freshwater Inflow to Texas Bays and Estuaries" (Longley 1994). The standards in that book were based on the relationship between fisheries harvest and inflow calculated using an optimization model and remained as one annual inflow volume per estuary. The 2010s saw a dramatic shift in the state approach where the primary management goal was to protect ecosystems by adopting standards for environmental flow regimes.

**Acknowledgements** This publication was funded in part through Contract No. 21-155-007-C879 from the Texas General Land Office (GLO) with Gulf of Mexico Energy Security Act of 2006 funding made available to the State of Texas and awarded under the Texas Coastal Management Program. The views contained herein are those of the authors and should not be interpreted as representing the views of the GLO or the State of Texas. Partial funding was also provided by the National Oceanic and Atmospheric Administration (NOAA), Office of Education, Educational Partnership Program (EPP) award (NA16SEC4810009). Its contents are solely the responsibility of the award recipient and do not necessarily represent the official views of the U.S. Department of Commerce, National Oceanic and Atmospheric Administration.

# References

Anchor QEA (2021) Evaluating the attainment of environmental flow standards. Final report to the Texas Water Development Board. Anchor QEA, LLC, Austin, TX, p 1239

Bao Y, Tung Y-K, Mays LW, Ward GH (1989) Analysis of the effect of freshwater inflows on estuary fishery resources. Report to Texas Water Development Board, by Center for Research in Water Resources, University of Texas, Austin, TX. Technical Memorandum 89-2. 49 p. http://www.twdb.texas.gov/publications/reports/contracted_reports/doc/9483707.pdf

Brazos BBASC [Brazos River and Associated Bay and Estuary System Basin and Bay Area Stakeholders Committee] (2012) Environmental flow standards and strategies recommendations report. https://www.tceq.texas.gov/assets/public/permitting/watersupply/water_rights/eflows/brazos_bbasc_report_8_22_2012_bbasc.pdf

Brazos BBEST [Brazos River Basin and Bay Expert Science Team] (2012) Environmental flow regime recommendations report. https://www.tceq.texas.gov/assets/public/permitting/watersupply/water_rights/eflows/brazos_bbest_complete_document.pdf

Chen GF (2010) Freshwater inflow recommendation for the Mission-Aransas estuarine system. Texas Parks and Wildlife Department, Ecosystem Resources Program, Coastal Fisheries Division, Austin, TX. 131 p. https://www.twdb.texas.gov/surfacewater/bays/major_estuaries/mission_aransas/index.asp

Childress RE, Hagen E, Williamson S (1975) The effects of freshwater inflows on hydrological and biological parameters in the San Antonio Bay system. Coastal Fisheries Branch, Texas Parks and Wildlife Department, Austin, TX. 190 p. http://hdl.handle.net/1969.3/26942

Colorado BBASC [Colorado and Lavaca Basin and Bay Area Stakeholder Committee] (2011) Environmental flows recommendation report. https://www.tceq.texas.gov/assets/public/permitting/watersupply/water_rights/eflows/collavbbascreport_82011.pdf

Colorado BBEST [Colorado and Lavaca Rivers and Matagorda and Lavaca Bays Basin and Bay Expert Science Team] (2011) Environmental flows regime recommendations report. https://www.tceq.texas.gov/assets/public/permitting/watersupply/water_rights/eflows/20110301clbbest_enviroflowreport.pdf

Copeland BJ (1966) Effects of decreased river flow on estuarine ecology. J Water Pollut Control Fed 38(11):1831–1839

Delrosario-Gomaa EA (2022) Texas 2007 Senate Bill 3 environmental flows process: a review and analysis for the priority coastal basins in Texas, USA. Dissertation, Texas A&M University-Corpus Christi

Elliott M, McLuskey DS (2002) The need for definitions in understanding estuaries. Est Coast Shelf Sci 55:815–827

Graves J (1971) Of plans and planning. In: Boyle RH, Graves J, Watkins TH (eds) The water hustlers. Sierra Club, San Francisco, CA, pp 38–62

Gray WG (1987) FLEET: Fast linear element explicit in time triangular finite element models for tidal circulation, User's Manual. University of Notre Dame, Notre Dame, Indiana

Guadalupe BBASC [Guadalupe, San Antonio, Mission, and Aransas Rivers and Mission, Copano, Aransas, and San Antonio Bays Basin and Bay Area Stakeholders Committee] (2011) Recommendations report. https://www.tceq.texas.gov/assets/public/permitting/watersupply/water_rights/eflows/20110901gsabbasc_report.pdf

Guadalupe BBEST [Guadalupe, San Antonio, Mission, and Aransas Rivers and Mission, Copano, Aransas, and San Antonio Bays Basin and Bay Expert Science Team] (2011) Environmental flows recommendations report. https://www.tceq.texas.gov/assets/public/permitting/watersupply/water_rights/eflows/20110301guadbbest_transmission.pdf

Gunter G, Hildebrand HH (1954) The relation of rainfall of the state and catch of the marine shrimp (*Penaeus setiferus*) in Texas Waters. Bull Mar Sci Gulf Carib 4:95–103

Hardy T, Winemiller K, Buskey E, Guillen G, Trungale J, Opdyke D, Annear T, Locke A, Estes C (2021) Statewide synthesis of environmental flow studies (2014-2017). Final report for TWDB Contract #1900012284

Hildebrand HH, Gunter G (1953) Correlation of rainfall with Texas catch of white shrimp, Penaeus setiferus (Linnaeus). Trans Am Fish Soc 82:151–155

Kuhn NL, Chen G (2005) Freshwater inflow recommendation for the Sabine Lake Estuary of Texas and Louisiana. Texas Parks and Wildlife Department, Coastal Fisheries Division, Coastal Studies Program, Austin, TX. 75 p. https://www.twdb.texas.gov/surfacewater/bays/major_estuaries/sabine_neches/index.asp

Lasdon LS (1980) GRG2 User's Guide. Graduate School of Business, University of Texas at Austin, Austin, TX

Lee W, Buzan D, Eldridge P, Pulich W (2001) Freshwater inflow recommendation for the Trinity-San Jacinto Estuary of Texas. Texas Parks and Wildlife Department, Coastal Studies Program, Resources Protection Division, Austin, TX. 66 p. https://www.twdb.texas.gov/surfacewater/bays/major_estuaries/trinity_san_jacinto/index.asp

Lockwood, Andrews & Newman, Inc (1967) A new concept-water for preservation of bays and estuaries. Texas Water Development Board Report 43. 39 p

Longley WL (ed) (1994) Freshwater inflows to Texas bays and estuaries: ecological relationships and methods for determination of needs. Texas Water Development Board and Texas Parks and Wildlife Department, Austin, TX. 386 p. https://repositories.lib.utexas.edu/bitstream/handle/2152/6728/FreshwaterInflows-+Ecological+Relationships+and+Methods+for+Determination+of+Needs+-+1994.pdf?sequence=2

Martin Q (1987) Estimating freshwater inflow needs for Texas estuaries by mathematical programming. Water Resource Res 23:230–238

Martin Q, Mosier D, Patek J, Gorham-Test C (1997) Freshwater needs of the Matagorda Bay system. Lower Colorado River Authority. 227 p. https://www.twdb.texas.gov/surfacewater/flows/freshwater/

Masch FD & Associates (1971) Tidal hydrodynamic and salinity models for San Antonio and Matagorda bays, Texas. Report to the Texas Water Development Board. 130 p

Matsumoto J, Powell G, Brock D (1994) Freshwater-inflow need of estuary computed by Texas estuarine MP model. J Water Resour Plan Manag 120(5):693–714

McLuskey DS, Elliott M (2007) Transitional waters: A new approach, semantics or just muddying the waters? Estuar Coast Shel Sci 71:359–363

Montagna PA, Palmer TA, Beseres Pollack J (2013) Hydrological changes and estuarine dynamics. SpringerBriefs in Environmental Sciences, New York, p 94

Nueces BBASC [Nueces River and Corpus Christi and Baffin Bay Basin and Bay Area Stakeholders Committee] (2012) Environmental flow standards and strategies recommendations report. https://www.tceq.texas.gov/assets/public/permitting/watersupply/water_rights/eflows/nuecesbbasc_recommendationsreport.pdf

Nueces BBEST [Nueces River and Corpus Christi and Baffin Bays Basin and Bay Expert Science Team] (2011) Environmental flows recommendations report. https://www.tceq.texas.gov/assets/public/permitting/watersupply/water_rights/eflows/20111028nuecesbbest_recommendations.pdf

Opdyke DR, Oborny EL, Vaugh SK, Mayes KB (2014) Texas environmental flow standards and the hydrology-based environmental flow regime methodology. Hydrol Sci J 59:820–830

Orlando SP Jr, Rozas LP, Ward GH, Klein CJ (1993) Salinity characteristics of Gulf of Mexico estuaries. National Oceanic and Atmospheric Administration, Office of Ocean Resources Conservation and Assessment, Silver Spring, MD, 209 p

Perillo GME (1995) Definitions and geomorphologic classifications of estuaries. In: Perillo GME (ed) Geomorphology and sedimentology of estuaries, Developments in sedimentology, vol 53, pp 17–47

Powell GL, Matsumoto J, Brock DA (2002) Methods for determining minimum freshwater inflow needs of Texas bays and estuaries. Estuaries 25(6B):1262–1274

Pritchard DW (1967) What is an estuary: physical viewpoint. In: Lauff GH (ed) Estuaries. American Association for the Advancement of Science, Washington, DC, pp 52–63. http://www.ecologyandsociety.org/vol12/iss1/art12/

Pulich W Jr, Lee WY, Loeffler C, Eldridge P, Hinson J, Minto M, German D (1998) Freshwater inflow recommendation for the Guadalupe Estuary of Texas. Texas Parks and Wildlife Department, Coastal Studies Technical Report No. 98-1. 102 p. https://www.twdb.texas.gov/surfacewater/bays/major_estuaries/guadalupe/index.asp

Pulich W Jr, Tolan J, Lee WY, Alvis W (2002) Freshwater inflow recommendation for the Nueces Estuary. Texas Parks and Wildlife Department, Resource Protection Division, Coastal Studies Program, Austin, TX. 99 p. https://www.twdb.texas.gov/surfacewater/bays/major_estuaries/nueces/index.asp

Rio Grande BBEST [Rio Grande, Rio Grande Estuary, and Lower Laguna Madre Basin and Bay Expert Science Team] (2012) Environmental flows recommendations report. https://www.tceq.texas.gov/assets/public/permitting/watersupply/water_rights/eflows/lowerrgbbest_finalreport.pdf

Sabine BBASC [Sabine and Neches Rivers and Sabine Lake Bay Basin and Bay Area Stakeholder Committee] (2009) Recommendation report. https://www.tceq.texas.gov/assets/public/permitting/watersupply/water_rights/eflows/2010snbbasc_finalrecommendations.pdf

Sabine BBEST [Sabine and Neches Rivers and Sabine Lake Bay Basin and Bay Expert Science Team] (2009) Environmental flows recommendations report. https://www.tceq.texas.gov/assets/public/permitting/watersupply/water_rights/eflows/sn_bbest_recommendationsreport.pdf

SAC [Science Advisory Committee] (2009a) Methodologies for establishing a freshwater inflow regime for Texas estuaries within the context of the Senate Bill 3 Environmental Flows Process. Report # SAC-2009-03-Rev1. Available at http://www.tceq.state.tx.us/assets/public/permitting/watersupply/water_rights/eflows/fwi20090605.pdf

SAC [Science Advisory Committee] (2009b) Use of hydrologic data in the development of instream flow recommendations for the environmental flows allocation process and the Hydrology-Based Environmental Flow Regime (HEFR) Methodology. Report # SAC-2009-01-Rev1

Tagliapietra D, Sigovini M, Ghirardini AV (2009) A review of terms and definitions to categorize estuaries, lagoons and associated environments. Mar Freshw Res 60:497–509

TCEQ [Texas Commission on Environmental Quality] (2011a) Chapter 298 – Environmental flow standards for surface water. Subchapter A: general provisions. https://www.tceq.texas.gov/assets/public/legal/rules/rules/pdflib/298a.pdf

TCEQ [Texas Commission on Environmental Quality] (2011b) Chapter 298 – Environmental flow standards for surface water. Subchapter B: Trinity and San Jacinto Rivers, and Galveston Bay. https://www.tceq.texas.gov/assets/public/legal/rules/rules/pdflib/298b.pdf

TCEQ [Texas Commission on Environmental Quality] (2012a) Chapter 298 – Environmental flow standards for surface water. Subchapter C: Sabine and Neches Rivers, and Sabine Lake Bay. https://www.tceq.texas.gov/assets/public/legal/rules/rules/pdflib/298c.pdf

TCEQ [Texas Commission on Environmental Quality] (2012b) Chapter 298 – Environmental flow standards for surface water. Subchapter D: Colorado and Lavaca Rivers, and Matagorda and Lavaca Bays. https://www.tceq.texas.gov/assets/public/legal/rules/rules/pdflib/298d.pdf

TCEQ [Texas Commission on Environmental Quality] (2012c) Chapter 298 – Environmental flow standards for surface water. Subchapter E: Guadalupe, San Antonio, Mission, and Aransas Rivers, and Mission, Copano, Aransas, and San Antonio Bays. https://www.tceq.texas.gov/assets/public/legal/rules/rules/pdflib/298e.pdf

TCEQ [Texas Commission on Environmental Quality] (2014a) Chapter 298 – Environmental flow standards for surface water. Subchapter F: Nueces River and Corpus Christi and Baffin Bays. https://www.tceq.texas.gov/assets/public/legal/rules/rules/pdflib/298f.pdf

TCEQ [Texas Commission on Environmental Quality] (2014b) Chapter 298 – Environmental flow standards for surface water. Subchapter G: Brazos River and Its Associated Bay and Estuary System. https://www.tceq.texas.gov/assets/public/legal/rules/rules/pdflib/298g.pdf

TCEQ [Texas Commission on Environmental Quality]. (2014c) Chapter 298 – Environmental flow standards for surface water. Subchapter H: Rio Grande, Rio Grande Estuary, and Lower Laguna Madre. https://www.tceq.texas.gov/assets/public/legal/rules/rules/pdflib/298h.pdf

TCEQ [Texas Commission on Environmental Quality] (2022) History of the TCEQ and its predecessor agencies. Date accessed 3NOV2022 https://www.tceq.texas.gov/agency/organization/tceqhistory.html

TDWR [Texas Department of Water Resources] (1980a) Lavaca-Tres Palacios Estuary: A study of the influence of freshwater inflows. LP-106. 348 p. https://www.twdb.texas.gov/surfacewater/flows/freshwater/

TDWR [Texas Department of Water Resources] (1980b) Guadalupe Estuary: A study of the influence of freshwater inflows. LP-107. 337 p. https://www.twdb.texas.gov/surfacewater/flows/freshwater/

TDWR [Texas Department of Water Resources] (1981a) Nueces and Mission-Aransas Estuaries: A study of the influence of freshwater inflows. LP-108. 378 p. https://www.twdb.texas.gov/surfacewater/flows/freshwater/

TDWR [Texas Department of Water Resources] (1981b) Trinity-San Jacinto Estuary: A study of the influence of freshwater inflows. LP-113. 411 p. https://www.twdb.texas.gov/surfacewater/flows/freshwater/

TDWR [Texas Department of Water Resources] (1981c) Sabine-Neches Estuary: A study of the influence of freshwater inflows. LP-116. 322 p. https://www.twdb.texas.gov/surfacewater/flows/freshwater/

TDWR [Texas Department of Water Resources] (1982) The influence of freshwater inflows upon the major bays and estuaries of the Texas Gulf coast. LP-115 Second Edition. 53 p + 6 appendices. https://www.twdb.texas.gov/surfacewater/flows/freshwater/

TDWR [Texas Department of Water Resources] (1983) Laguna Madre Estuary: A study of the influence of freshwater inflows. LP-182. 284 p. https://www.twdb.texas.gov/surfacewater/flows/freshwater/

Texas State Legislature. Water Code, Chapter 11. Date accessed 10 Apr 2020. https://statutes.capitol.texas.gov/Docs/WA/htm/WA.11.htm

Texas Water Development Board (1968) The Texas Water Plan. 249 p

TLWP [Texas Living Waters Project] (2017) The SB3 Environmental Flows Process. Texas Living Waters Project, Austin, Texas. https://texaslivingwaters.org/environmental-flows/sb3-environmental-flows-process/. Date accessed 6/21/2022.

Tolan JM, Lee WY, Chen G, Buzan D (2004) Freshwater inflow recommendation for the Laguna Madre Estuary. Texas Parks and Wildlife Department, Coastal Fisheries Division, Coastal Studies Program, Austin, TX. 115 p. https://www.twdb.texas.gov/surfacewater/flows/freshwater/

Tracor Inc. (1974) Galveston Bay Project Hydraulic Model User's Manual. by Hembree LA et al. Submitted to the Texas Water Quality Board. 101 p.

Trinity BBASC [Trinity and San Jacinto Rivers and Trinity Bay and Basin Stakeholders Committee] (2010a) Conditional approach report. https://www.tceq.texas.gov/assets/public/permitting/watersupply/water_rights/eflows/tsjbbasc2finalreport_conditional.pdf

Trinity BBASC [Trinity and San Jacinto Rivers and Trinity Bay and Basin Stakeholders Committee] (2010b) Regime approach report. https://www.tceq.texas.gov/assets/public/permitting/watersupply/water_rights/eflows/tsjbbasc1finalreport_regime.pdf

Trinity BBEST [Trinity and San Jacinto Rivers and Galveston Bay Basin and Bay Expert Science Team] (2009) Environmental flows recommendation report. https://www.tceq.texas.gov/assets/public/permitting/watersupply/water_rights/eflows/trinity_sanjacinto_bbestrecommendationsreport.pdf

Upper Rio Grande BBEST [Upper Rio Grande Basin and Bay Expert Science Team] (2012) Environmental flows recommendations report. https://www.tceq.texas.gov/assets/public/permitting/water-supply/water_rights/eflows/urgbbest_finalreport.pdf

Water Resources Engineers (1975) Computer program and documentation for the Dynamic Estuary Model with application to Sabine Lake estuarine system. Final Report to Texas Water Development Board.

Wood PJ, Hannah DM, Sadler JP (eds) (2007) Hydroecology and ecohydrology. Wiley

# Historical Perspective and Context of Freshwater Inflow Policy and Law in Texas

Myron J. Hess

## Abstract

This chapter provides an overview of water rights management in Texas, emphasizing policy and legal aspects of environmental flow protection, particularly freshwater inflows. It seeks to summarize the regulatory setting in which freshwater inflow science has been, and is being, applied. Rights to use state-owned surface water have been recognized since the 1800s, with regular consideration of environmental flow beginning only in 1985. Because older rights have the first claim to water, that legacy of older, perpetual rights lacking flow protections creates challenges for maintaining adequate flows. In 2007, the Texas Legislature enacted Senate Bill 3 seeking a comprehensive approach for flow protection. That approach, which remains only partially implemented, includes setting aside for flow protection reasonable amounts of water not already authorized for use, adopting flow standards with flow-protection requirements for new permits and targets for proactive protection efforts, and pursuing proactive strategies for voluntary efforts to move some existing rights to flow protection. It also provides for adaptive management to continue refining those elements. Following an initial flurry of activity through 2014, implementation has largely stalled. Recent developments may lead to renewed activity and a chance to address important shortcomings in flow protection.

## Keywords

Environmental standards · Surface water regulation · Texas 2007 Senate Bill 3 · Texas water code · Water law

## Abbreviations

| | |
|---|---|
| af | acre-feet |
| BBASC | Bay and Basin Area Stakeholder Committee |
| BBEST | Bay and Basin Expert Science Team |
| Brazos BBASC | Brazos River Basin and Associated Bay and Estuary System BBASC |
| Brazos BBEST | Brazos River Basin and Associated Bay and Estuary System BBEST |
| CL BBASC | Colorado and Lavaca Rivers and Matagorda and Lavaca Bays BBASC |
| CL BBEST | Colorado and Lavaca Rivers and Matagorda and Lavaca Bays BBEST |
| cfs | cubic feet per second |
| cms | cubic meters per second |
| COA | Certificate of Adjudication |
| DFC | Desired future condition |
| EFAG | Environmental Flows Advisory Group |
| GMA | Groundwater management area |
| GSA BBASC | Guadalupe, San Antonio, Mission, and Aransas Rivers and Mission, Copano, Aransas, and San Antonio Bays BBASC |
| GSA BBEST | Guadalupe, San Antonio, Mission, and Aransas Rivers and Mission, Copano, Aransas, and San Antonio Bays BBEST |
| HB | House Bill |
| LNRA | Lavaca-Navidad River Authority |
| LCRA | Lower Colorado River Authority |
| $m^3$ | Cubic meters |
| MBHE | Matagorda Bay Health Evaluation |
| NEAC | Nueces Estuary Advisory Council |
| Nueces BBASC | Nueces River and Corpus Christi and Baffin Bays BBASC |
| Nueces BBEST | Nueces River and Corpus Christi and Baffin Bays BBEST |
| SB 3 | Senate Bill 3 (Article 1) |

M. J. Hess (✉)
Law Office of Myron Hess PLLC, Tributary Consulting LLC, Austin, TX, USA
e-mail: myron@myronhess.com

© The Author(s) 2025
P. A. Montagna, A. R. Douglas (eds.), *Freshwater Inflows to Texas Bays and Estuaries*, Estuaries of the World, https://doi.org/10.1007/978-3-031-70882-4_2

| | |
|---|---|
| SN BBASC | Sabine and Neches Rivers and Sabine Lake Bay BBASC |
| SN BBEST | Sabine and Neches Rivers and Sabine Lake Bay BBEST |
| TAC | Texas Administrative Code |
| TCEQ | Texas Commission on Environmental Quality |
| SAC | Texas Environmental Flows Science Advisory Committee |
| TPWD | Texas Parks and Wildlife Department |
| TWC | Texas Water Code |
| TWDB | Texas Water Development Board |
| TSJ BBASC | Trinity and San Jacinto Rivers and Galveston Bay BBASC |
| WAM | Water Availability Model |
| WMP | Water Management Plan |
| yr | Year |

## Introduction

This chapter provides a broad overview of water rights management in Texas with an emphasis on policy and legal issues related to environmental flow protection and particularly to protection of freshwater inflows. The aim is to provide a basic understanding of the extent to which environmental-flow protection has been addressed previously in Texas, which sets the baseline for current efforts, along with an overview of the current regulatory approach and the status of its implementation. To help inform that understanding, the chapter begins with a broad overview of water quantity management in Texas. Rights to use state-owned surface water have been recognized in Texas since the 1800s, but environmental flow needs only began to be considered during the permitting process on a regular basis in 1985. Because older rights, which generally are perpetual, have the first claim to the available water during times of shortage, that legacy of older rights without flow-protection requirements creates a challenging setting for achieving effective flow protection. In 2007, the Texas Legislature enacted legislation, referred to as Senate Bill 3 (SB 3), intended to establish a comprehensive system for flow protection, while also allowing for regional variation in aspects of the implementation approaches taken.

SB 3 does not provide for reopening water rights recognized prior to September 1, 2007—the effective date of the legislation—to impose new flow-protection requirements. Instead, it relies on establishing set-asides of unappropriated flow, protections applied to new permits consistent with newly adopted environmental flow standards, and voluntary approaches for implementing proactive flow-protection strategies. The SB 3 process includes a critically important adaptive management component that is multi-faceted. That component contemplates ongoing monitoring and study efforts to refine knowledge of flow needs and to identify options for implementing proactive strategies to address impaired flow levels. Adaptive management also provides for periodic consideration, generally every 10 years, of revisions to adopted flow standards and of adjustments to flow protection provisions included in permits issued after September 1, 2007. Those adjustments allow for increasing, within prescribed limits, the specific protection levels included in the permit when initially issued if increases are appropriate to help achieve compliance with flow standards that are revised to be more protective.

However, implementation of SB 3 is incomplete. Some basins still lack flow standards. No unappropriated flow has been set aside for flow protection. No flow standards—the first of which were adopted in 2011—have been considered for revision, even though protection levels, especially for freshwater inflows, are far below the levels identified by scientists as adequate to protect a sound ecological environment. The adopted flow standards generally address both protection of instream flows—flow levels in rivers and streams at inland locations—and of freshwater inflows—flows of fresh water into coastal bays and estuaries. However, for some bays and estuaries, there are no separate freshwater inflow criteria. The components of the flow standards that do establish separate protection levels for freshwater inflows rely solely on a long-term modeling exercise, which assumes a recurrence of historical climate conditions, rather than on inflow protection conditions included in individual permits.

The absence of specific inflow-protection language in new permits impairs implementation of the authority to adjust such protections to increase flow protections through adaptive management. A legislatively mandated state-level review of approaches for facilitating voluntary conversions of reasonable levels of existing water rights to flow protection—a key component of proactive strategies—has not taken place. The challenge climate change poses for inflow protection is heightened because of the lack of environmental flow set-asides and of quantified inflow protection conditions in new permits, resulting in the risk of climate change impacts falling inordinately on bays and estuaries.

Fortunately, there are signs of some progress. Efforts by conservation groups to implement various types of proactive flow-protection strategies, although mostly on a small scale as a learning exercise, are underway. Ongoing state funding, although very limited, has allowed for continued studies that can help inform potential revisions to the adopted environmental flow standards and identification of potential proactive strategies. As reflected in the other chapters in this volume, important advances continue in understanding freshwater inflow needs. Finally, a 2022 review process for the Texas Commission on Environmental Quality (TCEQ)—the state agency charged with key aspects of SB 3 implementation—has resulted in recommendations to the Texas Legislature that could help invigorate the adaptive management component and overall implementation.

## Overview of Texas Water Law

Although the reality of the connection between surface water and groundwater is broadly understood, surface water and groundwater are managed separately in Texas. Certainly, progress has been made in the regulatory structure to begin acknowledging that those interconnections exist. However, limited ability to model and quantify the interactions, along with challenging legal precedents and regulatory structures, has constrained the realization of a cohesive management approach. The discussion in this chapter addresses water quantity regulation—regulation of rights to pump groundwater and to divert and impound surface water—rather than on laws or processes directly addressing water quality issues.

## Surface Water versus Groundwater

As a general proposition, water that is found in any type of surface watercourse is defined as state water. State water is recognized and managed as a publicly owned resource (TWC § 11.021). By contrast, percolating groundwater—water found below the surface of the ground that is not the underflow of a surface stream or an underground river—is recognized as being privately owned and managed separately from surface water (TWC § 36.001 (5), Eckstein and Hardberger 2020 § 1.10–1.11, Canseco 2020 § 5.9–5-11).

## Groundwater Management

There are varying degrees of regulation of groundwater pumping in Texas, ranging from governance solely under the common-law rule of capture, allowing virtually unlimited pumping for any use (and sometimes characterized as a rule that the biggest pump gets the water), to governance by a groundwater conservation district (GCD) that has authority to set pumping limits (Canseco 2020 § 5.75.16). The extent of authority and ability of GCDs to enforce pumping limits remains unclear, in significant part because of court rulings establishing groundwater as a property right that is privately owned in-place. Those rulings result in significant uncertainty about the extent to which regulatory bodies limiting pumping may be held liable for a taking of private property (Canseco 2020 § 5.11–5.13). Thus, the extent of regulation of pumping varies by location based on whether a specific area is located within a groundwater conservation district and, if so, on the individual district's regulatory approach

operating within the constraints of state law that is continuing to evolve.

Because the boundaries of GCDs usually are not co-extensive with the boundaries of aquifers[1] and because of interconnections between groundwater resources, Texas law provides for a joint planning approach for groundwater management. For that joint planning process, groundwater management areas (GMAs) are designated by the Texas Water Development Board (TWDB), a state agency. The GCDs within a particular GMA work together to identify quantified desired future conditions (DFCs). The DFCs identify target water levels in the aquifers within the GMA for future time periods.[2] Environmental impacts, including impacts on spring flow and on surface water, are among the factors that are recognized for consideration in establishing DFCs (TWC § 36.108 (d)(4), Norman and Hutchinson 2020 § 21.10–21.11). Although variable across the state, groundwater has been estimated to have provided, on average, about 30% of surface water baseflows[3] under historical conditions prior to the development of large surface water impoundments (Brun et al. 2016, p. 30, 34 Table 3–3). However, for various reasons, including the limited capacity of groundwater availability models to predict the impacts of pumping on springs and on surface water,[4] specific limits on pumping to protect spring flow and surface water flow remain relatively rare (Sugg et al. 2020 § 6.6). The DFC serves to guide the management of groundwater resources and the development, by the individual GCDs, of management plans that govern permitting (Norman and Hutchinson 2020 § 21.10–21.11).

### Submarine Flow

Traditionally, when assessing needs for freshwater inflows to Texas estuaries, the assumption has been that all inflows arrive as surface water flow and direct rainfall. However, as discussed in chapter "Groundwater-Surface Water Interactions in the Coastal Zone", there are also contributions of fresh, or brackish, groundwater that emerge directly into coastal bays and estuaries. As noted there, the extent of those direct submarine contributions along the Texas coast is mostly unquantified. However, increasing knowledge of

---

[1] Generally, the term aquifer refers to a geologic formation that has the capacity to hold a significant amount of water and to allow the recovery of that water in useful amounts through wells or springs (Eckstein and Hardberger 2020 § 1.11:1).

[2] The term "desired future condition" is defined in Section 36.001 (30) of the Texas Water Code (TWC § 36.001 (30)). General information about the GMA process is available on the TWDB website (TWDB GMA).

[3] Surface water baseflow, in the referenced study, refers to sustained flow levels not associated with specific, direct rainwater runoff (Brun et al. 2016, p. 29).

[4] The limitation comes primarily from issues such as the scale of the models and inadequate calibration of groundwater and surface–water interactions (Brun et al. 2016, p. 29).

such contributions illustrates the complexity of these natural systems and of the importance of retaining flexibility to adapt management as our understanding grows. Measured bay salinities used in developing inflow–salinity relationships, as discussed in chapter "Hydrology, Circulation, and Salinity", have played a major role in many of the efforts to develop freshwater inflow recommendations and resulting environmental flow standards (SAC Inflow Regime 2009, p. 72). Those measured salinities reflect, among other things, the impacts of both the surface water inflows, which generally are quantified, and submarine inflows, which generally are unquantified. If the amount of submarine flows changes significantly from historical levels, the amount of surface water inflows that would be required to achieve a particular salinity level also would change.

## Surface Water Management

Surface water in Texas is managed primarily through a prior-appropriation system, much like other western states. Although water flowing in water courses is recognized as state water, which is a publicly owned resource, water rights,[5] which generally are perpetual in duration, authorize the holder of the right to use and consume an identified quantity of surface water. Appropriation refers to the processes through which a right to use state water is acquired.[6] Some water rights authorize only non-consumptive uses, such as hydroelectric power generation. Other than for limited exempt uses, anyone wishing to divert or impound surface water must obtain a water right authorizing that diversion or impoundment for a defined use or combination of uses.[7] Water rights administration, including enforcement and amendment of existing rights and issuance of new permits, is overseen by the Texas Commission on Environmental Quality (TCEQ), a state agency.

Within an individual basin, the water right with the oldest priority date—priority date is based on when the application for the right was determined to be complete or, for older

rights that were initially based on use filings,[8] when the water was first put to a recognized beneficial use—generally has the first claim to the available water during times of shortage (TWC § 11.027). The major exception to the priority-based system is water rights management in the middle and lower Rio Grande. For those areas, water rights are managed through an equity-based system established by court order. That system manages water rights within three prioritization categories based on type of use, with all rights in an individual category having equal priority (Jarvis 2020 § 4.5).

Another exception to the priority system, and to the requirement for obtaining a specific water right, is an exemption from permitting recognized for domestic and livestock purposes (Caroom and Maxwell 2020, §10.14). Although limited in the amount and nature of use authorized, those exempt domestic and livestock rights are recognized as having a superior claim to even the most senior of the priority-based water rights (Jarvis 2020 § 4.7.1–4.7.2). At least in theory, the widespread existence of such domestic and livestock rights along rivers and streams should help to maintain flows. With adequate enforcement, the existence of those rights should result, when upstream flow is otherwise available, in some limited amount of water being allowed to pass downstream of priority-based water right holders regardless of whether the priority-based rights are subject to environmental flow protection provisions.

Because of the complexity of the various types of rights in surface water recognized over the centuries under Texas law and previously applicable Spanish and Mexican law, the state implemented a legislatively mandated adjudication process. Undertaken between 1967 and 2007, that adjudication process covered the portions of the state not included in the equity-based system for portions of the Rio Grande. In very general terms, the process sought to quantify and standardize a disparate collection of existing water rights arising under a variety of legal theories to allow for a more cohesive management approach (Jarvis 2020 § 4.6). The provisions of older water rights, as refined through the adjudication process, are reflected in Certificates of Adjudication (COA).

## Evaluation of Water Availability

For about the last 20 years, applications for surface water rights have been evaluated using the state's water availability model (WAM) for the relevant basin to assess if unappropriated water is available to support the application and to assess potential adverse effects on existing water rights.[9]

---

[5]The term "water right" is broad and includes any right acquired under the laws of Texas "to impound, divert, or use state water" (TWC § 11.002 (5)).

[6]The rules of the Texas Commission on Environmental Quality (TCEQ) define "appropriations," (30 TAC § 297.1 (3)) and "appropriative right" (30 TAC § 297.1 (4)). Generally, an appropriative right is a legal right to use a specified amount of state water for certain uses at identified locations that is acquired through a defined legal process.

[7]In addition to the use authorization, a permit also includes information including the source of supply, permittee, amount authorized for diversion or impoundment, priority date, and date issued (TWC § 11.135). As discussed in section "Overview of Environmental Flow Protection in Texas", recent water rights also may include provisions for protection of environmental flow.

[8]Many early rights were established before the state had an application and review process. The priority and extent of those early rights were initially based on filings of reported use (Jarvis 2020 § 4.4). As part of the state's water rights adjudication process, which is discussed later in this section, many of those rights were refined and limited in scope.

[9]Although limited modeling efforts were initiated in the 1970s and 1980s, early this century, with a commitment of state funding, compre-

Many older rights were recognized without consideration of the availability of unappropriated water or with limited understanding of that availability. In general terms, unappropriated water refers to water that has not already been authorized for use under an existing water right (Caroom and Maxwell 2020 § 10.5:1). New water rights may be granted even if unappropriated water adequate to satisfy the right is only available some of the time.[10] The combination of older rights issued without evaluation of availability of unappropriated water and newer rights issued for water available only some of the time means that, at various locations and times, authorizations to take water exceed the amount of water available.

In evaluating permit applications, WAM modeling is done using a monthly time-step: assessing water availability in monthly increments. Evaluations done with the WAM assume a recurrence of naturalized flows over the period of record used in developing the individual WAM. Naturalized flow is calculated by adjusting measured flows to approximate flows predicted to have occurred without major human impacts such as impoundment, diversion, or return flows (Alexander and Henderson 2020 §12.4). Although variable by river basin, the period of record for the WAMs generally is at least 50 years in duration and begins prior to the historical drought of record, which, for most of the state, includes a period beginning in the mid-1940s (Alexander and Henderson 2020 § 12.4). The WAM evaluations do not account for climate change impacts beyond the extent to which those impacts are reflected in flow and evaporation levels measured within the relevant period of record and captured in the naturalized flow calculations.

If environmental flow protections are included in a permit, such as an amount of flow that must be allowed to pass downstream of the diversion or impoundment location, those protections are represented in the WAM modeling as a monthly flow volume. For example, if permit conditions limit diversions under an existing water right to times when a specific flow level is being passed downstream for flow protection, the monthly volume corresponding to that flow level is included in modeling the water demand that must be met to protect the existing right (Alexander and Henderson 2020 §12.7). An application for a new water right would be evaluated to ensure it would avoid impairing the ability to meet the water demand of the existing right including compliance with the flow protection requirement applicable to the existing right. WAM evaluations also play a key role in the implementation of environmental flow protection, particularly for protection of freshwater inflows, under the SB 3 process, which is discussed in Section "An Overview of the Senate Bill 3 Process".

## Right of Full Consumption

Water rights in Texas authorize the full consumption of the amount of water authorized for diversion and use unless the water right imposes explicit limits on consumption or a quantified requirement for the return to the stream of water that is used but not consumed (TWC § 11.046 (c)). Such quantified return flow requirements and limits on consumption are very rare in Texas water rights. As a result, most water right holders are authorized to fully consume water authorized for diversion, including through multiple cycles of reuse[11] of the water. If water that is diverted is not consumed by use under the water right, it must be returned to the watercourse when feasible (TWC § 11.046 (a)). Because of that broad authorization of consumption, availability of unappropriated water for a new water right is assessed based on the assumption that all existing water rights will be fully exercised—that is, by assuming that the full authorized amount will be diverted and, unless the permit explicitly limits consumption or includes a quantified return flow requirement, fully consumed. In the state's WAM modeling, that is referred to as the full authorization simulation or WAM Run3 (Alexander and Henderson 2020 §12.7).

## Enforcement

The combination of water rights granted without comprehensive water availability analyses and of rights granted for water that is predicted to be available only some of the time can create significant enforcement challenges. For some areas of the state, particularly where demand commonly exceeds the available supply, watermaster operations, managed by TCEQ, have been established to provide real-time oversight of diversion and impoundment.[12] In other areas, where conflicts about competing rights to water have been less common, there is less oversight over water use. For those non-watermaster areas, use levels are self-reported on an annual basis and enforcement is undertaken through a complaint-based approach when a water right holder contacts TCEQ alleging an impairment of the water right

---

hensive WAMs were developed for almost all river basins in Texas. Those WAMs are updated, including by extending the period of record, periodically as funding is available (Alexander and Henderson 2020 § 12.3). Information about the WAM models is available on the TCEQ website.

[10]Water availability varies greatly between wet periods and dry periods. TCEQ's rules require that a "sufficient amount of unappropriated water" be available for a "sufficient amount of the time" to make the proposed use viable (30 TAC § 297.42 (a)). The rules also provide some guidance on requisite availability—addressing, in terms of percentage of time, how frequently some or all the requested volume of water must be available, for various types of uses (30 TAC § 297.42 (c)–(h)).

[11]Reuse is a complex issue in Texas water law, but, in simple terms, it generally refers to the deliberate use of treated wastewater for water supply (Gooch et al. 2020 § 24.2).

[12]General information about watermaster operations is available on the TCEQ website.

(Martinez et al. 2020 §13.11–13.13). However, without real-time information about which water rights are diverting or impounding water, the identification of out-of-priority diversions can be challenging.

## Overview of Environmental Flow Protection in Texas

Although surface water right authorizations in Texas date back to the 1800s (Jarvis 2020 § 4.1), routine imposition of permit provisions to address the environmental flow implications of those rights did not begin until 1985. At that point, the Texas Legislature first enacted broadly applicable requirements for inclusion of flow-protection measures when granting new water rights, discussed in section "Initial Efforts to Formalize Environmental Flow Consideration for Surface Water Permits". Because the older water rights, which lack environmental flow protections, have senior priority—giving them the first claim to the water, after exempt rights, during periods of shortage—the relatively recent efforts to protect environmental flows started out facing a major challenge. Most of the surface water in Texas that is reliably available during drought periods had been authorized for use and consumption before environmental flow impacts began to be addressed.

As reflected in Fig. 1, only a small percentage of the water authorized for diversion is subject to permit conditions providing protection of environmental flows. In addition, the rights with flow protection requirements are relatively recent rights with low, or junior, priority. Total statewide average annual streamflow under historical conditions—in this instance, conditions prior to significant impoundments being built—is estimated to have been about $3.77 \times 10^{10} \, \mathrm{m}^3$ ($3.05 \times 10^7$ af)[13] (Brun et al. 2016, p. 34 Table 3–3). As Fig. 1 shows, by 1985 about two-thirds of that annual average total, almost $2.5 \times 10^{10} \, \mathrm{m}^3$ ($2.02 \times 10^7$ af), had already been authorized for diversion and consumption under water rights without flow protection provisions. Although the values are statewide averages and conditions in individual basins and at specific locations vary, on a big picture level, the bulk of the reliably available surface water flow had already been allocated before flow protections began to be imposed on a regular basis. Those estimates reflect average flow values, so, even without adjusting for the impacts of climate change, during many years flow and water availability will be well below average levels.

As a result, especially during dry periods, environmental flow protection provisions applied only to new water rights

often have limited ability to ensure sufficient levels of continued flow. That is particularly true for the most common type of flow-protection provision, which merely requires a portion of the upstream flow to be allowed to pass downstream of an authorized diversion point or dam. That type of provision is referred to in this chapter as a "pass-through requirement." Diversion or impoundment pursuant to senior rights without flow protections that are located upstream of a new right can greatly reduce the level of flow arriving at the location of the new right to be passed downstream.[14]

In addition, for river and stream protection at inland locations, water rights that are not subject to flow-protection requirements and are located downstream of a new right may impound or divert some or all the flow that is passed downstream pursuant to the pass-through provision. The size and location of such older rights will greatly affect how far downstream the environmental flow benefit extends, including if those flows reach a bay or estuary. Despite those key limitations, well-constructed pass-through requirements provide important benefits by reducing the extent of additional flow reduction caused by the new water right. However, because of the cumulative impacts of existing rights that were issued without flow protections, pass-through requirements applied to new rights are not, alone, an adequate approach for protecting environmental flows.

As discussed in section "Examples of Strong Flow Protection Approaches in Texas Water Rights", other types of environmental flow-protection approaches, such as quantified return flow requirements,[15] quantified releases from storage, and allocations of a portion of project yield to flow protection, have been applied previously in some Texas water rights. Those approaches have the potential to improve artificially induced low flow conditions and, if used more widely, could help to mitigate the adverse impacts of senior water rights that lack flow-protection provisions. The Senate Bill 3 (SB 3) environmental flow-protection process, discussed in section "An Overview of the Senate Bill 3 Process", does not limit the use of such approaches in issuing new permits. That 2007 legislation also acknowledges the importance of pursuing various types of voluntary proactive flow-protection strategies, such as transactions with holders of existing rights. Proactive strategies could involve a variety

---

[13] An acre-foot (af) refers to the amount of water that would cover an acre of land to a depth of one foot and corresponds to about $3.26 \times 10^5$ gallons.

[14] It also is true that the presence of return flows from upstream rights or releases from storage for downstream delivery can increase flow over natural conditions at various times for some locations. Water discharged, or released from storage, specifically to support an authorized use downstream generally is protected from diversion by other water right holders regardless of priority date (TWC § 11.042, Caroom and Maxwell 2020 §10.7).

[15] Quantified return flow requirement refers to a requirement that can be imposed pursuant to Section 11.046 (b) of the Texas Water Code, for a specific amount or percentage of the water diverted to be returned to a watercourse, including for flow protection purposes (TWC § 11.046 (b)). Such requirements are very uncommon in Texas water rights.

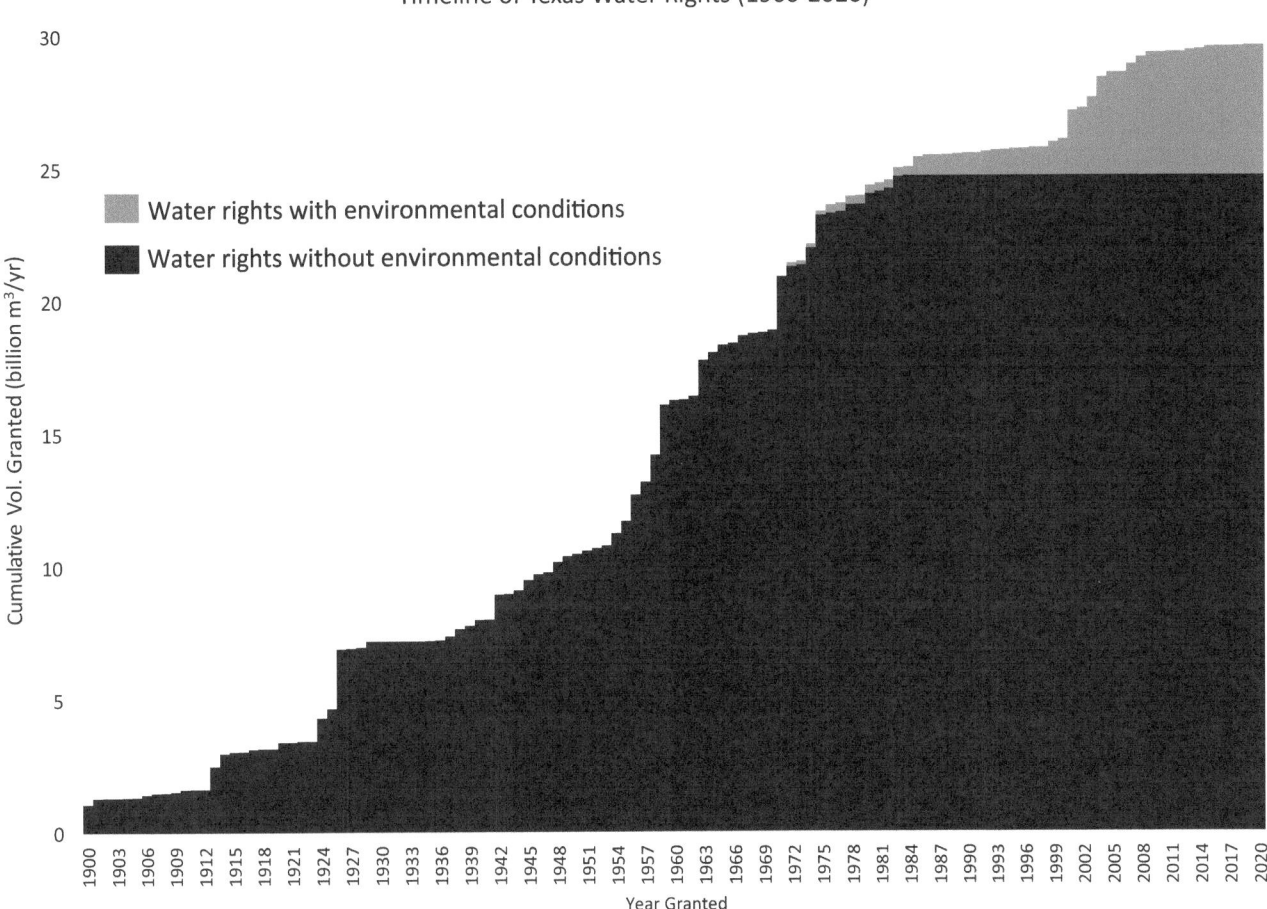

**Fig. 1** Chart, based on Texas Commission on Environmental Quality water right files, showing cumulative consumptive water rights by date from 1900 to 2020. Older water rights, which lack environmental flow protections, generally have the first claim to the available water during dry periods. Graphic developed by the National Wildlife Federation through its Texas Living Waters Project for its "Bays in Peril" and "Delivering on the Unrealized Potential of Senate Bill 3" publications. Used, with modifications, by permission

of approaches, including the conversion of portions of authorized diversions to flow protection or commitments for a portion of return flows from existing rights to continue to be discharged to help meet environmental flow needs.

## Examples of Strong Flow Protection Approaches in Texas Water Rights

The state's approach to environmental flow protection has changed over time. As reflected in Fig. 1, most water rights authorizing consumptive use,[16] and particularly the rights with the first claim on the limited flow available during dry

periods, lack permit conditions designed to limit adverse impacts on environmental flows. Legislation passed in 1985 began the process of making the inclusion of flow-protection provisions the norm rather than the exception. The state's approach for assessing and protecting environmental flows has evolved significantly since then. Some early examples of flow-protection provisions included in Texas permits, even predating those 1985 statutory changes, envisioned a more comprehensive approach than typically has been implemented in more recent permits and programs.

## Lake Texana and Palmetto Bend Reservoir
One of the earliest water rights that includes provisions for protection of freshwater inflows has a 1972 priority date and authorizes the construction and operation of two reservoirs, one each on the Lavaca and Navidad rivers, a short distance upstream of Lavaca Bay. The permit, now held solely by the Lavaca-Navidad River Authority (LNRA), initially was issued jointly to LNRA and the Texas Water Development

---

[16]Some water rights only authorize non-consumptive use, such as for hydroelectric power generation. As discussed in section "Right of Full Consumption", unless the water right specifically provides otherwise, water rights in Texas authorize full consumption of the amount authorized for use.

Board (TWDB).[17] The permit provisions were subsequently incorporated into a Certificate of Adjudication (COA), which has been amended multiple times. The COA, as initially amended, includes a broad, but vaguely defined, provision for protection of freshwater inflows generally stating that the authorization is subject to the release of water for the maintenance of the Lavaca-Matagorda Bay and Estuary System as may be determined by the agency that is now the Texas Commission on Environmental Quality (TCEQ) (COA 16-2095A, p. 4).[18]

Converting that broad statement into specific flow protections remains a challenging proposition. Specific freshwater inflow protections, in the form of pass-through requirements that vary by month based on reservoir storage, applicable to the existing Lake Texana on the Navidad River (referred to in the COA as Stage 1), were incorporated into the Certificate of Adjudication through a 1994 amendment (COA 16-2095B § 4.A.). The amended COA provisions applicable for the unbuilt second reservoir, Stage 2 (Palmetto Bend on the Lavaca River), include an explicit allocation of at least 2.234 $\times$ 10$^7$ m$^3$ (1.812 $\times$ 10$^4$ af) of water per year of the reservoir yield for maintenance of bay inflows. Those provisions also note that the entire Stage 2 authorization, which totals 5.932 $\times$ 10$^7$ m$^3$ (4.812 $\times$ 10$^4$ af) per year, is subject to release of water for the estuary. Finally, the COA provides that the freshwater inflow protections applicable to Lake Texana can be revisited if the second reservoir is built (COA 16-2095B § 1.B., 4.B.).

This early inflow protection effort, by allocating a portion of project yield and referencing a release requirement, provides for more robust protection of freshwater inflows than simply ensuring a pass-through of a portion of inflows available from upstream.[19] However, the extent to which those additional protections will be implemented remains uncertain. In 2020, LNRA filed an application for a new permit to authorize diversions from the Lavaca River and the construction of an off-channel reservoir. The application also signals an intention to abandon the existing authorization to construct the Palmetto Bend reservoir if the new application is granted (LNRA Application p. 5). Any permit issued based on the application would be subject to the applicable environmental flow standards adopted pursuant to Senate Bill 3, discussed in section "Senate Bill 3, A Revised Legislative Approach to Flow Protection", that currently only impose pass-through requirements.[20]

## Choke Canyon Reservoir

Another older Texas water right that includes significant provisions for protection of freshwater inflows authorizes the impoundment and use of water from the Choke Canyon Reservoir. That reservoir is located on the Frio River and is operated as a system in conjunction with the older Lake Corpus Christi. The Frio flows into the Nueces River, which flows into Lake Corpus Christi and then Nueces Bay and Corpus Christi Bay. The water right permit for Choke Canyon Reservoir was issued to the City of Corpus Christi and the Nueces River Authority with a 1976 priority date. The permit conditions were subsequently incorporated into a COA in 1981. The relevant provision of the COA addressing protection of freshwater inflows provides, in critical part:

> Owners shall provide not less than 151,000 acre-feet of water per annum for the estuaries by a combination of releases and spills from the reservoir system at Lake Corpus Christi Dam and return flows to Nueces and Corpus Christi Bays and other receiving estuaries.

(COA 21-3214 ¶ 5.B.). The provision establishes a minimum annual inflow volume of 1.863 $\times$ 10$^8$ m$^3$ (1.51 $\times$ 10$^5$ af) to be provided to the estuary system that includes Nueces and Corpus Christi bays through some combination of deliberate releases, spills, and return flows.[21] Additional language in the

---

[17]TWDB is a state agency involved primarily in water supply planning, flood planning, collection and analysis of water data, and water finance. TWDB is authorized to participate in water supply projects to support the maximum development of potential reservoir sites, with the state, through TWDB, temporarily retaining a share of ownership to produce the maximum potential yield from a project even if the project proponent initially did not need the full yield or could not afford the full cost (TWC § 15.301–.331, TWC § 16.131–.142).

[18]There are five amendments to the original Certificate, which are designated as COA 16-2095A through 16-2095E, all of which are available at the same location on the TCEQ website. Except to the extent specifically stated, an amendment does not change or supersede the terms of the original permit or previous amendments.

[19]As discussed in section "Overview of Environmental Flow Protection in Texas", pass-through requirements provide for a portion of the upstream flow occurring at a particular time to be allowed to pass downstream. A release requirement or a commitment of a portion of project yield can provide for a portion of the water that has been stored during wetter periods to be released and passed downstream to supplement flows at critical times.

[20]One apparent complication is that the environmental flow standards governing the LNRA application prohibit a new permit from reducing the attainment frequency of key inflow volumes below the levels predicted in the applicable WAM with the full exercise of all existing rights (30 TAC § 298.330 (c)). To assess the potential for such a reduction in attainment frequency, the full exercise of those existing rights must be modeled with any applicable environmental flow protections in effect. Because the details for applying the environmental flow protections in the current COA, which is an existing right, for the Palmetto Bend Reservoir currently are undetermined, it is unclear how the WAM evaluation will be undertaken to compare protection levels and assess compliance for the new application. The general approach used for implementing attainment frequency requirements is discussed in section "Adoption of Specific Freshwater Inflow Standards".

[21]This water right is one of the rare ones, as referenced in section "Right of Full Consumption", that includes a quantified return flow requirement, although the specific quantity of inflows required to be provided from return flows is not separately stated in the water right.

COA defers, for later consideration by what is now TCEQ,[22] thorny operational issues for implementing the requirement.

Beginning in 1992, TCEQ has issued a series of orders,[23] guided by the recommendations of the Nueces Estuary Advisory Council (TCEQ NEAC), which was created for that purpose by TCEQ. Those orders provide operational procedures for providing water to the estuary system, which include translating the requirement for releases from storage into a pass-through requirement that varies based on reservoir storage levels and on bay salinity levels. As explained in the report of the Nueces River and Corpus Christi and Baffin Bays Expert Science Team (Nueces BBEST),[24] the orders also establish incentives, in the form of decreased levels of required pass-through volumes, for delivering freshwater to strategic locations in the estuary system where it is likely to provide heightened benefit (Nueces BBEST 2011 § 2.1.5, p. 2–5).

## Lower Colorado River Authority Water Management Plan

The Lower Colorado River Authority (LCRA) operates two major water supply reservoirs (lakes Buchanan and Travis) on the Colorado River—the one in Texas, not the larger Colorado River in the western U.S.—as a system pursuant to a Water Management Plan (WMP). The WMP is required to be updated periodically, with the latest version having been approved in 2020 (LCRA WMP 2020). Pursuant to the WMP, LCRA uses both pass-throughs of reservoir inflows and releases from storage to help meet environmental flow needs. The requirement for LCRA to develop and periodically revise a WMP is set out in the COAs for Lake Buchanan and for Lake Travis issued in 1989 (COA 14-5478, COA 14-5482). The COA provisions make the WMP subject to approval by TCEQ and require environmental flow protection in the Colorado River below Austin and in the portion of the Matagorda Bay estuary system affected by inflows from the Colorado River.[25]

As authorized by those COAs, LCRA is allowed to manage, consistent with the WMP, the water supply from those two reservoirs in two categories: firm water and interruptible water. The firm water—water that is managed to be reliably available including during a recurrence of historical drought of record conditions under the WAM Run3 assumptions discussed in section "Right of Full Consumption"—is sold to customers with firm demands, such as municipal and industrial customers (LCRA WMP 2020, pp. 1-1–1-2). In the versions of the WMP developed so far, LCRA also has committed an allocation of firm water—most recently $4.125 \times 10^7$ m$^3$ ($3.344 \times 10^4$ af) per year—to help maintain environmental flows, including through releases from storage (LCRA WMP 2020, pp. 4–13).

Interruptible water—water that is predicted to be available somewhat regularly but not on a drought-reliable basis—is available to water customers, primarily for rice irrigation, who can accept a less dependable supply. Interruptible water also is used to help meet environmental flow needs, particularly outside of drought periods. Among other things, the WMP establishes the parameters for efforts to help meet environmental flow needs and for managing the interruptible supply to avoid impairing LCRA's ability to meet the demand for firm water. The supply of interruptible water decreases over time as the demands of firm-water customers grow to levels close to the full amount of the available firm water, which means less water will be available to be passed downstream for environmental flow protection. In addition, the amount of both interruptible and firm water available likely will decrease as the period-of-record used in water availability modeling is extended to include more recent periods that are drier than the historical drought of record.[26] To help ensure protection of firm water, the amount of interruptible water available for use in any year, including for flow protection, also is subject to adjustment on a seasonal basis based on reservoir storage levels (LCRA WMP 2020, pp. 4–13–4-20).

In 2008, LCRA completed extensive environmental flow studies and developed new recommendations for protection of instream flows in the river and of freshwater inflows to Matagorda Bay. For freshwater inflows, the revised recommendations, provided in the 2008 Matagorda Bay Health Evaluation (MBHE) report (LCRA 2008 MBHE), incorporate multiple levels of freshwater inflow targets. The targets range from a monthly threshold level, designed to be met each month to maintain an area of nursery habitat in the river delta during extreme drought conditions, up through four levels of increasing annual inflow amounts, referred to as the MBHE 1 through MBHE 4 Regimes. Those annual inflow amounts, each of which has an associated target attainment frequency reflecting the percentage of years in which the inflow volumes are recommended to be met, are flexibly dis-

---

[22] TCEQ is the successor agency to the Texas Water Commission and other agencies.

[23] Two such orders are available on the TCEQ website (TCEQ Order 1992, 2001).

[24] The role of such expert science teams is discussed in section "The Regional Actors: Bay and Basin Area Stakeholder Committees and Bay and Basin Expert Science Teams".

[25] As noted in those COAs, the adjudication process included a court review component. As part of that court review, various agreements were reached to resolve disputes leading to the unusual management approach discussed in this section, including flow protection aspects. The adjudication process did not usually address flow protection.

[26] A recent update to the relevant WAM extending the period of record through 2016 resulted in the identification of a new drought-of-record with a reduced combined firm yield of the reservoirs (LCRA WMP 2020, Chap 3).

tributed across seasonal targets (referred to as spring, fall, and intervening seasons).[27] The inflow regimes with smaller volumes have higher associated attainment frequencies, reflecting the goal that inflow volumes should be that high or higher in most years.

## Initial Efforts to Formalize Environmental Flow Consideration for Surface Water Permits

### Legislative Directives for Study and Consideration of Inflow Needs

Beginning in 1975, the Texas Legislature initiated steps towards formalizing consideration of environmental flow impacts in water rights permitting.[28] Senate Bill 137, passed that year, added maintenance of a "proper ecological environment" in the bays and estuaries to the statement of public policy goals for management of natural resources[29] and directed that the effects on the state's bay and estuaries had to be assessed as part of the permitting process for water rights. It also directed and provided funding for studies of the effects of freshwater inflows on bays and estuaries, while imposing a rather optimistic deadline for completion of "comprehensive studies" by the end of 1979 (Acts 1975 64th Leg. ch. 344).

In 1985, the Legislature built on those earlier efforts, directing that permits issued within 322 river kilometers (200 river miles) of the coast must include "to the extent practicable when considering all public interests" conditions considered necessary to maintain beneficial inflows (Acts 1985, 69th Leg. ch. 133, § 4.01, pp. 20–21).[30] That legislation also added a requirement that impacts to instream flows—flows in rivers and streams—must be assessed. Later amendments added the directive that, in addition to assessing impacts to instream flows, new water right permits must include conditions considered necessary to "maintain existing instream uses and water quality" (TWC § 11.147 (d)).

### Authorization for Potential Suspension of Flow Protection

That 1985 legislation imposing the requirement for permit conditions to maintain beneficial inflows also added authority to suspend those inflow protections upon a finding of an emergency for which no practical resolution, other than suspension, is available (Acts 1985, 69th Leg. ch. 133, § 4.02, p. 21). That language is codified in two separate provisions of the Water Code (TWC § 5.506, 11.148). A 2007 amendment to those provisions provides that set-asides of unappropriated water made pursuant to SB 3, discussed in section "Senate Bill 3, A Revised Legislative Approach to Flow Protection", also are subject to suspension under emergency conditions. Those provisions ensure that environmental flow protections may be limited or paused entirely to meet other types of human water needs on a temporary basis.

### Potential Allocation of Portion of Project Yield for Flow Protection

The same 1985 legislation also directed that 5% of the annual firm yield of reservoir projects built within 322 river kilometers (200 river miles) of the coast using a particular type of state funding was to be appropriated to the Texas Parks and Wildlife Department (TPWD), the state's fish and wildlife agency, specifically for releases to bays and estuaries and for instream uses (TWC § 15.3041, 16.1331). Those provisions, although still in effect, have yet to result in an appropriation to TPWD.[31] As discussed in section "Lake Texana and Palmetto Bend Reservoir", the water right authorizing Lake Texana and Palmetto Bend Reservoir, although predating this provision, does allocate a portion of the yield of the unbuilt Palmetto Bend Reservoir to freshwater inflow protection, but without an appropriation of water to TPWD.

### Creation of Texas Water Trust

The Texas Water Trust (TWC § 15.7031) was created in 1997 as part of the Texas Water Bank, overseen by the TWDB. The Texas Water Bank was envisioned as a mechanism for facilitating voluntary water transactions. Water rights initially issued for another use can be placed in the Texas Water Trust, either permanently or for a specified term, where they are dedicated to environmental flow protection purposes. As of early 2023, three rights are held in the Trust: two perpetual dedications on the Rio Grande from 2003 (totaling about

---

[27] The study approach and results, which form the basis of the Matagorda Bay inflow recommendations developed as part of the SB 3 process, discussed in section "Senate Bill 3, A Revised Legislative Approach to Flow Protection", are summarized in the report of the Colorado and Lavaca Rivers and Matagorda and Lavaca Bays Basin and Bay Expert Science Team (CL BBEST) (CL BBEST 2011, pp. 2-222–2-234). Those BBEST recommendations are summarized in Table 2.

[28] This was not the first formal recognition of the importance of environmental flows. In 1968, the first statewide water plan for Texas was developed. That plan, referencing an extensive data collection effort initiated in 1967, acknowledged the importance of freshwater inflows to bay and estuary health, particularly for maintaining commercial and sport fishing (TWDB 1968 State Water Plan Summary, pp. 33, 35–38).

[29] That goal is currently codified in Section 1.003 of the Texas Water Code (TWC § 1.003 (6)).

[30] The legislation added a definition of "beneficial inflows" referencing a "salinity, nutrient, and sediment regime adequate to maintain a sound ecological environment in the receiving bay and estuary system …." (TWC § 11.147 (a)). It also extended the 1979 timeframe for completing the ongoing bay and estuary studies.

[31] As discussed in section "Surface water management", appropriation refers to the legal process through which a water right is issued. In 2007, SB 3 added an express prohibition on the issuance of new permits for the purpose of environmental flow protection. That prohibition might complicate any potential for implementing such an appropriation to TPWD if the issue were to arise (TWC § 11.0237 (a)).

$1.524 \times 10^6$ m³/year ($1.236 \times 10^3$ af/year)) and one perpetual dedication on the Guadalupe River from 2006 (totaling about $4.081 \times 10^7$ m³/year ($3.311 \times 10^4$ af/year)) (TWDB Water Trust). Responding to the slow progress in getting rights placed into the Trust, legislation passed in 2021 defines a proactive role for TPWD in facilitating placement of additional rights in the Trust and, upon agreement by the holder of the underlying right, in managing rights held there (Acts 2021, 87th Leg. ch. 689).[32]

## Senate Bill 3, A Revised Legislative Approach to Flow Protection

In 2007, the Texas Legislature enacted an omnibus water bill, Senate Bill 3,[33] addressing a wide variety of water issues. Article 1 of Senate Bill 3 (SB 3) creates an extensive process for defining, and attempting to meet, environmental flow needs for Texas streams, rivers, and coastal waters (Acts 2007, 80th Leg ch 1430, Article 1). SB 3 focuses on four primary flow-protection components:

- defining environmental flow needs through the adoption of environmental flow standards (TWC § 11.1471 (a)(1));
- establishing set-asides of unappropriated water[34] to help meet those needs (TWC § 11.1471 (a)(2));
- implementing flow-protection conditions in new water right authorizations consistent with the adopted flow standards and any set-asides (TWC § 11.147 (e-3)); and
- facilitating implementation of proactive strategies to help meet the flow standards by, for example, pursuing voluntary transactions to redirect some previously appropriated water to flow-protection purposes (TWC § 11.02362 (o)).

That legislation does not provide for involuntarily reopening water rights issued prior to September 1, 2007, to impose new or changed environmental flow protections.[35]

In recognition of the challenge involved in assessing flow needs and in achieving protection of environmental flows, SB 3 includes an ongoing adaptive management component

through which flow-protection approaches are to be reviewed and refined on an ongoing basis (TWC § 11.02362 (p)). Those refinements can include periodic revisions to previously adopted flow standards and, if needed to meet revised flow standards, adjustments to strengthen, within defined limits, flow protection conditions in permits issued after September 1, 2007 (TWC § 11.1471 (a)(3)).

## An Overview of the Senate Bill 3 Process

The Senate Bill 3 process is complex and starts with appointments by the Environmental Flows Advisory Group (EFAG) and ends with regulations by Texas Commission on Environmental Quality (TCEQ) (Table 1). While specific groups were appointed, other entities, including the Texas Water Development Board (TWDB) and the Texas Parks and Wildlife Department (TPWD), also made important contributions.

**Table 1** Overview of Senate Bill 3 environmental flows process showing key actors with summary of roles played by each

| Senate Bill 3 Process Overview | |
| --- | --- |
| Statewide Actors | Regional Actors |
| **Environmental Flows Advisory Group (EFAG):**<br>• appoint SAC and BBASC members,<br>• oversee process schedule,<br>• study improved enforcement approaches,<br>• study ways to encourage conversion of existing rights to flow protection, and<br>• approve adaptive management work plans | **Bay and Basin Area Stakeholder Committee (BBASC):**<br>• appoint BBEST members;<br>• recommend, based on BBEST flow regime recommendations and other factors, flow standards and proactive strategies to help meet standards; and<br>• develop adaptive management work plans for monitoring, studies, and future reviews of recommendations, standards, and strategies |
| **Texas Environmental Flows Science Advisory Committee (SAC):**<br>provide scientific guidance to EFAG, BBESTs, BBASCs, and state agencies on implementation | |
| **Texas Commission on Environmental Quality (TCEQ):**<br>• after considering BBEST and BBASC recommendations and other factors, establish environmental flow set-asides and adopt flow standards;<br>• ensure new water rights comply with flow standards and set-asides;<br>• periodically consider revising standards; and<br>• as appropriate, adjust post-2007 rights to achieve compliance with revised standards | **Bay and Basin Expert Science Team (BBEST):**<br>• recommend, based solely on available science, flow regimes adequate to protect sound ecological environment; and<br>• advise and assist BBASC in its undertakings |

---

[32] Management of rights in the Trust could include, for example, determining when, for a water right with storage capacity, releases should be made to provide the greatest environmental benefit.

[33] House Bill 3, which also passed during that session, included the same provisions for environmental flow protection.

[34] Basically, unappropriated water means state-owned water that has not already been authorized for use under the state water rights system (Caroom and Maxwell 2020 §10.5.1).

[35] SECTION 1.27 of SB 3 provides that the statutory changes requiring compliance with flow standards and set-asides only apply to applications that were pending before TCEQ on the effective date of the legislation or are filed after that date (Acts 2007 80th Leg ch 1430, SECTION 1.27). The effective date of SB 3 was September 1, 2007.

## The Regional Actors: Bay and Basin Area Stakeholder Committees and Bay and Basin Expert Science Teams

SB 3 uses a regional approach in developing initial recommendations for flow-protection levels with a state-level oversight and decision process. The regional approach provides for the creation of a basin and bay area stakeholder committee (BBASC) and basin and bay expert science team (BBEST) for specific river basins and any associated bay or estuary systems. Those regional bodies develop recommendations for flow protection that are then considered by the state-level actors. Each BBEST, made up of technical experts appointed by the relevant BBASC, is directed solely to consider the best available science in formulating environmental flow regime recommendations that are to be developed through a process designed to achieve consensus (TWC § 11.02362 (m)). SB 3 directs that those environmental flow regime recommendations are to consist of a schedule of flow quantities reflecting both seasonal and yearly fluctuations as well as geographic variation and are to be set at a level adequate to protect a "sound ecological environment" and to maintain the "productivity, extent, and persistence of key aquatic habitats" (TWC § 11.002 (16)). The legislation does not include a definition of "sound ecological environment," leaving that issue to be addressed during the implementation process.

BBASCs are made up of stakeholders representing a variety of statutorily identified interest groups in the basin and bay area (TWC § 11.02362 (f)). The BBASCs are directed to consider the BBEST recommendations, along with other factors, including competing needs for water, in developing, through consensus if possible, recommendations for environmental flow standards and proactive strategies for helping to meet the standards (TWC § 11.02362 (o)). Each BBASC is also directed to develop, with the assistance of the relevant BBEST, a work plan with recommendations for studies and monitoring to help address knowledge gaps. Work plans also must provide for a periodic review—to occur on a default 10 year-cycle[36]—of the environmental flow regime recommendations, the environmental flow standards, and the strategies for meeting the flow standards (TWC § 11.02362 (p)).

## The State-Level Actors: Environmental Flows Advisory Group, Science Advisory Committee, and Texas Commission on Environmental Quality

The state-level entities with primary roles in the SB 3 process include the Environmental Flows Advisory Group (EFAG), the Texas Environmental Flows Science Advisory

Committee (SAC), and the TCEQ.[37] The EFAG, which is made up of legislators and state agency leaders (TWC § 11.0236), has general oversight authority over the SB 3 process, including through setting the boundaries of the bay and basin areas and the schedule for completion of various process components (TWC § 11.02362 (a)–(e)). The EFAG also appoints the members of the BBASCs to represent the various interest groups identified in statute (TWC § 11.02362 (f)), as well as the members of the SAC (TWC § 11.02361 (a) & (b)). Finally, the EFAG is charged with studying approaches for encouraging voluntary conversions of reasonable amounts of existing water rights to flow protection purposes as well as approaches for improving water rights administration and enforcement to help ensure ecologically sound rivers and bays (TWC § 11.0236 (i)).

The SAC, made up of technical experts knowledgeable about environmental flow science, is charged with developing technical guidance for the BBESTs and providing input to the EFAG on various technical issues related to flow protection, such as comments on flow standard recommendations and work plans (TWC § 11.02361 (e)). One example of technical guidance provided by the SAC is a reference to a definition of "sound ecological environment" for consideration in implementing the SB 3 process. That definition focuses on four key components: sustaining the full complement of native species, sustaining key habitats for those species, retaining key features of the natural flow regime needed by those species to complete their life cycle, and sustaining key ecosystem processes and services (SAC Lessons Learned 2010, pp. 2–3). Generally, the various BBESTs adopted, or referenced, that definition, or some variation thereof, in developing their recommendations.[38] TCEQ is the state's

---

[36] SB 3 directs that flow standards and set-asides may not be revised more frequently than once every 10 years unless the relevant adaptive management work plan provides for more frequent review (TWC § 11.1471 (f)). The legislation also directs that work plans must provide for reviews at least once every 10 years (TWC § 11.02362 (p)). Accordingly, a 10-year review is the default frequency, but shorter review cycles can be established.

[37] Both the Texas Parks and Wildlife Department (TPWD) and the Texas Water Development Board (TWDB) also have played key roles in providing data and expertise. In addition, TWDB has administered state funding to support the work of the SAC and BBESTs as well as funding to support various studies related to adaptive management.

[38] For example, the Sabine and Neches Rivers and Sabine Lake BBEST (SN BBEST) adopted the same definition (SN BBEST 2009, p. 9), the Trinity and San Jacinto and Galveston Bay BBEST (TSJ BBEST), although failing to reach agreement on a set of environmental flow recommendations—instead developing two sets of competing recommendations—also adopted that definition (TSJ BBEST 2009, pp. 4–5), as did the Brazos River and Bay BBEST (Brazos BBEST 2012, pp. 1–3). The Colorado and Lavaca Rivers and Matagorda and Lavaca Bays BBEST (CL BBEST) referenced the definition cited by the SAC but qualified its application by noting that the BBEST sought to maintain, through its flow recommendations, an "acceptably" sound ecological environment (CL BBEST 2011, pp. 1–3). Similarly, the Guadalupe, San Antonio, Mission, and Aransas Rivers and Mission, Copano, Aransas, and San Antonio Bays BBEST (GSA BBEST) added a qualifier of "to some reasonable level." (GSA BBEST 2011, pp. 1.5–1.8). The Nueces BBEST added similar qualifications of acceptable soundness. A noteworthy aspect is the determination by the Nueces BBEST that the "Nueces Bay and Delta region" represented an "unsound ecological environment" because of alterations in freshwater inflows (Nueces BBEST 2011, pp. 1-5–1-6).

primary environmental agency and has responsibility for managing surface water rights. Under SB 3, TCEQ, after considering the recommendations from the respective BBESTs and BBASCs along with other input, has responsibility for adopting rules addressing the key substantive components for flow protection that guide the agency's review of applications for new and amended water rights. TCEQ also has responsibility for issuing and administering water rights consistent with SB 3.

## Key Substantive Flow Protection Components under Senate Bill 3

SB 3 addresses four primary flow-protection components. TCEQ acts as the primary decisionmaker on three of those components after considering input from the other actors and the public. Responsibility for the fourth component is less clear. First, TCEQ is charged with defining environmental flow needs for the various river basin and bay systems through the adoption of "environmental flow standards adequate to support a sound ecological environment, to the maximum extent reasonable considering other public interests" and a variety of other factors (TWC § 11.1471 (a)(1) & (b)). Second, TCEQ is to establish set-asides of unappropriated water[39] "to satisfy the environmental flow standards to the maximum extent reasonable when considering human water needs" (TWC § 11.1471 (a)(2)). Next, as TCEQ evaluates new water right authorizations, the agency is charged with identifying flow-protection conditions to be incorporated into those new permits or amendments to help meet the adopted flow standards and comply with set-asides (TWC § 11.147 (e-3)).

A fourth major component of flow protection under SB 3 is the identification and implementation of proactive strategies to help meet the flow standards (TWC § 11.02362 (o)). Proactive strategies could involve a variety of approaches, including voluntary transactions to redirect some previously appropriated water to flow-protection purposes or commitments by water right holders to continue the discharge of a specific quantity of treated return flows to help sustain inflows. BBASCs are charged with identifying potential strategies to help meet the flow standards.[40] However, there is no clear assignment of responsibility or authority for the actual implementation of such strategies. Similarly, there is no funding identified. As noted in section "The State Level

Actors: Environmental Flows Advisory Group, Science Advisory Committee, and Texas Commission on Environmental Quality", the EFAG is charged with studying methods for encouraging the conversion of existing rights to flow protection (TWC § 11.0236 (i)). Presumably, that study, when undertaken, will include recommendations for addressing the various uncertainties associated with implementation of proactive strategies.

In recognition of the challenge involved in assessing flow needs and in achieving protection of environmental flows, SB 3 provides for an ongoing adaptive management component, on a 10-year default cycle, for continued review and refinement of flow protection efforts (TWC § 11.02362 (p)). Those refinements can include periodic revisions to previously adopted flow standards (TWC § 11.1471 (f)). If appropriate to help meet revised flow standards, TCEQ can strengthen, within legislatively defined limits, flow protection conditions included in permits issued after September 1, 2007—the date SB 3 became effective (TWC § 11.1471 (a) (3)). The adaptive management component also includes provisions for periodic work plan updates to identify studies to address unanswered questions about flow needs and to help identify proactive strategies for meeting flow needs (TWC § 11.02362 (p)).

## Current Status of Senate Bill 3 Implementation

### Adoption of Flow Standards

TCEQ has adopted initial environmental flow standards for most stream and river basins that flow to the major bay systems in Texas. The first such flow standards were adopted in 2011 and the most recent ones in 2014. As of early 2023, no environmental flow standards have been adopted for the Canadian, Red, or Sulphur River basins or for the Cypress Creek Basin, which do not flow to the Texas coast. Similarly, there are no applicable flow standards for various smaller coastal basins in Texas.

The adopted flow standards rely solely on a pass-through approach for flow protection: requiring some portion of flows from upstream to be allowed to pass downstream of new water rights. Although that approach will limit the extent to which new water rights subject to the flow standards further reduce the already altered flow patterns resulting from the effects of existing water rights, TCEQ has discretion to implement more protective approaches. As discussed in section "Examples of Strong Flow Protection Approaches in Texas Water Rights", consistent with Texas law, some existing water rights include more comprehensive flow-protection measures, such as quantified return flow requirements and required releases of stored water, that can help mitigate artificially induced periods of very low flows.

[39] Basically, unappropriated water means state-owned water that has not already been authorized for use under the state's water rights system (30 TAC § 297.1 (56), Caroom and Maxwell 2020 §10.5.1).

[40] Most of the BBASC reports include a list of potential proactive strategy approaches, although the approaches are conceptual in nature: describing types of approaches rather than recommendations for specific transactions or actions (CL BBASC 2011, pp. 129–133; GSA BBASC 2011, pp. 131–133; Brazos BBASC 2012, pp. 48–51; Nueces BBASC 2012, pp. 97–100).

## No Set-Asides Established

SB 3 directs the establishment of reasonable amounts of set-asides of unappropriated water—basically identifying quantities of water as off-limits from permitting for other uses—to remain available for environmental flow protection. One key advantage to establishing set-asides for flow protection is that set-asides would have enforceable priority dates and enjoy protection much like a water right.[41] The significance of that aspect is discussed in section "Adoption of Specific Freshwater Inflow Standards" under the Climate Change Risk heading. Despite the SB 3 directive to set aside an amount of unappropriated water, if available, to the maximum extent reasonable considering human water needs, TCEQ declined to establish any set-asides. Since that time, TCEQ has continued to determine unappropriated water was available in continuing to issue new water rights for various uses. The agency did note, when adopting flow standards in 2011 and again in 2012, that it might consider establishing set-asides later.[42]

## Freshwater Inflow Protection Approaches Taken in Adopted Flow Standards

The adopted flow standards generally use one of two basic approaches in providing protection of freshwater inflows. The first approach relies solely on standards for protection of instream flows at upstream locations in the larger streams and rivers that contribute flow to a major bay or estuary without specific criteria for inflow protection. Examples of this approach include Sabine Lake, which, despite the name, is a bay, and the Brazos River estuary. The second approach includes, in addition to criteria for instream flow protection at upstream locations, specific freshwater inflow criteria identifying inflow volumes and associated attainment frequencies specifying how frequently the volumes should be met. Examples of this approach include Galveston Bay, Matagorda and Lavaca Bays, San Antonio Bay, and the Nueces Bay and Delta.

### Reliance Solely on Instream Flow Criteria to Address Freshwater Inflows

The Sabine Lake estuary system is relatively small in areal extent, with the bay covering about $2.428 \times 10^4$ hectares ($6.0 \times 10^4$ acres), particularly when considered in relation to the average volume of river flows reaching the bay, about $1.726 \times 10^{10}$ m$^3$/yr ($1.400 \times 10^7$ af/yr) (SN BBEST 2009, p. 143). The Sabine and Neches Rivers and Sabine Lake Bay BBEST (SN BBEST), after developing its recommendations for instream flow protection, undertook an analysis of the expected adequacy of those recommendations to maintain a sound ecological environment in Sabine Lake. Concluding that its instream flow recommendations would be adequate for that purpose, the BBEST declined to recommend separate freshwater inflow criteria (SN BBEST 2009, pp. 18–19).

The instream flow criteria recommended by the SN BBEST rely on pass-through requirements that vary by location and season. Those criteria include a seasonally variable subsistence flow level applicable only during dry hydrological conditions and three different levels of seasonally variable base-flows[43] applicable under different hydrological conditions: dry, in effect during the driest 25% of seasons; average, in effect during the 50% of seasons that are neither dry nor wet; and wet, in effect during the wettest 25% of seasons.[44] The BBEST also recommended two different levels of seasonally variable pulse flows with the protected level determined based on hydrological condition (SN BBEST 2009, p. 105). Pulse flows are high-volume, short-duration flows that occur in response to rainfall events and, in most cases, are defined by a pulse trigger flow level, pulse volume, and pulse duration. The size and number of protected pulses may vary by season as well as hydrologic condition. As implemented in the adopted flow standards for the Sabine and Neches river basins, when a pulse trigger level is reached for a protected pulse, diversion or impoundment is only allowed when flow passing downstream exceeds the trigger level until either the pulse volume has been allowed to pass downstream or the pulse duration has been satisfied (30 TAC § 298.275 (d)). Many of the other BBESTs recommended

---

[41] The legislature directed that the priority date for an environmental flow set aside would correspond to the date that the relevant BBEST provided its environmental flow regime recommendations to TCEQ (TWC § 11.1471 (e)). The Legislature authorized TPWD to ensure the protection of set-asides in a manner similar to how the holder of a water right protects that right (TWC § 11.0841 (c)). As discussed in Section "Surface water management", priority dates are not used in portions of the Rio Grande basin, so if set-asides were established there, no priority date would apply.

[42] In responding to comments about the failure to adopt environmental flow set-asides during the rulemaking process for adoption of flow standards, TCEQ signaled a willingness to consider doing so in the future (Texas Register 2011, p. 2924, Texas Register 2012, p. 6652).

[43] Subsistence flow levels are focused primarily on maintaining adequate water quality conditions and sufficient habitat during drought periods to allow aquatic life to persist and recover when flow conditions improve. Base flow levels are intended to represent more normal flow conditions (SN BBEST 2009, p. 45).

[44] The use of hydrological condition to trigger changes in the levels of flow protections was designed to reflect the variation in natural flow levels in response to seasonal precipitation patterns, with higher flows during wetter periods, and the understanding that different groupings of species benefit from periods of high, average, and low flow conditions (SN BBEST 2009, pp. 16–17). The BBEST recommended that hydrological condition for a season should be determined based on cumulative reservoir storage at the beginning of the season (SN BBEST 2009, p. 105).

some variation on this basic approach for protection of instream flows at inland locations.

TCEQ, relying upon that inflow analysis by the SN BBEST, also declined to incorporate separate freshwater inflow components in the adopted flow standards for Sabine Lake (Texas Register May 6, 2011, p. 2982). However, the instream flow standards adopted by TCEQ generally protect much smaller volumes of flow than were recommended for protection by the BBEST.[45] Relying solely on instream flow criteria for bay inflow protection also can be problematic because flows protected at the most downstream instream flow measurement point, particularly for measurement points located far from the coast, can be diverted or impounded under existing water rights between the measurement point and the bay.

### Adoption of Specific Freshwater Inflow Standards

For protection of freshwater inflows to Galveston Bay, Matagorda Bay, Lavaca Bay, San Antonio Bay, and the Nueces Bay and Delta, the flow regime recommendations from the respective BBESTs include specific freshwater inflow criteria. Although varying in various details, each BBEST recommended freshwater inflow standards generally consisting of a set of flow volumes, or range of volumes, to be met as bay inflow during an identified season or other defined time-period, along with accompanying attainment frequencies. The attainment frequency component addresses how often, stated in percentage of months, seasons, or years, the associated volume criterion should be met. Where the relevant BBEST and BBASC provided consistent recommendations for the overall structure of freshwater inflow criteria,[46] the flow standards adopted by TCEQ generally reflect that structure. However, the attainment frequency component adopted by TCEQ for use in evaluating future

permit applications provides much lower levels of protection than the attainment frequency recommended by the relevant BBEST as being adequate to support a sound ecological environment. As explained in section "Key Substantive Flow Protection Components under Senate Bill 3", in adopting flow standards, TCEQ is directed to consider various factors in addition to adequacy to protect a sound ecological environment, which is the fundamental consideration for the BBESTs. In addition, as discussed in the following paragraphs, the TCEQ attainment frequencies are based on an assumption of full use of all existing water rights.

**Reliance Solely on Model Predictions of Long-term Attainment Frequency** For each of these bay systems, attainment frequency for various inflow volumes is assessed solely through water availability modeling and evaluated as a long-term value over the WAM period-of-record. For example, the Colorado and Lavaca Rivers and Matagorda and Lavaca Bays BBEST (CL BBEST) recommended that the highest volume inflow regime, referred to in the BBEST report as MBHE 4[47] and intended to represent very favorable conditions in the portion of Matagorda Bay most affected by Colorado River inflows, should be met in 35% of future years (CL BBEST 2011, pp. 2–234). With the exception of the flow standards for inflows to Galveston Bay, the adopted flow standards do not list the specific attainment frequency values used for evaluating new permits, instead stating, for example, that future water rights subject to the standards may not decrease the "modeled annual frequency" of an inflow regime (30 TAC § 298.330 (a)(2)). The "modeled annual frequency" is defined as the frequency at which specific inflow levels occur in TCEQ's WAM Run3 at the time the first application for a new water right subject to the flow standards is evaluated by TCEQ (30 TAC § 298.305 (12)).

Stated another way, those flow standards provide that the full exercise of any new water right granted may not reduce the frequency at which the key freshwater inflow volumes are predicted to occur below the frequency predicted for those volumes with only existing rights fully exercised.[48] For

---

[45]TCEQ adopted flow standards that include only a subsistence flow level and a single baseflow level for each season at each measurement point without any consideration of hydrological condition. The adopted subsistence flow levels match the BBEST recommendations. The adopted base flow levels are set 10% above the relevant dry base flow level recommended by the BBEST to be in effect only during the driest 25% of time. The flow standards also only protect the lower level of pulse flows recommended by the BBEST for protection. The difference in protected flow level is quite large. For example, at the most downstream measurement point on the Sabine River, the adopted flow standards for the spring season protect a subsistence flow level of 12.4 cubic-meters/second (cms) (436 cubic-feet/second (cfs)) and a base flow level of 37.6 cms (1329 cfs) (30 TAC § 298.280 (5)). By comparison, the SN BBEST recommended protection of that same subsistence flow level, a dry-condition base flow level of 34.2 cms (1208 cfs), an average-condition base flow level of 50.8 cms (1795 cfs), and a wet-condition base flow level of 99.2 cms (3505 cfs) (SN BBEST 2009, p 128).

[46]For Galveston Bay, neither the BBEST nor the BBASC was able to reach consensus on recommendations (TSJ BBEST 2009, TSJ BBASC Version 1 2010, TSJ BBASC Version 2 2010). As a result, TCEQ developed the approach that appears in the adopted flow standards.

[47]MBHE refers to the Matagorda Bay Health Evaluation study, discussed in section "Lower Colorado River Authority Water Management Plan". MBHE 4 is one of the four recommended annual inflow patterns, each of which consists of a seasonal regime of inflow volumes and associated attainment frequencies. Table 2 provides additional information about the MBHE inflow regimes.

[48]The general concept is that authorizations of additional diversion or impoundment should be limited to avoid causing an increase in how often the bay system fails to receive inflow volumes identified as particularly important for maintaining bay health. Those inflow standards for Matagorda Bay also include a criterion for the annual average inflow volume to the bay from the Colorado River that defines an allowable level of decrease in that parameter as evaluated through WAM modeling (30 TAC § 298.330 (a)(1)).

the MBHE 4 criterion, that WAM-calculated frequency with the full exercise of all existing rights was identified as 8.5% at the time the Colorado and Lavaca Rivers and Matagorda and Lavaca Bays BBASC (CL BBASC) report was developed (CL BBASC 2011, p. 119). The WAM calculations also assume a recurrence of historical hydrology. As reflected in Table 2, the differences between those BBEST attainment frequency recommendations, identified as adequate to protect a sound ecological environment, and the WAM Run3 calculated frequencies for Matagorda Bay are very substantial for all identified inflow regime levels.

For all bay systems using an attainment frequency approach, the WAM Run3 calculations reflect the amount of diversion, impoundment, and consumption that has been

**Table 2** Comparison of attainment frequencies for freshwater inflow volumes to Matagorda Bay from the Colorado River in Texas recommended by the CL BBEST as adequate to support a sound ecological environment to attainment frequencies for those same volumes protected by adopted Senate Bill 3 environmental flow standards. Attainment frequencies protected by flow standards assume full exercise of all existing water rights, including full authorized consumptive use (WAM Run3). Threshold Regime has a monthly volume and frequency, and other regimes have seasonally distributed volumes with annual attainment frequencies. WAM Run3 attainment frequencies are from Table 7.7-1 in CL BBASC report (2011)

| Freshwater Inflows to Matagorda Bay from Colorado River: Comparison of BBEST Recommendations to Flow Standards | Attainment Frequency (% of months or years) | |
| --- | --- | --- |
| Volumes and Seasons Consistent in BBEST Recommendations and Flow Standards (Flow Volume/Associated Time Period) | BBEST | Flow Standards (WAM Run3) |
| Threshold Regime (drought level inflows as minimum monthly volume) $1.85 \times 10^7$ m$^3$ ($1.5 \times 10^4$ af)/month | 100% | 65.5% |
| MBHE 1 Regime Spring/Fall/Intervening Season $1.406 \times 10^8$ m$^3$/$9.99 \times 10^7$ m$^3$/$1.295 \times 10^8$ m$^3$ $1.14 \times 10^5$ af/$8.1 \times 10^4$ af/$1.05 \times 10^5$ af | 90% | 35.6% |
| MBHE 2 Regime Spring/Fall/Intervening Season $2.081 \times 10^8$ m$^3$/$1.479 \times 10^8$ m$^3$/$1.917 \times 10^8$ m$^3$ $1.68 \times 10^5$ af/$1.199 \times 10^5$ af/$1.554 \times 10^5$ af | 75% | 16.9% |
| MBHE 3 Regime Spring/Fall/Intervening Season $3.037 \times 10^8$ m$^3$/$2.159 \times 10^8$ m$^3$/$2.798 \times 10^8$ m$^3$ $2.462 \times 10^5$ af/$1.75 \times 10^3$ af/$2.268 \times 10^5$ af | 60% | 11.9% |
| MBHE 4 Regime Spring/Fall/Intervening Season $5.343 \times 10^8$ m$^3$/$3.797 \times 10^8$ m$^3$/$4.922 \times 10^8$ m$^3$ $4.332 \times 10^5$ af/$3.078 \times 10^5$ af/$3.99 \times 10^5$ af | 35% | 8.5% |

authorized in existing water rights. No adjustments to historical hydrology are included in the model runs to reflect anticipated impacts of a changing climate or, in most instances, changes in groundwater contributions to surface flow. It is not likely that all existing water rights will be fully used, with full authorized consumption, every year, as is assumed in the Run3 calculation. However, it is unclear how much solace to take from that consideration because it also is not likely that the historical hydrology assumed for the WAM runs will be repeated.

The freshwater inflow standards for San Antonio Bay and for the Nueces Bay and Delta incorporate a similar approach for calculating and protecting freshwater inflow attainment frequencies, relying solely on a modeling exercise over the historical WAM period-of-record. However, those adopted flow standards, instead of prohibiting any change in predicted attainment frequencies from the levels corresponding to full use of existing permits (WAM Run3), prescribe specific limits on the allowable amount of predicted change[49] for different aspects of the freshwater inflow standards (30 TAC § 298.380 (a) San Antonio Bay, 30 TAC § 298.430 (a) Nueces Bay and Delta).

Because the attainment frequencies used for evaluating applications for new permits are not listed in the flow standards for most of the bay systems with specific inflow criteria (Matagorda Bay, Lavaca Bay, San Antonio Bay, and Nueces Bay and Delta), it is not possible to understand the protection levels provided by the standards from reviewing the standards themselves.[50] The standards for inflows to Galveston Bay from the Trinity and San Jacinto rivers differ

---

[49] For example, for one of the inflow regimes for inflows to San Antonio Bay, referred to by TCEQ as Spring 6, the flow standards provide that the modeled permitting frequency—the WAM Run3 value for existing permits—shall not be increased by more than 8% as a result of the issuance of new permits (30 TAC § 298.380 (a)(3)(B)). The Spring 6 inflow regime, which the BBEST referred to as the G1-D level, identifies a range of very low inflow volumes during spring months for which the BBEST identified a frequency of occurrence of no more than 9% of years as being adequate to support a sound ecological environment (GSA BBEST 2011 Table 6.1–18 p 6.23). The flow standards provide that the 8% value is to be added to the modeled result (30 TAC § 298.380 (a)(2)). If the modeled permitted frequency is 20% of years, the adopted flow standards allow that to be increased to 28% of years (20 + 8) as the result of the issuance of new permits.

[50] This aspect of the rules can be particularly confusing because, depending on which bay system is being reviewed, an annual strategy frequency, strategy target frequency, or a target frequency is sometimes stated. But, as explained in the definitions in the rules, except for Galveston Bay, those frequency values are used solely as a target for informing the use of proactive strategies to increase inflows. Specific attainment frequencies are not listed in the rules for WAM Run3 results for the other bay systems, which is how new applications are evaluated. One stated rationale for not listing those frequencies is that the WAM calculation may change somewhat, even without new water rights being authorized, due to factors such as minor changes in existing water rights, underlying modeling code, or the period-of-record for the WAM (CL BBASC 2011, pp. 119–120).

in that respect. For that bay system, the annual and seasonal attainment frequencies used in evaluating new permits, which are derived from WAM Run3 but generally include an allowance for some additional reduction in attainment frequency from the Run3 value to accommodate issuance of new permits, are listed in the flow standards (Fig. 30 TAC § 298.225 (a)).[51] Although labeled in the flow standards as "target frequencies," those are the values used in evaluating new permits affecting inflows to Galveston Bay (30 TAC § 298.225 (a), TCEQ SB 3 Permitting Guidelines § 6.1 p. 7). The inflow standards for Galveston Bay do not include strategy targets for informing implementation of proactive strategies.

**No Real-time Enforcement of Modeling-Based Inflow Standards** Because compliance with these types of freshwater inflow standards is assessed solely through a modeling exercise covering the full WAM period-of-record,[52] there is no real-time mechanism for enforcing a requirement that flows be allowed to pass downstream to the bay.[53] For example, a multi-month, or even multi-year, period of near-zero inflows to Matagorda Bay would not indicate the standards had been violated. Using the example modeling results from Table 1, because the flow standards are met if the Threshold Regime level is predicted to be met in at least 65.5% of months over that period-of-record, any combination of up to 34.5% of future months can be predicted to fall below the Threshold Regime level without violating the standards. How far inflows are predicted to fall below the Threshold Regime level during those months is not considered. In addition, the number of months that actual, as opposed to predicted, inflows are below that volume is irrelevant to the evaluation. As discussed in the following paragraph, permits developed using this approach lack any provisions referencing protection of freshwater inflows. As a result, the flow standards would not prevent the holder of a water right granted after adoption of the flow standards from continuing to divert or impound water during months when actual

inflows to the bay were at or close to zero regardless of how often that had already occurred.[54]

**No Permit Language to Inform Adaptive Management Adjustment of Inflow Protections** For the aspects of freshwater inflow standards that rely on inflow volumes and associated attainment frequencies, neither the volumes nor the attainment frequencies are being incorporated into, or even referenced in, new water right permits (TCEQ SB 3 Permitting Guidelines § 6.0). Unfortunately, in addition to limiting enforcement approaches, that absence creates a serious impediment to adjusting protection levels through adaptive management. SB 3 directs that flow protections applicable to permits governed by its provisions are subject to being increased within specific limits if more protective environmental flow standards are subsequently adopted. The statutory limits on increased protection are defined as a percentage increase (12.5%) of the protection requirement included in the initial permit (TWC § 11.147 (e-1)). Because no protection requirement for freshwater inflows is included in such permits, there is no value that can be increased and that important aspect of adaptive management is undermined.

**Climate Change Risk Falls Fully on Bay Inflows** Relying solely on the modeling approach for inflow protection also places the full risk of climate change, discussed in chapter "Climate Effects on Inflows", on freshwater inflows. Rather than providing for some level of shared risk with water rights issued after adoption of the flow standards, the modeling approach puts inflow protection last in line for claiming a right to water. Without any inflow-protection conditions in permits, actual shortfalls in future flows below the model projections will all come at the expense of freshwater inflows. Even the most junior water right issued many years after adoption of the applicable inflow standards will, in practical effect, be senior in priority to inflow protection because inflow protections based solely on a modeling exercise have no priority date.[55] As a result, if there is not sufficient water to meet both inflow protections and demands under the junior

---

[51] Unfortunately, adding to the potential for confusion, those attainment frequencies, which establish limits for permitting, are labeled as "target frequencies."

[52] As discussed in section "Evaluation of Water Availability", the WAM period of record for most basins begins in the 1940s and extends over a period of about 50 to 60 years (Alexander and Henderson 2020 § 12.7–12-8). More information about the WAM models and modeling is available on the TCEQ website (TCEQ WAMs).

[53] If, when assessing a water right application, the WAM modeling showed an impairment of the freshwater inflow standards, presumably the application either would be denied or would be modified to eliminate that impairment (TCEQ SB 3 Permitting Guidelines § 6.1). However, because any modification would be made to address a modeling prediction rather than a real-time freshwater inflow deficiency, it is not clear if any permit issued after such modifications necessarily would include provisions addressing protection of actual inflows on a real-time basis.

[54] Depending on location, diversion, or impoundment under a new permit may be restricted to avoid impairing downstream senior water rights during dry periods, but any water passed downstream for that purpose likely would be captured for those senior rights before reaching the bay.

[55] As discussed in section "Evaluation of Water Availability", by contrast, flow-protection conditions included in a permit that limit when the permittee can divert or impound water have, in practical effect, the same priority date as the permit. If inflow limits were included in permits, any water right that is junior in priority to a permit with such limits would not be allowed to impair the ability to divert or impound water under the permit by reducing freshwater inflows below those limits.

permit, the junior permit has the right to take the water. Environmental flow set-asides, if any were to be established as discussed in section "No Set-Asides Established", would provide for inflow protection on a real-time basis because they would be assigned a priority date that could be enforced against junior water rights. (TWC § 11.1471 (e)). In recognition of that, SB 3 assigns the Texas Parks and Wildlife Department responsibility to assert the priority date of set-asides (TWC § 11.0841 (c)). It also establishes a mechanism for temporarily suspending a set-aside to make the water available for other uses in the event of emergency conditions that cannot be resolved in some other way (TWC § 11.148 (a-1)).

### Target Frequencies for Improving Inflows Through Proactive Strategies

The adopted flow standards prescribing freshwater inflow protections for several bay systems include separate attainment frequencies intended to guide implementation of proactive strategies (Matagorda Bay (Fig. 30 TAC § 298.330 (a) (2)); Lavaca Bay (Fig. 30 TAC § 298.330 (c)); San Antonio Bay (Fig. 30 TAC § 298.380 (a)(3) spring, Fig. 30 TAC § 298.380 (a)(4) summer); Mission and Aransas Bays (Fig. 30 TAC § 298.380 (a)(5)); and Nueces Bay (Fig. 30 TAC § 298.430 (a)(3)). Those values reflect the attainment frequencies that the respective BBESTs identified as being adequate to support a sound ecological environment. As discussed in section "Adoption of Specific Freshwater Inflow Standards", those BBEST-recommended attainment frequencies call for much more frequent achievement of the recommended inflow volumes than is provided for in the attainment frequencies used in evaluating future water right applications under the TCEQ-adopted flow standards.

Those strategy, or target, frequencies serve solely to guide the use of voluntary approaches for increasing inflow levels (30 TAC § 298.305 (2), 30 TAC § 298.355 (8), and 30 TAC § 298.405 (9)). Like the other aspects of those inflow standards, the strategy targets are stated as attainment frequencies to be assessed through modeling over the full WAM period-of-record. Importantly, the flow standards for those bay systems provide that if proactive strategies are implemented in a manner that improves the attainment frequency reflected in the WAM modeling, those improved attainment frequencies are protected against impairment by subsequent permit applications (30 TAC § 298.330 (b) & (d) Matagorda & Lavaca bays, 30 TAC § 298.380 (b) San Antonio-Mission-Aransas bays, 30 TAC § 298.430 (b) Nueces Bay).[56]

A settlement agreement[57] the City of Houston executed in 2011 with several conservation organizations to resolve protests of two water right applications by the City illustrates one type of approach that can help maintain essential inflows and could improve attainment frequencies reflected in WAM modeling (Settlement Agreement Houston).[58] One of those applications sought authorization for the indirect reuse[59] of up to $7.161 \times 10^8$ m$^3$ ($5.809 \times 10^5$ af) per year of treated wastewater return flows from the City's water supply sources. The City agreed to ensure, subject to limited exceptions, that at least 50% of the total wastewater authorized for treatment at the City's wastewater plants from its current supply sources would continue to be discharged and allowed to flow to Galveston Bay. That represents a long-term commitment of about $3.580 \times 10^8$ m$^3$ ($2.905 \times 10^5$ af) per year of drought reliable inflows to the bay that otherwise could have been reused and consumed. The agreement contemplates further efforts to formalize recognition of the dedication of those return flows to flow protection. Formal recognition could result in a change in the attainment frequencies for freshwater inflows reflected in the WAM.[60]

Proactive strategy commitments that large and with that level of potential to change attainment frequencies, although extremely beneficial, currently are rare. Relatively small-scale proactive strategies designed to provide strategically targeted incremental benefits also may be important contributors to bay health, particularly during serious drought periods. For example, three nonprofit organizations, the Galveston Bay Foundation and the Texas chapter of the Nature Conservancy, working with Texas Water Trade,[61]

---

[56] Without this type of provision, implementation of proactive strategies that resulted in a change in inflow attainment frequencies in WAM modeling—above the levels protected in the flow standards—could create the potential for new permits being issued that would deprive the bay of the inflows otherwise provided through implementation of the strategies.

[57] The broad outlines of the settlement are summarized in a 2011 article in the Houston Chronicle (Houston Chronicle 3/8/2011).

[58] Unfortunately, the current environmental flow standards for inflows to Galveston Bay do not include strategy targets or language about protecting attainment frequencies improved through such strategies. However, the adaptive management component of SB 3, discussed in section "Adaptive Management", provides a potential mechanism for revising the standards to add strategy targets.

[59] Water rights in Texas authorize, in the absence of specific permit language to the contrary, complete consumption of the water diverted prior to its return to a watercourse (TWC § 11.046 (c)), known as direct reuse. However, additional authorization is required to implement "indirect reuse," which involves the discharge of return flows, discussed in section "Right of Full Consumption", to a watercourse and subsequent diversion downstream (TWC § 11.042 (b)–(c)).

[60] Without the settlement agreement, the City had the right to directly reuse and consume all its return flows. As discussed in section "Right of Full Consumption", the full authorization run of the WAM (WAM Run3), which is what the attainment frequencies in the flow standards are based on, assumes such full consumption. Because the terms of the agreement limiting the City's right to direct reuse are not currently reflected in its water rights, the relevant WAM Run3 does not reflect changed attainment frequencies based on the agreement.

[61] Texas Water Trade is working in Texas, with various partners, to facilitate a variety of market transactions for flow protection (Texas Water Trade Website).

have begun implementing smaller inflow-protection transactions involving purchases of water rather than purchases of water rights.[62] One of those transactions involves the purchase of $6.167 \times 10^6$ m$^3$ ($5.0 \times 10^3$ af) of water for wetland and estuary habitat that can be used to provide freshwater to specific portions of Galveston Bay at key times. Another involves a water transaction benefiting 121 hectares (300 acres) of coastal marsh habitat along Matagorda Bay (Texas Water Trade Flows Fund). Although such smaller transactions likely have the potential to provide important individual and cumulative environmental benefits, they are unlikely, especially because the duration of the individual transactions may be brief, to result in a change in WAM attainment frequencies.

## Adaptive Management

Adaptive management is a key component of SB 3. In adopting environmental flow standards, TCEQ acknowledged the importance of the adaptive management aspects for ensuring protection of a sound ecological environment in the state's surface waters.

> HB3/SB3 is an adaptive management process. As such, the determination of whether a sound ecological environment exists at some future date, and how that sound ecological environment should be protected, is a topic that can be considered by future science team and stakeholder groups when considering recommendations for revisions to the adopted rules.

(Texas Register 2012, p. 6649). Unfortunately, the adaptive management component currently appears to lack momentum. As discussed in section "The Regional Actors: Bay and Basin Area Stakeholder Committees and Bay and Basin Expert Science Teams", SB 3 contemplates a review at least once every 10 years for the environmental flow recommendations by the BBASCs and BBESTs, for environmental flow standards, and for proactive strategies (TWC § 11.02362 (p)). Flow standards were adopted for various bay and basin areas by TCEQ in 2011, 2012, and 2014 (Texas Register 2011, p. 2908; Texas Register 2012, p. 6629; Texas Register 2014, p. 1416). Various adaptive management studies have been funded to help inform potential revisions to aspects of the standards and to evaluate some proactive strategy options.[63]

However, as of early 2023, there is no announced plan for formally initiating the adaptive management process. The issue of lost momentum in adopting and revisiting environmental flow standards has been acknowledged in the review, in anticipation of the 2023 Session of the Texas Legislature, by the Texas Sunset Advisory Commission[64] of programs implemented by TCEQ. The uncompleted work in implementing SB 3 has been recognized and recommendations for helping to re-energize those efforts have been adopted and forwarded for consideration by the Texas Legislature during its 2023 legislative session (Sunset Advisory Commission Staff Report with Commission Decisions 2022, pp. 43–46).

## Recommendations for Research and Analysis to Inform Senate Bill 3 Implementation

Various statutory and funding refinements are needed to help get effective implementation of Senate Bill 3 back on track,[65] some of which may arise from the Sunset Advisory Commission process discussed in section "Adaptive Management". Regardless of when those refinements are put in place, additional analyses and information will be needed to identify the best path forward. The recommendations provided here focus on potential studies and analyses that could help to inform efforts to address key shortcomings through a reinvigorated SB 3 process.

**Analysis of Approaches for Implementing Proactive Flow Protection Strategies** As discussed in section "The State Level Actors: Environmental Flows Advisory Group, Science Advisory Committee, and Texas Commission on Environmental Quality", SB 3 calls for a state-level study, overseen by the Environmental Flows Advisory Group, of methods for encouraging proactive strategies to convert reasonable amounts of existing water rights to flow protection purposes. Unfortunately, that study has not been conducted and progress on implementing such strategies has been quite limited. Because of the legacy of perpetual water rights that lack flow-protection provisions and that have the first claim to available water during dry periods, implementation of such strategies on a significant level will be necessary for

---

[62] The purchase of a water right would allow conversion of the authorized amount to flow protection purposes for the duration of the right, which usually is perpetual. By contrast, the purchase of water involves obtaining a specific quantity of water over a defined time-period, which might range from a month to multiple years. A short-term purchase of water is likely to be more readily available and much less expensive than the purchase of the underlying water right.

[63] Although not entirely current as of early 2023, the TWDB website provides a listing of, and links to, many of the studies undertaken, with funding overseen by TWDB, to help inform potential revisions to flow standards and evaluation of potential affirmative strategies to improve flow conditions (TWDB Statewide Environmental Flows (SB 3)).

[64] Texas agencies undergo a periodic review, normally on a 12-year cycle, by the Texas Sunset Advisory Commission to determine if the agency should be continued in existence and, if so, whether adjustments to agency procedures or programs may be appropriate. The Sunset Commission, made up of legislators and legislative appointees, makes recommendations that are then considered by the Texas Legislature (Texas Sunset Advisory Commission 2021 Sunset in Texas).

[65] A brief discussion of potential statutory refinements can be found in a 2021 document developed by the author for the Texas Living Waters Project (Delivering on the Unrealized Potential of Senate Bill 32021 p 47–51).

ensuring adequate freshwater inflows for the long-term. Analysis of potential approaches for incentivizing such conversions could provide significant value in informing that statewide study when it does occur and in assisting Bay and Basin Area Stakeholder Committees with the identification of potential proactive strategies for consideration during the adaptive management process.

**Identification of Inflow Targets for Proactive Strategies to Supplement Drought Period Inflows** In addition to identifying ways to encourage the implementation of proactive strategies, identification of the most important locations and times for delivery of freshwater inflows is needed. Particularly during drought conditions, when inflows will be especially low, targeted amounts of relatively small quantities of freshwater strategically delivered to key habitats have the potential to provide important benefits (Montagna et al. 2021, pp. 134–135). Analysis of the potential value of this approach in specific bay systems and of the recommended timing, volume, and location for delivery of inflows to provide the greatest benefit could advance the development and implementation of the types of proactive strategies referenced in section "Target Frequencies for Improving Inflows Through Proactive Strategies".

**Analysis of Options for Quantifiable Permit Conditions for Inflow Protection** As discussed in section "Adoption of Specific Freshwater Inflow Standards", SB 3 requires a flow-protection adjustment provision to be included in new permits and certain permit amendments approved after September 1, 2007. If applicable flow standards are made more protective, that provision allows for adjusting flow protections by up to a 12.5% increase in the annualized total of specific pass-through or release requirements included in the permit or amendment. Currently, protections for instream flows at inland locations on rivers and streams are being implemented through specific permit requirements, but protections for freshwater inflows, as applied for most bay and estuary systems, are being implemented solely through a modeling exercise undertaken at the time of permit review. Reliance solely on a modeling approach results in permits that lack any quantified requirements for freshwater inflow protections to which that flow-protection adjustment provision can be applied.

The ability to adjust protections to reflect improved understanding of inflow needs over time is a critical component of the flow-protection approach of SB 3. Reliance solely on a modeling approach also means that freshwater inflow protections are, in effect, treated as junior in priority even to permits issued after the flow standards were adopted. As also discussed in section "Adoption of Specific Freshwater Inflow Standards", that unduly places the risk from climate change on inflow protections. Accordingly, there is a need to identify

approaches for supplementing the current modeling approach by including quantified freshwater inflow protection language in new permits.

**Informing More Refined Criteria for Attainment Frequency Evaluation** For many bay systems, compliance with freshwater inflow provisions currently is assessed solely through evaluation of long-term attainment frequencies evaluated over the full WAM period-of-record—generally 50 years or longer. In addition to creating problems for implementing the flow-protection adjustment provision during adaptive management, that evaluation approach fails to provide a readily available tool for assessing, and addressing, potentially problematic shorter-term periods of extremely low inflows during drought periods.[66] For example, the adopted flow standards for inflows to Galveston Bay from the Trinity and San Jacinto rivers have no inflow criteria in effect, even on an annual basis, during the driest 25 percent of the time.[67] Development of permit criteria and other indicators for identifying problematic drought-level inflows on a real-time, or near real-time, basis could provide important benefits both for implementing permit requirements and for informing implementation of proactive strategies. The Threshold Inflow Regime level of a minimum monthly inflow of $1.85 \times 10^7$ m$^3$ ($1.5 \times 10^4$ af) from the Colorado River to Matagorda Bay, discussed in section "Adoption of Specific Freshwater Inflow Standards" and summarized in Table 2, is an example of one such approach. Analyses of potential approaches for use in other bay systems could provide beneficial insights.

**Approaches for Establishing Set-Asides for Flow Protection** Although, as discussed in section "No Set-Asides Established", TCEQ declined to establish any set-asides of unappropriated water for environmental flow protection, the agency has acknowledged its willingness to consider doing so in the future. Unfortunately, the amount of unappropriated water available to be set aside continues to decrease as new permits are granted. Indeed, for many bay systems, there may be little, if any, unappropriated water available on a fully reliable basis—that is, water available even during future serious drought conditions under WAM Run3 assumptions—to be set aside for freshwater inflow protection. However, the failure to set aside any unappropri-

---

[66] The Nueces BBASC Report identifies a related research priority focused on duration of no-flow periods (Nueces BBASC 2012, p. 107).

[67] The lowest volume targets for annual inflow amounts have an annual attainment frequency of 75%, which means there are no minimum inflow levels set for the remaining 25% of years. Similarly, the lowest volume targets for seasonal inflows have an attainment frequency of 60%, which means there are no minimum inflow levels for 40% of seasons (Figure 30 TAC § 298.225 (a)).

ated flow, even as other water rights continue to be issued, appears to be inconsistent with the directives of SB 3 and unnecessarily allows for the worsening of an already highly challenging flow-protection situation.

Having water set aside for flow protection, even if that water would not be reliably available during extreme drought conditions, would help to limit the risk to bay inflows from climate change and to help make water available for creative management approaches. For example, if water is set aside for flow protection, it may be possible to implement approaches for storing some of that water during years of more average rainfall for release for bay inflows during drought years. Analyses of the availability, by basin, of unappropriated water for set-asides and of approaches for characterizing set-asides to maximize inflow benefits consistent with balancing impacts on future permitting decisions could provide important insights for the adaptive management process.

**Acknowledgement** This publication was funded in part through Contract No. 21-155-007-C879 from the Texas General Land Office (GLO) with Gulf of Mexico Energy Security Act of 2006 funding made available to the State of Texas and awarded under the Texas Coastal Management Program. The views contained herein are those of the author and should not be interpreted as representing the views of the GLO or the State of Texas.

# References

Agreement for settlement of disputes regarding water rights applications numbers 5826 and 5827 by the City of Houston (2011), on file with the author

Alexander Martin K, Henderson R (2020) Determining surface water availability. In Sahs M (ed) Essentials of Texas water resources, 6. State Bar of Texas, ENRLS Austin, TX, pp 12-1–12-11

Brazos River and Associated Bay and Estuary System Basin and Bay Area Stakeholders Committee Environmental Flow Standards and Strategies Recommendations Report (2012). Available via https://tamucc-ir.tdl.org/handle/1969.6/94337. Accessed 16 Jan 2023

Brazos River Basin and Bay Expert Science Team Environmental Flow Regime Recommendations Report (2012). Available via https://tamucc-ir.tdl.org/handle/1969.6/94735. Accessed 24 Feb 2023

Brun B, Jackson K, Lake P, Walker J (2016) Texas aquifers study: groundwater quantity, Quality, Flow and Contributions to Surface Water. Available via https://texashistory.unt.edu/ark:/67531/metapth1114570/m2/1/high_res_d/UNT-0002-0007.pdf. Accessed 9 Jan 2023

Canseco S (2020) Groundwater law and regulation. In Sahs M (ed) Essentials of Texas water resources, 6th edn. State Bar of Texas, ENRLS Austin, TX, pp 5-1–5-19

Caroom D, Maxwell S (2020) Surface water rights permitting. In Sahs M (ed) Essentials of Texas water resources, 6th edn. State Bar of Texas, ENRLS Austin, TX pp 10-1–10-18

Certificate of Adjudication 14-5478, Texas Commission on Environmental Quality website. https://gisweb.tceq.texas.gov/WRRetrieveRights/?ID=ADJ5478. Accessed 13 Feb 2023

Certificate of Adjudication 14-5482, Texas Commission on Environmental Quality website. https://gisweb.tceq.texas.gov/WRRetrieveRights/?ID=ADJ5482. Accessed 13 Feb 2023

Certificate of Adjudication 16-2095, Texas Commission on Environmental Quality website. https://gisweb.tceq.texas.gov/WRRetrieveRights/?ID=ADJ2095. Accessed 13 Feb 2023

Certificate of Adjudication 21-3214, Texas Commission on Environmental Quality website. https://gisweb.tceq.texas.gov/WRRetrieveRights/?ID=ADJ3214. Accessed 13 Feb 2023

Colorado and Lavaca Basin and Bay Area Stakeholder Committee Environmental Flows Recommendation Report (2011). Available via https://tamucc-ir.tdl.org/handle/1969.6/94359. Accessed 10 March 2023

Colorado and Lavaca Rivers and Matagorda and Lavaca Bays Basin and Bay Area Expert Science Team Environmental Flows Regime Recommendations Report (2011). Available via https://tamucc-ir.tdl.org/handle/1969.6/94332. Accessed 16 Jan 2023

Eckstein G, Hardberger A (2020) Scientific, legal, and ethical foundations for Texas water law. In Sahs M (ed) Essentials of Texas water resources, 6th edn. State Bar of Texas, ENRLS Austin, TX, pp 1-1–1-29

Gooch T, Sloan D, Acevedo A (2020) Water reuse. In Sahs M (ed) Essentials of Texas water resources, 6th edn. State Bar of Texas, ENRLS Austin, TX, pp 24-1–24-38

Guadalupe, San Antonio, Mission and Aransas Rivers and Mission, Copano, and Aransas and San Antonio Bays Basin and Bay Expert Science Team Environmental Flows Recommendations Report (2011). Available via https://tamucc-ir.tdl.org/handle/1969.6/94333. Accessed 16 Jan 2023

Guadalupe, San Antonio, Mission and Aransas Rivers and Mission, Copano, Aransas, and San Antonio Bays Bay and Basin Area Stakeholders Committee Recommendations Report (2011). Available via https://tamucc-ir.tdl.org/handle/1969.6/94367. Accessed 21 Feb 2023

Hess M (2021) Delivering on the unrealized potential of Senate Bill 3. Available via https://texaslivingwaters.org/wp-content/uploads/2021/05/The-Unrealized-Potential-of-SB-3.pdf. Accessed 21 Feb 2023

Houston Chronicle, Water pact a win-win for Houston-Galveston Bay (March 8, 2011). https://www.chron.com/news/houston-texas/article/Water-pact-a-win-win-for-Houston-Galveston-Bay-1689281.php. Accessed 13 Feb 2023

Jarvis G (2020) Historical development of Texas surface water law: background of the appropriation and permitting system and management of surface water resources. In Sahs M (ed) Essentials of Texas water resources, 6th edn. State Bar of Texas, ENRLS Austin, TX, pp 4-1–4-36

Lavaca-Navidad River Authority water right application number 13728. Available via https://www.tceq.texas.gov/downloads/permitting/water-rights/pending/admin-complete/lnra-13728-adcpackage.pdf. Accessed 27 Feb 2023

LCRA – SAWS Water Project (2008) Matagorda Bay Inflow Criteria (Colorado River) Matagorda Bay Health Evaluation, on file with the author

Lower Colorado River Authority (2020) Lakes Buchanan and Travis water management plan and drought contingency plan. Available via https://www.lcra.org/download/2020-water-management-plan/?wpdmdl=11923. Accessed 26 Feb 2023

Martinez R, Smith R, Tadema D, Groetsch I (2020) Water rights enforcement. In Sahs M (ed) Essentials of Texas water resources, 6th edn. State Bar of Texas, ENRLS Austin, TX p 13-7–13-12

Montagna PA, McKinney L, Yoskowitz D (2021) Focused flows to maintain natural nursery habitats. Tx Water J 12(1):129–139. https://doi.org/10.21423/twj.v12i1.7123

Norman M, Hutchinson W (2020) Groundwater management area joint planning. In Sahs M (ed) Essentials of Texas water resources, 6th edn. State Bar of Texas, ENRLS Austin, TX pp 21-1–21-22

Nueces River and Corpus Christi and Baffin Bay Basin and Bay Area Stakeholder Committee Environmental Flow Standards and Strategies Recommendation Report (2012). Available via https://tamucc-ir.tdl.org/handle/1969.6/94369. Accessed 16 Jan 2023

Nueces River and Corpus Christi and Baffin Bays Basin and Bays Expert Science Team Environmental Flows Recommendation Report (2011). Available via https://tamucc-ir.tdl.org/handle/1969.6/94335. Accessed 16 Jan 2023

Sabine and Neches Rivers and Sabine Lake Bay Basin and Bay Expert Science Team Environmental Flows Recommendations Report (2009) Available via https://tamucc-ir.tdl.org/handle/1969.6/94336. Accessed 10 March 2023

Senate Bill 3 Science Advisory Committee for Environmental Flows (2009) Methodologies for establishing a freshwater inflow regime for Texas estuaries within the context of the Senate Bill 3 environmental flows process. Available via https://tamucc-ir.tdl.org/handle/1969.6/94344. Accessed 17 Jan 2023

Senate Bill 3 Science Advisory Committee for Environmental Flows (2010) Lessons learned from initial SB3 BBEST Activities. Available at https://tamucc-ir.tdl.org/handle/1969.6/94351. Accessed 3 Feb 2023

Senate Bill 3 Science Advisory Committee for Environmental Flows Guidance and discussion papers (2009–2011). Available via https://www.tceq.texas.gov/permitting/water_rights/wr_technical-resources/eflows/resources.html. Accessed 13 Feb 2023

Sugg Z, Ziaja S, Schlager E (2020) Conjunctive management of surface water and groundwater resources. In Sahs M (ed) Essentials of Texas water resources, 6th edn. State Bar of Texas, ENRLS Austin, TX pp 6-1–6-16

Texas Administrative Code, Title 30 § 297.1 (17). https://texreg.sos.state.tx.us/public/readtac$ext.TacPage?sl=R&app=9&p_dir=&p_rloc=&p_tloc=&p_ploc=&pg=1&p_tac=&ti=30&pt=1&ch=297&rl=1. Accessed 10 Oct 2022

Texas Administrative Code, Title 30, § 297.42. https://texreg.sos.state.tx.us/public/readtac$ext.TacPage?sl=R&app=9&p_dir=&p_rloc=&p_tloc=&p_ploc=&pg=1&p_tac=&ti=30&pt=1&ch=297&rl=42. Accessed 10 Oct 2022

Texas Administrative Code, Title 30 § 298.225. https://texreg.sos.state.tx.us/public/readtac$ext.TacPage?sl=R&app=9&p_dir=&p_rloc=&p_tloc=&p_ploc=&pg=1&p_tac=&ti=30&pt=1&ch=298&rl=225

Texas Administrative Code, Title 30 Figure § 298.225 (a). https://texreg.sos.state.tx.us/fids/201101539-1.html. Accessed 10 Oct 2022

Texas Administrative Code, Title 30 § 298.275 (d). https://texreg.sos.state.tx.us/public/readtac$ext.TacPage?sl=R&app=9&p_dir=&p_rloc=&p_tloc=&p_ploc=&pg=1&p_tac=&ti=30&pt=1&ch=298&rl=275. Accessed 10 Oct 2022

Texas Administrative Code, Title 30 Figure § 298.280 (5). https://texreg.sos.state.tx.us/fids/201101540-5.html. Accessed 10 Oct 2022

Texas Administrative Code, Title 30 § 298.305. https://texreg.sos.state.tx.us/public/readtac$ext.TacPage?sl=T&app=9&p_dir=P&p_rloc=158574&p_tloc=&p_ploc=1&pg=6&p_tac=&ti=30&pt=1&ch=298&rl=330. Accessed 10 Oct 2022

Texas Administrative Code, Title 30 § 298.330. https://texreg.sos.state.tx.us/public/readtac$ext.TacPage?sl=R&app=9&p_dir=&p_rloc=&p_tloc=&p_ploc=&pg=1&p_tac=&ti=30&pt=1&ch=298&rl=330. Accessed 10 Oct 2022

Texas Administrative Code, Title 30 Figure § 298.330 (a)(2). https://texreg.sos.state.tx.us/fids/201204234-4.html. Accessed 10 Oct 2022

Texas Administrative Code, Title 30 Figure § 298.330(c). https://texreg.sos.state.tx.us/fids/201204234-5.html. Accessed 10 Oct 2022

Texas Administrative Code, Title 30 § 298.355 (8). https://texreg.sos.state.tx.us/public/readtac$ext.TacPage?sl=T&app=9&p_dir=N&p_rloc=158477&p_tloc=&p_ploc=1&pg=15&p_tac=&ti=30&pt=1&ch=298&rl=330. Accessed 10 Oct 2022

Texas Administrative Code, Title 30 § 298.380(a). https://texreg.sos.state.tx.us/public/readtac$ext.TacPage?sl=R&app=9&p_dir=&p_rloc=&p_tloc=&p_ploc=&pg=1&p_tac=&ti=30&pt=1&ch=298&rl=380

Texas Administrative Code, Title 30 Figure § 298.380(a)(3). https://texreg.sos.state.tx.us/fids/201204235-1.html. Accessed 10 Oct 2022

Texas Administrative Code, Title 30 Figure § 298.380(a)(4). https://texreg.sos.state.tx.us/fids/201204235-2.html. Accessed 10 Oct 2022

Texas Administrative Code, Title 30 Figure § 298.380(a)(5). https://texreg.sos.state.tx.us/fids/201204235-3.html. Accessed 10 Oct 2022

Texas Administrative Code, Title 30 § 298.405 (9). https://texreg.sos.state.tx.us/public/readtac$ext.TacPage?sl=T&app=9&p_dir=N&p_rloc=166351&p_tloc=&p_ploc=1&pg=24&p_tac=&ti=30&pt=1&ch=298&rl=330. Accessed 10 Oct 2022

Texas Administrative Code, Title 30 § 298.430 (b). https://texreg.sos.state.tx.us/public/readtac$ext.TacPage?sl=T&app=9&p_dir=N&p_rloc=166355&p_tloc=&p_ploc=1&pg=18&p_tac=&ti=30&pt=1&ch=298&rl=330. Accessed 10 Oct 2022

Texas Administrative Code, Title 30 Figure § 298.430(a)(3). https://texreg.sos.state.tx.us/fids/201400706-1.html. Accessed 10 Oct 2022

Texas Commission on Environmental Quality Agreed Order Choke Canyon Reservoir (1992). https://www.tceq.texas.gov/downloads/permitting/water-rights/nueces-estuary-advisory-council/1992-agreedorder.pdf. Accessed 10 Oct 2022

Texas Commission on Environmental Quality Agreed Order Choke Canyon Reservoir (2001). https://www.tceq.texas.gov/downloads/permitting/water-rights/nueces-estuary-advisory-council/2001-neac-agreedorder.pdf. Accessed 13 Feb 2023

Texas Commission on Environmental Quality Nueces Estuary Advisory Council. https://www.tceq.texas.gov/permitting/water_rights/neac-advisory.html. Accessed 13 Feb 2023

Texas Commission on Environmental Quality, Senate Bill 3 Revised Permitting Guidelines (undated draft). Available via https://wayback.archive-it.org/414/20210528093937/https://www.tceq.texas.gov/assets/public/permitting/watersupply/water_rights/eflows/revised_draft_sb3_implementation_guidelines.pdf. Accessed 13 Feb 2023

Texas Commission on Environmental Quality, water availability modeling information. https://www.tceq.texas.gov/permitting/water_rights/wr_technical-resources/wam.html. Accessed 30 Jan 2023

Texas Commission on Environmental Quality, watermaster information. https://www.tceq.texas.gov/permitting/water_rights/wmaster. Accessed 30 Jan 2023

Texas Legislature, Acts 1975, ch. 344, 64th Leg., SB 137. https://lrl.texas.gov/scanned/sessionLaws/64-0/SB_137_CH_344.pdf. Accessed 17 Jan 2023

Texas Legislature, Acts 1985, ch. 133, 69th Leg., HB 2. https://lrl.texas.gov/scanned/sessionLaws/69-0/HB_2_CH_133.pdf. Accessed 17 Jan 2023

Texas Legislature, Acts 1985, ch. 795, 69th Leg., SB 249. https://lrl.texas.gov/scanned/sessionLaws/69-0/SB_249_CH_795.pdf. Accessed 17 Jan 2023

Texas Legislature, Acts 1987, ch. 419, 70th Leg., SB 683. https://lrl.texas.gov/scanned/sessionLaws/70-0/SB_683_CH_419.pdf. Accessed 17 Jan 2023

Texas Legislature, Acts 2007, ch. 1351, 80th Leg., SB 3. https://capitol.texas.gov/tlodocs/80R/billtext/pdf/SB00003F.pdf#navpanes=0. Accessed 13 Feb 2023

Texas Legislature, Acts 2021, ch. 689, 87th Leg., HB 2225. https://capitol.texas.gov/tlodocs/87R/billtext/pdf/HB02225F.pdf#navpanes=0. Accessed 13 Feb 2023

Texas Register Vol. 36 No. 18, 36 TexReg 2908 (May 6, 2011). Available via https://texashistory.unt.edu/ark:/67531/metapth176618/m2/1/high_res_d/0506is.pdf. Accessed 13 Feb 2023

Texas Register Vol. 37 No. 34, 37 TexReg 6629 (Aug. 24, 2012). Available via https://texashistory.unt.edu/ark:/67531/metapth253226/m2/1/high_res_d/0824is.pdf. Accessed 13 Feb 2023

Texas Register Vol. 39 No 9, 39 TexReg 1416 (Feb. 28, 2014). Available via https://texashistory.unt.edu/ark:/67531/metapth396818/m2/1/high_res_d/0228is.pdf. Accessed 13 Feb 2023

Texas Sunset Advisory Commission, Staff Report with Commission Decisions, Texas Commission on Environmental Quality, Texas Low-Level Radioactive Waste Disposal Compact Commission (2022–23). Available via https://www.sunset.texas.gov/public/uploads/2022-12/TCEQ%20Staff%20Report%20with%20Commission%20Decisions_11-22-22.pdf. Accessed 13 Feb 2023

Texas Sunset Advisory Commission, Sunset in Texas (Sept. 2021). Available at https://www.sunset.texas.gov/public/uploads/files/reports/Sunset%20in%20Texas%202022-23.pdf. Accessed 13 Feb 2023

Texas Water Code § 1.003. https://statutes.capitol.texas.gov/Docs/WA/htm/WA.1.htm#1.003

Texas Water Code § 5.506. https://statutes.capitol.texas.gov/Docs/WA/htm/WA.5.htm#5.506

Texas Water Code § 11.002. https://statutes.capitol.texas.gov/Docs/WA/htm/WA.11.htm#11.002

Texas Water Code § 11.021. https://statutes.capitol.texas.gov/Docs/WA/htm/WA.11.htm#11.021

Texas Water Code § 11.0236. https://statutes.capitol.texas.gov/Docs/WA/htm/WA.11.htm#11.0236

Texas Water Code § 11.02361. https://statutes.capitol.texas.gov/Docs/WA/htm/WA.11.htm#11.02361

Texas Water Code § 11.02362. https://statutes.capitol.texas.gov/Docs/WA/htm/WA.11.htm#11.02362

Texas Water Code § 11.0237. https://statutes.capitol.texas.gov/Docs/WA/htm/WA.11.htm#11.0237

Texas Water Code § 11.027. https://statutes.capitol.texas.gov/Docs/WA/htm/WA.11.htm#11.027

Texas Water Code § 11.042. https://statutes.capitol.texas.gov/Docs/WA/htm/WA.11.htm#11.042

Texas Water Code § 11.046. https://statutes.capitol.texas.gov/Docs/WA/htm/WA.11.htm#11.046

Texas Water Code § 11.0841. https://statutes.capitol.texas.gov/Docs/WA/htm/WA.11.htm#11.0841

Texas Water Code § 11.135. https://statutes.capitol.texas.gov/Docs/WA/htm/WA.11.htm#11.135

Texas Water Code § 11.138. https://statutes.capitol.texas.gov/Docs/WA/htm/WA.11.htm#11.138

Texas Water Code § 11.1381. https://statutes.capitol.texas.gov/Docs/WA/htm/WA.11.htm#11.1381

Texas Water Code § 11.147. https://statutes.capitol.texas.gov/Docs/WA/htm/WA.11.htm#11.147

Texas Water Code § 11.1471. https://statutes.capitol.texas.gov/Docs/WA/htm/WA.11.htm#11.1471

Texas Water Code § 11.148. https://statutes.capitol.texas.gov/Docs/WA/htm/WA.11.htm#11.148

Texas Water Code § 15.301–.331. https://statutes.capitol.texas.gov/Docs/WA/htm/WA.15.htm#15.301

Texas Water Code § 15.3041. https://statutes.capitol.texas.gov/Docs/WA/htm/WA.15.htm#15.3041

Texas Water Code § 15.7031. https://statutes.capitol.texas.gov/Docs/WA/htm/WA.15.htm#15.7031

Texas Water Code § 16.131–.142. https://statutes.capitol.texas.gov/Docs/WA/htm/WA.16.htm#16.131

Texas Water Code § 16.1331. https://statutes.capitol.texas.gov/Docs/WA/htm/WA.16.htm#16.1331

Texas Water Code § 36.001. https://statutes.capitol.texas.gov/Docs/WA/htm/WA.36.htm#36.001

Texas Water Code § 36.108. https://statutes.capitol.texas.gov/Docs/WA/htm/WA.36.htm#36.108

Texas Water Development Board, 1968 State Water Plan. https://www.twdb.texas.gov/waterplanning/swp/1968/index.asp. Accessed 15 Jan 2023

Texas Water Development Board, groundwater management area information. https://www.twdb.texas.gov/groundwater/management_areas/index.asp. Accessed 13 Feb 2023

Texas Water Development Board, groundwater availability models information. https://www.twdb.texas.gov/groundwater/models/index.asp. Accessed 13 Feb 2023

Texas Water Development Board, Senate Bill 3 studies information. http://www.twdb.texas.gov/surfacewater/flows/environmental/index.asp. Accessed 13 Feb 2023

Texas Water Development Board, Texas water trust information. https://www.twdb.texas.gov/waterplanning/waterbank/trust/index.asp. Accessed 18 Jan 2023

Texas Water Trade. https://texaswatertrade.org/. Accessed 13 Feb 2023

Texas Water Trade, Texas Flows Fund priority projects. https://texaswatertrade.org/our-solutions/our-priorities/texas-flows-fund/. Accessed 18 Jan 2023

Trinity and San Jacinto and Galveston Bay Area Stakeholder Committee, Report of the Trinity-San Jacinto-Trinity bay and basin stakeholders committee (2010). Available via https://tamucc-ir.tdl.org/handle/1969.6/94330. Accessed 10 March 2023

Trinity and San Jacinto and Galveston Bay Area Stakeholder Committee, Recommended environmental flow standards and strategies for the Trinity and San Jacinto river basins and Galveston Bay (2010). Available via https://tamucc-ir.tdl.org/handle/1969.6/94329. Accessed 10 March 2023

Trinity and San Jacinto and Galveston Bay Expert Science Team Environmental Flows Recommendations Report (2009). Available at https://tamucc-ir.tdl.org/handle/1969.6/94338. Accessed 27 Feb 2023

# Climate Effects on Inflows

John W. Nielsen-Gammon ⓘ and Alison A. Tarter ⓘ

### Abstract

Climate has a profound influence on the most fundamental properties of Texas bays and estuaries: their water temperatures and salinities, their circulation, and even their locations and spatial extents. With climate change, these influences will evolve and possibly change the character of bays and estuaries. Key climate changes include warmer temperatures, changes in amount and distribution of precipitation, and sea level rise. Inflows to Texas bays and estuaries are not altered solely by climate change, but it is possible to infer the impacts attributable to climate change. Specifically, hydrologic model simulations driven by downscaled climate model projections tend to predict roughly equal chances of increases or decreases in total inflows, but with a clear tendency for more erratic inflows as the wettest months get wetter and the driest months get drier. Projections of most hydrological variables have largely spread across models, indicating that change is likely, but the specific magnitude of change cannot be discerned at this time. Both increasing variability and a long-term trend imply that estuarine ecosystems must be able to adapt to a broader range of conditions than those found historically. Management decisions should allow for a range of possible future climate impacts.

### Keywords

Climate, climate change, runoff, extreme rainfall, downscaling, hydrologic model

J. W. Nielsen-Gammon (✉) · A. A. Tarter
Department of Atmospheric Sciences, Texas A&M University,
College Station, TX, USA
e-mail: n-g@tamu.edu; tartera@exchange.tamu.edu

## Abbreviations

| | |
|---|---|
| CMIP5 | Coupled Model Intercomparison Project phase 5 |
| ET | Evapotranspiration |
| GCM | Global Climate Model |
| LOCA | Localized Constructed Analog downscaling method |
| RCP | Representative Concentration Pathway |
| TCR | Transient Climate Response |
| VIC | Variable Infiltration Capacity hydrologic model |

## Introduction

The importance of present-day climate can be seen in the dramatic change in properties of estuaries along the Texas coast from southwest to northeast (Montagna et al. 2018). The differences in inflows are related to upstream basin sizes and precipitation, while differences in evaporation rates are driven by local temperatures, wind, and sunlight. These important influences have never been static, but in addition to fluctuations caused by natural factors such as the El Niño-Southern Oscillation (ENSO) (Tolan 2007; Kim et al. 2014), some are also changing rapidly as a result of large-scale climate change (Nielsen-Gammon 2011; Nielsen-Gammon et al. 2021). Through simulations of future climate and understanding of the physical processes that are involved, it is possible to get a sense of the sorts of changes that are likely or possible to occur.

This chapter will address Texas climate, climate change, and impacts on inflows to bays and estuaries through the lens of the hydrologic model simulations. The simulations are driven by downscaled global climate model simulations and projections as described by Vano et al. (2020). These projections include the direct effects of climate changes as well as indirect effects, such as through climate-driven changes in ENSO. They are, however, simulations, and as such only represent possible scenarios of what may happen. Consistency

© The Author(s) 2025
P. A. Montagna, A. R. Douglas (eds.), *Freshwater Inflows to Texas Bays and Estuaries*, Estuaries of the World,
https://doi.org/10.1007/978-3-031-70882-4_3

among simulations provides evidence that the projected changes may be robust, but it is possible that the actual changes may even lie outside the envelope of simulated changes. Also, many other factors besides climate change will impact inflows over time, including changes in residential and industrial water use, agriculture and land use, channelization, and streamflow management policies.

The discussions will focus on river basins that can serve as reference points for the wide range of Texas inflows. Along-coast differences in climate will be represented by differences in small coastal drainage basins surrounding Houston (NE) and Kingsville (SW). Along-coast differences in riverine inflows will be represented by the Sabine-Neches Rivers (NE) and the Nueces River (SW). Differences related to the size of drainage basins will be represented by the Lavaca River and the Brazos River, both of which discharge along the central Texas coast.

Most information on future trends is based on global climate model (GCM) output from the Coupled Model Intercomparison Project phase 5 (CMIP5) that has been downscaled using the Localized Constructed Analog (LOCA) method (Pierce et al. 2014, 2015). This statistical downscaling adjusts the GCM output to agree with local historical observations and also estimates climate patterns at high spatial resolution from the relatively coarse GCM output. In this case, the local observations are the Livneh gridded dataset (Livneh et al. 2013, 2015) of daily maximum and minimum temperatures and precipitation over the period 1976–2005. The use of historic weather patterns means, for example, that downscaled temperatures during simulated onshore flow days will have smooth variations at the coastline while temperatures during offshore flow days will have a sharp change at the coastline, consistent with what actually happens. The downscaled temperature and precipitation during the historical period will have similar long-term average values as well as similar temporal variability to what actually happened during 1976–2005.

The comparisons in this chapter use the simulations from downscaled GCM historical output as a reference for the historical climate, rather than actual historical inflows. This ensures that the downscaled climate model projections represent the simulated change from current to future climate. If an observed historical data source was used, the difference from projected downscaled GCM output would partly be due to differences between the historical data source and the downscaled historical GCM simulations. Using simulations for both past and future conditions provide a reasonable representation of current climate and isolates the simulated climate changes.

The downscaled GCM data consists solely of temperatures, precipitation, and wind. These data have been used as input to the Variable Infiltration Capacity (VIC) hydrologic model (Liang et al. 1996) to simulate processes such as evaporation, runoff, and baseflow at the same spatial resolution as the downscaled LOCA information. The VIC simulations are not bias-corrected, but they are based on the LOCA bias-corrected temperatures and precipitation. The historical VIC output should not be regarded as equivalent to historical observations. Much of what VIC simulates is unobserved.

The VIC simulations do not include reservoir storage, so the simulations are a modeled estimate of what would have happened in the absence of human storage and diversions of water. They also represent outflow at the mouths of the rivers, which due to channelization may have been deposited into estuaries in different locations and to differing extents over time. Human influences can be substantial; reservoir storage and release tend to reduce high flows and, depending on the circumstances, can increase or decrease low flows. This would be a challenge for quantitative modeling of bay and estuary behavior, but the purpose of this chapter is to isolate climate influences, which the VIC simulations do.

## Data and Analyses

Model output from individual runs of 32 GCM models were downscaled to produce sets of LOCA-VIC output for RCP 4.5 and RCP 8.5 projections, where RCP stands for Representative Concentration Pathway. The GCMs are listed in Table 1. From among those models, results are presented mainly from nine highlighted GCMs identified by Kothari et al. (2021) for being especially realistic in their representation of various aspects of Texas climate relevant to droughts.

All meteorological and hydrological variables are aggregated to a monthly time scale and averaged or summed over each drainage basin. Time lags in runoff from basin to discharge point should be small on a monthly basis and are therefore ignored.

Model output is displayed as annual time series, smoothed time series, or time series percentiles. Output from each model should be interpreted as a plausible evolution of the climate system under the influence of greenhouse gases, aerosols, and other human-caused changes. The annual time series show the sequence of annual average or annual total values from individual climate model simulations. Their noisiness reflects the variability of the weather from year to year as represented by the simulations.

The smoothed time series are an estimate of the climate change with weather variability removed. They differ from each other because each model has its own temperature sensitivity to changes in the Earth's radiative balance and also its own reconfigured weather and climate patterns in response to the changing temperatures. A sense of the differences in temperature sensitivity across models is given by the last column in Table 1, which shows the Transient Climate Response

**Table 1** List of models whose downscaled output was used in this chapter, along with their designations in figures including ensemble members. Highlighted models are listed in bold. If different ensemble member numbers were used from RCP 4.5 and RCP 8.5, the respective RCPs are listed in parentheses. Transient Climate Response (TCR) values were compiled and reported in Meehl et al. (2020)

| Model | Designation | Institution | TCR |
|---|---|---|---|
| ACCESS1.0<br>ACCESS1.3 | access1-0.1<br>access1-3.1 | Commonwealth Scientific and Industrial Research Organisation, Australia and Bureau of Meteorology, Australia | 1.9<br>1.6 |
| **BCC-CSM1.1**<br>BCC-CSM1.1(m) | **bcc-csm1-1.1**<br>bcc-csm1-1-m.1 | Beijing Climate Center, China Meteorological Administration | **1.7**<br>2.1 |
| CanESM2 | canesm2.1 | Canadian Centre for Climate Modelling and Analysis | 2.3 |
| **CCSM4** | **ccsm4.6** | National Center for Atmospheric Research | **1.7** |
| CESM1(BGC)<br>CESM1(CAM5) | cesm1-bgc.1<br>cesm1-cam5.1 | National Science Foundation, Department of Energy, National Center for Atmospheric Research | –<br>– |
| CMCC-CM<br>CMCC-CMS | cmcc-cm.1<br>cmcc-cms.1 | Centro Euro-Mediterraneo per I Cambiamenti Climatici | –<br>– |
| **CNRM-CM5** | **cnrm-cm5.1** | Centre National de Recherches Meteorologiques/Centre Europeen de Recherche et Formation Avancees en Calcul Scientifique | **2.0** |
| **CSIRO-Mk3.6.0** | **csiro-mk3-6-0.1** | Commonwealth Scientific and Industrial Research Organization in collaboration with the Queensland Climate Change Centre of Excellence | **1.7** |
| EC-EARTH | ec-earth.8 (4.5)<br>ec-earth.2 (8.5) | EC-EARTH consortium | – |
| FGOALS-g2 | fgoals-g2.1 | LASG, Institute of Atmospheric Physics, Chinese Academy of Sciences; and CESS, Tsinghua University | 1.4 |
| GFDL-CM3<br>GFDL-ESM2G<br>**GFDL-ESM2M** | gfdl-cm3.1<br>gfdl-esm2g.1<br>**gfdl-esm2m.1** | Geophysical Fluid Dynamics Laboratory | 1.9<br>1.1<br>**1.4** |
| GISS-E2-H | giss-e2-h.6 (4.5)<br>giss-e2-h.2 (8.5) | NASA Goddard Institute for Space Studies | 1.7 |
| GISS-E2-M | giss-e2-m.6 (4.5)<br>giss-e2-m.2 (8.5) | NASA Goddard Institute for Space Studies | 1.5 |
| HadGEM2-AO | hadgem2-ao.1 | National Institute of Meteorological Research / Korea Meteorological Administration | – |
| HadGEM2-CC<br>HadGEM2-ES | hadgem2-cc.1<br>hadgem2-es.1 | Met Office Hadley Centre | –<br>2.5 |
| INM-CM4 | inmcm4.1 | Institute for Numerical Mathematics | 1.3 |
| **IPSL-CM5A-LR**<br>IPSL-CM5A-MR | **ipsl-cm5a-lr.1**<br>ipsl-cm5a-mr.1 | Institut Pierre-Simon Laplace | **2.0**<br>2.0 |
| MIROC-ESM<br>MIROC-ESM-CHEM<br><br>**MIROC5** | miroc-esm.1<br>miroc-esm-chem.1<br><br>**miroc5.1** | Japan Agency for Marine-Earth Science and Technology, Atmosphere and Ocean Research Institute (The University of Tokyo), and National Institute for Environmental Studies | 2.2<br>–<br><br>**1.4** |
| MPI-ESM-LR<br>MPI-ESM-MR | mpi-esm-lr.1<br>mpi-esm-mr.1 | Max Planck Institute for Meteorology (MPI-M) | 2.0<br>2.0 |
| **MRI-CGCM3** | **mri-cgcm3.1** | Meteorological Research Institute | **1.6** |
| **NorESM1-M** | **noresm1-1.1** | Norwegian Climate Centre | **1.4** |

(TCR), a measure of how much the simulated global temperature changes after several decades of steadily rising carbon dioxide concentrations. Larger values imply more rapid simulated changes, with differences varying by as much as a factor of two. The highlighted model TCR ranges from 1.4 to 2.0, with an average of about 1.65. This is slightly below the assessed value from the latest Intergovernmental Panel on Climate Change (IPCC) report, which asserted a best estimate of 1.8 and a likely range of 1.4 to 2.2 (Forster et al. 2021). Thus, these model simulations are just slightly conservative regarding the magnitude of projected global temperature change.

The projections from individual models are also aggregated and summarized for the historical period (1950–2009) and the future period (2025–2084). Because these are single-model aggregations, the spread in a single aggregation excludes differences in climate sensitivity. However, because they span six decades, part of the spread corresponds to the changing average conditions over the period rather than to year-to-year variations in the weather. The information is presented in this manner because the historical period gives a sense of the range of conditions experienced in recent decades and the projection period gives a sense of the range of conditions possible in the coming several decades. Both increasing variability and an increasing long-term trend would imply that estuarine ecosystems must be able to adapt to a broader range of conditions than those found historically.

Some aggregated projections are presented as monthly values to illustrate the approximate historical climate as well as future conditions. The historical conditions are indicated by lines at the 10th, 50th, and 90th percentile for each high-lighted model, while the downward-pointing triangles, circles, and upward-pointing triangles represent the projected values. Comparison of lines and circles in particular months indicate the direction and magnitude of projected changes in typical and more extreme values.

The full range of monthly conditions are depicted by boxen plots (also known as letter-value plots) and violin plots. The boxen plots represent ranges of distributions, with the thickest box corresponding to the 25–75% range, the next thickest box corresponding to the 12.5–87.5% range, and so forth, with half of the remaining distribution incorporated with each box increment. Violin plots represent the likelihood of different values by showing smoothed empirical distributions.

An RCP is a scenario for future concentrations of greenhouse gases and aerosols driven by changes in emissions from such processes as the burning of fossil fuels. Projections are shown here for both RCP 4.5 and RCP 8.5. RCP 8.5 is a high-end business-as-usual projection, with emissions unlikely to be so large even without further climate action (Hausfather and Peters 2020). RCP 8.5 is useful both as something approaching a worst-case scenario and as a tool for identifying the climate change signal. Because the magnitude of the change is so large, it is relatively easy to distinguish it from random distributional changes over short periods caused by the weather. Conversely, RCP 4.5 assumes substantial climate action and is a plausible best-case scenario. Neither RCP includes changes in volcanic and solar activity, so they are not comprehensive forecasts of the climate system, just simulations of the possible anthropogenic contributions to climate change. For more on RCPs, see van Vuuren et al. (2011).

For most parameters, the greatest uncertainty in projected future conditions in a Texas river basin is from not knowing the particular weather events for that year in advance. Averaging over historical and future 60-year periods greatly reduces the uncertainty due to the weather and leaves uncertainties associated with future greenhouse gas emissions and future climate responses. Of these, generally the greatest unknown is the particular climate response, which can vary greatly from model to model (Lehner et al. 2020). Hence, the problem of knowing future climate-driven conditions in Texas bays and estuaries comes down to the fact that we know that the climate of the future will be different from the climate of the past, but we cannot tell by how much exactly.

It is crucial to keep in mind that these are "just" model projections, and reality could turn out quite a bit differently, with changes likely to be generally less severe but possibly more severe in some aspects compared to the RCP 8.5 projections.

## Temperature

Climatological temperatures in Texas increase from north to south across Texas in most seasons, though they are more uniform in the summer months and locally higher altitudes can be cooler. The Gulf of Mexico exerts a moderating influence, keeping coastal temperatures cooler during spring and summer days and warmer during fall and winter nights. Under climate change, temperatures are projected to increase by a few degree Celsius.

The solid and dashed lines in Fig. 1 show the seasonal variation in temperature in the vicinity of Houston. Climatological average temperatures in July and August, the hottest months of the year, are around 28°C, while average temperatures in December and January, the coldest months of the year, are around 11°C. Monthly average temperatures are much more variable in the winter than in the summer, with the range from the 10th to 90th percentile being around 3°C in summer and 7°C in winter.

The solid and dashed lines in Fig. 2 shows the seasonal variation in historical temperatures for the Kingsville area. While Houston is representative of the East Texas coastal area, Kingsville is representative of the South Texas coastal area. The same seasonal pattern is evident in Kingsville as in Houston but, consistent with Texas statewide climate, the seasonal range of temperatures is larger in the northern location and temperatures in general are higher in the southern location.

All nine models give similar annual cycles for the 1950–2009 period in Figs. 1 and 2 because they include the 1976–2005 downscaling calibration period where agreement with observations is enforced mathematically. Cross-model differences are much larger in projected temperatures, as shown using symbols for the 2025–2084 period in the RCP 8.5 scenario. Nonetheless, under this high-end scenario, models consistently project that both maximum and minimum temperatures in 2025–2084 will average about 3°C warmer than temperatures in 1950–2009. This represents a trend that is similar to observed historical trends in Texas since 1975 (Nielsen-Gammon et al. 2021).

There is not much interannual variability in Texas summertime temperatures. The projected 2025–2084 temperatures are warm enough that the 10th percentile summer month in 2025–2084 is projected to be similar to the 90th percentile summer month in 1950–2009. In other words, summer heat that used to occur only 10% of the time is pro-

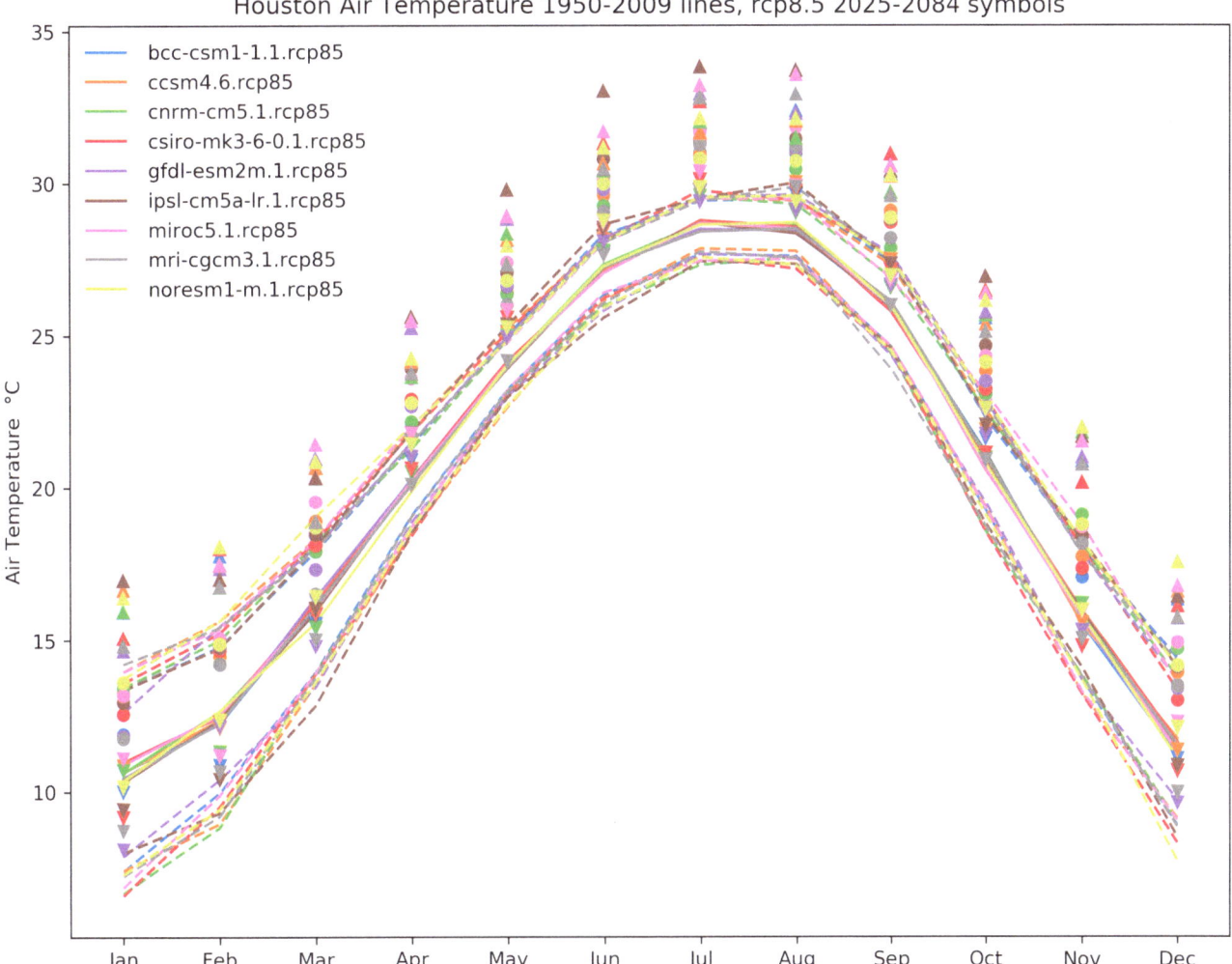

**Fig. 1** Monthly temperatures from nine highlighted climate models for historical (1950–2009, lines) and future (2025–2084, symbols) periods in the Houston area, using RCP 8.5 projections. Solid lines and circles show the median values, while dashed lines and triangles show the 10th and 90th percentiles of monthly values over each period for each model

jected to occur 90% of the time over the next several decades under RCP 8.5.

Figure 3 shows the evolution of annual average temperatures around Kingsville under RCP 8.5. Trends are similar for other basins. Amidst year-to-year variability, temperatures continue to rise under this scenario, and indeed the rate of temperature rise itself increases. Figure 4 shows the evolution of 30-year average annual temperatures for the same area under RCP 4.5. The rate of temperature increase slows down dramatically during the second half of the twenty-first century. These characteristics are generally true for all RCP 4.5 and RCP 8.5 variables: climate change accelerates during the latter part of the twenty-first century under RCP 8.5 but is subdued under RCP 4.5. The 30-year averages in Fig. 4 highlight the differences across models in the rate of climate change. Those differences are much larger on a local scale than a global scale as simulated changes in weather patterns affect climate locally.

Changes in means and extremes over the entire year are more clearly depicted in Fig. 5. The bulges near the top and bottom of each distribution represent summer and winter, when temperatures tend to stay within the same range for a prolonged period. Figure 5 highlights the dramatic change in heat extremes under RCP 8.5. Summertime temperatures that were nearly unheard of are projected to become commonplace. Cold extremes decrease in frequency, though not as rapidly as hot extremes increase in frequency because the range of possibilities during the cold season is much broader.

Similar patterns of temperature change are found in other coastal and near-coastal basins in Texas. Temperatures increase in all seasons, with the summer season temperatures moving partly outside the realm of historic temperatures and into virtually unprecedented territory.

Overall, temperatures increase in all seasons at all locations, with unusually hot temperatures becoming consider-

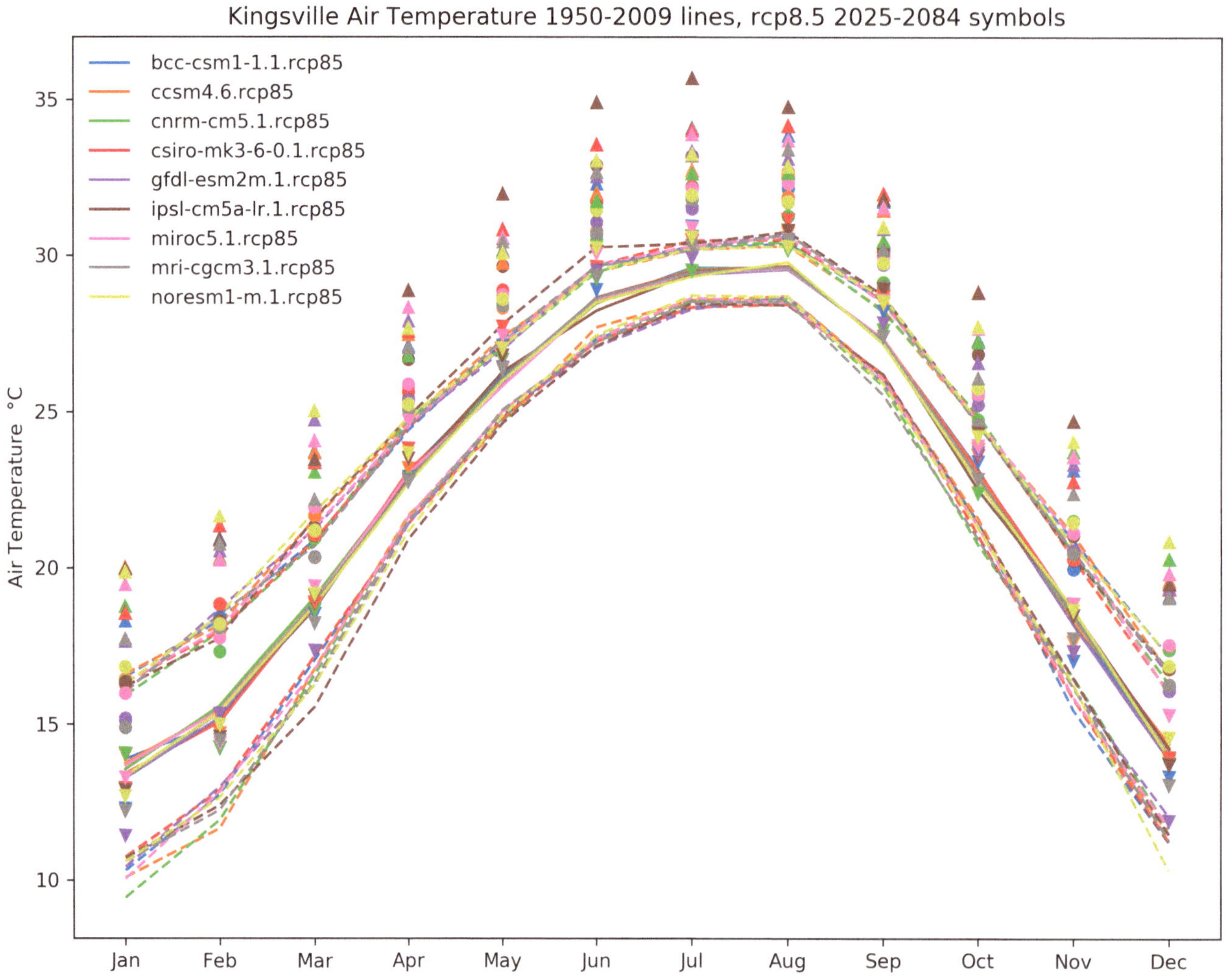

Kingsville Air Temperature 1950-2009 lines, rcp8.5 2025-2084 symbols

**Fig. 2** Monthly temperatures, as in Fig. 1, but for the Kingsville area

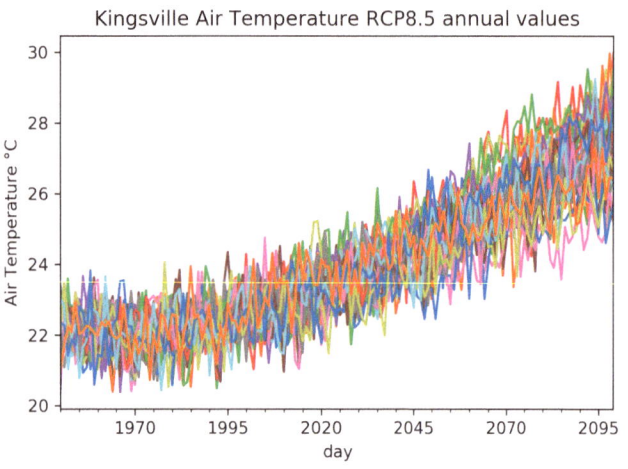

**Fig. 3** Annual average temperatures, all models, RCP 8.5, for the Kingsville area, showing magnitude and range of projected temperature increase in the context of interannual variability

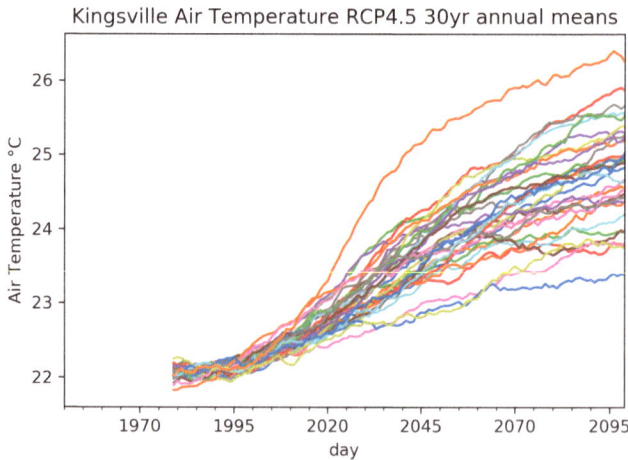

**Fig. 4** Thirty-year running average temperatures, all models, RCP 4.5, for the Kingsville area, showing differences in projections of average climate conditions

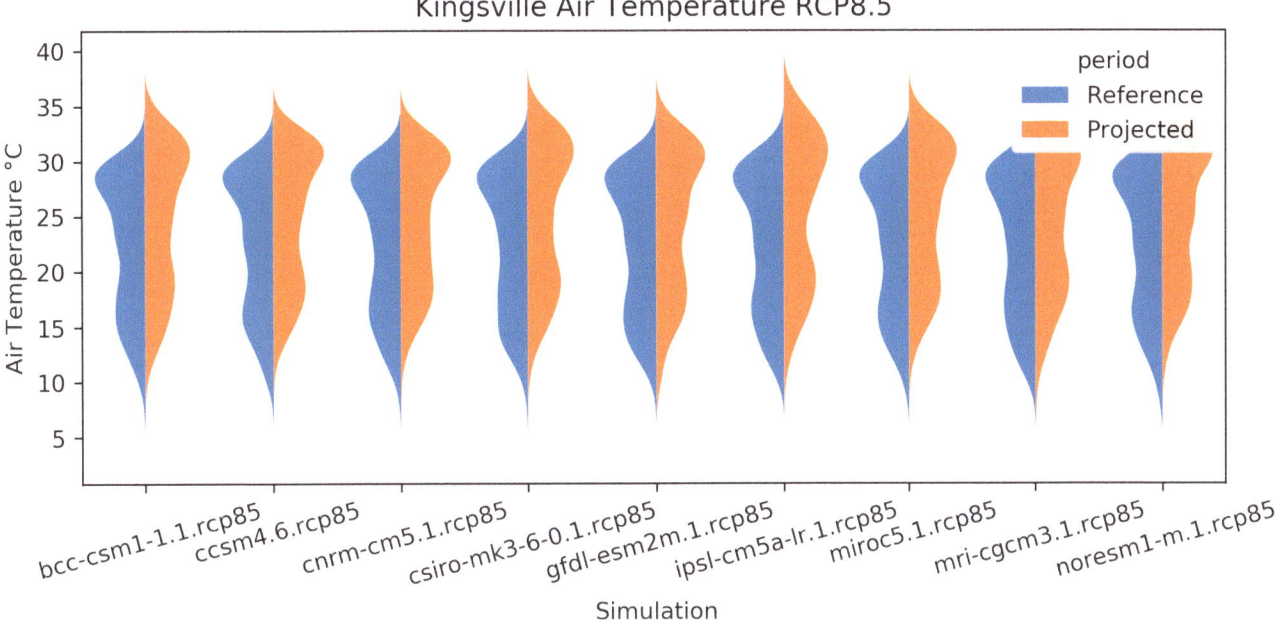

**Fig. 5** Violin plot of distribution of monthly average temperatures in Kingsville drainage area in highlighted models for historic and future conditions under RCP 8.5. In this and other violin plots, the thickness of the colors shows the relatively likelihood of temperatures in the reference and projected time periods. Here the predominant change is an upward shift in the temperature distribution

ably more common and unusually cold temperatures becoming somewhat less common. Temperatures rise steadily until the middle of the twenty-first century, at which point they could accelerate (RCP 8.5), begin to level off (RCP 4.5), or maintain their pace.

## Humidity and Water Surface Evaporation

The relative humidity in the eastern coastal area is climatologically fairly constant throughout the year, averaging about 65% (Fig. 6). In the southern coastal area, by contrast, conditions are drier, and there is a prominent seasonal cycle, with relative humidity reaching a minimum in spring and a maximum in fall (Fig. 7). With climate change, relative humidity trends are unclear, but potential evapotranspiration is projected to increase.

In Houston (Fig. 6), model projections of average relative humidity range from a slight increase to a slight decrease. At the 10th percentile, several of the models project much more frequent occurrences of dry conditions during spring, summer, and early fall. But even if relative humidity were to stay the same, the rate of evaporation from the surface of bays and estuaries would increase due to the rising temperatures and resulting increase in the vapor pressure deficit, the difference between the amount of water vapor in the air and the maximum capacity for water vapor in the air. The only ways such an increase in evaporation could be avoided would be through an increase in cloud cover, which would cut back on

the amount of sunlight available to drive evaporation, or a decrease in wind speed.

This expectation is confirmed by Fig. 8, which shows potential evapotranspiration (ET) for the Houston drainage area. Potential evapotranspiration (ET) is the amount of evaporation that would occur if the surface was not energy-limited or water-limited. The potential ET increases in all model projections by between 5% and 20% over the course of the twenty-first century.

In Kingsville (Fig. 7), by contrast, there are substantial declines in relative humidity projected by all models between January and July. This coincides with the already dry period in South Texas. Humidity conditions are influenced by the presence of dry air aloft during the springtime from the higher elevations of northern Mexico that often depletes moisture from low-level air. With northeastern Mexico generally projected to see declines in precipitation, particularly during the wintertime (Colorado-Ruiz et al. 2018), the consequence for South Texas is projected to be drier atmospheric conditions during winter, spring, and early summer, with consequent increases in surface evaporation.

There are no long-term observations of surface evaporation from Texas bays and estuaries, although estimates based on pan evaporation are available from the Texas Water Development Board. Evaporation from lakes has increased over the period 1985–2014 across most of Texas (Zhao and Gao 2019), and lake evaporation is projected to increase by a few percent across Texas over the next few decades (Zhao et al. 2023).

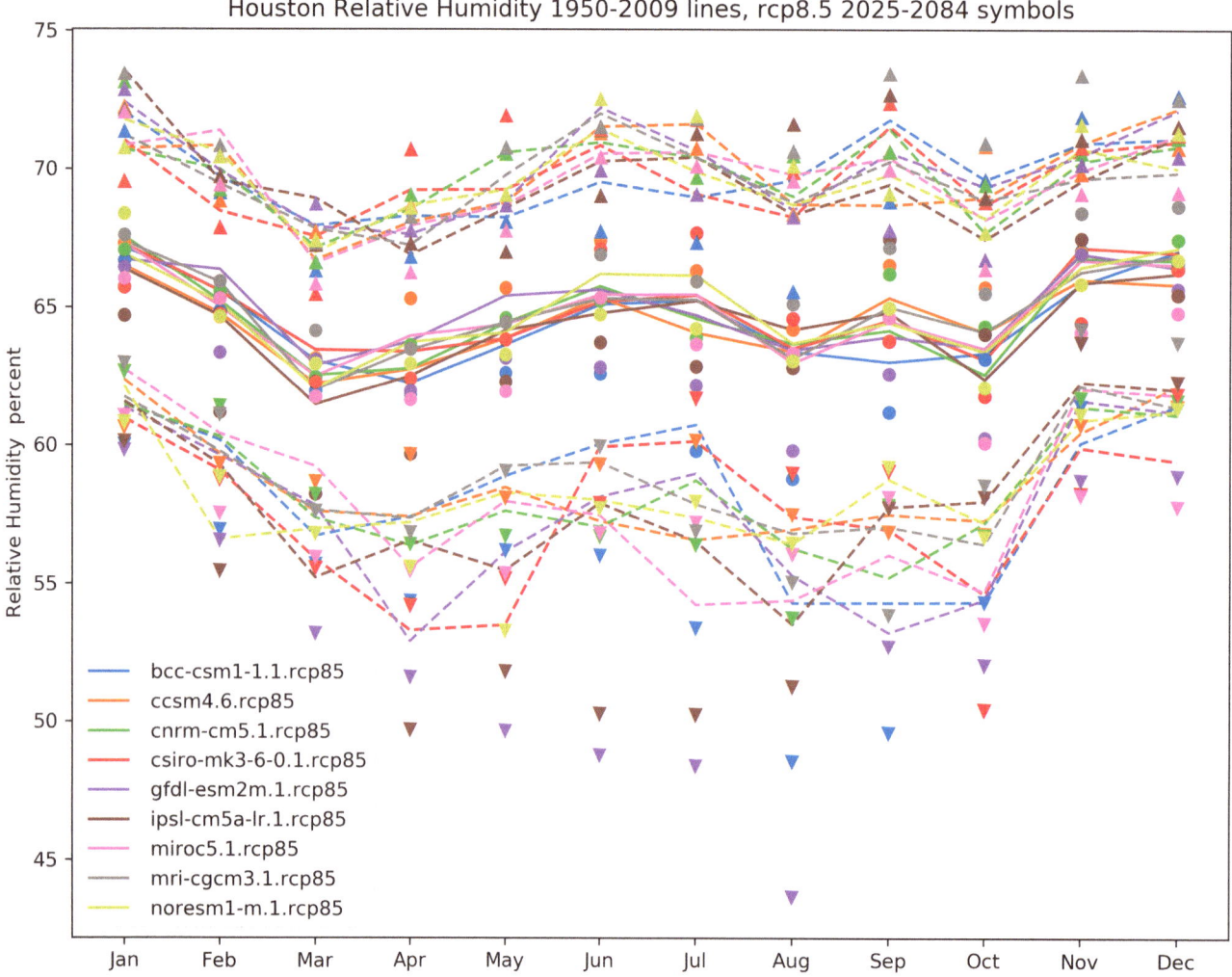

**Fig. 6** Monthly mean 10th, 50th, and 90th percentile historical and RCP 8.5 projected relative humidity values from nine highlighted models for the Houston drainage area, with little seasonal cycle in relative humidity

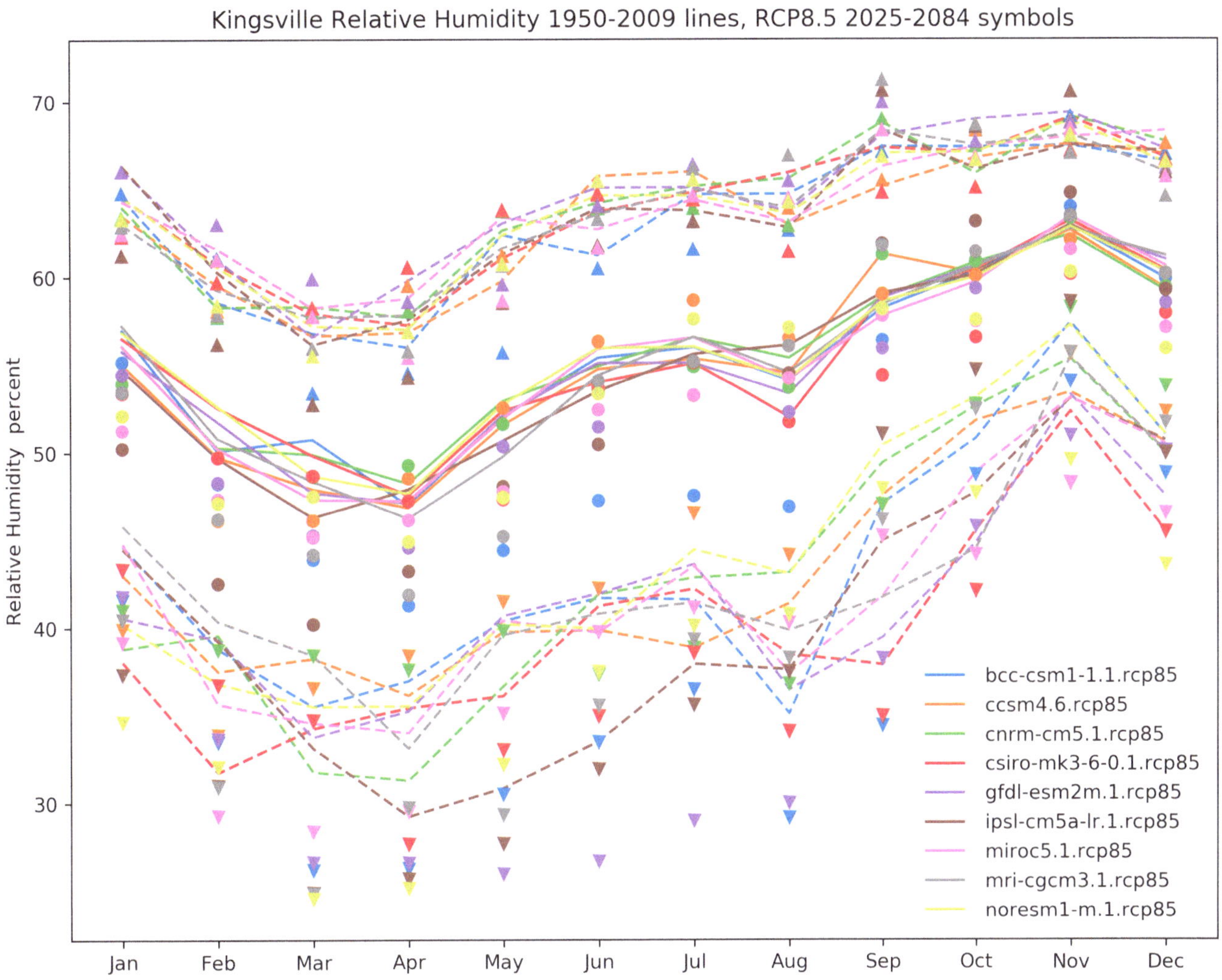

## Kingsville Relative Humidity 1950-2009 lines, RCP8.5 2025-2084 symbols

legend:
- bcc-csm1-1.1.rcp85
- ccsm4.6.rcp85
- cnrm-cm5.1.rcp85
- csiro-mk3-6-0.1.rcp85
- gfdl-esm2m.1.rcp85
- ipsl-cm5a-lr.1.rcp85
- miroc5.1.rcp85
- mri-cgcm3.1.rcp85
- noresm1-m.1.rcp85

**Fig. 7** Monthly mean 10th, 50th, and 90th percentile historical and RCP 8.5 projected relative humidity values from nine highlighted models for the Kingsville drainage area, showing a substantial seasonal cycle

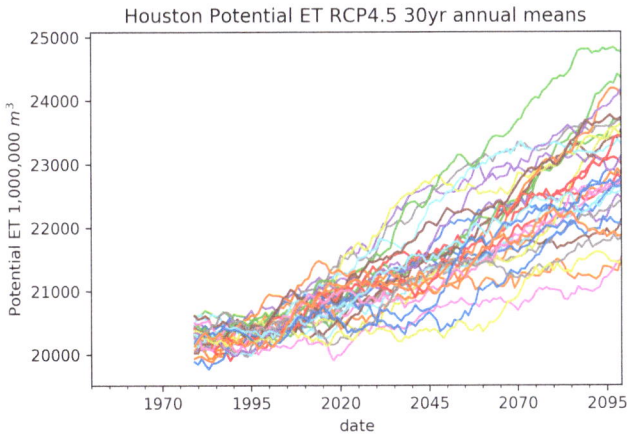

**Fig. 8** Potential evapotranspiration (ET), 30-year running means for all 32 RCP 8.5 model projections for the Houston area, showing robust increase

## Rainfall

Rainfall in Texas generally increases from west to east. In most of the state, there are climatological peaks in rainfall in late spring and early fall, but East Texas has nearly uniform rainfall throughout the year and far West Texas sees the bulk of its rain during the summertime. Snowfall is an extremely minor contributor to inflows to Texas bays and estuaries and is ignored here. Computer models feature widely diverging projections of future average precipitation, but all project an increase in the frequency of extremely dry months.

Figure 9 shows the rainfall for the combined Sabine-Neches basin. While the basin receives around $10 \times 10^9 \, m^3$ of water every month on average, the wettest month tends to be March and the driest tends to be August. The steadiness of rainfall is also reflected in the relative (for Texas) lack of

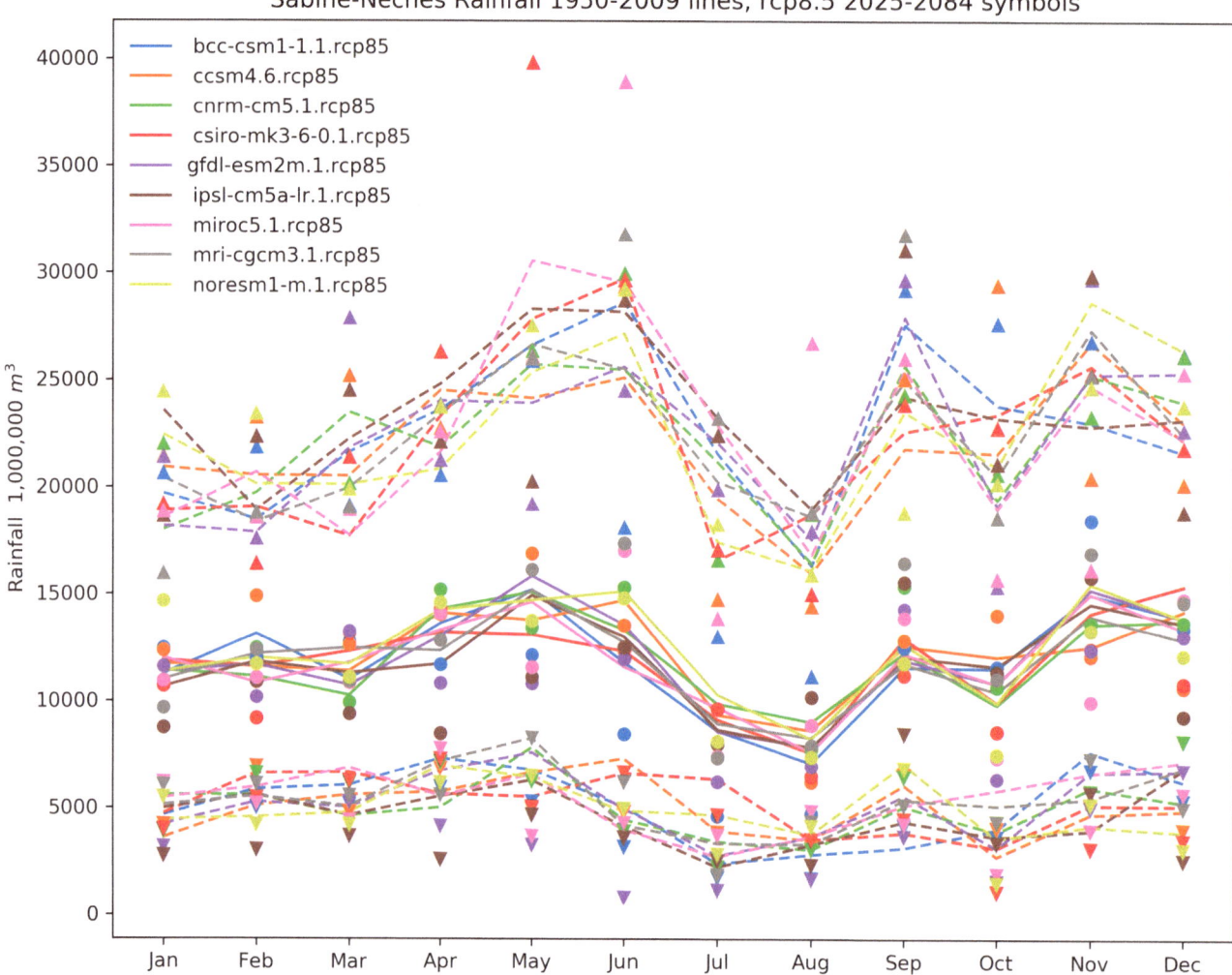

**Fig. 9** Monthly total 10th, 50th, and 90th percentile historical and RCP 8.5 projected rainfall values from nine highlighted models for the combined Sabine-Neches drainage area showing weak seasonal cycle and inconsistent changes across models

interannual variability, with the 10th percentile monthly rainfall around $5 \times 10^9$ m³ and the 90th percentile monthly rainfall in excess of $20 \times 10^9$ m³.

Farther west and south, there is more seasonality in the precipitation. The Nueces drainage basin (Fig. 10) shows the two-peak structure common to most of Texas, with December through March being the driest months and a secondary minimum of rainfall appearing in July and August. Also, the fractional differences across percentiles are much larger for the Nueces basin than the Sabine-Neches basin, despite the two basins having similar sizes. In the Nueces basin, dry months have almost no precipitation, while wet months are as much as three times as wet as the normal median value. Overall rainfall accumulation in the Nueces basin is about a third that of the Sabine-Neches basin, reflecting the west-east precipitation gradient across Texas.

There are no noticeable seasonal patterns to the rainfall trends in the two basins. The cumulative annual trends for Sabine-Neches (Fig. 11) highlight model disagreements, with five models projecting a decline ranging up to 15% and four projecting almost no change in the median. Nonetheless, there is model agreement in the extremes, with eight out of nine models projecting an increase in the frequency of very dry months as well as an increase in the frequency of very wet months. On the whole, the model average standard deviation in the Sabine-Neches basin (not shown) increases by about 20% from 1950–1979 to 2070–2099.

Broadly speaking, Texas lies close to a GCM projection hinge point, with increased precipitation projected to the northeast and decreased precipitation projected to the southwest. While the Sabine-Neches basin has little overall projected change in rainfall on average, the model results for

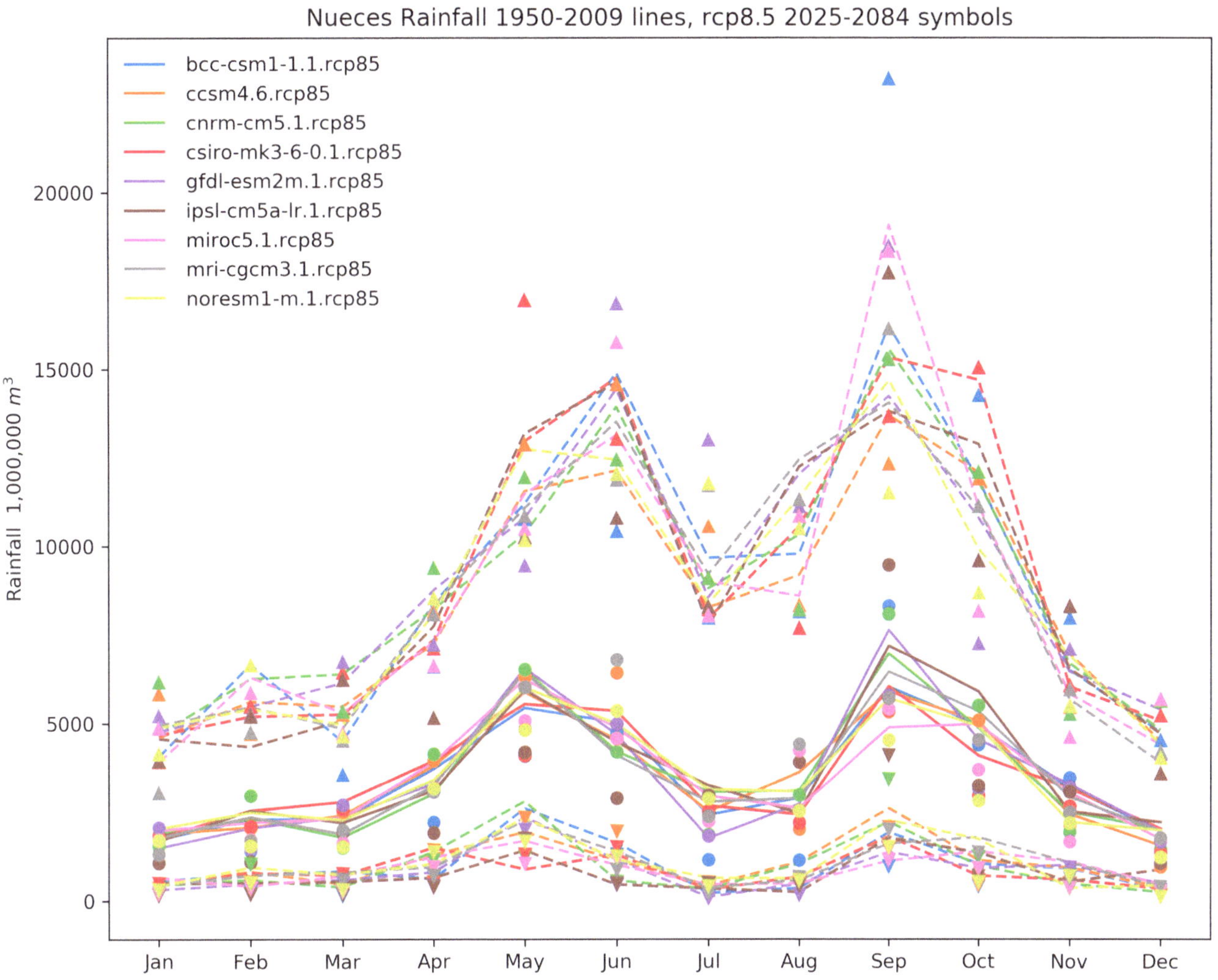

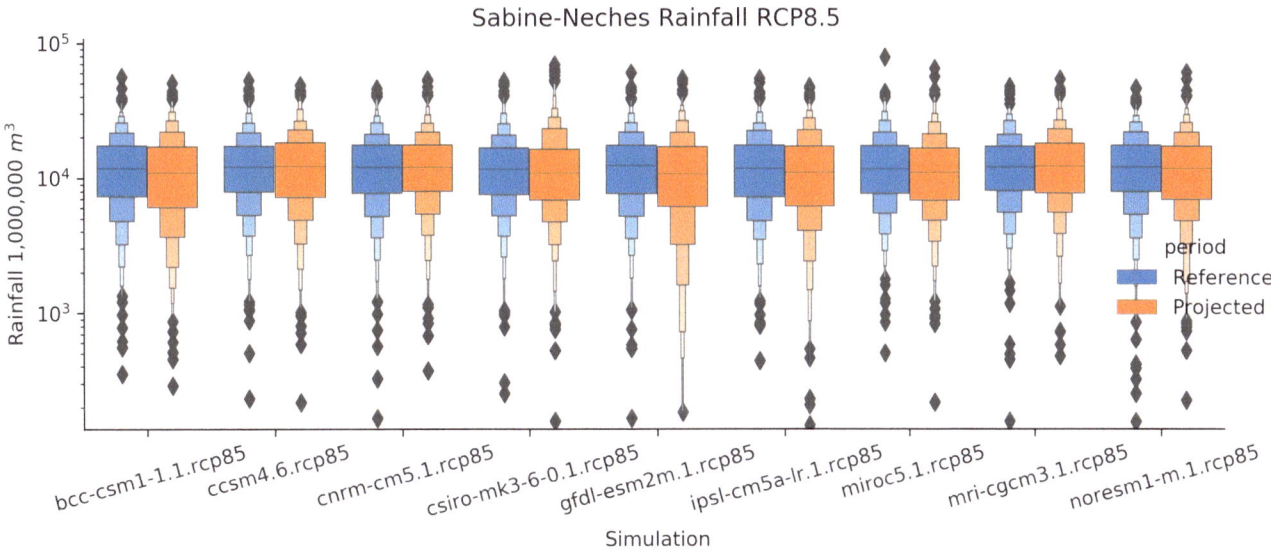

**Fig. 10** Monthly total 10th, 50th, and 90th percentile historical and RCP 8.5 projected rainfall values from nine highlighted models for the Nueces drainage area, showing strong seasonal cycle and inconsistent trends across models

**Fig. 11** Boxen plot of the distribution of historical and RCP 8.5 projected rainfall values from nine highlighted models for the combined Sabine-Neches drainage area showing projected increase in extremes

the Nueces basin in South Texas project much more of a tendency for decreased rainfall (Fig. 12). The median decline in monthly median precipitation is around 10%. Only two of the nine models project an increase in extremely wet months, while all project an increase in extremely dry months. The change in rainfall standard deviation tends to be slightly positive but varies wildly among model simulations.

Figure 13 further shows the model disagreement among all 32 GCMs for Nueces basin rainfall. Though the majority of GCMs show a decline, the changes by the year 2100 range from a 40% increase to a 40% decrease. This large variability means that it will be challenging to make a robust estimate of the underlying climate trend for at least several more decades. Historically, precipitation has increased by about 10%, mostly in central and eastern Texas (Nielsen-Gammon et al. 2021), which could mean that the models are generally wrong or that Texas has received extra rainfall due to natural variability.

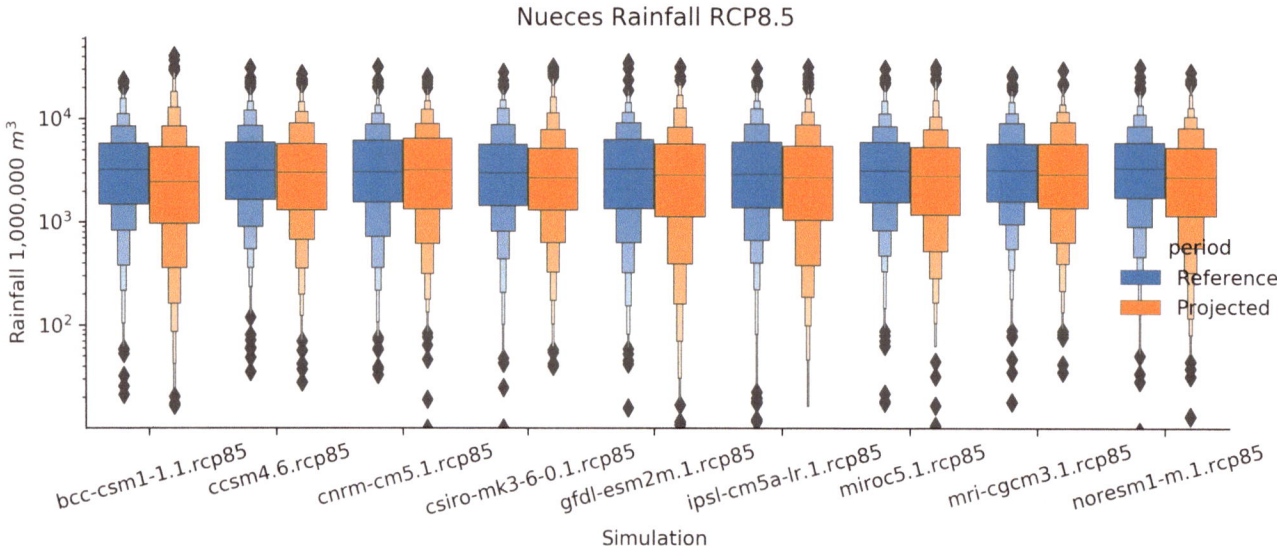

**Fig. 12** Boxen plot of the distribution of historical and RCP 8.5 projected rainfall values from nine highlighted models for the Nueces drainage area showing tendency for decreased precipitation

**Fig. 13** Thirty-year running means of rainfall in the Nueces River drainage for all 32 RCP 8.5 climate simulations showing large uncertainty among projections

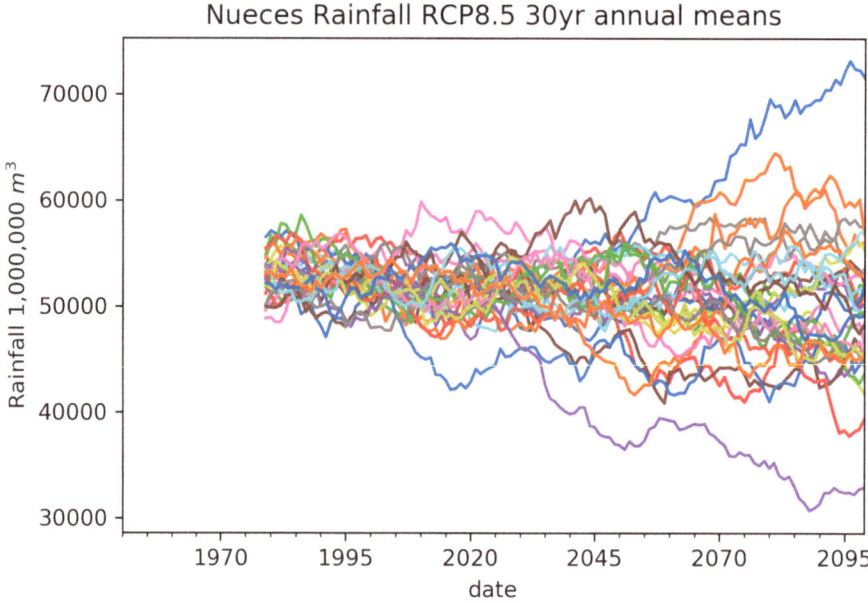

## Soil Moisture

The higher temperatures, greater potential ET, and possible decrease in rainfall all suggest a decrease in soil moisture. Such a decrease has already been documented for both east and west Texas (Nielsen-Gammon et al. 2020). The VIC model projections consistently indicate decreasing soil moisture in the future.

The VIC-simulated top layer soil moisture for the Sabine-Neches basin (Fig. 14) indicates that August is typically the driest month, and the wettest months are January and February. The overall pattern is consistent with soil moisture observations in Oklahoma (Illston et al. 2004).

The VIC-simulated top layer soil moisture for the drier Nueces basin (Fig. 15) is more complex. In addition to the summer soil moisture minimum, there is a secondary mini-

mum in April when relative humidity values are low (see Fig. 7) prior to the May rains.

The change in soil moisture in RCP 8.5 for the Sabine-Neches basin (Fig. 16) shows a slight general decrease in the typical, fairly high soil moisture values. At the drier values of soil moisture, there is an enhanced tendency for occurrence in the projections compared to the historical simulations.

Drier conditions prevail in the Kingsville drainage area. There, the violin plots (Fig. 17) show that dry and wet conditions are nearly equally common, but that climate models project a substantial increase in the fraction of soil moisture values considered dry.

The VIC simulations of soil moisture do not include a dynamic vegetation model. Since soil moisture is sensitive to transport via roots into plants with subsequent ET, this shortcoming may lead to a bias in the soil moisture simulations.

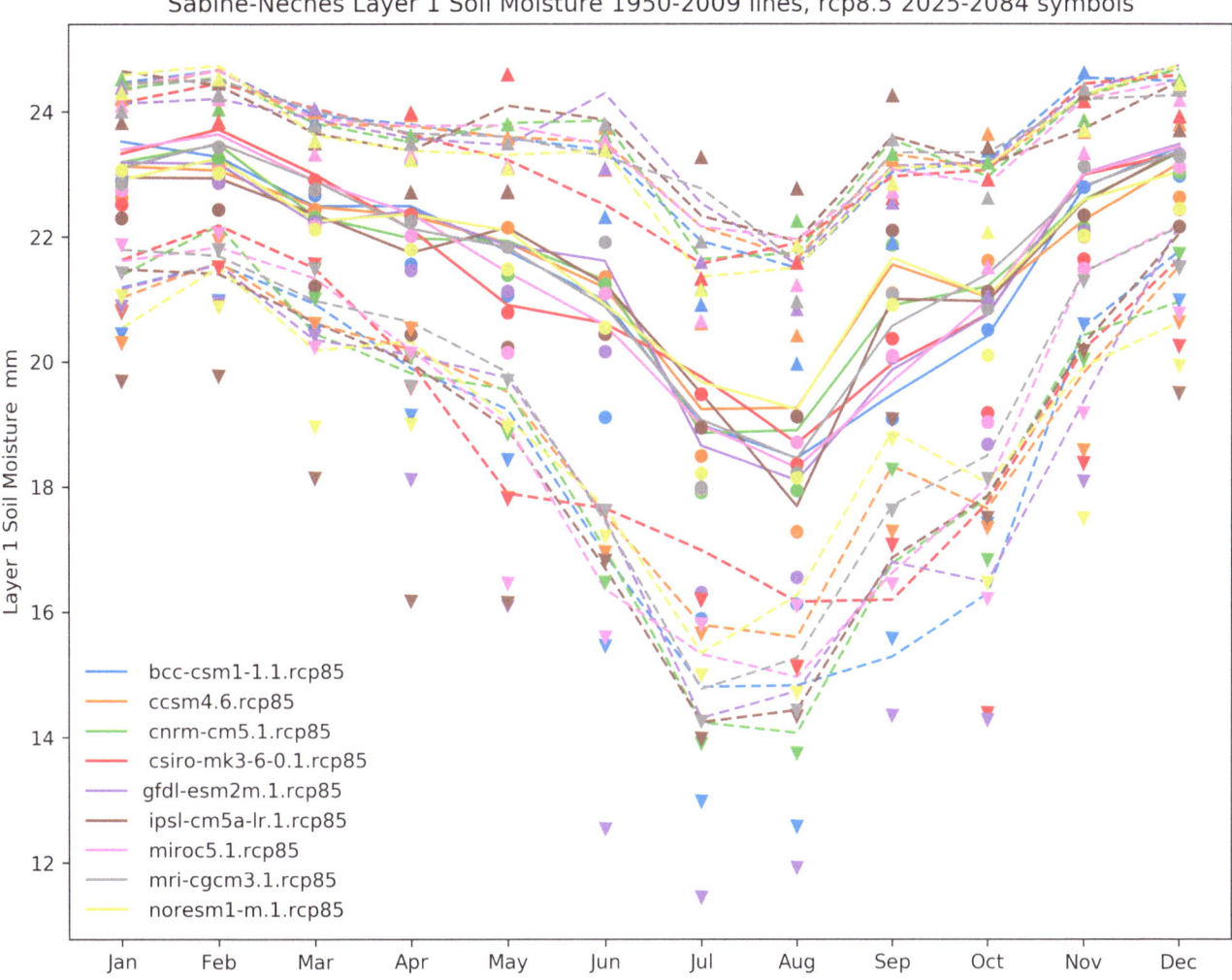

**Fig. 14** Monthly 10th, 50th, and 90th percentile historical and RCP 8.5 projected layer 1 soil moisture values from nine highlighted models for the Sabine-Neches drainage area showing seasonal cycle and projected decline

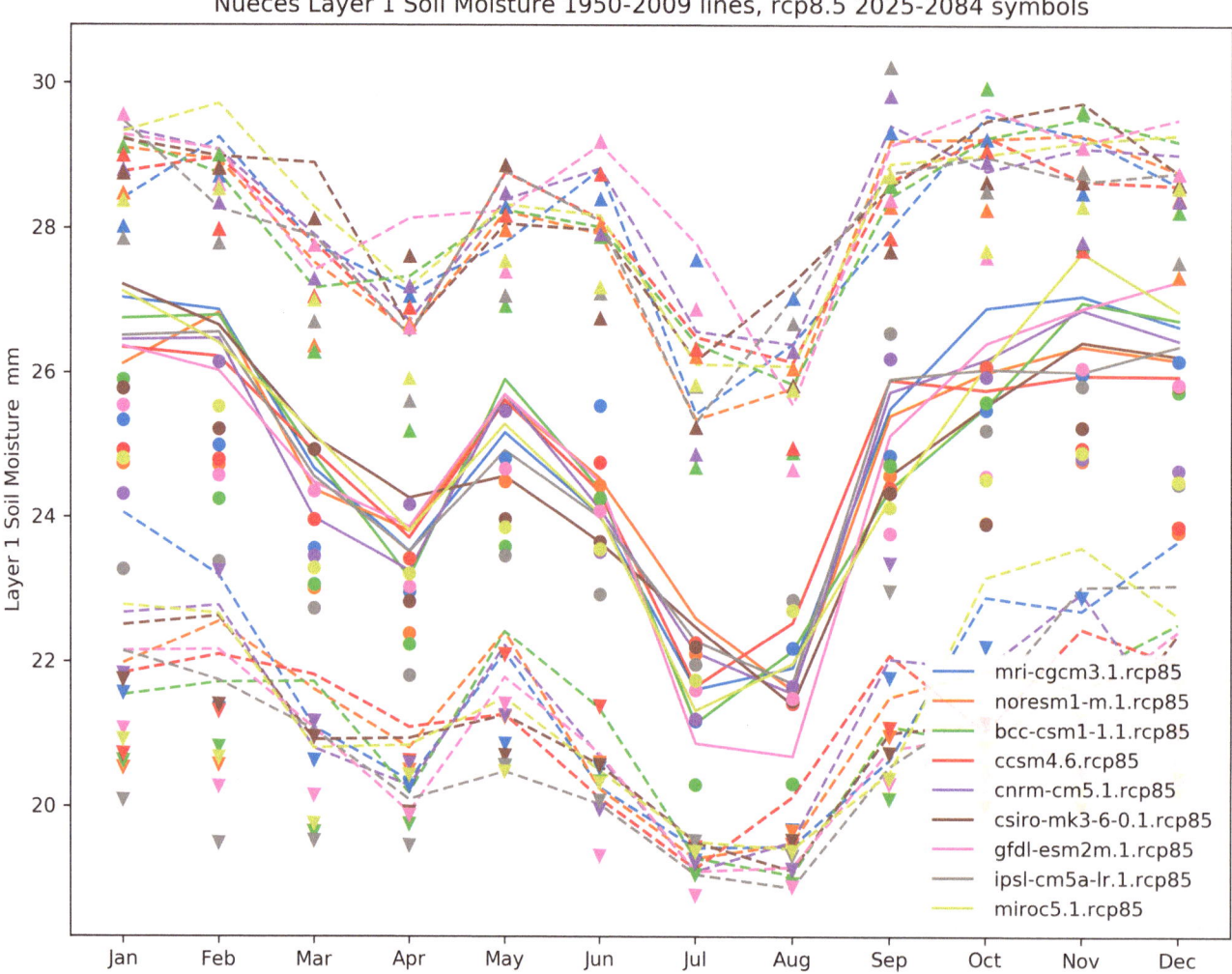

**Fig. 15** Monthly 10th, 50th, and 90th percentile historical and RCP 8.5 projected layer 1 soil moisture values from nine highlighted models for the Nueces drainage area showing seasonal cycle and projected decline

Specifically, with carbon dioxide increasing, plants become more water-efficient as their stomata need not open as far or as long to take in a certain amount of water vapor. Therefore, the VIC model may be underestimating soil moisture if this process is not included.

Whatever the magnitude of decrease in soil moisture, the consequences for inflows to bays and reservoirs may be severe. Decreased soil moisture implies that there should be less baseflow (the water soaking through the soil and joining the water table) and there should also be less runoff (because soil can take up more water when it starts out drier). Conversely, the tendency for rainfall to be more extreme suggests that a greater percentage of rainfall may run off.

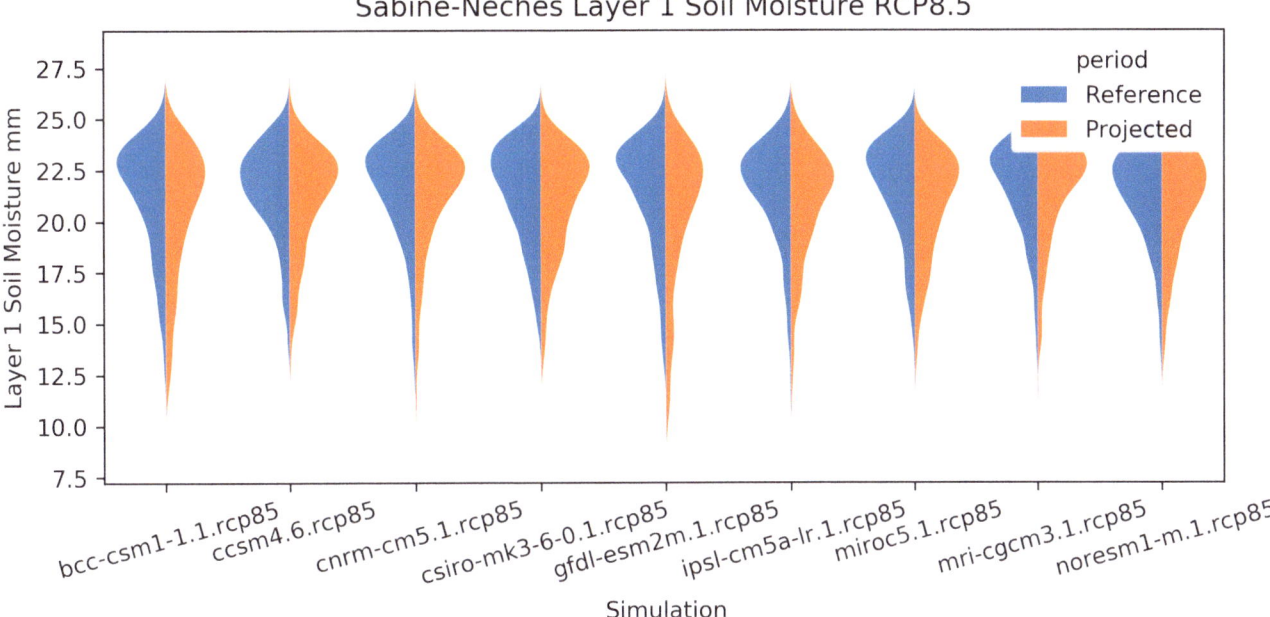

**Fig. 16** Violin plot of distribution of monthly soil moisture values in the Sabine-Neches drainage area in highlighted models for historic and future conditions under RCP 8.5, showing projected slight soil moisture decrease

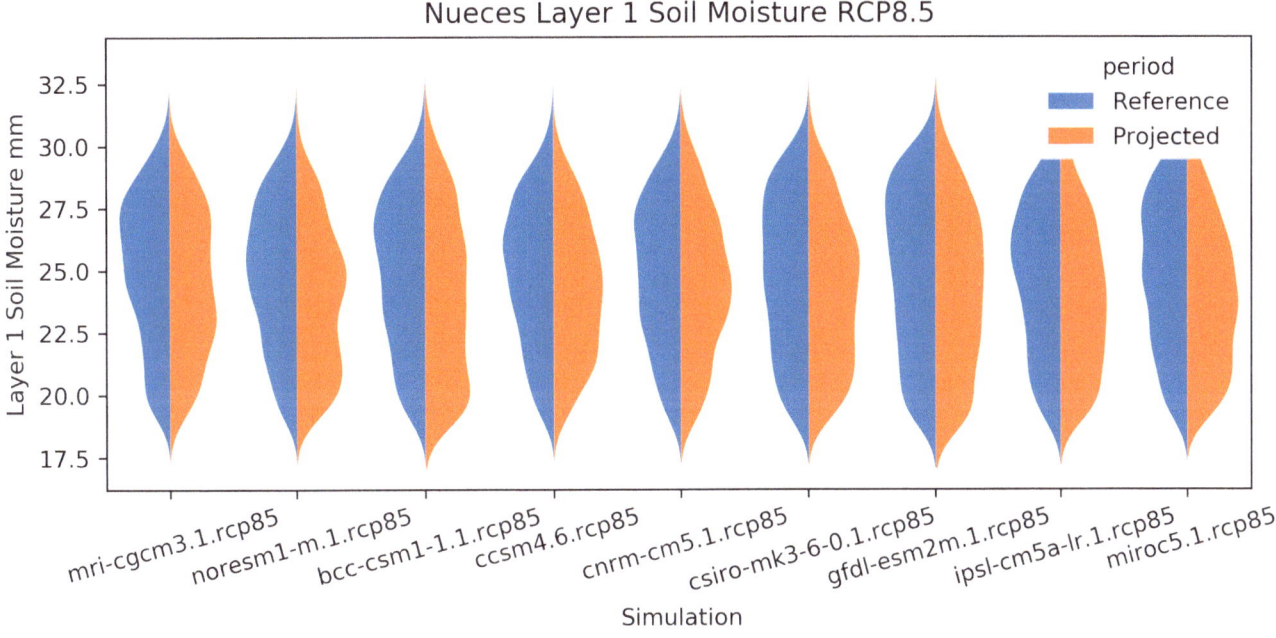

**Fig. 17** Violin plot of distribution of monthly soil moisture values in the Nueces drainage area in highlighted models for historic and future conditions under RCP 8.5, showing substantial projected increase in drier soils

## Baseflow and Runoff

Projections of runoff tend to indicate a future decrease in median values of runoff but with an increase in runoff during high-runoff months.

Baseflow in the Sabine-Neches basin peaks in February and March (Fig. 18) with typical simulated monthly values of $1000–2000 \times 10^6$ m³. During late summer and early fall, baseflow reaches a minimum of about a tenth its February-March value. Runoff is typically a bit less than $1000 \times 10^6$ m³, but it persists through the year, declining to about $300 \times 10^6$ m³ during July and August (Fig. 19).

Median baseflow (Fig. 20) is simulated to decrease in seven out of nine VIC RCP 8.5 model runs and only increase

slightly in the other two, consistent with the projected decline in soil moisture. Similar statistics apply to runoff (Fig. 21), with the same seven models projecting a decrease. However, at the high end of monthly values, the runoff increases, while at the low end of monthly values, the runoff decreases. So, inflows become more erratic, with some model runs indicating an overall increase and others indicating an overall decrease.

In the Nueces River basin, where precipitation generally decreased, median inflows also decreased (Fig. 22). Despite this decrease, the majority of model simulations showed an increase in the frequency of the most extreme heavy precipitation. Also very consistent was the increase in frequency of extreme dry months.

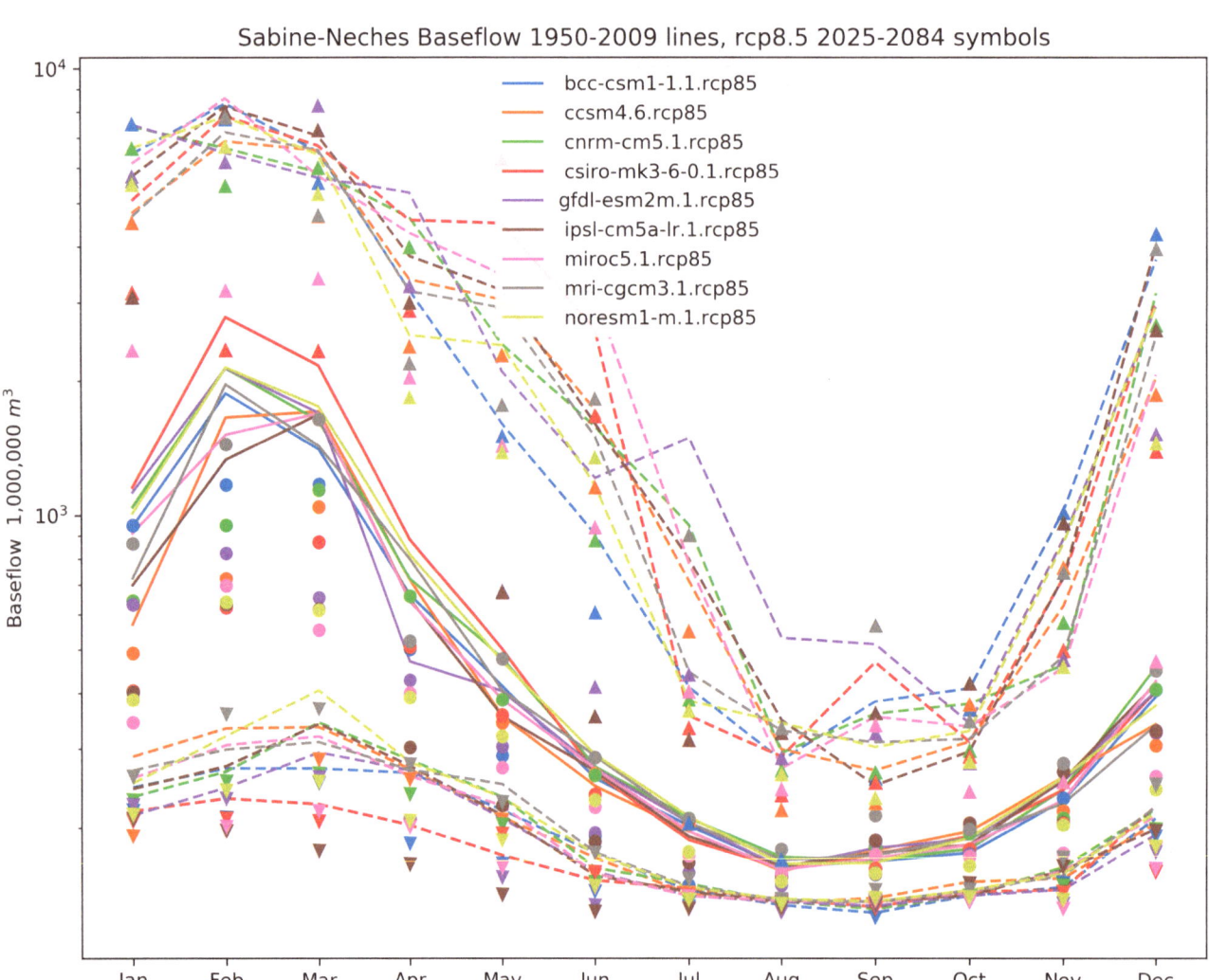

**Fig. 18** Monthly 10th, 50th, and 90th percentile historical and RCP 8.5 projected baseflow values from nine highlighted models for the Sabine-Neches drainage area, showing strong seasonality and general projected decline

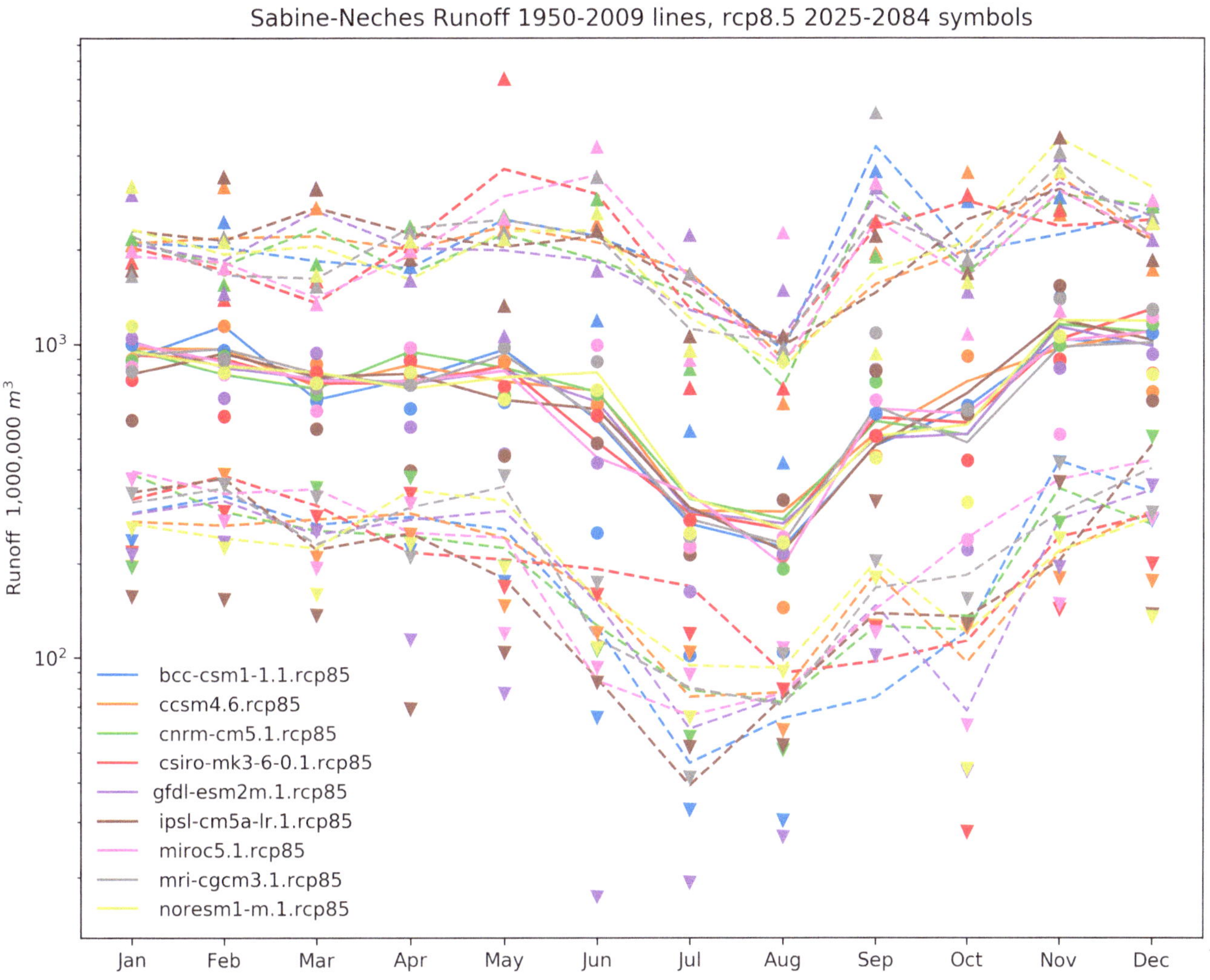

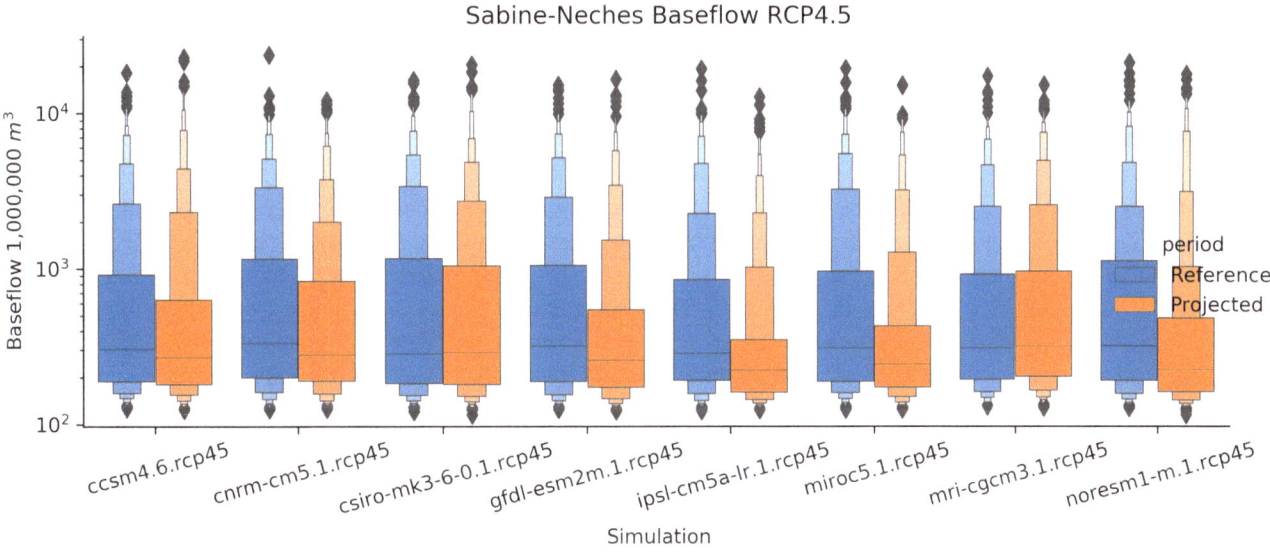

**Fig. 19** Monthly 10th, 50th, and 90th percentile historical and RCP 8.5 projected runoff values from nine highlighted models for the Sabine-Neches drainage area, showing weak seasonality and weak projected trends

**Fig. 20** Boxen plot of the distribution of historical and RCP 8.5 projected annual baseflow values from nine highlighted models for the Sabine-Neches drainage area, showing general decrease

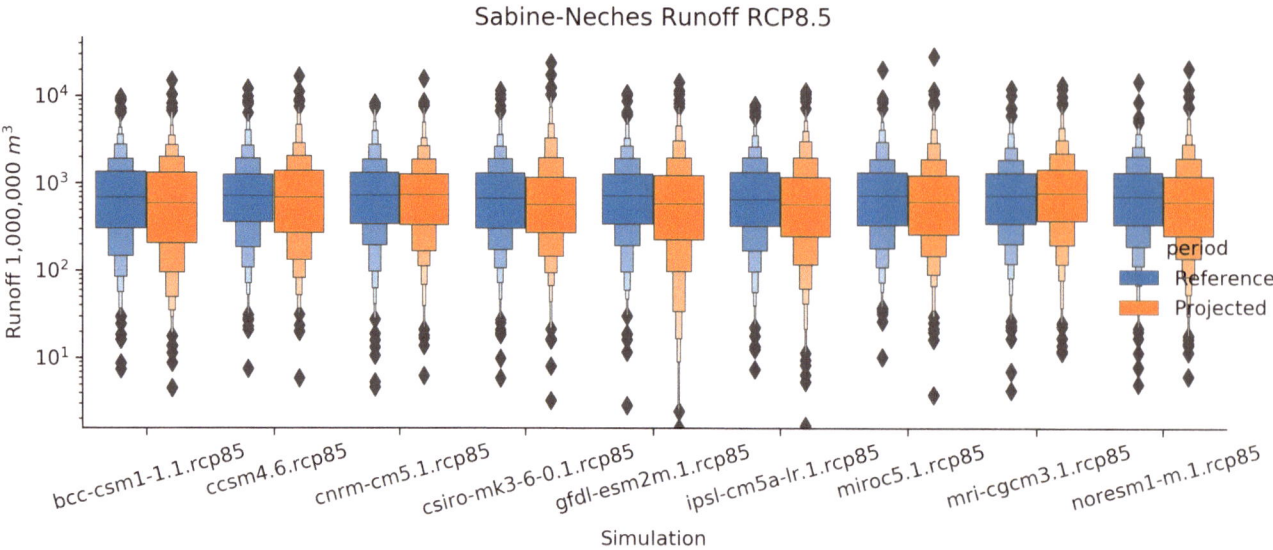

**Fig. 21** Boxen plot of the distribution of historical and RCP 8.5 projected annual runoff values from nine highlighted models for the Sabine-Neches drainage area, showing projected increase in the range of runoff values

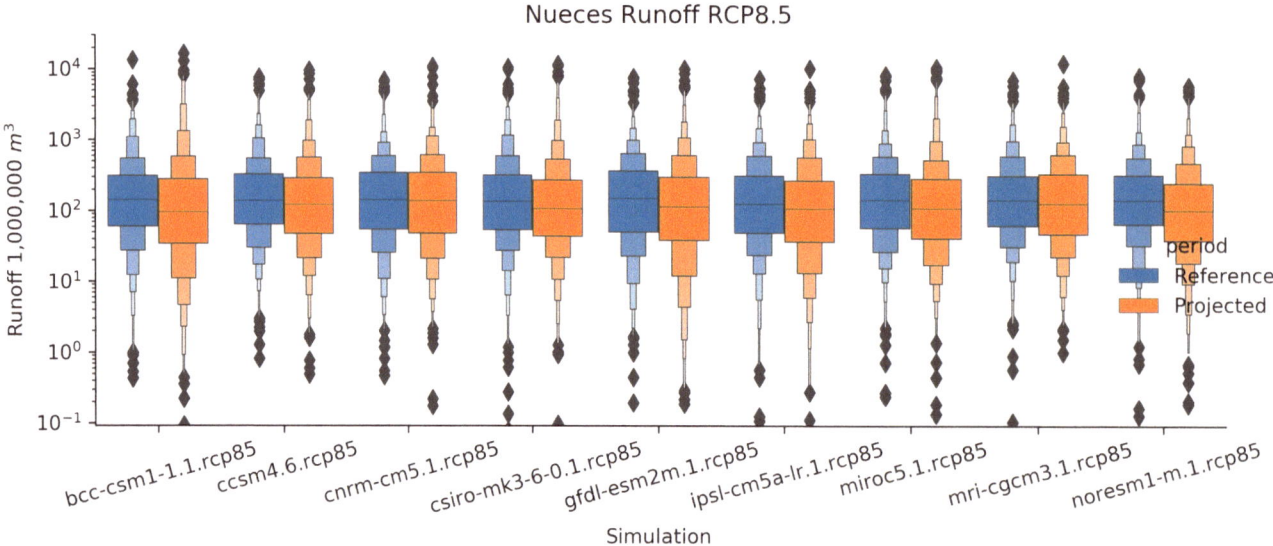

**Fig. 22** Boxen plot of the distribution of historical and RCP 8.5 projected annual runoff values from nine highlighted models for the Nueces drainage area, showing general decrease except for largest runoff values

At the high end of drainage basin size, the Brazos River basin shows an expansion of both the higher values of runoff and the lower values of runoff (Fig. 23). The 12 largest simulated monthly runoff values are all after the year 2045. The standard deviation of runoff (Fig. 24) increases in most models, with an average increase of about 40%. Baseflow decreases in five models, with little or no change in the other four.

The much smaller Lavaca River basin is simulated to have a decreasing median of runoff in all nine RCP 8.5 simulations (Fig. 25). Nonetheless, the frequency of large extreme precipitation values increases in eight out of the nine RCP 8.5 simulations. The overall runoff change is again unclear, with the median simulation implying a slight increase in runoff.

**Fig. 23** Annual total runoff, all models, for the Brazos River basin, using RCP 8.5, showing projected increase in largest monthly runoff and decrease in smallest monthly runoff

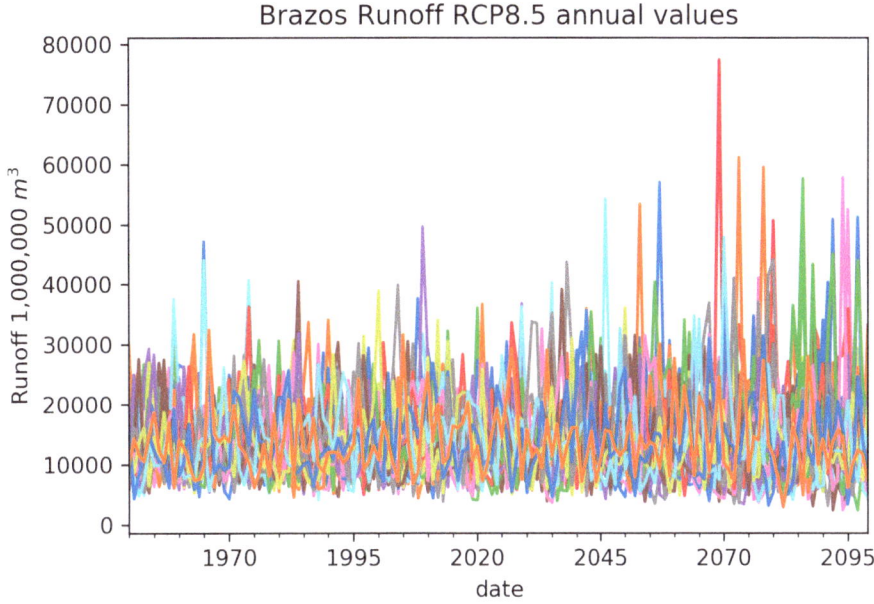

**Fig. 24** Standard deviation of annual total runoff, all models, for the Brazos River basin, using RCP 8.5, over 30-year rolling windows, showing projected erratic increase in variability

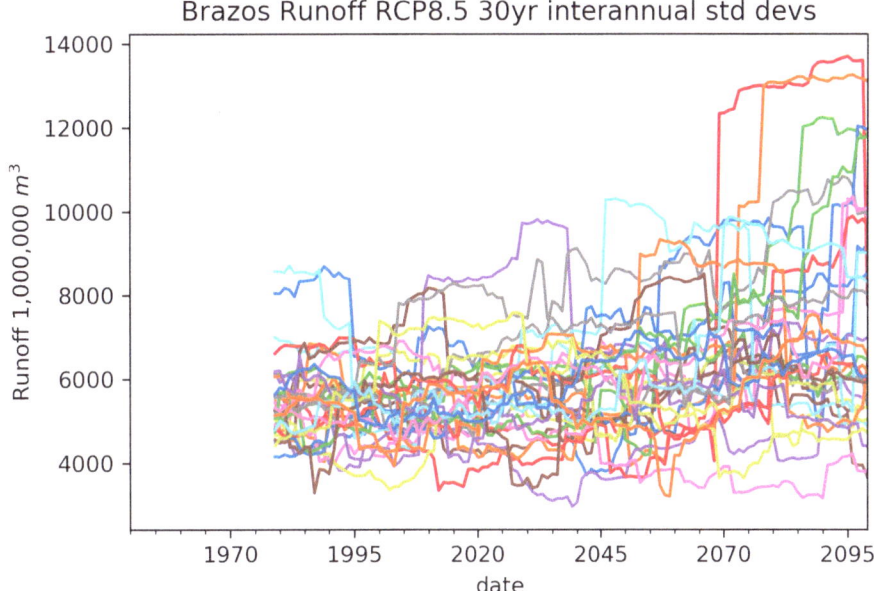

Across the board, projections are similar: little change to a substantial decrease in median inflows, while the majority of projections indicate a slight increase in runoff overall. The large range of runoff projection values means that both large increases and large decreases are possible. However, it is comforting that the large projected decrease in soil moisture does not translate into a large projected decrease of inflows.

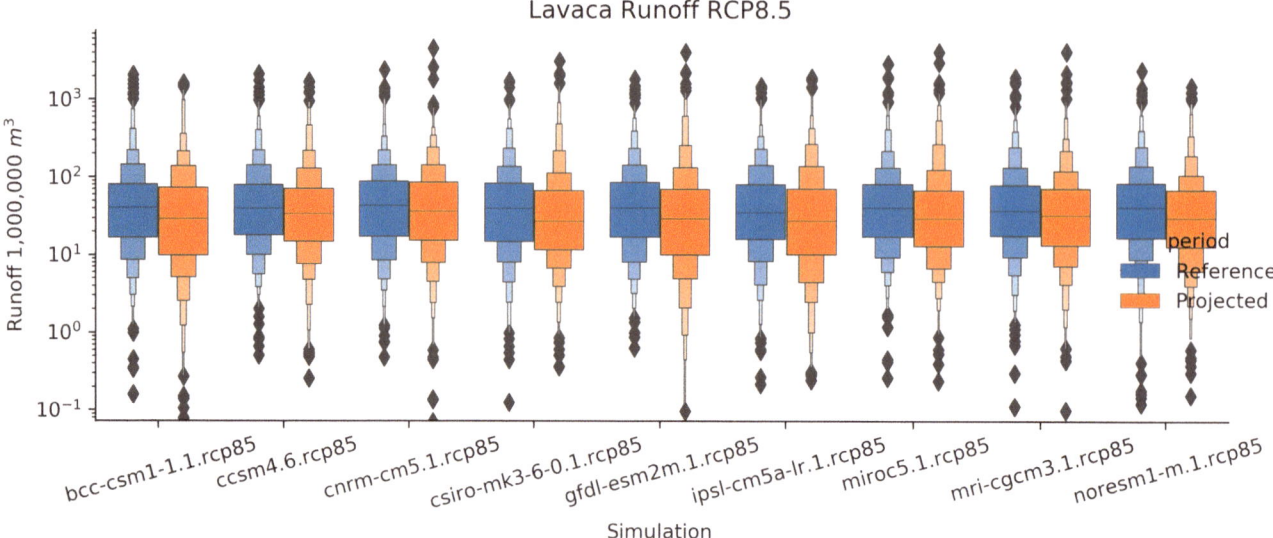

**Fig. 25** Boxen plot of the distribution of historical and RCP 8.5 projected runoff values from nine highlighted models for the Lavaca drainage area, showing projected decrease in median flows but mostly increases in the largest flows

## Total Runoff

Further insight into inflows can be obtained by combining baseflow and runoff and examining total runoff on a month-by-month basis. Here results are shown from three models that collectively illustrate the differing climate change outcomes simulated by the VIC model from different down-scaled GCM projections. All include declines in median monthly streamflow but differ in the magnitude of increase in 95th percentile streamflow and in the future seasonality of peak streamflow.

Linear trends are shown for the period 1975–2100 for the RCP 8.5 projections; over this period, the magnitude of change increases slightly (Fig. 3) but a steady trend is a reasonable approximation. As before, remember that the RCP 8.5 projection is a likely overestimate of the total magnitude of climate change.

Figure 26 shows the VIC projections from the CCSM4 model. The quantile regression lines are shown at the 95th, 50th, and 5th percentiles. The 50th percentile is the median, while the 95th percentile line should be exceeded about six times per decade and the 5th percentile line should fail to be exceeded about six times per decade. These two extremes have slightly diverging trends in the CCSM4-VIC projection, with both extremes becoming somewhat more extreme. In this simulation, the low flows are mostly in the cold sea-

son (C, November through April), while high flows are distributed across the cold season as well as the peak rainfall periods of spring (S, May and June) and fall (F, September and October).

The BCC-CCSM1.1-VIC projection (Fig. 27) features a more dramatic divergence of trends, with extreme low flows prominent in both the November through April cold season and the July-August hot season, and hot-season extreme low flows become relatively more common over time. The extreme high flows occur mainly in the fall and the cold half of the year by mid-century. However, the largest trend takes place in the median flows, declining by about 25% over the period. This means that typical flows are projected to become lower and somewhat more consistent, despite the increasing trend for occasional large discharge events.

The MIROC5-VIC RCP8.5 projection has no increase in extreme total monthly runoff values (Fig. 28). Nonetheless, there is still a tendency for runoff to become more erratic, as both median and extreme low flows decline. In this simulation, low flows become almost exclusively the province of the cold season, while high flows during the cold season become extremely rare.

These three-model runs agree on the tendency for extreme low and median flows to decline but disagree on the trend for extreme high flows. They also agree that many of the extreme low flows should occur during the cold season but differ on

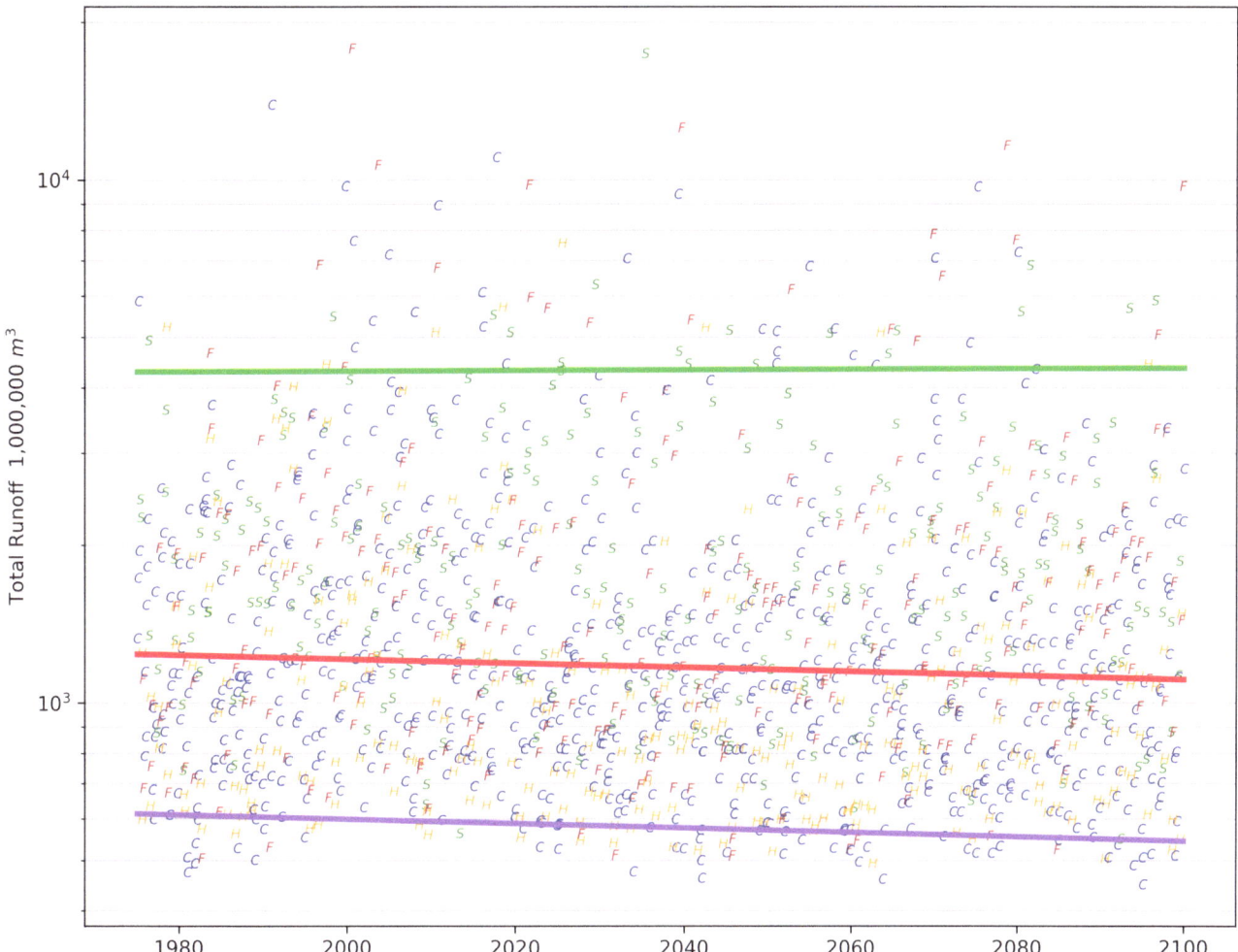

**Fig. 26** Monthly values of total runoff, Brazos River, VIC driven by downscaled CCSM4 RCP 8.5 ensemble member 6. Lines (top to bottom, colored green, red, and purple) are 95th, 50th, and 5th percentile quantile regression trends, while characters show individual monthly total runoff values. *C* represents the months of November through April, *S* May and June, *H* July and August, and *F* September and October

the likelihood of extreme low flows during the summertime. The model runs also disagree on the seasons in which extreme high flows are most likely.

The Lavaca River drainage basin, being much smaller than the Brazos River drainage basin, should be relatively more susceptible to trends in smaller-scale rainfall events such as tropical cyclones. The model runs considered here do not realistically simulate tropical cyclones such as hurricanes, though they do simulate tropical waves and other sorts of tropical disturbances. Given this shortcoming, the trends in extreme high inflows from smaller basins during the warm

seasons should be considered to be less reliable than the trends in extreme high inflows from larger basins. Of the three model runs examined closely here, the BCC-CSM1.1-VIC Lavaca River simulation (Fig. 29) has the largest differences compared with the corresponding Brazos River simulation, but all three have similar sorts of differences. Extreme low and median flows decline in the Lavaca as well as Brazos River, but the extreme low flows are more likely, and increasingly likely, in July and August. Also, the extreme high flows either increase less or even decrease. The results of simulations for all basins are summarized in Figs. 30, 31,

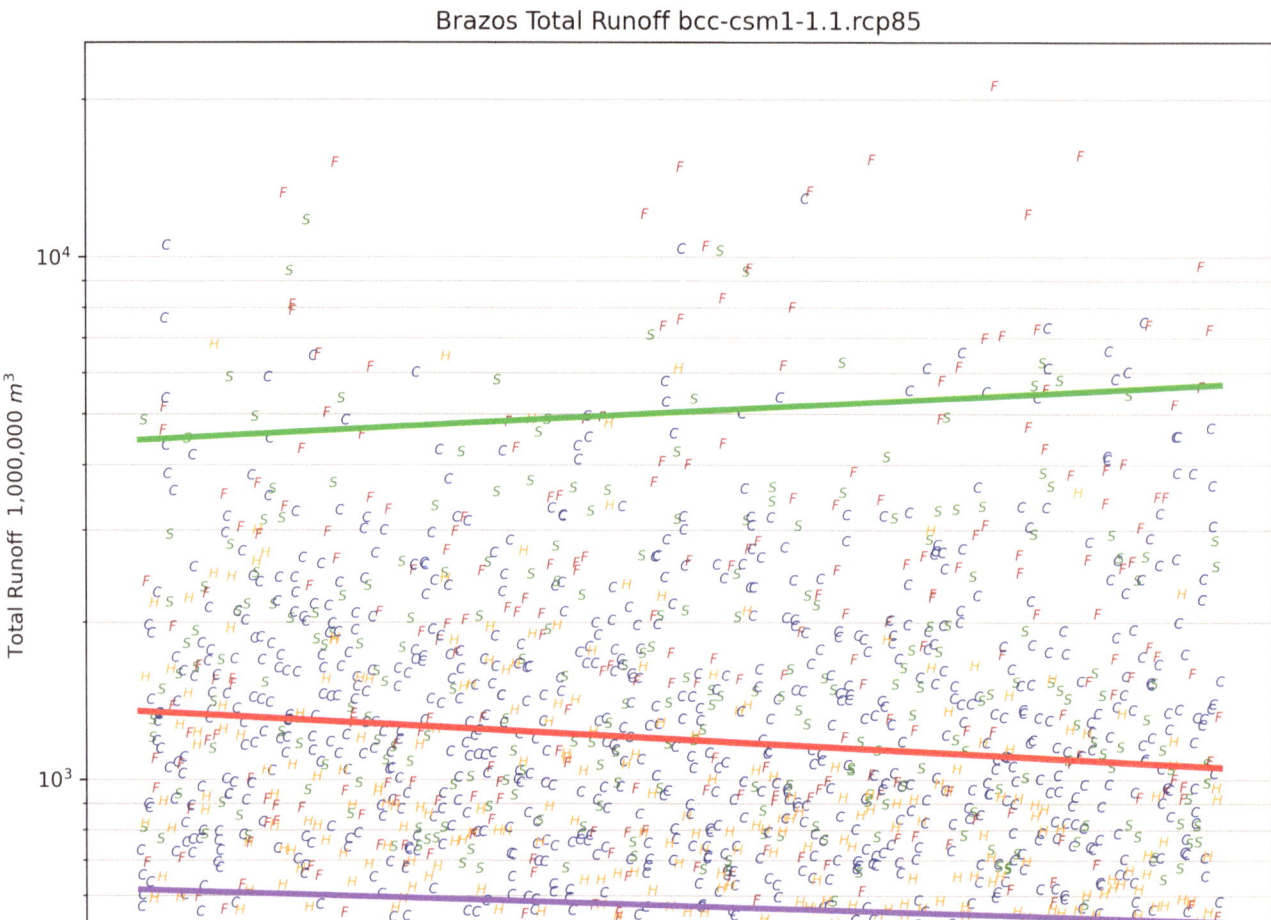

**Fig. 27** Monthly values of total runoff from VIC model driven by downscaled BCC-CSM1.1 RCP 8.5 projection, Brazos River. Lines and symbols as in Fig. 26

and 32. The drainage basins for which calculations were performed are plotted left to right from southwest to northeast, and the model runs highlighted in Figs. 26, 27, 28, and 29 are shown with solid markers. Several general patterns are apparent in Figs. 30, 31, and 32.

First, the trends in median flows tend to be smaller than the trends in extreme low flows, with trends in extreme high flows tending to be largest. Indeed, there is not a single model run for any basin that projects an increase in median flows, and most trends in extreme low flows are likewise negative. In every basin, at least one simulation projects an increase in extreme high flows, and in many basins a majority of simulations do so. Because extreme high flows are several times as large as median flows, the consistent projected decline in median discharge does not translate consistently into a projected decline in total discharge. Second, there is a clear tendency for the trends to decrease from southwest to northeast. In the Sabine/Neches, San Jacinto, and Houston area drainages, there are almost no projected positive trends in any of the three percentiles, and the majority of trends decreases more rapidly than 2% per decade. In the Nueces and drainages farther southwest, most decreases are less extreme than 2.5% per decade, with extreme high discharges generally projected to increase.

Third, there is a general tendency for larger basins to have more positive or less negative trends than smaller basins. The

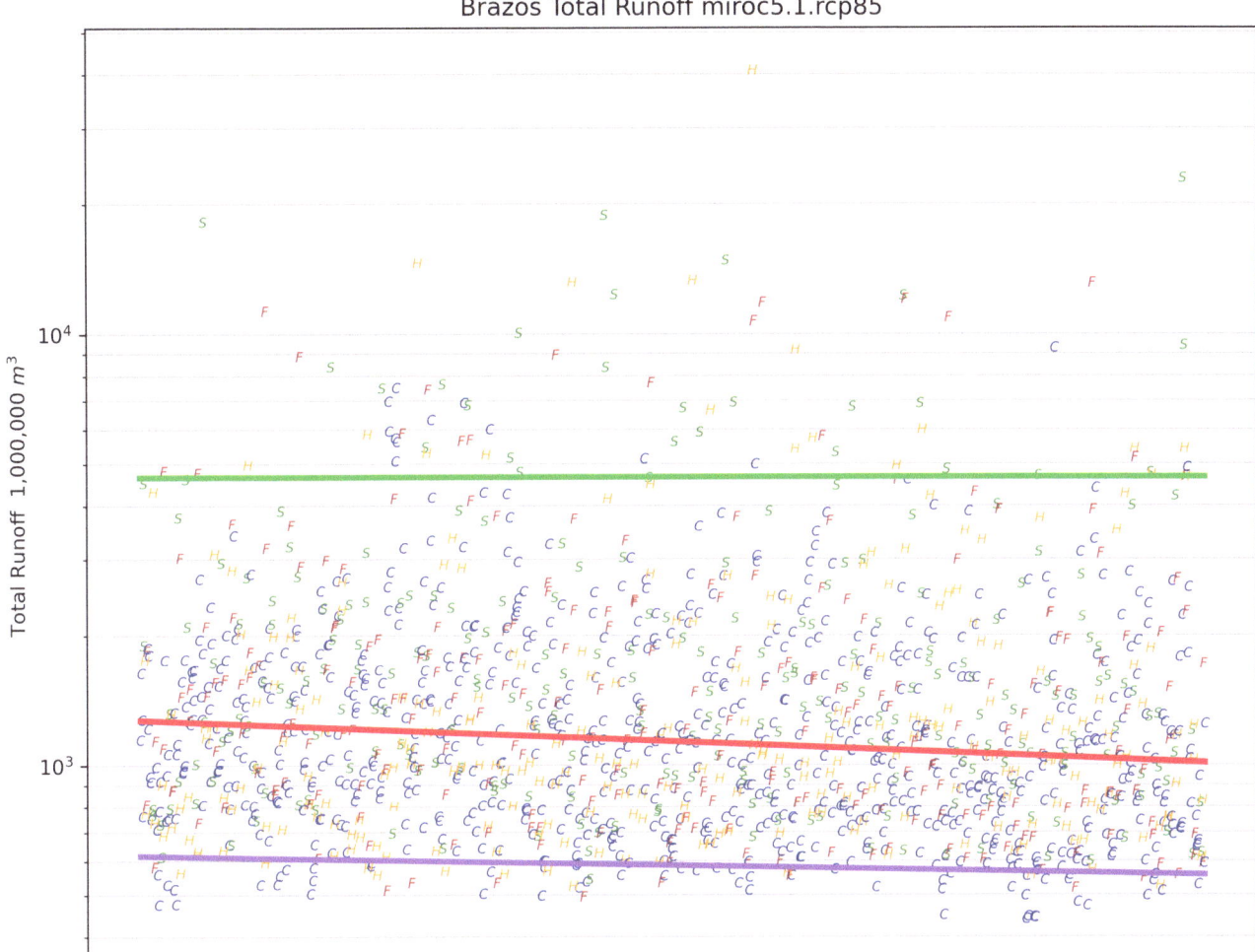

**Fig. 28** Monthly values of total runoff from VIC model driven by downscaled MIROC5 RCP 8.5 projection, Brazos River. Lines and symbols as in Fig. 26

two largest basins, Brazos and Colorado, have fewer projections of large decreases of inflows compared to their neighbors. This tendency, while dramatic in many model simulations, is not universal, with a couple of models showing more negative trends in the larger basins.

There are several sources of uncertainty in the trend projections. First, these are RCP 8.5 trends, and while we do not know the future concentrations of greenhouse gases, they will probably increase less rapidly than in the RCP 8.5 projection. A much more optimistic projection such as RCP 4.5 would have changes about half as large by the end of the twenty-first century, though there would be little difference between them in the next few decades.

Second, as is apparent in Figs. 26, 27, 28, and 29, each model simulates a random sequence of weather, and that randomness introduces noise in the trends estimated from each model's simulations. The magnitude of the uncertainty due to weather noise can be estimated using a technique known as annual block bootstraps. The 90% confidence intervals for the trends span a typical range of about +/− 3.5% per decade for extreme high inflows, about +/− 1.2% per decade for median inflows, and about +/− 0.6% per decade for extreme low inflows. This is the uncertainty of each individual model's projection. However, this uncertainty is reduced by about 67% for the overall projections because each model simulates a different random sequence of weather, and much of the random error cancels out.

Finally, these are modeled trends, and models do not represent the local implications of global climate change perfectly. The magnitude of that error is not necessarily

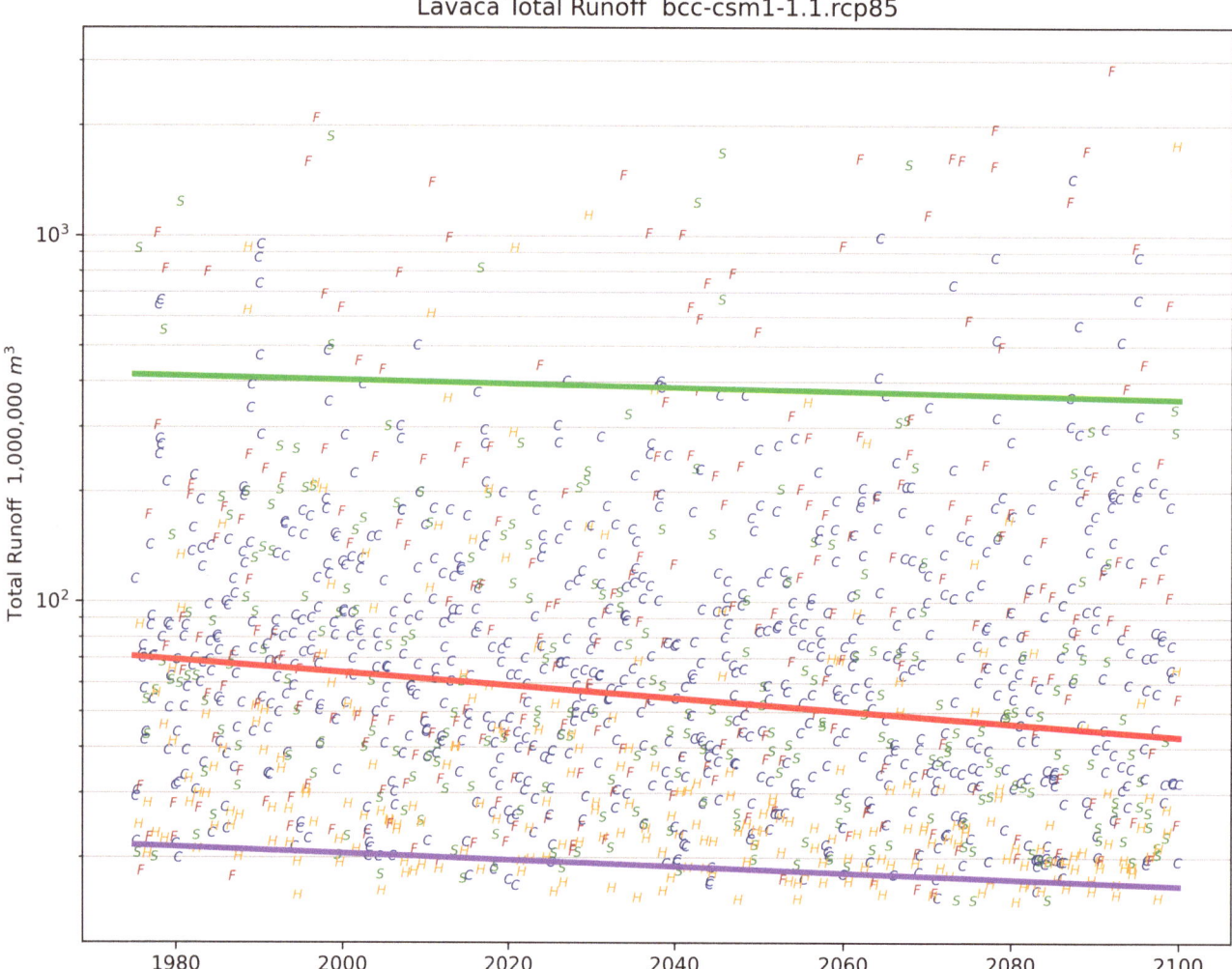

**Fig. 29** Monthly values of total runoff from VIC model driven by a downscaled BCC-CSM1.1 RCP 8.5 projection, Lavaca River. Lines and symbols are as in Fig. 26

correlated with the models' accuracy in representing the baseline climate, but some sense of the magnitude of that uncertainty is given by the range of different trends simulated by the different models, to the extent that their spread is larger than the spread caused by the weather randomness.

Overall, these three sources of error in the projected climate-driven trends seem similar in magnitude, and together imply that one should view the RCP 8.5 projections as indicative of the direction of change and representing a soft upper limit on the magnitude of change.

One other point to keep in mind is that the same uncertainty associated with the randomness of the weather that affects each model's projected trend also will affect the actual future trends, which themselves are subject to the randomness of the weather. In other words, even if we knew exactly how the climate will change in the future, we would still only be able to predict the long-term trend in median inflows to within about 1.2% per decade, with the weather-driven variability from one decade to another being even larger.

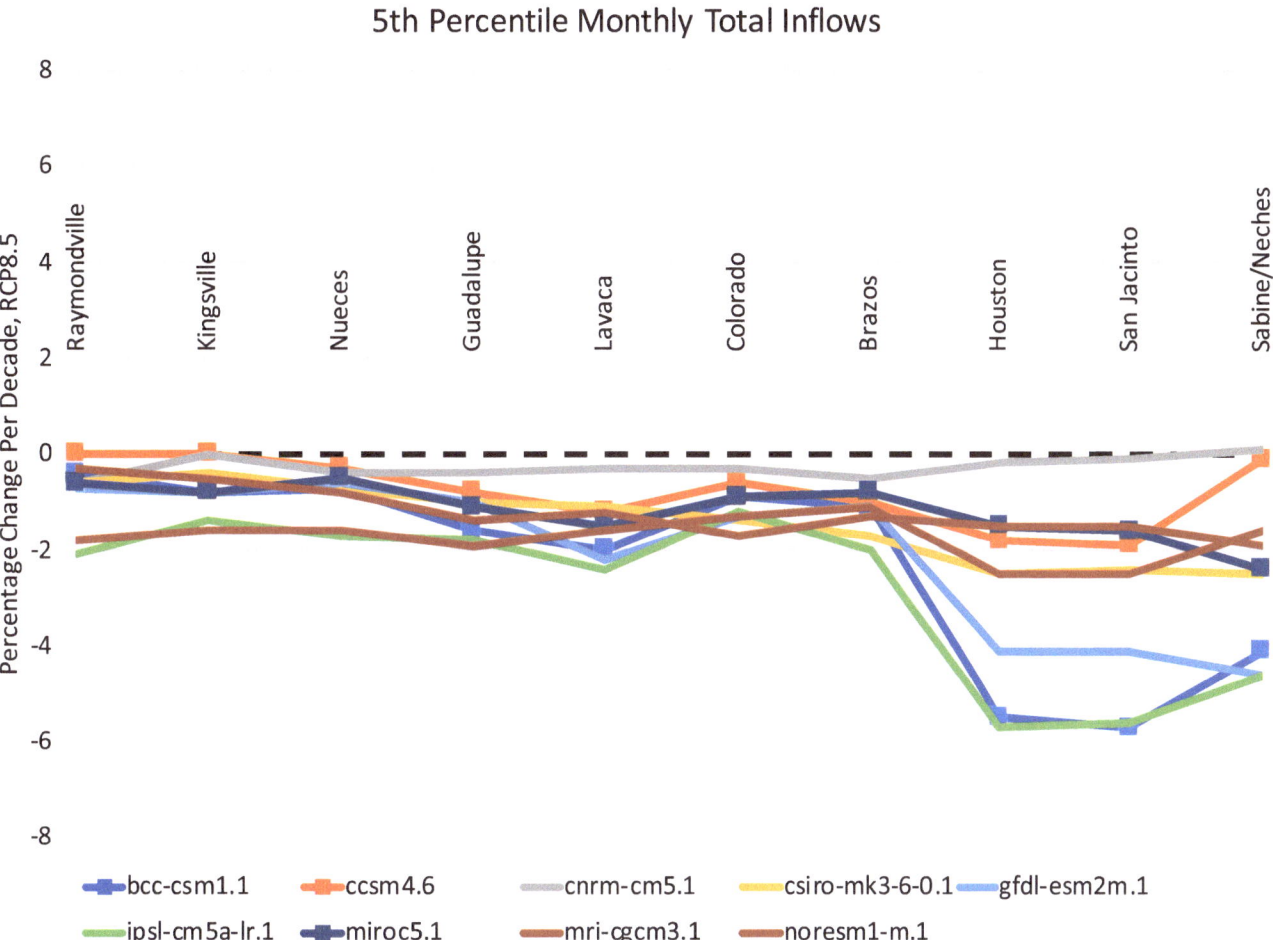

**Fig. 30** Projected decadal trends in low flows (5th percentile of monthly inflows) for each Texas drainage basin analyzed, using the RCP 8.5 projections. Basins are ordered from southwest to northeast

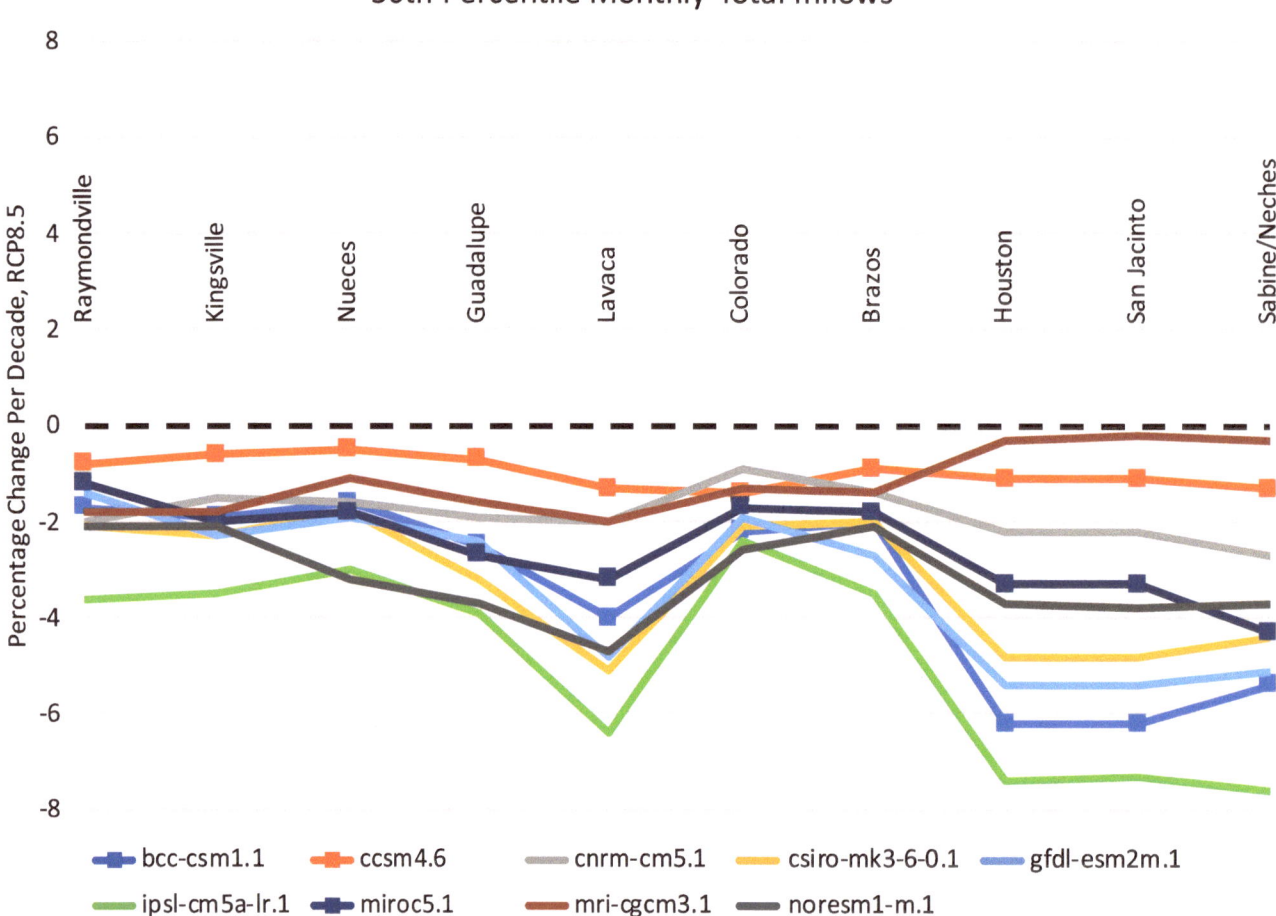

**Fig. 31** Projected decadal trends in median flows (50th percentile of monthly inflows) for each Texas drainage basin analyzed, using the RCP 8.5 projections. Basins are ordered from southwest to northeast

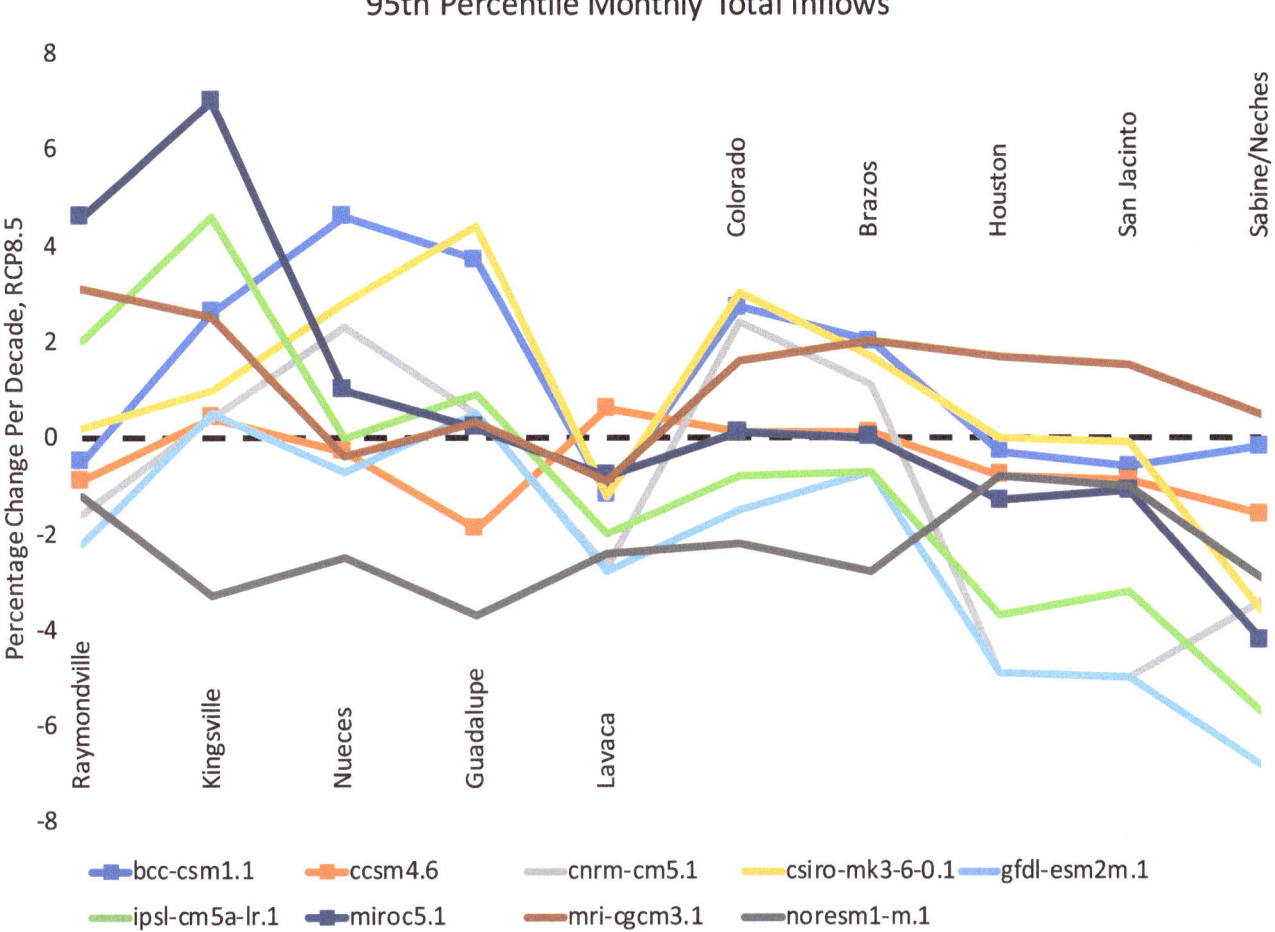

**Fig. 32** Projected decadal trends in high flows (95th percentile of monthly inflows) for each Texas drainage basin analyzed, using the RCP 8.5 projections. Basins are ordered from southwest to northeast

## Sea Level Rise and Hurricanes

Sea level rise is a consequence of climate change that has had profound impacts on Texas bays and estuaries over geologic time. Presently, the rate of relative sea level rise along the Texas coast is between about 3 and 7 mm/yr. This rate of sea level rise is smaller than the most rapid rises during the transition out of the last interglacial period 10,000–20,000 years ago, but it is larger than any known sea level changes over the past 5000 years (Davis 2011).

Sea level rise is not included in the LOCA VIC downscaled projections. The latest National Oceanic and Atmospheric Administration projections (Sweet et al. 2022) indicate a relative sea level rise for the western Gulf of Mexico of 0.3–0.5 m between 2020 and 2050, which would represent an acceleration over past trends.

It is beyond the scope of this chapter to examine how relative sea level rise will alter the currents and exchanges of fresh and salt water in Texas bays and estuaries. However, relative sea level rise and hurricanes interact to produce changing extreme weather impacts.

Hurricanes and climate change are the subject of much interest and much misinformation. The best available summary is a recent multiple-expert review paper (Knutson et al. 2020), which presented relatively high confidence that sea level rise would lead to greater inundation from hurricanes, that hurricane precipitation rates would increase, that average tropical cyclone intensity would increase, and that the proportion of tropical cyclones that become very intense will increase. There was relatively low confidence that the overall number of tropical cyclones will change in any particular way, that total numbers of very intense tropical cyclones would increase, or that tropical cyclones would slow down.

The combination of greater average intensity and sea level rise increase the odds that storms will create new channels through barrier islands and thereby alter the salinity and currents of individual bays and estuaries. As sea level continues to rise, the probability of this will continue to increase unless the barrier islands can respond to gradually rising seas by retreating inland and gaining elevation. This is more easily done in places where the barrier islands are not already developed.

## Conclusion

The climate of Texas contributes to the diversity of its bays and estuaries. From Brownsville to Port Arthur, average rainfall increases, average inflow basin size increases, and average temperature decreases. These factors lead most directly to the large differences of salinity from Laguna Madre to Sabine Lake.

Climate change is but one of many man-made and natural factors affecting Texas bays and estuaries. Given the growing demand for fresh water by Texas, it is not clear that climate change is even the dominant change factor, since managed and unmanaged streams have substantially different flow trends (Rodgers et al. 2020). Nonetheless, climate change is happening, and some effects are clear. Rising air temperatures are leading to increased water temperatures and increased potential evapotranspiration, which along with increased water diversions contribute to increased salinity (Bugica et al. 2020).

Less easy to identify are climate-driven changes in inflows. According to climate models, extremes as well as means ought to be changing, and the erratic nature of extreme events makes underlying trends challenging or impossible to identify with a limited observational dataset. Variability in Texas bay and estuary inflows is substantial (Ward 1980), driven by the seasonal cycle and the erratic nature of Texas weather, enhanced by larger-scale modes of climate variability such as ENSO.

Climate models, and hydrologic simulations driven by their downscaled output, are not clear regarding whether inflow volume overall will increase or decrease. However, there is a clear simulated trend toward more erratic inflows, as represented by increases in variability, decreases in the magnitude of low flows and median flows, and possible increases in the magnitude of high flows. Ecosystems in estuaries are presently adapted to large inflow variability (Pollack et al. 2011), but bays and estuaries will be at increasing risk of inflows outside the recent historical envelope. Fortunately, super-extreme inflow events such as Harvey (Steichen et al. 2020; Thyng et al. 2020) will not become the norm.

In general, all inflow trends are more negative toward the Louisiana border than toward the Mexico border, and there is also some tendency for smaller drainage basins to have more negative trends than larger drainage basins. Given the uncertain nature of future climate impacts, management decisions regarding bays and estuaries should focus on resilience. The future state of the bays and estuaries can neither be predicted nor determined through intentional action. No matter what, due to the increasing temperatures, the future state of Texas bays and estuaries will be unlike the past state of Texas bays and estuaries.

**Acknowledgements** This publication was funded in part through Contract No. 21-155-007-C879 from the Texas General Land Office (GLO) with Gulf of Mexico Energy Security Act of 2006 funding made available to the State of Texas and awarded under the Texas Coastal Management Program. The views contained herein are those of the authors and should not be interpreted as representing the views of the GLO or the State of Texas.

The basin-aggregated VIC model projections presented in this chapter are archived in the Gulf of Mexico Research Initiative Information & Data Cooperative (GRIIDC) under doi 10.7266/qhkhp60n.

We acknowledge the World Climate Research Program's Working Group on Coupled Modeling, which is responsible for CMIP, and we thank the climate modeling groups (listed in Table 1) for producing and making available their model output. For CMIP, the U.S. Department of Energy's Program for Climate Model Diagnosis and Intercomparison provides coordinating support and led development of software infrastructure in partnership with the Global Organization for Earth System Science Portals.

## References

Bugica K, Sterba-Boatwright B, Wetz MS (2020) Water quality trends in Texas estuaries. Mar Poll Bull 152:110903. https://doi.org/10.1016/j.marpolbul.2020.110903

Colorado-Ruiz G, Cavazos T, Salinas JA, De Grau P, Ayala R (2018) Climate change projections from Coupled Model Intercomparison Project phase 5 multi-model weighted ensembles for Mexico, the North American monsoon, and the mid-summer drought region. Int J Climatol 38:5699–5716. https://doi.org/10.1002/joc.5773

Davis RA Jr (2011) Sea-level change in the Gulf of Mexico. Texas A&M University Press, College Station, p 172

Forster P, Storelvmo T, Armour K, Collins W, Dufresne JL, Frame D, Lunt DJ, Mauritsen T, Palmer MD, Watanabe M, Wild M, Zhang H (2021) The earth's energy budget, climate feedbacks, and climate sensitivity. In: Masson-Delmotte V, Zhai P, Pirani A, Connors SL, Péan C, Berger S, Caud N, Chen Y, Goldfarb L, Gomis MI, Huang M, Leitzell K, Lonnoy E, Matthews JBR, Maycock TK, Waterfield T, Yelekçi O, Yu R, Zhou B (eds) Climate change 2021: The physical science basis. Contribution of working group I to the sixth assessment report of the intergovernmental panel on climate change. Cambridge University Press, pp 923–1054. doi: https://doi.org/10.1017/9781009157896.009

Hausfather Z, Peters GP (2020) Emissions – the 'business as usual' story is misleading. Nature 577:618–620. https://doi.org/10.1038/d41586-020-00177-3

Illston BG, Basara JB, Crawford KC (2004) Seasonal to interannual variations of soil moisture measured in Oklahoma. Int J Climatol 24:1883–1896. https://doi.org/10.1002/joc.1077

Kim HC, Son S, Montagna P, Spiering B, Nam J (2014) Linkage between freshwater inflow and primary productivity in Texas estuaries: Downscaling effects of climate variability. J Coastal Res 68:65–73

Knutson T, Camargo SJ, Chan JCL, Emanuel K, Ho C-H, Kossin J, Mohapatra M, Satoh M, Sugi M, Walsh K, Wu L (2020) Tropical cyclones and climate change assessment, Part II: Projected response to anthropogenic warming. Bull Amer Meteorol Soc 101:E303–E322. https://doi.org/10.1175/BAMS-D-18-0194.1

Kothari K, Ale S, Bordovsky JP, Munster CL, Singh VP, Nielsen-Gammon J, Hoogenboom G (2021) Potential genotype-based climate change adaptation strategies for sustaining cotton production in the Texas High Plains. Field Crops Res 271:108261. https://doi.org/10.1016/j.fcr.2021.108261

Lehner F, Deer C, Maher N, Marotzke J, Fischer EM, Brunner L, Knutti R, Hawkins E (2020) Partitioning climate projection uncertainty with multiple large ensembles and CMIP5/6. Earth Syst Dyn 11:491–508. https://doi.org/10.5194/esd-11-491-2020

Liang X, Wood EF, Lettenmaier DP (1996) Surface soil moisture parameterization of the VIC-2L model: evaluation and modification. Glo Plan Change 13:195–206. https://doi.org/10.1016/0921-8181(95)00046-1

Livneh B, Rosenberg EA, Lin C, Nijssen B, Mishra V, Andreadis KM, Maurer EP, Lettenmaier DP (2013) A long-term hydrologically based dataset of land surface fluxes and states for the conterminous United States: update and extensions. J Climate 27:477–486. https://doi.org/10.1175/JCLI-D-12-00508.1

Livneh B, Bohn TJ, Pierce DW, Munoz-Arriola F, Nijssen B, Vose R, Cayan DR, Brekke L (2015) A spatially comprehensive, hydrometeorological data set for Mexico, the US, and southern Canada, 1950-2013. Sci Data 2:150042. https://doi.org/10.1038/sdata.2015.42

Meehl GA, Senior CA, Eyring V, Flato G, Lamarque J-F, Stouffer RJ, Taylor KE, Schlund M (2020) Context for interpreting equilibrium climate sensitivity and transient climate response from the CMIP6 Earth system models. Sci Adv 6:eaba1981. https://doi.org/10.1126/sciadv.aba1981

Montagna PA, Hu X, Palmer TA, Wetx M (2018) Effect of hydrological variability on the biogeochemistry of estuaries along a regional climatic gradient. Limnol and Oceanog 63:2465–2478

Nielsen-Gammon JW (2011) The changing climate of Texas. In: Schmandt J, North GR, Clarkson J (eds) The impact of global warming on Texas. University of Texas Press, pp 39–68

Nielsen-Gammon JW, Banner JL, Cook BI, Tremaine DM, Wong CI, Mace RE, Gao H, Yang Z-L, Flores Gonzalez M, Hoffpauir R, Gooch T, Kloesel K (2020) Unprecedented drought challenges for Texas water resources in a changing climate: What do researchers and stakeholders need to know? Earth's. Future 8:e2020EF001552. https://doi.org/10.1029/2020EF001552

Nielsen-Gammon JW, Holman S, Buley A, Jorgensen S, Escobedo J, Ott C, Dedrick J, Van Fleet A (2021) Assessment of historic and future trends in extreme weather in Texas, 1900-2036: 2021 Update. Document OSC-202101, Office of the State Climatologist, Texas A&M University. Available at https://climatexas.tamu.edu/products/texas-extreme-weather-report/index.html Accessed 1 Oct 2022

Pierce DW, Cayan DR, Thrasher BL (2014) Statistical downscaling using localized constructed analogs (LOCA). J Hydromet 15:2558–2585. https://doi.org/10.1175/JHM-D-14-0082.1

Pierce DW, Cayan DR, Maurer EP, Abatzoglou JT, Hegewisch KC (2015) Improved bias correction techniques for hydrological simulations of climate change. J Hydromet 16:2421–2442. https://doi.org/10.1175/JHM-D-14-0236.1

Pollack JB, Kim HC, Morgan EK, Montagna PA (2011) Role of flood disturbance in natural oyster (Crassostrea virginica) population maintenance in an estuary in South Texas, USA. Estuaries and Coasts 34:187–197

Rodgers K, Roland V, Hoos A, Crowley-Ornelas E, Knight R (2020) An analysis of streamflow trends in the Southern and Southeastern US from 1950-2015. Water 12:3345. https://doi.org/10.3390/w12123345

Steichen JL, Labonté JM, Windham R, Hala D, Kaiser K, Setta S, Faulkner PC, Bacosa H, Yan G, Kamalanathan M, Quigg A (2020) Microbial, physical, and chemical changes in Galveston Bay following an extreme flooding event, Hurricane Harvey. Front Marine Sci 7:l. https://doi.org/10.3389/fmars.2020.00186

Sweet WV, Hamlington BD, Kopp RE, Weaver CP, Barnard PL, Bekaert D, Brooks W, Craghan M, Dusek G, Frederikse T, Garner G, Genz AS, Krasting JP, Larour E, Marcy D, Marra JJ, Obeysekera J, Osler M, Pendleton M, Roman D, Schmied L, Veatch W, White KD, Zuzak C (2022) Global and regional sea level rise scenarios for the United States: up- dated mean projections and extreme water level probabilities along U.S. Coastlines. NOAA Technical Report NOS 01. National Oceanic and Atmospheric Administration, National Ocean Service, Available from https://oceanservice.noaa.gov/hazards/sealevelrise/noaa-nos-techrpt01-global-regional-SLR-scenarios-US.pdf. Accessed 1 Oct 2022

Thyng KM, Hetland RD, Socolofsky SA, Fernando N, Turner EL, Schoenbaechler C (2020) Hurricane Harvey caused unprecedented freshwater inflow to Galveston Bay. Estuaries Coasts 43:1836–1852

Tolan JM (2007) El Niño-Southern Oscillation impacts translated to the watershed scale: Estuarine salinity patterns along the Texas Gulf Coast, 1982 to 2004. Estuarine, Coastal and Shelf Sci 72:247–260

van Vuuren DP, Edmonds J, Kainuma M, Riahi K, Thomson A, Hibbard K, Hurtt GC, Kram T, Krey V, Lamarque J-F, Masui T, Meinshausen M, Nakicenovic N, Smith SJ, Rose SK (2011) The representative concentration pathways: an overview. Clim Ch 109:5. https://doi.org/10.1007/s10584-011-0148-z

Vano J, Hamman J, Gutmann E, Wood A, Mizukami N, Clark M, Pierce D, Cayan D, Wobus C, Nowak K, Arnold J (2020) Comparing downscaled LOCA and BCSD CMIP5 climate and hydrology projections - release of downscaled LOCA CMIP5 hydrology. US Bureau of Reclamation. Available at: https://gdo-dcp.ucllnl.org/downscaled_cmip_projections/techmemo/LOCA_BCSD_hydrology_tech_memo.pdf. Accessed 1 Oct 2022

Ward, G. H. (1980). Hydrography and circulation processes of Gulf estuaries. In Estuarine and wetland processes: with emphasis on modeling. Boston, MA: Springer US. pp. 183–215

Zhao G, Gao H (2019) Estimating reservoir evaporation losses for the United States: Fusing remote sensing and modeling approaches. Remote Sens Env 226:109–124. https://doi.org/10.1016/j.rse.2019.03.015

Zhao B, Kao S-C, Zhao G, Gangrade S, Rastoki D, Ashfaq M, Gao H (2023) Evaluating enhanced reservoir evaporation losses from CMIP6-based future projections in the contiguous United States. Earth's. Future 11:e2022EF002961. https://doi.org/10.1029/2022EF002961

# Hydrology, Circulation, and Salinity

Daniel Opdyke ⓘ, Josef Hoffmann, Paul A. Montagna ⓘ, and Joseph F. Trungale ⓘ

**Abstract**

Patterns of freshwater inflow and salinity are presented for the Sabine-Neches, Trinity-San Jacinto, Colorado-Lavaca, Guadalupe, Mission-Aransas, Nueces, and Laguna Madre estuaries of Texas. There is a strong precipitation gradient from west to east, which translates into a strong freshwater inflow gradient where estuary inflows generally increase from west to east. There is a strong correlation between inflows of adjacent estuaries. Another driver of inflow is a cycle of extreme droughts and floods. Inflows to each estuary are highly variable, with the standard deviation of monthly inflows exceeding the mean in all estuaries except the Sabine-Neches. Seasonal patterns of inflows exist, with the easternmost estuaries (Sabine-Neches and Trinity-San Jacinto) exhibiting high inflows during winter and spring, mid-coast estuaries (Colorado-Lavaca and Guadalupe) having high spring and fall inflows, and lower coast estuaries (Mission-Aransas, Nueces, and Laguna Madre) having higher fall inflows. Freshwater inflow patterns create similar patterns in salinity, although salinity is complicated by the estuaries having differing volumes, tidal passes, and tidal prisms. In general, the Sabine-Neches Estuary is often oligohaline, whereas the Nueces Estuary and the Laguna Madre are often euhaline to hyperhaline, and the intervening estuaries are mesohaline and polyhaline. In most cases, salinity variability is highest where the salinity average is moderate in locations that swing between fresh and salt water. Near the Gulf passes, and near freshwater sources, the average salinity is high, and low, respectively, but the standard deviation is relatively low. More salinity monitoring and a more modern approach to salinity modeling is needed. Long-term trends are uncertain because of high spatial and temporal variability of inflow and salinity.

**Keywords**

Circulation · Hydrodynamic modeling · Salinity · Drought · Flood · Precipitation

**Supplementary Information** The online version contains supplementary material available at https://doi.org/10.1007/978-3-031-70882-4_4.

D. Opdyke (✉)
Anchor QEA, LLC, Austin, TX, USA
e-mail: dopdyke@anchorqea.com

J. Hoffmann · J. F. Trungale
The Nature Conservancy, Colorado River Program,
Boulder, CO, USA
e-mail: joseph.trungale@tnc.org; joseph.trungale@tnc.org

P. A. Montagna
Texas A&M University-Corpus Christi, Harte Research Institute,
Corpus Christi, TX, USA
e-mail: Paul.Montagna@tamucc.edu

## Abbreviations

| | |
|---|---|
| ac-ft | Acre-feet |
| ac-ft/d | Acre-feet per day |
| ac-ft/mo | Acre-feet per month |
| cfs | Cubic feet per second |
| ENSO | El Niño–southern oscillation |
| ft³/s | Cubic feet per second |
| GRIIDC | Gulf of Mexico Research Initiative Information and Data Cooperative |
| HDR | HDR Engineering, Inc. |
| HRI | Harte Research Institute |
| km | Kilometer |
| km² | Square kilometer |
| m³ | Cubic meter |
| m³/mo | Cubic meters per month |
| m³/s | Cubic meters per second |
| mi² | Square mile |
| NOAA | National Oceanic and Atmospheric Administration |
| psu | Practical salinity unit |

© The Author(s) 2025
P. A. Montagna, A. R. Douglas (eds.), *Freshwater Inflows to Texas Bays and Estuaries*, Estuaries of the World, https://doi.org/10.1007/978-3-031-70882-4_4

| | |
|---|---|
| psu/yr | Practical salinity units per year |
| SI | International System of Units |
| TAMU | Texas A&M University |
| TCEQ | Texas Commission on Environmental Quality |
| TDWR | Texas Department of Water Resources |
| TWDB | Texas Water Development Board |
| TxRR | Texas Rainfall-Runoff |
| USEPA | U.S. Environmental Protection Agency |
| USGS | U.S. Geological Survey |

## Introduction

In Texas, precipitation and river flows are characterized by extremes, both spatially and temporally. Dramatic variations occur from the swamps of east Texas to the deserts of west Texas, and from "the time it never rained" (Kelton 1973) in the 1950s to the back-to-back-to-back 500-year floods of 2015, 2016, and 2017 in Houston (Lotz 2017; Ingraham 2017). As a result of these widely varying river flows, Texas has equally variable estuaries, from the Sabine-Neches estuary in the northeast, which is often called Sabine Lake, to the Laguna Madre in the southwest coast of Texas, part of the largest hyper-saline estuary in the world (NPS 2022; Tweedley et al. 2019).

The purpose here is to describe Texas watersheds, explore the spatial and temporal variations in precipitation and hydrology, and describe how freshwater combines with tides and circulation patterns to create equally dramatic variations in estuarine salinity. Salinity is a focus because it has been described as a key characteristic of estuaries that influences species abundance and diversity (Smyth and Elliott 2016), despite the former—widely held but now understood as misguided—perception in Texas that water reaching the Gulf of Mexico was wasted (Ward 2005 and citations therein). The remaining chapters in this book will highlight the importance, and limitations, of salinity in controlling estuarine biota. Accordingly, this chapter serves as a foundation for the chapters to come. Data supporting the tables and figures of this chapter are available online (Trungale et al. 2023).

There are seven bar-built estuaries along the Texas coast (Fig. 1). The estuary names, which result from the river source, will be used here as opposed to the major or primary bay names and include the following (ordered from northeast to southwest):

**Sabine-Neches Estuary:** This estuary is also referred to as Sabine Lake.

**Trinity-San Jacinto Estuary:** The upstream portion of this estuary, near the terminus of the Trinity River, is referred to as Trinity Bay, and the larger estuary is referred to as the Galveston Bay system.

**Colorado-Lavaca Estuary:** The upstream portion of this estuary, near the terminus of the Lavaca River, is referred to as Lavaca Bay, and the larger estuary is referred to as the Matagorda Bay system. The Colorado River flows into the eastern portion (or arm) of Matagorda Bay.

**Guadalupe Estuary:** This estuary is also referred to as San Antonio Bay.

**Mission-Aransas Estuary:** The upstream portion of this estuary, near the terminus of the Mission River, is referred to as Mission Bay, the adjacent larger estuary is referred to as Copano Bay, the waterbody closer to the Gulf is Aransas Bay, and together these bays are referred to as the Aransas Bay system.

**Nueces Estuary:** The upstream portion of this estuary, near the terminus of the Nueces River, is referred to as Nueces Bay, and the adjacent larger estuary is referred to as the Corpus Christi Bay system.

**Laguna Madre Estuary:** The Upper Laguna Madre is the estuary to the north of the coastal land mass commonly known as the Landcut, and the Lower Laguna Madre is to the south of the Landcut. This estuary will be discussed as two estuaries (upper and lower) where data permit. There are no major rivers flowing into Baffin Bay nor Laguna Madre.

Although the mouths of the Brazos River, San Bernard River, and the Rio Grande are technically estuaries, these are drowned river valleys, or riverine estuaries, lacking enclosed bay systems. There are also minor bay systems that lack direct drainage from a freshwater inflow source. The largest minor bays along the Texas coast (from northeast to southwest) are Christmas Bay, Cedar Lakes, East Matagorda Bay, and South Bay. River estuaries and minor bays will not be examined here, but they have been the subject of a previous study (Palmer et al. 2011) and the data are available online (Montagna 2022; Montagna and Kalke 2023).

## Bathymetry

Each bay system was filled with sediments within recent geologic time scales. The result is that these bay systems are very shallow (Figs. 2, 3, 4, 5, 6, 7, 8, and 9). Figures 2, 3, 4, 5, 6, 7, 8, and 9 are based on the Continuously Updated Digital Elevation Model data from National Oceanic and

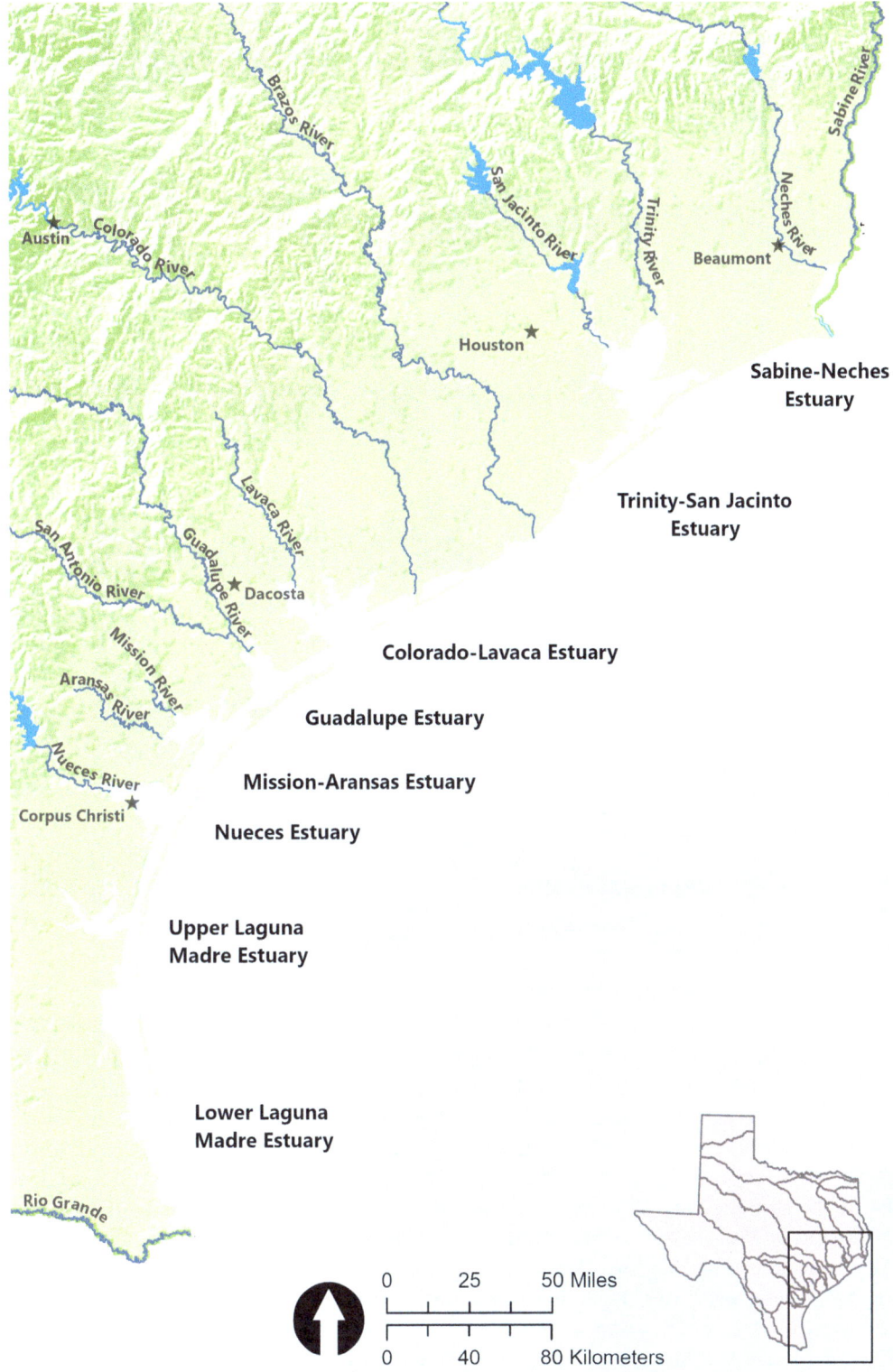

**Fig. 1** Map of Texas estuaries and major rivers that discharge into each estuary

**Fig. 2** Bathymetry for the
Sabine-Neches Estuary

Atmospheric Administration (NOAA) (Amante et al. 2023; NOAA 2023). Another source of bathymetric data are the NOAA nautical charts available at https://www.nautical-charts.noaa.gov/. The depth ranges from 1 to 5 m in the bays. San Antonio Bay is the shallowest, mostly less than 2 m (Fig. 5). Laguna Madre is also shallow (Figs. 8 and 9). Corpus Christi Bay is the deepest reaching near 5 m (Fig. 7). The secondary bay areas closet to the river mouths in all bays are the shallowest, typically 1–2 m. The primary bays connected to the Gulf of Mexico are typically the deepest at 3–5 m. The Intracoastal Waterway runs along the entire coast, parallel to the mainland and barrier islands, and is typically dredged to 12 ft (~4 m). Major ship channels in Galveston Bay, Matagorda Bay, Corpus Christi Bay, and the Brazos Santiago Pass in the Lower Laguna Madre are dredged to between 38 to 54 ft (~12 to 16m). The channel through Sabine Pass (dredged to 48 ft [15m]) connects directly to ports and does not cross the bay. Some bays have been deepened by dredging for oyster shell, which was used as a construction material. The shape, size, and depth of the bays modify tides and circulation.

**Fig. 3** Bathymetry for the
Trinity-San Jacinto Estuary

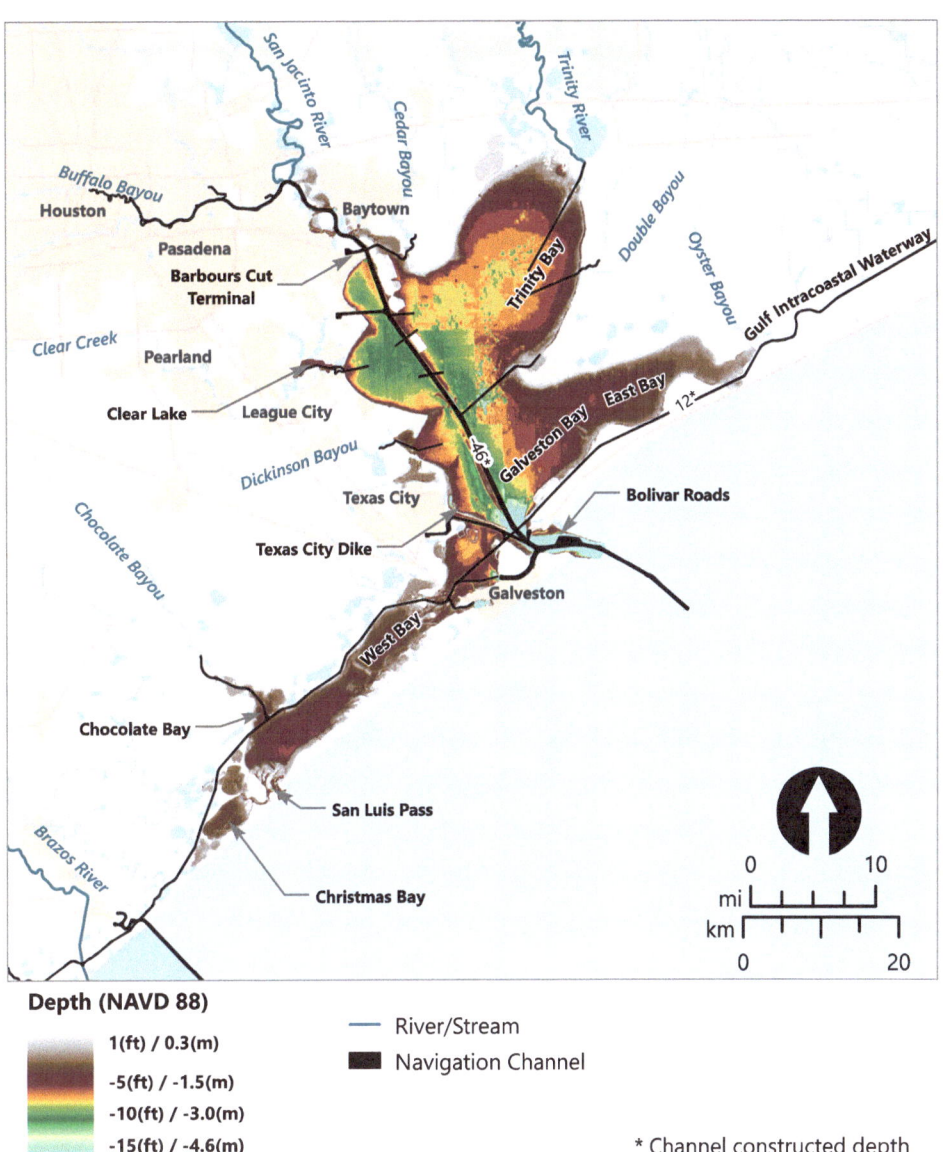

**Fig. 4** Bathymetry for the
Colorado-Lavaca Estuary

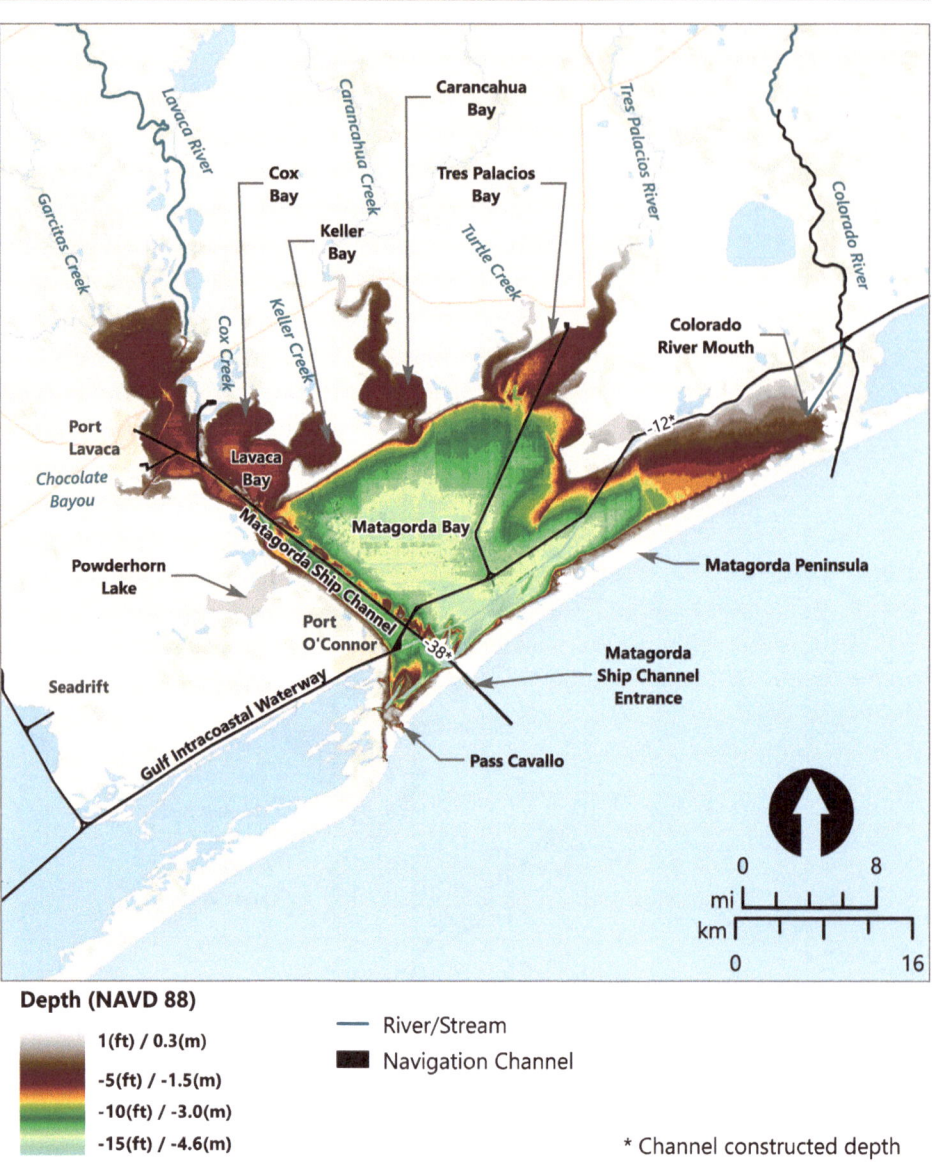

**Fig. 5** Bathymetry for the
Guadalupe Estuary

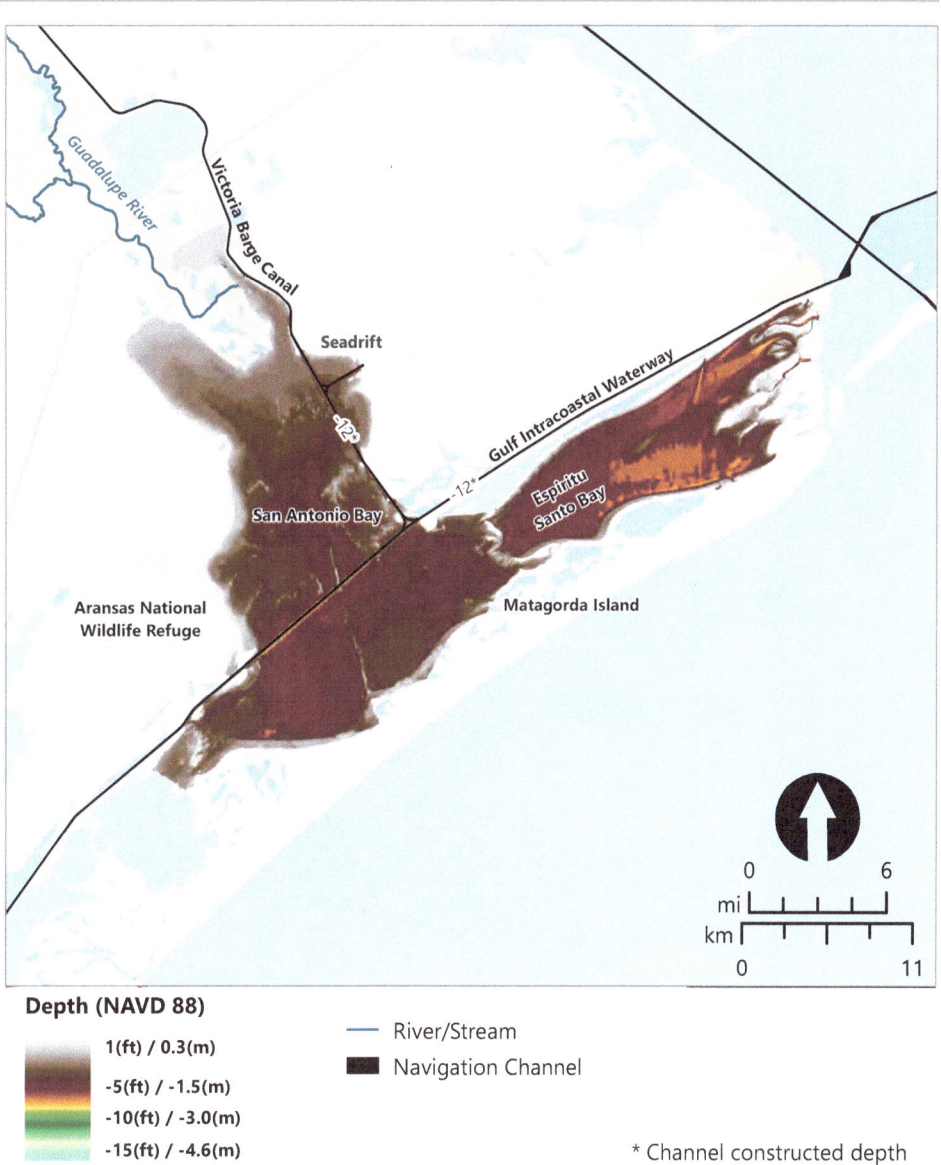

Depth (NAVD 88)
1(ft) / 0.3(m)
-5(ft) / -1.5(m)
-10(ft) / -3.0(m)
-15(ft) / -4.6(m)

River/Stream
Navigation Channel

* Channel constructed depth

**Fig. 6** Bathymetry for the
Mission-Aransas Estuary

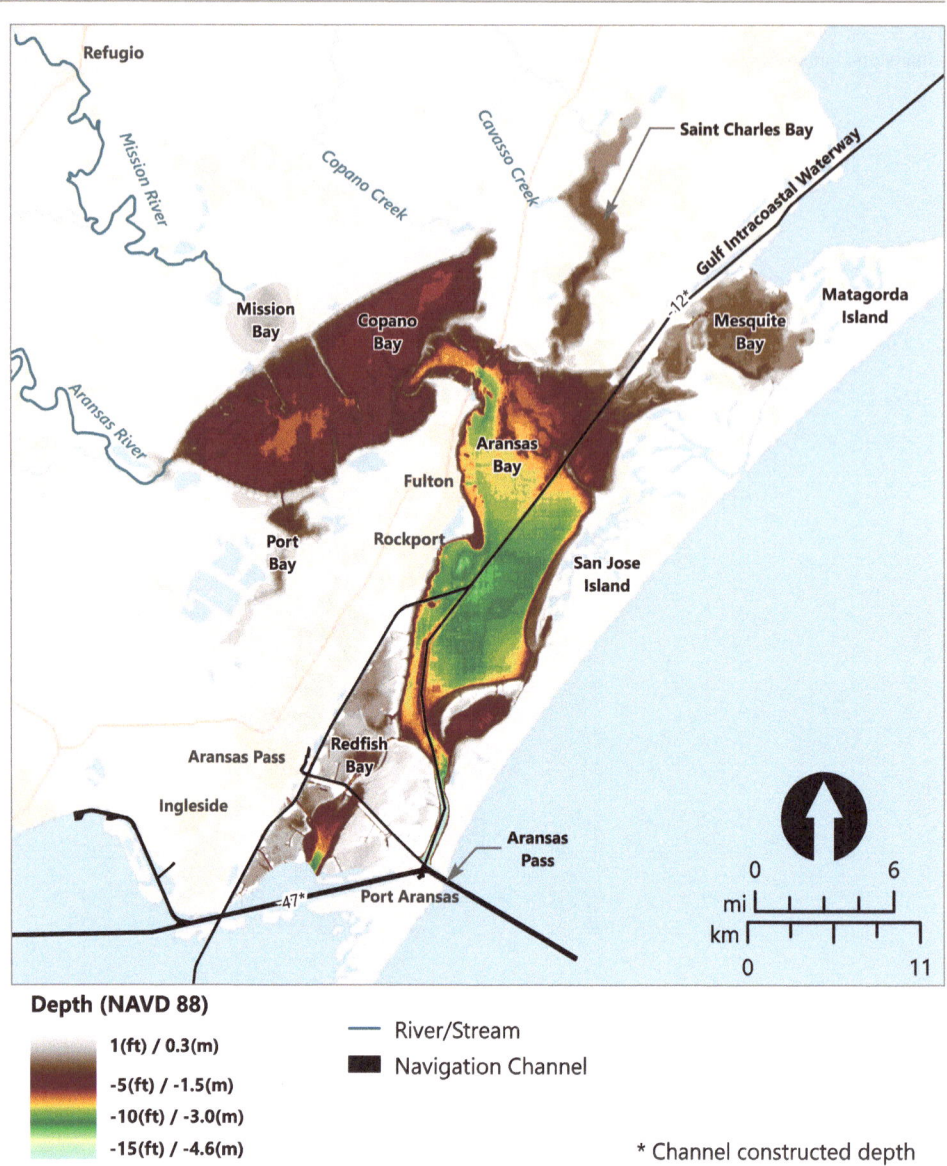

**Fig. 7** Bathymetry for the
Nueces Estuary

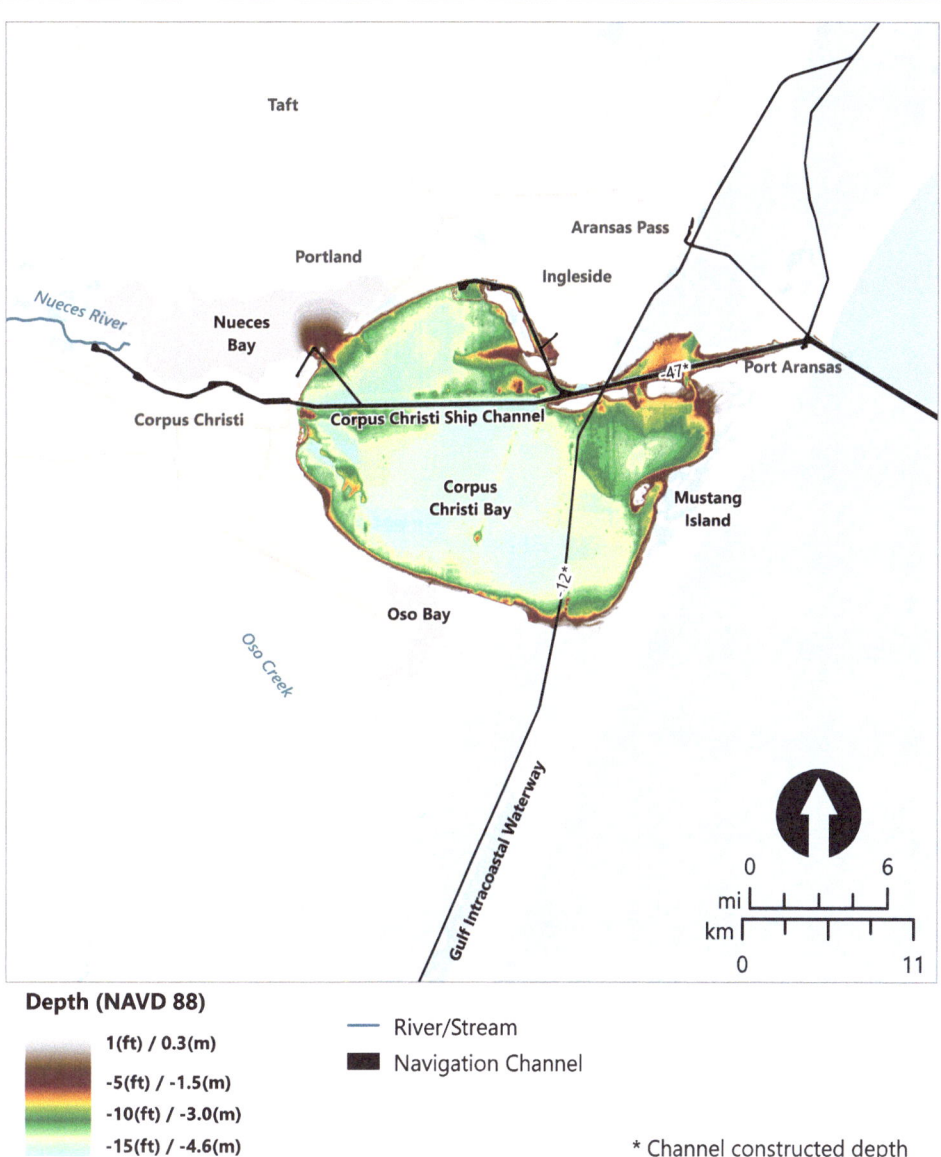

**Fig. 8** Bathymetry for the upper Laguna Madre Estuary

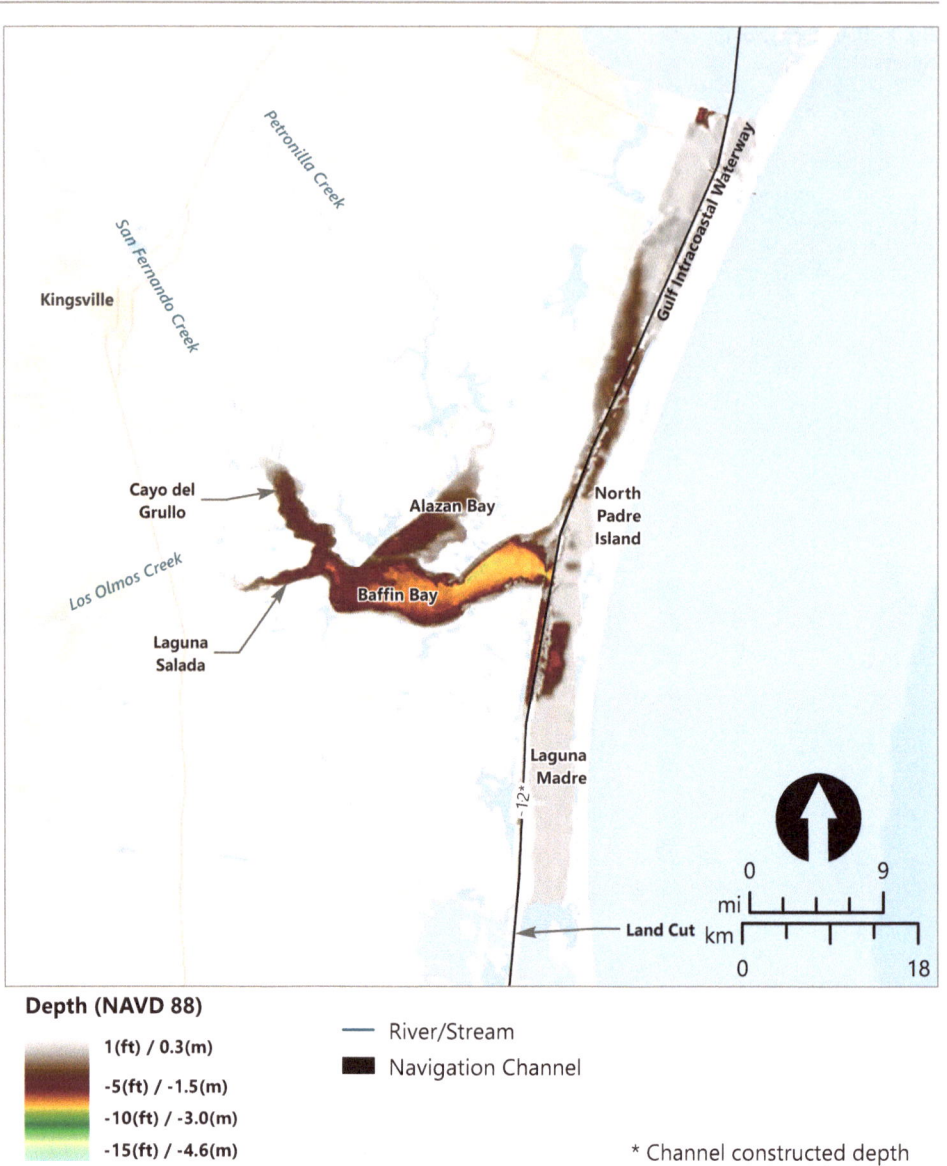

**Depth (NAVD 88)**

| | |
|---|---|
| | 1(ft) / 0.3(m) |
| | -5(ft) / -1.5(m) |
| | -10(ft) / -3.0(m) |
| | -15(ft) / -4.6(m) |

—— River/Stream

▨ Navigation Channel

\* Channel constructed depth

**Fig. 9** Bathymetry for the lower Laguna Madre Estuary

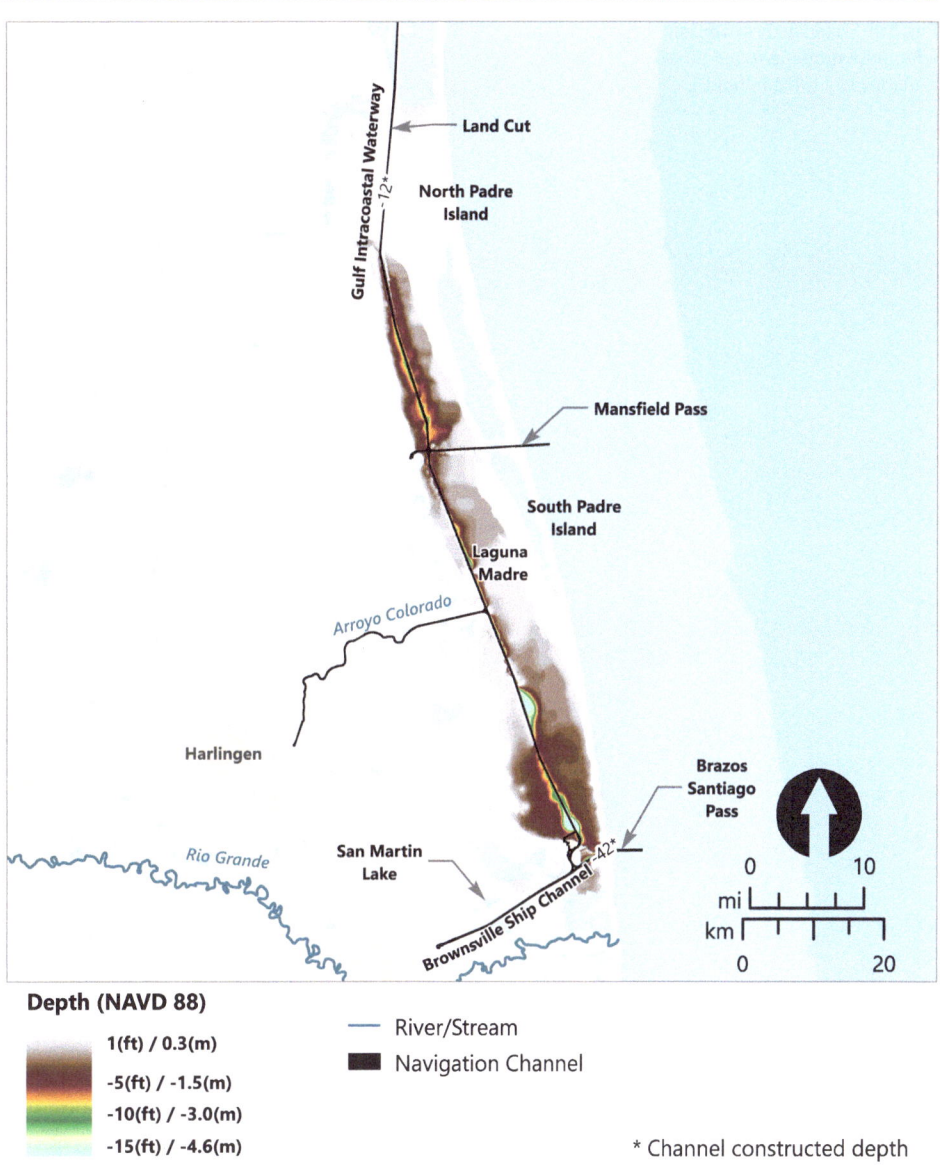

**Depth (NAVD 88)**

1(ft) / 0.3(m)
-5(ft) / -1.5(m)
-10(ft) / -3.0(m)
-15(ft) / -4.6(m)

—— River/Stream
■■ Navigation Channel

\* Channel constructed depth

## Watersheds

A watershed (sometimes referred to as a watershed area, drainage area, catchment, or basin) refers to the area of land where precipitation, and resulting runoff, drains to a specific location (Maidment 1993). When focusing on estuaries, the watershed is defined as the entire land area that contributes freshwater to the estuary.

Elevations in Texas are generally highest in the western and northern portions of the state and lowest along the shoreline with the Gulf of Mexico. Accordingly, most rivers start in the north and west and drain southeast (Fig. 10). Of the seven Texas estuaries, most have all of their watershed in Texas. The two exceptions include the Sabine-Neches

Estuary, which has some watershed areas in Louisiana, and the Colorado-Lavaca Estuary, which has some watershed areas in New Mexico but is considered to be non-contributing to the coast (TWDB 2022a).

The total contributing watershed area for each of the seven Texas estuaries varies (Table 1). Among the estuaries, the Laguna Madre is unique in that it consists of two distinct portions: the northern portion is referred to as the Upper Laguna Madre, and the southern portion is referred to as the Lower Laguna Madre. These two portions are partially separated by a coastal land mass commonly known as the Landcut, which does not contribute runoff to the estuary and is therefore not considered a part of the Laguna Madre watershed (Schoenbaechler et al. 2011a). The Gulf Intracoastal

**Fig. 10** Watersheds in Texas, with those draining to the estuaries identified in bold

**Table 1** Areas for the watershed and each estuary. These areas are based on watershed delineations published by the Texas Water Development Board downstream of U.S. Geological Survey (USGS) stream gages, combined with the upstream drainage areas reported by USGS for each stream gage. Estuary area and volume are from Engle et al. (2007)

| Estuary | Watershed area | | Estuary area | Estuary volume |
|---|---|---|---|---|
| | Square miles (mi²) | Square kilometers (km²) | Square kilometers (km²) | billion Cubic meters (10⁹ m³) |
| Sabine-Neches | 20,749 | 53,740 | 264.7 | 0.660 |
| Trinity-San Jacinto | 24,220 | 62,730 | 1456.4 | 2.707 |
| Colorado-Lavaca | 35,402 | 91,691 | 1114.9 | 2.430 |
| Guadalupe Estuary | 10,526 | 27,262 | 587.0 | 0.728 |
| Mission-Aransas | 2778 | 7195 | 532.0 | 0.870 |
| Nueces | 17,388 | 45,035 | 570.7 | 1.280 |
| Laguna Madre | 6040 | 15,644 | 2137.8 | 1.755 |

Waterway through the Landcut connects Upper and Lower Laguna Madre. The Laguna Madre watershed area listed in Table 1 represents the combined drainage areas for the Upper and Lower portions of the estuary but does not include the non-contributing area. Similarly, the watershed area shown for the Colorado-Lavaca estuary does not include the non-contributing area. Although there is variation in the size of watersheds draining to Texas estuaries, significant portions of the watershed areas in west and south Texas are arid and hence contribute little runoff.

These watersheds have been, in certain locations, modified for human purposes. For example, floodways carry high flows from the Rio Grande to the Laguna Madre, to reduce flooding in communities downstream along the Rio Grande. This effectively increases the size of the Laguna Madre watershed area during flood conditions. As another example, the City of Corpus Christi (in the Nueces watershed) imports water from the Colorado-Lavaca watershed via the Mary Rhodes pipeline, ultimately resulting in discharges to the Nueces estuary. The watershed areas in Table 1 do not account for these anthropogenic activities, but at the scale of this analysis, they can be considered to be minor.

## Precipitation

### Spatial Variations in Precipitation

Precipitation exhibits a strong increasing gradient from west to east across Texas, with modest irregularities that are largely caused by topographic variations, such as the mountains of west Texas and the hill country of central Texas (Fig. 11, the annual precipitation map is based on mean annual precipitation data from the PRISM Climate Group (2022) using a data period of 1991 through 2020). Annual precipitation decreases from 142 cm (56 in) in east Texas to 15 cm (6 in) in far west Texas. Therefore, the wettest areas of Texas receive approximately seven times more rainfall than the driest, which is tied with California for the largest range in precipitation in the contiguous United States (Ward 2005).

**Fig. 11** Average annual precipitation across Texas (English and SI units). Labels for contour lines around the state are provided in English units only

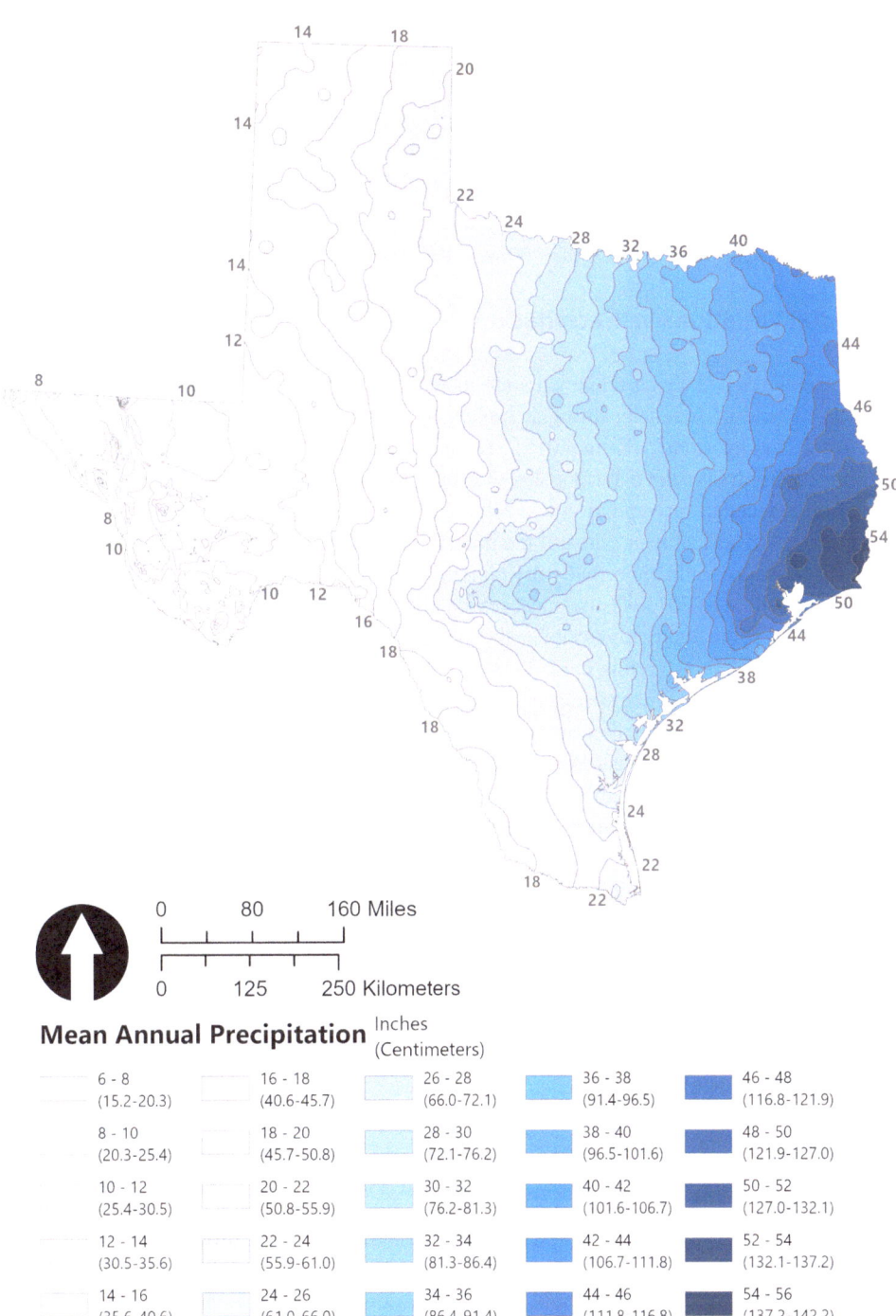

**Mean Annual Precipitation** Inches (Centimeters)

| | | | | |
|---|---|---|---|---|
| 6 - 8 (15.2-20.3) | 16 - 18 (40.6-45.7) | 26 - 28 (66.0-72.1) | 36 - 38 (91.4-96.5) | 46 - 48 (116.8-121.9) |
| 8 - 10 (20.3-25.4) | 18 - 20 (45.7-50.8) | 28 - 30 (72.1-76.2) | 38 - 40 (96.5-101.6) | 48 - 50 (121.9-127.0) |
| 10 - 12 (25.4-30.5) | 20 - 22 (50.8-55.9) | 30 - 32 (76.2-81.3) | 40 - 42 (101.6-106.7) | 50 - 52 (127.0-132.1) |
| 12 - 14 (30.5-35.6) | 22 - 24 (55.9-61.0) | 32 - 34 (81.3-86.4) | 42 - 44 (106.7-111.8) | 52 - 54 (132.1-137.2) |
| 14 - 16 (35.6-40.6) | 24 - 26 (61.0-66.0) | 34 - 36 (86.4-91.4) | 44 - 46 (111.8-116.8) | 54 - 56 (137.2-142.2) |

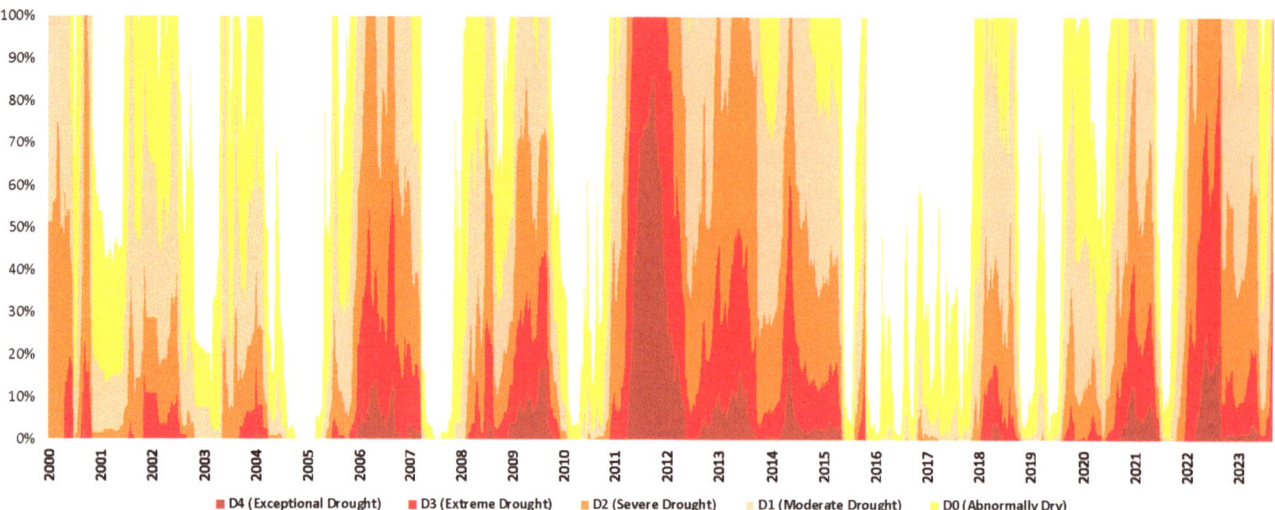

**Fig. 12** Texas Drought Monitor showing the percent of Texas in various stages of drought from 2000 to 2023. The exceptional drought of 2011 is discernible in this figure and will be evident in many other figures in this chapter

## Temporal Variations in Precipitation

Texas was described by an unnamed state meteorologist in 1927 as "a land of perennial drought, broken by the occasional devastating flood" (King 2015). The U.S. Drought Monitor (2023) tracks the current status of drought in Texas (based on antecedent precipitation and related metrics) and also provides historical data. Most of the time, at least some locations in Texas are in abnormally dry conditions or worse (Fig. 12). Drought in 2011 was the most severe in recent history, and, by some metrics, exceeded the previous record drought of the 1950s (Hoerling et al. 2013).

At the opposite end of the spectrum are floods. Hurricane Harvey, in 2017, delivered an entire year's worth of rainfall to some areas (e.g., an area near Port Arthur received over 60 inches) and set multiple rainfall records (TAMU 2018; TWDB 2019). Although rainfall events as large as Hurricane Harvey are rare, significant rainfall events are not uncommon in Texas. Indeed, Central Texas has been referred to as "flash flood alley" due to the occurrence of intense storms coupled with steep topography and thin soils (Wasson 2021); the rest of Texas is not immune to floods either. Indeed, every county in Texas has experienced flooding (Perry 2019; TWDB 2019) and a helpful compendium of significant floods is found in Burnett (2008).

## Freshwater Inflows

Rainfall begets runoff. The spatial and temporal variations in precipitation lead to even larger variations in river flows. Keeping with the focus of this book, this evaluation will explore the spatial and temporal variations in freshwater inflows to Texas estuaries.

## Methodology for Calculating Freshwater Inflows

Each estuary, except Laguna Madre, has one or more major riverine inflows. The Upper Laguna Madre has freshwater inflows, but due to its relatively small watershed area and arid conditions, only creeks and arroyos are present. The Lower Laguna Madre is similarly limited to creeks and arroyos, with the exception of the Arroyo Colorado which does carry some floodwaters from the Rio Grande as well as wastewater and irrigation return flows and hence is an important source of freshwater inflows at times. Each estuary also has a local watershed area that drains to the estuary without first discharging to a river. Each riverine inflow has a flow gauge, typically operated by the U.S. Geological Survey (USGS 2022), a short distance upstream of the estuary; therefore, the majority of the inflows to each estuary are directly measured. Flows from the watershed area downstream of the gage, including adjacent local watershed areas, are not directly measured and must be estimated using a model. The modeled downstream flows are then added to the measured riverine flows to estimate the total freshwater inflows to each estuary.

Long-term estimates of historical freshwater inflows into each estuary have been calculated by the Texas Water Development Board (TWDB) for the period from January 1941 through December 2018 (TWDB 2022b). The earliest estimates, covering the period 1941 through 1976, were calculated on a monthly basis and presented in a series of Texas Department of Water Resources (TDWR) studies published between 1980 and 1983 (TDWR 1980a, b, 1981a, b, c, and 1983). These early estimates were subsequently extended to cover the period 1977 through 1987 in support of additional studies of the Texas estuaries; the subsequent estimates were calculated on a daily basis and are presented by Longley (1994). The latest estimates covering the period 1987 through 2018 have been developed on a daily basis by TWDB in support of continued research and planning studies over the past 30 years; the latest estimates for each estuary are periodically published in coastal hydrology reports issued by TWDB (Neupane and Schoenbaechler 2021; Schoenbaechler et al. 2011b; Guthrie and Lu 2010; Schoenbaechler et al. 2011b, c, d, e, f). From 1941 to 1976, the monthly freshwater inflow estimates are only available as net total inflow for each estuary, whereas the 1977–2018 daily estimates include separate values for each major river and tributary inflow.

For all time periods, the basic formula for estimating freshwater inflows is the same and includes the net contributions of four components: (1) flows measured at upstream USGS stream gages; (2) modeled runoff from ungaged portions of the estuary's watershed; (3) flows diverted from ungaged watershed areas for agricultural, municipal, and industrial use; and (4) unconsumed flows returned to streams in ungaged watershed areas by agricultural, municipal, and industrial users (Neupane and Schoenbaechler 2021; Schoenbaechler et al. 2011b; Guthrie and Lu 2010; Schoenbaechler et al. 2011b, c, d, e, f). The net freshwater inflow resulting from these four components is calculated as follows:

$$\text{Net Inflow} = \text{Gaged Flow} + \text{Modeled Runoff} - \text{Diverted Flow} + \text{Returned Flow}$$

Methods for determining modeled runoff, diverted flows, and returned flows varied somewhat between the time periods of 1941–1976 and 1977–2018.

Modeled runoff

For the 1941–1976 estimates, modeled runoff in ungaged areas was calculated using a water yield model developed by TWDB, which was based on methods of the Soil Conservation Service and calibrated to gaged portions of each estuary's watershed. In the 1977–2018 estimates, modeled runoff was calculated using the TWDB Texas Rainfall-Runoff (TxRR) model (Matsumoto 1992a).

Diverted flow

In the 1941–1976 estimates, diverted flows were obtained from the TDWR reported water usage system. In the 1977–2018 estimates, diverted flows were obtained from various sources, including the following:

Data collected by the Texas Commission on Environmental Quality (TCEQ) and the U.S. Environmental Protection Agency (USEPA)

Data collected by local water authorities

Data compiled by HDR Engineering, Inc. (HDR), based on individual reports from water right holders

Returned flows

In the 1941–1976 estimates, returned flows from municipal and industrial users were obtained from the TDWR self-reporting system; returned flows from agricultural users were calculated based on data collected through previous rice irrigation return flow studies (TDWR 1974, 1975). In the 1977–2018 estimates, returned flows were obtained from various sources, including the following:

Data collected by TCEQ and USEPA

TWDB Irrigation Water Use estimates, for agricultural returned flows prior to 2008

Data collected by local water authorities

Data compiled by HDR based on individual reports from water right holders

In the 1977–2018 estimates, the net inflow obtained from the formula above is calculated for each of the major rivers and tributaries that discharge into each estuary. Modeled runoff from the surrounding watersheds is applied to the nearest major river or tributary after being adjusted for any diverted or returned flow (Matsumoto et al. 2014; Guthrie et al. 2012; Schoenbaechler et al. 2011a, g, h).

Using this methodology, diverted and returned flows are generally aggregated on a watershed level and applied to the nearest major river or tributary. However, there are four power plants adjacent to Texas estuaries where these flows have been identified as separate inflow and withdrawal estimates by TWDB. These include the diversion of large amounts of water for cooling purposes from one location followed by return flows to a different location (Guthrie et al. 2014; Schoenbaechler et al. 2011a). However, there is little water consumed through this process. For that reason, although these flows are important for circulation and salinity modeling, they have no meaningful effect on net freshwater inflow volumes for the estuaries.

The following subsections present analyses of the long-term freshwater inflow estimates.

## Freshwater Inflow Estimates

The historical freshwater inflow estimates can be visualized in several ways. Examples are provided below and in the online supplement to this book. For annual and monthly flow totals, units are acre-feet (ac-ft) or cubic meters (m$^3$). For daily flow rates, cubic feet per second (ft$^3$/s) or cubic meters per second (m$^3$/s) are used. The conversion is 1 ft$^3$/s = 1.98 acre-feet per day (ac-ft/d) = 0.0283 m$^3$/s. Some figures show flows from individual tributaries. When all such inflows are summed, the result is termed "total inflows." Note that, unless otherwise specified, this "total inflows" calculation does not include precipitation or evaporation. The presentation for Figs. 13, 14, 15, 16, 17, 18, 19, and 20 shows individual inflows broken out by color for each inflow source.

Estimated total annual inflows to the Sabine-Neches Estuary start in 1977 because, as previously described, that is the first year for which estimates of individual inflows are available from TWDB (Fig. 13). For this estuary, the majority of the inflows are from the Sabine and Neches Rivers, with a lesser contribution from other inflows. This pattern is common for Texas estuaries, which typically have one or two major riverine inflows, as well as contributions from a few, much smaller, coastal drainages. As with other estuaries a sharp decline during the drought of 2011 is visible.

For the Trinity-San Jacinto Estuary, the Trinity and San Jacinto rivers constitute the majority of the inflows, generally followed by Buffalo Bayou, which drains significant portions of Houston (Fig. 14). Much like for the Sabine-Neches Estuary, the influence of the notable floods in 2015, 2016, and 2017 are shown.

For the Colorado-Lavaca Estuary, the Colorado River provides the majority of the inflows, followed by the Lavaca River and local creeks (Fig. 15). The standout inflow occurred as a result of the Christmas flood of 1991, with significant inflows well into 1992.

For the Guadalupe Estuary, the San Antonio River joins the Guadalupe River shortly before the head of the estuary, so the flows are combined herein. Also, because of the shape

of this estuary, there are no other significant inflows (Fig. 16). Similar to the Colorado-Lavaca Estuary, the Christmas flood of 1991, followed by a wet winter, resulted in high inflows in 1992. Equally notable is the high flows of 2002. The rain was so heavy in early July upstream of Canyon Lake (the largest reservoir in the basin) that the reservoir overtopped its spillway for the first time since the dam was constructed in 1964. As a result, 67,000 cubic feet per second (cfs) carved a new canyon downstream of the lake (Canyon Lake Gorge Preservation Society 2022).

For the Mission-Aransas Estuary, the primary inflows are generally from the Mission and Aransas Rivers, although Cavasso Creek sometimes contributes significant inflows (Fig. 17).

For the Nueces Estuary, the majority of the inflows are contributed by the Nueces River (Fig. 18). There is no other river that drains to this estuary, although Oso Creek provides notable inflows in some years. This watershed suffered from the same July 2002 flood as the Guadalupe watershed, resulting in the highest inflows in the period of record. Hurricane Harvey contributed significant inflows in 2017, and a series of tropical disturbances led to high inflows in the later half of 2018.

For the Upper Laguna Madre, the primary inflow is San Fernando Creek, and the high flow events generally follow a different pattern than watersheds to the north (Fig. 19). Here the largest inflow occurred in 2010 as a result of Hurricane

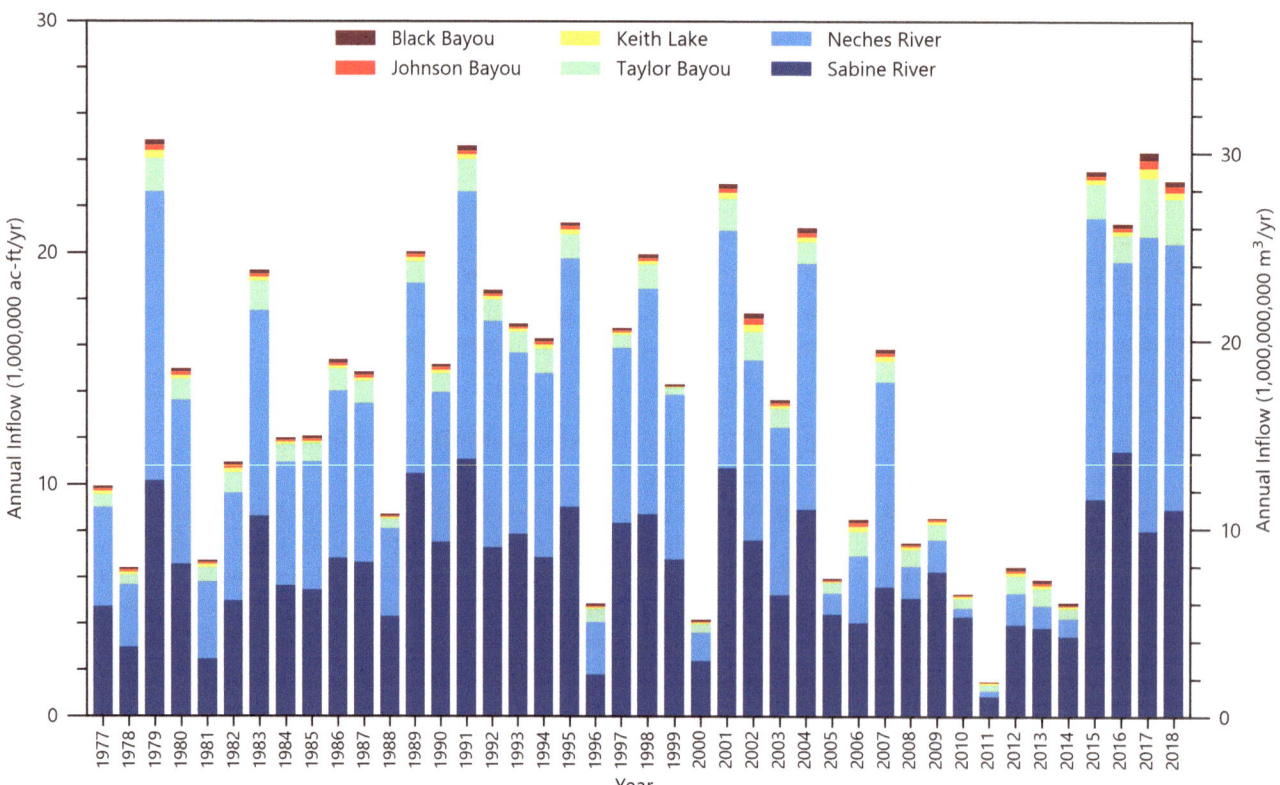

**Fig. 13** Annual inflows to the Sabine-Neches Estuary, including riverine inflows and local drainages

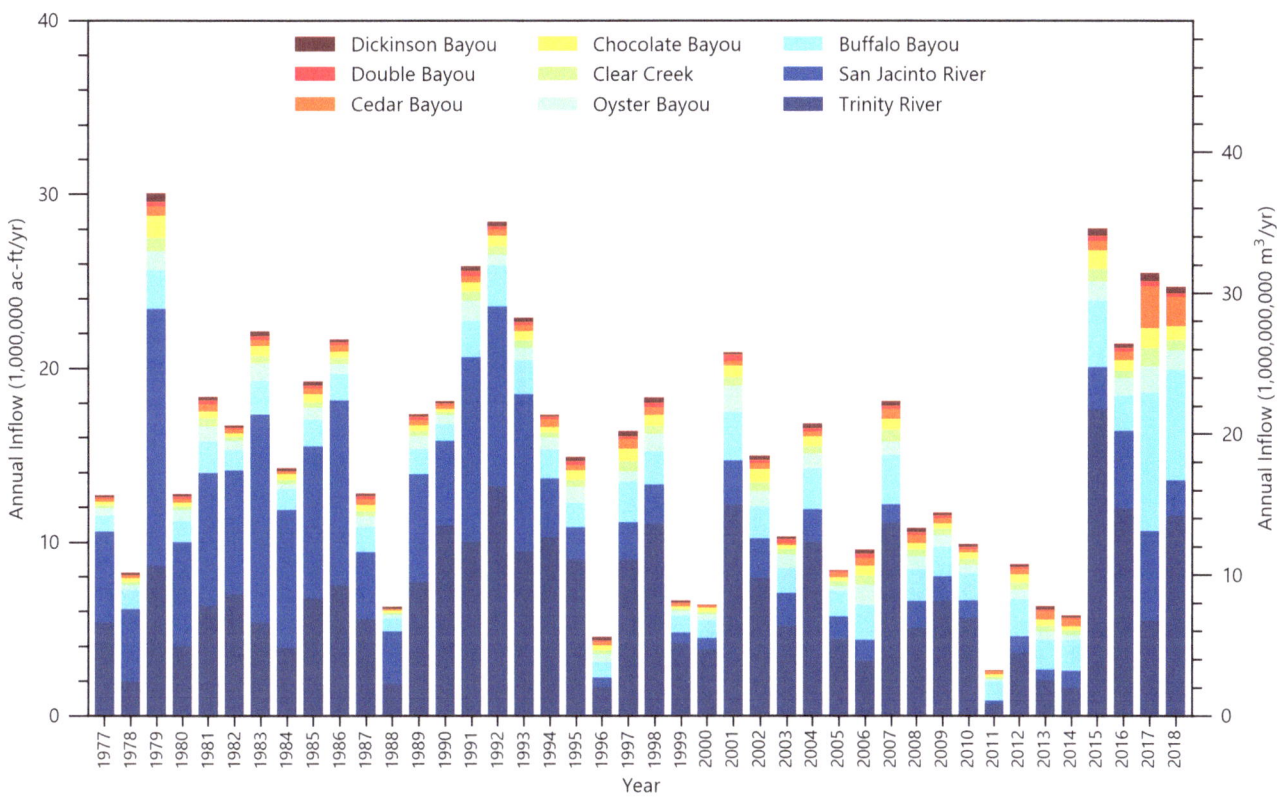

**Fig. 14** Annual Inflows to the Trinity-San Jacinto Estuary, including riverine inflows and local drainages

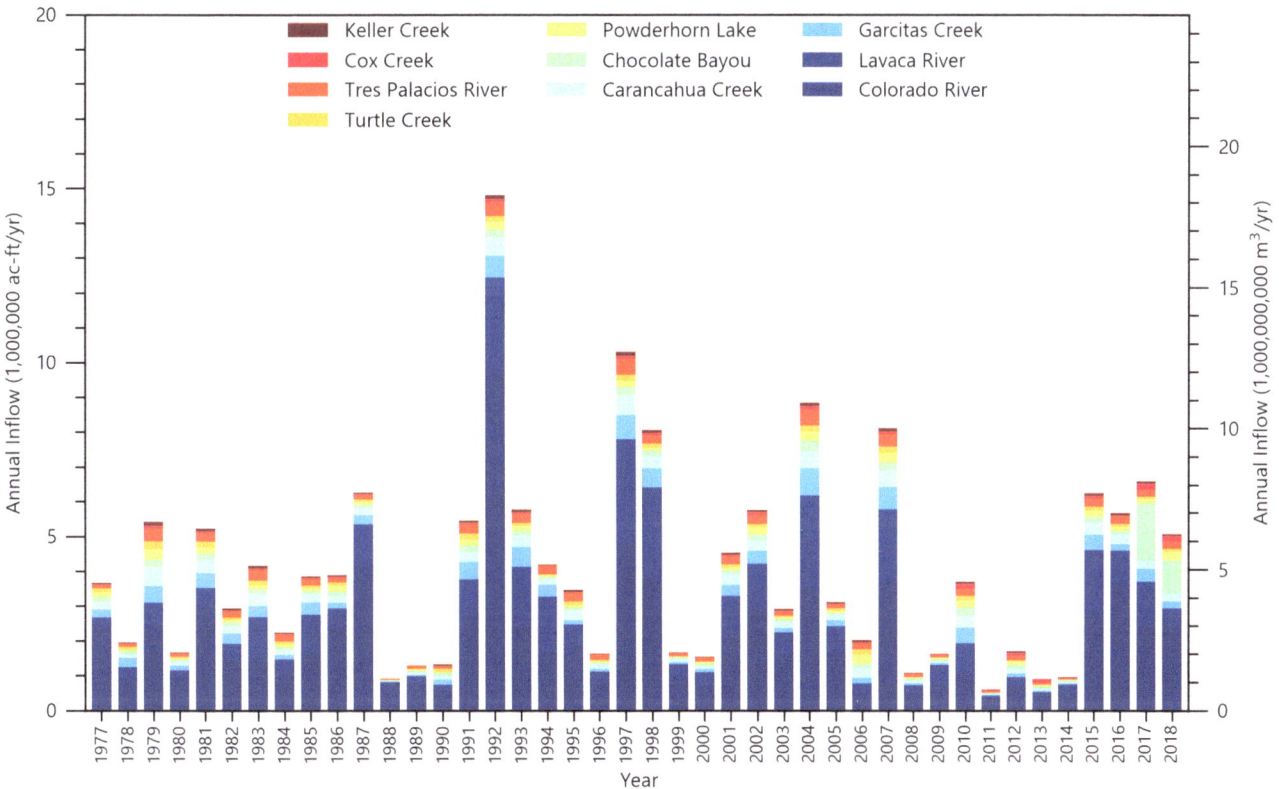

**Fig. 15** Annual inflows to the Colorado-Lavaca Estuary, including riverine inflows and local drainages

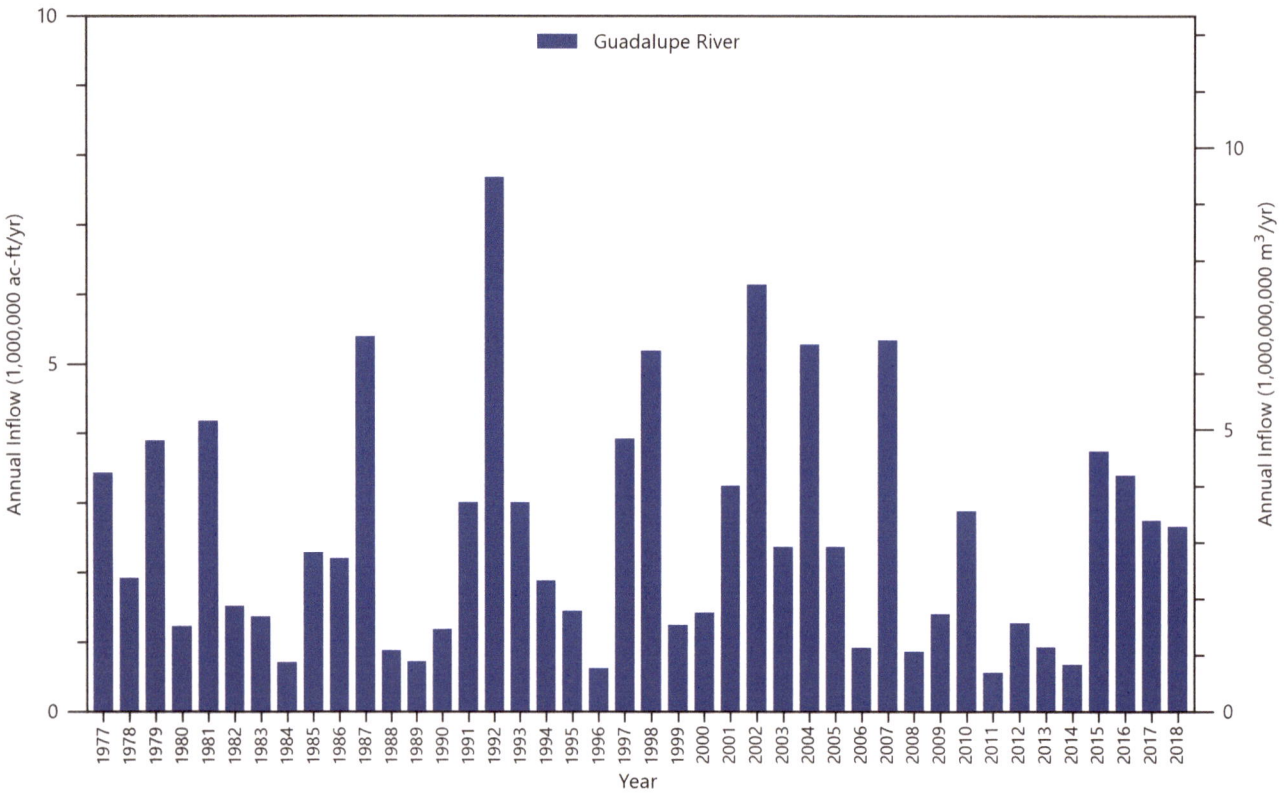

**Fig. 16** Annual inflows to the Guadalupe Estuary, including riverine inflows and local drainages

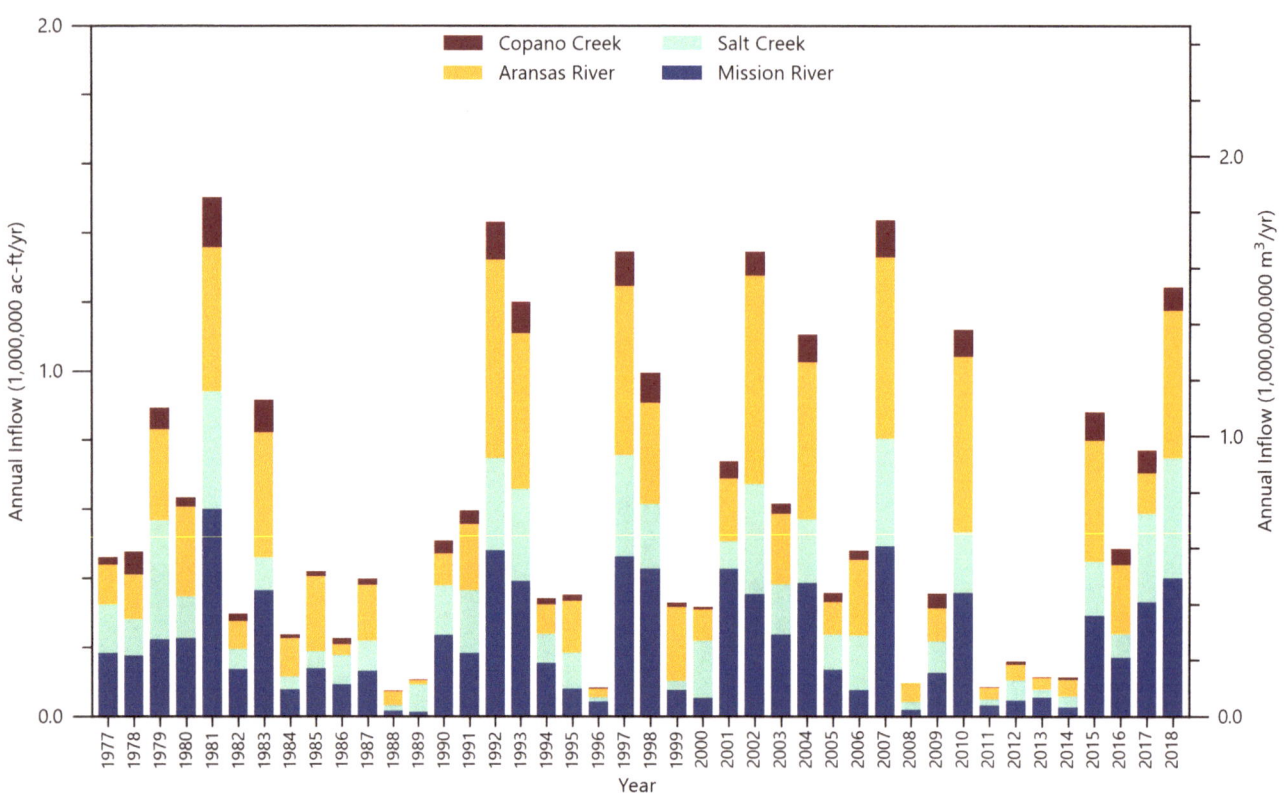

**Fig. 17** Annual inflows to the Mission-Aransas Estuary, including riverine inflows and local drainages

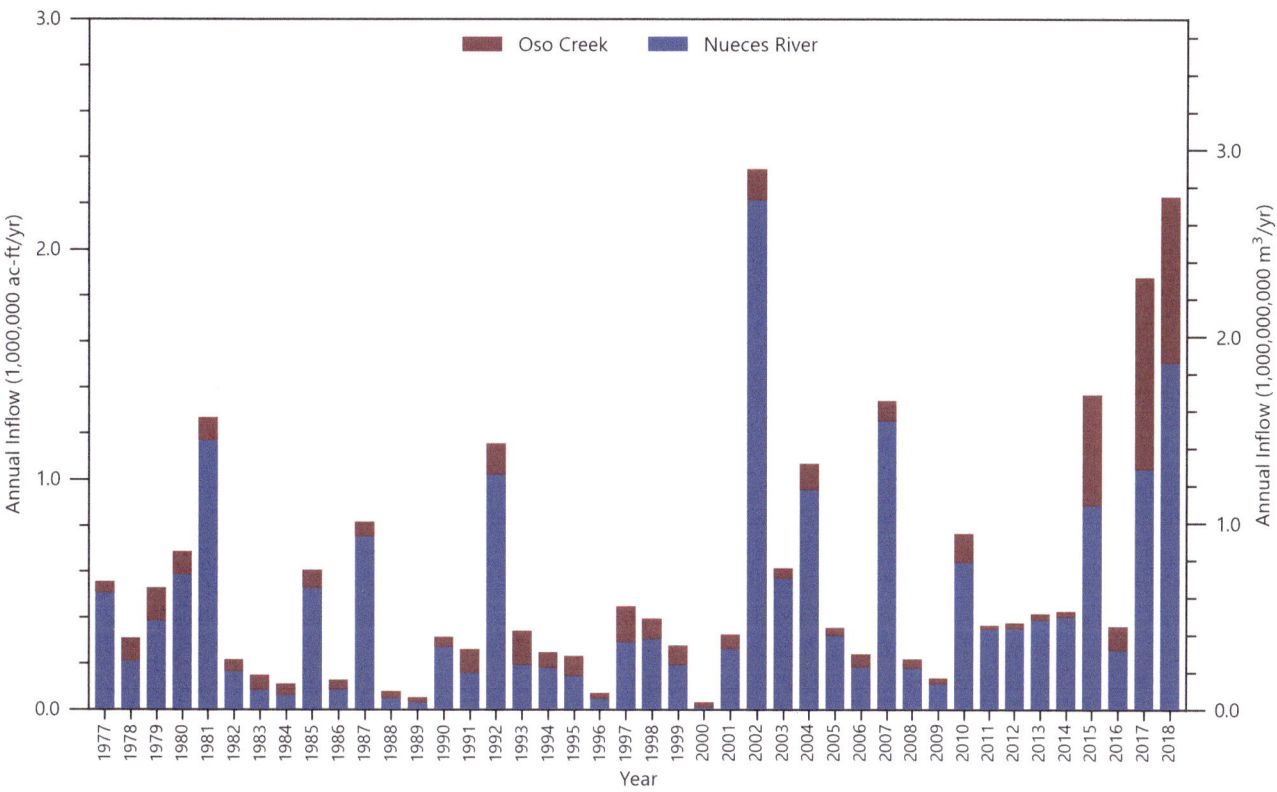

**Fig. 18** Annual inflows to the Nueces Estuary, including riverine inflows and local drainages

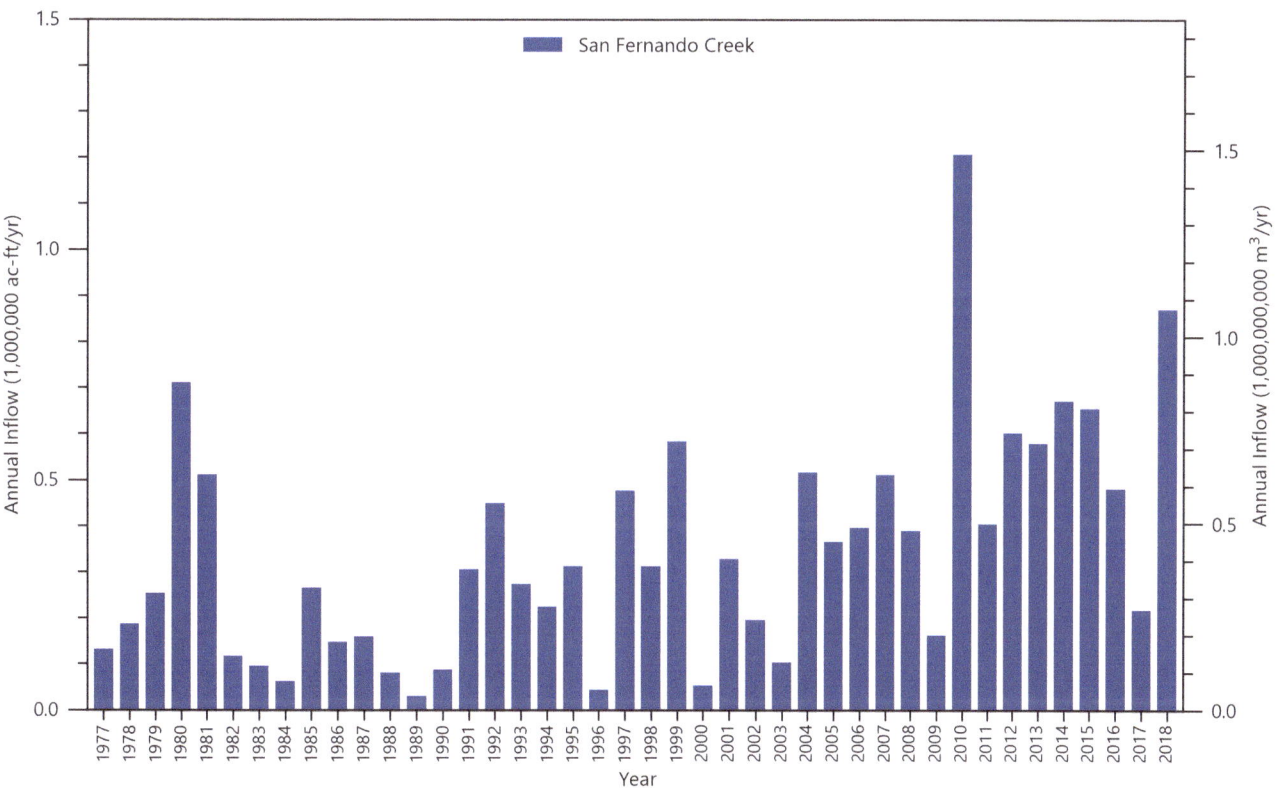

**Fig. 19** Annual inflows to the Upper Laguna Madre, including local drainages

Alex in July (NWS 2022a), tropical storm Hermine (September; Tompkins 2010) and Hurricane Karl (NWS 2022b).

For the Lower Laguna Madre, the primary inflow is the Arroyo Colorado (Fig. 20), which, during flood events, carries some flow from the Rio Grande. The largest inflows match the pattern of the Upper Laguna Madre, with 2010 in this basin dominated by Hurricane Alex.

Figure 21 illustrates monthly total inflows (i.e., all riverine and tributary inflows) to each estuary. This figure begins in 1941, which is the first year of estimates of total inflows (separate data for the Upper and Lower Laguna Madre are not available back to 1941, so all inflows to the Laguna Madre are combined in this figure). A version of Fig. 21 using logarithmic y axis scales is available in the online supplement. The patterns from 1977 to 2018 are similar in this figure as in the figures above (where individual tributaries are shown). Additional patterns are evident in the older period of record, most notably the extreme drought of the 1950s and the somewhat less severe drought of the 1960s, both of which are, remarkably, evident in every estuary. In the mid-coast to South Texas, the impacts of Hurricane Beulah in September 1967 are visible.

Similarities, especially between adjacent panels, are evident in Fig. 21. To quantify the similarity, the Pearson correlation coefficient between each pair of estuaries was calculated and is shown in Table 2. Each adjacent pair has a correlation coefficient of at least 0.57, indicating that there is a fairly strong relationship between adjacent estuaries. The correlation coefficients uniformly decrease with distance but remain positive, except for the pair of the Sabine-Neches Estuary and the Laguna Madre, which is very weakly negative. This analysis shows that nearly all the estuaries trend together and that the effect is strongest for adjacent estuaries. Daily inflow estimates to each estuary are also available from 1977 to 2018. These are presented in a series of raster hydrographs (Koehler 2004) in the online supplement.

To explore patterns across the Texas coast, Fig. 22 illustrates the mean and median monthly freshwater inflow for each estuary, based on the 1941–2018 data. The estuaries are ordered from west to east, and the figure illustrates how inflows generally mirror the west-to-east gradient in precipitation in Texas. The Sabine-Neches Estuary has the highest inflows, with the Trinity-San Jacinto Estuary a relatively close second. The two mid-coast estuaries (Colorado-Lavaca and Guadalupe) also have relatively similar inflows. The three lower coast estuaries have the lowest inflows. Accordingly, estuaries along the Texas coast can be thought of as forming these three groups. Some deviations in the overall west-to-east gradient do occur. The Mission-Aransas has the lowest inflows, in part because it has the smallest watershed area. The Nueces Estuary inflows are lower, in general, than those

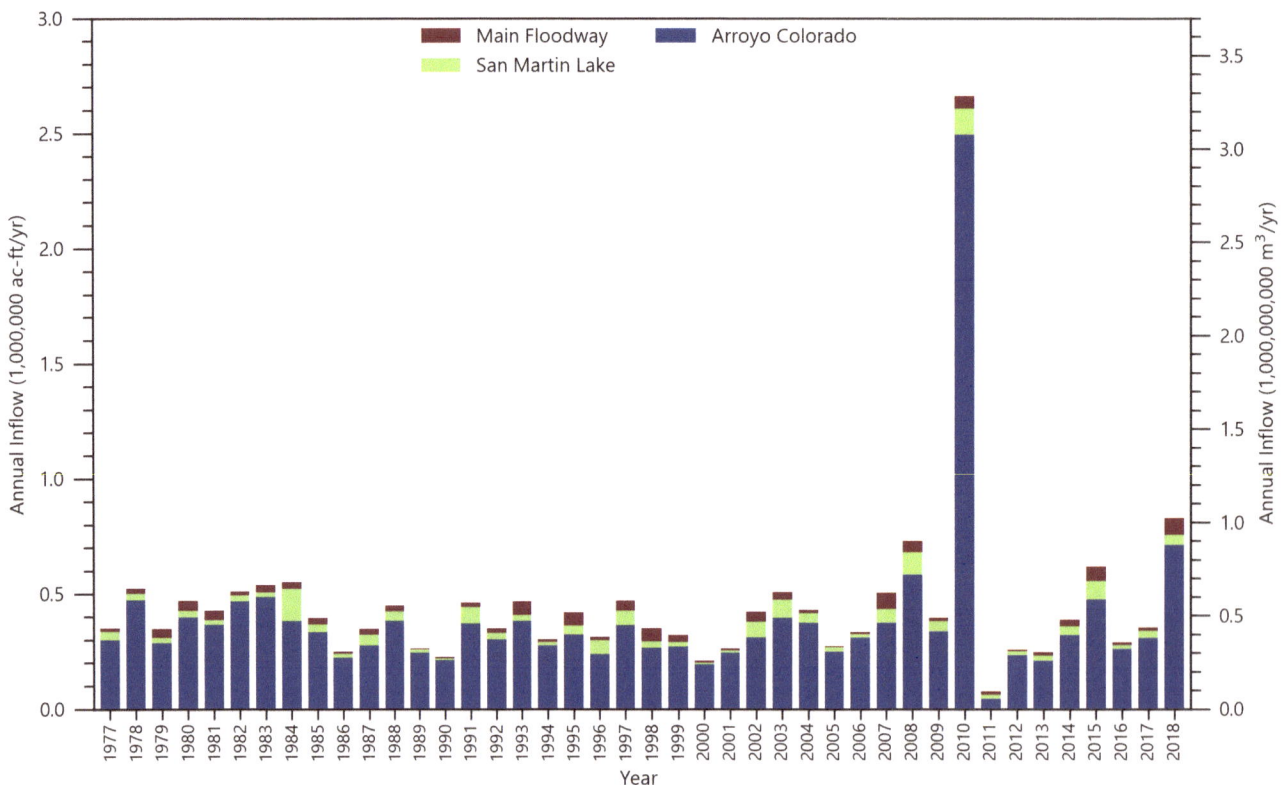

**Fig. 20** Annual inflows to the Lower Laguna Madre, including local drainages

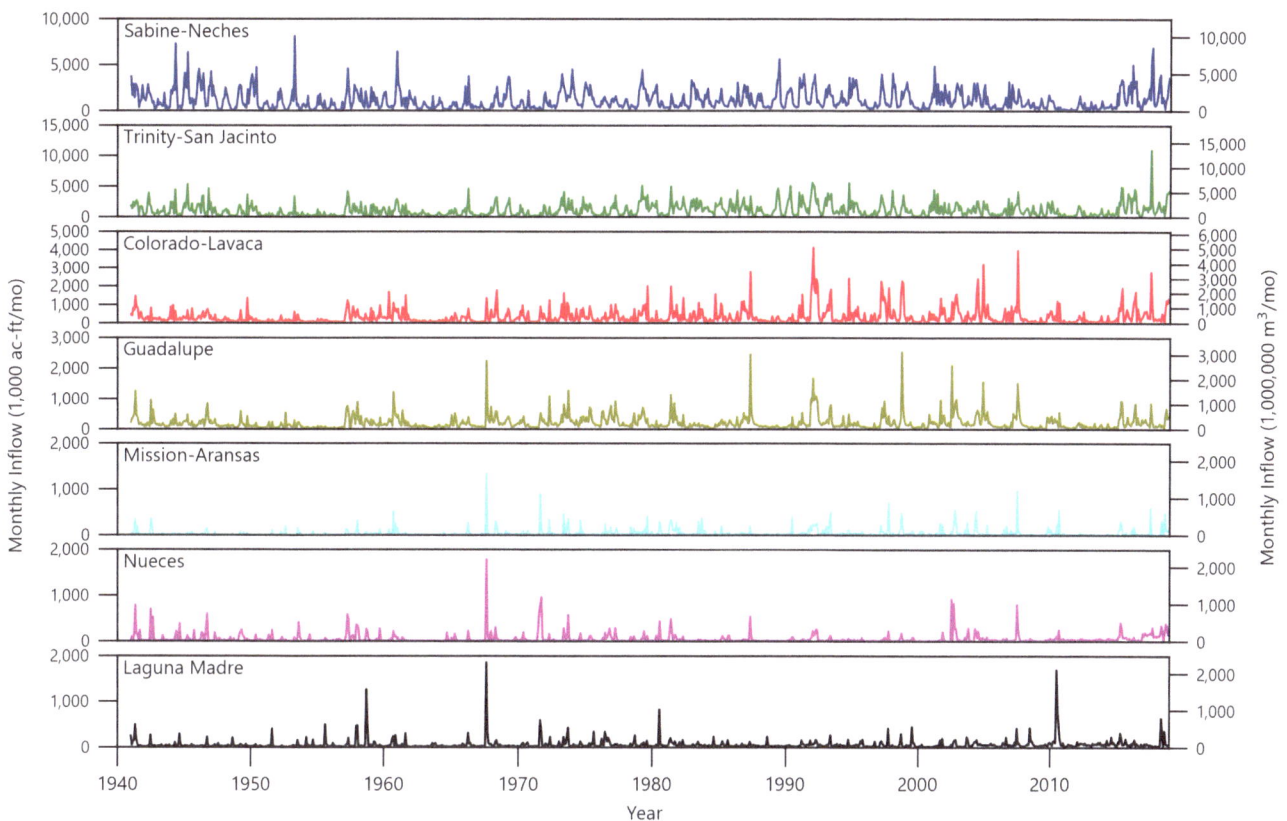

**Fig. 21** Time series of total monthly inflows to each estuary from 1941 to 2018. Note the changing y axis scale from panel to panel

**Table 2** Correlation matrix for monthly inflows between each pair of estuaries. Color bins: 0.01–0.2 dark red, 0.21–0.4 red, 0.41–0.6 pink, 0.61–0.8 light blue

| Estuary | SN | TJ | CL | GE | MA | NC | LM |
|---|---|---|---|---|---|---|---|
| Sabine-Neches (SN) | 1.00 | | | | | | |
| Trinity-San Jacinto (TJ) | 0.76 | 1.00 | | | | | |
| Colorado-Lavaca (CL) | 0.42 | 0.67 | 1.00 | | | | |
| Guadalupe (GE) | 0.31 | 0.50 | 0.79 | 1.00 | | | |
| Mission-Aransas (MA) | 0.13 | 0.32 | 0.62 | 0.63 | 1.00 | | |
| Nueces (NC) | 0.09 | 0.20 | 0.42 | 0.63 | 0.66 | 1.00 | |
| Laguna Madre (LM) | - 0.01 | 0.07 | 0.28 | 0.36 | 0.58 | 0.57 | 1.00 |

in the Laguna Madre, despite having a larger watershed area. This is likely due to a combination of more significant reservoir storage and human uses in the Nueces watershed, coupled with a substantial portion of the Nueces watershed providing recharge to the Edwards Aquifer (which primarily discharges via springs in the Guadalupe watershed). For each estuary, the mean inflow is substantially greater than the median. This is due to infrequent, large floods elevating the mean while having no impact on the median.

The values shown in Fig. 22 (representing 1941–2018) are generally similar to those shown in Fig. 4.1.2 of Longley (1994; representing 1941–1987). However, the mean and median inflows to the Trinity-San Jacinto Estuary are notably higher in the extended period of record (by 46% for the mean and 64% for the median). This may be a result of increased rainfall in recent years (Nielsen-Gammon et al. 2019) and is likely also a result of increased wastewater discharges from the major cities in the watershed (including

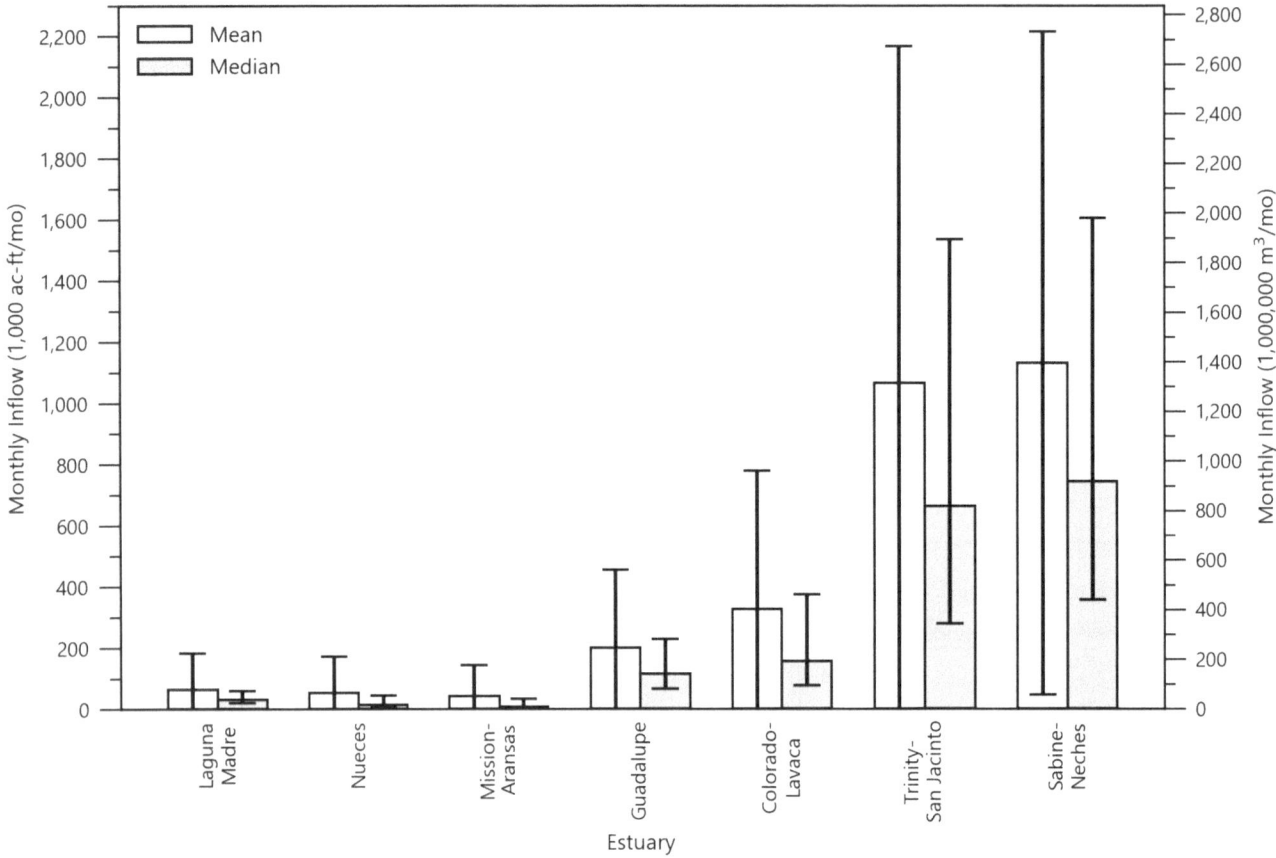

**Fig. 22** Mean and median monthly freshwater inflows for each estuary. Error bars overlain on the mean values represent one standard deviation of the data. It is important to recognize that this is not the standard deviation of instantaneous flows but rather the standard deviation of monthly flows (only monthly flows are available from 1941 to 2018). Therefore, it does not represent all the variability in actual flows. In all cases except the Sabine-Neches estuary, the standard deviation exceeds the mean, which highlights the extraordinary temporal variability of freshwater inflows to Texas estuaries. Error bars on the median represent the 25th and 75th percentiles. These are not symmetric, and the 75th percentile is always farther from the median than the 25th percentile is, which highlights the positive skew of the inflows

Dallas, Fort Worth, and Houston), which obtain some of their water supplies from groundwater and inter-basin transfers.

Table 3a (with units 1000 acre-feet per month [ac-ft/mo]) and Fig. 3b (with units 1,000,000 cubic meters per month [m³/mo]) provide flow statistics that mirror Fig. 22. These tables show that not only is the standard deviation of the monthly flow values greater than the mean in all estuaries except the Sabine-Neches, but the coefficient of variation (defined as the standard deviation divided by the mean) increases to the south, reflecting the very flashy nature of runoff in south Texas.

Another perspective on monthly inflows across all estuaries is provided in Fig. 23. Here the 1941–2018 monthly inflow data for each estuary are ranked and plotted from largest to smallest, in what is sometimes called an exceedance probability curve. Again, the Sabine-Neches and Trinity-San Jacinto estuaries stand out as having the highest inflows. Indeed, the 90% exceedance level for each of these estuaries is at approximately the 5% exceedance level for each of the Mission-Aransas and Nueces Estuaries and the Laguna Madre. During wet periods, the inflows from the Mission-Aransas and Nueces estuaries and Laguna Madre are similar, but during

dry periods, the Mission-Aransas has the lowest inflows, with the Nueces Estuary being somewhat higher and the Laguna Madre higher still (albeit still low compared to east Texas). For the Nueces Estuary, some of the reduction in inflows during dry periods is due to storage of river flows by Lake Corpus Christi and Choke Canyon Reservoir followed by withdrawals for consumptive use. The Mission-Aransas estuary and Laguna Madre do not have similarly large reservoirs or human uses. These total inflow data are also presented as cumulative inflows versus time in the online supplement.

## Freshwater Inflow Balance with Precipitation and Evaporation

As an added perspective on freshwater in each estuary, the total inflows (which, in this context, are the total riverine and tributary inflows) can be combined with precipitation and evaporation estimates to calculate a freshwater inflow balance. The precipitation estimates are based on data obtained from the National Weather Service and TWDB archived records,

**Table 3** Monthly total inflow statistics for each estuary. The exceedance probability is the probability that the given flow is equaled or exceeded. Abbreviations: SD = Standard Deviation, CV = Coefficient of Variation, Min = Minimum, Max = Maximum. (A) 1000 ac-ft/mo. (B) 1,000,000 m³/mo)

| (A) Estuary | Mean | SD | CV | Min | Exceedance probability | | | | | Max |
|---|---|---|---|---|---|---|---|---|---|---|
| | | | | | 90% | 75% | 50% | 25% | 10% | |
| Sabine-Neches | 1130 | 1085 | 96% | 26 | 176 | 357 | 743 | 1608 | 2653 | 8091 |
| Trinity-San Jacinto | 1066 | 1100 | 103% | 16 | 136 | 277 | 664 | 1540 | 2540 | 10,943 |
| Colorado-Lavaca | 326 | 453 | 139% | 16 | 41 | 77 | 156 | 376 | 791 | 4108 |
| Guadalupe | 201 | 254 | 126% | 5 | 40 | 66 | 116 | 229 | 449 | 2520 |
| Mission-Aransas | 42 | 102 | 241% | 0 | 1 | 2 | 7 | 32 | 117 | 1340 |
| Nueces | 52 | 120 | 231% | 0 | 3 | 5 | 13 | 43 | 138 | 1779 |
| Laguna Madre | 62 | 121 | 195% | 2 | 14 | 19 | 30 | 59 | 124 | 1855 |
| (B) Estuary | Mean | SD | CV | Min | Exceedance Probability | | | | | Max |
| | | | | | 90% | 75% | 50% | 25% | 10% | |
| Sabine-Neches | 1394 | 1338 | 96% | 31 | 218 | 440 | 917 | 1983 | 3273 | 9981 |
| Trinity-San Jacinto | 1315 | 1357 | 103% | 19 | 168 | 342 | 819 | 1899 | 3133 | 13,497 |
| Colorado-Lavaca | 402 | 559 | 139% | 20 | 51 | 95 | 192 | 464 | 976 | 5067 |
| Guadalupe | 248 | 313 | 126% | 6 | 49 | 82 | 143 | 283 | 554 | 3108 |
| Mission-Aransas | 52 | 126 | 241% | 0 | 1 | 2 | 9 | 39 | 145 | 1653 |
| Nueces | 64 | 148 | 231% | 0 | 3 | 6 | 16 | 53 | 170 | 2195 |
| Laguna Madre | 77 | 149 | 195% | 2 | 17 | 23 | 37 | 73 | 153 | 2288 |

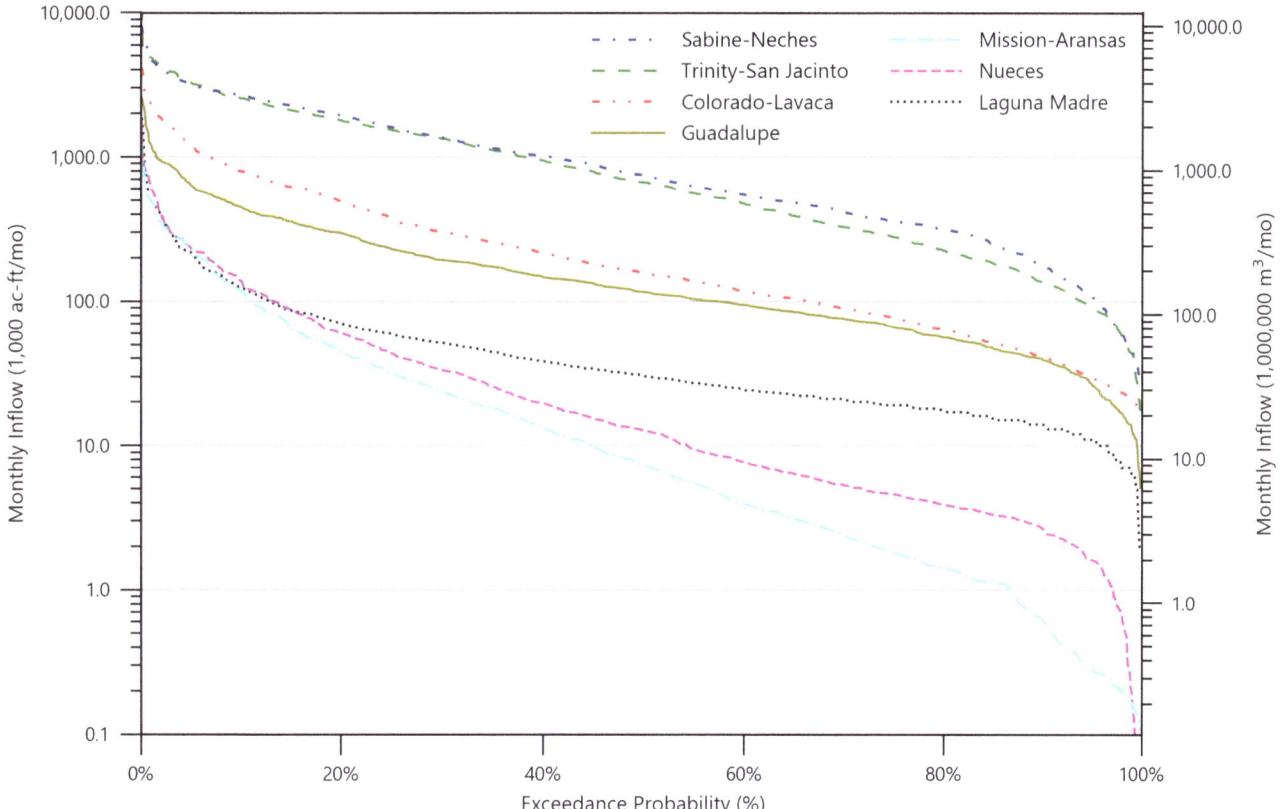

**Fig. 23** Exceedance probability curves of monthly total inflows for the period of 1941–2018

which collectively span the period 1900–2020 (TWDB 2022c). Evaporation estimates are based on 1954–2021 data compiled from multiple organizations and spatially processed by TWDB (2022d). For the calculation of the freshwater inflow balance, the period 1954–2018 was used because it was the longest period for which the total inflow estimates, precipitation data, and evaporation data were all simultaneously available.

The estimates of precipitation and evaporation depths (in units of inches per day [in/day] and inches per month [in/mo], respectively) are multiplied by the surface area of each estuary to get a volume of water per unit time. The net freshwater inflow balance is then the sum of total inflows and precipitation minus evaporation. This calculation further illustrates the three groups of Texas estuaries, with the Sabine-Neches and Trinity-San Jacinto estuaries having the highest net freshwater, the Colorado-Lavaca and Guadalupe estuaries having intermediate values, and the South Texas estuaries having the lowest net freshwater (Fig. 24). Indeed, the Laguna Madre has a net negative balance, which is why this estuary is so often hypersaline. A similar net negative balance was calculated for this estuary by Schoenbaechler et al. (2011e, f).

## Seasonal Patterns

Seasonal patterns in total inflows are demonstrated by the average inflow by month for each estuary (Fig. 25). The y axis uses a log scale to better highlight three seasonal patterns.

The Sabine-Neches and Trinity-San Jacinto Estuaries (East Texas): These estuaries display one seasonal peak (winter and spring), with notably lower summer and fall inflows. In these systems, consistent rainfall during the winter and spring tends to overshadow the influence that occasional tropical storms have on summer and fall flows.

Colorado-Lavaca and Guadalupe Estuaries (Mid-Coast): These estuaries display two seasonal peaks (spring and fall), with the fall peak being lower than the spring peak. In these systems, reduced winter and spring rainfall depresses flows in those seasons (all relative to east Texas), while the effects of tropical storms are apparent, contributing to the fall peak.

Mission-Aransas and Nueces Estuaries and Laguna Madre (South Texas): These estuaries display two seasonal peaks (late spring and fall), with the fall peak being higher than the spring peak. In these systems, reduced winter and spring rainfall depresses flows in those seasons (all relative

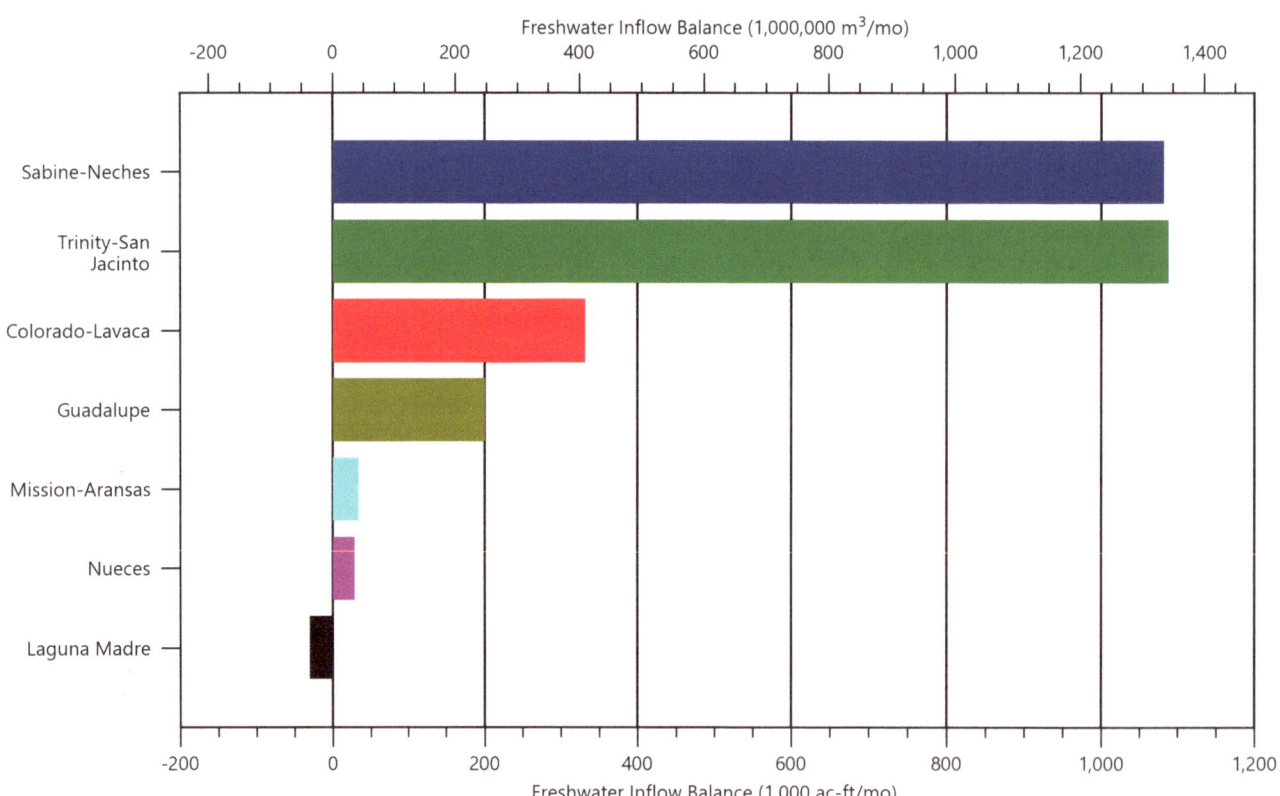

**Fig. 24** Freshwater inflow balance for each estuary

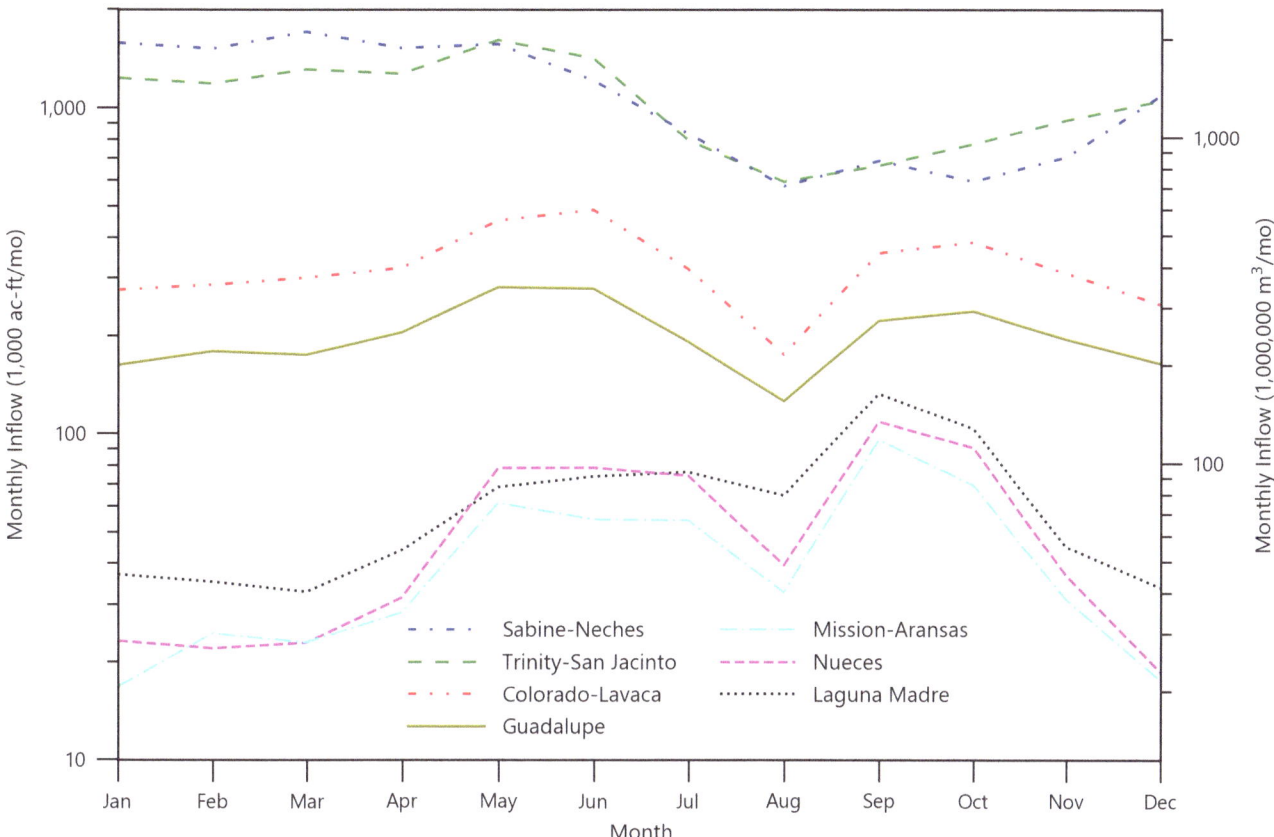

**Fig. 25** Average inflow by month for each Texas estuary. A log y axis is used to better visualize patterns

to the mid-coast and east Texas), while the effects of tropical storms is even more apparent, contributing to the fall peak.

These seasonal patterns are consistent with the precipitation patterns illustrated in Ward (2005) and reinforce the grouping of estuaries by total inflow, where the Sabine-Neches and Trinity-San Jacinto have the highest inflows, the Colorado-Lavaca and Guadalupe have intermediate inflows, and the Mission-Aransas, Nueces, and Laguna Madre have the lowest inflows.

## Long-Term Temporal Trends in Freshwater Inflows

In Texas, long-term shifts in freshwater inflows occur slowly, if at all, whereas short- and intermediate-term inflows exhibit significant variability. As a result, long-term trend analyses are plagued by periods of record that are too short, and periods of record that start or end during abnormally wet or dry periods. Wurbs and Zhang (2014) perceptively state, "The appearance of trends can change significantly with a several-year lengthening or shortening of the period-of-analysis." Indeed, the current effort performed similar numerical tests and generated similarly mercurial results. Additionally,

Wurbs (2021) points out that some locations have experienced long-term increases in mean flows, whereas others have experienced decreases, but these patterns may be different for low flows or high flows, and some of these trend observations are dependent on the period of analysis and even the time interval used in the analysis (e.g., daily versus monthly flows).

Another concern relates to the concept of statistical significance. Temporal trends are often reported as statistically significant, or not, based on arbitrary decisions, such as the choice of type I error rate. In linear regression, this affects the p-value that is deemed to be statistically significant. Furthermore, the p-value decreases with sample size, and it is common for large datasets to have statistically significant trends because of the large sample size, even when the trend itself is very small (Smith 2020; Armstrong 2019; Sullivan and Feinn 2012). This behavior has led authors to focus on the effect size, i.e., the magnitude of trend, as opposed to simply its statistical significance (Sullivan and Feinn 2012). Hydrologic data are relatively easy to record, which often results in large sample sizes. The effects of such large sample sizes on p-values are important to recognize.

A number of authors have evaluated temporal trends in flow data close to Texas estuaries. Broad observations include the following:

- Wurbs and Zhang (2014) performed a linear regression on observed flows at multiple locations throughout Texas. The following list indicates the stations near estuaries and the trend identified:
  - Increasing trend in runoff
    - Trinity River at Romayor
    - Colorado River near Bay City
    - Lavaca River near Edna
    - Guadalupe River at Victoria
    - San Antonio River at Goliad
  - Decreasing trend in runoff
    - Sabine River near Ruliff
    - Neches River at Evadale
    - West Fork San Jacinto River near Conroe
    - Nueces River near Mathis
  - Of the nine stations identified above, the increasing and decreasing trends are essentially evenly split, and no pattern along the coast is evident.
  - These authors take pains to explain that long-term changes cannot be accurately described by linear regression and that their conclusions instead serve as an "indication of possible change."
- Harwell et al. (2020) evaluated temporal trends in streamflow using a nonparametric approach for several basins in Texas, including the Neches, Trinity, Colorado, and Guadalupe. The following list indicates the stations near estuaries and the annual trend identified:
  - Neches River at Evadale, negative
  - Trinity River at Romayor, positive
  - Colorado River near Bay City, positive
  - Guadalupe River at Victoria, positive
  - None of these trends were identified as statistically significant.
  - All four of these stations exhibited the same trend direction as identified in Wurbs and Zhang (2014).
- Asquith et al. (1997) used a nonparametric approach to study trends in inflows to the Mission-Aransas, Nueces, and Laguna Madre estuaries and concluded, "No evidence for trends in estimated annual inflow volumes for any bay system for 1977–1993 is indicated."
- Wurbs and Yang (2022) evaluated the temporal trend in annual flows at the Trinity River at Romayor gage using linear regression and indicated an increasing trend but did not comment on the statistical significance of the trend.
- Joseph et al. (2013) used a nonparametric approach to look at streamflow trends using 1950–2009 data at downstream gages in the Nueces, San Antonio, and Guadalupe River basins. They concluded that the San Antonio and Guadalupe Rivers exhibited a statistically significant increase in annual flows, whereas the Nueces River exhibited a non-significant decrease in annual flows.
- Longley (1994) used a nonparametric approach to evaluate trends in inflows from 1941 to 1987 for six Texas estuaries and observed a significant trend for only one estuary: a 1.2% increase in inflow per year for the Mission-Aransas Estuary.

The current study used linear regression to evaluate long-term temporal trends in monthly total inflows to each estuary. Graphical results are provided in the online supplement. In all cases, the p-values of the temporal trends were less than 0.05 (not shown), but this apparent statistical significance of the temporal trends is largely based on the extraordinary number of data points. Of the seven estuaries, only one indicated a decreasing trend (Sabine-Neches), whereas the other six indicated a positive trend.

Wurbs (2021) notes that for precipitation and evaporation, "Permanent long-term trends, if they exist, are hidden by the great continuous variability." The same could be said for freshwater inflows. Conversely, Nielsen-Gammon et al. (2019) demonstrated long-term increasing trends in precipitation across much of Texas, except for west Texas. Many of the inflow trends identified above are positive, although some inconsistencies occur in the literature, and it is difficult to draw definitive conclusions.

This analysis of trends in inflows focused on long-term trends. The presence or absence of such trends does not inform how climate change (which is often predicted to increase variability), factors that influence climate variability such as El Niño-Southern Oscillation and the Pacific Decadal Oscillation, and human uses might affect shorter-term patterns. For example, human uses may have little impact on annual flows during most years but may have a substantial impact on flows during dry years.

## Inflow Summary

Based on these inflow data, several patterns emerge:

- Freshwater inflow variations across estuaries are primarily a result of their locations on the coast and the corresponding precipitation rates in their watersheds, with the sizes of the watersheds and human activities being secondary factors.

There is a strong west-to-east gradient in inflows, with lower inflows in the western estuaries and higher inflows in the eastern estuaries. This pattern matches the west-to-east gradient in precipitation.

The flows in adjacent pairs of estuaries are correlated, with the correlation between pairs of estuaries that are not adjacent weakening with distance.

Temporal variations in each estuary are significant, with the standard deviation of monthly flows exceeding the mean in all estuaries except the Sabine-Neches.

There are seasonal patterns to inflows, with the eastern-most estuaries exhibiting high inflows during winter and spring, mid-coast estuaries having high spring and fall inflows, and lower coast estuaries having higher fall inflows.

Long-term trends in inflows suggest the possibility of increasing inflows more often than decreasing, but, as noted by Wurbs (2021), it is very difficult to extract subtle long-term trends from data with such high variability.

## Tides

Gulf of Mexico saltwater enters Texas estuaries by tides, winds, and density currents. The tidal component is called tidal pumping (Day et al. 1989). Tides result from the gravitational interaction between the moon, the sun, and the earth. This astronomical tide is made up of the sum of sinusoidal waves that represent the rise and fall of water level at different periodicities, like a musical tone is made up of harmonics. The tidal constituents are forced directly by gravitational accelerations from relative movements within the earth-moon-sun system, by the mutual interaction of constituents, and/or modified by the water depth and size and shapes of landforms beneath the sea level. Ocean tidal dynamics is complex and is better explained in textbooks (Britton & Morton 1989; Thurman and Burton 2001) or treatises devoted to the subject (Hicks 2006; Parker 2007).

Ideally, there are three types of tides: diurnal, semidiurnal, and mixed tides. Diurnal tides have a single high and low water mark each day with a tidal period of one lunar day (i.e., 24.8 h). Semidiurnal tides have two high and low water marks each day with a period of one-half lunar day (i.e., 12.4 h). Mixed tides have characteristics of both and are common along the Texas coast. Two other tidal constituents are common to the Texas coast: the fortnightly tide (13.6 days) and the secular semiannual or seasonal tide (which actually is comprised of both annual and semiannual constituents) (Ward 1997). The secular semiannual tide is less well understood and varies considerably from year-to-year.

## Tidal Variation in Texas

The tidal variations within Texas estuaries are forced by the tidal cycles in the Gulf of Mexico. Generally, tides in the Gulf of Mexico are diurnal with some mixed tides around the northeast and northwest of the Gulf (Espino and Rojas 2004). Mean tidal ranges based on tide measurements provided by the National Oceanic and Atmospheric Administration (NOAA) tide gages (NOAA 2022) average 0.86 ft. (0.26 m) (Table 4). Tidal ranges less than 6 feet are commonly referred to as micro-tidal, and all the tidal ranges shown in the table are substantially less than 6 feet (~2 m). Tidal ranges at the entrance (i.e., Gulf passes, Fig. 26) are higher than within estuaries, i.e., Port O'Connor, Aransas Wildlife Refuge, Rockport, USS Lexington, and South Padre Island Coast Guard Station. The relatively low tidal ranges experienced by Texas estuaries contribute to the relatively low energy of the Texas coastal environment, including relatively modest tidal flushing of these estuaries.

The Texas coast experiences peak high tides in May and October; and lowest tides in January and July (Fig. 27). Unfortunately, hurricanes are also common during the fall high tides, making coastal flooding more problematic. The middle coast locations (Galveston, Matagorda, and Port Aransas) exhibit similar ranges, with lower ranges occurring towards the north at Sabine Pass, and lower still towards the south at Brazos Santiago. There is also a shift in the peak of the first high tide to April in the south.

**Table 4** Tidal variations in Texas estuary entrances and bays based on data from selected NOAA gages

| Estuary | Tide gage and Location | Established | Mean diurnal tidal range (ft) | (m) |
|---|---|---|---|---|
| Sabine-Neches | 8770822 Texas Point, Sabine Pass Entrance, TX | Jun 2011 | 1.93 | 0.59 |
| Sabine-Neches | 8770475 Port Arthur, TX | Apr 1996 | 1.04 | 0.32 |
| Trinity-San Jacinto | 8771341 Galveston Bay Entrance, North Jetty, TX | Apr 2000 | 1.67 | 0.51 |
| Trinity-San Jacinto | 8771972 San Luis Pass, TX | Jan 1998 | 1.17 | 0.36 |
| Trinity-San Jacinto | 8770613 Morgans Point, Barbours Cut, TX | Jun 1992 | 1.33 | 0.41 |
| Colorado-Lavaca | 8773767 Matagorda Bay Entrance Channel, TX | Jun 1973 | 1.24 | 0.38 |
| Colorado-Lavaca | 8773259 Port Lavaca, TX | Jan 1989 | 0.90 | 0.27 |
| Guadalupe | 8774230 Aransas Wildlife Refuge, TX | Oct 2012 | 0.33 | 0.10 |
| Mission-Aransas | 8774770 Rockport, TX | Feb 1937 | 0.36 | 0.11 |
| Mission-Aransas/Nueces | 8775241 Aransas, Aransas Pass, TX | Aug 2016 | 1.37 | 0.42 |
| Nueces | 8775296 USS Lexington, Corpus Christi Bay, TX | Jan 2004 | 0.61 | 0.19 |
| Lower Laguna Madre | 8779749 South Padre Island Brazos Santiago Entrance, TX | Aug 2016 | 1.47 | 0.45 |
| Lower Laguna Madre | 8779280 Realitos Peninsula, TX | Feb 2009 | 0.56 | 0.17 |

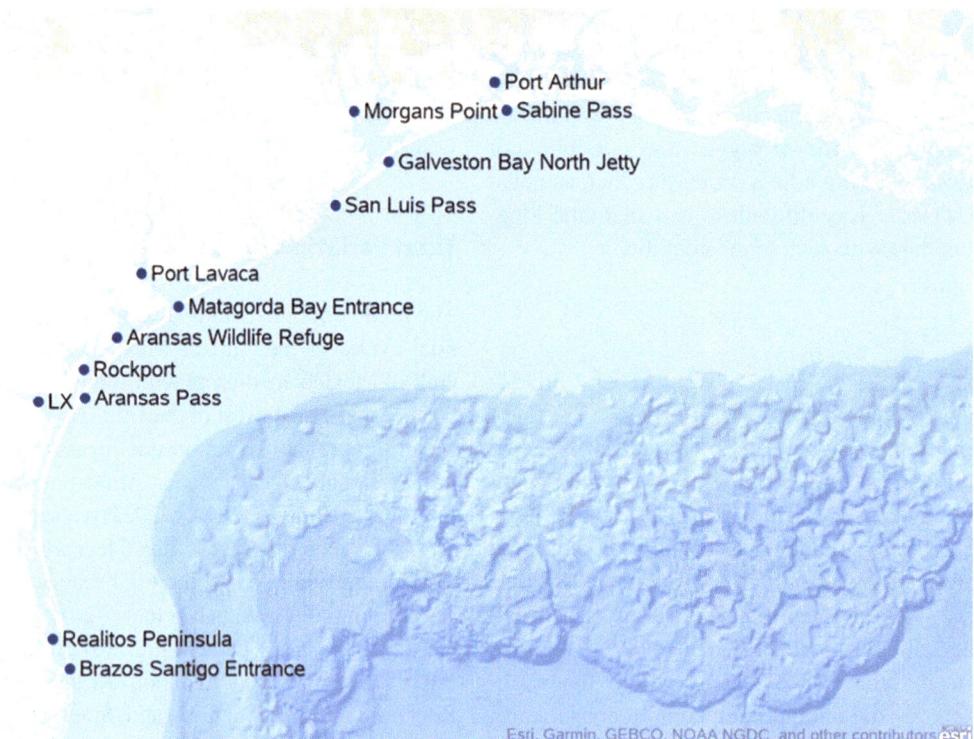

**Fig. 26** Locations of tide gages along the Texas coast (Table 4, NOAA 2022). Abbreviation: LX = USS Lexington

**Fig. 27** Mean monthly tides (NAVD88) at Gulf of Mexico passes (Fig. 26)

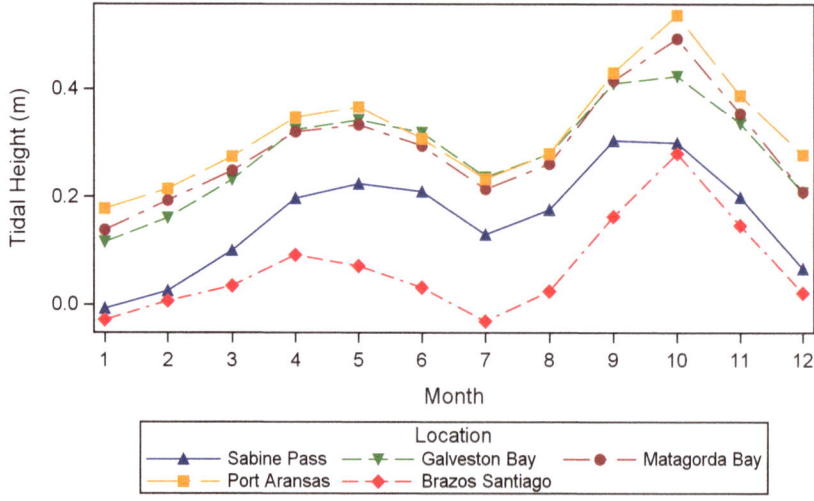

An examination of daily tides by hour for the semiannual tides demonstrates the three main components of tidal signals along the Texas coast (Fig. 28). The daily tide is composed of both diurnal and mixed tides. While most of the signals are diurnal, a mixed tide can be seen from October 11–16, 27–29, and mid-January. The effect of a strong cold front on October 9, 2020, depressed water levels for a short time. The phases of the moon can be seen especially in January with expansions of the range from 9 to 15 and contractions from 17 to 20. The October ranges are nearly all positive, while the January ranges are nearly all negative. For the most part, Texas has high tides in spring and fall, and low tides in winter and summer, the daily tidal range is small, and high winds associated with fronts and storms can change tidal range dramatically.

The effect of a strong frontal passage over Sabine Pass on October 9, 2020 demonstrates the importance of meteorological effects on tidal level along the Texas coast (Fig. 28). The water levels are depressed to −0.5 m for a short time during a period when tide levels should be high due to the secular tide. While both barometric pressure and wind play a role in the depressed tidal level, wind stress is far more

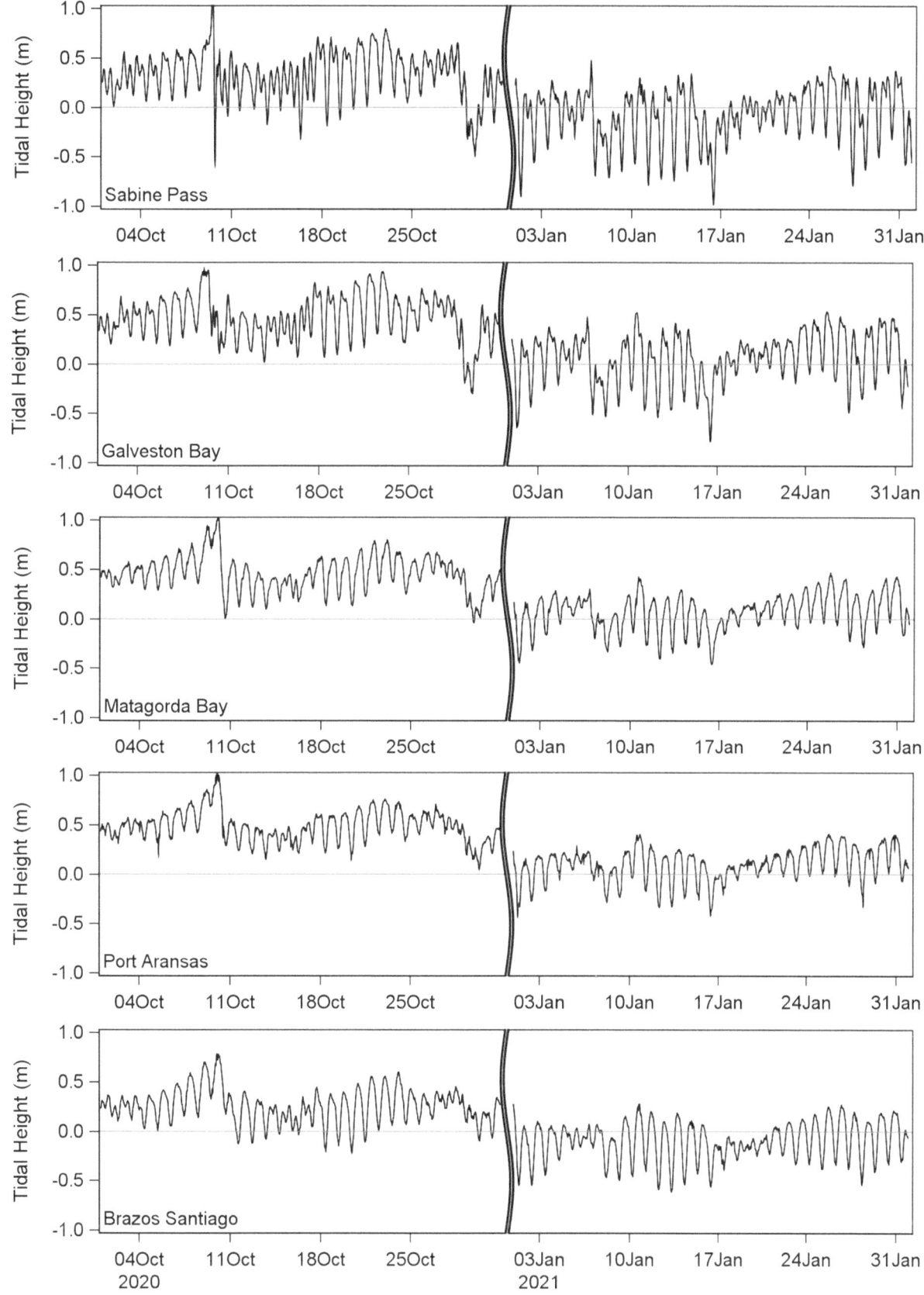

**Fig. 28** Mean hourly tides (NAVD88) during October 2020 and January 2021

important in driving tidal level during frontal passages (Ward 1997). Winds during frontal approach set-up water levels and reverse after frontal passage leading to lower water levels.

## Tidal Variation in Bays

Salt water enters Texas estuaries from the Gulf of Mexico through the tidal passes by tidal pumping. However, the exchange is greatly influenced by the narrow inlet into a much larger basin. Texas estuaries are mostly bar-built estuaries with sandy barrier islands facing the Gulf of Mexico. The exchange of Gulf water is within narrow tidal passes (or inlets). The narrow entrance acts as a filter for higher frequency (shorter period) oscillations, resulting in lower frequency (longer period) oscillations allowed to pass more readily than higher frequencies (shorter periods). The dampening of the tidal signal is also called the stilling well effect of co-oscillating basins because the large ocean basin and the bay are connected by a narrow inlet. The net effect is a reduction in amplitude of the tidal constituents. For example, the short-term daily amplitude of a semidiurnal tide traversing Aransas Pass to Corpus Christi Bay is reduced by about 90% (Ward 1997). In contrast, the fortnightly tide and the secular tide are nearly unaffected.

Tidal flood is the incoming tide that supplies salt water to the bays. Tidal ebb is the outgoing tide that drains the bays. The fortnightly and secular tides are the most important tidal constituents driving these dynamics. These longer-term filling and depletion events can remain in place for days to weeks (Fig. 28). In contrast, the shorter-term semidiurnal and diurnal tidal influxes are filtered by the narrow constriction of the tidal inlet by the stilling effect.

"Tidal prism is the amount of water that flows into and out of an estuary or bay with the flood and ebb of the tide, excluding any contribution from freshwater inflows" (Hume 2005). Thus, tidal prism is an important indicator of the hydrodynamic processes in an estuary. Because Texas bays are shallow, the tidal prism can contribute to a high amount of flushing of the bays during the secular tides. However, as shown in Fig. 28, meteorological events such as frontal passages, can provide larger volumes of water exchange than tidal prisms.

## Circulation and Salinity

Fresh water enters estuaries at the landward side of each estuary. By contrast, salt water enters from the Gulf of Mexico. This leads to the definition of an estuary: a location where fresh water and saltwater mix. Gulf water enters estuaries largely as a result of tides, wind, and density-driven currents. Because the density of water increases with salin-

ity, Gulf water tends to underflow freshwater, which can lead to vertical stratification, especially in deeper waters and in areas of low turbulence. Furthermore, the difference in density between salt and fresh water creates a driving force that intrudes salt water into the bay and forces fresher, surface, water out of the bay. This pattern, termed a density current (or gravitational circulation) causes water movement even in the absence of wind and tides, as long as there are horizontal gradients in salinity between the bay and Gulf of Mexico. These forces, combined with the physiography of the bay and the Coriolis force, result in complex circulation patterns and hence complex salinity patterns.

Salinity has been described as a key characteristic of estuaries that influences species abundance and diversity (Smyth and Elliott 2016). Because the important Texas estuaries are enclosed bay systems, their salinity is primarily controlled by the balance between freshwater inflows, tidal exchange, wind, and longshore currents. Freshwater inflows have zero salinity, whereas tides bring in Gulf water, which has an average salinity of 36 practical salinity units (psu) (Espino and Rojas 2004). Circulation causes these flows to mix, resulting in widespread areas of intermediate salinity.

Historically, Texas estuaries have experienced significant variability in salinity, but organisms in each have adapted to "typical" conditions for that estuary. The conditions that an estuary has exhibited in the past may be different in the future, due to the following human influences:

- Reduction in freshwater inflows due to human diversion and consumptive use
- Increase in freshwater inflows due to transfers of water into a watershed from outside of the watershed
- Increases in tidal exchange due to relative sea level rise (i.e., a combination of land subsidence and eustatic sea level rise)
- Increases in tidal exchange due to construction or expansion of ship channels
- Increases in salinity due to increased evaporation caused by global warming

Most of these human influences, including the most important ones, will result in higher salinity in Texas estuaries. Many of these influences, and their effects on salinity, can be understood through the development of hydrodynamic and salinity numerical models (also referred to as circulation models). Such models combine data on water flows, precipitation, evaporation, bathymetry, and meteorological conditions to predict water elevations, water velocities, and salinity concentrations.

Although a variety of organizations and researchers have built hydrodynamic and salinity models of individual Texas estuaries, the only modeling framework that has been consistently applied to all Texas estuaries is TxBLEND. TxBLEND is a two-dimensional (depth-averaged), finite-element estuary model that has been used by TWDB for many years to simulate water elevation, velocity, and salinity in Texas estuaries (TWDB 2022c). The TxBLEND models used in this chapter were obtained from TWDB.

TxBLEND is an expansion of the salinity model BLEND developed by Dr. William Gray at Notre Dame University, which has been modified by TWDB to accommodate additional forcing processes (such as tides, riverine inflow, wind, and evaporation), initial salinity concentrations, and various utility routines for simulation of Texas estuaries (Matsumoto 1992b). TxBLEND solves the continuity and momentum equations for hydrodynamics and the advection-diffusion equation for salinity transport (TWDB 2022c); these equations are solved on a numerical grid consisting of linear triangular elements. Inputs into TxBLEND include the numerical grid and associated bathymetry; daily average flow rates for riverine inflows; 2-h tide data at the offshore grid boundaries; time-series wind magnitudes and directions as either hourly data, 3-h averages, or daily averages; daily total precipitation; and daily total evaporation (Matsumoto 1992a, b; Guthrie et al. 2012). Outputs from TxBLEND include daily- and monthly-average salinities at every grid node; daily-average current velocities at every grid node; and time-series water levels, current velocities, and salinities at user-specified grid nodes and user-specified time intervals.

Calibration and validation of the TxBLEND models for each of the Texas estuaries are documented in reports issued by TWDB (Matsumoto et al. 2014; Guthrie et al. 2010, 2012; Schoenbaechler et al. 2011a, g, h). The TxBLEND models and input files used in this chapter were the latest versions available through TWDB for each estuary as of September 2021. The start and end dates of the simulation files obtained from TWDB varied somewhat among the models. TWDB's model start dates were not modified for this chapter. However, the end date for each model was uniformly extended for this chapter through December 31, 2018, to coincide with the end date of the TWDB freshwater inflow estimates that were used as inputs into the models. The resulting simulation periods for the estuary models presented in this chapter are shown in Table 5.

**Table 5** TxBLEND simulation periods for each estuary

| Estuary | TxBLEND Simulation Period |
| --- | --- |
| Sabine-Neches | January 1, 1990 – December 31, 2018 |
| Trinity-San Jacinto | January 1, 1982 – December 31, 2018 |
| Colorado-Lavaca | January 1, 1987 – December 31, 2018 |
| Guadalupe | January 1, 1987 – December 31, 2018 |
| Mission-Aransas | January 1, 1987 – December 31, 2018 |
| Nueces | January 1, 1987 – December 31, 2018 |
| Upper Laguna Madre | January 1, 1987 – December 31, 2018 |
| Lower Laguna Madre | January 1, 1987 – December 31, 2018 |

Maps of the TxBLEND model domains in the vicinity of each estuary are shown in Figs. 29, 30, 31, 32, 33, 34, 35, and 36. Locations of modeled tributary inflows for each estuary are also plotted on these maps, along with the locations of salinity datasondes established by various organizations (such as the Lower Colorado River Authority, NOAA, Texas A&M University (TAMU), Texas Parks and Wildlife Department (TPWD), and TWDB and the locations of NOAA tide stations. The locations and time-series data for the salinity datasondes shown in these maps were obtained from TWDB (2022e). Each datasonde is classified as either a live station (a station that is currently active at the time of this writing and posts its measurements in real time to a publicly available web page), an active station (a station that is currently active at the time of this writing but does not post its measurements in real time to a publicly available web page), or an inactive station (a station for which historical data are available but which no longer collects data).

As with any numerical model, the TxBLEND models represent the conditions specified by the model developer, which usually reflect the estuary conditions at the time of model development. Because Texas estuaries are not static, some changes have occurred since the development of these TxBLEND models, including the following:

In the Trinity-San Jacinto Estuary model, a former pass called Rollover Pass is included in the model. This pass was closed in 2019.

In the Nueces Estuary model, the mouth of Cedar Bayou was simulated as closed. This bayou has alternated between closed and open over the years.

In the Colorado-Lavaca Estuary model, the Colorado River discharges into the eastern arm of Matagorda Bay, which did not occur in the real world until 1993.

## Circulation

TxBLEND predicts water velocity and direction throughout the modeled period of record. Details regarding these outputs are sometimes important to evaluate ecosystem responses (e.g., drift of algal blooms or oyster larvae), but most organisms are either fixed (e.g., adult oysters) or independently motile (e.g., finfish) and therefore are not strongly affected by particular currents. Accordingly, detailed outputs are not presented here. Rather, Figs. 37, 38, 39, 40, 41, 42, 43, and 44 illustrate the net (or "residual") water velocities and directions, averaged over several tidal cycles during typical conditions, in each estuary. Each of these figures was constructed as follows:

- For each day in the modeled period of record, the area-weighted average bay salinity was calculated.
- The median of the modeled period of record time series of average daily bay salinities was calculated.

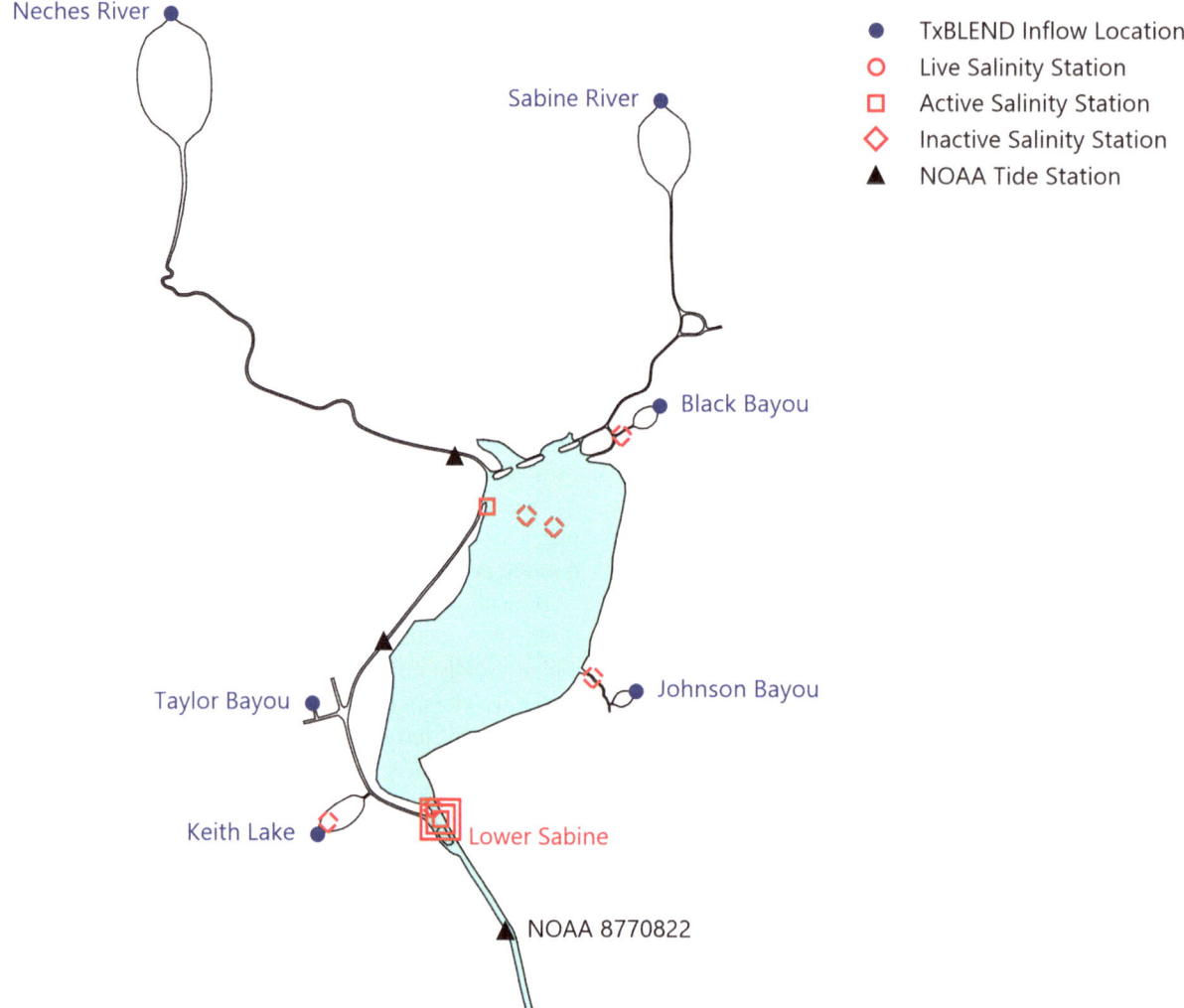

**Fig. 29** Locations of TxBLEND inflows, salinity stations, and tide stations for the Sabine-Neches Estuary. Salinity stations analyzed for temporal trends in later sections of this chapter (Table 7) are identified with station labels. Tide stations presented in Table 4 are identified with station labels. Salinity stations represented by a single shape with a dashed border have a period of record spanning less than 10 years. Salinity stations represented by a single shape with a solid border have a period of record spanning more than 10 years but less than 20 years. Salinity stations represented by two concentric shapes have a period of record spanning more than 20 years but less than 30 years. Salinity stations represented by three concentric shapes have a period of record spanning more than 30 years but less than 40 years. Blue shading denotes the portion of the estuary over which TxBLEND salinity results were extracted, and in some cases spatially averaged, for analyses presented in subsequent sections of this chapter

- The 28-day window that had a median salinity most closely equal to the long-term median salinity was found.
- The net water velocity and direction for the 28-day window was calculated for each model grid cell.
- These net velocities and directions were combined into flow paths.

This approach is an adaptation of that described by Matsumoto et al. (2005). This approach describes a typical circulation pattern, but it should not be construed as the only circulation pattern for each estuary. Different tides, wind, freshwater inflows, and other factors can dramatically change the circulation.

For all estuaries, the net velocity vectors generally go from the freshwater inputs to the Gulf passes. Velocities are often highest in the vicinity of the major freshwater inputs, in the vicinity of the Gulf passes, and along ship channels (where present). In the Trinity-San Jacinto Estuary, local increases in velocity are visible at the location of return flows from the P.H. Robinson Power Plant (in the southwestern lobe of upper Galveston Bay) and the Cedar Bayou Power Plant (in upper Trinity Bay). In the Nueces Estuary, return flow from the Nueces Bay Power Plant (discharged near the southern shoreline of Nueces Bay) is notably conspicuous and appears responsible for driving water movement from the eastern half of Nueces Bay into Corpus Christi Bay. In

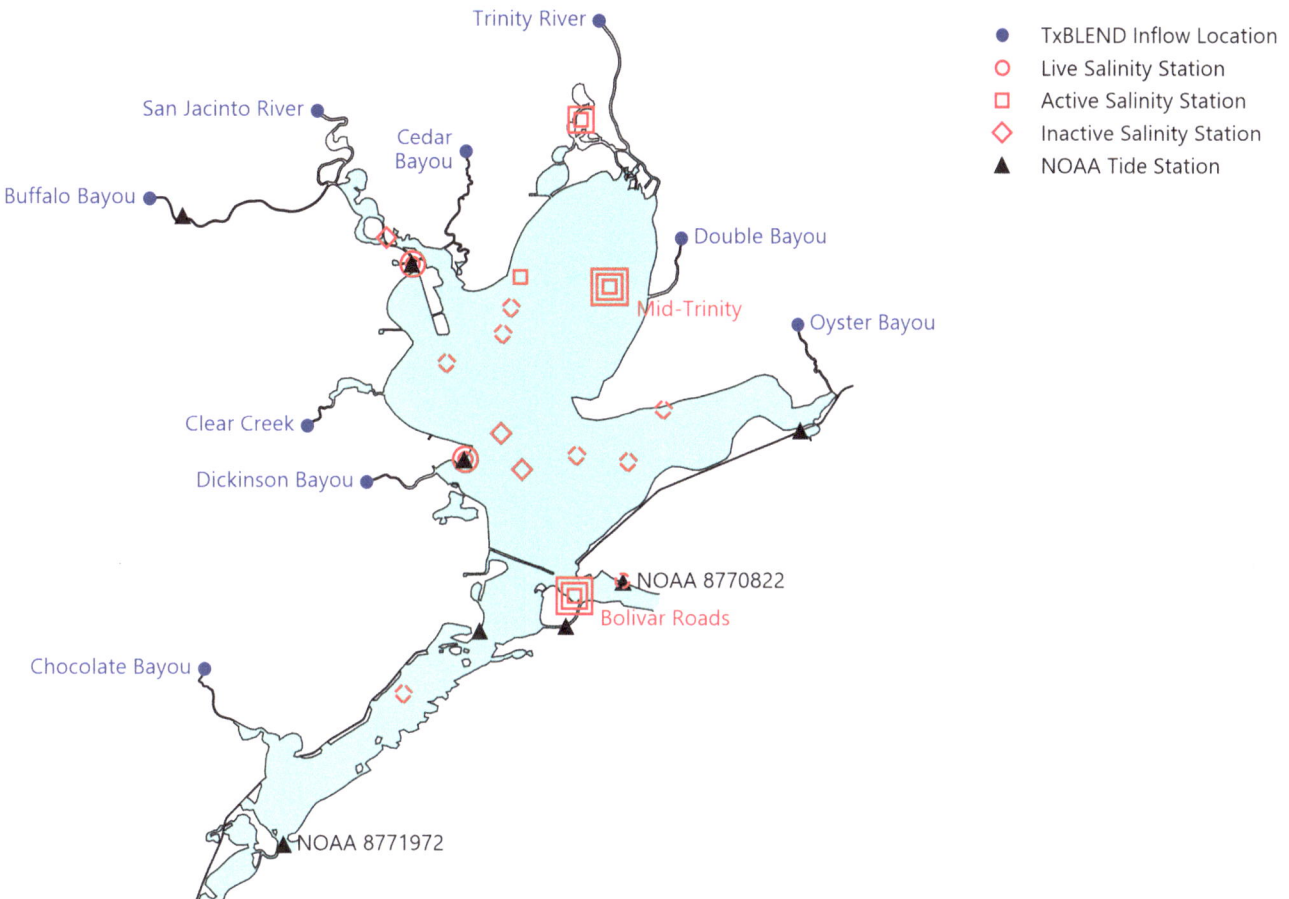

**Fig. 30** Locations of TxBLEND inflows, salinity stations, and tide stations for the Trinity-San Jacinto Estuary. Colors, symbols, and labels as in Fig. 29

the Upper Laguna Madre, a small, isolated increase in velocity is visible at the diversion site for the Barney Davis Power Plant (near the western shoreline in the upper part of the estuary). With the exception of the Lower Laguna Madre, gyres are observed in various locations throughout each estuary (e.g., the eastern side of Trinity Bay in the Trinity-San Jacinto Estuary and the center of the Colorado-Lavaca Estuary).

## Salinity

Circulation in each estuary mixes freshwater from rivers with salt water from the Gulf of Mexico to create spatial and temporal variations in salinity. Because organisms have preferences, and often requirements, for certain salinity levels, these variations help to define organism presence and abundance.

Because salinity monitoring data are only available for limited locations and times, a holistic view of salinity for each estuary is better accomplished using model results. In the following paragraphs, TxBLEND salinity model outputs are discussed and illustrated, including temporal variations in salinity for each estuary, spatial variations within each estuary, and spatial variations across the estuaries.

Figures 45, 46, 47, 48, 49, 50, 51, and 52 illustrate the percentage of each estuary in salinity ranges over time. These figures use color to depict the percent of each estuary's area in ranges of 5 psu. Superimposed on the figure is the total monthly estuary inflow. High flow periods are associated with greater percentages of low salinity (e.g., purple and blue), whereas low flow periods are associated with higher salinity (yellow and red). To best visualize the salinity ranges within each individual estuary, the color ramp for the salinity scale is different for each figure, varying between 0 psu and the maximum monthly average salinity value calculated by the model within the estuary, rounded up to the nearest 5 psu.

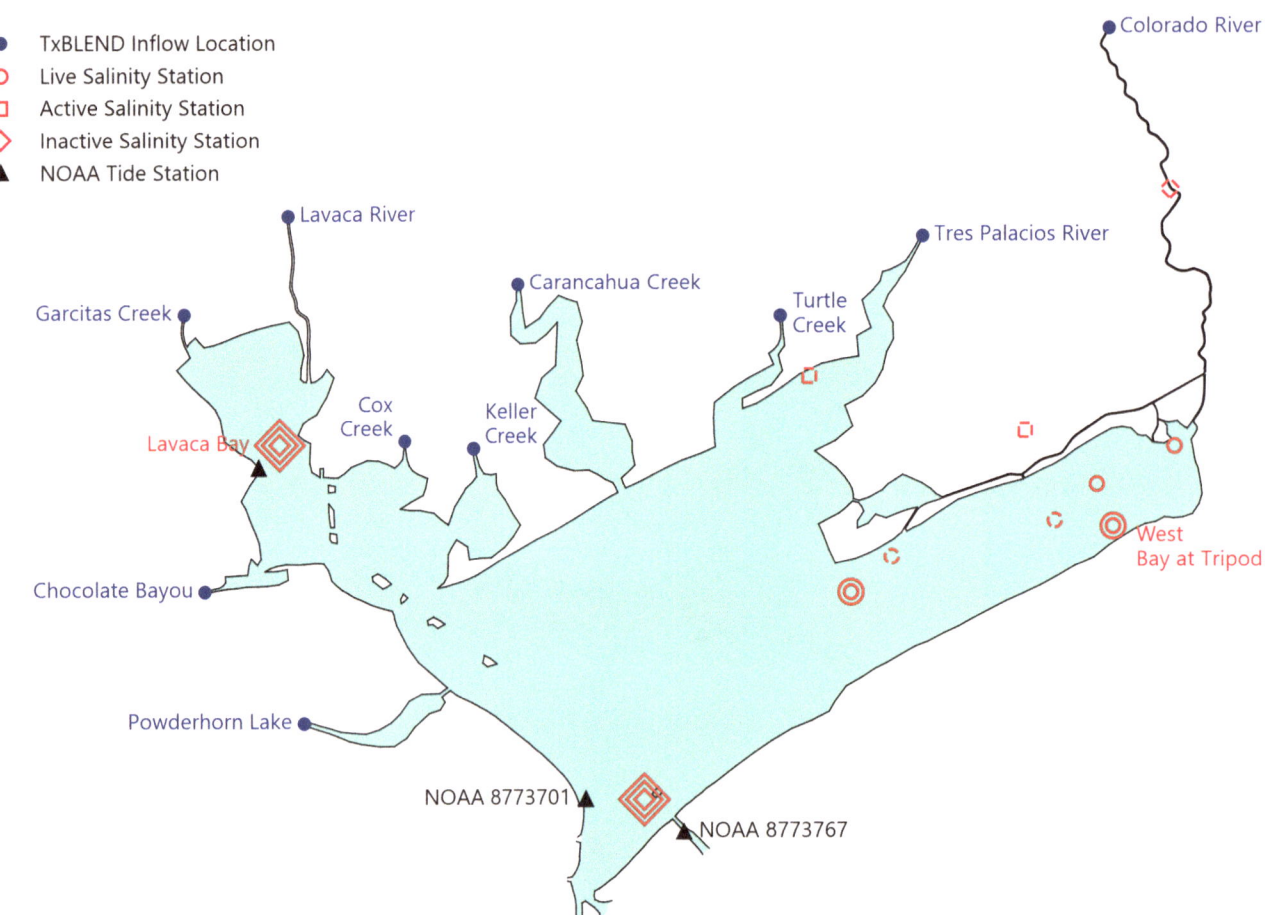

**Fig. 31** Locations of TxBLEND inflows, salinity stations, and tide stations for the Colorado-Lavaca Estuary. Colors, symbols, and labels as in Fig. 29

In all estuaries, there is the expected inverse relationship between inflow and salinity. In most estuaries, the effects of the 2011 drought are seen, with some of the highest sustained salinity values in the modeled period of record. High flow events (e.g., the Christmas 1991 flood in the Colorado-Lavaca and Guadalupe watersheds, the July 2002 flood in the Guadalupe watershed, Hurricane Alex in July 2010 in the Upper and Lower Laguna Madre, and other events discussed above) result in the largest area of low salinities. For the Trinity-San Jacinto Estuary, Hurricane Harvey brought in freshwater equivalent to three times the entire bay volume and depressed salinities for 2 months (Du et al. 2019).

The small tidal range in the Texas estuaries allows freshwater to remain in the bay systems for extended time periods, creating mesohaline habitats for organisms in many of the estuaries. For the Laguna Madre, the limited Gulf passes and net negative freshwater balance, coupled with the small tidal range, contribute to the frequency of hypersaline conditions. Similar time series figures of monthly total inflow and

bay average salinity (as opposed to percent of the bay in salinity ranges) are provided in the online supplement. Raster plots of bay daily average salinity values are also provided in the online supplement.

The long-term average modeled salinity in each estuary is summarized in Table 6. As expected from the net freshwater balance results, there is generally a gradient of low salinity to high salinity from east to west (north to south). Some deviations from this pattern occur, likely because some estuaries have significant Gulf passes (e.g., Colorado-Lavaca) whereas others do not (e.g., Guadalupe and Mission-Aransas).

The modeled data are compared with measurements taken by Texas Parks and Wildlife Department (TPWD), Coastal Fisheries monitoring data. TPWD measures salinity in at least 20 stations every month, in a randomized design, so the data were averaged by month to obtain an estuary-wide average each month, and then the mean was calculated. The only estuary that matches perfectly is Trinity-San Jacinto. The largest difference between modeled and measured was in

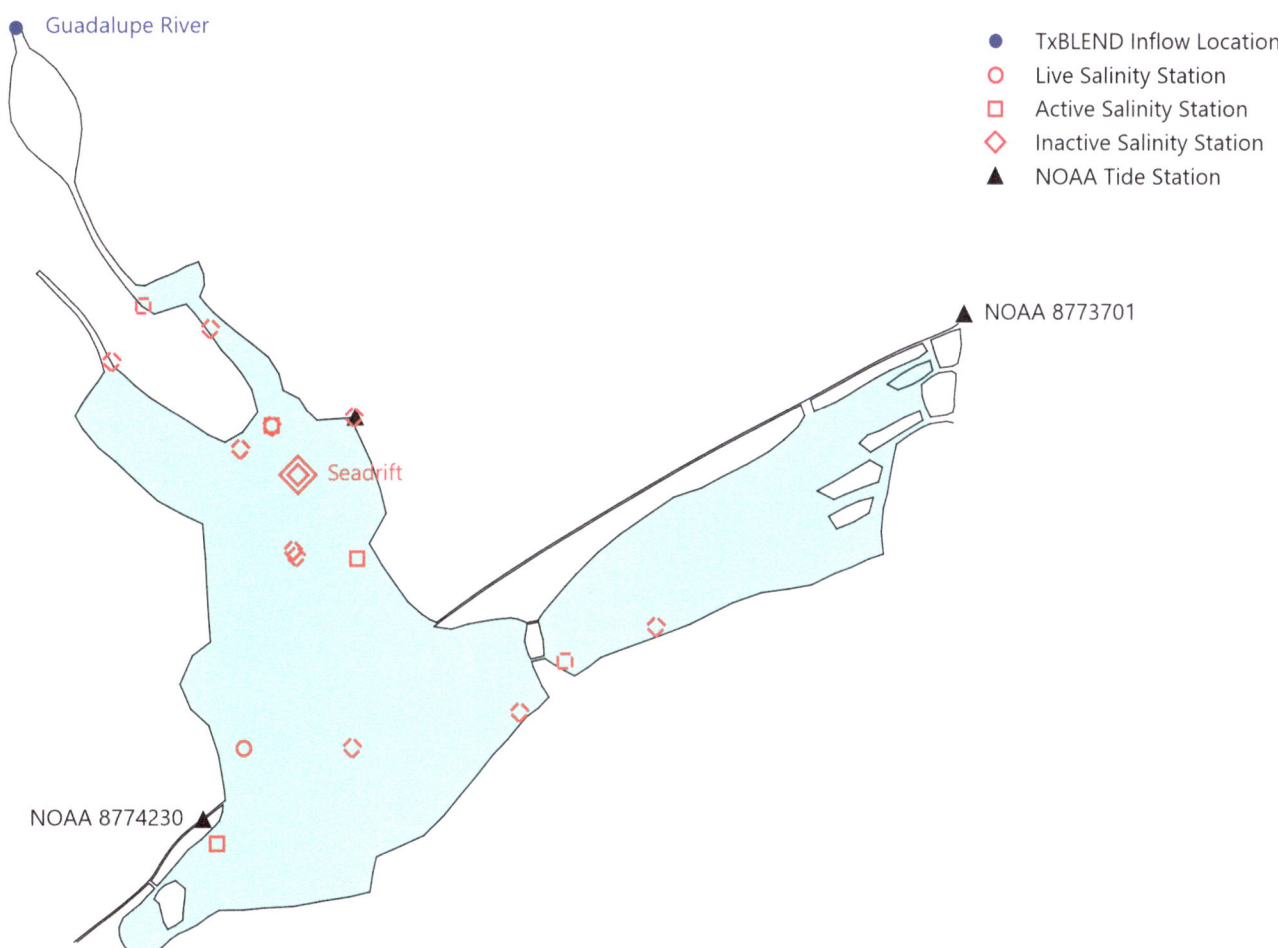

**Fig. 32** Locations of TxBLEND inflows, salinity stations, and tide stations for the Guadalupe Estuary. Colors, symbols, and labels as in Fig. 29

between upper and lower Laguna Madre. While the model is calculating consistently across space and time, the TPWD data are sparser and randomly sampled. Also, the modeled salinity is uniform with depth (because the model is 2D) while the TPWD data are collected near the surface. These differences account for some of the discrepancy between the model and data.

To focus on seasonal salinity patterns, Fig. 53 shows the average salinity by month for each estuary. These patterns have an annual peak in salinity in late summer for all estuaries except the Sabine-Neches which shows no pattern of seasonality.

In general, the Sabine-Neches Estuary is often oligohaline (salinity between 0.5 and 5 psu) whereas the Nueces Estuary and Laguna Madre are often euhaline (salinity between 30 and 40 psu) or even hyperhaline (salinity greater than 40 psu; also referred to as hypersaline) (Anon 1958). The intervening estuaries are generally mesohaline (salinity

between 5 and 18 psu) or polyhaline (salinity between 18 and 30 psu) (Anon 1958). However, due to the high level of temporal variability across seasons and years, and spatial variability within each estuary, these classifications do not fully capture the nature of each estuary.

Figures 54, 55, 56, 57, 58, 59, 60, and 61 illustrate the long-term average of daily average salinity and long-term standard deviation of daily average salinity in each estuary. Some organisms are highly adapted to variable salinity conditions and would be expected to be abundant where salinity is variable. Indeed, Attrill (2002) showed that species diversity decreased with salinity range (which is a measure of variability) in the Thames Estuary and further argued that salinity variation, rather than absolute salinity tolerance, drives the distribution of organisms in an estuary (Attrill 2002). Therefore, understanding the spatial extent of highly variable salinity conditions may provide clues as to species' distributions.

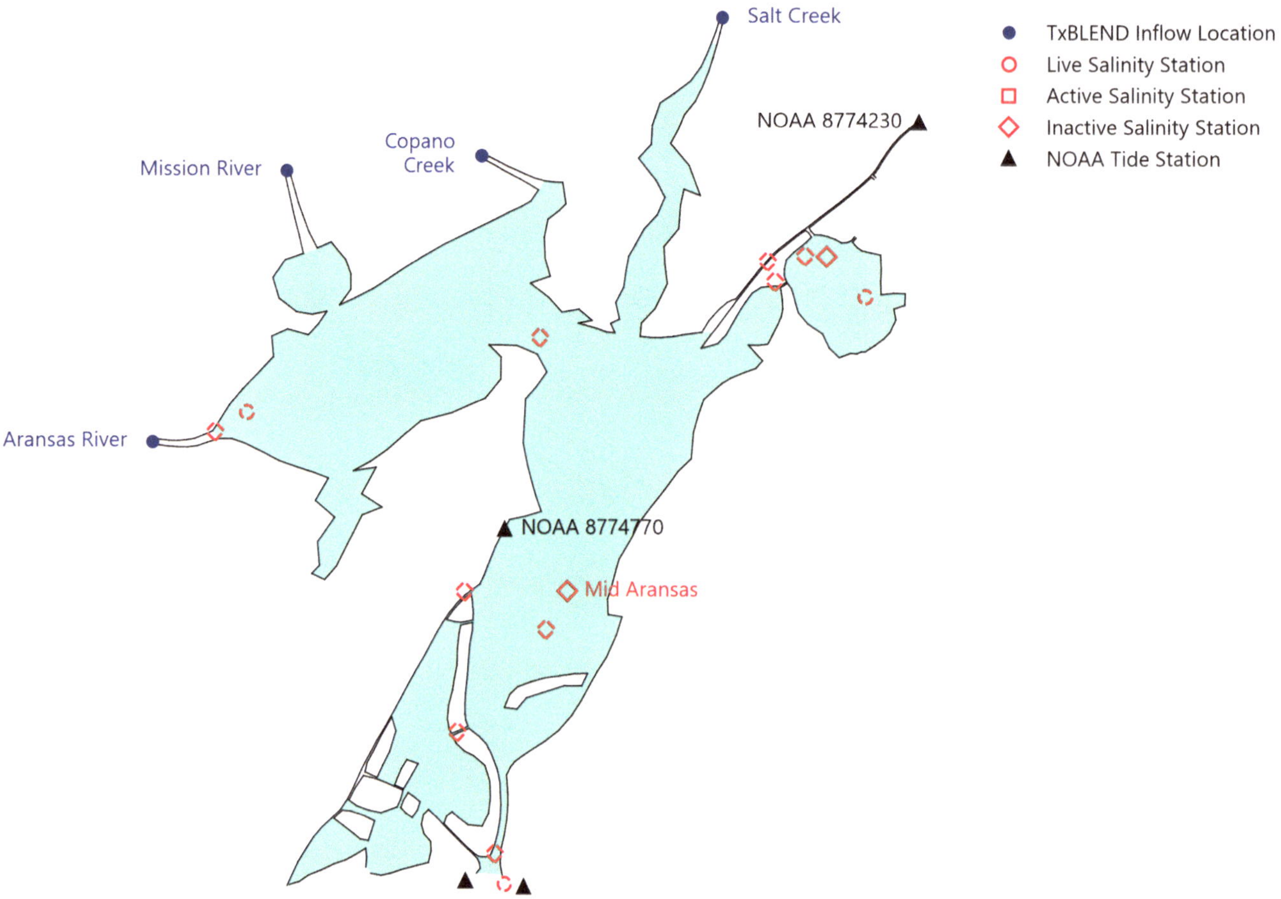

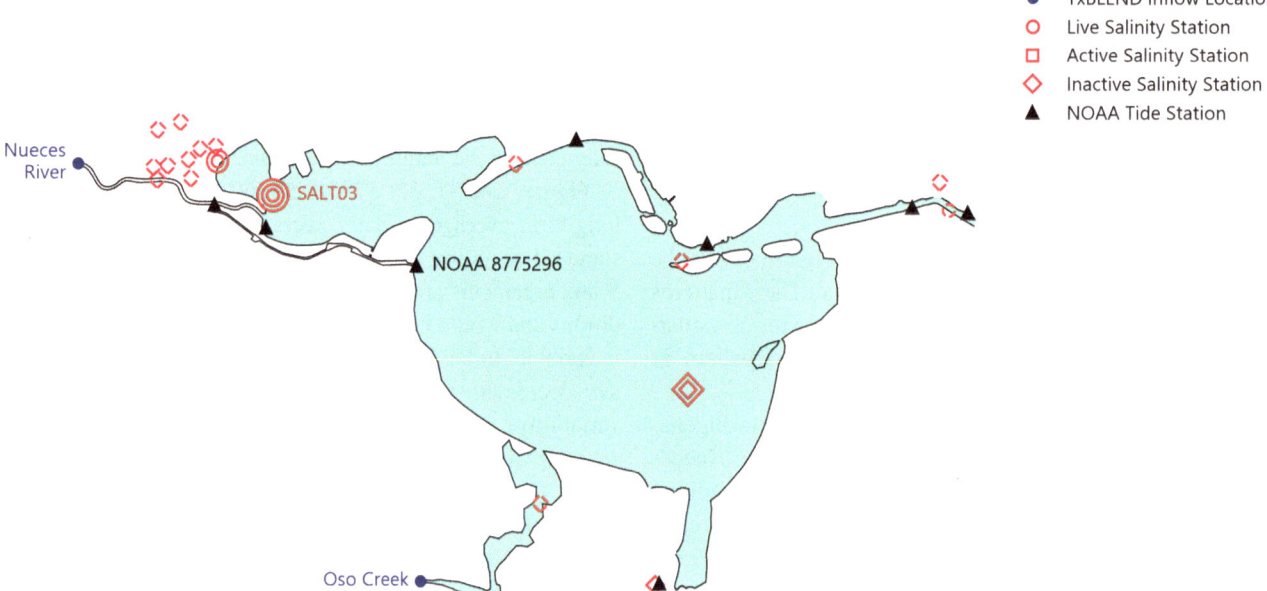

**Fig. 33** Locations of TxBLEND inflows, salinity stations, and tide stations for the Mission-Aransas Estuary. Colors, symbols, and labels as in Fig. 29

**Fig. 34** Locations of TxBLEND inflows, salinity stations, and tide stations for the Nueces Estuary. Colors, symbols, and labels as in Fig. 29

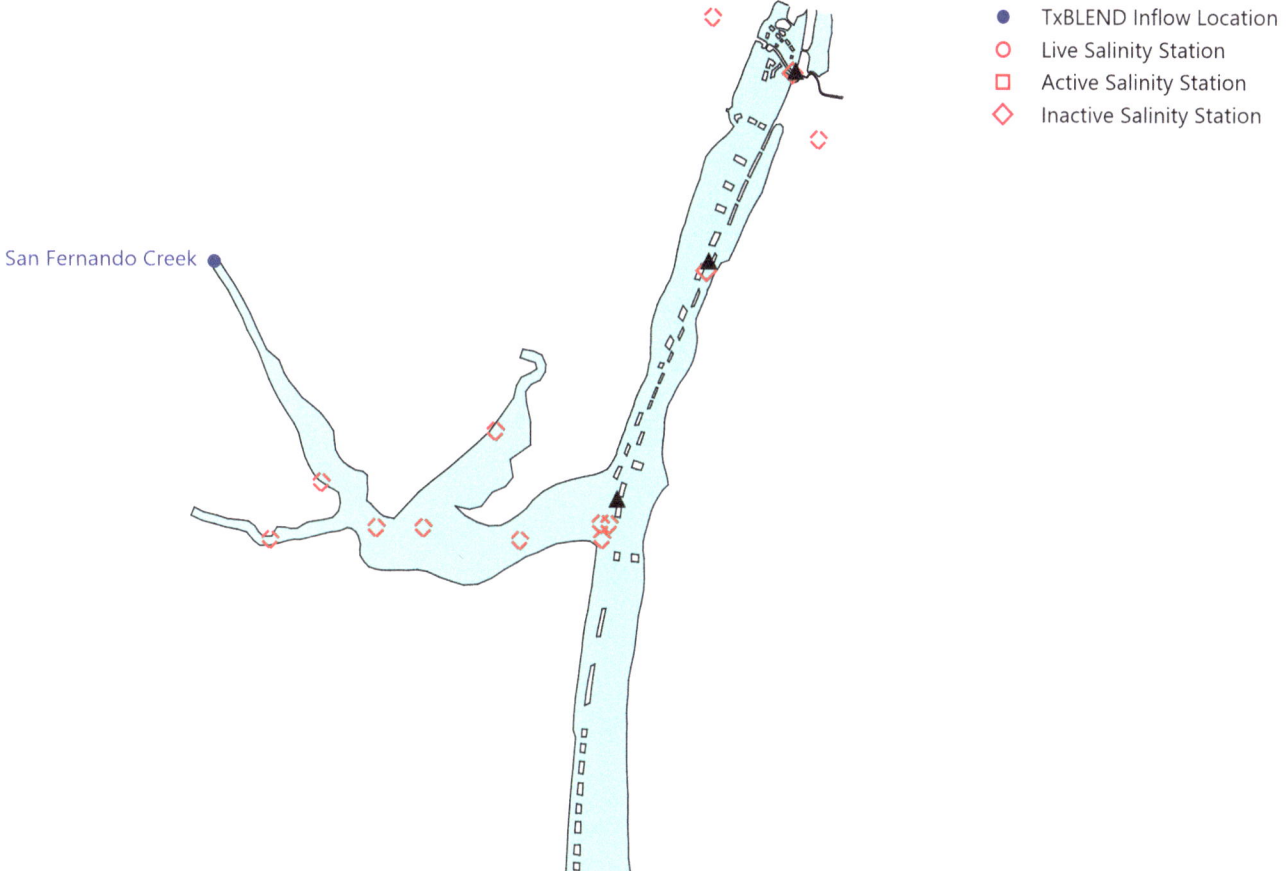

**Fig. 35** Locations of TxBLEND inflows, salinity stations, and tide stations for the Upper Laguna Madre Estuary. Colors, symbols, and labels as in Fig. 29

In most cases, the standard deviation is highest where the salinity average is moderate. These are locations that swing between fresh and salt water, exhibiting high variability. Near the Gulf passes, and near the freshwater inflows, the average salinity is high, and low, respectively, but the standard deviation is relatively low in both types of locations due to the relatively constant nature of salinity at these boundaries. The Lower Laguna Madre has a different pattern due to the frequent presence of hypersalinity (i.e., salinity higher than seawater). In this system, the standard deviation is highest furthest from the Gulf passes, where hypersalinity is most common.

In the Sabine-Neches and Nueces estuaries, the main body of the estuary also has relatively low salinity variability (e.g., light blue or green comprising the majority of the main body of each estuary:), whereas others exhibit somewhat higher variability (primarily yellow: Colorado-Lavaca and Lower Laguna Madre), or the highest (primarily orange and red: Trinity-San Jacinto, Guadalupe, Mission-Aransas, and Upper Laguna Madre [particularly Baffin Bay]). The gradients of absolute salinity and salinity variability across the Texas estu-aries will help provide context to the chemical and biological discussions in subsequent chapters. Maps of the monthly average salinity are provided in the online supplement.

## Salinity Versus Freshwater Inflow

As a case study of the relationship between salinity and freshwater inflow in one of the Texas estuaries, Fig. 62 shows a plot of monthly average salinity (measured at salinity station SALT03) versus cumulative monthly inflow (measured at USGS gage 08211500, Nueces River at Calallen, Texas) in the Nueces Estuary. The period of record for this plot, based on the simultaneous availability of salinity and inflow data, is October 1992 through July 2022. A log-scale is used for the x-axis to better visualize trends across the wide range of monthly inflow values.

To demonstrate the relationship between the measured salinities and inflows, Fig. 51 includes a regression curve of the data. This curve is a piecewise function that consists of two segments.

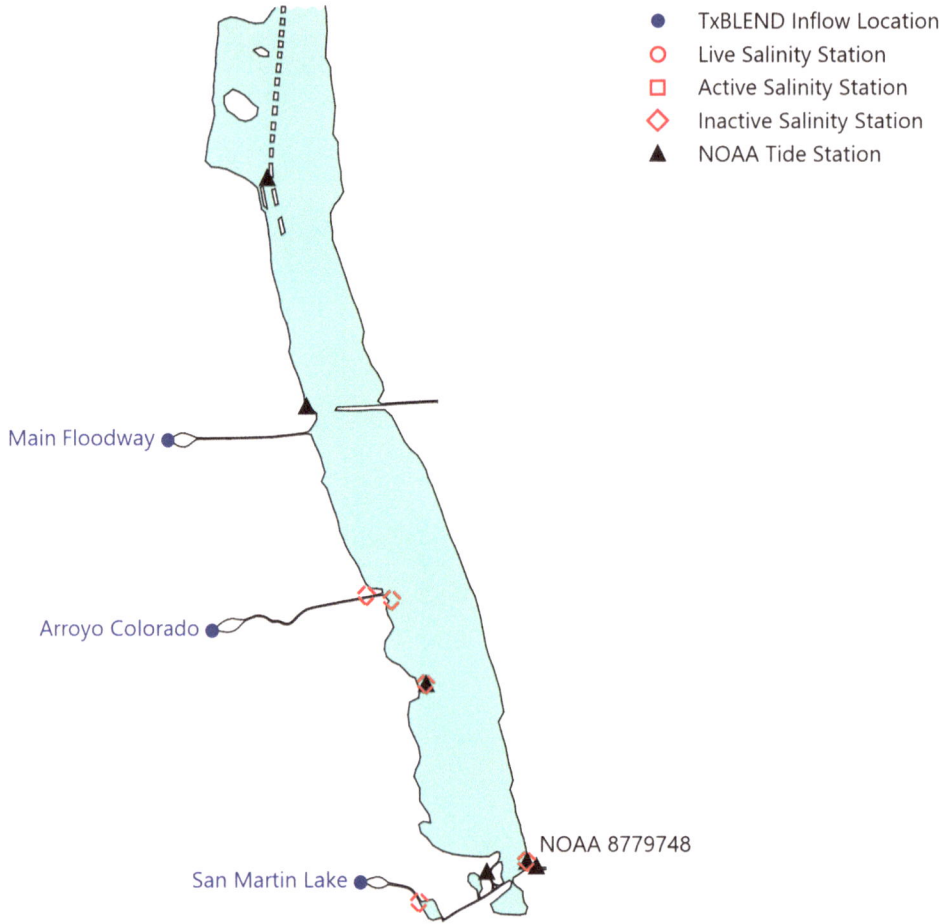

**Fig. 36** Locations of TxBLEND inflows, salinity stations, and tide stations for the Lower Laguna Madre Estuary. Colors, symbols, and labels as in Fig. 29

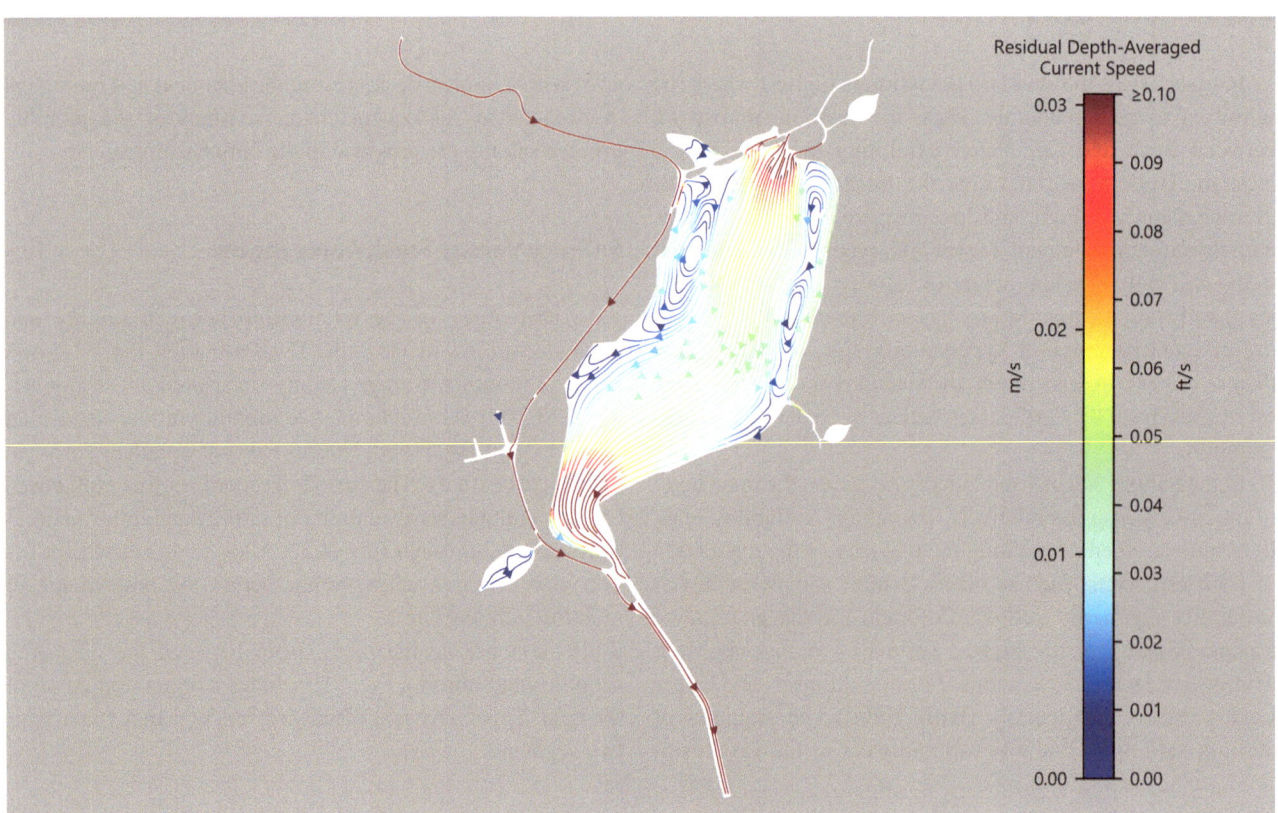

**Fig. 37** Average net velocity flowlines for the Sabine-Neches Estuary. Flowlines illustrate a typical circulation pattern, with the color of the line scaled to the average net velocity

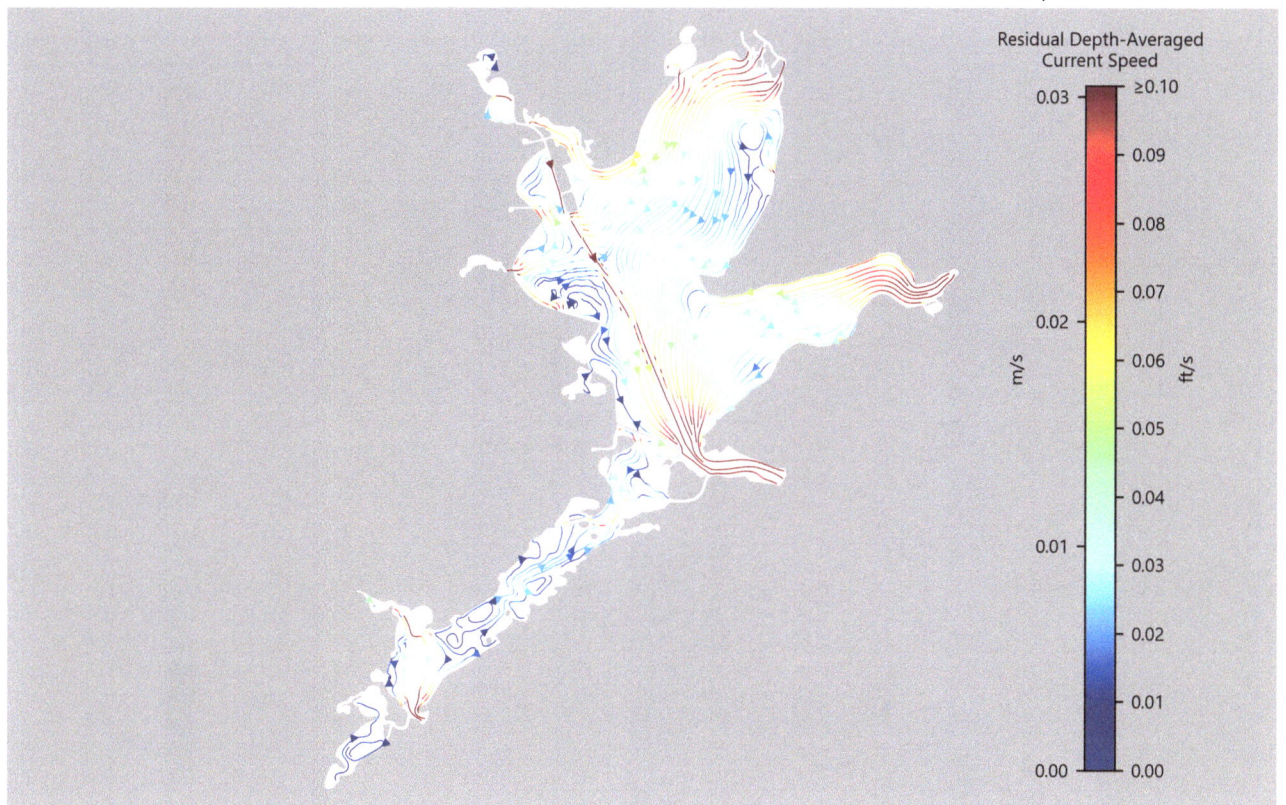

**Fig. 38** Average net velocity flowlines for the Trinity-San Jacinto Estuary. Flowlines color-scaled to the average net velocity

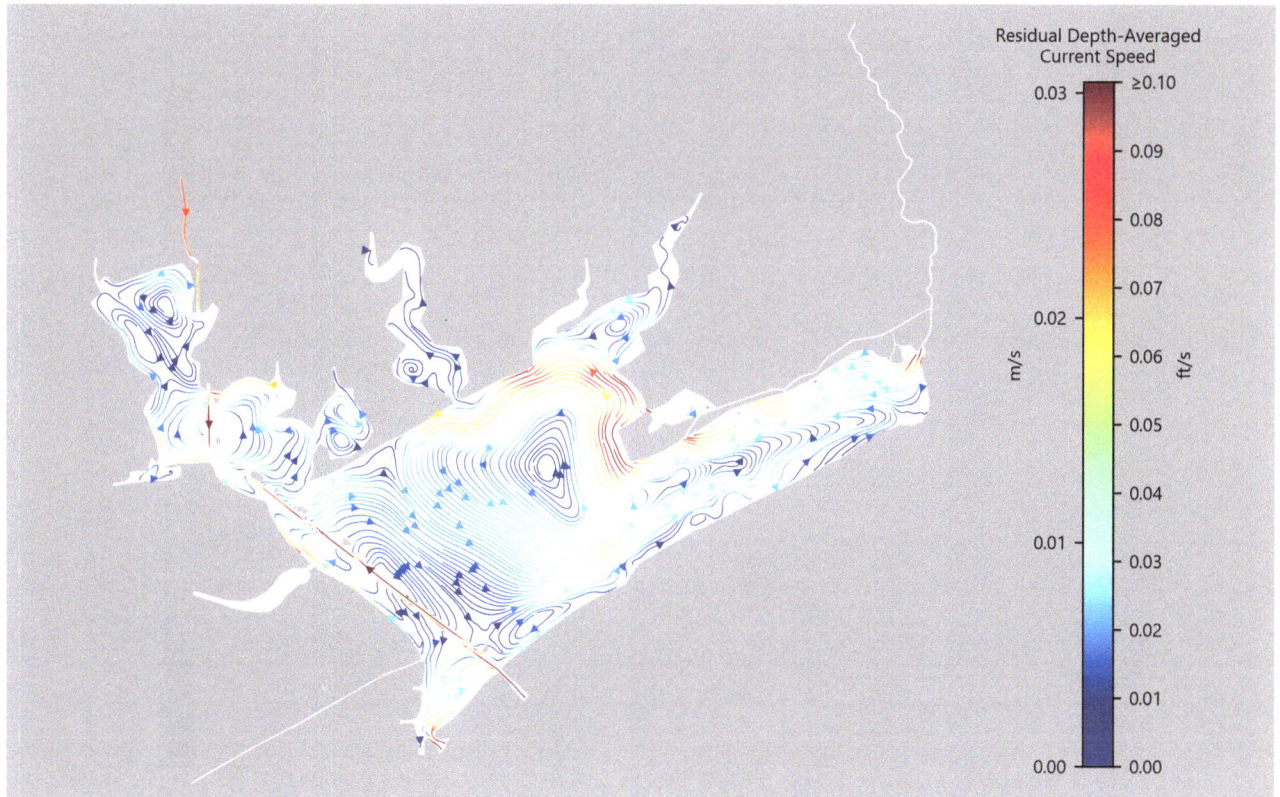

**Fig. 39** Average net velocity flowlines for the Colorado-Lavaca Estuary. Flowlines color-scaled to the average net velocity

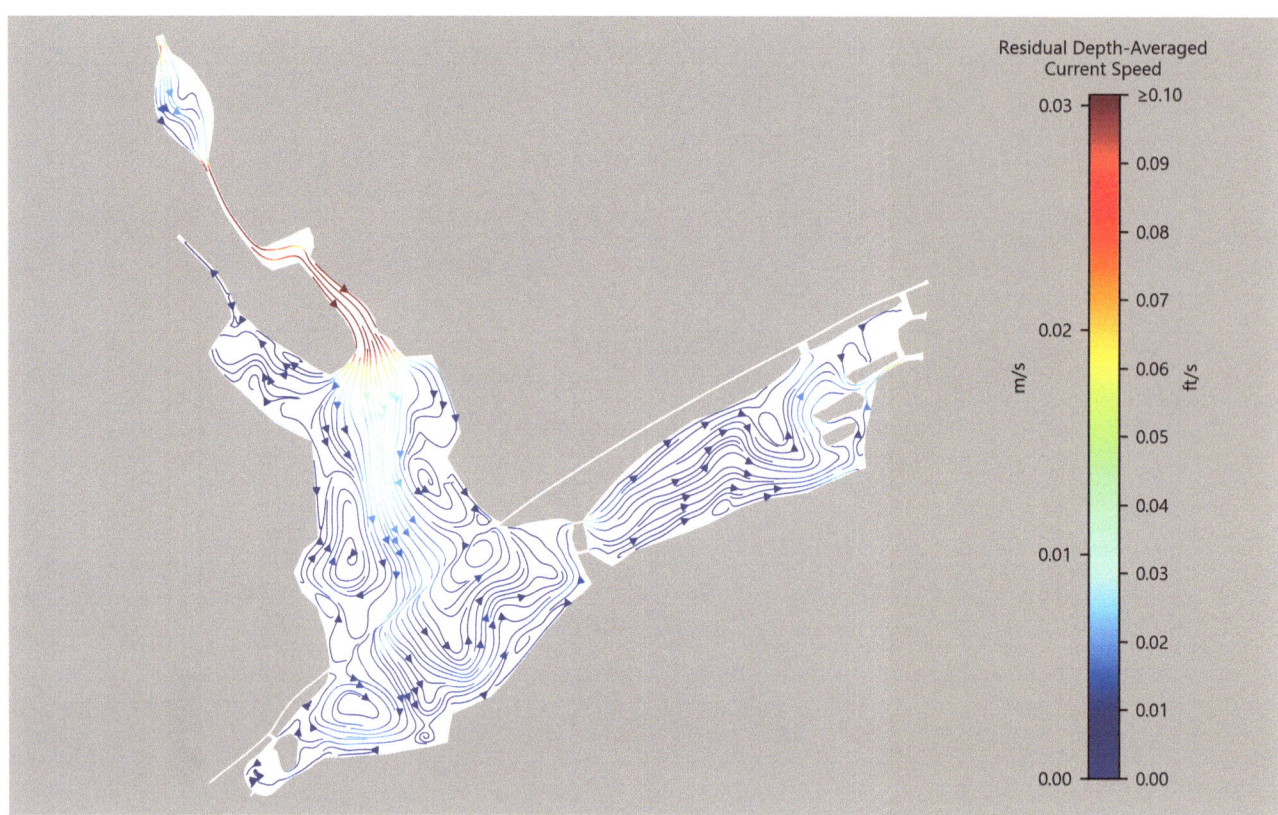

**Fig. 40** Average net velocity flowlines for the Guadalupe Estuary. Flowlines color-scaled to the average net velocity

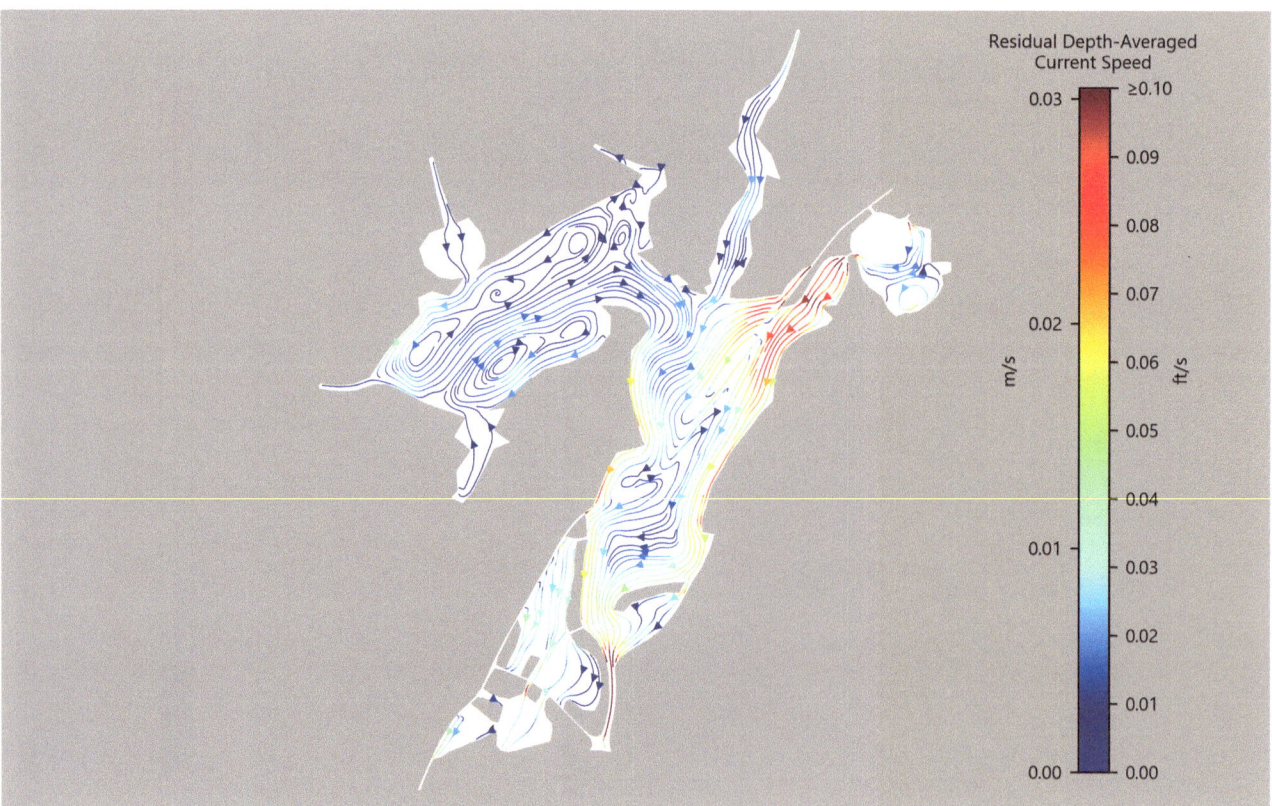

**Fig. 41** Average net velocity flowlines for the Mission-Aransas Estuary Flowlines color-scaled to the average net velocity

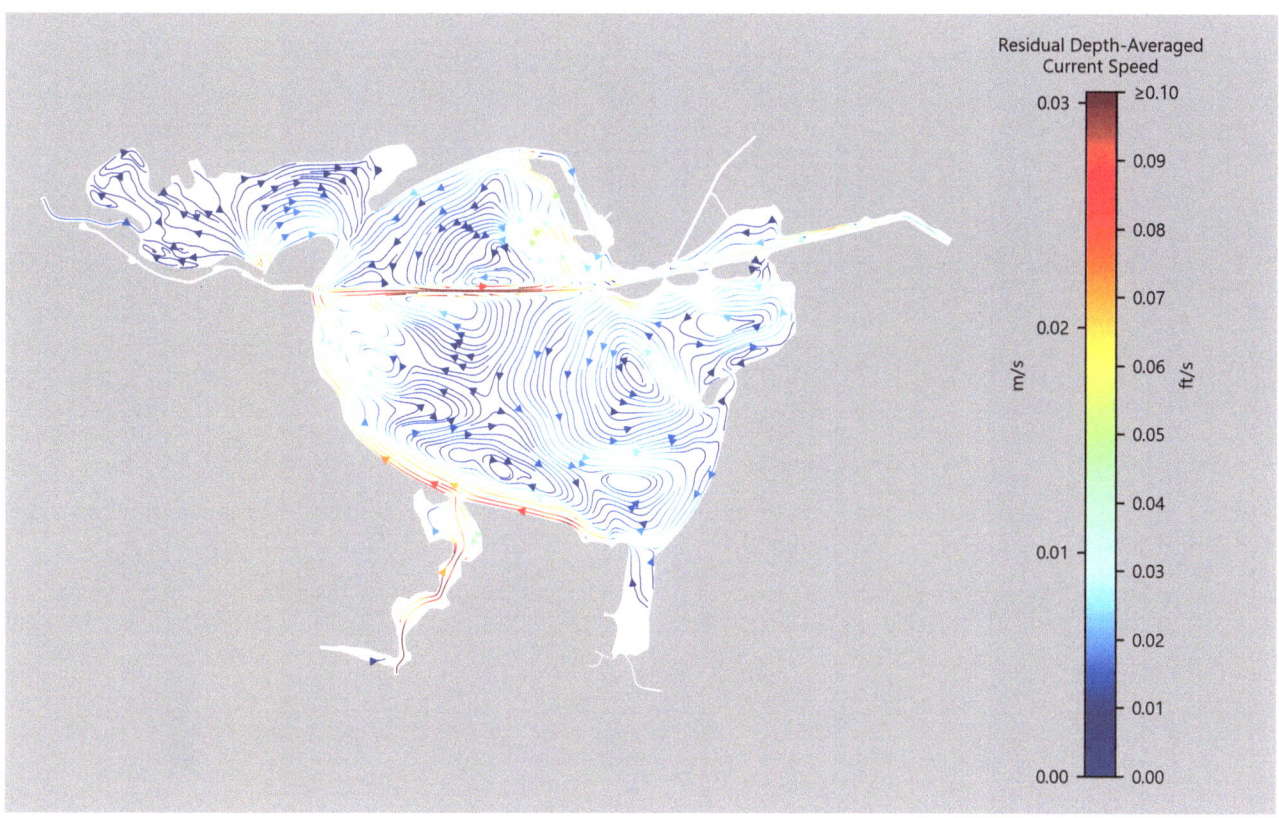

**Fig. 42** Average net velocity flowlines for the Nueces Estuary. Flowlines color-scaled to the average net velocity

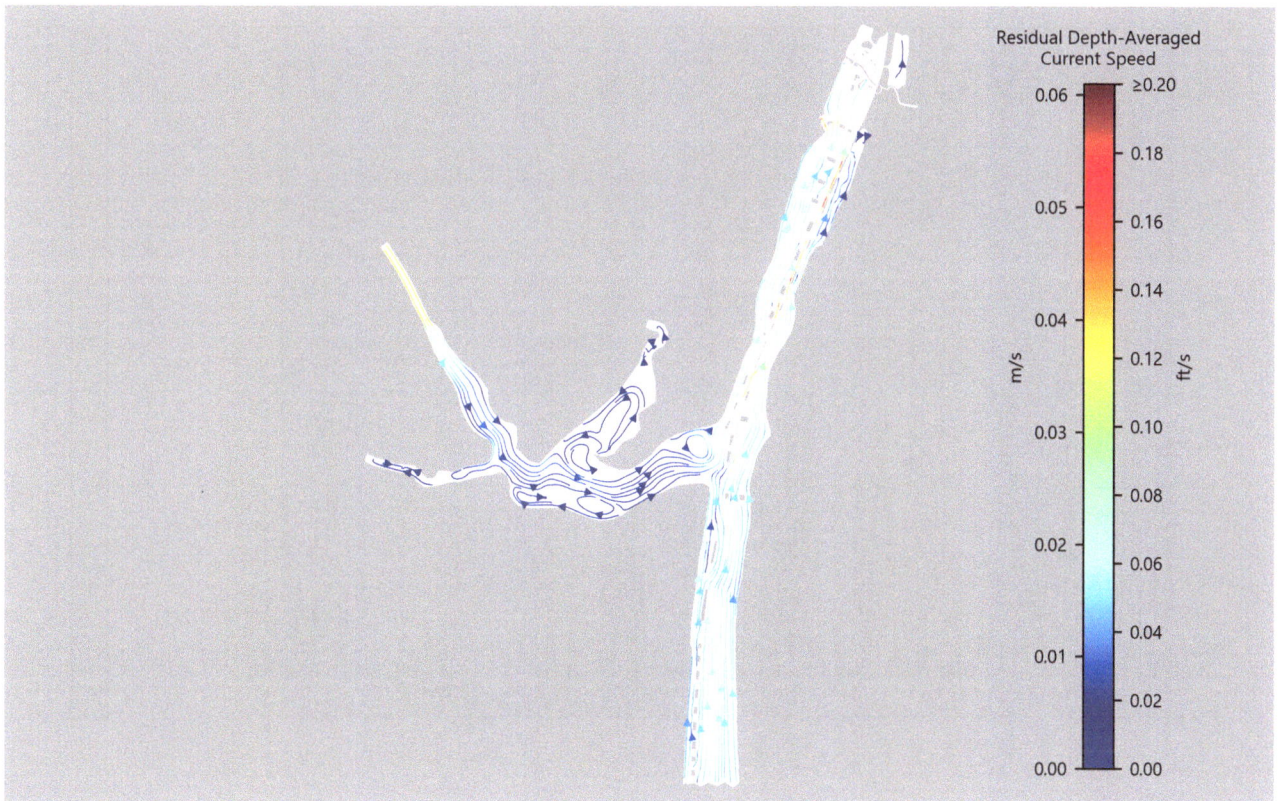

**Fig. 43** Average net velocity flowlines for the Upper Laguna Madre Estuary. Flowlines color-scaled to the average net velocity

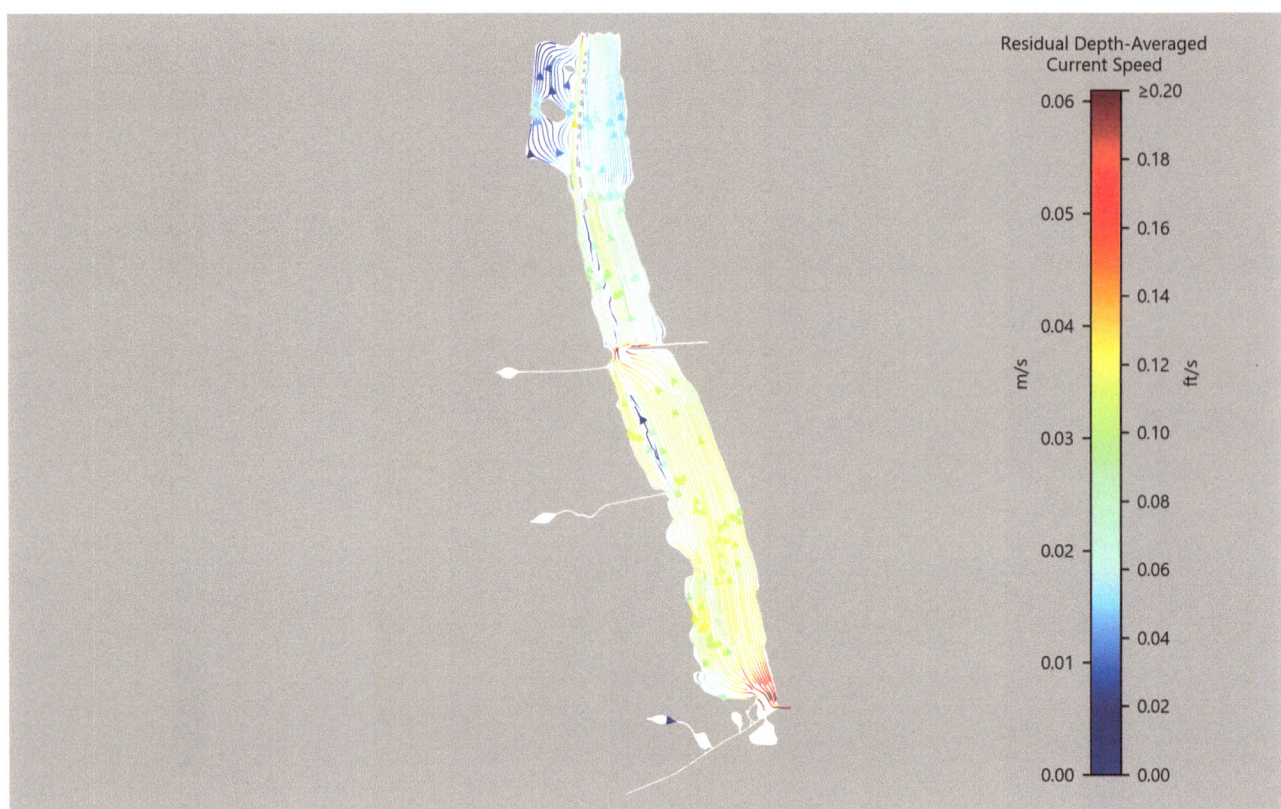

**Fig. 44** Average net velocity flowlines for the Lower Laguna Madre Estuary. Flowlines color-scaled to the average net velocity

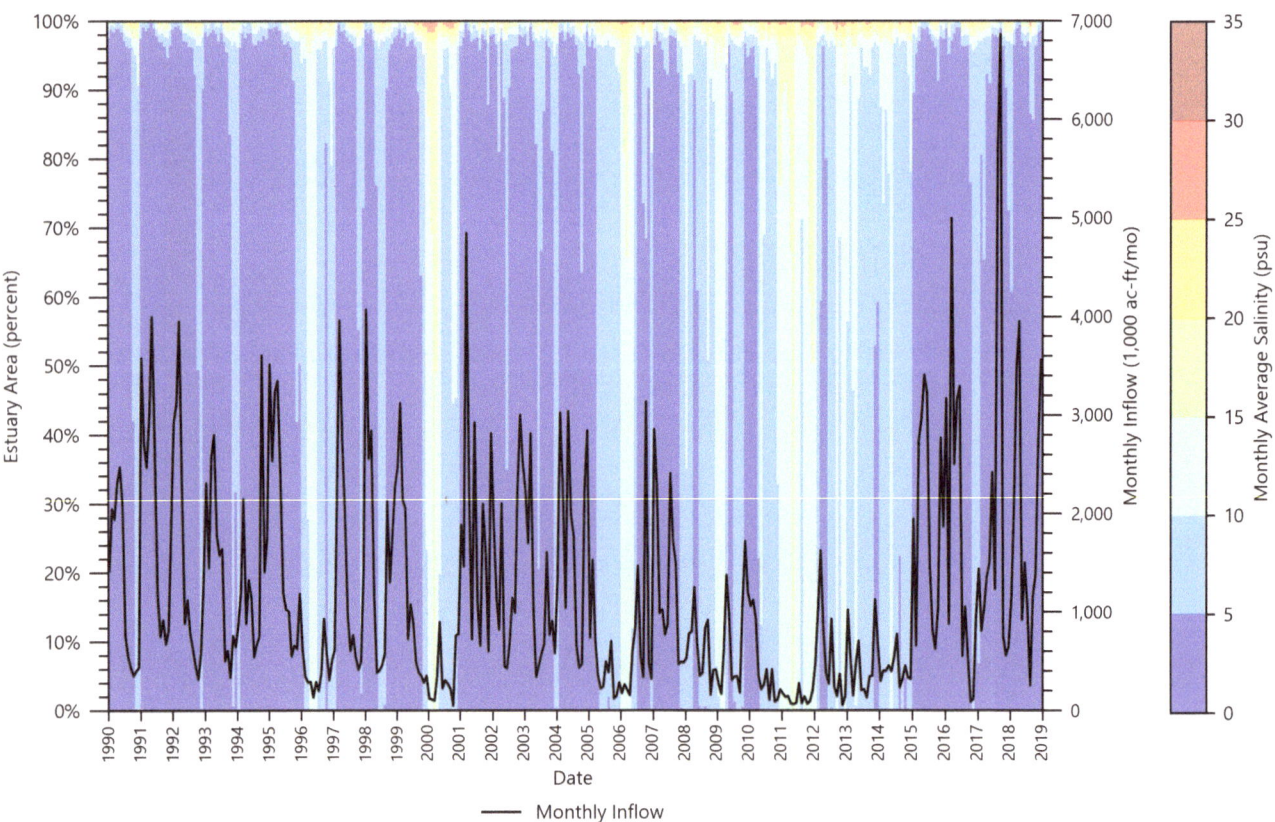

**Fig. 45** Percent of Sabine-Neches Estuary area in salinity ranges. Superimposed on the colors denoting salinity is a black line showing the monthly total inflow to the estuary

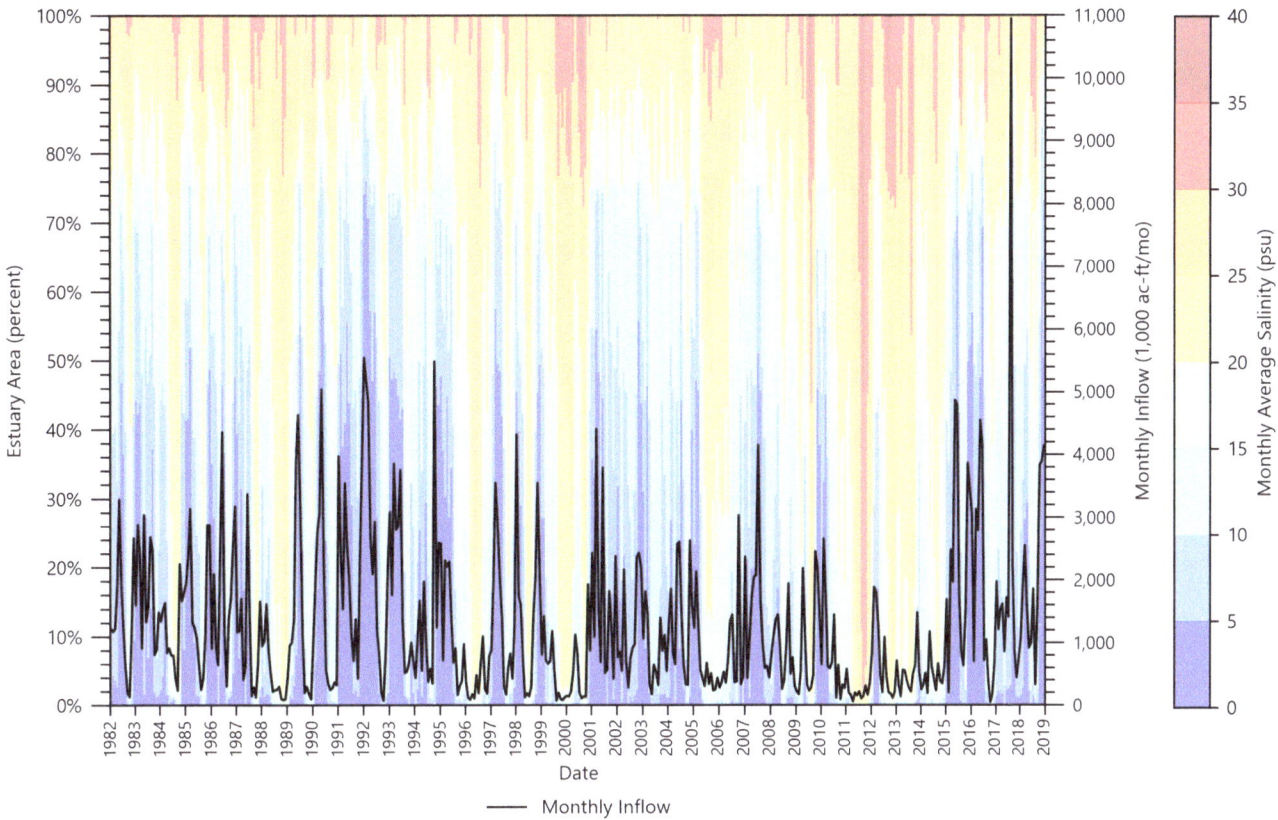

**Fig. 46** Percent of Trinity-San Jacinto Estuary area in salinity ranges. Superimposed on the colors denoting salinity is a black line showing the monthly total inflow to the estuary

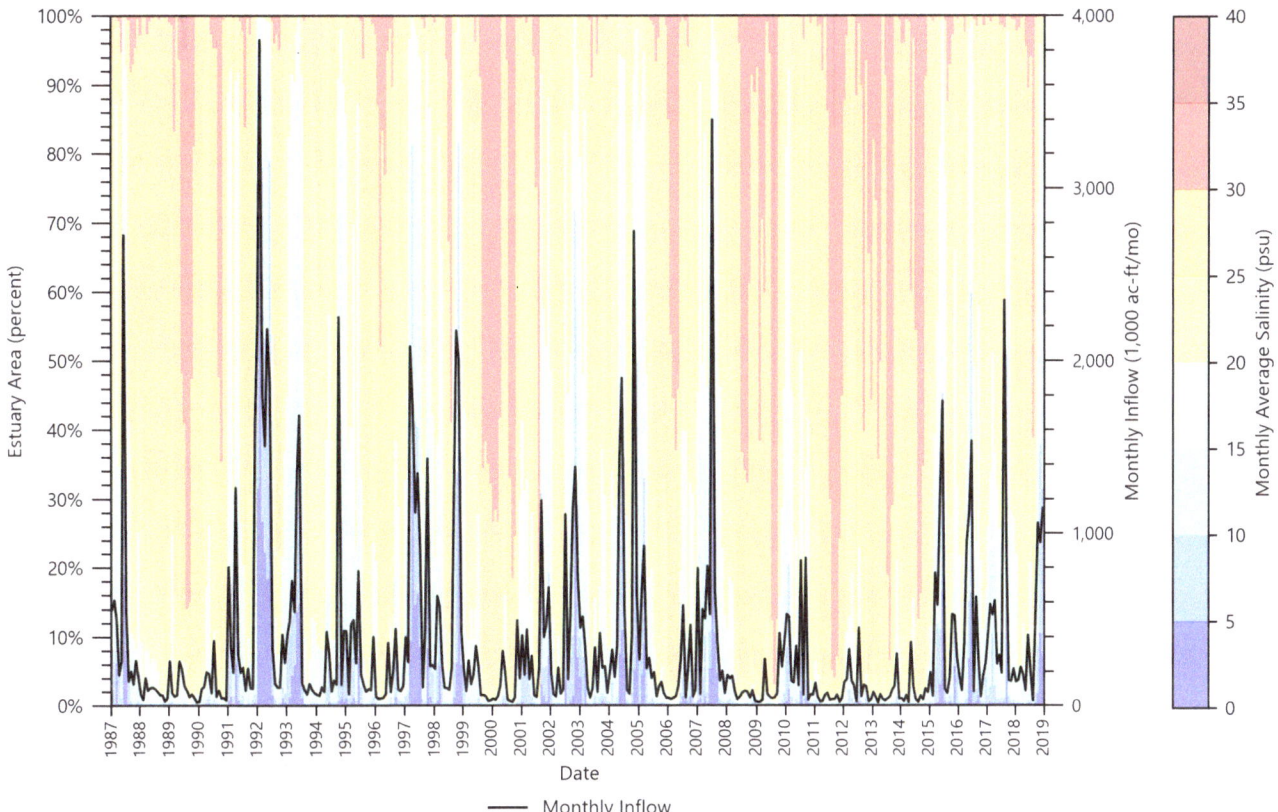

**Fig. 47** Percent of Colorado-Lavaca Estuary area in salinity ranges. Superimposed on the colors denoting salinity is a black line showing the monthly total inflow to the estuary

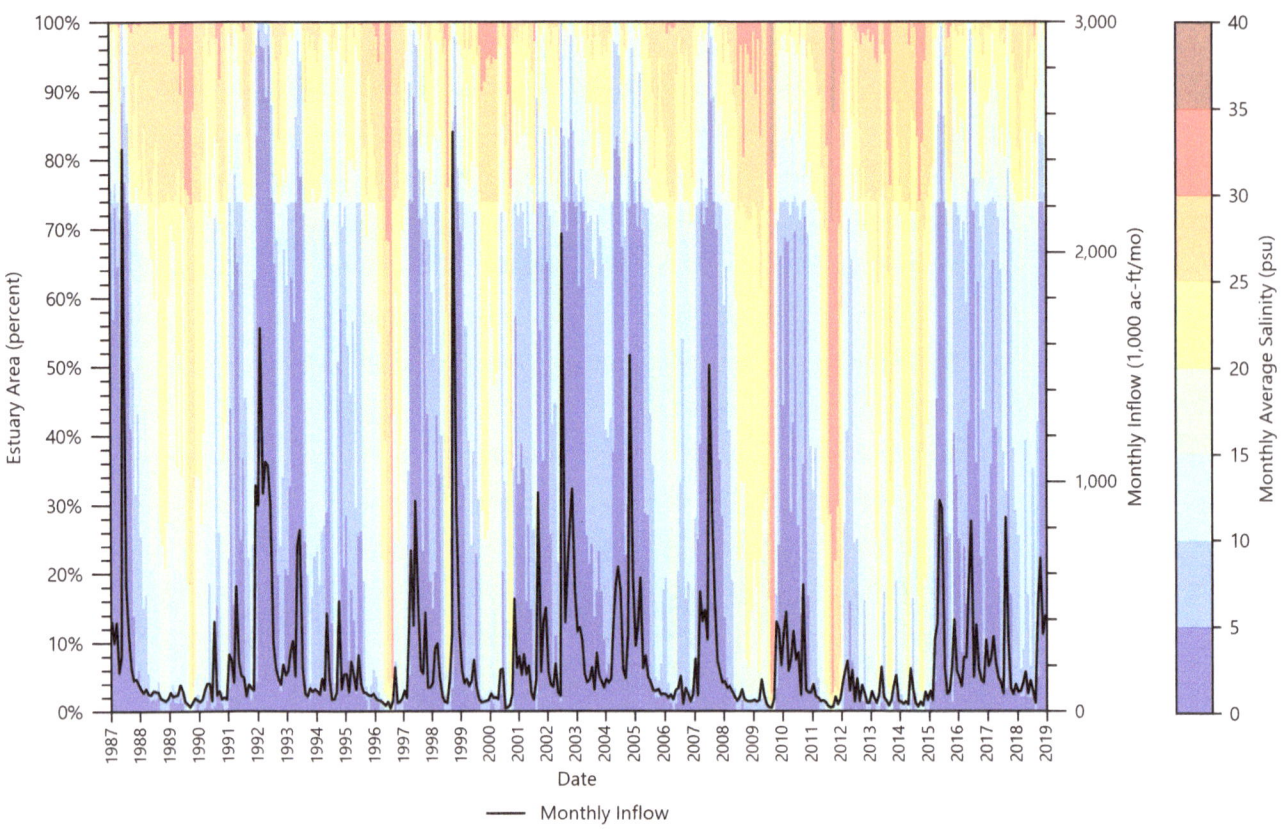

**Fig. 48** Percent of Guadalupe Estuary area in salinity ranges. Superimposed on the colors denoting salinity is a black line showing the monthly total inflow to the estuary

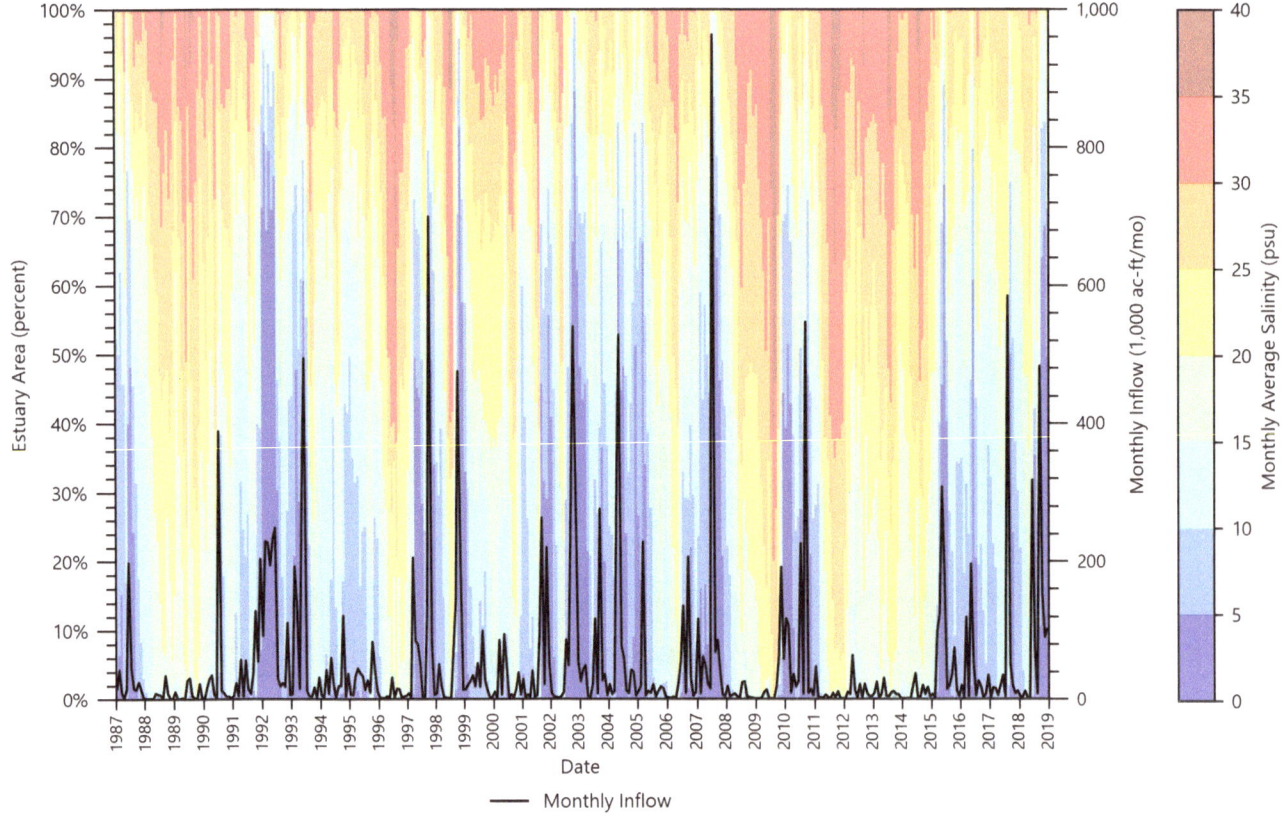

**Fig. 49** Percent of Mission-Aransas Estuary area in salinity ranges. Superimposed on the colors denoting salinity is a black line showing the monthly total inflow to the estuary

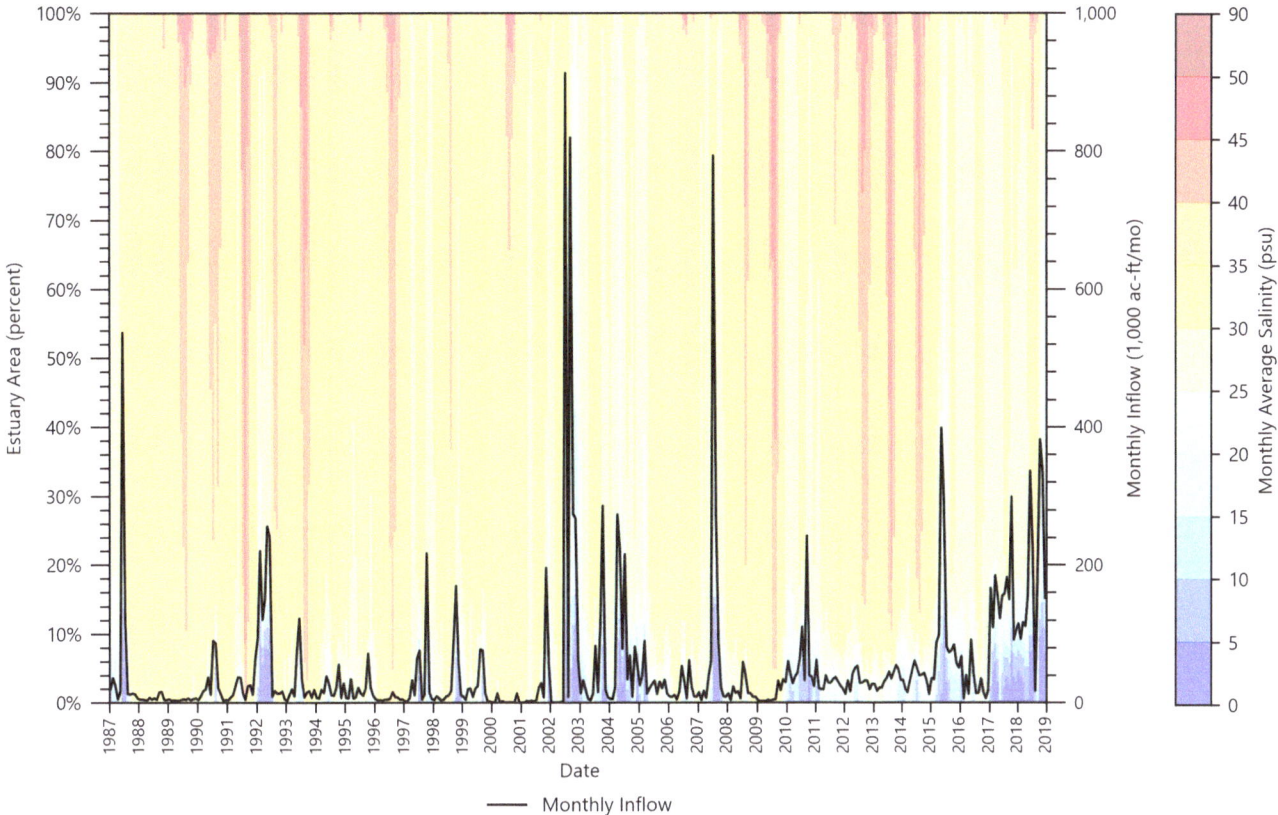

**Fig. 50** Percent of Nueces Estuary area in salinity ranges. Superimposed on the colors denoting salinity is a black line showing the monthly total inflow to the estuary. There is a small area of this estuary that approached 90 psu for a short period of time, resulting in the wide range of the scale bar

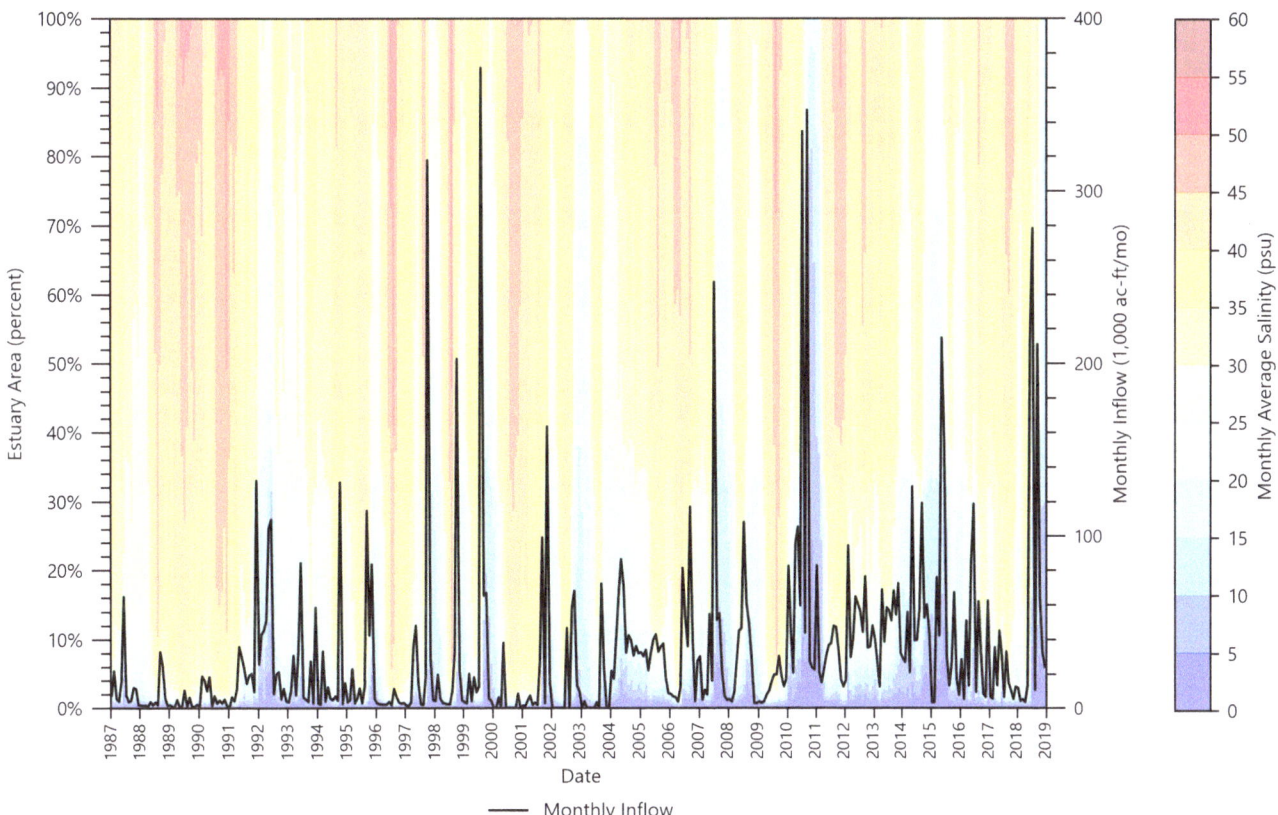

**Fig. 51** Percent of Upper Laguna Madre Estuary area in salinity ranges. Superimposed on the colors denoting salinity is a black line showing the monthly total inflow to the estuary

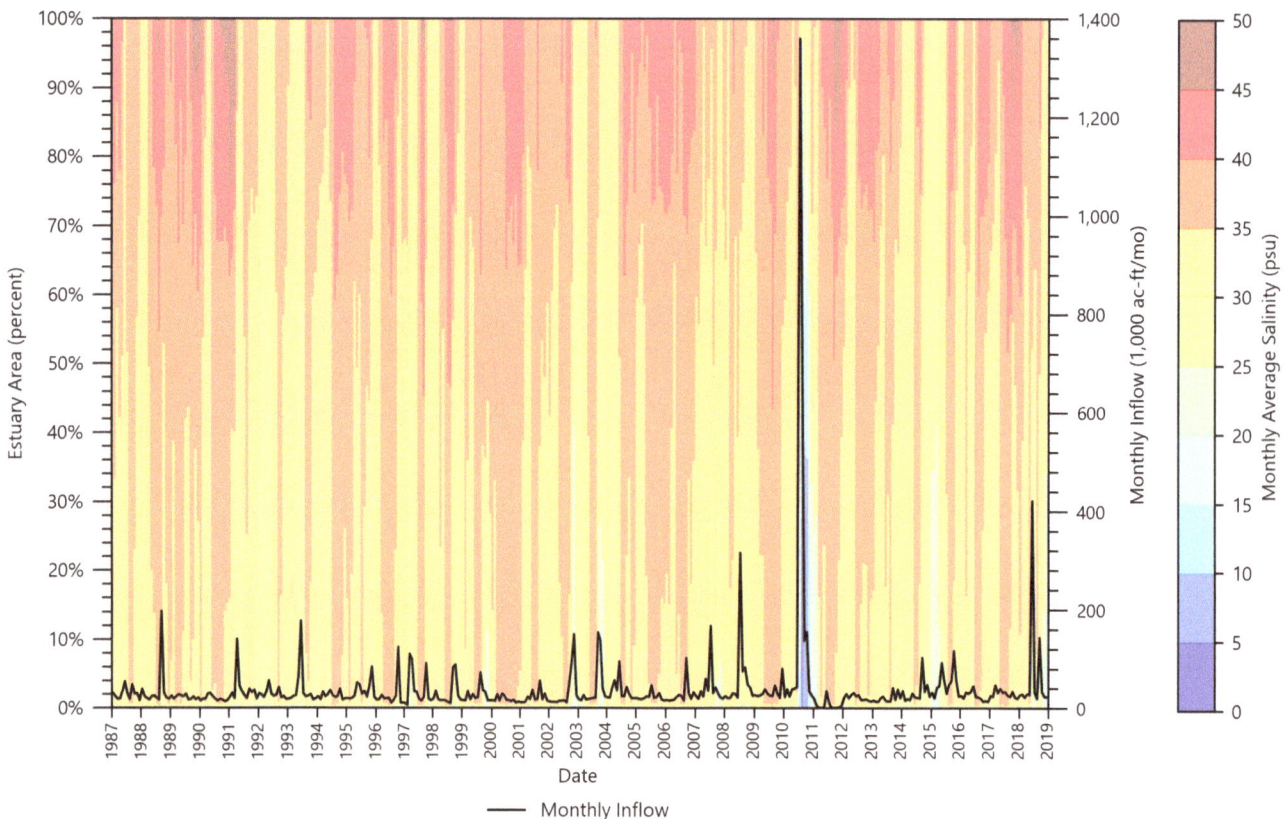

**Fig. 52** Percent of Lower Laguna Madre Estuary area in salinity ranges. Superimposed on the colors denoting salinity is a black line showing the monthly total inflow to the estuary

**Table 6** Long-term (1987–2019) average of daily average salinity values for each estuary based on TxBLEND model outputs and Texas Parks and Wildlife Department monitoring (1977–2020, average estuary-wide by month)

| Estuary | Long-term average salinity (psu) | | | |
|---|---|---|---|---|
| | TxBLEND | TPWD | | |
| | | Mean | STD | n |
| Sabine-Neches | 5 | 8 | 5 | 423 |
| Trinity-San Jacinto | 16 | 16 | 6 | 536 |
| Colorado-Lavaca | 23 | 21 | 7 | 536 |
| Guadalupe | 14 | 18 | 7 | 536 |
| Mission-Aransas | 18 | 19 | 8 | 536 |
| Nueces | 32 | 30 | 5 | 535 |
| Upper Laguna Madre | 32 | 37 | 9 | 535 |
| Lower Laguna Madre | 35 | 32 | 5 | 535 |

1. The first segment of the regression curve, comprising the upper range of inflow values, is a second-degree polynomial function fitted to the entire dataset but truncated to the range of log-inflow values between 0.3 and 6.0.
2. To prevent the regression curve from producing an unrealistic decreasing salinity trend for the lower range of inflow values, the regression curve is a constant horizontal line (at 37.3 psu) for the range of log-inflow values between −2.1 and 0.3.

Much of the unexplained variance in Fig. 62 (i.e., the residuals between the data and the regression line) is likely due to antecedent inflows. For example, if the antecedent month had a high inflow, the regression will likely overpre-

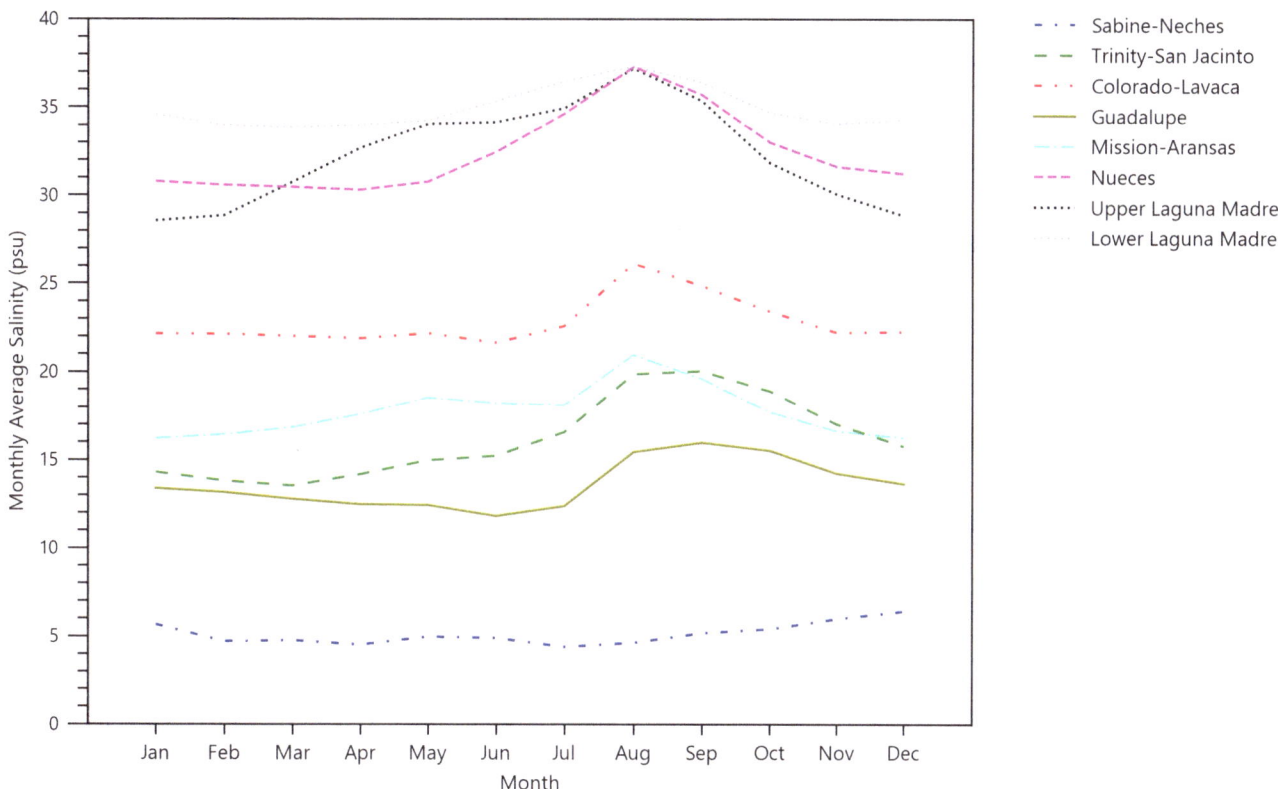

**Fig. 53** Average of monthly average salinity by month for each estuary

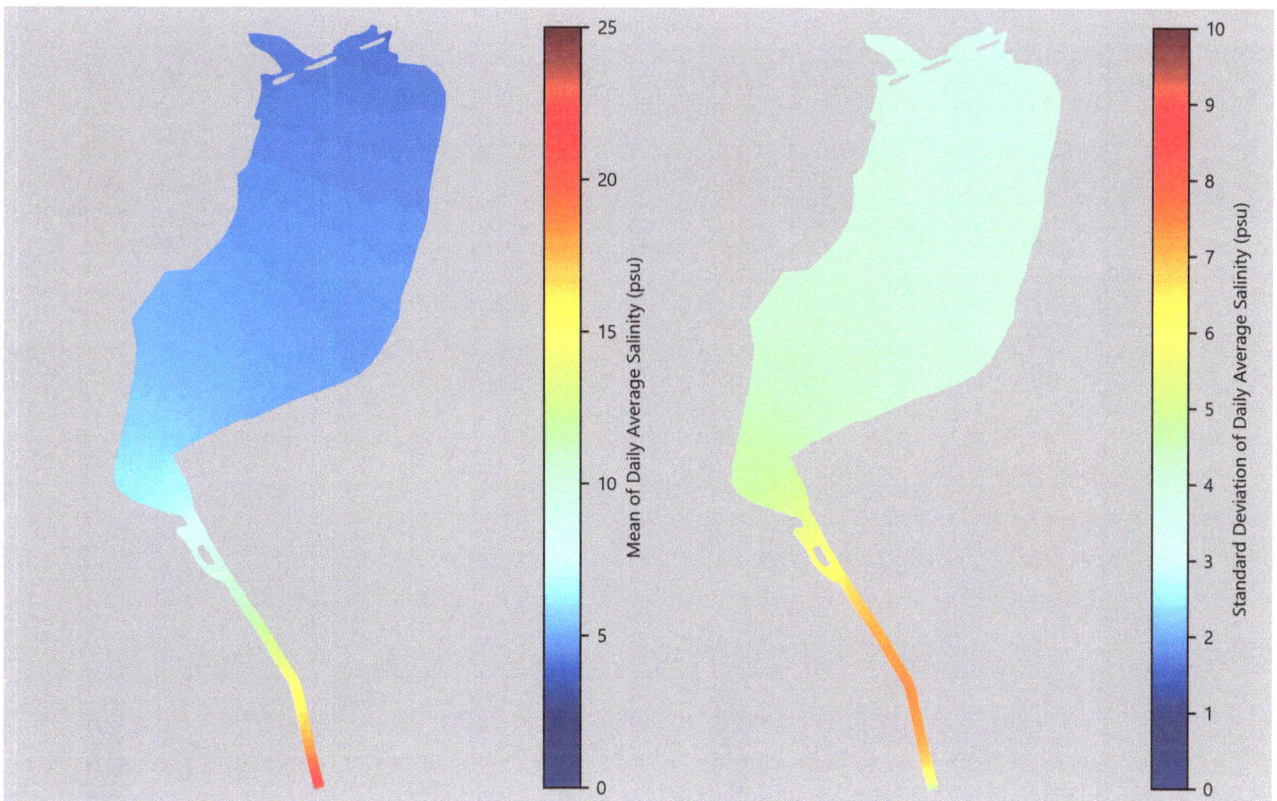

**Fig. 54** Salinity for the Sabine-Neches Estuary, the long-term average of daily average salinity values on the left, and the standard deviation of the daily average on the right

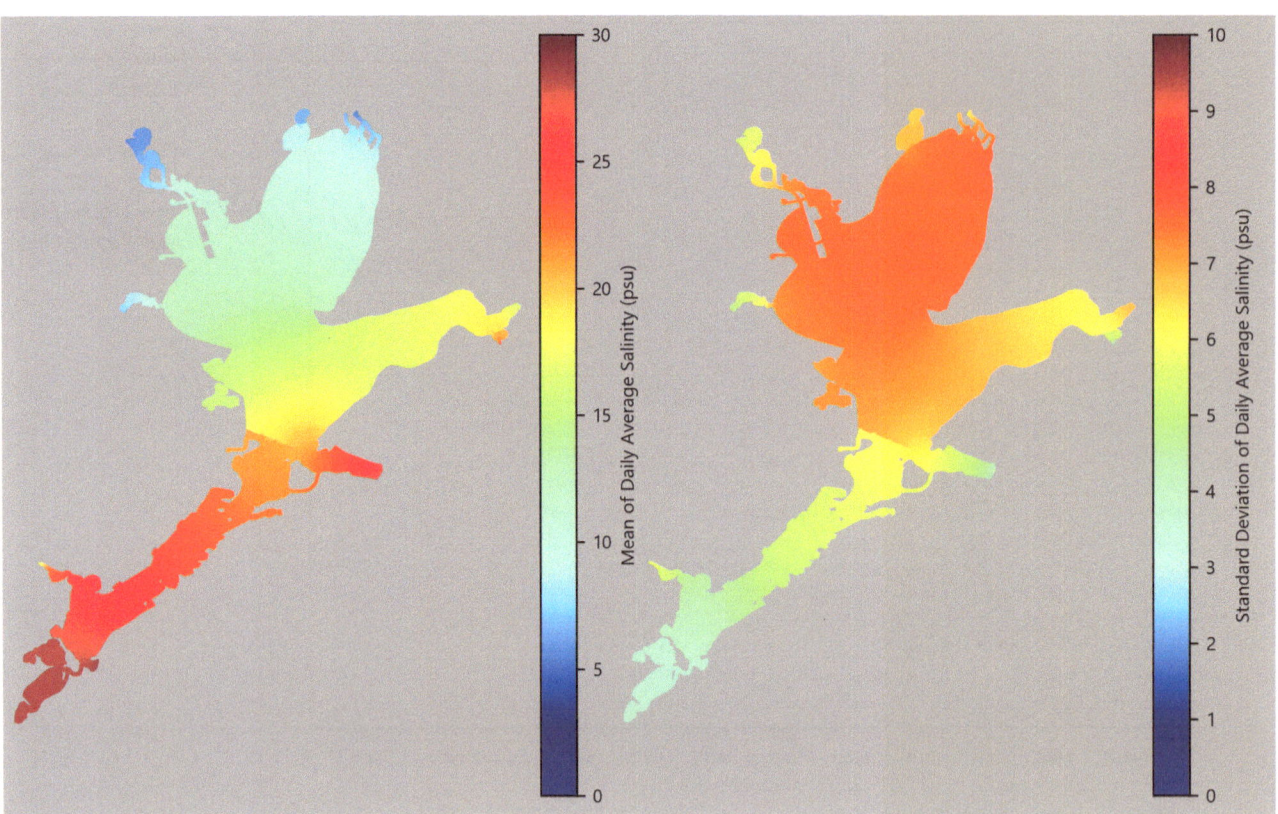

**Fig. 55** Salinity for the Trinity-San Jacinto Estuary, the long-term average of daily average salinity values on the left, and the standard deviation of the daily average on the right

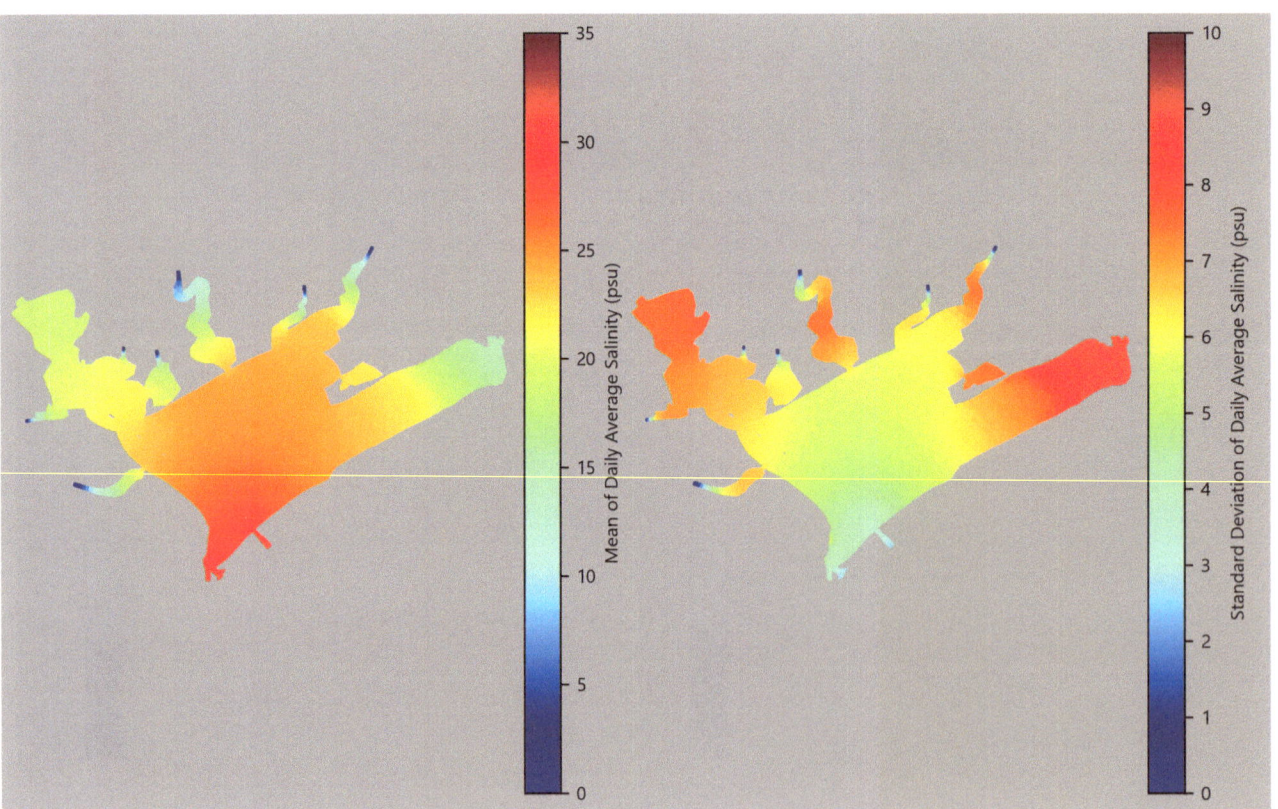

**Fig. 56** Salinity for the Colorado-Lavaca Estuary, the long-term average of daily average salinity values on the left, and the standard deviation of the daily average on the right

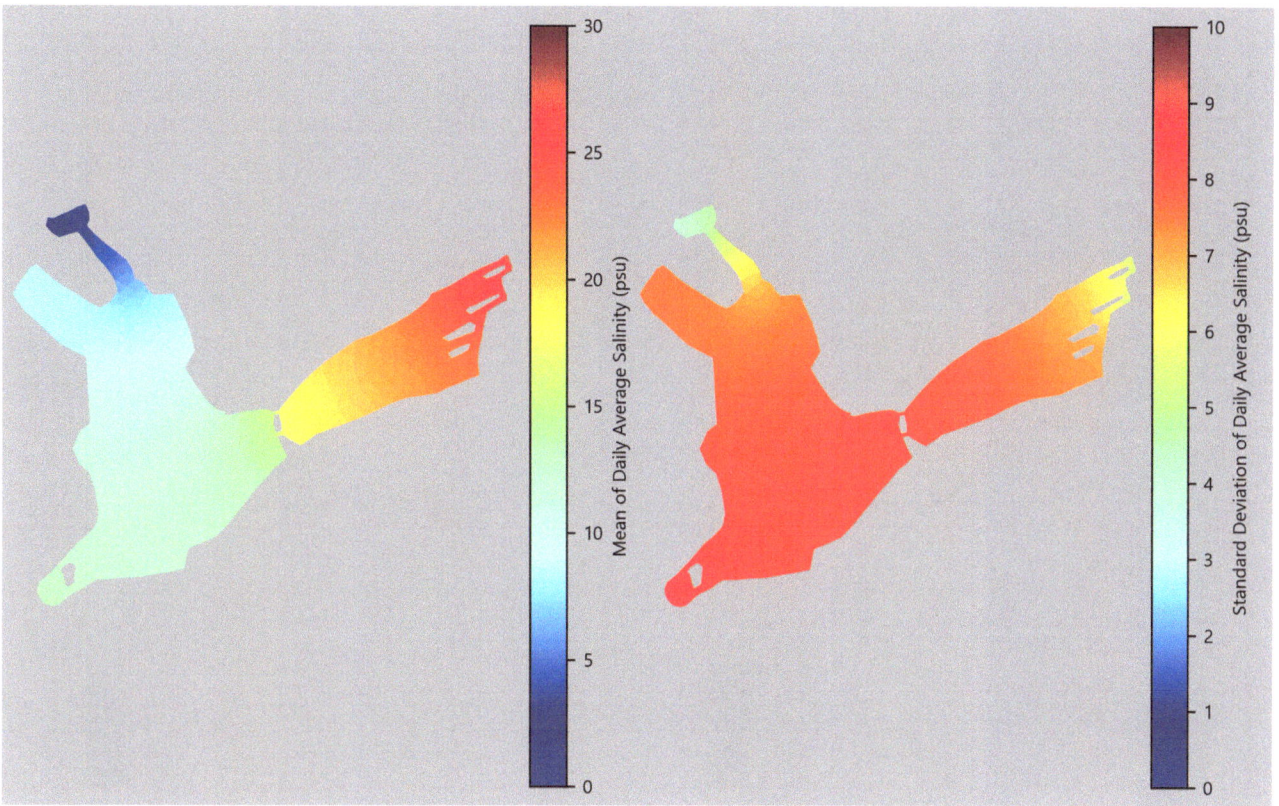

**Fig. 57** Salinity for the Guadalupe Estuary, the long-term average of daily average salinity values on the left, and the standard deviation of the daily average on the right

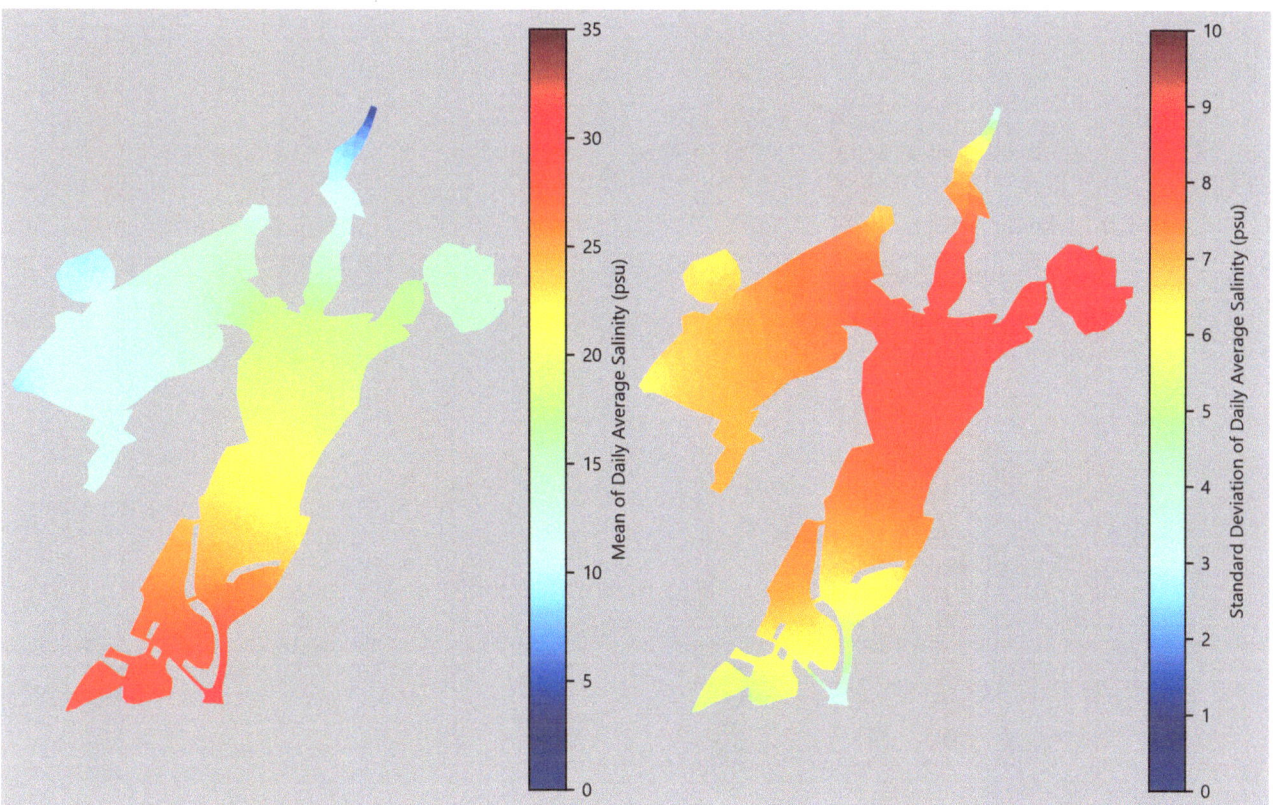

**Fig. 58** Salinity for the Mission-Aransas Estuary, the long-term average of daily average salinity values on the left, and the standard deviation of the daily average on the right

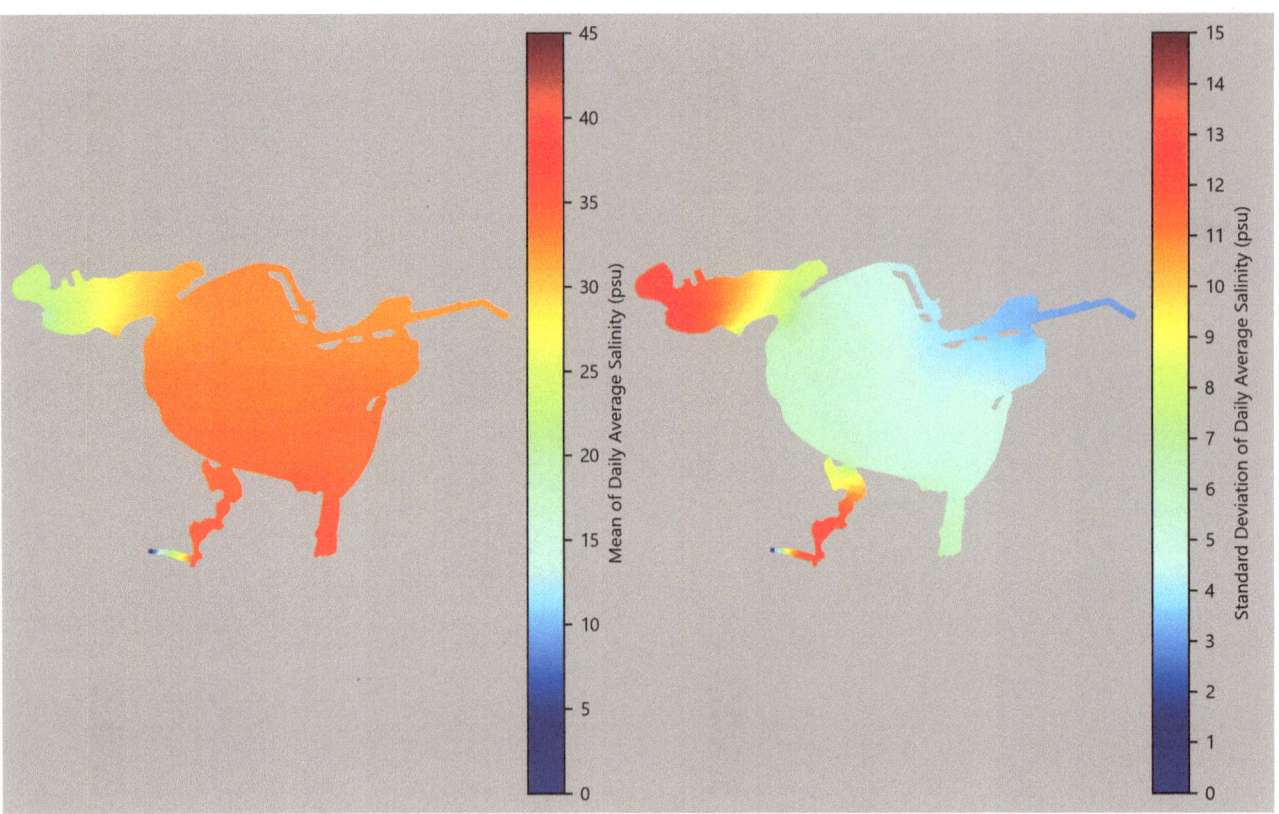

**Fig. 59** Salinity for the Nueces Estuary, the long-term average of daily average salinity values on the left, and the standard deviation of the daily average on the right

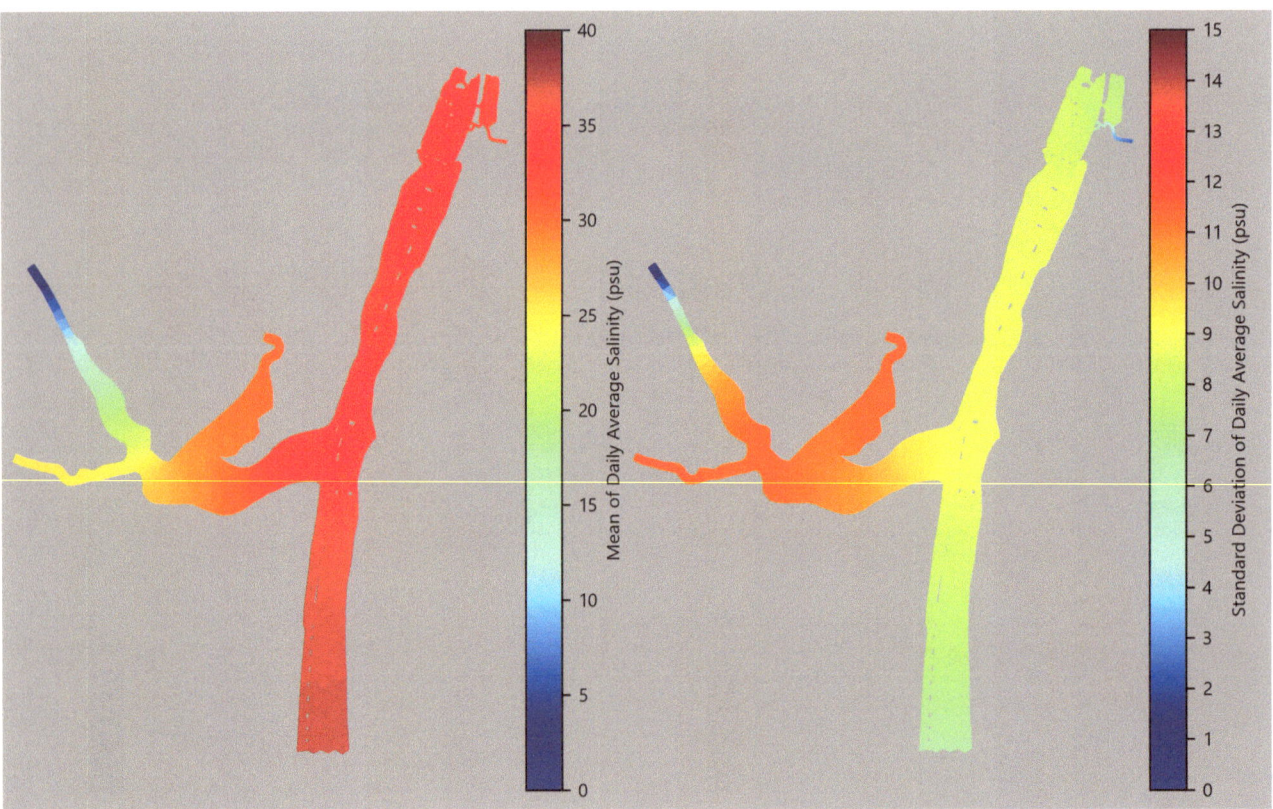

**Fig. 60** Salinity for the Upper Laguna Madre Estuary, the long-term average of daily average salinity values on the left, and the standard deviation of the daily average on the right

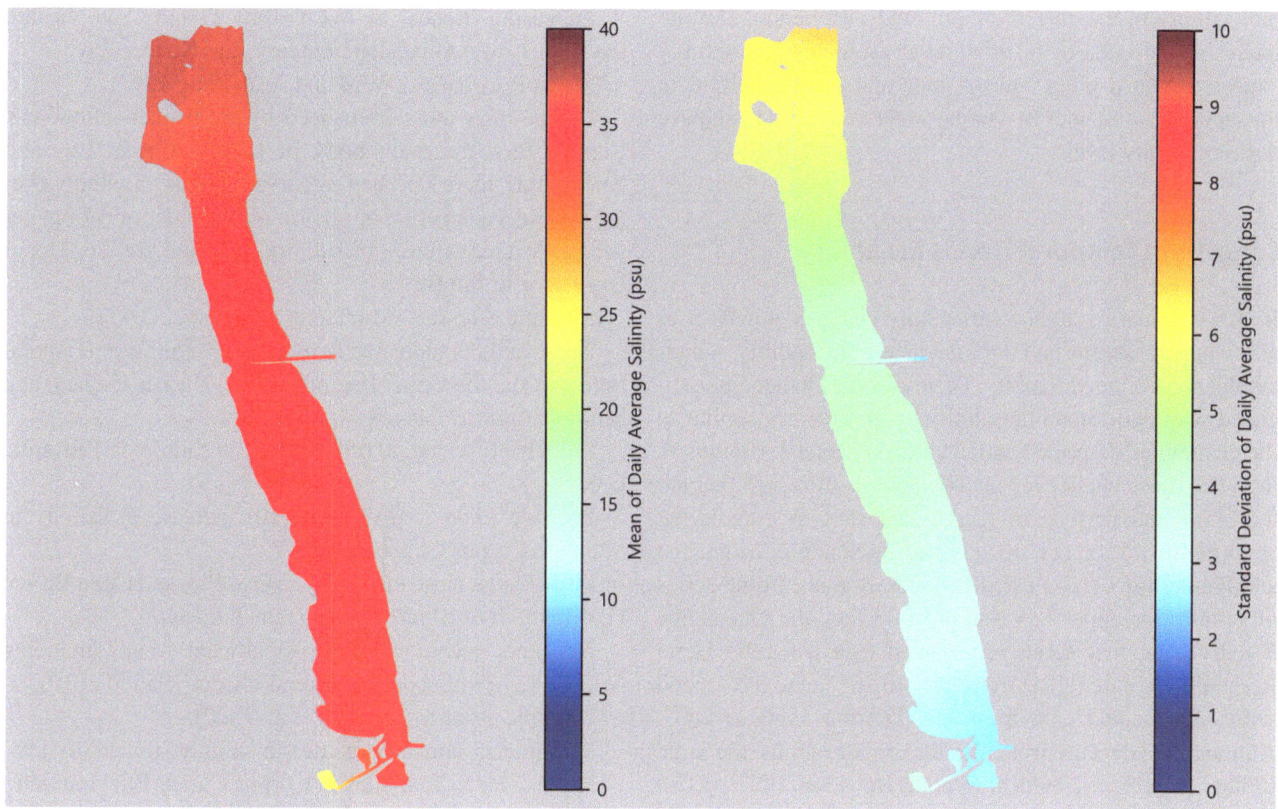

**Fig. 61** Salinity for the Lower Laguna Madre Estuary, the long-term average of daily average salinity values on the left, and the standard deviation of the daily average on the right

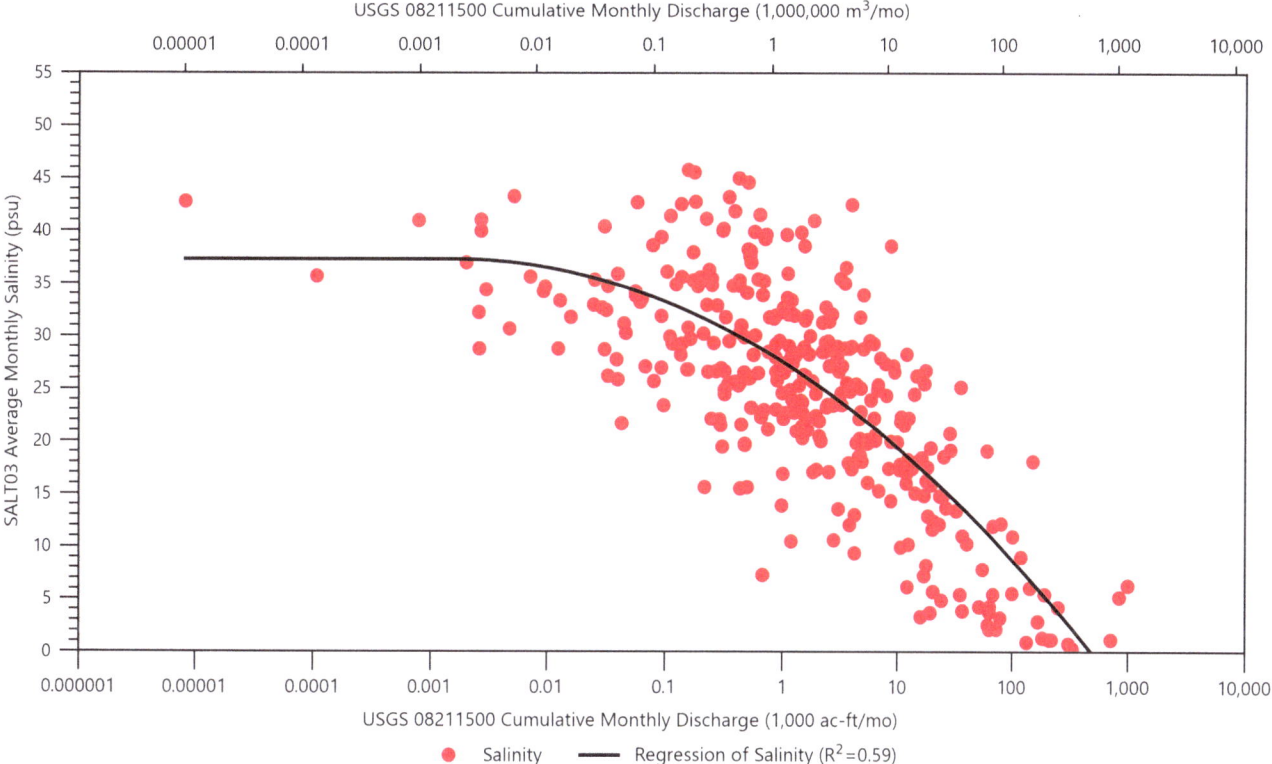

**Fig. 62** Regression curve of monthly average salinity (measured at salinity station SALT03) versus cumulative monthly discharge (measured at USGS gage 08211500, Nueces River at Calallen, Texas) in the Nueces Estuary

dict salinity in the current month (and vice versa). Despite the residual noise, Fig. 62 provides an estimate of the salinity expected with a given inflow value and can be helpful in management strategies that target water deliveries to achieve desired salinity levels.

## Long-Term Temporal Trends in Salinity

In Texas estuaries, long-term salinity changes slowly, if at all, whereas short- and intermediate-term salinity values exhibit significant variability. Therefore, the challenges associated with understanding salinity trends are very similar to the challenges described earlier in this chapter for trends in inflows. These challenges include periods of record that are too short and periods of record that start or end during abnormally wet or dry periods. As an example, Montagna and Palmer (2014) evaluated two closely-related time series of salinity for upper San Antonio Bay (i.e., the Guadalupe Estuary); the first series represented data from the Harte Research Institute (HRI) from 1987 to 2012, and the second series represented data from TPWD from 1986 to 2009. Although the data are from different organizations, the same salinity patterns are evident in both. However, 1987 was an abnormally wet year, and 2011 (into 2012) was abnormally dry. Accordingly, their statistical analysis of the HRI dataset indicated a significant increasing trend in salinity. Conversely, 1986 was an average year, and although 2009 was dry, it was not as dry as 2011 and 2012. Their statistical analysis of the TPWD data showed no significant trend in salinity. This example illustrates how conclusions may be largely dependent on climate at the beginning and end of the salinity time series.

Much like streamflows, salinity data are relatively easy to record, which often results in large sample sizes. A number of authors have evaluated salinity trends in Texas estuaries. Broad observations include the following:

Galveston Bay (TAMU 2018)

"Salinity shows no distinctive trend throughout the study time period [ostensibly 1973–2014, but the actual period varied by location]."

Guadalupe Estuary (Montagna and Palmer 2014)

Significantly increasing salinity in HRI data (1987–2012) from upper San Antonio Bay (based on p-value <0.05)

No significant trends in TPWD data (1986–2009) from upper San Antonio Bay or from HRI (1987–2012) or TPWD data in lower San Antonio Bay

Multiple estuaries (Longley 1994)

The statistical significance of salinity temporal trends was tested at each of 22 locations across six estuaries (based on p-value <0.05). Four locations were identified as significant, as follows:

Decreasing trend in lower Sabine Lake

Increasing trends in West Bay (Trinity-San Jacinto Estuary), Lower Guadalupe Estuary, and Nueces Bay

Multiple estuaries (Ward and Armstrong 1997)

"In the bays more influenced by freshwater inflow, viz. Copano Bay, the main body of Corpus Christi Bay and Nueces Bay, there has been a general increase in salinity over the three-decade period of record, on the order of 0.1 ppt per year...No clear trends in salinity emerged for the Upper Laguna or Baffin Bay"

Multiple estuaries (Montagna and Palmer 2012)

These authors indicated four areas of significant temporal trends across the Upper Laguna Madre, Nueces Estuary, and Mission-Aransas Estuary:

Probable (p-value <0.05) decrease in salinity in Petronilla Creek

Possible ($0.05 \leq$ p-value <0.10) increase in salinity in Southern Corpus Christi Bay

Probable increase in salinity in Oso Bay and Conn Brown Harbor [both of which represent small areas]

No other locations had a significant trend, including Nueces Bay (with a period of record from 1959 to 2010).

Multiple estuaries (Bugica et al. 2020)

"Significant annual increases in salinity [p-value < 0.05] were observed at four sites in Corpus Christi Bay, four sites in Nueces Bay, three sites in Lavaca Bay, and one site in each of the following locations: Matagorda Bay, Copano Bay, Aransas Bay, Upper Laguna Madre, San Antonio Bay, and Keller Bay."

However, no changes were observed at other stations in these estuaries.

No changes were observed in the Trinity-San Jacinto and Sabine-Neches estuaries.

The authors did attempt to account for climate, by comparing against El Niño–Southern Oscillation (ENSO) index, precipitation, and evaporation, and concluded that "increasing salinity trends are likely not due to natural climate variability … but instead are symptomatic of other factors such as growing human water demand in watersheds."

Multiple estuaries (Montagna et al. 2011)

Using data from 1976 to 2007, the authors conclude "there is no evidence of salinity change over time."

The above literature used a variety of data sources and statistical techniques, but all had the common interest of identifying statistically significant trends in salinity in Texas estuaries. Despite the common goal, these sources present few commonalities and more than one inconsistency. Unfortunately, this is not unexpected when trying to identify subtle shifts in noisy data that are influenced by important short- and intermediate-term shifts (such as caused by intense wet and dry climatic periods).

For the current effort, temporal trends in salinity at eight fixed datasonde locations were evaluated using linear regression. These datasets have robust periods of record (25 years or more) and few, if any, apparent quality control concerns (note that stations in the Laguna Madre were omitted from this analysis due to never reporting salinities above 50 psu and often reporting very low salinity values in the absence of rainfall). These data sondes represent fixed locations, which results in less variability than would occur if the regression was performed on data compiled from multiple locations in each estuary.

Regression figures are provided in the online supplement (Figs. 4.S.42, 4.S.43, 4.S.44, 4.S.45, 4.S.46, 4.S.47, 4.S.48, and 4.S.49), and a summary of the results is shown in Table 7. In all cases, the p-values of the temporal trends were less than 0.05 (not shown), but this apparent significance of the temporal trends is largely based on the very large number of data points. Of the eight stations tested, all eight showed increasing salinity with time. Of the eight, the strongest effect is at the SALT03 station in Nueces Bay, which is consistent with increasing water demands in this arid part of Texas.

**Table 7** Temporal trends of salinities measured at fixed locations in the Texas estuaries

| Estuary | Salinity station | Period of record | % of record complete | Mean (psu) | Standard deviation (psu) | Rate of change (psu/yr) |
|---|---|---|---|---|---|---|
| Sabine-Neches | Lower Sabine | May 1990–Jun 2022 | 59% | 13.2 | 8.0 | 0.05 |
| Trinity-San Jacinto | Mid-Trinity | Dec 1986–Aug 2022 | 54% | 10.6 | 7.4 | 0.05 |
| | Bolivar Roads | May 1990–Jun 2022 | 55% | 22.7 | 6.5 | 0.09 |
| Colorado-Lavaca | Lavaca Bay | Dec 1986–Feb 2017 | 57% | 17.7 | 9.0 | 0.11 |
| | West Bay at Tripod | Dec 1992–Jul 2018 | 79% | 19.8 | 8.4 | 0.02 |
| Guadalupe | Seadrift | Dec 1986–Mar 2016 | 50% | 13.4 | 8.6 | 0.05 |
| Mission-Aransas | Mid Aransas | Dec 1986–Oct 2000 | 40% | 24.1 | 7.8 | 0.11 |
| Nueces | SALT03 | Dec 1991–Aug 2022 | 92% | 24.5 | 10.9 | 0.28 |

This analysis suggests that salinities are increasing with time in Texas estuaries. The evidence for trends in inflows was equivocal but suggested possible increases in inflows in some areas. However, increases in temperature have been documented (Nielsen-Gammon et al. 2019), which increases evaporation. For all of these time series, longer periods of record will be helpful in updating our understanding in the coming years.

## Conclusions and Recommendations

Texas estuaries exhibit strong gradients in inflows and salinity from east to west (north to south). This provides a useful natural experiment, in which the presence and abundance of various species can be evaluated in light of each estuary's specific characteristics. The average inflow to the Sabine-Neches Estuary is nearly 20 times the inflow to the Laguna Madre, and, as a result, the average salinity is less than one-sixth of that in the Laguna Madre. Intervening estuaries are generally between these bounds, with each adjacent set exhibiting a relatively high correlation in inflows; smaller correlations exist between estuaries that are not adjacent.

Based on the magnitude of inflows, seasonal patterns in inflows, and spatial arrangement, it is convenient to assign each estuary into one of three groups, as follows:

Sabine-Neches and Trinity-San Jacinto Estuaries (East Texas)

High inflows and one seasonal peak (winter and spring), with notably lower summer and fall inflows

Colorado-Lavaca and Guadalupe Estuaries (Mid-Coast)

Moderate inflows and two seasonal peaks (spring and fall), with the fall peak being lower than the spring peak

Mission-Aransas and Nueces Estuaries and Laguna Madre (South Texas)

Low inflows and two seasonal peaks (late spring and fall), with the fall peak being higher than the spring peak

These inflow patterns create similar spatial patterns in salinity, although salinity is complicated by the estuaries having differing volumes, tidal passes, and tidal prisms. In general, the Sabine-Neches Estuary is often oligohaline, whereas the Nueces Estuary and Laguna Madre are often euhaline to hyperhaline (also referred to as hypersaline), and the intervening estuaries are mesohaline and polyhaline.

Subsequent chapters will show that although some species are found in all Texas estuaries, most species are much more prevalent in some estuaries than others. Salinity is an important driver of these gradients, but temperature, nutrients, and other factors certainly contribute and will be discussed.

This evaluation has benefitted from a rich and relatively long-term dataset of inflows and salinity coupled with consistent modeling throughout the Texas coast. However, certain limitations of the data and modeling exist, as follows:

- Temporal trends
  - Long-term temporal trends in highly variable data are difficult to discern. However, they are critically important to management of water resources because management decisions are often based on an assumption of stationarity (e.g., consistent long-term climate), and if that assumption does not hold, ecological outcomes will be different, and quite possibly worse, than expected.
- Salinity modeling
  - The periods of record for the TxBLEND models are updated rarely, and the calibrations and numerical grids are updated even more rarely. As a result, these models have not benefitted from recently collected data. In some cases, the model grids no longer accurately reflect important characteristics of the estuary (e.g., Rollover Pass in the Trinity-San Jacinto Estuary was closed in 2019, and the ship channels in many estuaries are currently being deepened and widened).
- Submarine groundwater discharge
  - A component of the freshwater inflow balance that is largely unquantified is groundwater discharge directly to each estuary. This is the subject of Chap. 5.

Based on this evaluation, we recommend the following actions:

- Continuation and expansion of long-term monitoring of inflows (primarily performed by USGS) and salinity (performed by TWDB and others)
  - Such data are critical to evaluating short-term, intermediate-term, and long-term trends. They are also invaluable to setting up and calibrating numerical models. As such, these data collection efforts should be maintained and expanded where possible.
  - All salinity data should be accompanied by a discussion of the quality control procedures followed when collecting and compiling the data. Any data that do not meet quality control requirements should be flagged or removed from the dataset, with an explanation provided.
- Periodic updating of trend analyses for both inflows and salinity.
  - Long term trend analyses are confounded by highly variable data and multi-year climate patterns, but as datasets get longer, these analyses will become more robust
  - For locations where inflows may be increasing over time, examine possible causative factors, including increases in precipitation, increases in impervious cover, and return flows from groundwater-sourced water supplies.
  - For locations where inflows may be decreasing over time, examine possible causative factors, including decreases in precipitation, increases in temperature, and human uses.
- Replacement or enhancement of modeling tools
  - TxRR and TxBLEND can no longer be considered state of the art. In particular, TxBLEND has been subject to significant criticism for years (e.g., Monismith 2005), and it is generally recognized that superior modeling platforms now exist. Unfortunately, (re) building models throughout the Texas coast requires significant resources, and although efforts are being made to accomplish this, progress is slow due to limited funding.
  - These modeling tools should be replaced with state-of-the-art tools, calibrated to recent data, and made available in more user-friendly formats. The proliferation of proposed ship channel deepening and widening projects throughout Texas provides just one example of the utility of these models. It is imperative that stakeholders have a common and confident understanding of the impacts of these projects on salinity (and subsequently, biota).
  - Updated models should be built to reflect current conditions, e.g., the opening or closure of passes, large interbasin water transfers, deepened ship channels, and other significant features.
  - Ideally, updated models would accurately depict conditions near the head of each estuary and in adjoining marshes. These are the areas that are most suitable for delivery of freshwater inflows, beneficial placement of dredged material, or other activities to restore habitat.
  - As will be described in subsequent chapters, nutrients, suspended sediment, and other constituents also influence the ecosystem of each estuary. Updated models would likely have the capability to add these components.

**Acknowledgement** This publication was funded in part through Contract No. 21-155-007-C879 from the Texas General Land Office (GLO) with Gulf of Mexico Energy Security Act of 2006 funding made available to the State of Texas and awarded under the Texas Coastal Management Program. The views contained herein are those of the authors and should not be interpreted as representing the views of the GLO or the State of Texas.

## References

Amante CJ, Love M, Carignan K, Sutherland MG, MacFerrin M, Lim E (2023) Continuously updated digital elevation models (CUDEMs) to support coastal inundation modeling. Remote Sens (Basel) 15:1702. https://doi.org/10.3390/rs15061702

Anon (1958) The Venice system for the classification of marine waters according to salinity. Limnol Oceanogr 3:346–347

Armstrong RA (2019) Is there a large sample size problem? Ophthal Physiol Optics 39:129–130

Asquith et al (1997) Status, trends, and changes in freshwater inflows to bay systems in the Corpus Christi Bay National Estuary Program Study Area. Corpus Christi Bay National Estuary Program, CCBNEP-17 September 1997. https://www.cbbep.org/publications/virtuallibrary/CC17.pdf

Attrill MJ (2002) A testable linear model for diversity trends in estuaries. J Anim Ecol 71(2):262–269

Britton JC, Morton B (1989) Shore ecology of the Gulf of Mexico. University of Texas Press, Austin, p 387

Bugica K, Sterba-Boatwright B, Wetz MS (2020) Water quality trends in Texas estuaries. Mar Pollut Bull 152:110903. https://doi.org/10.1016/j.marpolbul.2020.110903

Burnett J (2008) Flash floods in Texas. Texas A&M University Press, College Station, TX. 350 p

Canyon Lake Gorge Preservation Society (2022). History—the 2022 flood. Accessed 3 Sept 2022. https://gorgepreservationsociety.org/history

Day JW Jr, Hall CAS, Kemp WM, Yáñez-Arancibia A (1989) Estuarine ecology. Wiley

Du J, Park K, Dellapenna TM, Clay JM (2019) Dramatic hydrodynamic and sedimentary responses in Galveston Bay and adjacent inner shelf to Hurricane Harvey. Sci Total Environ 653:554–564

Engle VD, Kurts JC, Smith LM, Chancy C, Bourgeois P (2007) A classification of U.S. estuaries based on physical and hydrologic attributes. Environ Monit Assess 129:397–412. https://doi.org/10.1007/s10661-006-9372-9

Espino G, Rojas J (2004) Physical and chemical characteristics of the Gulf of Mexico. Environmental analysis of the Gulf of Mexico, vol 1. DF Harte Research Institute for Gulf of Mexico Studies Special Publication Series, Mexico, pp 41–61

Guthrie CG, Lu Q (2010) Coastal hydrology for the Guadalupe Estuary: updated hydrology with emphasis on diversion and return flow data for 2000–2009. Texas Water Development Board, Austin, TX, p 28

Guthrie CG, Matsumoto J, Lu Q (2010) TxBLEND model calibration and validation for the Guadalupe and Mission-Aransas Estuaries. Texas Water Development Board, Austin, TX. 46 p

Guthrie CG, Schoenbaechler C, Matsumoto J, Lu Q (2012) TxBLEND model calibration and validation for the Trinity-San Jacinto Estuary. Texas Water Development Board, Austin, TX. 56 p

Guthrie CG, Schoenbaechler C, Negusse S, Matsumoto J, McEwen T, Crockett D (2014) Development of a Galveston Bay TxBLEND hydrodynamic and salinity transport model through the historic drought of the 1950's. Texas Water Development Board, Austin, TX. Revised March 10, 2014, 110 p

Harwell GR, McDowell JS, Gunn CL, Garrett BS (2020) Precipitation, temperature, groundwater-level elevation, streamflow, and potential flood storage trends within the Brazos, Colorado, Big Cypress, Guadalupe, Neches, Sulphur, and Trinity River Basins in Texas Through 2017. Scientific Investigations Report 2019–5137. Version 1.1, April 2020

Hicks SD (2006) Understanding tides. US Department of Commerce, National Oceanic and Atmospheric Administration, National Ocean Service

Hoerling M, Kumar A, Dole R, Nielsen-Gammon JW, Eischeid J, Perlwitz J, Quan X-W, Zhang T, Pegion P, Chen M (2013) Anatomy of an extreme event. J Climate 26:2811–2832

Hume TM (2005) Tidal prism. In: Schwartz ML (ed) Encyclopedia of coastal science. Encyclopedia of earth science series. Springer, Dordrecht. https://doi.org/10.1007/1-4020-3880-1_320

Ingraham C (2017) Houston is experiencing its third '500-year' flood in 3 years. How is that possible? Washington Post August 29, 2017. https://www.washingtonpost.com/news/wonk/wp/2017/08/29/houston-is-experiencing-its-third-500-year-flood-in-3-years-how-is-that-possible/

Joseph JF, Falcon HE, Sharif HO (2013) Hydrologic trends and correlations in South Texas River basins: 1950-2009. J Hydrol Eng 18(12):1653–1662

Kelton E (1973) The time it never rained. Doubleday & Company, New York. 373 p

King M (2015) Point Austin: It's rainin' down in Texas: marking time and history by the Texas floods. Austin Chronicle May 29, 2015. https://www.austinchronicle.com/news/2015-05-29/point-austin-its-rainin-down-in-texas/

Koehler RB (2004) Raster-based analysis and visualization of hydrologic time-series. University of Arizona. https://repository.arizona.edu/handle/10150/280516

Longley WL (1994) Freshwater inflows to Texas bays and estuaries: ecological relationship and methods for determination of needs. Texas Water Development Board, Austin, Texas. 386 p

Lotz A (2017) Cities, counties take steps to prepare for flood season. Community Impact. March 2, 2017. https://communityimpact.com/houston/tomball-magnolia/city-county/2017/03/02/cities-counties-take-steps-prepare-flood-season/

Maidment DR (1993) Handbook of hydrology. McGraw Hill. 1424 p

Matsumoto J (1992a) User's manual for the Texas Water Development Board's rainfall-runoff model, TxRR. Texas Water Development Board, Austin, TX. 28 p

Matsumoto J (1992b) User's manual for the Texas Water Development Board's circulation and salinity model: TxBLEND. Texas Water Development Board, Austin, TX. 268 p

Matsumoto J, Powell GL, Brock DA, Paternostro C (2005) Effects of structures and practices on the circulation and salinity patterns of Galveston Bay, Texas. Prepared for the Galveston Bay Estuary Program. Texas Water Development Board, Austin, TX. 138 p

Matsumoto J, Guthrie CG, Crockett D, McEwen T (2014) TxBLEND model extension and salinity validation for the Sabine-Neches Estuary: extending simulations through 2013. Texas Water Development Board, Austin, TX. 24 p

Monismith SG (2005) Letter to TWDB regarding the suitability of the hydrodynamics/transport model (code) TxBLEND. January 29, 2005

Montagna PA (2022) Rio Grande water and sediment quality 2000–2005. Distributed by: Gulf of Mexico Research Initiative Information and Data Cooperative (GRIIDC), Harte Research Institute, Texas A&M University–Corpus Christi. https://doi.org/10.7266/xzyn67pf

Montagna PA, Brenner J, Gibeaut J, Morehead S (2011) Chapter 4: coastal impact. In: Schmandt J, North GR, Clarkson J (eds) The impact of global warming on Texas, 2nd edn, pp 96–123

Montagna PA, Kalke RD (2023) Effect of freshwater inflow on macrobenthos in minor bay and river-dominated estuaries. Distributed by: Gulf of Mexico Research Initiative Information and Data Cooperative (GRIIDC), Harte Research Institute, Texas A&M University–Corpus Christi. https://doi.org/10.7266/btehjm5n

Montagna PA, Palmer TA (2012) Water and sediment quality status and trends in the coastal bend: phase 2: data analysis. Publication CBBEP—78 August 2012. https://www.cbbep.org/publications/publication1206updated.pdf

Montagna PA, Palmer TA (2014) Report 1: water quality, benthic Macrofauna, and Epibenthic Fauna (pp. 1-20). In: Stanzel KM, Dodson JA (eds) San Antonio Bay: status and trends reports, Coastal Bend Bays & Estuaries Program Publication CBBEP-92, pp 1–241

Nielsen-Gammon J, Escobedo J, Ott C, Dedrick J, Van Fleet A (2019) Assessment of historic and future trends of extreme weather in Texas, 1900–2036. Texas A&M University, Office of the Texas State Climatologist. Document OSC-202001. March 5, 2019

Neupane R, Schoenbaechler C (2021) Coastal hydrology for the Trinity-San Jacinto Estuary. Texas Water Development Board, Austin, TX. 20 p

NOAA (National Oceanic and Atmospheric Administration) (2022). Tides & currents. Accessed July 5, 2022. https://tidesandcurrents. noaa.gov/

NOAA (2023) Continuously updated digital elevation model (CUDEM)—1/9 arc-second resolution bathymetric-topographic tiles. https://www.ncei.noaa.gov/metadata/geoportal/rest/metadata/ item/gov.noaa.ngdc.mgg.dem:999919/html. Accessed 11 Sept 2023

NPS (National Park Service) (2022) The Laguna Madre. Accessed September 9, 2022. https://www.nps.gov/pais/learn/nature/ laguna.htm#:~:text=The%20Laguna%20Madre%20of%20 Texas,hypersaline%20lagoons%20in%20the%20world

NWS (National Weather Service) (2022a) Record flooding along the Rio Grande from Hurricane Alex July 8–9, 2010. Accessed 3 Sept 2022. https://www.weather.gov/crp/20100709_Flood

NWS (National Weather Service) (2022b) Very Heavy Rains and Flooding in the Coastal Bend September 17th–21st 2010. Accessed 3 Sept 2022. https://www.weather.gov/crp/sept2010flood

Palmer TA, Montagna PA, Pollack JB, Kalke RD, DeYoe HR (2011) The role of freshwater inflow in lagoons, rivers, and bays. Hydrobiologia 667:49–67

Parker BP (2007) Tidal analysis and prediction. National Oceanic and Atmospheric Administration, National Ocean Service. https:// repository.oceanbestpractices.org/handle/11329/632

Perry C (2019) Commentary: water fuels our future. Texas Water J 10:22–23

PRISM Climate Group (2022) 30-year normals. Accessed 28 July 2022. https://prism.oregonstate.edu/normals/

Schoenbaechler C, Guthrie CG, Matsumoto J, Lu Q (2011a) TxBLEND model calibration and validation for the Laguna Madre Estuary. Texas Water Development Board, Austin, TX. 60 p

Schoenbaechler C, Guthrie CG, Lu Q (2011b) Coastal hydrology for the Lavaca-Colorado Estuary. Texas Water Development Board, Austin, TX. 18 p

Schoenbaechler C, Guthrie CG, Lu Q (2011c) Coastal hydrology for the Mission-Aransas Estuary. Texas Water Development Board, Austin, TX. 15 p

Schoenbaechler C, Guthrie CG, Lu Q (2011d) Coastal hydrology for the Nueces Estuary: hydrology for version #TWDB201101 with updates to diversion and return data for 2000–2009. Texas Water Development Board, Austin, TX. 20 p

Schoenbaechler C, Guthrie CG, Lu Q (2011e) Coastal hydrology for the Laguna Madre Estuary, with emphasis on the upper Laguna Madre. Texas Water Development Board, Austin, TX. 28 p

Schoenbaechler C, Guthrie CG, Lu Q (2011f) Coastal hydrology for the Laguna Madre Estuary, with emphasis on the lower Laguna Madre. Texas Water Development Board, Austin, TX. 29 p

Schoenbaechler C, Guthrie CG, Matsumoto J, Lu Q, Negusse S (2011g) TxBLEND model calibration and validation for the Lavaca-Colorado Estuary and East Matagorda Bay. Texas Water Development Board, Austin, TX. 72 p

Schoenbaechler C, Guthrie CG, Matsumoto J, Lu Q, Negusse S (2011h) TxBLEND model calibration and validation for the Nueces Estuary. Texas Water Development Board, Austin, TX. 72 p

Smith EP (2020) Ending reliance on statistical significance will improve environmental inference and communication. Estuar Coasts 43:1–6. https://doi.org/10.1007/s12237-019-00679-y

Smyth K, Elliott M (2016) In: Solan M, Open NWN (eds) Effects of changing salinity on the ecology of the marine environment. Chapter 9 in stressors in the marine environment. University Press. https://doi.org/10.1093/acprof:oso/9780198718826.003.0009

Sullivan GM, Feinn R (2012) Using effect size–or why the p value is not enough. J Grad Med Educ 4:279–282. https://doi.org/10.4300/ JGME-D-12-00156.1

TAMU (Texas A&M University) (2018) Eye of the storm: report of the Governor's Commission to Rebuild Texas. November 2018

TDWR (Texas Department of Water Resources) (1974) Rice irrigation return-flow study. Texas Water Development Board, Austin, TX. 135 p

TDWR (1975) Inventories of irrigation in Texas 1958, 1964, 1969, and 1974. Report 196. Texas Water Development Board, Austin, TX. 265 p

TDWR (1980a) Lavaca-Tres Palacios Estuary: a study of the influence of freshwater inflows. Texas Department of Water Resources LP-106. Texas Department of Water Resources, Austin, TX

TDWR (1980b) Guadalupe Estuary: a study of the influence of freshwater inflows. Texas Department of Water Resources LP-107. Texas Department of Water Resources, Austin, TX

TDWR (1981a) Nueces and Mission-Aransas Estuaries: a study of the influence of freshwater inflows. Texas Department of Water Resources LP-108. Texas Department of Water Resources, Austin, TX

TDWR (1981b) Trinity-San Jacinto Estuary: a study of the influence of freshwater inflows. Texas Department of Water Resources LP-113. Texas Department of Water Resources, Austin, TX

TDWR (1981c) Sabine-Neches Estuary: a study of the influence of freshwater inflows. Texas Department of Water Resources LP-116. Texas Department of Water Resources, Austin, TX

TDWR (1983) Laguna Madre Estuary: a study of the influence of freshwater inflows. Texas Department of Water Resources LP-182. Texas Department of Water Resources, Austin, TX

Thurman HV, Burton EA (2001) Introduction to oceanography, 9th edn. Prentice Hall

Tompkins S (2010) Lower Laguna Madre's salinity lowered by storm runoff. Houston Chronicle. September 12, 2010. https://www. chron.com/sports/outdoors/article/Lower-Laguna-Madre-s-salinity-lowered-by-storm-1699418.php

Trungale J, Hoffmann J, Montagna PA, Opdyke D (2023) Freshwater inflows to Texas bays and estuaries: hydrology, circulation, and salinity data (1977-2018). Distributed by: Gulf of Mexico Research Initiative Information and Data Cooperative (GRIIDC), Harte Research Institute, Texas A&M University–Corpus Christi https:// doi.org/10.7266/8KR3TNDT

TWDB (Texas Water Development Board) (2019) State flood assessment. Report to the Legislature, 86th Legislative Session. January 2019. 54 p

TWDB (2022a) Lake Travis (Colorado River Basin). Accessed 9 Sept 2022. Available at: https://www.twdb.texas.gov/surfacewater/rivers/ reservoirs/travis/index.asp

TWDB (2022b) Coastal hydrology. Accessed 26 Aug 2022. Available at: https://www.twdb.texas.gov/surfacewater/bays/coastal_hydrology/index.asp

TWDB (2022c) Estuary circulation & salinity models. Accessed 4 Sept 2022. Available at: https://www.twdb.texas.gov/surfacewater/bays/ models/index.asp

TWDB (2022d) Lake evaporation and precipitation. Accessed 4 Sept 2022. Available at https://waterdatafortexas.org/ lake-evaporation-rainfall

TWDB (2022e) Texas bays & estuaries continuous water quality monitoring stations. Accessed 17 Aug 2022. Available at https://www. waterdatafortexas.org/coastal

Tweedley JR, Dittmann SR, Whitfield AK, Withers K, Hoeksema SD, Potter IC (2019) Hypersalinity: global distribution, causes, and present and future effects on the biota of estuaries and lagoons. In: Coasts and estuaries: the future. Elsevier, pp 523–546. https://doi. org/10.1016/B978-0-12-814003-1.00030-7

U.S. Drought Monitor (2023) Accessed 25 Aug 2023. https://drought-monitor.unl.edu/

USGS (U.S. Geological Survey) (2022) USGS Water Data for Texas. Accessed September 2, 2022. Available at: https://waterdata.usgs. gov/tx/nwis/

Ward G (1997) Processes and trends of circulation within the Corpus Christi Bay National Estuary Program Study Area. Report CCBNEP-21, Corpus Christi Bay National Estuary Program, Corpus Christi, TX. https://www.cbbep.org/publications/CCBNEP21.pdf

Ward GH (2005) Texas water at the Century's turn—perspectives, reflections and a comfort bag. In: Norwine J, Giardino J, Krishnamurthy S (eds) Water for Texas. Texas A&M University Press, College Station, pp 17–43

Ward GH, Armstrong NE (1997) Current status and historical trends of ambient water, sediment, Fish and Shellfish Tissue Quality in the Corpus Christi Bay National Estuary Program Study Area. Publication CCBNEP-13

Wasson M (2021) Why Texas is called flash flood alley. Spectrum News 1. https://spectrumlocalnews.com/tx/south-texas-el-paso/weather/2021/04/26/texas-is-called-flash-flood-alley

Wurbs RA (2021) Monthly river flows in Texas for natural and developed conditions. Water Cycle 2:1–14. https://doi.org/10.1016/j.watcyc.2020.10.001

Wurbs RA, Yang M-Y (2022) Statistical assessments of river flow alterations and environmental flow standards. J Water Manag Model. https://doi.org/10.14796/JWMM.C481

Wurbs RW, Zhang Y (2014) River system hydrology in Texas. Texas Water Resources Institute TR-461. Texas A&M University. August 2014

# Groundwater-Surface Water Interactions in the Coastal Zone

Audrey R. Douglas ⓘ and Dorina Murgulet ⓘ

## Abstract

Groundwater-surface water interactions are an important process in coastal and estuarine environments that influence water budgets and bring bioactive solutes, such as nutrients, gases, and trace metals, yet are often overlooked. Submarine groundwater discharge (SGD), or the exchange of water between coastal sediments and surface waters, is a critical component of the global hydrologic and biogeochemical cycles that link terrestrial waters to marine environments. Fresh SGD, which may account for up to 10% of total freshwater inflows to the ocean globally, is a source of new nutrients to a system, whereas saline SGD is considered a source of recycled nutrients from sediments. Like surface water estuaries, within the coastal aquifer terrestrial and marine waters meet in a subterranean estuary where groundwater from land drainage is measurably diluted by seawater recirculating through the aquifer and altered to become biogeochemically distinct. SGD has been quantified in three Texas estuaries with Nueces Estuary consistently having the highest fresh SGD rates and Mission-Aransas and Upper Laguna Madre Estuaries being dominated by saline SGD. SGD-derived nutrient fluxes have been shown to be substantial in Texas estuaries but the influence of these nutrients on coastal ecosystem functions and services still requires further investigation.

## Keywords

Submarine groundwater discharge · Subterranean estuary · Coastal aquifer · Coastal hydrology · Gulf coast aquifer

A. R. Douglas (✉)
Harte Research Institute, Texas A&M University-Corpus Christi, Corpus Christi, TX, USA
e-mail: Audrey.Douglas@taumcc.edu

D. Murgulet
Center for Water Supply Studies, Texas A&M University-Corpus Christi, Corpus Christi, TX, USA
e-mail: Dorina.Murgulet@tamucc.edu

## Abbreviations

| | |
|---|---|
| $CH_4$ | Methane |
| CHO | Carbon-hydrogen-oxygen (organic molecules) |
| $CO_2$ | Carbon dioxide |
| CWSS | Center for Water Supply Studies |
| DIC | Dissolved inorganic carbon |
| DIN | Dissolved inorganic nitrogen |
| DIP | Dissolved inorganic phosphorous |
| DO | Dissolved oxygen |
| DOC | Dissolved organic carbon |
| DOM | Dissolved organic matter |
| Fe | Iron |
| FWI | Freshwater inflow |
| GCA | Gulf Coast Aquifer |
| GW-SW | Groundwater-surface water |
| $HSiO_3^-$ | Hydrogen silicate |
| Mn | Manganese |
| N | Nitrogen |
| $NH_4^+$ | Ammonium |
| $NO_2^-$ | Nitrite |
| $NO_3^-$ | Nitrate |
| NOx | Nitrate plus nitrite ($NO_3^- + NO_2^-$) |
| P | Phosphorous |
| $PO_4^{3-}$ | Phosphate |
| S | Sulfur |
| SGD | Submarine groundwater discharge |
| Si | Silica |
| $SiO_4^{2-}$ | Silicate |
| STE | Subterranean estuary |
| TDN | Total dissolved nitrogen |
| TGLO | Texas General Land Office |
| TWDB | Texas Water Development Board |
| $^{222}Rn$ | Radon (half-life 3.8 days) |
| $^{223}Ra$ | Radium-223 (11.4 days) |
| $^{224}Ra$ | Radium-224 (half-life 3.6 days) |
| $^{226}Ra$ | Radium-226 (half-life 1600 years) |
| $^{228}Ra$ | Radium-228 (half-life 5.7 years) |

P. A. Montagna, A. R. Douglas (eds.), *Freshwater Inflows to Texas Bays and Estuaries*, Estuaries of the World, https://doi.org/10.1007/978-3-031-70882-4_5

## Introduction

Though not as obvious as river discharge, continental groundwaters also discharge directly into the sea as submarine groundwater discharge (SGD) wherever a hydrologic connection exists between aquifers and sea floor (Figs. 1 and 2). SGD is an important component of the global hydrologic and biogeochemical cycles that link terrestrial waters to marine environments. It is defined as any flow of water across the sediment-water interface, irrespective of its composition (e.g., fresh or saline), origin (e.g., terrestrial or oceanic), or driving force (e.g., terrestrial hydraulic gradients, tidal pumping, density- or wind-driven circulation, or bioturbation) (Fig. 2) (Moore 2010; Santos et al. 2012). Thus, SGD encompasses any and all flow of water on continental margins from the seabed to the coastal ocean and may occur over scales as large as many meters to kilometers (Fig. 2 1–4, 9) or smaller than a few meters or centimeters (Fig. 2 5–12). However, SGD does not include processes such as deep-sea hydrothermal circulation, deep fluid expulsion at convergent margins, and density-driven cold seeps on continental slopes

(Moore 2010). While submarine springs and seeps have been observed for many years in karst (Fleury et al. 2007; Murgulet et al. 2020; Pétré et al. 2020) and volcanic (Bokuniewicz et al. 2006; Lee and Kim 2007; Dimova et al. 2012; Cardenas et al. 2020) topography and have been documented since at least the Roman period (Kohout 1966; Rocha et al. 2021), these features have traditionally been perceived as hydrologic marvels rather than objects for serious investigation or concern. Moreover, submarine springs are an exception as most groundwater discharges to the sea along or near the shoreline or offshore on the continental shelf are diffusive and cannot be observed with the naked eye. Thus, these submarine groundwater discharges are inherently difficult to quantify. Nevertheless, since Kohout (1966) identified submarine springs as a "neglected phenomenon of coastal hydrology," the related fields of study have grown tremendously.

Over the last four decades, recognition has emerged that groundwater may be both volumetrically and chemically important to coastal ecosystem functions (i.e., physicochemical and biological processes) and ecosystem services (i.e.,

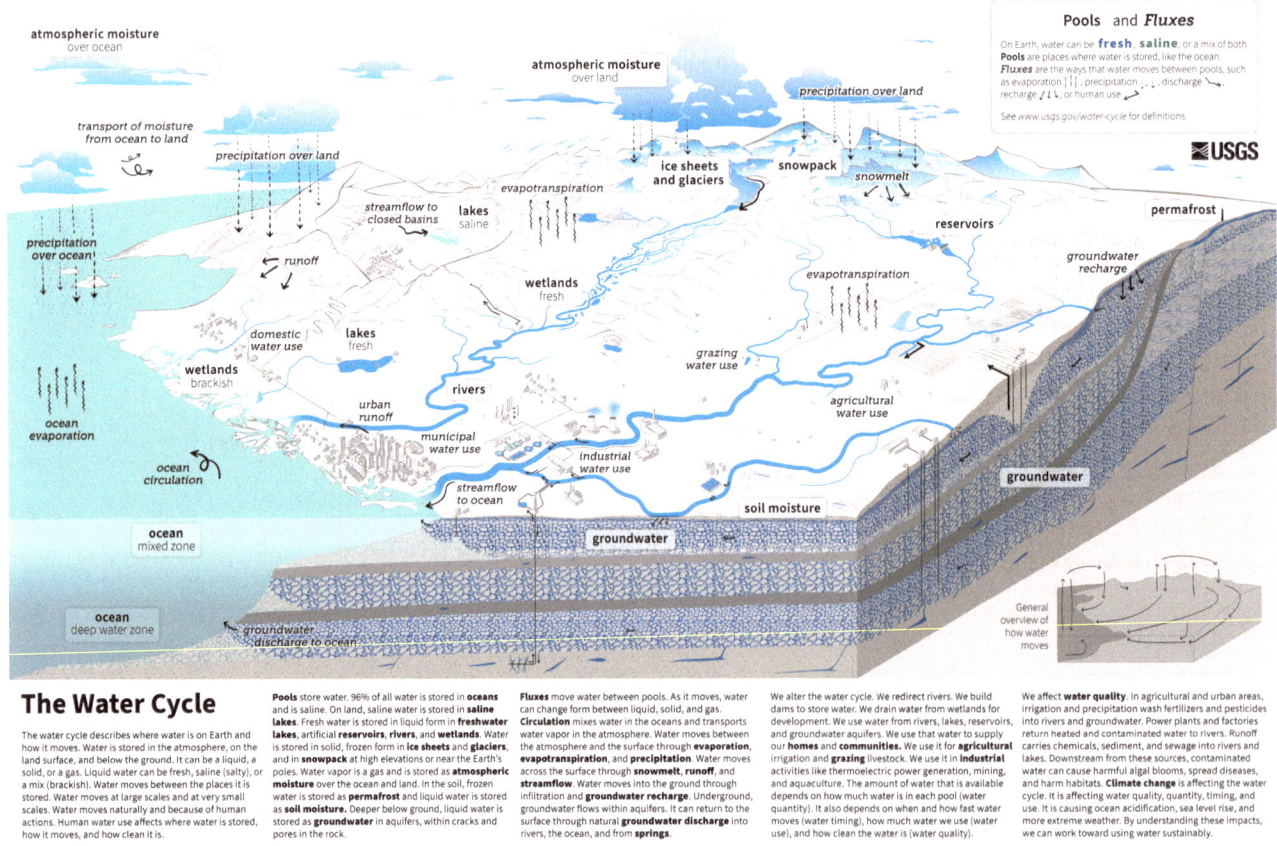

**Fig. 1** Diagram of the water cycle showing where water is on Earth, how water moves (fluxes), where water is stored (pools), the composition of water bodies (fresh, saline, or mixed), and the human water uses. The diagram was created by the USGS VizLab, in collaboration with the USGS Water Resources Mission Area Web Communications Branch, for the USGS Water Science School. (Corson-Dosch et al. 2022)

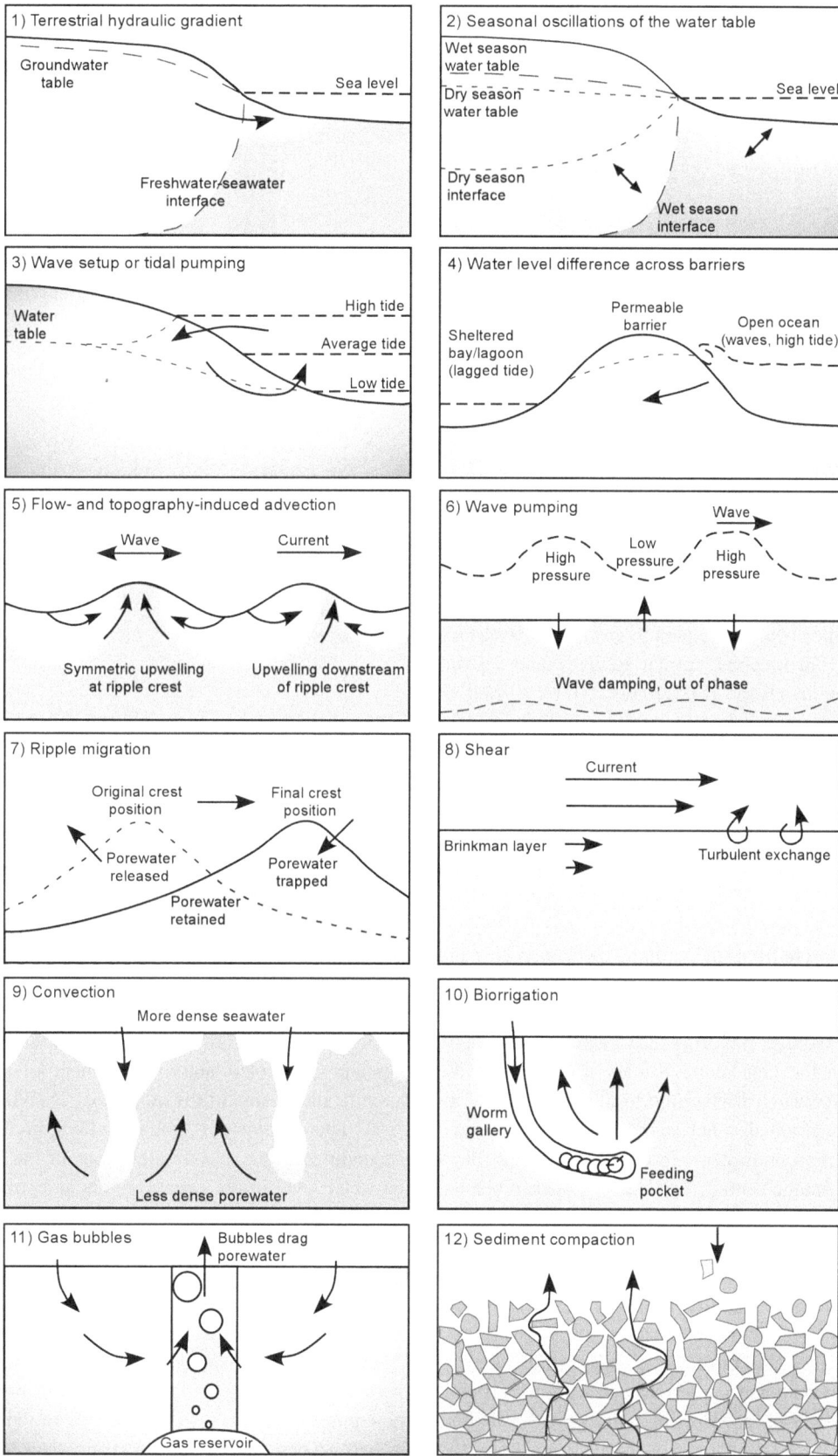

**Fig. 2** Conceptual diagrams of the drivers of porewater (or groundwater) advection in permeable sediments from large-scale processes occurring over areas from many meters to kilometers (1–4, 9) to small-scale processes occurring over areas smaller than a few meters (5–12). Figure reprinted from Santos et al. (2012) with permission from Elsevier

**Fig. 3** Published articles that use common keywords related to groundwater discharge in the coastal zone as indexed by the Web of Science through December 31, 2022

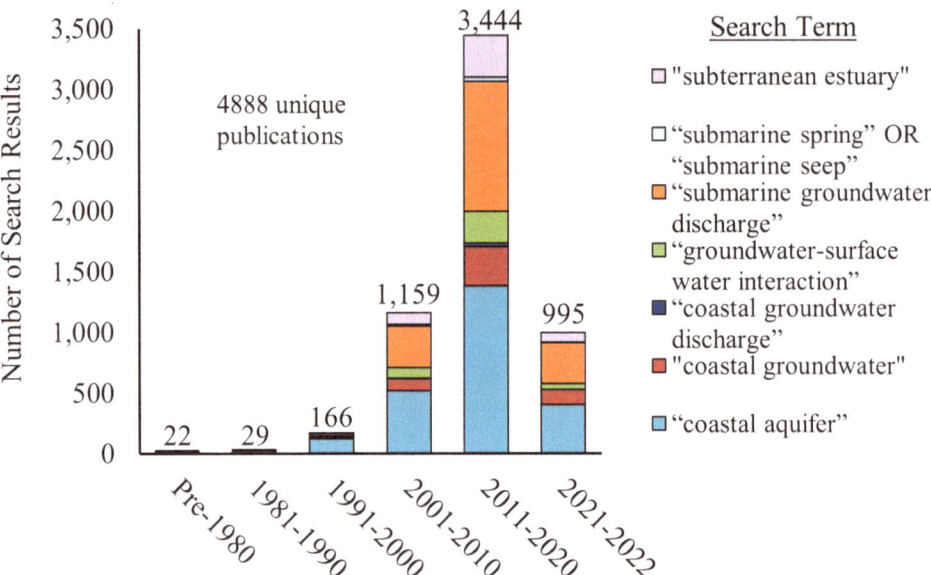

the benefits that are obtained from ecosystems that support, directly or indirectly, the survival and quality of life) leading to a proliferation of research related to groundwater-surface water (GW-SW) interactions, coastal aquifers, and submarine groundwater discharge (SGD) (Fig. 3). In a Web of Science search, the number of peer-reviewed research articles published between 2011 and 2020 for each keyword increased 2.6- to 5.8-fold over the previous decade. However, while scientific interest in GW-SW interactions in coastal areas has increased, data points are skewed and large swathes of coastline remain unstudied (Santos et al. 2021), such as the Texas coast. Of the 4702 unique peer-reviewed publications related to groundwater in the coastal zone on Web of Science, only 20 were from studies along the Texas coast and the majority investigated three estuary systems (i.e., Mission-Aransas Estuary, Nueces Estuary, and Upper Laguna Madre Estuary) with no studies exploring the Lower Laguna Madre Estuary or from the Guadalupe Estuary north. While the impacts of groundwater discharging to the Texas coast are still not well constrained, other studies have demonstrated the importance of groundwater in the coastal zone regardless of geologic or climatic setting. The USGS's revised version of the classic "The Water Cycle" diagram supports this chapter's suggestions and demonstrates the complexity of GW-SW interactions (i.e., fluxes between groundwater and surface water) and by expanding upon the role of groundwater as a source of water and solutes. For instance, the new cycle shows areas of aquifer recharge and groundwater discharge along a river course, depicts the pathways for groundwater discharge to the ocean from different aquifer formations that outcrop at varying depths and distances from shore, and

incorporates some of the potential coupled effects of human water use on groundwater and surface water (Fig. 1) (Corson-Dosch et al. 2022).

## Groundwater-Surface Water Interactions

The interaction between groundwater and surface water occurs at a range of spatial and temporal scales with highly variable magnitudes depending on meteorological, fluvial, hydrological, geological, and anthropogenic processes (Santos et al. 2012; Murgulet et al. 2016, 2022; Moosdorf et al. 2021). Nearly all surface water features (e.g., streams, lakes, wetlands, estuaries) interact with groundwater (Winter et al. 1998). Traditionally, groundwater and surface water have been considered separately in resource management practices, though it is well understood that they are hydrologically connected and development of one affects the quantity and quality of the other (Fig. 1) (Winter et al. 1998). Rivers and streams are strongly influenced by groundwater. Depending on the relative elevations of the water table and the water level in the stream, rivers and streams may alternate between gaining streams (i.e., groundwater supported, water table elevation > stream water level elevation) and losing streams (i.e., recharge groundwater/aquifer, water table elevation < stream water level elevation) along their watercourses (Winter et al. 1998; Fetter 2001). Furthermore, stream flow may be supported year-round from groundwater (perennial streams), fluctuate between receiving groundwater support (following wet periods or in the early stages of droughts when water table > stream water level) and losing water to an aquifer (following storm events and extended dry periods when water table < stream water level) seasonally or with hydroclimatic changes (intermittent streams), or are located above the water table year-round and only flow for a short duration following precipitation events and thus always recharge the groundwater (ephemeral streams). Therefore, the water quality and potential contamination of aquifers have "downstream" impacts through the rivers discharging to the coastal zone, whereas the water quality and potential contamination of surface waters have possible impacts on groundwater chemistry and uses. Furthermore, the same GW-SW interaction processes occurring within streams continue out into estuaries and the coastal zone in the form of fresh SGD, or saltwater intrusion depending on the hydraulic gradients and hydrometeorological conditions. The saline/recirculated SGD component, while still influenced by the land-sea hydraulic gradients, is mostly the result of oceanic forcings (e.g., wave setup or tidal pumping and salinity differences) (Fig. 2).

Only over the past couple decades has research started to explore the role of groundwater as a source of freshwater (Mulligan and Charette 2006; El-Gamal et al. 2012; Lopez et al. 2020), nutrients (Slomp and Van Cappellen 2004; Rodellas et al. 2015; Douglas et al. 2021b), and contaminants (Bone et al. 2007; McKenzie et al. 2020; Szymczycha et al. 2020) to the coastal zone. Thus, to date, assessments of groundwater-derived freshwater inflows and nutrient inputs and cycling in the coastal zone have not commonly been included in coastal management practices. However, a greater focus on GW-SW interactions in Texas by state agencies has arisen from the impacts on streamflow related to the increasing utilization of water resources and the greater frequency of droughts (Young et al. 2018 and references therein). Furthermore, GW-SW interactions have additional effects on regulatory issues related to freshwater inflows, such as implementation of environmental flow recommendations, the management of water rights along a river's course, compliance with the Endangered Species Act in both terrestrial and marine systems, and acquisition of bed and banks permits (Young et al. 2018). However, the most overlooked is the integration of SGD fluxes of solutes such as nutrients in the nonpoint source (NPS) pollution reduction management strategies.

## Submarine Groundwater Discharge

There is a growing body of literature applying direct and indirect measurements that provides evidence that SGD, or the exchange of water between coastal sediments and surface waters as previously described, can account for a substantial fraction of freshwater and solute inflows to a system (Moore 2010; Taniguchi et al. 2019; Santos et al. 2021). Though the specific discharge rates may typically be low (e.g., 1–30 cm $d^{-1}$), the total discharge can be high because the seepage areas involved are large (Burnett et al. 2003). Groundwater of varying salinities enters coastal waters through SGD bringing with it bioactive solutes, such as nutrients (e.g., N, P, Si, DOM), gases (e.g., methane, carbon dioxide), and trace metals (e.g., iron, nickel, zinc). Fresh SGD is a source of new nutrients to a system, whereas saline SGD is considered a source of recycled nutrients from sediments. Solute fluxes from SGD may exceed river inputs in some systems, with ~60% of studies reviewed by Santos et al. (2021) finding greater SGD-derived N fluxes than riverine input, which often resulted in N/P ratios greater than the Redfield ratio, thus counteracting N limitation. Depending on the system-specific conditions, coastal ecosystems may experience many possible positive or negative impacts of SGD, including enhanced coral calcification, primary productivity, fisheries, denitrification, and pollutant attenuation, or eutrophication, algal blooms, deoxygenation, and localized acidification, respectively.

Quantification of groundwater fluxes is difficult due to significant spatial and temporal heterogeneity and the inherent logistical complications of measuring a dispersed flux of water. Additionally, individual SGD measurement techniques may not account for all discharge components. Efforts to identify and quantify SGD have relied upon several techniques. Among the most common methods in the coastal zone are the application of elemental and isotopic geochemistry, such as naturally occurring radioisotopes of radon ($^{222}Rn$) (Burnett and Dulaiova 2003; Murgulet et al. 2018) and the radium quartet ($^{223}Ra$, $^{224}Ra$, $^{226}Ra$, $^{228}Ra$) (Garcia-Orellana et al. 2021), stable isotopes of oxygen ($\delta^{18}O$) and hydrogen ($\delta D$) (Lopez et al. 2020; Murgulet et al. 2022), and trace metals (Montluçon and Sañudo-Wilhelmy 2001; Mori et al. 2019; Spalt et al. 2020), and modeling groundwater flow, such as Darcy's law (Mulligan and Charette 2006; Douglas et al. 2020) and density-dependent flow and transport simulation codes (Guo and Langevin 2002; Murgulet and Tick 2016). Seepage meters have been deployed in shallow and calm conditions to quantify SGD flux (volume discharge normalized to area and time) by direct measurements of changing water volume in attached bags, dilution of dye in an attached chamber, or changes in temperature caused by groundwater convection (Lee 1977; Taniguchi and Fukuo 1993; Sholkovitz et al. 2003). Geophysical techniques, like electrical resistivity profiles, have been used to identify subsurface seepage faces and/or porewater salinity (Stieglitz et al. 2008; Bighash and Murgulet 2015; Spalt et al. 2018). Thermal imaging has been used to locate and map discharge plumes (Danielescu et al. 2009; Kelly et al. 2013). However, each of these techniques has varying spatial and temporal integration scales to consider in the final SGD estimates. For example, regional estimates based on Darcy's law account for terrestrial-derived SGD based on the hydraulic gradient between the aquifer and surface water but few account for recirculated seawater (Li et al. 2009), whereas estimates based on a radium mass balance approach (Charette et al. 2003; Breier and Edmonds 2007; Breier et al. 2010) may provide a reasonable estimate of saline, recirculated SGD but fail to represent lower salinity SGD due to the salinity dependence of radium sorption processes (e.g., radium largely remains particle-bound at salinities lower than ~15) (Mulligan and Charette 2006; Douglas et al. 2020). In contrast, a radon mass balance approach may account for total SGD (i.e., fresh and saline; terrestrial and recirculated seawater) as radon is a noble gas and does not react to different salinity regimes as drastically as radium, though radon activities may be overestimated by as much as 20% in cooler, more saline waters (Schubert et al. 2012). Thus, the application of multiple methods for measuring and/or modeling SGD is recommended for a more complete understanding of the processes controlling GW-SW interactions.

Whatever the method used to quantify SGD, it is a global geological process by which groundwater of varying salinities enters coastal waters (Fig. 1) and is known to transport bioactive solutes (Lecher and Mackey 2018). Consequently, SGD may influence biota through nutrient fluxes, principally N which is often growth limiting for phytoplankton and macrophytes, as groundwater is usually enriched in N compared to P (Slomp and Van Cappellen 2004). Thus, eutrophication from SGD-derived nutrient inputs may alter the balance of nutrients (i.e., the Redfield ratio) and fuel phytoplankton growth potentially leading to harmful algal blooms and other cascading effects. SGD-derived solute fluxes are most commonly determined by multiplying the solute concentration of the groundwater endmember by the calculated SGD rate. Thus, adequately constraining the groundwater endmember is key for both the solute concentrations and the groundwater flux.

## Subterranean Estuary

Traditionally, the study of groundwater in coastal areas has been approached from two separate practices: (1) oceanography and marine science and (2) hydrogeology (Moore 1999; Duque et al. 2020). The land-hydrology viewpoint has conventionally ignored processes beyond the coastline, largely due to a dearth of offshore hydrostratigraphic data and difficulties simulating variable-density flow and focused instead on freshwater as a water supply resource regarding saltwater

as a contaminant to be avoided. In contrast, the marine science perspective focuses on understanding ocean processes, where aquifers are considered a source of chemical components but habitually dismiss or minimize the role of terrestrial hydrogeologic processes with sparse groundwater endmember measurements. In truth, much like the surface estuaries, coastal aquifers exist at the intersection of land and sea and the processes of interest in coastal areas to both marine science and hydrogeology are not physically distinct but occur simultaneously in space and time.

In an attempt to integrate the hydrogeologic and oceanographic approaches to coastal groundwater, analogously with the definition for surface estuaries by Pritchard (1967), Moore (1999) coined the term subterranean estuary (STE) as "a coastal aquifer where groundwater from land drainage measurably dilutes seawater that has invaded the aquifer through a free connection to the sea." Duque et al. (2020) added to this original definition that the STE is "the part of the coastal aquifer that interacts actively with the ocean" and "includes all of the relevant physical, chemical, and biological aspects that excite researchers working at the land-sea intersection." The mixture of terrestrial freshwater and oceanic saltwater received in STEs is exposed to intense biogeochemical processes as it interacts with the aquifer solids that along with the different physical boundaries create clear distinctions between subterranean and surface estuaries (Moore and Joye 2021; Rocha et al. 2021). On its own, a given STE is not comprised of the mixing of fresh groundwater and seawater alone. Consequently, there has been some confusion in terminology between SGD, GW-SW interactions, and STEs as outflows from STEs constitute SGD and are evidence of GW-SW interactions but not all SGD is the product of STE outflow and some GW-SW interactions are not STEs in the formal sense. Thus, Rocha et al. (2021) proposed three criteria to define STEs that account for the reactive mixing of freshwater with seawater that create the salinity gradients that provide the characteristic estuarine structure of STEs in addition to creating biogeochemically distinct subterranean water bodies:

1. Hydrological—a net flow of freshwater to the ocean on time scales of management (i.e., < decades).
2. Geological—the presence of coastal permeable sediments or rocks that act as a conduit for advection.
3. Chemical—the occurrence of freshwater from a continental watershed sufficient to alter the chemical composition of seawater circulating through the coastal aquifer.

All estuarine water masses, both surface and subterranean, are products of mixing (Fig. 4) which is often non-conservative creating biogeochemical and physical distinctions between estuaries and adjacent water bodies. In STEs, mixing is primarily driven by ocean forcing (e.g.,

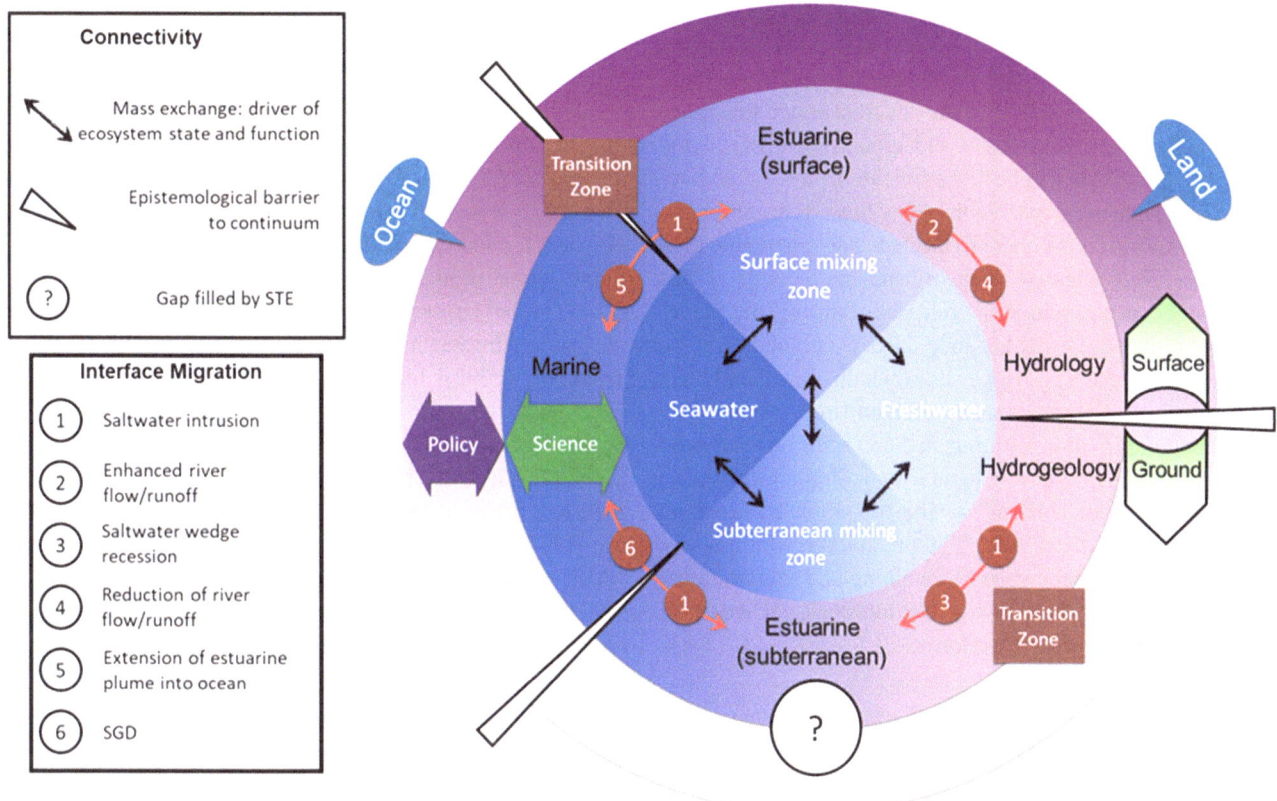

**Fig. 4** Representation of the place of the Subterranean Estuary in the coastal zone. Reprinted from Rocha et al. (2021) with permission from Elsevier

tides, waves, sea-level changes) and freshwater inflows that are regulated by the terrestrial hydraulic gradient and aquifer permeability. The key difference between STEs and surface estuaries is the reactive and transport timescales, which are generally much longer in STEs than surface systems, varying from hours to centuries.

Within Texas, the Texas Water Code (TWC) and Texas Commission on Environmental Quality (TCEQ) definitions of and rules pertaining to "state water," synonymous with surface water, and groundwater divide the STE and impede cohesive, holistic management of groundwater and surface water resources (Young et al. 2018) (see Chap. 2). However, the TWC Chap. 290, Subchapter D begins to bridge this divide in the definition of groundwater under the direct influence of surface water by recognizing the variability of groundwater quality and biogeochemistry resulting from climatological and geological influences on GW-SW interactions. The TWC Chap. 290, Subchapter D defines groundwater under the direct influence of surface water as:

"Any water beneath the surface of the ground with:

(A) significant occurrence of insects or other macroorganisms, algae, or large-diameter pathogens such as *Giardia lamblia* or *Cryptosporidium*;

(B) significant and relatively rapid shifts in water characteristics such as turbidity, temperature, conductivity, or pH which closely correlate to climatological or surface water conditions; or

(C) site-specific characteristics including measurements of water quality parameters, well construction details, existing geological attributions, and other features that are similar to groundwater sources that have been identified by the executive director as being under the direct influence of surface water."

The subterranean estuary encompasses multiple transition zones that encompass the full range of oxic (dissolved oxygen $[DO] > 2$ mg $L^{-1}$ $O_2$), dysoxic, suboxic, and anoxic (DO $<2$ mg $L^{-1}$) reactions included in the oxidation-reduction ladder and the specific organisms which mediate these reactions (Tyson and Pearson 1991; Slomp and Van Cappellen 2004; Moore and Joye 2021). For instance, in dysoxic to anoxic regions oxygen is consumed creating transition zones from oxic to reduced species that include: $NO_3^- — NO_2^- — NH_4^+$ (denitrifiers), $Mn^{4+} — Mn^{3+} — Mn^{2+}$ (manganese reducers), $Fe^{3+} — Fe^{2+}$ (iron reducers), $SO_4^{2-} — S^{2-}$ (sulfate reducers), $CO_2 — CH_4$ (methanogens). The rate of reactions in the STE is also moderated by organic matter substrates as C may be limiting. For example, Slomp and Van Cappellen (2004) explain that when anoxic groundwater meets oxic seawater, if denitrification is C-limited (i.e., limited organic substrate), high $NO_3^-$ anoxic groundwater may be transported to surface waters with little modification and groundwater $PO_4^{3-}$ and $NH_4^+$ concentrations will be low as release from organic matter will not have occurred. On the other hand, if organic matter decomposition is not C-limited and denitrification proceeds, $NH_4^+$ released from organic matter will be the dominant N species and $PO_4^{3-}$ released from organic matter and the reduced Fe-oxides may also accumulate in the groundwater of the STE. However, when this $PO_4^{3-}$ and $Fe^{2+}$-rich groundwater meets oxic seawater, an "iron curtain" may form removing P and resulting in SGD with an N/P ratio typically above Redfield. Charette and Sholkovitz (2002) described the "iron curtain" as the transition zone where soluble $Fe^{2+}$ is oxidized to $Fe^{3+}$ and precipitates where it then acts as a geochemical barrier by retaining and accumulating certain dissolved species, such as phosphorous, carried to the STE by groundwater or seawater.

## Submarine Groundwater Discharge Trends

### Global SGD Trends

In comparison with the global magnitude of gauged river discharge, the magnitude of SGD is poorly constrained though it is known for having a major influence on the environmental condition of many nearshore marine environments as an essential component of biogeochemical budgets. Though there remain serious data gaps from coastal zones of large portions of the world, especially in South America and Africa

(Taniguchi et al. 2002; Santos et al. 2021), a few attempts to estimate groundwater discharge rates to the coast on near-global scales (i.e., 60°S to 70°N) have been made. The fresh component of SGD has been estimated to be up to 10% of the river discharge to global oceans and comparable to riverine inputs for solutes such as C, Si, iron, and strontium (Taniguchi et al. 2002; Moore 2010). Using an observationally constrained radium isotope model, Kwon et al. (2014) found global total SGD, i.e., saline and fresh SGD, to be $(12 \pm 3) \times 10^4$ km³ year⁻¹, which is 3–4 times greater than the riverine freshwater fluxes to the ocean proposed by Dai and Trenberth (2002), likely due to the recirculated seawater included in the estimate that may account for >90% of total SGD. Zhou et al. (2019) applied a high-resolution water budget approach to estimate the total annual volume of fresh SGD to the near-global coast and found fresh SGD accounted for ~489 km³ year⁻¹, or roughly 1% of annual river discharge, which was in line with previous estimates (Zektser and Loaiciga 1993; Church 1996). Additionally, Zhou et al. (2019) found that wet, tropical coasts export more than 56% of all fresh SGD compared to the roughly 10% from arid, mid-latitude regions (23–40°) and that high-relief, tectonically active coastlines have significantly greater ratios of fresh SGD rates than passive margins (Fig. 5). These low fresh SGD rates in dry, low relief mid-latitude regions leave these regions vulnerable to groundwater salinization as (1) surface water is also generally limited, accounting for ~17% of global river discharge, and (2) rates of groundwater extraction increase with coastal population growth facilitating landward hydraulic gradients.

Compared to tropical and mid-latitude regions, less is known about SGD in polar regions even though groundwater is projected to become an increasing source of freshwater

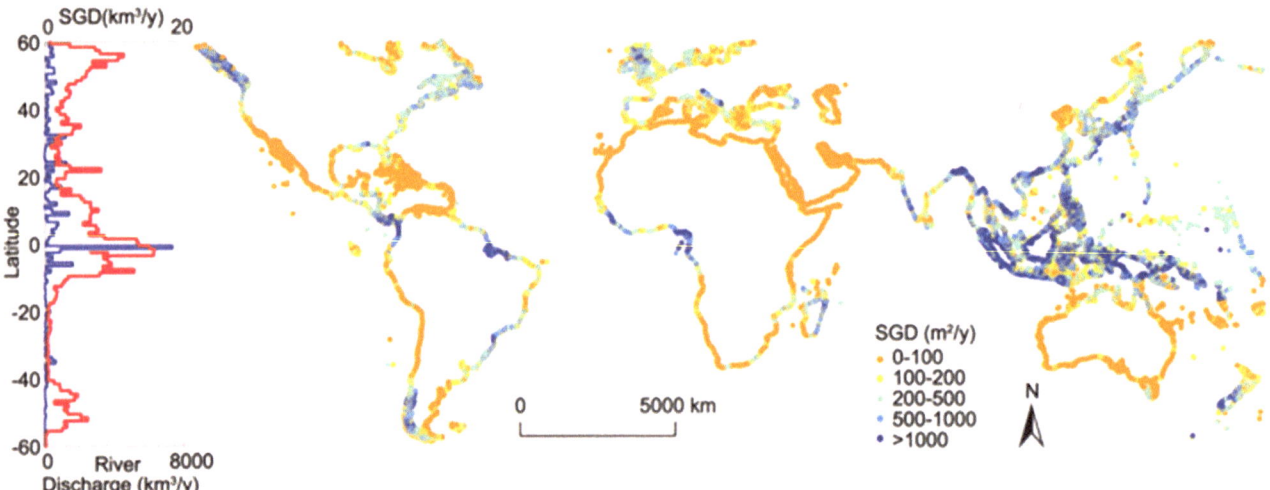

**Fig. 5** Estimates of fresh SGD rates (uncertainty estimated to be 177.0 m² year⁻¹) along the near-global coastline. (left) Comparison of fresh SGD, blue (Zhou et al. 2019), and river discharge, red (Dai and Trenberth 2002), by latitude. Note the different ranges of horizontal axes for river discharge and fresh SGD. Figure reprinted from Zhou et al. (2019) with permission from Wiley

and solutes to the Arctic Ocean as global temperatures rise, sea ice melts, and permafrost thaws. Based on a radium mass balance model, a case study in Russia's Buor-Khaya Gulf in the Laptev Sea found SGD rates in the submarine permafrost talik zone (i.e., a layer of year-round unfrozen ground) were $1.7 \times 10^6$ m$^3$ day$^{-1}$ which exceeds the typical April discharge of the Yana River (Charkin et al. 2017). Additionally, an assessment of the role of SGD as a source of nutrients and trace metals to the coastal Arctic Ocean and Gulf of Alaska found SGD was a source of $NO_3^-$ and Fe to the Arctic Ocean and conservative extrapolations indicate 1.5–17.5 times more $NO_3^-$ is supplied to the Gulf of Alaska from SGD than from rivers (Lecher et al. 2016). Thus, SGD is a global phenomenon that plays a critical role in water and solute budgets in the coastal zone.

## SGD in Texas

### Hydrogeologic Setting

The Texas estuaries overlie the Gulf Coast Aquifer (GCA), which is a major aquifer paralleling Texas' Gulf of Mexico coastline. The GCA spans an area of 108,702 km$^2$ (41,970 mi$^2$), which includes 56 counties and consists of several confined and unconfined aquifers composed of discontinuous Quaternary- and Tertiary-age sedimentary deposits of sand, silt, clay, and gravel beds (Bruun et al. 2016). Approximately 81% of the GCA falls under a groundwater conservation district (GCD) or other districts, such as the Harris-Galveston and Fort Bend subsidence districts, tasked with developing a groundwater management plan (see Chap. 2). From the land-surface downward, the GCA is comprised of the following hydrogeologic units: the Chicot aquifer, the Evangeline aquifer, the Burkeville confining unit, the Jasper aquifer, and the Catahoula confining unit (Baker 1979; Chowdury and Turco 2006; Kasmarek 2012; Ellis et al. 2023). All the sedimentary units dip and thicken toward the Gulf of Mexico. Thus, the Texas estuaries are in contact with the Chicot aquifer's alluvium and Beaumont Formations (Bruun et al. 2016). Cross-formational flow between the different aquifers and the confining units occurs and is generally upward with approximately $1.7 \times 10^6$ m$^3$ year$^{-1}$ (1400 acre-ft year$^{-1}$) flowing from the Jasper aquifer to the Burkeville confining unit, $7.4 \times 10^6$ m$^3$ year$^{-1}$ (6000 acre-ft year$^{-1}$) from the Burkeville confining unit to the Evangeline aquifer, and $24.7 \times 10^6$ m$^3$ year$^{-1}$ (20,000 acre-ft year$^{-1}$) from the Evangeline aquifer to the overlying Chicot aquifer (Bruun et al. 2016).

The TWDB studies of the GCA and modeled groundwater flow toward the Gulf of Mexico have found the freshwater part of the aquifer extends beneath the Gulf of Mexico to the southeast (Bruun et al. 2016). The GCA's freshwater saturated thickness averages about 305 m (1,000 ft). The maximum total sand thickness ranges from 396 m (1,300 ft)

in the northeast to 213 m (700 ft). The Lavaca-Matagorda estuary marks a transition point for groundwater salinity (Young et al. 2010, 2016; Bruun et al. 2016) where northeast of the estuary groundwater is predominantly fresh, whereas southwest of the estuary groundwater becomes more brackish to saline. The hydraulic conductivity (a saturated formation's ability to transmit fluid) of the aquifer also decreases from 2.13 m day$^{-1}$ (7 ft day$^{-1}$) in the northeast to 0.3 m day$^{-1}$ (1 ft day$^{-1}$) in the south. The transmissivity (the ability of a formation to transmit groundwater throughout its entire saturated thickness) of the aquifer ranges from over 1301 m$^2$ day$^{-1}$ (14,000 ft$^2$ day$^{-1}$) in the northeast to less than 93 m$^2$ day$^{-1}$ (1000 ft$^2$ day$^{-1}$) in the south (George et al. 2011). While all Texas aquifers contribute some groundwater to the baseflow in streams, on average an estimated $11.4 \times 10^9$ m$^3$ year$^{-1}$ (9.3 million acre-ft year$^{-1}$) of groundwater flows from aquifers to surface water in Texas accounting for approximately 30% of the average surface water flows (Bruun et al. 2016). The GCA discharges the most groundwater to surface water of all Texas aquifers, with an estimated $4.69 \times 10^9$ m$^3$ year$^{-1}$ (3.8 million acre-ft year$^{-1}$), and follows similar trends to groundwater salinity with greater estimated discharge rates in the northeast and the lowest estimated discharge rates in the southwest. A large portion of the water contained within the GCA is used for irrigation, municipal drinking water, and industrial purposes by the rapidly growing coastal communities (Chowdury and Turco 2006; George et al. 2011).

Of the GCA's hydrogeologic units, the shallowest, Chicot aquifer, is of most interest to coastal GW-SW interactions and the focal subunit of the GCA for the remainder of this chapter. The Chicot aquifer was deposited during the Holocene and Pleistocene and comprises the following stratigraphic units (youngest to oldest): the alluvium, the Beaumont Formation, the Lissie Formation which is subdivided into the Montgomery and Bentley Formations, and the Willis Formation. All the sedimentary units dip and thicken toward the Gulf of Mexico. Thus, the Texas estuaries are in contact with the shallowest of the Chicot aquifer's formations: the alluvium and Beaumont Formations. Additionally, the down-dip boundary for the GCA should allow groundwater discharge across a large area of the ocean bottom with two of the three groundwater availability models extending the regional flow system to ~10 miles beyond the coastline (Bruun et al. 2016). Thus, these models allow for flow between the Gulf of Mexico and the groundwater in the Chicot aquifer. In the coastal counties, the water table is typically 4.3–6.1 m (14–20 ft) below the surface and water table conditions (i.e., unconfined aquifer exposed to atmospheric pressure only) occur principally at depths less than 30.5 m (100 ft) (Hammond 1969; Scanlon et al. 2005).

The GCA geology is complex due to cyclic deposition of sedimentary facies created by changes in land-surface subsidence of the depositional basin and sea-level transgressions (rise) and regressions (decline) that produced the discontinuous beds of sand, silt, clay, and gravel (Kasmarek and Robinson 2004; Chowdury and Turco 2006) (see Chap. 6). Consequently, the aquifer facies alternate between the predominantly continental sediments that compose the aquifers and the predominantly marine sediments that compose the confining units and clay layers within aquifers. Thus, the Chicot aquifer and the GCA system as a whole has a high degree of heterogeneity in both lateral and vertical extents and groundwater may be locally confined by interbedded silt/clay lenses less than 1.8 m (6 ft) thick. Consequently, groundwater discharge to coastal waters may be associated with a complicated assortment of confined, semi-confined, and unconfined aquifers that results in often patchy, diffuse, and temporally variable SGD fluxes (Fig. 2) (Santos et al. 2012). Even though locally, deeper groundwater may have been depleted to where groundwater elevations are below sea level (e.g., Kingsville area), locally higher hydraulic heads than seawater still support flow toward estuaries and the Gulf of Mexico (e.g., north Nueces Bay) (Douglas et al. 2020; Murgulet et al. 2022).

## Spatial and Temporal SGD Trends

To date, the few studies quantifying SGD in Texas have focused on the three estuaries of the coastal bend (i.e., Mission-Aransas, Nueces, and Upper Laguna Madre Estuaries), and, outside of the groundwater modeling of the GCA conducted by the TWDB described above, this component of the hydrologic cycle has, to our knowledge, not been closely investigated in the estuaries of the upper and lower Texas coast (e.g., the Guadalupe, Lavaca-Colorado, Trinity-San Jacinto, Sabine-Neches, and Lower Laguna Madre Estuaries). Thus, there is a need for future assessments of SGD and its associated impacts on water inflows and salinity, nutrient, and contaminant budgets in the upper and lower Texas coast. The following studies are part of the Center for Water Supply Studies' (CWSS) efforts to fill these gaps through studies funded by the Texas General Land Office (TGLO) and the Texas Sea Grant. The studies included below have used radon as a tracer and all calculated fluxes included here have been updated to account for the influence of winds in the days prior to sample collection on the radon inventory. Thus, values reported in the published literature may be slightly different but do not deviate significantly from those reported herein. All CWSS applied some combination of nutrient, major ion, stable and radiogenic isotopes, and electrical resistivity measurements to quantify SGD rates and solute fluxes both spatially and temporally.

Considerable variation in SGD estimates and associated nutrient fluxes have been observed between and within the Mission-Aransas (Douglas et al. 2017; Spalt et al. 2018, 2020), Nueces (Murgulet et al. 2015, 2018; Douglas et al.

2020, 2021a, b), and Upper Laguna Madre estuaries (Lopez et al. 2018, 2020; Murgulet et al. 2021) (Figs. 6 and 7). Overall, the highest $^{222}$Rn-derived total SGD rates have consistently been observed in Nueces Estuary (12.0 ± 7.9 to 374.7 ± 71.4 cm day$^{-1}$, $\mu$ = 94.8 cm d$^{-1}$). The highest $^{222}$Rn-derived total SGD rates were observed in north Nueces Bay (57.1 ± 5.0 to 345.4 ± 51.5 cm day$^{-1}$, $\mu$ = 131.3 cm day$^{-1}$) where steep hydraulic gradients and regional groundwater flow toward the bay enhance SGD and at the mouth of Oso Bay (250.4 ± 47.6 to 374.7 ± 71.4 cm day$^{-1}$, $\mu$ = 312.6 cm day$^{-1}$) near to growth faults transecting the bay that may serve as conduits for vertical groundwater flow. The lower Nueces River was also observed to have consistently elevated total SGD rates (82.5 ± 9.2 to 141.9 ± 22.8 cm day$^{-1}$, $\mu$ = 108.4 cm day$^{-1}$) indicating groundwater support of the river downstream of the saltwater barrier dam at Calallen, diversions to Rincon Bayou, and municipal intakes. The lowest total SGD rates were observed at the mouth of Laguna Madre (13.4 ± 10.7 to 43.2 ± 15.2 cm day$^{-1}$) and in the middle of Nueces Bay after the spring 2015 flooding (12.0 ± 7.9 to 39.8 ± 8.1 cm day$^{-1}$, $\mu$ = 25.2 cm day$^{-1}$) where mid-bay muds likely act as a low hydraulic conductivity barrier inhibiting advective fluxes of groundwater. Interestingly, the middle of Nueces Bay had high total SGD rates (82.6 ± 26.8 to 213.1 ± 29.9 cm day$^{-1}$) during the dry period prior to the late spring 2015 flood event and radium activity ratios indicated a deeper groundwater source during this period compared to the shallow groundwater signature post flooding (Douglas et al. 2020). Furthermore, the average seasonal porewater activity ratios (avg. AR 2.0–10.4, overall avg. porewater AR 6.5) aligned with the range of shallow (~3–7 m bgs, $n$ = 4, avg. AR 14.4) and deep groundwater AR (~60–177 m bgs, $n$ = 6, avg. AR 1.1) and the average AR of all groundwater (6.4) indicating groundwater source mixing occur within the bay's subsurface. In comparison, the range of total SGD rates in Mission-Aransas (16.6 ± 7.1 to 87.4 ± 18.4 cm day$^{-1}$, $\mu$ = 39.6 cm day$^{-1}$) and Upper Laguna Madre (6.2 ± 1.6 to 58.0 ± 16.5 cm day$^{-1}$, $\mu$ = 21.4 cm day$^{-1}$) estuaries was smaller and less variable over space and time.

In Nueces Bay, Douglas et al. (2020) attempted to differentiate the fresh/terrestrial SGD and the saline recirculated SGD in Nueces Bay using Darcy's law (i.e., fresh/terrestrial SGD) and $^{222}$Rn (total SGD) and $^{226}$Ra (saline/recirculated SGD) isotope mass balances. Saline SGD fluxes (4–39 cm day$^{-1}$) were comparable to previous radium-based studies in Nueces Bay (Breier et al. 2005, 2010; Breier and Edmonds 2007) but orders of magnitude less than total SGD and Darcy estimates, which ranged over two orders of magnitude (12.0–345.4 and 9–828 cm day$^{-1}$, respectively). In the previous studies in Nueces Bay, the authors were concerned the Ra-derived SGD estimates were unrealistically high for such a clay-rich semi-arid environment and compared to river discharge, so they assessed the possibility that leaking

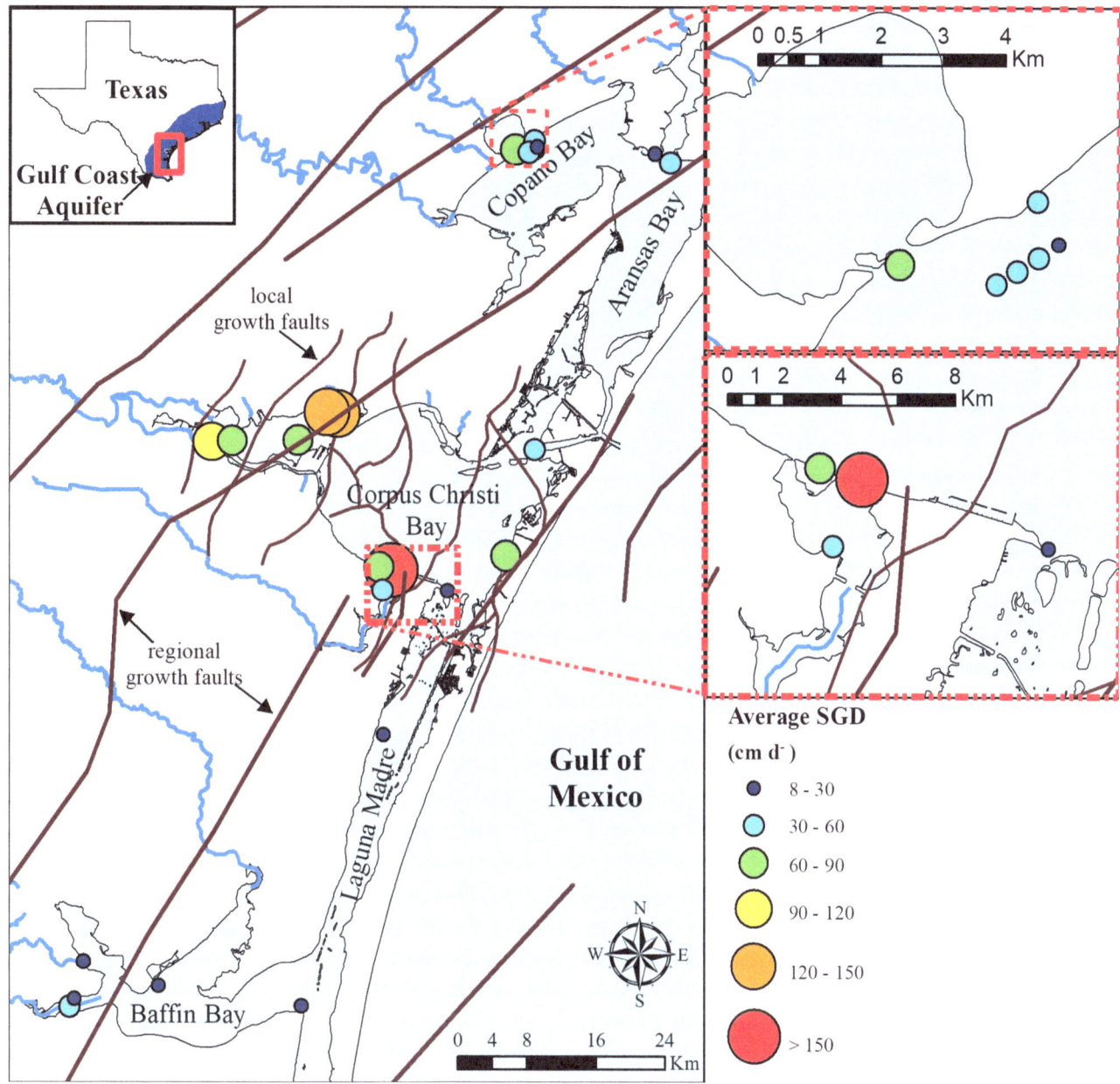

**Fig. 6** Overall average SGD rates (cm day$^{-1}$) from all Center for Water Supply Studies stations sampled between January 2014 and January 2018 for Mission-Aransas, Nueces, and Upper Laguna Madre Estuaries. Local growth faults as generalized by Brown et al. (2004) from 3D seismic and well-data analyses. Regional growth fault data from Texas Commission on Environmental Quality (Nicot et al. 2010)

oil/gas wells and pipelines could be the source of excess Ra in the system. However, these studies were not able to confirm this hypothesis. While the potential impacts of leakage of oil-field brines on these tracer measurements should not be ignored in the heavily developed estuary systems of the Texas coast, such as Nueces Bay (Fig. 8), particularly as the aging infrastructure increases the risks, the much greater total SGD estimates observed by Douglas et al. (2020) and Murgulet et al. (2022) support the occurrence of even greater SGD rates within this bay system due to a combination of a steeper shoreline and positive hydraulic gradients north of

the bay, natural heterogeneities (e.g., sediment type, depositional environment, growth faults), and anthropogenic disturbances (e.g., oil/gas drilling, dredging, oil-field brine discharges) that facilitate groundwater flow to the bay. Furthermore, any excess $^{226}$Ra, the parent of $^{222}$Rn, should be accounted for in the radon mass balance as part of the $^{226}$Ra-supported radon component where $^{226}$Ra is subtracted from the total $^{222}$Rn measurements to calculate an excess $^{222}$Rn (Douglas et al. 2020). The Nueces Bay studies (Murgulet et al. 2016; Douglas et al. 2020, 2021a, b) captured the end of a prolonged drought (fall—winter 2014) fol-

**Fig. 7** Seasonal average SGD (cm day⁻¹) by estuary system for each monitoring site. Note the different scale for Nueces Estuary.
7AB = Station 7 Aransas Bay (Mission Bay Mouth),
13AB = Station 13 Aransas Bay (St. Charles Bay Mouth),
19AB = Station 19 Aransas Bay (Redfish Bay),
GI = Goose Island State Park,
LMo = Laguna Madre Mouth,
OS = Oso Bay Mouth,
SI = Shamrock Island,
UB = University Beach,
OB = Oso Bay, 7NB = Station 7 Nueces Bay (north bay),
8NB = Station 8 Nueces Bay (Nueces River Mouth),
12NB = Station 12 Nueces Bay (mid-bay),
14NB = Station 14 Nueces Bay (lower Nueces River),
9BB = Station 9 Baffin Bay (Cayo de Grullo),
10BB = Station 10 Baffin Bay (Alazan Bay), 11BB = Station 11 Baffin Bay (Laguna Salada), 12BB = Station 12 Baffin Bay (Baffin Bay mouth), LM = Laguna Madre

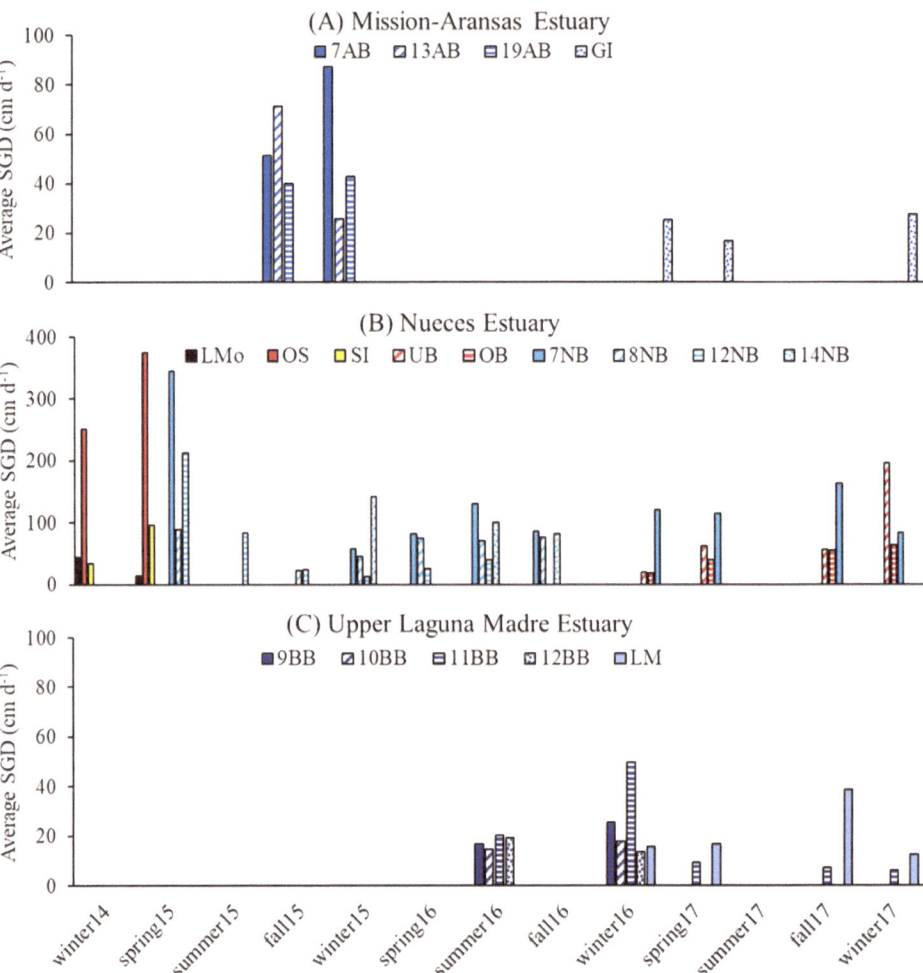

lowed by a flooding period (spring—summer 2015) and return to baseflow conditions (winter 2015—summer 2015) and demonstrated that changing hydroclimatic conditions impact: (1) groundwater source mixing within the sediment as the dominant groundwater source with the deeper groundwater signature dominated during dry periods and the shallow during wet periods, (2) SGD as a significant source of inorganic and organic nutrients in semi-arid environments with saline/recirculated SGD providing old N and fresh/terrestrial SGD supplying new N, and (3) DOM composition as increased SGD during flood recession reduced C:N ratios and brought in more N- and S-containing compounds, while the riverine inputs decreased within 6 months from the flood peak.

Similarly in Baffin Bay, Lopez et al. (2020) used ²²²Rn, long-lived ²²⁶Ra, and short-lived ²²³Ra mass balances to differentiate the fresh/terrestrial SGD and the saline/recirculated SGD. This study found the ²²²Rn and ²²⁶Ra mass balance models produced similar SGD estimates within the range of uncertainty, while the ²²³Ra produced substantially higher SGD rates. As Baffin Bay is a hypersaline system, the impacts of salinity on Ra activities and SGD estimates

are expected to be negligible. Thus, the larger SGD rates obtained from short-lived isotopes imply higher saline/recirculated SGD rates over short time scales, likely due to the influence of wind-driven recirculation. Overall, bay-wide SGD rates for the three tracers ranged from 1.6 to 41.3 cm day⁻¹. The negligible fresh/terrestrial SGD inputs to this system are likely due to a combination of the limited annual rainfall in the watershed, high evaporation rates, and groundwater flow patterns. Unlike Nueces Estuary, deep groundwater input is not expected to occur in Baffin Bay due to hydraulic gradients dipping inland away from the bay in the Kingsville area where drawdowns may exceed 46 m (151 ft) (Chowdhury et al. 2004). A more recent study conducted between February 2020 and January 2021 used time-series ²²²Rn measurements at two locations along the Laguna Salada northern shore and groundwater activities from four nearshore monitoring wells and four nearshore porewater samples as endmembers (Murgulet et al. 2021). The SGD estimates ranged from 3 to 40 cm day⁻¹ ($\bar{x}$: 17 cm day⁻¹, $n = 21$). During spring (March–May), SGD decreased at both locations, similar to Darcy's law SGD

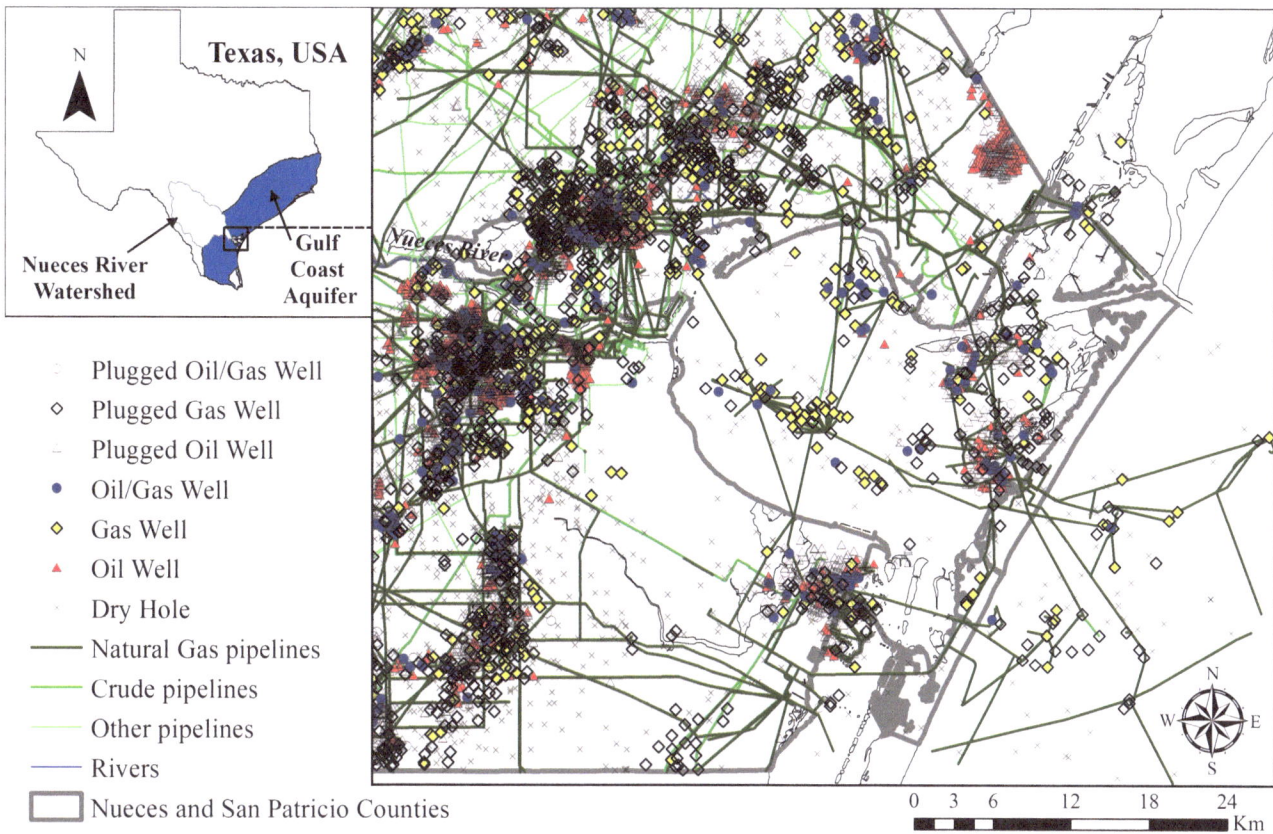

**Fig. 8** Locations of oil and gas wells and pipelines in and around Nueces Estuary. All oil and gas well and pipeline data for Nueces and San Patricio counties were provided by the Railroad Commission of Texas and are accessible through the Public GIS Viewer (RRC 2018)

estimate trends. In fall and winter, SGD increased at both locations.

The three south Texas estuaries (Mission-Aransas, Nueces, and Upper Laguna Madre) were studied in conjunction for 1 year (Winter 2017—Winter 2018) to assess how SGD magnitudes and composition (i.e., fresh or saline/recirculated) vary in nearshore low inflow environments (Fig. 7) (Murgulet et al. 2022). Time-series $^{222}$Rn mass balance models found much higher SGD inputs to the Nueces Estuary (Nueces, Corpus Christi, and Oso Bays $\bar{x}$ = 120, 83, and 44 cm day$^{-1}$, respectively) and lower inputs to the Mission-Aransas and Upper Laguna Madre Estuaries (Aransas Bay, Upper Laguna Madre, and Baffin Bay $\bar{x}$ = 23, 21, and 18 cm day$^{-1}$, respectively). Consistent with the findings of the previous studies, these greater groundwater inputs were attributed to anthropogenically disturbed sediments and potentially surfacing growth faults throughout Nueces Estuary (Fig. 6); however, the connections between SGD and disturbed sediments or geologic features have not been confirmed and require further investigation. The lower groundwater inputs to the Upper Laguna Madre Estuary were explained by the drier climate, less anthropogenic disturbance, and the neighboring groundwater cone of depression around Kingsville. Comparison of $^{222}$Rn-derived

SGD rates with short-lived $^{224}$Ra- and long-lived $^{226}$Ra-derived SGD rates indicated saline/recirculated SGD contributions were greater in Upper Laguna Madre Estuary, Aransas Bay, and Oso Bay where the two methods were in close agreement. However, the two methods were markedly different in Nueces Estuary where $^{222}$Rn-derived rates were substantially higher than rates from $^{224}$Ra and $^{226}$Ra indicating a greater terrestrial source than recirculating seawater. Furthermore, stable isotopes of water ($\delta^{18}$O and $\delta$D) supported greater recirculation in Laguna Madre Estuary and Aransas Bay where the effects of evaporation/salinization were more pronounced, whereas greater terrestrial/$^{222}$Rn-derived SGD inputs showed less evaporation/salinization.

**Porewater Chemistry**

Characterizing the solute concentrations of the groundwater endmember is incredibly important to estimating SGD-derived fluxes in estuarine systems. Shallow porewater (0.3–2 m below the sediment surface) solute concentrations are generally considered the best endmember as it is deep enough to reduce surface water influences and shallow enough to realistically represent the solute concentrations that may reach the surface in terms of flux rates (i.e., time); however, it should be noted that depending on the specific characteristics of the STE the porewater passes through, the solute composition of the discharging water may be altered as discussed above (see Section "Subterranean Estuary"). The CWSS measured porewater nutrient concentrations at 52 unique stations collecting 170 samples by Date/Time-Station (averaging replicates) from Mission-Aransas ($n = 28$), Nueces ($n = 78$), and Upper Laguna Madre ($n = 64$) Estuaries between January 2014 and January 2018. However, because porewater could not be consistently collected at all SGD time-series monitoring locations, the event average porewater concentration from the spatial study was used in the flux calculations for SGD locations without a porewater sample.

The CWSS found that the nutrient porewater concentrations measured varied spatially and temporally (Murgulet et al. 2015, 2019; Douglas et al. 2017, 2021b; Lopez et al. 2018). Overall, porewater had low DO (0.1–10.9 mg $L^{-1}$, $\mu = 1.8$ mg $L^{-1}$) and high $NH_4^+$ (0.14–5531 $\mu$mol $L^{-1}$, $\mu = 262.4$ $\mu$mol $L^{-1}$), $NO_3^- + NO_2^-$ (0.01–281.1 $\mu$mol $L^{-1}$, $\mu = 7.2$ $\mu$mol $L^{-1}$, referred to hereafter as NOx), $PO_4^{3-}$ (0.01–45 $\mu$mol $L^{-1}$, $\mu = 7.8$ $\mu$mol $L^{-1}$), $SiO_4^{4-}$ (0.86–576 $\mu$mol $L^{-1}$, $\mu = 176.4$ $\mu$mol $L^{-1}$), and DOC (69.3–15,243.5 $\mu$mol $L^{-1}$, $\mu = 616.3$ $\mu$mol $L^{-1}$) (Table 1). The high ammonium concentrations in the porewater samples interfered with the TDN measurements in several studies; thus, TDN and DON concentrations and fluxes will not be discussed. Porewater salinities were observed to be like sea-water or greater in the Upper Laguna Madre Estuary (31.7–67.1) in all samples indicating a general lack of fresh-water input from groundwater. In comparison, porewater in Mission-Aransas and Nueces Estuaries exhibited signs of fresh groundwater inputs and freshening events (i.e., salinities <30) in the STE were observed at 8 different stations across 6 different seasons in Mission-Aransas Estuary and 6 different stations across 9 different seasons in Nueces Estuary. Spatially, on average, Nueces and Baffin Bays pore-water had the highest DOC ($\mu = 768.6$ and 750.6 $\mu$mol $L^{-1}$, respectively) and $PO_4^{3-}$ ($\mu = 11.3$ and 12.3 $\mu$mol $L^{-1}$, respectively), while Baffin Bay had the highest $NH_4^+$ ($\mu = 726.6$ $\mu$mol $L^{-1}$), DIN ($\mu = 723.3$ $\mu$mol $L^{-1}$), and $SiO_4^{4-}$ ($\mu = 267.7$ $\mu$mol $L^{-1}$) and Oso Bay had the highest NOx ($\mu = 101.0$ $\mu$mol $L^{-1}$). In contrast, Aransas Bay had low DOC and NOx ($\mu = 304.5$ and 0.92 $\mu$mol $L^{-1}$, respectively), Oso Bay had low DOC and $NH_4^+$ ($\mu = 334.6$ and 38.0 $\mu$mol $L^{-1}$, respectively), Corpus Christi Bay had low NOx ($\mu = 0.52$ and 101.5 $\mu$mol $L^{-1}$, respectively), and Laguna Madre had low NOx, and $PO_4^{3-}$ ($\mu = 0.74$ and 2.45 $\mu$mol $L^{-1}$, respectively).

The relative concentrations of DOC, $NH_4^+$, and NOx in these anoxic sediments have implications for the processes controlling elemental cycling such as whether N is converted to $N_2$ and lost from system through denitrification or N is conserved in the system as $NH_4^+$ through dissimilatory nitrate reduction to ammonium (DNRA) (Giblin et al. 2013). While denitrification and DNRA have similar requirements

**Table 1** Summary statistics of porewater chemistry by estuary

| | | Temp (°C) | DO (mg $L^{-1}$) | Sal | pH | DOC ($\mu$mol $L^{-1}$) | $NH_4^+$ | $NO_3^- + NO_2^-$ | DIN | $PO_4^{3-}$ | $SiO_4^{4-}$ |
|---|---|---|---|---|---|---|---|---|---|---|---|
| Mission-Aransas Estuary | min | 12.5 | 0.3 | 8.0 | 6.9 | 124.0 | 0.1 | 0.06 | 0.6 | 0.4 | 24.1 |
| | max | 35.1 | 9.4 | 40.8 | 8.3 | 545.7 | 809.3 | 6.6 | 809.6 | 11.2 | 393.5 |
| | avg | 22.9 | 2.1 | 25.3 | 7.7 | 304.5 | 111.4 | 0.9 | 111.0 | 5.2 | 155.4 |
| | stdev | 6.5 | 2.8 | 9.5 | 0.4 | 115.2 | 181.6 | 1.4 | 181.9 | 3.4 | 89.3 |
| | n | 27 | 22 | 27 | 27 | 24 | 27 | 27 | 27 | 27 | 27 |
| Nueces Estuary | min | 11.6 | 0.1 | 0.3 | 6.6 | 69.3 | 0.8 | 0.02 | 0.1 | 0.01 | 0.9 |
| | max | 33.3 | 10.9 | 49.0 | 8.5 | 15,243.5 | 1,660.0 | 281.1 | 1,660.0 | 31.8 | 576.0 |
| | avg | 22.8 | 1.8 | 33.4 | 7.5 | 637.1 | 125.7 | 13.2 | 131.2 | 7.8 | 144.9 |
| | stdev | 6.1 | 1.8 | 7.1 | 0.4 | 1,861.4 | 237.7 | 46.1 | 238.7 | 7.6 | 110.1 |
| | n | 68 | 65 | 66 | 68 | 75 | 79 | 71 | 80 | 80 | 78 |
| Upper Laguna Madre Estuary | min | 9.4 | 0.2 | 31.7 | 6.3 | 197.8 | 0.8 | 0.01 | 1.0 | 0.01 | 0.9 |
| | max | 38.1 | 10.9 | 67.1 | 8.6 | 4,542.8 | 5,531.3 | 34.2 | 5,538.5 | 45.0 | 576.0 |
| | avg | 24.1 | 1.7 | 50.5 | 7.1 | 632.8 | 416.7 | 2.4 | 414.0 | 7.5 | 200.9 |
| | stdev | 6.4 | 2.2 | 7.9 | 0.6 | 635.9 | 772.9 | 4.9 | 775.7 | 10.9 | 135.4 |
| | n | 58 | 49 | 51 | 58 | 69 | 75 | 70 | 75 | 75 | 73 |
| Overall | min | 9.4 | 0.1 | 0.3 | 6.3 | 69.3 | 0.1 | 0.01 | 0.1 | 0.01 | 0.9 |
| | max | 38.1 | 10.9 | 67.1 | 8.6 | 15,243.5 | 5,531.3 | 281.1 | 5,538.5 | 45.0 | 576.0 |
| | avg | 23.5 | 1.8 | 37.8 | 7.4 | 616.3 | 262.4 | 7.2 | 262.9 | 7.8 | 176.4 |
| | stdev | 6.2 | 2.0 | 12.6 | 0.5 | 1,369.2 | 564.5 | 31.8 | 564.2 | 8.9 | 115.8 |
| | n | 149 | 132 | 141 | 149 | 153 | 166 | 155 | 167 | 167 | 165 |

(i.e., anoxic conditions, available nitrate, and organic substrates) and have been shown to compete in estuarine systems, DNRA is thought to be favored in high-carbon low-nitrate systems, such as Baffin Bay (An and Gardner 2002), while denitrification is favored under high-nitrate low-carbon conditions (Burgin and Hamilton 2007).

## SGD-Derived Nutrient Fluxes

With consistently high SGD rates and moderately high pore-water nutrient concentrations in Nueces Estuary, it is no surprise that SGD-derived nutrient fluxes are orders of magnitude greater than either Mission-Aransas or Upper Laguna Madre Estuary (Table 2). However, it is a testament to how much greater the porewater $NH_4^+$ concentrations are in Upper Laguna Madre Estuary (i.e., Baffin Bay) that the overall average $NH_4^+$ and DIN fluxes in Nueces Estuary are only a little over twice those in Upper Laguna Madre Estuary, whereas Mission-Aransas Estuary fluxes are sixfold less even with SGD rates in Upper Laguna Madre Estuary on average a third of those in Nueces Estuary and half those in Mission-Aransas Estuary.

### Mission-Aransas Estuary

SGD-derived nutrient fluxes exhibited spatio-temporal differences in Mission-Aransas Estuary (Douglas et al. 2017). The average SGD-derived fluxes in late summer 2015 were 182 mmol m$^{-2}$ day$^{-1}$ of DOC, 1.12 mmol m$^{-2}$ day$^{-1}$ of DIN,

77.6 mmol m$^{-2}$ day$^{-1}$ $SiO_4^{4-}$, and 2.31 mmol m$^{-2}$ day$^{-1}$ of $PO_4^{3-}$. By late fall 2015, the average SGD-derived nutrient fluxes for DIN (16.4 mmol m$^{-2}$ day$^{-1}$) had increased by an order of magnitude, largely due to the $NH_4^+$ component, while the SGD fluxes of DOC (215 mmol m$^{-2}$ day$^{-1}$) and $SiO_4^{4-}$ (77.2 mmol m$^{-2}$ day$^{-1}$) remained similar and $PO_4^{3-}$ (1.53 mmol m$^{-2}$ day$^{-1}$) decreased by approximately one quarter. In late summer, the mouth of St. Charles Bay had the lowest DIN fluxes and the highest $PO_4^{3-}$ and $SiO_4^{4-}$ fluxes. By late fall, the mouth of St. Charles Bay had the lowest DOC and $SiO_4^{4-}$, DIN was lowest at the mouth of Mission Bay and highest in Redfish Bay, and $PO_4^{3-}$ was similar throughout the system. DOC was highest at the mouth of Mission Bay in both seasons indicating consistent organic matter input to the Mission Bay area.

In Copano Bay at the mouth of Mission Bay, Spalt et al. (2020) used a combination of spatial and temporal geochemistry of porewater and surface water combined with SGD-derived solute fluxes and turnover times to examine the significance of SGD in delivering nutrients to a paleovalley system aligned with productive oyster reefs (see Section "Impacts of Extreme Events"). Solute fluxes of DIN, DOC, total alkalinity, and dissolved inorganic carbon (DIC) were slightly greater at the oyster reef than elsewhere in the bay system. Mooney and McClelland (2012) found Copano Bay receives $2.29 \times 10^5$ mol N year$^{-1}$ in a dry year and approximately two orders of magnitude more ($2.37 \times 10^7$ mol N year$^{-1}$)

**Table 2** Summary statistics of the average Rn-derived submarine groundwater discharge (SGD) rates and the corresponding nutrient fluxes by estuary and overall

| | | Average SGD (cm day$^{-1}$) | Fluxes (mmol m$^{-2}$ day$^{-1}$) | | | | | |
| --- | --- | --- | --- | --- | --- | --- | --- | --- |
| | | | $NH_4^+$ | $NO_3^- + NO_2^-$ | DIN | $PO_4^{3-}$ | $SiO_4^{4-}$ | DOC |
| Mission-Aransas Estuary | Minimum | 16.6 | 1.0 | 0.4 | 5.0 | 0.7 | 95.6 | 475.4 |
| | Maximum | 87.4 | 626.2 | 19.9 | 630.2 | 36.7 | 1,357.1 | 4,442.5 |
| | Mean | 39.6 | 137.9 | 6.3 | 144.3 | 17.2 | 587.3 | 1,549.2 |
| | Stdev | 19.3 | 189.9 | 6.0 | 187.5 | 8.7 | 387.2 | 1,241.5 |
| | n | 14 | 13 | 13 | 13 | 13 | 13 | 9 |
| Nueces Estuary | Minimum | 12.0 | 58.5 | 0.7 | 75.0 | 8.0 | 144.0 | 395.6 |
| | Maximum | 374.7 | 7,848.7 | 343.2 | 7,896.6 | 604.0 | 12,235.3 | 526,449.5 |
| | Mean | 94.8 | 1,651.7 | 31.9 | 1,683.6 | 122.4 | 1,894.8 | 32,233.6 |
| | Stdev | 84.2 | 2,042.2 | 73.4 | 2,032.2 | 144.2 | 2,463.8 | 107,104.8 |
| | n | 38 | 36 | 36 | 36 | 36 | 36 | 33 |
| Upper Laguna Madre Estuary | Minimum | 6.2 | 10.6 | 0.2 | 11.1 | 0.9 | 117.7 | 281.3 |
| | Maximum | 58.0 | 3879.1 | 33.4 | 3889.8 | 91.5 | 2001.4 | 4145.4 |
| | Mean | 21.4 | 689.7 | 6.4 | 696.1 | 15.9 | 471.5 | 1355.9 |
| | Stdev | 14.8 | 1183.0 | 8.8 | 1183.8 | 25.5 | 493.3 | 1208.8 |
| | n | 16 | 14 | 14 | 14 | 14 | 14 | 14 |
| Overall | Minimum | 6.2 | 1.0 | 0.2 | 5.0 | 0.9 | 117.7 | 281.3 |
| | Maximum | 374.7 | 7848.7 | 343.2 | 7896.6 | 604.0 | 12,235.3 | 526,449.5 |
| | Mean | 66.2 | 1189.5 | 21.8 | 1211.2 | 81.4 | 1364.1 | 19,582.7 |
| | Stdev | 71.6 | 1790.1 | 58.6 | 1787.5 | 124.0 | 2046.2 | 83,118.4 |
| | n | 68 | 59 | 59 | 59 | 59 | 59 | 56 |

during wet years from the Mission River sub-watershed. Spalt et al. (2020) found that SGD could supply $0.41 \times 10^7$ to $4.1 \times 10^7$ mol DIN year$^{-1}$ for the 1 km$^2$ study area which far exceeds the riverine inputs to Copano Bay from Mission River during dry conditions and is approximately equivalent during wet conditions. While the average flushing time for Mission-Aransas Estuary is approximately 38 days (Breier et al. 2010), Spalt et al. (2020) found that the turnover times (i.e., the ratio of the water column analyte inventory (mmol m$^{-2}$) to the SGD-derived solute rate (mmol m$^{-2}$ day$^{-1}$)) for all nutrients were < 11 days which is substantially shorter than the flushing time. The overall low turnover times could be explained by rapid assimilation under the study conditions, loss to the atmosphere (i.e., NH$_4^+$), or overestimation of the SGD-derived flux due to geochemical reactions within the sediment that occur between the depth at which the porewater was sampled and the sediment-water interface (i.e., porewater endmember concentrations of NH$_4^+$ vs NO$_3^-$, Fe, Mn, and more).

## Nueces Estuary

The seasonal variability of water column nutrient concentrations was assessed in Nueces Bay by Douglas et al. (2021b) at 15 sites and SGD-derived nutrient fluxes from groundwater and porewater at four sites across hydroclimatic conditions from September 2014 through July 2016, including drought and flood conditions (note: SGD was not measured in September 2014). This study found that, on average, 67% of the variance in the water column water quality was due to temporal differences, while 22% was explained by spatial differences demonstrating greater temporal control on discharge and geochemical perturbations. Additionally, a principal component analysis revealed three principal components accounting for a total of 55.5% of the variability with surface water in all seasons: Freshwater inflow as salinity and nutrients have an inverse relationship and low salinity is expected with high inflows (PC1 28.8%), saline/recirculated SGD and recycled nitrogen as Ra and reduced N species (i.e., NH$_4^+$ and DON) are positive and pH is negative (PC2 15.6%), and total SGD and "new" nitrogen as Rn and NOx are positively correlated (PC3 11.2%) (Fig. 9). Temporally, the June 2015 flood event had the highest PC1 scores followed by the September 2015 flood recession. In contrast, spatially the Nueces River and samples near the river mouth (i.e., west bay) had the highest PC1 scores, while the Corpus Christi Bay samples had the lowest PC1 scores indicating that freshwater inflow had the greatest effect in the tidal river and during flooding. Samples from the drought period before the flood event (i.e., September 2014, December 2014) had higher PC2 scores compared to samples following the flood event indicating greater contribu-

tions of "old" N and saline/recirculated SGD before flooding. Winter and summer samples had the highest PC3 scores, particularly in Nueces River and the north bay, compared to spring and fall samples indicating that the total SGD and "new" N are greater in the river and nearer shore than further offshore in the bay and demonstrating that groundwater support to rivers should not be ignored. This study shows that combining geochemical tracers of groundwater discharge with nutrient studies can reveal the potential influence of SGD on surface water nutrient trends.

In general, Douglas et al. (2021b) found SGD fluxes from the average porewater endmember were greater than the average watershed groundwater endmember for DOC, NH$_4^+$, and PO$_4^{3-}$, while porewater fluxes were less than groundwater fluxes for NOx and SiO$_4^{4-}$ across all seasons (Fig. 10). The one exception being June 2015, when NOx fluxes from porewater exceeded fluxes from groundwater indicative of either NOx accumulation or production in the porewater, or a groundwater source not accounted for, such as from the agricultural area to the north. Overall, total SGD nutrient fluxes were greater than saline nutrient fluxes and both were orders of magnitude greater than nutrient fluxes calculated from the TWDB's FWI estimates to Nueces Bay (Fig. 10). Of note, this study used the TWDB FWI estimates instead of the USGS stream gauge at Calallen for comparison because the stream gauge is upstream of municipal diversions and a saltwater barrier dam and outside of the flood event and the flood recession period observed stream flow was negligible as water did not flow over the dam. Total SGD fluxes (saline+fresh) for DOC, NH$_4^+$, PO$_4^{3-}$, and SiO$_4^{4-}$ were highest in December 2014 (i.e., drought) and lowest in June 2015 (i.e., flood). In contrast, the highest total SGD flux of NOx occurred in June 2015 when SGD rates were lowest but porewater NOx concentrations were highest. Elevated NOx fluxes in June 2015 and December 2014 parallel Nueces River surface water samples from those events that load positively on PC3 indicating a substantial groundwater flux of new N to the tidal river. Following precipitation events, groundwater recharge within the watershed to the surficial aquifer increases leading to enhanced groundwater discharge to the river as the hydraulic gradients are steeper along riverbanks. In a companion study, Douglas et al. (2021a) assessed the changes in DOM molecular composition from the summer 2015 flood event through flood recession and return to baseflow and semi-wet conditions and found the flood recession DOM composition was substantially different from the other seasons due to increased N- and S-containing compounds likely from enhanced SGD.

Previous nitrogen budgets for Nueces Bay and Nueces Estuary generated net negative balances indicating more N was leaving the system through diversions, tidal exchange,

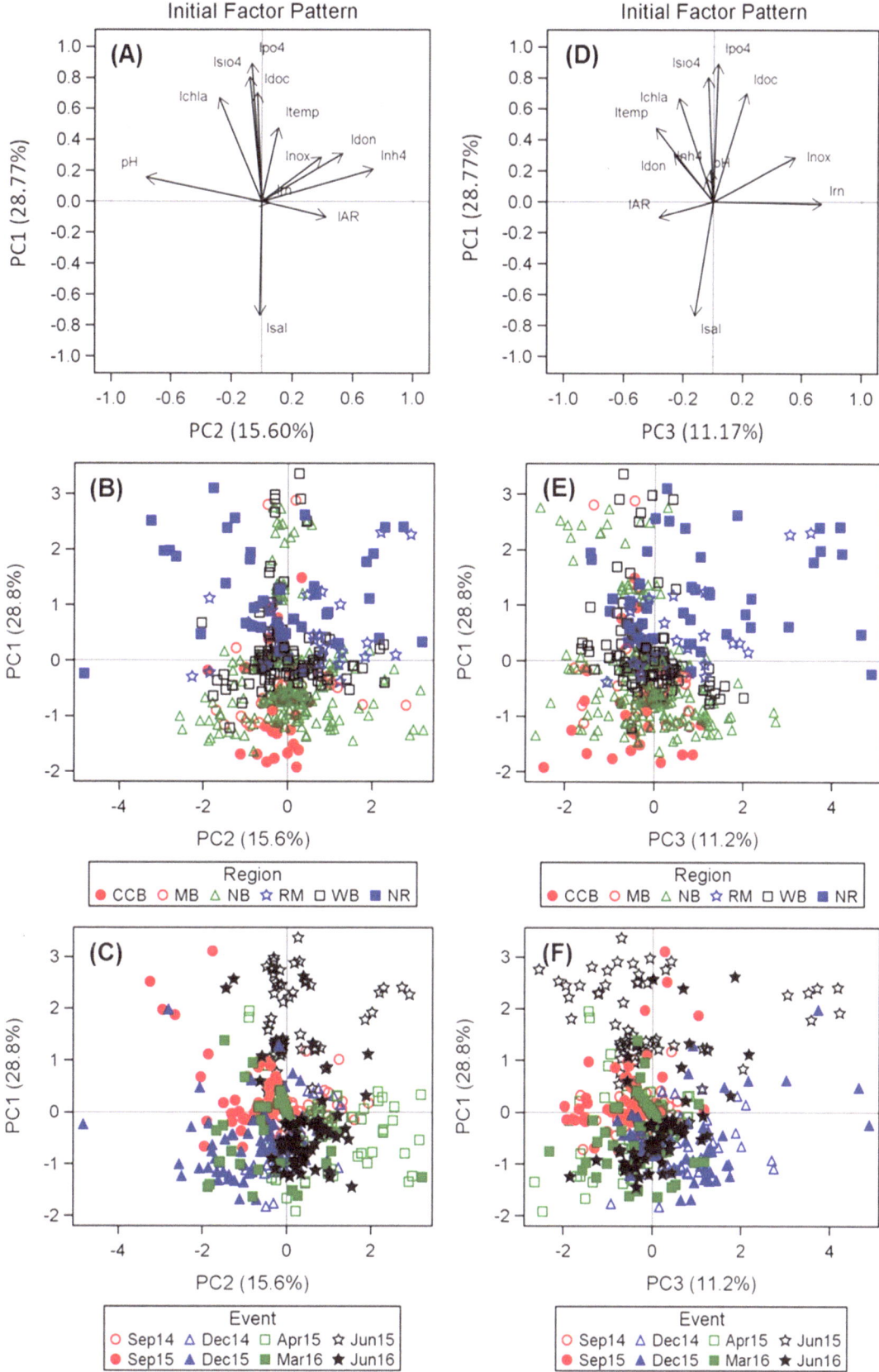

**Fig. 9** Principal components (PC) analysis of water quality and SGD tracers. Variable loads for PC1 and PC2 (**a, d**). Sample scores by region for PC1 and PC2 (**b**) and PC1 and PC3 (**e**). Sample scores by sampling region for PC1 and PC2 (**c**) and PC1 and PC3 (**f**). Abbreviations: temp = temperature; sal = salinity; chl-a = chlorophyll-α; doc = dissolved organic carbon; don = dissolved organic nitrogen; NH4 = ammo-nium; NOx = nitrate+nitrite; PO4 = phosphate; SiO4 = silicate; Rn = radon-222; AR = radium 224:226 activity ratio; CCB = Corpus Christi Bay; MB = mid-bay; NB = north bay; WB = west bay; RM = river mouth; NR = Nueces River. Reprinted from Douglas et al. (2021b) with permission from Elsevier

**Fig. 10** Nutrient fluxes for surface runoff (gray area), total SGD (black/white bars), and saline SGD (gray/white bars) over time. Color patterns reflect the greater fluxes are from average porewater endmember (PW, white with black/gray dots) or average groundwater endmember (GW, gray/black with white dots). (**a**) Dissolved organic carbon fluxes. (**b**) Nitrate+nitrite fluxes. (**c**) Ammonium fluxes. (**d**) Phosphate fluxes. (**e**) Silicate fluxes. Modified and reprinted from Douglas et al. (2021b) with permission from Elsevier

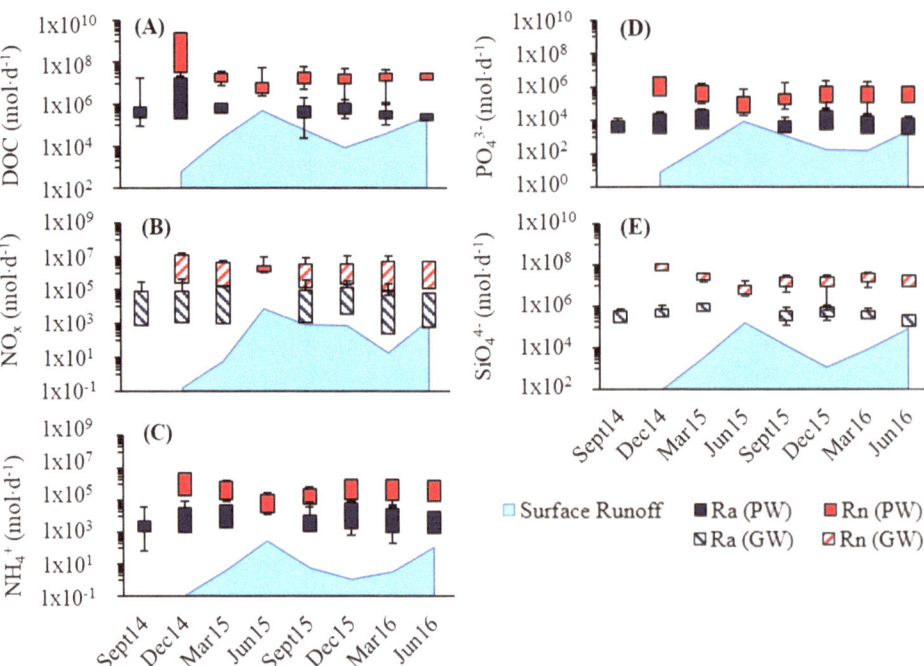

denitrification, and nitrogen burial than was entering the system from known sources (i.e., gauged streams, ungauged streams, wastewater treatment plants, other return flows, wet/dry deposition, nitrogen fixation, and SGD) (Brock 2001; Anchor QEA 2017). However, these budgets either did not account for SGD-derived N (Brock 2001) or estimated SGD-derived N from $NO_3^-$, and thus likely underestimated the contribution of total N from groundwater (Anchor QEA 2017). Douglas et al. (2021b) updated the SGD variable in the most recent N budget from Anchor QEA, for years 1986–2015, by applying the bay-wide TDN porewater fluxes for June 2015—March 2016 from Douglas et al. (2021a), which removed ammonium prior to measurement to reduce interference and reduce error. The original SGD N flux estimate was $75 \times 10^6$ g N year$^{-1}$ from regional groundwater $NO_3^-$; however, the Rn-derived N flux averaged $46 \times 10^9$ g N year$^{-1}$ and the Ra-derived N flux averaged $1.4 \times 10^9$ g N year$^{-1}$. Thus, the Ra-derived annual average TDN fluxes were up to an order of magnitude more than those used by Anchor QEA (2017), while the Rn-derived TDN fluxes were two orders of magnitude more than the N fluxes Anchor QEA (2017) used. Thus, SGD supplies more N to the system than accounted for by all estimated inputs and sinks creating an excess of $45.2 \times 10^9$ g N year$^{-1}$, on average. Even the most conservative total fluxes, calculated using the highest groundwater endmember in the SGD calculation, generate a substantial surplus of N to the system which may have considerable

impacts on the microbial and planktonic communities within the larger Nueces Estuary.

**Upper Laguna Madre Estuary**

In Upper Laguna Madre Estuary, SGD-derived nutrient fluxes were believed to be dependent on spatial and temporal nutrient concentrations in the porewater rather than hydroclimatic conditions (Lopez et al. 2018). For instance, as a result of porewater DIN concentration differences, average SGD-derived DIN and $HPO_4^{2-}$ fluxes (178.1 mmol m$^{-2}$ day$^{-1}$ and 5.0 mmol m$^{-2}$ day$^{-1}$, respectively) were larger in summer than in late fall (92.3 mmol m$^{-2}$ day$^{-1}$ and 1.2 mmol m$^{-2}$ day$^{-1}$, respectively), while $HSiO_3^-$ fluxes were similar between the seasons (63.9 mmol m$^{-2}$ day$^{-1}$ and 73.6 mmol m$^{-2}$ day$^{-1}$) and DOC fluxes increased from summer (156.7 mmol m$^{-2}$ day$^{-1}$) to fall (277.4 mmol m$^{-2}$ day$^{-1}$). The SGD-derived DIN fluxes to Baffin Bay were mainly in the form of $NH_4^+$. Lopez et al. (2018) extrapolated the average solute fluxes from SGD to the area of Baffin Bay (i.e., SGD-derived nutrient flux multiplied by the area of Baffin Bay) and found SGD supplied almost five orders of magnitude more DIN ($1029 \times 10^5$ mol day$^{-1}$ and $235 \times 10^5$ mol day$^{-1}$, respectively) than the surface runoff in both summer and late fall (750 mol day$^{-1}$ and 470 mol day$^{-1}$, respectively). Similarly, bay-wide average SGD inputs of DOC, $HSiO_3^-$, and $HPO_4^{2-}$ fluxes were three to five orders of magnitude larger than the surface runoff. The study showed that the nutrient input associated with SGD, regardless of if derived from groundwater inputs or from seawater recirculation within bay sediments, is significant in this shallow bay system during normal dry conditions.

In the Laguna Salada study (Murgulet et al. 2021), nutrient fluxes were calculated as the product of radon-derived SGD rates using the average groundwater endmember and the average of local groundwater and porewater nutrient concentrations. The largest NOx flux was derived using the average well concentrations, referred to as terrestrial, because groundwater was always about two orders of magnitude higher than porewaters. Terrestrial NOx fluxes were the highest in February 2021 ($25.6 \times 10^3$ µmol day$^{-1}$) and January 2020 ($25.6 \times 10^3$ µmol day$^{-1}$), while the lowest occurred in August. Generally, the terrestrial fluxes decreased from February to June 2020 at both sites, with peaks in July at both and in April at RP. The deeper porewater ($\sim$1 m) endmember concentrations resulted in NOx fluxes about one order of magnitude lower than the terrestrial ones, likely due to transformation of $NO_3^-$ to reduced forms like $NH_4^+$ along the transport path of groundwater within the anoxic, hypersaline subterranean estuary. In general, shallow porewater fluxes were slightly lower than those from deeper porewater, but both showed a similar pattern with increasing magnitudes after September. Like NOx, the largest $NH_4^+$ flux was derived using the average well concentrations, but when compared to those derived from porewaters, they were similar in range, except for January 2021 and February 2020. The terrestrial $NH_4^+$ fluxes exceeded those from porewater up to July, after which shallow porewater-derived fluxes were slightly higher indicating production of $NH_4^+$ in sediments from sources other than groundwater. Compared to the inorganic N species, fluxes of DON were overall highest in the first half of 2020 and peaked in June (shallow and deep porewater endmembers derived fluxes: $29.2 \times 10^3$ and $27.9 \times 10^3$ µmol day$^{-1}$, respectively) after which DON fluxes decreased significantly regardless of the source (terrestrial or pore). The terrestrial DON inputs slightly exceeding those from porewater in colder months (e.g., October—January). Fluxes of $HPO_4^{2-}$ and $HSiO_3^-$ varied by site and month during the study period. For instance, $HPO_4^{2-}$ ranged between $6.7 \times 10^3$ µmol day$^{-1}$ (using the shallow porewater as the source) in July and 81 µmol day$^{-1}$ in August (using both deep and groundwater sources). Overall, shallow porewater-derived fluxes exceeded the deeper and local groundwater sources, indicating that there is an additional source of $HPO_4^{2-}$ from the sediment. Fluxes of $HSiO_3^-$ were within similar ranges as $NH_4^+$, but interestingly followed a pattern like DON where $HSiO_3^-$ fluxes were highest in the first half of 2020 with peaks in early spring and summer, decreased significantly after July regardless of the source

(terrestrial or pore), and started increasing again in October. Terrestrial $HSiO_3^-$ inputs slightly exceed those from porewater throughout the entire monitoring period. The highest $HSiO_3^-$ flux was estimated in June from both shallow and deep porewater endmembers ($190 \times 10^3$ and $177.5 \times 10^3$ µmol day$^{-1}$, respectively) and the lowest in August from deep porewater ($398.1 \times 10$ µmol day$^{-1}$), although September and December match closely. Thus, the results of this study reveal much lower solute fluxes nearshore in Laguna Salada when compared with the bay-wide fluxes measured in 2016. This is mainly the result of lower concentrations measured in porewater nearshore. It is thus inferred that although SGD fluxes may be similar among nearshore and offshore in the bay, groundwater solute concentrations are driving lower nutrient fluxes to nearshore waters in Laguna Salada.

## Other Considerations

### Influence of Geologic Setting

The flow of groundwater is influenced by the geologic heterogeneity of the system, including but not limited to the porosity and permeability of the sediments and the depositional environment. For example, it is expected that groundwater discharge rates will be higher in unconsolidated sediments with a higher sand content because they have a greater intrinsic permeability and hydraulic conductivity (see Chap. 6). Consequently, SGD studies that focus on systems or include stations with high clay content and finer grains are not common.

In a high-resolution investigation of a 1 km$^2$ area along the northern shoreline of Copano Bay, Spalt et al. (2018, 2020) approached for the first time the role of depositional environments as hydrologic controls on SGD and nutrient supply to oyster reefs. The studies focused on Copano Bay near the mouth of Mission Bay and revealed higher SGD rates in proximity to oyster reefs that paralleled the paleovalley margins ($63 \pm 11$ cm day$^{-1}$ in January and $29 \pm 11$ cm day$^{-1}$ in July) when compared to the paleovalley, i.e., relict fluvial channels incised during a glacial maximum and subsequently filled or buried by younger, finer grained sediments as energy levels within the channel decreased ($32 \pm 9$ cm day$^{-1}$ in January and $20 \pm 7$ cm day$^{-1}$ in July). Furthermore, SGD from the 1 km$^2$ area supplied twofold to one order of magnitude more DIN than riverine inputs to Copano Bay during dry periods and DIN inputs were equivalent during wet periods. Thus, geologic features like paleovalleys may create preferential flow paths for SGD and create areas within the estuary that provide favorable conditions for estuarine biota, like primary producers and filter feeders such as oysters.

Similarly, in Nueces Bay, simultaneous continuous resistivity surveys and continuous $^{222}$Rn measurements compared to previous studies of sediment textural class distributions (Shideler et al. 1981) found greater SGD rates near the transitions from the higher clay content mid-bay muds to the sandier sediments nearshore (Douglas et al. 2020). Additionally, during porewater collection in the middle of Nueces Bay, soft, unconsolidated, and saturated silty-clay sediments (or "fluffy" mud) were repeatedly observed to ~3 m below the sediment-water interface indicating the integrity of the bottom clay-rich layer as a reliable confining unit had been altered to favor vertical advective flow (Barbour and Fredlund 1989; Gerla 1992; Silliman et al. 2002), similar to terrestrial occurrences of quicksand or soap holes (Toth 1971). However, the vertical groundwater flow is slowed by the low intrinsic permeability of the silty-clay mid-bay muds and redirected along the edges of the mid-bay muds (i.e., follows the path of least resistance) resulting in the observed increase in SGD near the sediment transitions. Meanwhile, in Baffin Bay, continuous resistivity surveys and high spatial and temporal resolution Rn measurements revealed enhanced SGD at relic serpulid reefs (Lopez et al. 2020) where the sandy substrates serpulid worms prefer (Simms et al. 2010) potentially provide a preferential groundwater flow path.

## Impacts of Extreme Events

Extreme hydrometeorological and hydroclimatic events, such as hurricanes, droughts, and floods, have been predicted to increase in frequency and intensity due to climate change (Webster et al. 2005). Coastal aquifers and SGD, influenced by both terrestrial and marine forces, are increasingly affected by climate variations and sea-level rise. Hydroclimatic conditions within watersheds are important and could significantly alter the hydraulic gradients in coastal areas. For instance, it is expected that during wet periods with net aquifer recharge, the aquifer's freshwater-seawater interface will shift seaward as the water table (aquifer) height increases and the hydraulic gradient increases. However, during dry periods, the water table drops due to limited recharge and decreases the hydraulic gradient resulting in the freshwater-seawater boundary shifting landward. The impacts of these hydroclimatic shifts are generally greater and the response times faster in the shallower unconfined and semi-confined aquifers than in the deeper confined aquifers. Thus, greater SGD rates with a greater freshwater component are expected to accompany or follow wet periods, while

decreasing rates with an increasing saline component are expected as dry periods carry on over weeks, months, and years. Over shorter time periods of days, major storm events, like hurricanes, may alter the hydraulic gradients in the impacted coastal areas causing landward encroachment of seawater where there is positive storm surge (Xiao et al. 2019) and enhanced hydraulic gradients and SGD where there is a negative storm surge (Douglas et al. 2022). The relatively short-term storm events may continue to impact the surficial aquifer's water quality for days to months following the storm's passage (Douglas et al. 2022) and the impacts of these natural perturbations on Texas' coastal aquifers and SGD fluxes require further investigation over the long term.

During the Murgulet et al. (2022) inter-bay comparison study discussed above, Hurricane Harvey made landfall in the northern extent of the study area during the late summer sampling event and proceeded to deliver record amounts of precipitation over the northern watersheds in aquifer outcrop areas resulting in a 1000-year flood event. Overall, regional average SGD rates were greater immediately following ($\bar{x}$ = 58.3 cm day$^{-1}$) and 3 months after Harvey's passage ($\bar{x}$ = 68.3 cm day$^{-1}$) than in the preceding winter ($\bar{x}$ = 41.5 cm day$^{-1}$) and spring ($\bar{x}$ = 43 cm day$^{-1}$) demonstrating that extreme precipitation events can enhance terrestrial groundwater inputs by increasing coastal aquifer recharge and steepening hydraulic gradients toward the coast. Furthermore, in Nueces Estuary, the impact of Hurricane Harvey manifested through negative storm surges and limited surface runoff which enhanced tidal pumping mechanisms caused by sudden drops in sea level and steepening of the hydraulic gradient between the water table and the sea. Using pre- and post-landfall surface-, pore-, and groundwater nutrient measurements and dissolved organic matter (DOM) molecular characterization analyses, found inorganic and organic nutrient compositions and inputs to vary substantially after landfall nearshore Corpus Christi Bay. The enhanced flushing of porewater resulting from the sudden decreases in sea level corresponded with pulses of NH$_4^+$ and lower proportions of N-containing DOM, likely from carbon-rich groundwater or a benthic source. Recovery to mean sea level associated with inputs of lower DIN and greater proportions of N-containing DOM. A month after landfall a higher terrestrial CHO- and S-containing (or sulfurized) DOM composition mirrored nearshore groundwater signatures, in which the CHO- and S-containing DOM are more abundant when compared to surface water. The study points at the enhanced groundwater inputs and flushing of porewater due to considerable drops in sea level and steepening of hydraulic gradients toward the coast as causes for the sulfurization of DOM. Additionally, the increased presence of

S-containing DOM indicates the mobilization of terrigenous DOM which is then exported to the estuary via SGD, including both recirculated seawater through bottom sediments and terrestrial groundwater fluxes. These limited works show the need for future work to better understand the effects of both negative and positive surges on SGD and the associated nutrient fluxes for improved awareness of the physicochemical and biological changes following a large storm event within Texas estuaries.

## Saltwater Intrusion and Submarine Groundwater Discharge

Saltwater intrusion, the landward excursion of seawater in coastal aquifers, and subsequent groundwater salinization are accelerating along many coastlines and expanding the STE (Moore and Joye 2021). Saltwater intrusion may result from groundwater extraction, sea-level rise, reduced groundwater recharge, and human disturbances to confining layers. Over-pumping of fresh groundwater from coastal aquifers enhances groundwater intrusion by reducing the hydrostatic pressure and drawing saltwater into the aquifer from the ocean, an adjacent aquifer, or relic saltwater from deeper in the aquifer. In addition to over-pumping, sea-level rise also alters the hydrostatic balance between the coastal aquifer and the coastal ocean. On a coastal plain with low relief (for example, a slope of 1:1000), a 10 cm rise in sea level can lead to a 100 m inland intrusion of seawater into the aquifer (Werner and Simmons 2009). Additionally, seawater intrusion may be exacerbated by processes that direct precipitation into surface runoff and impede groundwater recharge, such as covering permeable surfaces with impermeable surfaces (e.g., roads, parking lots, buildings), sewer and canal construction, and draining freshwater coastal wetlands (Moore and Joye 2021).

As little as 1%, seawater will render groundwater nonpotable by taste. Intrusion of saltwater into freshwater coastal aquifers may cause profound changes in the biogeochemistry of the affected aquifers and STE as increased saltwater intrusion alters the ionic strength and oxidative capacity of these systems (Moore and Joye 2021). The increased ionic strength causes some particle-bound, surface-active ions to be released to solution thus creating elevated concentrations of some dissolved chemical species, such as nutrients (e.g., $NH_4^+$, $PO_4^{3-}$), trace metals and gases (e.g., Fe, Mn, Ba, Hg, As), sulfides, and radionuclides (e.g., Ra), in groundwater. Additionally, the oxidation capacity of the STE increases with increased salinity from saltwater intrusion because once dissolved oxygen is depleted in saltwater, sulfate becomes the primary electron acceptor for organic matter oxidation and saltwater generally has a high sulfate concentration. Moore and Joye (2021) found that the introduction of sulfate ion, in effect, transforms aquifers from oxidant-limited to nearly oxidant unlimited thus stimulating "the biogeochemical reactions that degrade organic matter via anaerobic metabolisms, leading to increased concentrations of nutrients, dissolved carbon (DIC and DOC), hydrogen sulfide and metals in the subterranean estuary." Consequently, STE outflows as SGD in areas of saltwater intrusion are highly enriched in nutrients, carbon, sulfide, and metals and have been linked to harmful algal blooms, eutrophication, and hypoxia (Hwang et al. 2005; Lee et al. 2009; Peterson et al. 2016).

## Effects of Submarine Groundwater Discharge on Marine Biota

Though studies of solute fluxes associated with SGD have become more common around the globe in recent years, the actual impacts of SGD on marine biota are still not well constrained. While various methods have been used to assess effects of SGD on organisms, including field campaigns measuring SGD in situ along with enumeration of organisms and their physiologies (Blanco et al. 2008; Su et al. 2014; Honda et al. 2018), mesocosm and bottle incubation experiments seeking to understand the dose response of organisms to groundwater additions (Garcés et al. 2011; Lecher et al. 2015, Lecher et al. 2017), and modeling organism response to set concentrations of nutrient additions (Wang et al. 2018), the studies all found that SGD is a key source of nutrients to coastal waters and may cause obvious changes in primary production and nutrient structure. For instance, in Daya Bay, China, SGD was found to be the primary source of $NO_3^-$ N and Si contributing up to 68% and 42% of all considered sources, respectively (Wang et al. 2018), and in Dongshan Bay, China, fresh SGD accounted for 13% of total SGD but delivered 7.4–80.5% of the total SGD-derived nutrients, thus providing an important source of new nutrients with a high N/P ratio (585) to a fish and shellfish mariculture bay (Yu et al. 2022). Furthermore, in Dongshan Bay, SGD provided 2.8–21.2 times more nutrients than local rivers to the mariculture bay suggesting that SGD may be important to the success of the mariculture industry in this system. We refer the reader to Lecher and Mackey (2018) for a more comprehensive review of the effects of SGD on marine macrophytes (e.g., seagrasses and kelp), macrofauna (e.g., coral and meiofauna), and ecosystems than is provided here.

The most common organisms studied with respect to SGD are phytoplankton and bacteria as they provide the base of many aquatic ecosystems and are supported by the N, P, and Si brought into the system with SGD. Several studies have shown correlations and covariance between $^{222}$Rn and chlorophyll-α indicating that SGD relieves nutrient limitations and fuels primary production (Blanco et al. 2008; Lecher et al. 2015, 2017). For instance, diatoms have positive responses to SGD inputs as SGD is generally enriched in DIN, DIP, and dissolved Si, which the diatoms need for their frustules, due to the rock-water interactions in the coastal aquifer dissolving aquifer substrate. In the Gulf of Mexico, salinity-loving diatom species like *Pseudo-nitzschia* spp. have been shown to respond positively to SGD, likely due to the increase in nitrate from brackish or recirculated SGD without a corresponding decrease in salinity, as would be expected from increased riverine inputs (Liefer et al. 2009). While the positive and negative influences of SGD on marine biota are being investigated elsewhere, this remains an unanswered question in Texas coastal ecosystems. However, nutrient data from Murgulet et al. (2019), collected in conjunction with the Murgulet et al. (2022) study, show that groundwater and porewater in Mission-Aransas, Nueces, and Upper Laguna Madre Estuaries are often elevated in N and Si nutrients relative to Redfield ratios meaning that SGD is a potential year-round source of Si and N for phytoplankton (Fig. 11). Similarly, Santos et al. (2021) looked at the DIN:DIP versus DIN:DSi ratios of fresh and/or saline SGD from 239 study cases from 31 countries and found that 58% of the compiled studies were enriched in Si, 36% were enriched in N, and 6% were enriched in P relative to the Redfield ratio. Furthermore, like the groundwater endmembers in Oso Bay, Laguna Madre, and Baffin Bay in Murgulet et al. (2019) (Fig. 11), the DIN:DIP ratios in SGD were com-

monly >16, even at Si-enriched locations, demonstrating that SGD may counter N limitation in most coastal waters and fuel phytoplankton blooms.

## Impacts of Anthropogenic Disturbances to Bay Sediments on Groundwater and Hydrogeology

Existing ship channels, such as the Intracoastal Waterway, the Houston, Corpus Christi, and the Matagorda ship channels, and ongoing drilling and dredging projects may potentially impact local hydrology and groundwater quality in the shallow unconfined and semi-confined aquifers of the GCA. For instance, in the upper 30.5 m (100 ft) of the Beaumont Formation near Lavaca Bay in Guadalupe estuary, three primary saturated sand and silt zones occurring at relatively consistent depths were identified with intervening clay layers in the EPA's site assessment of the ALCOA Superfund site (Scanlon et al. 2005) (EPA 2001). Zone A, at an elevation of 1.5 to −0.3 m mean sea level (5 to −1 ft msl), discharges to the bay system at the shoreline where it crops out. Zone B, at −6.1 to −9.1 m msl (−20 to −30 ft msl), discharges to the bay system in the existing ship channel and turning basin where the deep channel intercepts Zone B. Zone C, at −14.9 to −20 m msl (−49 to −66 ft msl), does not currently outcrop in the bay system. Groundwater discharges upward through strata offshore from both Zone B and Zone C. Assuming the depth to Zone C remains consistent under the bay, the currently proposed deepening of the Matagorda Ship Channel to ~13 m (~44 ft) with an allowable over depth of 0.61 m (2 ft) would leave ~1 m (~3.3 ft)

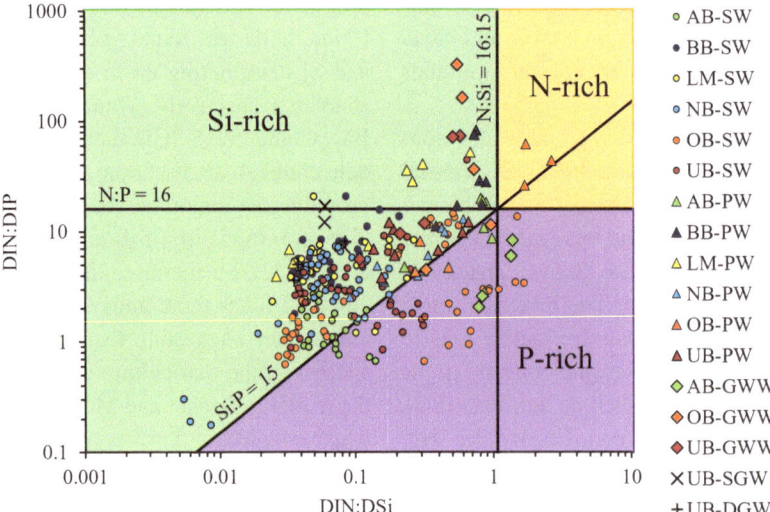

**Fig. 11** Potential surface water nutrient enrichment relative to the Redfield ratio and support from submarine groundwater discharge as shown by dissolved inorganic nitrogen (DIN):dissolved inorganic phosphorous (DIP) versus DIN:dissolved silicate (DSi) ratios in surface water and porewater/groundwater endmembers. AB = Aransas Bay,

BB = Baffin Bay, LM = Laguna Madre, NB = Nueces Bay, OB = Oso Bay, UB = University Beach, SW = surface water, PW = Porewater, GWW = dug well, SGW = shallow groundwater, DGW = deep groundwater

of confining sediment in place and, thus, should not intersect this groundwater-bearing layer (Montagna et al. 2021 and references therein). However, thinning the confining layer may also enhance groundwater discharge to the bay (von Ahn et al. 2021). Thus, dredging bay sediments and placement of oil/gas and water wells risks breaching underlying confining layers or reducing their reliability as confining layers which may disturb natural pressure and flow fields in the shallow subsurface or allow increased communication between previously confined groundwater lenses and bay surface waters (i.e., short-circuit confining layers) (Duncan 1972; Lautier 1998; Moore 1999; Teatini et al. 2017). Furthermore, interruption of the hydraulic connection, and the associated nutrient inputs, to the seafloor further offshore could negatively impact benthic communities at the base of the food chain (Pinckney 2018). However, given the quantity of dredging and drilling projects in the coastal zone, the potential impacts of these projects on GW-SW interactions and coastal hydrology are still poorly constrained.

The few studies of the impacts of dredging and ship channels on coastal hydrology reveal that impacts may vary widely from relatively minor and smaller in scale (Jorgensen 1976; Lautier 1998; Bellino and Spechler 2013) to potentially major and larger in scale (Duncan 1972; Wang 1988; Murgulet and Tick 2008; Santos et al. 2008; Teatini et al. 2017; von Ahn et al. 2021). These studies observed that navigation channels and dredging for irrigation canals may enhance SGD or serve as conduits for deeper groundwater discharge in areas where the confining clay layer is thinned or breached, ship channels may serve as the source of saltwater intrusion into shallow groundwater, saltwater intrusion into coastal aquifers is greater from larger and deeper channels than from smaller and shallower channels, and ships passing through navigation channels may enhance recirculated SGD through ship-wake effects. Thus, the potential impacts of anthropogenic disturbances to bay sediments on GW-SW interactions will greatly depend on the local conditions (e.g., geology, climate) and should be assessed for existing and future projects. Baseline studies of SGD should be conducted prior to the port expansion projects and further deepening and widening of navigation channels along the Texas coastline.

## SGD and Ecosystem Services

To help facilitate incorporation of SGD into management planning and community discussions, we recommend the development of a framework that places SGD within the larger discussion of environmental flows and ecosystem services. For example, Alorda-Kleinglass et al. (2021) used the ecosystem services approach based on the Millennium Ecosystem Assessment (2005) to develop a common and interdisciplinary framework to assess the broader social implications of SGD. The framework (Fig. 12a) is based on four broad ecosystem service categories identified as:

1. Supporting: indirect services provided by SGD that do not directly affect human well-being but sustain the existence of other ecosystem services.
2. Provisioning: products that SGD provides to society.
3. Regulating: services that control crucial processes for habitats and coastal ecosystems influenced by SGD.
4. Cultural: non-material benefits provided by SGD that contribute to human values and influence behavior. Note that the subjectivity of the observer plays a big role in the perception of this service and varies across stakeholders and communities.

These broad categories are further subdivided into outcomes that reflect the different services of each category from SGD and relate to each other or can be prioritized through synergies (i.e., mutually beneficial relationship between two or more outcomes) and trade-offs (i.e., societal choice to prioritize one outcome in exchange of one or more of the other outcomes) in order to achieve coastal society well-being (i.e., the capacity of an ecosystem service to provide the conditions for physical, social, psychological, and spiritual fulfillment). Because trade-offs related to SGD-ecosystem services can become especially complex when economic, cultural, or political interests are at stake (Fig. 12b), policies and management strategies need to be developed that consider the synergies and trade-offs between SGD and environmental flows or ecosystem services that achieve or maintain the well-being of our coastal communities. For example, apply the framework to a community that decides SGD is a freshwater resource to be collected and used for their own consumption resulting in the reduced flux of water and solutes to the coastal ecosystem (Fig. 12b) (Alorda-Kleinglass et al. 2021). The trade-offs may be that the provisioning of freshwater resources reduces the regulat-

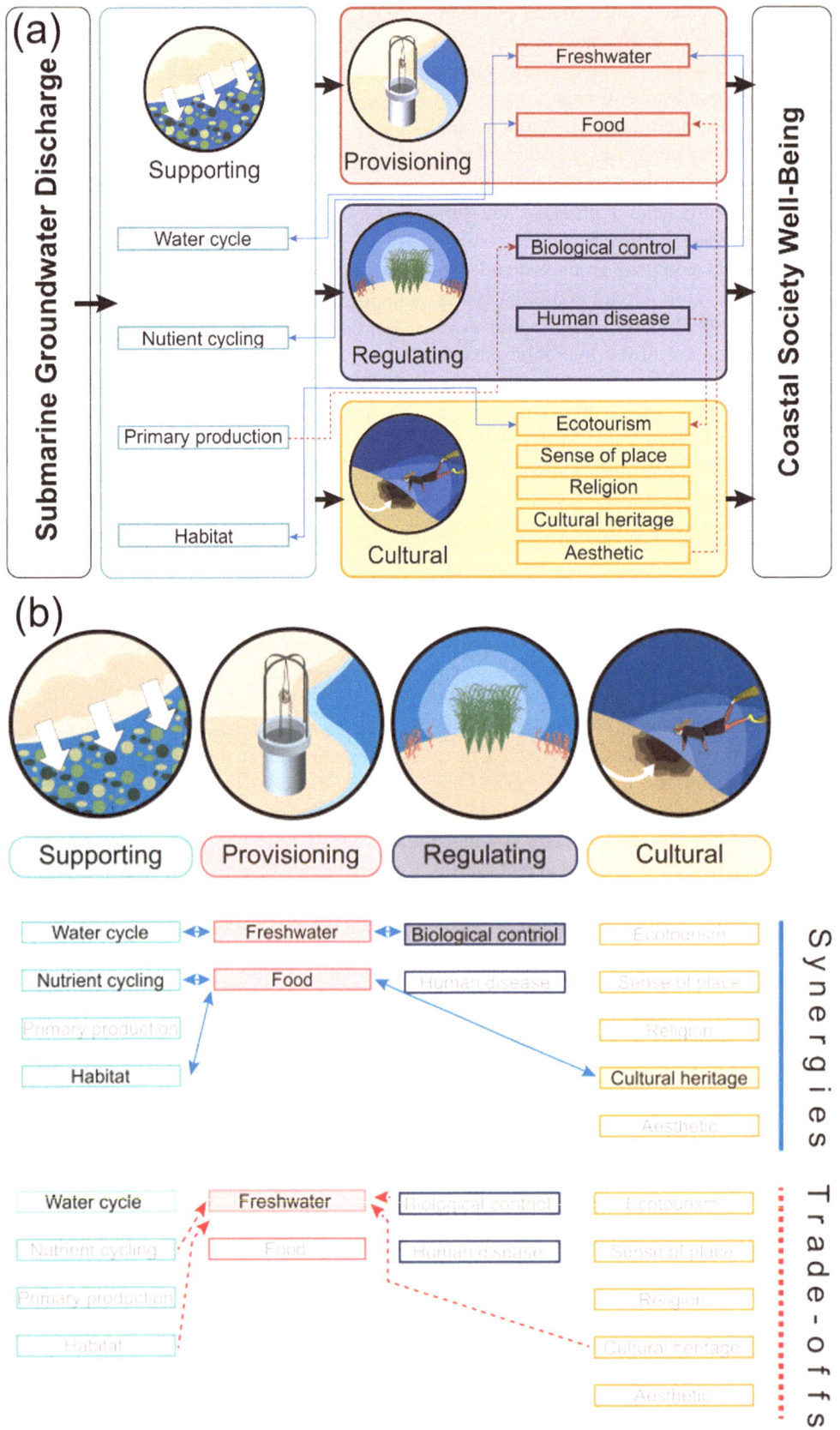

**Fig. 12** (**a**) Conceptual framework for SGD-ecosystem services. Outcomes (squared boxes) for the four ecosystem services, synergies (blue arrows), and trade-offs (red arrows) are shown. (**b**) Example of the application of the framework to a community that decides SGD is a freshwater resource to be collected and used for their own consumption resulting in reduced flux of water and solutes to the coastal ecosystem. Trade-offs = provision of freshwater resource in exchange for reduced regulating and supporting services in the coastal ecosystem and detriment to nearby coastal communities. Figures arranged and reprinted from Alorda-Kleinglass et al. (2021) under Creative Commons license CC BY 4.0 (https://creativecommons.org/licenses/by/4.0/)

ing and supporting services from the supply of water and nutrients to the coastal ecosystem and detriment to nearby coastal communities whose cultural heritage may depend on consumption of SGD-dependent species. Applying this scenario to Texas, pumping of groundwater for municipal or industrial uses, would reduce the flow of groundwater to the coast and the Gulf of Mexico, potentially limiting the supply of essential nutrients and freshwater inputs to N- or P-depleted waters.

## Summary and Recommendations

Despite the growing body of evidence supporting the role of SGD and the STE in coastal ecosystem functions and ecosystem services around the globe, these components of the hydrologic cycle remain absent from most environmental management frameworks, particularly from water and nutrient budgets. The innate separation of surface and groundwater systems at institutional, legislative, and political levels generally excludes GW-SW interactions from consideration in governance practice. Thus, SGD and the STE remain largely overlooked in coastal monitoring and management plans worldwide and in Texas. Consequently, a potentially ecologically significant source of freshwater and solutes, such as nutrients in NPS pollution, to bays and estuaries is missing from management strategies. For instance, the current and ongoing nutrient reduction efforts should incorporate the contribution from terrestrial groundwater inputs for a more successful outcome. Thus, we agree with Rocha et al. (2021) and Moore and Joye (2021) who both "strongly advocate for the inclusion of subterranean estuaries as essential components of coastal ecosystem function research, monitoring, policy, and management" and further suggest considering groundwater and surface water as one interconnected water system in future research, monitoring, policy, and management decisions.

In Texas, efforts have begun to quantify SGD rates and associated solute fluxes, but sizable portions of the coast remain unstudied, including four of the major estuarine systems (i.e., Trinity-San Jacinto, Lavaca-Colorado, Guadalupe, and Lower Laguna Madre). For the bays and estuaries that have been studied, a substantial portion of the SGD is saline/recirculated (e.g., Baffin Bay and Aransas Bay); however, SGD may provide a substantial input of fresh/terrestrial water and solutes of terrestrial origin to the rivers and streams flowing into the bays as well as to the bays directly (e.g., Nueces Estuary). These SGD fluxes have been shown to vary spatially within a bay system due to geologic setting and anthropogenic disturbances of the bay sediments and temporally due to dry/wet periods and major storm events raising or lowering the water table height or sea level and thus alter-

ing the hydraulic gradient between the aquifer and the bay. Additionally, the SGD-derived nutrient fluxes have been shown to exceed riverine nutrient inputs in some systems that have previously unbalanced nutrient budgets or have been deemed impaired by previous assessments (i.e., Nueces Bay) demonstrating the need for GW-SW interactions to be included in future monitoring and management plans. Furthermore, the existing studies hint at large differences in SGD inputs among climatic extremes and hydroclimatic gradients and call for long-standing monitoring of both SGD inputs and STE chemical composition to better prepare long-term water and solute budgets. We recommend estuarine ecologists, environmental chemists, and coastal hydrogeologists collaborate to co-develop research questions and projects that begin to connect the subterranean and the surface estuaries.

To incorporate SGD into solute budgets and management strategies, groundwater inputs need to be monitored similarly to surface water inputs (e.g., stream gauges and data buoys). However, with the current methods and technology, such efforts would be incredibly time-intensive and costly. Thus, a more limited application of these methods within each estuary system (i.e., biweekly, monthly, or quarterly 24-hour monitoring at a select few sites) to determine a range of SGD rates or an average annual SGD rate would be a most immediate and simplistic solution. In contrast, monitoring of shallow groundwater is less time-intensive and can aid valuable information regarding nutrient loading and aquifer recharge that can help determine temporal and spatial changes in terrestrial groundwater inputs and source determination. From these groundwater monitoring efforts, groundwater inputs can then be added to surface water budgets and incorporated into management decisions and strategies. For instance, shallow groundwater that is high in nutrients and near agricultural areas may indicate excessive fertilizer usage and will inform nutrient reduction efforts by estimating excess application. An evaluation of potential nutrient sources based on land use/land cover, hydrologic inputs (surface and subsurface runoff), bay residence times, among others, will inform a holistic mass balance model, leading to a greater understanding of nutrient sources to Texas estuaries, which is an important step in meeting the overarching goal of the Coastal Zone Act Reauthorization Amendments (CZARA or §6217) to "protect coastal waters" by "control" of NPS pollution. Finally, the adoption and adaptation of the SGD-ecosystem services framework for management discussions could further aid the inclusion of SGD in the discussion and protection of environmental flows and ecosystem services leading to better-informed decisions that consider and account for the trade-offs between different stakeholders and communities.

**Acknowledgments** This publication was funded in part through Contract No. 21-155-007-C879 from the Texas General Land Office (GLO) with Gulf of Mexico Energy Security Act of 2006 funding made available to the State of Texas and awarded under the Texas Coastal Management Program. The views contained herein are those of the authors and should not be interpreted as representing the views of the GLO or the State of Texas.

**Data Sources** All Center for Water Supply Studies data are publicly available or available by request to Dorina Murgulet (dorina.murgulet@tamucc.edu). The CWSS data used and presented in this chapter are available on GRIIDC: https://doi.org/10.7266/SJACB00R.

# References

Alorda-Kleinglass A, Ruiz-Mallén I, Diego-Feliu M, Rodellas V, Bruach-Menchén JM, Garcia-Orellana J (2021) The social implications of submarine groundwater discharge from an ecosystem services perspective: a systematic review. Earth Sci Rev 221:103742. https://doi.org/10.1016/j.earscirev.2021.103742

An S, Gardner WS (2002) Dissimilatory nitrate reduction to ammonium (DNRA) as a nitrogen link, versus denitrification as a sink in a shallow estuary (Laguna Madre/Baffin Bay, Texas). Mar Ecol Prog Ser 237:41–50. https://doi.org/10.3354/meps237041

Anchor QEA (2017) Nutrient budget for Nueces Bay. Texas Water Development Board, Austin, TX. 83 p

Baker ET (1979) Stratigraphic and hydrogeologic framework of part of the coastal plain of Texas. 77–712. Survey USG, 34 p. https://doi.org/10.3133/ofr77712

Barbour S, Fredlund D (1989) Mechanisms of osmotic flow and volume change in clay soils. Can Geotech J 26:551–562. https://doi.org/10.1139/t89-068

Bellino JC, Spechler RM (2013) Potential effects of deepening the St. Johns River navigation channel on saltwater intrusion in the surficial aquifer system, Jacksonville, Florida. 2013–5146. Survey USG, Reston, VA., 46 p. https://doi.org/10.3133/sir20135146

Bighash P, Murgulet D (2015) Application of factor analysis and electrical resistivity to understand groundwater contributions to coastal embayments in semi-arid and hypersaline coastal settings. Sci Total Environ 532:688–701. https://doi.org/10.1016/j.scitotenv.2015.06.077

Blanco AC, Nadaoka K, Yamamoto T (2008) Planktonic and benthic microalgal community composition as indicators of terrestrial influence on a fringing reef in Ishigaki Island, Southwest Japan. Mar Environ Res 66:520–535. https://doi.org/10.1016/j.marenvres.2008.08.005

Bokuniewicz H, Rapaglia J, Beck A (2006) Submarine groundwater discharge (SGD) from a volcanic Island: a case study in Mauritius Island. Int J Oceans Oceanogr 1:319–336

Bone SE, Charette MA, Lamborg CH, Gonneea ME (2007) Has submarine groundwater discharge been overlooked as a source of mercury to coastal waters? Environ Sci Technol 41:3090–3095. https://doi.org/10.1021/es0622453

Breier JA, Breier CF, Edmonds HN (2005) Detecting submarine groundwater discharge with synoptic surveys of sediment resistivity, radium, and salinity. Geophys Res Lett 32. https://doi.org/10.1029/2005GL024639

Breier JA, Breier CF, Edmonds HN (2010) Seasonal dynamics of dissolved Ra isotopes in the semi-arid bays of south Texas. Mar Chem 122:39–50. https://doi.org/10.1016/j.marchem.2010.08.008

Breier JA, Edmonds HN (2007) High Ra-226 and Ra-228 activities in Nueces Bay, Texas indicate large submarine saline discharges. Mar Chem 103:131–145. https://doi.org/10.1016/j.marchem.2006.06.015

Brock DA (2001) Nitrogen budget for low and high freshwater inflows, Nueces Estuary, TX. Estuaries 24:509–521. https://doi.org/10.2307/1353253

Brown LF Jr, Loucks RG, Trevio RH, Hammes U (2004) Understanding growth-faulted, intraslope subbasins by applying sequence-stratigraphic principles: examples from the south Texas Oligocene Frio formation. AAPG Bull 88:1501–1522. https://doi.org/10.1306/07010404023

Bruun B, Jackson K, Lake P, Walker J (2016) Texas aquifers study: groundwater quantity, quality, flow, and contributions to surface water. Board TWD, Austin, TX. 336 p

Burgin AJ, Hamilton SK (2007) Have we overemphasized the role of denitrification in aquatic ecosystems? A review of nitrate removal pathways. Front Ecol Environ 5:89–96. https://doi.org/10.1890/1540-9295(2007)5[89:HWOTRO]2.0.CO;2

Burnett WC, Bokuniewicz H, Huettel M, Moore WS, Taniguchi M (2003) Groundwater and pore water inputs to the coastal zone. Biogeochemistry 66:3–33. https://doi.org/10.1023/B:BIOG.0000006066.21240.53

Burnett WC, Dulaiova H (2003) Estimating the dynamics of groundwater input into the coastal zone via continuous radon-222 measurements. J Environ Radioact 69:21–35. https://doi.org/10.1016/S0265-931X(03)00084-5

Cardenas MB, Rodolfo RS, Lapus MR, Cabria HB, Fullon J, Gojunco GR, Breecker DO, Cantarero DM, Evaristo J, Siringan FP (2020) Submarine groundwater and vent discharge in a volcanic area associated with coastal acidification. Geophys Res Lett 47:e2019GL085730. https://doi.org/10.1029/2019GL085730

Charette MA, Sholkovitz ER (2002) Oxidative precipitation of groundwater-derived ferrous iron in the subterranean estuary of a coastal bay. Geophys Res Lett 29:85–81. https://doi.org/10.1029/2001GL014512

Charette MA, Splivallo R, Herbold C, Bollinger MS, Moore WS (2003) Salt marsh submarine groundwater discharge as traced by radium isotopes. Mar Chem 84:113–121. https://doi.org/10.1016/j.marchem.2003.07.001

Charkin AN, Rutgers van der Loeff M, Shakhova NE, Gustafsson Ö, Dudarev OV, Cherepnev MS, Salyuk AN, Koshurnikov AV, Spivak EA, Gunar AY, Ruban AS, Semiletov IP (2017) Discovery and characterization of submarine groundwater discharge in the Siberian Arctic seas: a case study in the Buor-Khaya Gulf, Laptev Sea. Cryosphere 11:2305–2327. https://doi.org/10.5194/tc-11-2305-2017

Chowdhury AH, Wade S, Mace RE, Ridgeway C (2004) Groundwater availability model of the central gulf coast aquifer system: numerical simulations through 1999. Model Report. Texas Water Development Board, Austin, TX, USA, 14 p.

Chowdury AH, Turco MJ (2006) Chapter 2 geology of the Gulf coast aquifer, Texas. In: Mace RE, Davidson SC, Angle ES, Mullican WF (eds) Aquifers of the Gulf coast of Texas. Texas Water Development Board, Austin, TX, pp 23–50

Church TM (1996) An underground route for the water cycle. Nature 380:579–580. https://doi.org/10.1038/380579a0

Corson-Dosch H, Nell C, Volentine R, Archer AA, Bechtel E, Bruce JL, Felts N, Gross TA, Lopez-Trujillo D, Riggs CE, Read EK (2022) The water cycle. Water Data Visualizations Team, U.S. Geological Survey https://www.usgs.gov/media/images/water-cycle-png

Dai A, Trenberth KE (2002) Estimates of freshwater discharge from continents: latitudinal and seasonal variations. J Hydrometeorol 3:660–687. https://doi.org/10.1175/1525-7541(2002)003<0660:EOFDFC>2.0.CO;2

Danielescu S, MacQuarrie KT, Faux RN (2009) The integration of thermal infrared imaging, discharge measurements and numerical simulation to quantify the relative contributions of freshwater inflows to small estuaries in Atlantic Canada. Hydrol Process 23:2847–2859. https://doi.org/10.1002/hyp.7383

Dimova NT, Swarzenski PW, Dulaiova H, Glenn CR (2012) Utilizing multichannel electrical resistivity methods to examine the dynamics of the fresh water-seawater interface in two Hawaiian groundwater systems. J Geophys Res Oceans 117. https://doi.org/10.1029/2011JC007509

Douglas A, Murgulet D, Wetz MS, Spalt N, Lopez C, Wang H (2017) Evaluating groundwater inflow and nutrient transport to Texas coastal embayments. TGLO Contract No. 15-047-000-8392 Texas General Land Office, Austin, TX, 109 p

Douglas AR, Murgulet D, Abdulla HA (2021a) Impacts of hydroclimatic variability on surface water and porewater dissolved organic matter in a semi-arid estuary. Mar Chem 235:104006. https://doi.org/10.1016/j.marchem.2021.104006

Douglas AR, Murgulet D, Greige M, Das K, Felix JD, Abdulla HAN (2022) Organic matter composition and inorganic nitrogen response to Hurricane Harvey's negative storm surge in Corpus Christi Bay, Texas. Front Mar Sci. https://doi.org/10.3389/fmars.2022.961206

Douglas AR, Murgulet D, Montagna PA (2021b) Hydroclimatic variability drives submarine groundwater discharge and nutrient fluxes in an anthropogenically disturbed, semi-arid estuary. Sci Total Environ 755:142574. https://doi.org/10.1016/j.scitotenv.2020.142574

Douglas AR, Murgulet D, Peterson RN (2020) Submarine groundwater discharge in an anthropogenically disturbed, semi-arid estuary. J Hydrol 580:124369. https://doi.org/10.1016/j.jhydrol.2019.124369

Duncan DA (1972) High resolution seismic survey. Port Royal Sound Environ Stud:85–106

Duque C, Michael HA, Wilson AM (2020) The subterranean estuary: technical term, simple analogy, or source of confusion? Water Resour Res 56:e2019WR026554. https://doi.org/10.1029/2019WR026554

El-Gamal AA, Peterson RN, Burnett WC (2012) Detecting freshwater inputs via groundwater discharge to Marina Lagoon, Mediterranean Coast, Egypt. Estuaries Coast 35:1486–1499. https://doi.org/10.1007/s12237-012-9539-2

Ellis J, Knight JE, White JT, Sneed M, Hughes JD, Ramage JK, Braun CL, Teeple A, Foster LK, Rendon SH, Brandt JT (2023) Hydrogeology, land-surface subsidence, and documentation of the Gulf Coast Land Subsidence and Groundwater-Flow (GULF) model, southeast Texas, 1897–2018. Survey USG, Reston, VA. https://doi.org/10.3133/pp1877

Environmental Protection Agency (EPA) (2001) Region 6 Superfund Division. Record of Decision: Alcoa (Point Comfort)/Lavaca Bay Site Point Comfort, Texas. CERCLIS# TXD 008123168. United States Environmental Protection Agency, Washington, DC, USA. https://semspub.epa.gov/work/06/908161.pdf.

Fetter CW (2001) Applied hydrogeology. Prentice-Hall, Inc, Englewood Cliffs, p 598

Fleury P, Bakalowicz M, de Marsily G (2007) Submarine springs and coastal karst aquifers: a review. J Hydrol 339:79–92. https://doi.org/10.1016/j.jhydrol.2007.03.009

Garcés E, Basterretxea G, Tovar-Sánchez A (2011) Changes in microbial communities in response to submarine groundwater input. Mar Ecol Prog Ser 438:47–58. https://doi.org/10.3354/meps09311

Garcia-Orellana J, Rodellas V, Tamborski J, Diego-Feliu M, van Beek P, Weinstein Y, Charette M, Alorda-Kleinglass A, Michael HA, Stieglitz T, Scholten J (2021) Radium isotopes as submarine groundwater discharge (SGD) tracers: review and recommendations. Earth Sci Rev 220:103681. https://doi.org/10.1016/j.earscirev.2021.103681

George PG, Mace RE, Petrossian R (2011) Aquifers of Texas. TWDB Report 380. Board TWD, Austin, TX. 182 p

Gerla PJ (1992) The relationship of water-table changes to the capillary fringe, evapotranspiration, and precipitation in intermittent wetlands. Wetlands 12:91–98. https://doi.org/10.1007/bf03160590

Giblin AE, Tobias CR, Song B, Weston N, Banta GT, Rivera-Monroy VH (2013) The importance of dissimilatory nitrate reduction to ammonium (DNRA) in the nitrogen cycle of coastal ecosystems. Oceanography 26:124–131

Guo W, Langevin CD (2002) User's guide to SEAWAT; a computer program for simulation of three-dimensional variable-density groundwater flow. Techniques of Water-Resources Investigations 06-A7. USGS, 77 p. https://doi.org/10.3133/twri06A7

Hammond WW Jr (1969) Ground-water resources of Matagorda County, Texas. TWDB report 91. Board TWD, Austin, TX. 163 p

Honda H, Sugimoto R, Kobayashi S (2018) Submarine groundwater discharge and its influence on primary production in Japanese coasts: case study in Obama Bay. The water-energy-food nexus: human-environmental security in the asia-pacific ring of fire, pp 101-115. https://doi.org/10.1007/978-981-10-7383-0_8

Hwang D-W, Kim G, Lee Y-W, Yang H-S (2005) Estimating submarine inputs of groundwater and nutrients to a coastal bay using radium isotopes. Mar Chem 96:61–71. https://doi.org/10.1016/j.marchem.2004.11.002

Jorgensen DG (1976) Salt-water encroachment in aquifers near the Houston Ship Channel, Texas. 76-781. Survey USG, 58 p. https://doi.org/10.3133/ofr76781

Kasmarek MC (2012) Hydrogeology and simulation of groundwater flow and land-surface subsidence in the northern part of the Gulf coast aquifer system, Texas, 1891-2009 (ver. 1.1 December 2013). 55 p. https://doi.org/10.3133/sir20125154

Kasmarek MC, Robinson JL (2004) Hydrogeology and simulation of ground-water flow and land-surface subsidence in the northern part of the Gulf coast aquifer system, Texas. 2004–5102. Survey USG, 103 p. https://doi.org/10.3133/sir20045102

Kelly JL, Glenn CR, Lucey PG (2013) High-resolution aerial infrared mapping of groundwater discharge to the coastal ocean. Limnol Oceanogr Methods 11:262–277. https://doi.org/10.4319/lom.2013.11.262

Kohout F (1966) Submarine springs: a neglected phenomenon of coastal hydrology. Hydrology 26:391–413

Kwon EY, Kim G, Primeau F, Moore WS, Cho HM, DeVries T, Sarmiento JL, Charette MA, Cho YK (2014) Global estimate of submarine groundwater discharge based on an observationally constrained radium isotope model. Geophys Res Lett 41:8438–8444. https://doi.org/10.1002/2014GL061574

Lautier JC (1998) Hydrogeologic assessment of the proposed deepening of the Wilmington Harbor shipping channel, New Hanover and Brunswick counties, North Carolina. NC Department of Environment, Health and Natural Resources, Division of …

Lecher AL, Chien C-T, Paytan A (2016) Submarine groundwater discharge as a source of nutrients to the North Pacific and Arctic coastal ocean. Mar Chem 186:167–177. https://doi.org/10.1016/j.marchem.2016.09.008

Lecher AL, Mackey K, Kudela R, Ryan J, Fisher A, Murray J, Paytan A (2015) Nutrient loading through submarine groundwater discharge and phytoplankton growth in Monterey Bay, CA. Environ Sci Technol 49:6665–6673. https://doi.org/10.1021/acs.est.5b00909

Lecher AL, Mackey KR (2018) Synthesizing the effects of submarine groundwater discharge on marine biota. Hydrology 5:60. https://doi.org/10.3390/hydrology5040060

Lecher AL, Mackey KRM, Paytan A (2017) River and submarine groundwater discharge effects on diatom phytoplankton abundance in the Gulf of Alaska. Hydrology 4:61. https://doi.org/10.3390/hydrology4040061

Lee DR (1977) A device for measuring seepage flux in lakes and estuaries. Limnol Oceanogr 22:140–147. https://doi.org/10.4319/lo.1977.22.1.0140

Lee JM, Kim G (2007) Estimating submarine discharge of fresh groundwater from a volcanic Island using a freshwater budget of the coastal water column. Geophys Res Lett 34. https://doi.org/10.1029/2007GL029818

Lee Y-W, Hwang D-W, Kim G, Lee W-C, Oh H-T (2009) Nutrient inputs from submarine groundwater discharge (SGD) in Masan Bay, an embayment surrounded by heavily industrialized cities,

Korea. Sci Total Environ 407:3181–3188. https://doi.org/10.1016/j.scitotenv.2008.04.013

Li X, Hu BX, Burnett WC, Santos IR, Chanton JP (2009) Submarine ground water discharge driven by tidal pumping in a heterogeneous aquifer. Ground Water 47:558–568. https://doi.org/10.1111/j.1745-6584.2009.00563.x

Liefer JD, MacIntyre HL, Novoveská L, Smith WL, Dorsey CP (2009) Temporal and spatial variability in Pseudo-nitzschia spp. in Alabama coastal waters: a "hot spot" linked to submarine groundwater discharge? Harmful Algae 8:706–714. https://doi.org/10.1016/j.hal.2009.02.003

Lopez C, Murgulet D, Douglas A, Murgulet V (2018) Impacts of temporal and spatial variation of submarine groundwater discharge on nutrient fluxes to Texas coastal Embayments, phase III (Baffin Bay). TGLO Contract No. 16-060-000-9104. Texas General Land Office, Austin, TX. 106 p

Lopez CV, Murgulet D, Santos IR (2020) Radioactive and stable isotope measurements reveal saline submarine groundwater discharge in a semiarid estuary. J Hydrol 590:125395. https://doi.org/10.1016/j.jhydrol.2020.125395

McKenzie T, Holloway C, Dulai H, Tucker JP, Sugimoto R, Nakajima T, Harada K, Santos IR (2020) Submarine groundwater discharge: a previously undocumented source of contaminants of emerging concern to the coastal ocean (Sydney, Australia). Mar Pollut Bull 160:111519. https://doi.org/10.1016/j.marpolbul.2020.111519

Montagna PA, Gibeaut J, Dotson M, Douglas AR, Magolan J, Palacios JM, Rener L, Subedee M, Trevino K (2021) Evaluation of the proposal for widening and deepening the Matagorda Ship Channel. Final report to the Matagorda Bay Mitigation Trust, Contract Number 016. Texas A&M University-Corpus Christi, Harte Research Institute, Corpus Christi, Texas, USA, 63 p

Montluçon D, Sañudo-Wilhelmy SA (2001) Influence of net groundwater discharge on the chemical composition of a coastal environment: Flanders Bay, Long Island, New York. Environ Sci Technol 35:480–486. https://doi.org/10.1021/es9914442

Mooney RF, McClelland JW (2012) Watershed Export Events and Ecosystem Responses in the Mission–Aransas National Estuarine Research Reserve, South Texas. Estuar Coasts 35:1468–1485. https://doi.org/10.1007/s12237-012-9537-4

Moore WS (1999) The subterranean estuary: a reaction zone of ground water and sea water. Mar Chem 65:111–125. https://doi.org/10.1016/S0304-4203(99)00014-6

Moore WS (2010) The effect of submarine groundwater discharge on the ocean. Ann Rev Mar Sci 2:59–88. https://doi.org/10.1146/annurev-marine-120308-081019

Moore WS, Joye SB (2021) Saltwater intrusion and submarine groundwater discharge: acceleration of biogeochemical reactions in changing coastal aquifers. Front Earth Sci 9:600710. https://doi.org/10.3389/feart.2021.600710

Moosdorf N, Böttcher ME, Adyasari D, Erkul E, Gilfedder BS, Greskowiak J, Jenner A-K, Kotwicki L, Massmann G, Müller-Petke M (2021) A state-of-the-art perspective on the characterization of subterranean estuaries at the regional scale. Front Earth Sci 9:601293. https://doi.org/10.3389/feart.2021.601293

Mori C, Santos IR, Brumsack H-J, Schnetger B, Dittmar T, Seidel M (2019) Non-conservative behavior of dissolved organic matter and trace metals (Mn, Fe, Ba) driven by porewater exchange in a subtropical mangrove-estuary. Front Mar Sci 6:481. https://doi.org/10.3389/fmars.2019.00481

Mulligan AE, Charette MA (2006) Intercomparison of submarine groundwater discharge estimates from a sandy unconfined aquifer. J Hydrol 327:411–425. https://doi.org/10.1016/j.jhydrol.2005.11.056

Murgulet D, Douglas A, Lopez C, Gyawali B, Murgulet V (2019) Impacts of temporal and spatial variation of submarine groundwater discharge on nutrient fluxes to Texas coastal embayments. TGLO Contract No. 17-182-000-9819 Texas General Land Office, Austin, TX, 76 p

Murgulet D, Douglas AR, Herrera Silveira JA, Mariño Tapia I, Valle-Levinson A (2020) Submarine groundwater discharge along the northern coast of the Yucatán Peninsula. https://doi.org/10.5038/9781733375313.1060

Murgulet D, Felix JD, Lopez C, Qui Y (2021) Nonpoint source nutrient pollution study in Baffin Bay Texas, phase I. GLO Contract No. 20-036-000-B744. Texas General Land Office, Austin, TX. 71 p

Murgulet D, Lopez CV, Douglas AR (2022) Radioactive and stable isotopes reveal variations in nearshore submarine groundwater discharge composition and magnitude across low inflow northwestern Gulf of Mexico estuaries. Sci Total Environ 823:153814. https://doi.org/10.1016/j.scitotenv.2022.153814

Murgulet D, Murgulet V, Spalt N, Douglas A, Hay RG (2016) Impact of hydrological alterations on river-groundwater exchange and water quality in a semi-arid area: Nueces River, Texas. Sci Total Environ 572:595–607. https://doi.org/10.1016/j.scitotenv.2016.07.198

Murgulet D, Tick G (2008) The extent of saltwater intrusion in southern Baldwin County, Alabama. Environ Geol 55:1235–1245. https://doi.org/10.1007/s00254-007-1068-0

Murgulet D, Tick GR (2016) Effect of variable-density groundwater flow on nitrate flux to coastal waters. Hydrol Process 30:302–319. https://doi.org/10.1002/hyp.10580

Murgulet D, Trevino M, Douglas A, Spalt N, Hu X, Murgulet V (2018) Temporal and spatial fluctuations of groundwater-derived alkalinity fluxes to a semiarid coastal embayment. Sci Total Environ 630:1343–1359. https://doi.org/10.1016/j.scitotenv.2018.02.333

Murgulet D, Wetz MS, Douglas A, McBee W, Spalt N, Linares K (2015) Evaluating groundwater inflow and nutrient transport to Texas coastal Embayments. TGLO Contract No. 14-081-000-7949. Texas General Land Office, 277 p

Nicot J, Scanlon B, Yang C, Gates J (2010) Geological and geographical attributes of the South Texas Uranium Province. The University of Texas at Austin, Bureau of Economic Geology: Austin, TX, USA

Peterson RN, Moore WS, Chappel SL, Viso RF, Libes SM, Peterson LE (2016) A new perspective on coastal hypoxia: the role of saline groundwater. Mar Chem 179:1–11. https://doi.org/10.1016/j.marchem.2015.12.005

Pétré M-A, Ladouche B, Seidel J-L, Hemelsdaël R, de Montety V, Batiot-Guilhe C, Lamotte C (2020) Hydraulic and geochemical impact of occasional saltwater intrusions through a submarine spring in a karst and thermal aquifer (Balaruc peninsula near Montpellier, France). Hydrol Earth Syst Sci 24:5655–5672. https://doi.org/10.5194/hess-24-5655-2020

Pinckney JL (2018) A mini-review of the contribution of benthic microalgae to the ecology of the continental shelf in the South Atlantic bight. Estuar Coasts 41:2070–2078. https://doi.org/10.1007/s12237-018-0401-z

Pritchard DW (1967) What is an estuary: physical viewpoint. In: Lauff GH (ed) Estuaries. American Association for the Advancement of Science, Washington, pp 52–63

Rocha C, Robinson CE, Santos IR, Waska H, Michael HA, Bokuniewicz HJ (2021) A place for subterranean estuaries in the coastal zone. Estuar Coast Shelf Sci 250:107167. https://doi.org/10.1016/j.ecss.2021.107167

Rodellas V, Garcia-Orellana J, Masque P, Feldman M, Weinstein Y (2015) Submarine groundwater discharge as a major source of nutrients to the Mediterranean Sea. Proc Natl Acad Sci U S A 112:3926–3930. https://doi.org/10.1073/pnas.1419049112

RRC (Railroad Commision of Texas) (2018) Public GIS viewer (map). Austin, TX, 2018: https://www.rrc.texas.gov/resource-center/research/gis-viewer/

Santos IR, Chen X, Lecher AL, Sawyer AH, Moosdorf N, Rodellas V, Tamborski J, Cho H-M, Dimova N, Sugimoto R, Bonaglia S, Li H, Hajati M-C, Li L (2021) Submarine groundwater discharge impacts on coastal nutrient biogeochemistry. Nat Rev Earth Environ 2:307–323. https://doi.org/10.1038/s43017-021-00152-0

Santos IR, Eyre BD, and Huettel M (2012). The driving forces of porewater and groundwater flow in permeable coastal sediments: a review. Estuar Coast Shelf Sci 98: 1–15. http://dx.doi.org/https://doi.org/10.1016/j.ecss.2011.10.024

Santos IR, Niencheski F, Burnett W, Peterson R, Chanton J, Andrade CF, Milani IB, Schmidt A, Knoeller K (2008) Tracing anthropogenically driven groundwater discharge into a coastal lagoon from southern Brazil. J Hydrol 353:275–293. https://doi.org/10.1016/j.jhydrol.2008.02.010

Scanlon BR, Tachovsky JA, Reedy R, Nicot J-P, Keese K, Raymon MS, Merwade V, Howard MT, Wells GL, Mullins GJ, Ortiz DM (2005). Groundwater-surface water interactions in Texas. Austin, Texas, USA, 240 p

Schubert M, Paschke A, Lieberman E, Burnett WC (2012) Air–water partitioning of 222Rn and its dependence on water temperature and salinity. Environ Sci Technol 46:3905–3911. https://doi.org/10.1021/es204680n

Shideler GL, Stelting CE, McGowen JH (1981) Maps showing textural characteristics of benthic sediments in the Corpus Christi Bay estuarine system, South Texas. US Geological Survey, Reston, VA

Sholkovitz E, Herbold C, Charette M (2003) An automated dye-dilution based seepage meter for the time-series measurement of submarine groundwater discharge. Limnol Oceanogr Methods 1:16–28. https://doi.org/10.4319/lom.2003.1.16

Silliman SE, Berkowitz B, Simunek J, van Genuchten MT (2002) Fluid flow and solute migration within the capillary fringe. Ground Water 40:76–84. https://doi.org/10.1111/j.1745-6584.2002.tb02493.x

Simms AR, Aryal N, Miller L, Yokoyama Y (2010) The incised valley of Baffin Bay, Texas: a tale of two climates. Sedimentology 57:642–669. https://doi.org/10.1111/j.1365-3091.2009.01111.x

Slomp CP, Van Cappellen P (2004) Nutrient inputs to the coastal ocean through submarine groundwater discharge: controls and potential impact. J Hydrol 295:64–86. https://doi.org/10.1016/j.jhydrol.2004.02.018

Spalt N, Murgulet D, Abdulla H (2020) Spatial variation and availability of nutrients at an oyster reef in relation to submarine groundwater discharge. Sci Total Environ 710:136283. https://doi.org/10.1016/j.scitotenv.2019.136283

Spalt N, Murgulet D, Hu X (2018) Relating estuarine geology to groundwater discharge at an oyster reef in Copano Bay, TX. J Hydrol 564:785–801. https://doi.org/10.1016/j.jhydrol.2018.07.048

Stieglitz T, Taniguchi M, Neylon S (2008) Spatial variability of submarine groundwater discharge, Ubatuba, Brazil. Estuar Coast Shelf Sci 76:493–500. https://doi.org/10.1016/j.ecss.2007.07.03

Su N, Burnett WC, MacIntyre HL, Liefer JD, Peterson RN, Viso R (2014) Natural radon and radium isotopes for assessing groundwater discharge into little lagoon, AL: implications for harmful algal blooms. Estuar Coasts 37:893–910. https://doi.org/10.1007/s12237-013-9734-9

Szymczycha B, Borecka M, Białk-Bielińska A, Siedlewicz G, Pazdro K (2020) Submarine groundwater discharge as a source of pharmaceutical and caffeine residues in coastal ecosystem: Bay of Puck, southern Baltic Sea case study. Sci Total Environ 713:136522. https://doi.org/10.1016/j.scitotenv.2020.136522

Taniguchi M, Burnett WC, Cable JE, Turner JV (2002) Investigation of submarine groundwater discharge. Hydrol Process 16:2115–2129. https://doi.org/10.1002/hyp.1145

Taniguchi M, Dulai H, Burnett KM, Santos IR, Sugimoto R, Stieglitz T, Kim G, Moosdorf N, Burnett WC (2019) Submarine groundwater discharge: updates on its measurement techniques, geophysical drivers, magnitudes, and effects. Front Environ Sci 7:141. https://doi.org/10.3389/fenvs.2019.00141

Taniguchi M, Fukuo Y (1993) Continuous measurements of groundwater seepage using an automatic seepage meter. Ground Water 31:675–679. https://doi.org/10.1111/j.1745-6584.1993.tb00601.x

Teatini P, Isotton G, Nardean S, Ferronato M, Mazzia A, Da Lio C, Zaggia L, Bellafiore D, Zecchin M, Baradello L, Cellone F, Corami F, Gambaro A, Libralato G, Morabito E, Volpi Ghirardini A, Broglia R, Zaghi S, Tosi L (2017) Hydrogeological effects of dredging navigable canals through lagoon shallows. A case study in Venice. Hydrol Earth Syst Sci 21:5627–5646. https://doi.org/10.5194/hess-21-5627-2017

Toth J (1971) Groundwater discharge: a common generator of diverse geologic and morphologic phenomena. Hydrol Sci J 16:7–24. https://doi.org/10.1080/02626667109493029

Tyson RV, Pearson TH (1991) Modern and ancient continental shelf anoxia: an overview. Geol Soc Lond Spec Publ 58:1–24. https://doi.org/10.1144/GSL.SP.1991.058.01.01

von Ahn CME, Scholten JC, Malik C, Feldens P, Liu B, Dellwig O, Jenner A-K, Papenmeier S, Schmiedinger I, Zeller MA (2021) A multi-tracer study of fresh water sources for a temperate urbanized Coastal Bay (southern Baltic Sea). Front Environ Sci 406. https://doi.org/10.3389/fenvs.2021.642346

Wang FC (1988) Dynamics of saltwater intrusion in coastal channels. J Geophys Res Oceans 93:6937–6946. https://doi.org/10.1029/JC093iC06p06937

Wang X, Li H, Zheng C, Yang J, Zhang Y, Zhang M, Qi Z, Xiao K, Zhang X (2018) Submarine groundwater discharge as an important nutrient source influencing nutrient structure in coastal water of Daya Bay, China. Geochim Cosmochim Acta 225:52–65. https://doi.org/10.1016/j.gca.2018.01.029

Webster PJ, Holland GJ, Curry JA, Chang H-R (2005) Changes in tropical cyclone number, duration, and intensity in a warming environment. Science 309:1844–1846. https://doi.org/10.1126/science.1116448

Werner AD, Simmons CT (2009) Impact of sea-level rise on sea water intrusion in coastal aquifers. Groundwater 47:197–204. https://doi.org/10.1111/j.1745-6584.2008.00535.x

Winter TC, Harvey JW, Franke OL, Alley WM (1998) Ground water and surface water: a single resource. 1139. Survey USG, 79 p. https://doi.org/10.3133/cir1139

Xiao H, Wang D, Medeiros SC, Bilskie MV, Hagen SC, Hall CR (2019) Exploration of the effects of storm surge on the extent of saltwater intrusion into the surficial aquifer in coastal east-Central Florida (USA). Sci Total Environ 648:1002–1017. https://doi.org/10.1016/j.scitotenv.2018.08.199

Young SC, Budge T, Knox PR, Kalbouss R, Baker E, Hamlin S, Galloway B, Deeds N (2010) Hydrostratigraphy of the Gulf coast aquifer from the Brazos River to the Rio Grande. Final Report Contract# 0804830795. Texas Water Development Board, Austin, TX, USA, 213 p

Young SC, Jigmond M, Deeds N, Blainey J, Ewing TE, Banerj D, Piemonti D, Jones T, Griffith C, Martinez G, Hudson C, Hamlin S, Sutherland J (2016) Final report: identification of potential brackish groundwater production areas-Gulf Coast aquifer system. Contract No. 1600011947. Texas Water Development Board, Austin, TX, USA, 636 p

Young SC, Mace RE, Rubinstein C (2018) Surface water-groundwater interactions issues in Texas. Texas Water J 9:129–149. https://doi.org/10.21423/twj.v9i1.7084

Yu X, Liu J, Chen X, Huang D, Yu T, Peng T, Du J (2022) Submarine groundwater-derived inorganic and organic nutrients vs. mariculture discharge and river contributions in a typical mariculture bay. J Hydrol 613:128342. https://doi.org/10.1016/j.jhydrol.2022.128342

Zektser I, Loaiciga HA (1993) Groundwater fluxes in the global hydrologic cycle: past, present and future. J Hydrol 144:405–427. https://doi.org/10.1016/0022-1694(93)90182-9

Zhou Y, Sawyer AH, David CH, Famiglietti JS (2019) Fresh submarine groundwater discharge to the near-global coast. Geophys Res Lett 46:5855–5863. https://doi.org/10.1029/2019GL082749

# Influence of Inflows on Estuary Sediments

Audrey R. Douglas ⓘ, Paul A. Montagna ⓘ, and Timothy Dellapenna ⓘ

## Abstract

Rivers deliver sediments, organic matter/carbon, contaminants, and nutrients from watersheds. Sediments form delta, estuarine, marsh, and wetland habitats. Most of the sediments are derived from weathered rocks that are transported by rivers during floods as both bedload (sand and gravel) and suspended load (silts and clays, i.e., mud). Siliciclastic sand can enter the estuary naturally through four mechanisms, which are: (1) shoreline erosion; (2) inflow from rivers; (3) advection into the bay via the flood tide, through tidal inlets and cuts; and (4) barrier island overwash. River-mouth/saltwater wedge dynamics trap the bedload (coarse fraction) within bayhead deltas of estuaries, with only the suspended load (mud fraction) being delivered to the estuaries under most conditions. Sand entering the bay through the tidal inlets/cuts, are trapped in the flood tidal delta proximal to the bay mouth. Barrier island overwash sands are also trapped proximal to the overwash sites. As a result, the interior of most estuarine sediment is muddy, with sandy shorelines, shoreline shoals, bayhead, and flood tidal deltas. Carbonate sands also form in situ by the abrasion of shells, most notably, oyster shells, forming shell shoals and oyster reef. Over time, the greatest "natural" changes to sediment distribution within estuaries are caused by extreme storms and floods that can deliver the equivalent of decades worth of "average fluvial sediment load" to bays within a few days. Additionally, sediment distributions and loads within estuaries can be altered by anthropogenic activities, including dredging and formation of dredge spoil island, formation of mitigation oyster reefs and wetlands, alterations in shoreline protection, and alteration of bay configurations and size. The carbon content of sediments increases with decreasing inflow and increasing salinity across the state where Sabine Lake averages 1% and Laguna Madre averages 2.5%. The nitrogen content of sediments is generally low, always <0.1%, but can be as high as 0.25% in Laguna Madre. Sedimentation rates generally range from 0.07 to 2.2 cm year$^{-1}$. There has been sediment compaction and subsidence over time.

## Keywords

Mud · Sand · Sediment grain size · Sediment carbon · Sediment content · Sediment nitrogen · Sedimentation · Saltwater wedge dynamics · Barrier islands

## Introduction

The three main responses in estuaries to inflow are salinity changes, nutrient loads, and sediment loads and distribution. The changing salinity levels, nutrient loading, and sediment loading provided by freshwater inflow are important for overall productivity and geochemical cycling of the estuary. Sediment has many important roles in the estuarine environment and forms the substrate, on (and in) which many organisms live and feed. In addition, sediment can provide several biogeochemical functions. For example, modification of overlying water quality through denitrification and sequestration of sediment-bound contaminants and nutrients. Consequently, human activities and upstream changes in inflow, such as dams, diversions, or land use landcover changes, can affect the quantity, quality, and timing of their delivery to the estuary as well (Alber 2002). Thus, understanding and monitoring sediment quality, quantity, location, and transport is important for managing ecosystem health.

**Supplementary Information** The online version contains supplementary material available at https://doi.org/10.1007/978-3-031-70882-4_6.

A. R. Douglas (✉) · P. A. Montagna
Texas A&M University-Corpus Christi, Corpus Christi, TX, USA
e-mail: Audrey.Douglas@taumcc.edu

T. Dellapenna
Texas A&M University-Galveston, Galveston, TX, USA

© The Author(s) 2025
P. A. Montagna, A. R. Douglas (eds.), *Freshwater Inflows to Texas Bays and Estuaries*, Estuaries of the World,
https://doi.org/10.1007/978-3-031-70882-4_6

The idea that the climatic gradient along the Texas coast provides an ideal setting to study sedimentation as it relates to freshwater inflow was proposed by Shepard and Rusnak (1957). The difference between hypersaline environments in Laguna Madre and oligohaline environments near river and creek mouths provided a striking contrast. Clearly, there was a difference in sediment sources that required study. Texas bays also have limited exchange rates with the Gulf of Mexico because of the long, continuous, barrier islands and the bay sediments are thus more likely to be permanent rather than being constantly transported by waves. The semi-arid bay sediments also exhibited signs of excessive evaporation because of the presence of gypsum and carbonate aggregates. The sediments of Texas bays reflect the influence of the variability of freshwater inflow.

## What Are Sediments?

Sediments are assemblages of individual grains of weathered rocks and minerals, as well as plant and animal remains that are moved and deposited in new locations (Bianchi 2006; Fetter 2001). Sediments are unconsolidated with openings between the sediment grains called pore spaces. Four dominant processes control sediment dynamics in estuaries: (1) erosion of the bed, (2) transportation, (3) deposition, and (4) consolidation of deposited sediments (Nichols and Biggs 1985). Rocks are broken down into smaller particles, or clasts, by way of physical and chemical weathering. Through erosion, water, wind, ice, and gravity transport these particles elsewhere by dislodging, dissolving, or removing surface material. Erosion rates vary by orders of magnitude depending on several factors including the erosion process, soil type, topography, and land use land cover type (Longley 1994). The amount of sediment that reaches a watershed's outlet can be significantly less than what is eroded due to particle deposition along the way. Sediments in estuaries are derived from a diverse set of sources including stream inputs (fluvial), atmospheric inputs (aeolian), continental shelf (marine), bio-

logical activity (biogenic), and erosion of estuarine margins (shoreline). The types and quantity of sedimentary material available for deposition, hydrodynamic processes, and basin geometry determine the composition and distribution of sedimentary facies in an estuary (Bianchi et al. 1999). Thus, the natural gradients in estuarine sedimentary processes are primarily driven by changes in river inflow, tidal current, waves, and meteorological forces (Bianchi 2006).

Rivers and streams typically carry various types of sediment, including non-living and inorganic materials such as clay and organic colloids, organo-clay complexes, silt, sand, gravel, twigs, and more. The nutrients transported in/on these materials are vital to the well-being of estuarine ecosystems. These nutrients may be absorbed or adsorbed to the materials, or the materials themselves may serve as nutrients once they break down. The organic matter content, particle size distribution, and whether the sediment is in suspension or part of the bed load can be used to characterize sediments transported by rivers and streams.

## Sediment Classification

An important property of sediment is its texture, which includes grain size, sorting, and maturity. Broadly, sediments may be classified by their grain size as gravel (coarsest), sand, silt, or clay (finest) (Table 1). According to the Wentworth-Lane size classification scale (Pettijohn 1975), sands are retained on a 63 µm mesh sieve (>63 µm but <2 mm), but may include gravel sized particles (>2 mm but <64 mm) and silts and clays are those particles that pass through a 63 µm mesh sieve. Combined, silt and clay are often referred to as mud. Sediments are transported to estuaries and mobilized within estuaries both in suspension and as bed load. Suspended load typically consists of the finer grained clay- to silt-sized particles that are dispersed in the water column by the upward component of turbulence or by colloidal suspension, whereas bed load refers to coarser sediment grain sizes and debris, such as sands, gravels, and

**Table 1** Sediment grain-size classification (modified from Fetter (2001) and Wentworth (1922))

| Name | Size range (mm) | Grain size | Energy conditions | Example |
|---|---|---|---|---|
| Boulder | >305 | Coarse-grained | High energy | Larger than a basketball |
| Cobbles | 76–305 | | | Grapefruit |
| Coarse Gravel | 19–76 | | | Lemon |
| Fine Gravel | 4.75–19 | | | Pea |
| Coarse Sand | 2–4.75 | | | Water softener salt |
| Medium Sand | 0.42–2 | | | Table salt |
| Fine Sand | 0.075–0.42 | | | Powdered sugar |
| Fines (mud) | <0.075 (<62 µm) | | | Talcum powder |
| Silt | 0.075–0.004 (2–62 µm) | | | Feels gritty in teeth |
| Clay | <0.004 (<2 µm) | Fine-grained | Low energy | Microscopic; feels sticky |

whole shells, that are moved along the seabed by saltation or traction (i.e., jumping, rolling, or sliding).

Turbulence keeps sediment in suspension, with coarser grains requiring greater turbulence than fine particles. As a result, greater energy (waves and tides) is required to transport sand and gravel than silts and clays, consequently, the coarser size fraction is generally only found in the shallower, more energetic portions of the estuaries and the finer grained fraction (muds) are found within the deeper, interior of the estuary. Sorting is a measure of the uniformity of grain sizes and indicates the selectivity of transportation processes. Samples that are uniform in particle size are considered "well-sorted," whereas samples consisting of a variety of different particle sizes are regarded as "poorly-sorted." Considering poorly sorted sediments in the coastal environment, wave or current action will transport coarser particles a shorter distance than finer particles. Since more time is available during transport of particles from their source to their point of deposition, sorting increases, clay content decreases (fine grains stay in suspension longer), less durable non-quartz minerals decrease, and sediment grains are rounded through abrasion.

Sediment composition, sorting, and depositional/erosional processes occurring within estuarine environments also influence the porosity, intrinsic permeability (i.e., the ability of a porous medium to allow a fluid to pass through it), and hydraulic conductivity (i.e., the ease with which a fluid can move through pore spaces, partially depends on intrinsic permeability) of the sediments. As a measure of the volume of the empty spaces in a rock or sediment, porosity is determined by sediment grain size, sorting, and the shape of sediment grains. Porosity will be greater in well-sorted sediments than poorly-sorted sediments because the more uniform grain size will not have finer grains filling the pore spaces. While clay-rich sediments may have higher porosities than sand-rich sediments, the pore spaces are smaller and water and other fluids do not move easily through them, resulting in lower permeability (Table 2).

## Estuary Geomorphology and Sediment Sources

The definition of an estuary as "a semi-enclosed coastal body of water with a free connection with the ocean and within which sea water is measurably diluted by freshwater from land drainage" provided by Pritchard (1967) covers the functional role of an estuary but does not describe how the estuary was formed. However, there are several other estuarine classification schemes that attempt to account for the processes forming and maintaining estuaries. For instance, Pritchard (1952) recognized four estuary typologies based on geomorphology: (1) drowned river valleys created by sea

**Table 2** Ranges of intrinsic permeabilities and hydraulic conductivities for unconsolidated sediments (modified from Fetter (2001))

| Material | Intrinsic permeability (cm$^2$) | Hydraulic conductivity (cm/s) |
|---|---|---|
| Clay | $10^{-6}$–0.001 | $10^{-9}$–$10^{-6}$ |
| Silt, sandy silts, clayey sands, till | 0.001–0.1 | $10^{-6}$–$10^{-4}$ |
| Silty sands, fine sands | 0.01–1 | $10^{-5}$–0.001 |
| Well-sorted sands, glacial outwash | 1–100 | 0.001–0.1 |
| Well-sorted gravel | 10–1000 | 0.01–1 |

level change or sediment starvation in coastal plains, (2) fjords formed by glaciation, (3) bar-built estuaries formed by sediment deposition by winds and tides, and (4) tectonic estuaries caused by faults in the coastal zone. Another scheme by Davies (1973) established a continuum of inlet types based on the wave energy expended on the coast with lagoons enclosed by sandy spits on the low energy end and muddy deltas formed by river processes at the high energy end. One more classification system by Fairbridge (1980) created seven basic physiographic types based on the overall relief of the region and the extent circulation is restricted at the mouth: (1) coastal plain (i.e., drowned river valley), (2) bar-built, (3) delta, (4) blind, (5) ria, (6) tectonic, and (7) fjord. This classification system was further modified to better encompass the spectrum and diversity of estuarine systems by considering the prior structure from which the estuary was formed (Perillo 1995). Thus, the classification of an individual estuarine system may change over geologic time scales as with periods of sea level rise and sea level fall, the configuration may change. Consequently, estuarine systems may have features of multiple typologies. The modern Texas coastline includes elements of each classification system, with drowned river valley, bar-built, blind estuaries, and lagoons being the primary estuarine types found. Thus, the Texas coast includes extensive coastal wetlands and bays, chenier plains, barrier islands and peninsulas, the Brazos wave-dominated delta, and several tidal inlet/delta complexes (Anderson et al. 2014).

In geologic time scales, estuaries are ephemeral coastal features that begin to fill in with sediments upon formation (Bianchi 2006 and references therein). Modern estuaries are recent surface features that formed during the stable interglacial periods of the Holocene epoch, 0–10,000 years before present (y BP), following extensive sea level rise at the end of the Pleistocene epoch (1.8 million years to 11,700 y BP) (Bianchi 2006, Kasmarek and Robinson 2004). During lowstands of sea level, when sea level was as much as 130 m lower than today and where the coast was at or near the continental shelf break, 100–200 km offshore of the present-day shoreline, the modern-day continental shelf was part of the coastal plain and rivers which flowed across it incised river valleys. As sea level rose, these valleys became inundated and transitioned to estuaries and today, the modern estuaries are situated over the incised valleys. Two of these river valleys, the Brazos and Rio Grande, had enough sediment to completely fill them, hence there are no large estuaries associated with these rivers. The remainder delivered a lower load of sediment and the estuaries have remained. As sea level rises, accommodation space is created, which can be filled with sediment. For an estuary to be maintained over Holocene timescales, there needs to be a close balance between the formation of accommodation space and sedimentation. If sedimentation is too high, the bay will fill and transition to a river delta with no estuarine embayment. If sedimentation rates are too low, the bay will deepen as sediments are transported out of the estuary through erosional and subtractive processes and ultimately transition into an open embayment. The Texas coast is classified as microtidal, with a tidal range normally ranging between 0.5 and 0.7 m (Armstrong 1982), and the presence of barrier islands along its coast classifies the bays as wave dominated (as opposed to tide-dominated; Dalrymple et al. 1992).

The barrier islands formed as wind- or water-deposited sediments accumulated to form a ridge immediately landward of the shoreline. The area landward of the ridge was flooded by the slow submergence during the Holocene, thus forming the barrier and lagoon system we see today (Hoyt 1967). In general, the larger bays and estuaries occupy the central and upper parts of the Texas coastal zone with bay size largely determined by the size of the Pleistocene rivers that incised the valleys they now occupy (McGowen and Morton 1979). Sediment is contributed to the estuaries by rivers that discharge at the heads of bays, and by the Gulf of Mexico through tidal inlets (mostly under normal sea conditions) and across barrier islands and peninsulas during storms. Sediment delivered to the coastal zone by Texas streams is chiefly a function of climate. Identifiable sediment delivered to bays, estuaries, and lagoons from the Gulf of Mexico is chiefly sand, shell, and rock fragments that are eroded from the inner shelf, shoreface, and beaches by tropical storms and hurricanes.

Siliciclastic sand can enter the estuary naturally through four mechanisms: (1) shoreline erosion; (2) inflow from rivers; (3) advection into the bay via the flood tide, through tidal inlets and cuts; and (4) barrier island overwash. Sediments are transported from watersheds through streams to estuaries and within estuaries in suspension (i.e., finer, lighter particles) and as bed load (i.e., coarser, heavier particles). Fluvial sediments may be derived from land runoff, bank erosion, and the internal reworking of in-channel deposits. Depending on the volume of water and its velocity, water can carry sediment downstream from a creek, into a river, and eventually to where the river discharges to the coastal zone. Sediment sources can vary considerably in the upper and lower regions

of the drainage basin and estuary (Christopherson 2004). The upstream tributaries in a drainage basin usually have small and irregular discharges, with the majority of the stream-energy expended in turbulent eddies. Consequently, hydraulic action, the work of water moving materials through the exertion of pressure and shearing force, is maximized in these upstream sections, though there is little coarse-textured load. However, the downstream portions of the river move much larger volumes of water and carry greater suspended sediment loads. Further downstream in the coastal zone, the lower regions of the estuary with higher salinity are expected to have more biological inputs (e.g., plankton deposition) whereas the upper regions with lower salinity are expected to be dominated by terrigenous inputs (Bianchi 2006).

In contrast, biogenic sediments grow in situ and are primarily composed of calcium carbonate ($CaCO_3$) originating from the remains of marine organisms, such as corals, foraminifera, shell-bearing organisms (like oysters, clams, and snails), and certain types of algae (like coccolithophores and calcareous red algae). For example, carbonate sands are often found in shallow marine environments and form by the abrasion of shells, most notably oyster shells along the Texas coast, after the organism dies forming shell shoals and reefs. The breakdown products of these organisms have mixed to various degrees with terrigenous material and siliciclastic sediments across continental shelves from sea-level lowstand to sea-level highstand. However, because biogenic "clasts" grow in situ, these sediments do not reflect the ambient energy of the environment or follow the normal energy considerations related to bedload transport as other types of sediment. The proportion of biogenic sediment may increase where river inflows are low (Flemming 2020).

The distribution of sediment size classes within estuaries is largely controlled by estuarine dynamics (Nichols and Biggs 1985). At the marine end, coarse grained sediment is advected in via tidal currents, through tidal inlets and cuts, and once the water is advected past the constrictions of the inlet, the currents rapidly decrease, causing the coarse bedload fraction to deposit, forming a flood tidal delta. At the fluvial end of the estuary, saltwater wedge (hereafter saltwedge) dynamics control the distribution of sediment (Nichols and Biggs 1985).

Salinity is a measure of the ionic concentration within the water column. In freshwater, the ionic concentration is very low, generally below 1 ppt, consequently the net charges on suspended sedimentary particles are negative and these particles repel each other. As the particles enter the lower range of the brackish part of the estuary, the ions in the water column neutralize the net negative charge on the suspended sediment particles, allowing for static forces to act on the particles, resulting in their flocculation. Although silt and clay sized particles cannot settle under these lower energy flow conditions, when they flocculate into sand sized particles, they begin to settle. Saltwater is denser than freshwater, and at least in a time averaged sense, a saltwedge forms at the head of estuaries often in the estuarine tributaries above the enclosed embayment portion of the estuary. As the fluvial water encounters the saltwedge, there is a separation of flow, with a freshwater layer flowing atop the saltwater layer (Nichols and Biggs 1985).

The point where the salt lens pinches out upstream is called the null point (Nichols and Biggs 1985). Above the null point, the entire water column is flowing downstream. However, at the null point, there is a separation of flow, with the freshwater lens flowing seaward, rising above the saltwater layer. At this point, bedload cannot be transported downstream, so it is trapped at and above the null point. The suspended load is transported downstream; however, as it encounters the brackish water, flocculation occurs and the particles sink into the saltwedge portion of the flow and is transported back upstream toward the null point, where,

when it encounters fresher water the flocs breakdown. This creates a feedback-loop, trapping a portion of the estuary's suspended sediment load toward the head of the estuaries, creating what is called the estuarine turbidity maximum. The null point is not fixed in time or space, it migrates up and downstream with the tides as well as with increasing or decreasing river discharge. During high discharge events, the null point can move into the open portion of the estuary, transporting sand with it. During Hurricane Harvey, the saltwedge in Galveston Bay was pushed all the way out of the bay into the Gulf of Mexico, as evidenced by the large sediment plume expanding out from the Bolivar Roads inlet (Fig. 1), depositing a sand layer across the bay (Dellapenna et al. 2022). The average position of the null point normally coincides with the position of the bayhead delta.

In general, high-flow periods have elevated sediment inputs that results in high estuarine turbidity. Sediments delivered to estuaries via freshwater inflow help to build and

**Fig. 1** NASA Earth Observatory image by Jesse Allen showing rivers and bays loaded with sediment following Hurricane Harvey, using data from the Land Atmosphere Near Real-time Capability for EOS (LANCE). Data acquired from the Moderate Resolution Imaging Spec-troradiometer (MODIS) on NASA's Terra satellite on August 31, 2017. https://visibleearth.nasa.gov/images/90866/texas-waters-run-brown-after-harvey

stabilize estuarine habitats, such as wetlands, tidal flats, and shoals (Olsen et al. 2006). Additionally, sediment transport has an important role in nutrient supply to the system and contaminant transport as transported sediments are a reservoir of organic and inorganic nutrients and contaminants bonded with sediment particles. These bonded nutrients and contaminants are then released in the water column through biogeochemical reactions. Thus, sediments and particulate matter carried by rivers provide the primary energy source for estuarine organisms (Day et al. 1989) and estuarine nutrient levels depend, in part, on the sediment carried to it.

Freshwater diversions from estuaries and upstream impoundments are decreasing the delivery of water and sediment to the coastal zone. Over the past 200 years, changes in sediment discharge are primarily due to anthropogenic factors including (a) deforestation and agriculture, (b) changes in land management, and (c) construction of dams, diversions, and levees (McKee and Baskaran 1999). Globally, reservoirs and water diversions have resulted in a net reduction of sediment delivery to estuaries by roughly 10% and prevented about 30% of the sediment from reaching the oceans (Syvitski et al. 2005; Vörösmarty et al. 2003). Riverbanks, deltas, and bays are common areas where water-borne sediment accumulates. There are only two rivers within the Atlantic Basin of the US, including the Gulf of Mexico, which consistently flow into the ocean, the Mississippi and the Brazos Rivers. The remainder either have so much freshwater removed that they no longer consistently flow into the ocean, for example, the Rio Grande, or they flow through estuaries. As a result, within the continental US, approximately 90% of the sediment eroded from land is stored somewhere between the river and the sea (Meade et al. 1990), i.e., the estuaries.

## Data and Methods

### Spatial Data

The Bureau of Economic Geology (BEG) performed a comprehensive investigation of state submerged lands in the 1970s to develop a baseline inventory of geological and biological data for future environmental monitoring (McGowen and Morton 1979; Wermund et al. 1989). The state-owned submerged lands of Texas encompass nearly 6000 square miles (15,540 km$^2$) and extend from Louisiana to Mexico. The area includes the bays, estuaries, lagoons, and the inner continental shelf 10.3 miles (16.6 km) seaward of the Gulf of Mexico shoreline split in to seven areas: Beaumont-Port Arthur, Houston-Galveston, Bay City-Freeport, Port Lavaca, Corpus Christi, Kingsville, and Brownsville-Harlingen (Fig. 2). The BEG study characterized the surficial sediments of the submerged lands of Texas with grab samples of the top 4–18 cm of sediment at 3346 sites. Total organic carbon (TOC) and multiple metals (boron, barium, calcium, chromium, copper, iron, manganese, nickel, lead, and strontium) were measured at a subset of these sites. The data from this effort were used to map and assess the spatial distribution of sediment composition and associated chemical characteristics in Texas bays, estuaries, and inner shelf. Data were interpolated in ArcMap 10.8 using kriging method. Sediment grain size data were plotted in ternary diagrams following the modified Folk sediment classification scheme (Folk 1980). Ternary plots were made with Python 3.9 standard libraries and python-ternary package (Harper et al. 2015). A principal component analysis (PCA) was performed on all grain size, total organic carbon, and selected trace metal data using SAS 9.4 statistical software with PROC FACTOR after variable standardization to a normal distribution with mean of zero and standard deviation of 1. The BEG submerged lands data were compared with Longley (1994) and a literature review.

### Temporal and Depth Data

The Harte Research Institute (and formerly the University of Texas Marine Science Institute) under the direction of Paul Montagna collected and analyzed surface sediment data from 1987 to 2019 in four estuaries. Additionally, 1-m long cores were collected to analyze carbon (C) and nitrogen (N) content change with sediment depth to estimate the role of inflow in elemental burial. The sediment grain size method is based upon practices in benthic ecology and sedimentology (Folk 1980; Lewis 1984; Plumb 1981). A 20 cm$^3$ sediment sample was mixed with 50 ml of hydrogen peroxide and 75 ml of deionized water to digest organic material in the sample. The sample was wet sieved through a 62 μm mesh stainless steel screen using a vacuum pump and a Millipore Hydrosol SST filter holder to separate gravel and sand from silt and clay. After drying, the gravel and sand were separated on a 125 μm screen. The silt and clay fractions were measured using pipette analysis. Percent contribution by weight was measured for four components: gravel, sand, silt, and clay. Texas estuary sediments do not have a source of rocks; thus, the gravel fraction is often composed of biogenic carbonates such as decomposing mollusk shells. A PCA was performed on surface (0–3 cm) sediment data using SAS 9.4 statistical software after transformation by standardization to a normal distribution with mean value of zero and standard deviation of 1.

For the last 27 years (1993–2022), the Taiwanese owned Formosa Plastics Corporation (FPC), Texas, Point Comfort Facility has been monitoring sediment and water quality, including sediment grain size and chemical data, around the plant's outfall discharging wastewater effluent into eastern

**Fig. 2** Spatial distribution of percent sand sediment characteristics for state waters of Texas from the Bureau of Economic Geology's (BEG) investigation. The seven areas covered by the submerged lands of Texas project are shown

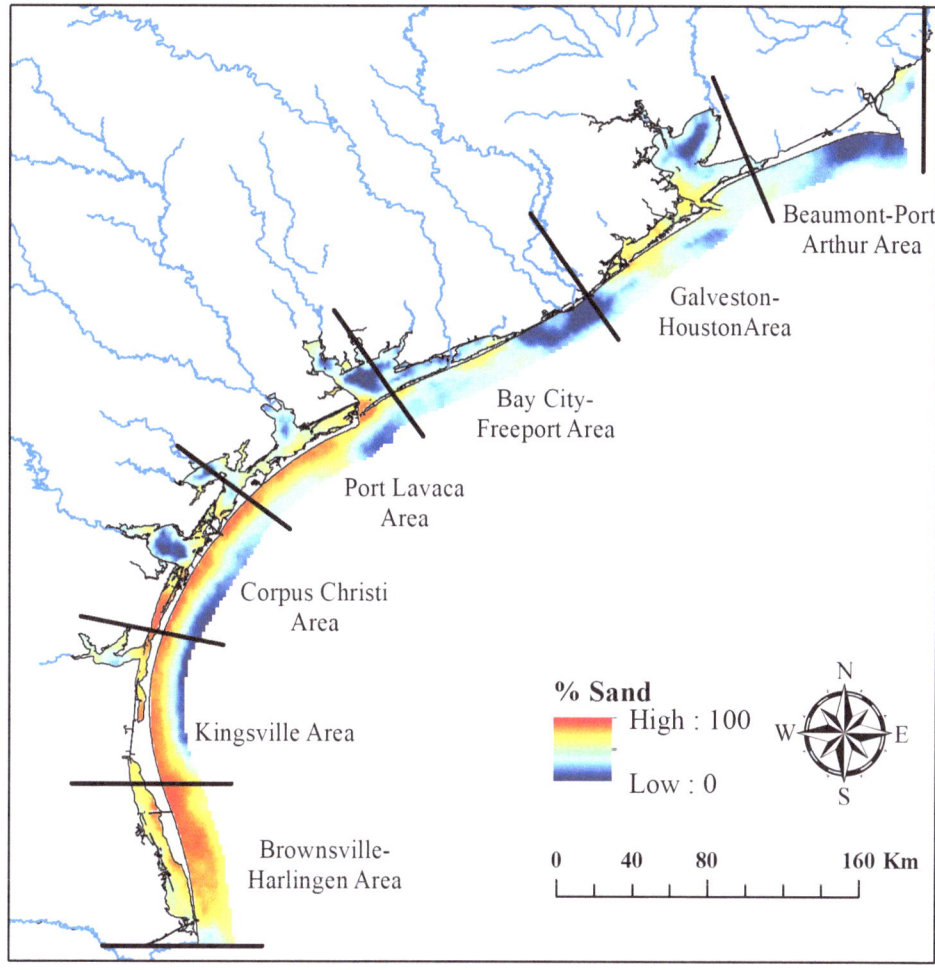

Lavaca Bay. The FPC primarily produces polyvinyl chloride, polyethylene, and polypropylene resins as well as other plastic products. The FPC erected a diffuser in Lavaca Bay to ensure adequate dilution of the plant's wastewater discharge. In May 1993, following the regulatory objectives of the Texas Commission on Environmental Quality (TCEQ) and the U.S. Environmental Protection Agency (EPA) as outlined in TCEQ Wastewater Permit #02436, EPA Permit #TX0085570, a Scope of Work, entitled *Receiving Water Monitoring Program, Scope of Work for the Formosa Plastics Corporation, TX, Point Comfort, Texas Facility*, was submitted to the Texas Water Commission (now the TCEQ) to satisfy the permit requirements. Following additional revisions and a final approval, monitoring trips began in May of 1993 and continued as of December 2023.

## Sedimentation Rates

Rivers transport sediments to the coast, so sedimentation rates are important for understanding effects of inflow. There are two ways in which sedimentation rates have been esti-

mated: by comparing bathymetry over long time scales or the profile of radioisotopes within sediments. Typically, radioisotopes from lithogenic sources (i.e., $^{226}$Ra, $^{228}$Ra, $^{228}$Th, $^{230}$Th, $^{232}$Th) or atomic bomb fallout ($^{137}$Cs, $^{210}$Pb) are used. In some cases, legacy pollution by heavy metals can be traced as it is buried over time. Estimates are taken from the literature.

## Results and Discussion

### Sediment Grain Size Distribution

Longley (1994) summarized the fraction of sand, silt, and clay in the suspended sediment load carried by seven Texas rivers (Fig. 4.4.2 in Longley 1994) from Sabine at Bon Wier south to Nueces River at Three Rivers. The data showed a decrease in suspended sand content and a corresponding increase in suspended clay content moving south, which corresponds with decreasing riverine discharge rates from north to south. However, some of the measurements were taken substantially upstream of the coast and do not necessarily reflect the sediment size distribution of the coast as further erosion and deposition in reservoirs and elsewhere would occur throughout transit.

The Submerged Land of Texas study assessed the sediment particle distribution within the bay and estuary systems and along the inner shelf of the Gulf of Mexico (Wermund et al. 1989). These data showed that the surficial sediments are predominantly sand and mud with few areas containing more shell/gravel (Fig. 3a). The occurrence of shell/gravel decreases from northeast to southwest while the percent sand increases. Contrary to the findings presented for suspended sediments in Texas rivers by Longley (1994), in the coastal zone, the sand content increases and the clay content decreases moving south. The greatest sand content occurs in the Kingsville and Brownsville-Harlingen areas (Figs. 2 and 3b) where inflows are the lowest, suggesting starvation of fine grain sediments to these low inflow estuaries and/or greater export of fine grain sediments than inputs. In the case of net export of fine grain sediments, or low effective sedimentation rates, the remaining sands and gravels are essentially lag deposits. One sediment core study in Baffin Bay, found sand/shell hash layers and erosional contacts were common throughout the bay system but more so in the stratigraphy of the shallower tributary bays where the bay bottom is more subject to higher energy events (Besonen et al. 2016). In contrast, the northern estuaries with greater inflows have much higher silt and clay content demonstrating a more continuous supply of fine grain sediments that exceeds within estuary export. The low sediment inputs in the southern estuaries have implications for the nutrient dynamics in these systems as the organic and inorganic nutrients that sorb to sediments would be greatly reduced.

Of the geochemical parameters, TOC ($r = 0.645$), iron ($r = 0.72$), and nickel ($r = 0.62$) had the strongest positive relationships with mud content. TOC is highest in the mid-bay muds in Trinity Bay, Copano and Aransas Bays, Corpus Christi Bay, and Baffin Bay. Nickel is highest on the shelf near the Sabine-Neches estuary and within Matagorda Bay. Boron ($r = 0.57$), chromium ($r = 0.49$), copper (0.54), manganese ($r = 0.51$), lead ($r = 0.33$), and zinc ($r = 0.48$) had weaker positive relationships with mud content and generally higher concentrations were observed in the north and lower concentrations were observed in the south. Barium ($r = 0.19$), calcium ($r = -0.11$), and strontium ($r = -0.16$) exhibited little to no relationship with mud content; however, elevated strontium was observed in the upper and lower Laguna Madre sediments, particularly in Baffin Bay, likely due to the arid environment leading to the formation of evaporites within the estuary.

A PCA performed on select sediment grain size and geochemical variables shows the difference in grain size and explains 44% of the variance with TOC and most of the metals associated with the higher mud content in the northern estuaries (Fig. 4a, b). Due to their small size, high surface area, and high surface reactivities, clay particles are particularly important when considering particle-particle and particle-dissolved constituent interactions in aquatic systems (Bianchi 2006, Hao et al. 2019). Clay minerals have been shown to have a strong affinity for trace metal adsorption (Gu and Evans 2007, Liu et al. 2018), particularly in the presence of humic substances (Hizal and Apak 2006), which are a major component of the dissolved organic carbon pool in aquatic environments that has a significantly higher concentration in estuaries near terrestrial inputs than in the open ocean (Opsahl and Benner 1997). Interestingly, sediment Ca and Sr do not follow trends in sediment type but instead separate out and drive PC2, explaining an additional 15% of the variance (Fig. 4a). The Ca and Sr in the sediment appear to

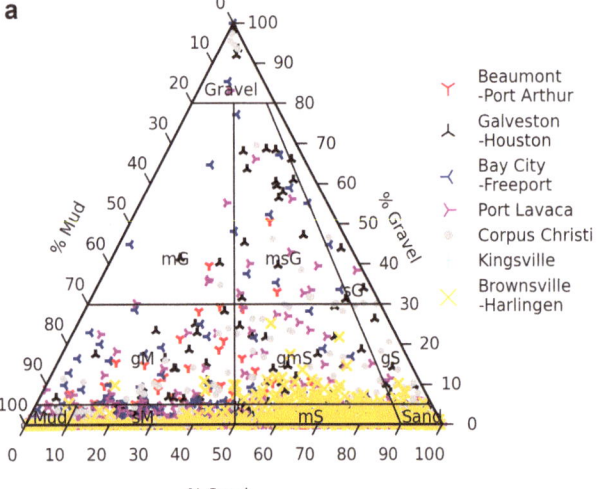
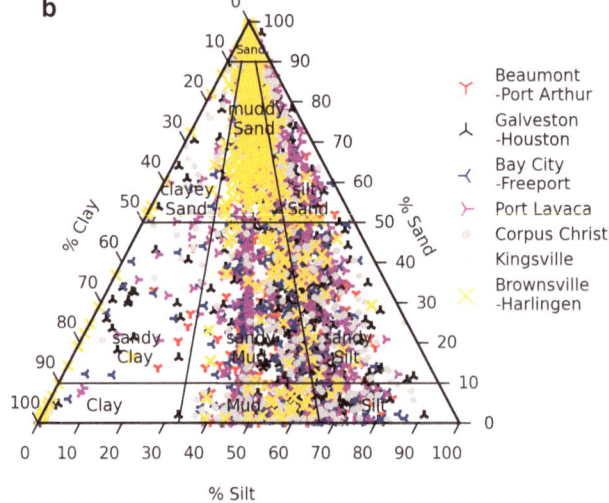

**Fig. 3** Modified folk classification for (**a**) percent gravel/shell-sand-mud and (**b**) percent sand-silt-clay of surficial sediments from the submerged lands of Texas. M, mud; m, muddy; S, sand; s, sandy; G, gravel; g, gravelly

be driven by proximity to and magnitude of freshwater inflows with higher Ca and Sr values in secondary bays than primary bays (Fig. 4b). Evidence of total alkalinity consumption in the water column via CaCO₃ precipitation was observed in Nueces Bay during a period with high freshwater inflows, spring 2015 (Murgulet et al. 2018). Thus, one explanation for the higher concentrations of Ca and Sr in secondary bay sediments may be due to precipitation of carbonate as calcite ($CaCO_3$) or strontianite ($SrCO_3$) where riverine and marine waters meet. As carbonate minerals are highly soluble, precipitation of $(Ca,Sr)CO_3$ seems more likely than the transport of highly soluble limestone from the watershed. However, formation of evaporites containing Ca (i.e., gypsum [$CaSO_4 \cdot 2H_2O$], anhydrite [$CaSO_4$]), and Sr (i.e., celestite [$SrSO_4$]) in and around the shallower secondary bays cannot be ruled out. Additional chemical information, such as Ca and Sr concentrations, total alkalinity, and isotopes of C and Sr, in the sediment, porewater, and water column would be necessary to tease out the process(es) responsible for these higher Ca and Sr concentrations in secondary bays in this dataset.

## Sediment Loading

Longley (1994) found that the average sediment concentrations in stream discharge, given by the slope of the double-mass curve, decreased for all TWDB monitoring stations. The greatest decrease occurred in Trinity River (1936–1946: 1.076 tons ac-ft⁻¹ [0.791 kg m⁻³], 1969–1986: 0.152 tons ac-ft⁻¹ [0.112 kg m⁻³]). The highest average sediment concentrations occurred in San Antonio River (0.728 tons ac-ft⁻¹, 0.535 kg m⁻³) and the lowest average sediment concentrations were in the Nueces River (0.034 tons ac-ft⁻¹, 0.025 kg m⁻³), measured just downstream of Lake Corpus Christi. The changes in sediment concentrations most often corresponded to the construction of reservoirs (Trinity River) or flooding/high inflow events (Lavaca, Guadalupe, and San Antonio rivers), though some changes had no identifiable cause. Additionally, the absolute sediment load in terms of tons of sediment per year also decreased. Though the Trinity River's absolute sediment load declined (1936–1946: 7.52 million tons year⁻¹ [6280 million kg year⁻¹], 1969–1986: 0.80 million tons year⁻¹ [7.26 million kg year⁻¹]), it still

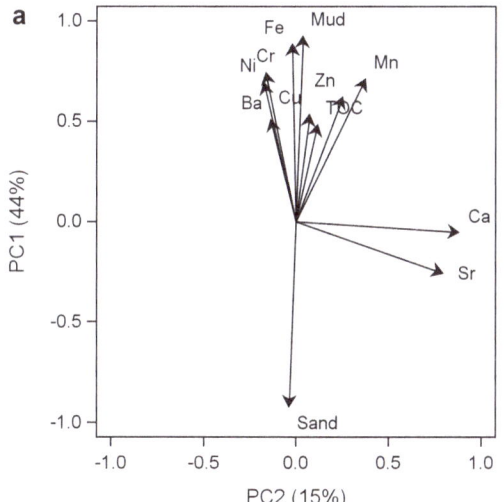

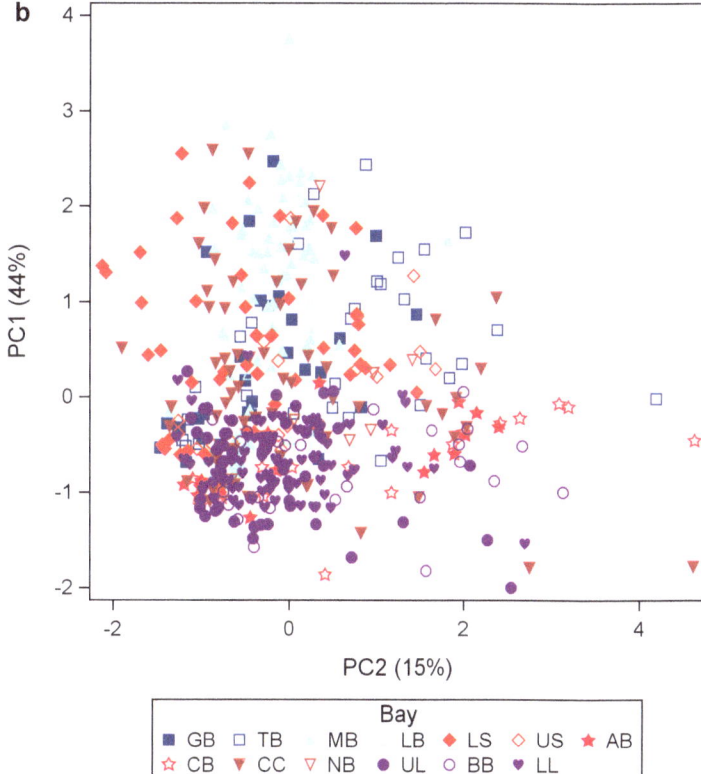

**Fig. 4** PCA for sediment grain size and select geochemical variables on BEG Submerged Lands data. (**a**) Vector loads for variables. (**b**) Sample scores for estuaries (colors) and bays (shapes) with filled symbols for primary bays and open symbols for secondary bays. Abbreviations: Fe = Percent Iron (%), Mn = Manganese (ppm), Ba = Barium (ppm), Cu = Copper (ppm), Cr = Chromium (ppm), Ni = Nickel (ppm), Zn = Zinc (ppm), Ca = Percent Calcium (%), Sr = Strontium (ppm), TOC = percent total organic carbon (%). Bays: GB = Galveston Bay, TB = Trinity Bay, MB = Matagorda Bay, LB = Lavaca Bay, LS = Lower San Antonio Bay, US = Upper San Antonio Bay, AB = Aransas Bay, CB = Copano Bay, CC = Corpus Christi Bay, NB = Nueces Bay, UL = Upper Laguna Madre, BB = Baffin Bay, and LL = Lower Laguna Madre

delivered the largest absolute sediment load while the Lavaca River delivered the smallest (0.147 million tons year$^{-1}$, 133 million kg year$^{-1}$).

## Elemental Burial

An important aspect of estuary C and N budgets is loss to sediments or burial (Brock 2001). There are allochthonous sources of C and N into estuaries with river flow and autochthonous sources produced and recycled by biological processes. The C and N content in sediments was measured by HRI. In each estuary, at least two replicates were collected in at least six stations and averaged by depth. Elemental content is expected to decrease with sediment depth (Fig. 5). The C content average by estuary varies from 0.5% to 3% and is different among the estuaries. The N content average by estuary is very low ranging from zero to 0.1%, although the surface sediments in Laguna Madre can be as high as 0.25% because of the presence of seagrasses. On average, there is 23 times more C than N in sediments. The C and N buried in sediments is refractory, meaning it is the fraction remaining after microbial degradation of freshly deposited material.

There is also an inverse influence of inflow as indicated by the average salinities when the 1-m long sediment samples were collected (Fig. 6). As salinity increases, there is an increase in C ($C$ = 0.00650 + Salinity*0.000377, $n$ = 51, $P$ = 0.0007) and N content ($N$ = 0.000482 + Salinity*0.00000940, $n$ = 51, $P$ = 0.0769), although the relationship with $N$ is >0.05. There is lower C content in the lowest salinity estuaries (Sabine-Neches and Trinity-San Jacinto) but high C content in three of the six stations in the Guadalupe Estuary that were closest to the San Antonio River mouth. The higher salinity estuaries (Nueces, and Upper and Lower Laguna

Madre) had higher C content. The differences between Upper and Lower Laguna Madre systems are explained by the presence of seagrass habitat in the Laguna portion and muddy habitat in the Baffin Bay portion. Turtle grass (*Thalassia testudinum*) dominates (64% of seagrasses in) Lower Laguna Madre and shoal grass (*Halodule wrightii*) dominates (92% of seagrasses in) Upper Laguna Madre (Onuf 2007). However, it is likely the large N content difference between Upper and Lower Laguna Madre explains the lack of correlation between salinity and N content.

## Temporal Dynamics of Sediments

The USGS has been collecting information on sediment transport in rivers and streams in the United States since 1889; however, the amount and location of sediment monitoring has changed over time. For example, between the 1970s and 2000s, the number of sites with discrete suspended-sediment concentration data declined by about 30%, the number of stations with at least 10 discrete suspended-sediment samples per decade decreased by ~40%, and the median number of daily-record sediment stations decreased by 60% (Lee and Glysson 2013). In contrast, the number of continuous turbidity sites and in situ acoustic Doppler sensors has increased since the 1990s, with a maximum of 139 turbidity sites in 2010 and an unknown number of acoustics sites. Thus, although there has been an increased prevalence of continuous turbidity and acoustic sites, the decreasing long-term discrete suspended-sediment concentration sites may make it more difficult to identify long-term changes in sediment transport, particularly across more recent decades (Lee and Glysson 2013).

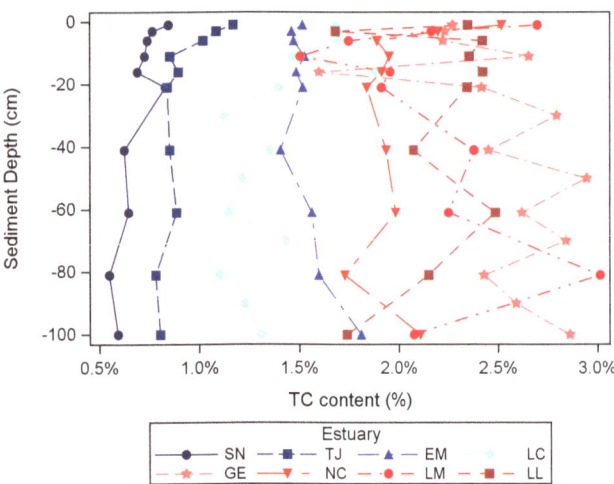
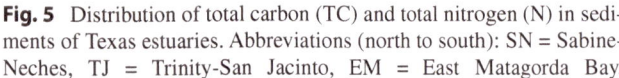
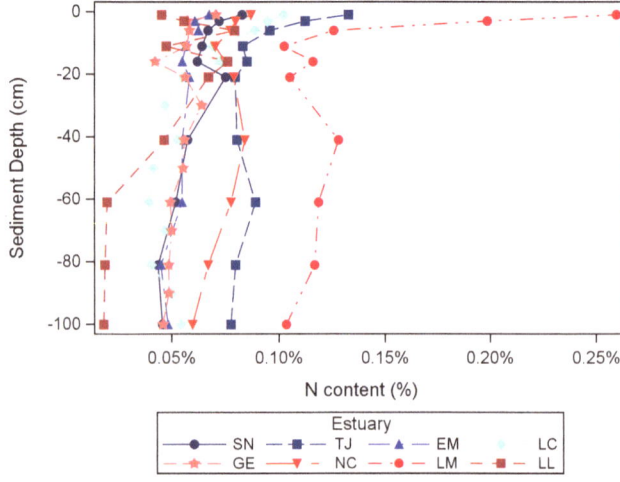

**Fig. 5** Distribution of total carbon (TC) and total nitrogen (N) in sediments of Texas estuaries. Abbreviations (north to south): SN = Sabine-Neches, TJ = Trinity-San Jacinto, EM = East Matagorda Bay,

LC = Lavaca-Colorado, GE = Guadalupe, NC = Nueces, LM = Upper Laguna Madre, and LL = Lower Laguna Madre

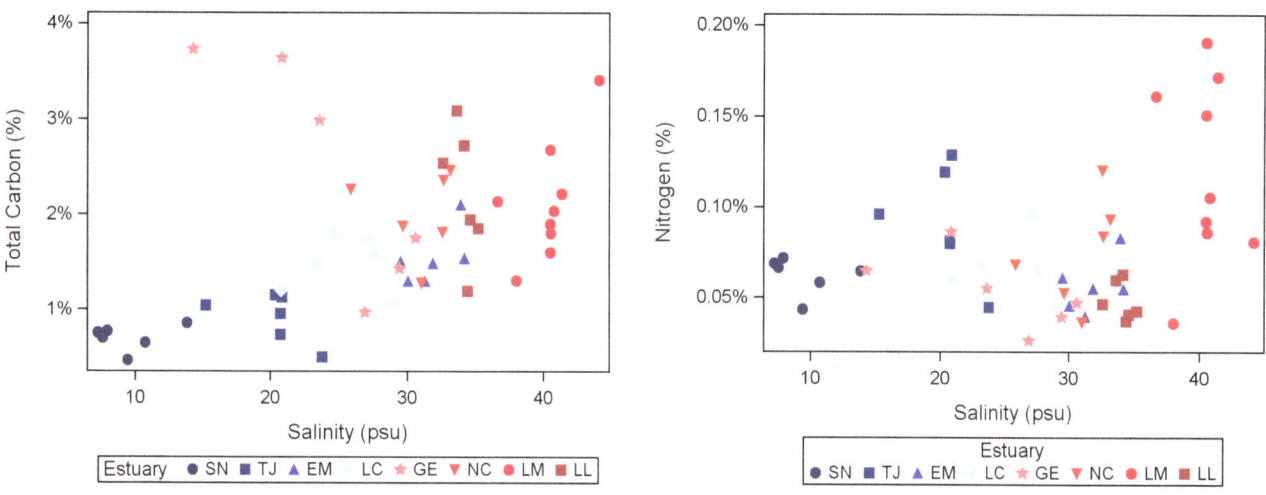

**Fig. 6** Elemental content by station within estuary averaged over all sediment depths (0–100 cm) versus salinity at time of collection. Estuary abbreviations as in Fig. 5

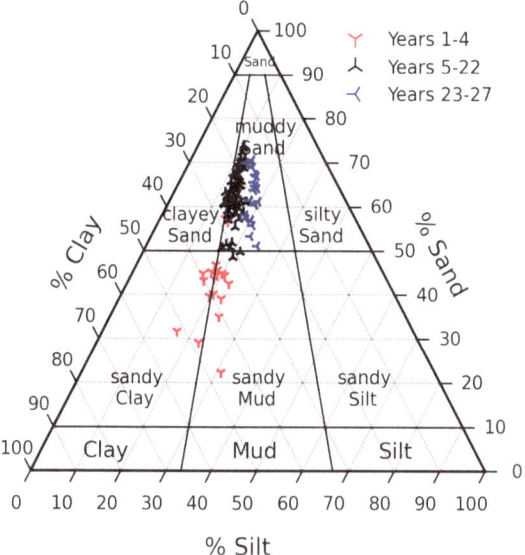

**Fig. 7** Modified Folk classification for Formosa trip average sand, silt, and clay

## Formosa Monitoring

The Formosa Plastic Corporation (FPC) monitored the sediment grain size and the chemistry of sediment and porewater in Lavaca Bay. Over time, the sand content in Lavaca Bay increased while the silt and clay content decreased (Fig. 7). Lavaca River and two creeks (Garcitas Creek and Placedo Creek) flow into Lavaca Bay. Decreased silt and clay inputs to Lavaca Bay may be due, in part, to the Palmetto Bend Dam and Lake Texana acting as a sediment trap for much of the watershed bedload and allowing time for some of the suspended load to settle. Consequently, the Lavaca River inflows, since the dam's completion in 1979 and the start of planned water impoundments in May 1980, would not be bringing in as much sediment and the higher clay content in

the first 4 years of the Formosa study (1993–1996) likely reflected residual existing fine grained sediment content of the bay bottom before it was eroded or resuspended and transported out of the system over time faster than it could be replenished. The smaller creeks flowing into Lavaca Bay may also be contributing less sediment to the system over time as the duration of dry periods has increased and while the frequency of extreme heavy precipitation has increased (see Chap. 3). Thus, the sediment supply from Garcitas Creek and Placedo Creek may fluctuate greatly with short periods of high sediment input and longer periods of reduced sediment input.

## HRI Data

Surface sediments (i.e., the top 3 cm) in five estuaries (GE, LC, LM, MA, and NC) were sampled annually between 1987 and 2018. Sampling and analyte methods are described in Montagna and Kalke (1992) and Palmer et al. (2011). Within each estuary is a primary bay connected to the Gulf of Mexico and a secondary bay receiving fresh water from a river, so the data are analyzed using a 3-way block model where year, estuary, and bay type are main effects. However, the three-way interaction is deleted because of an incomplete block design and a lack of replication (Table 3).

The only variables that are different among years are TOC content and the stable isotope values of TOC ($\delta^{13}$C TOC) and nitrogen ($\delta^{15}$N) (Table 3). Between 1995 and 2005 the TOC content and isotope values were roughly parallel indicating that higher carbon production was related to lighter carbon isotope values (Fig. 8). Typically, lighter isotope values represent a source from phytoplankton and heavier isotope values represent a source from aquatic plants or benthic microalgae (Fry 2006; Bouillon et al. 2011). Because these sediment samples are from 22 muddy stations and only 1 seagrass station, the carbon isotope values here likely repre-

**Table 3** Probability levels for sources of variation in ANOVA results for sediment variables. Highlighted cells are less than 0.05. TOC = total organic carbon, $N$ = Number of observations

| Variable | N | Mean | Year | Estuary | Bay | Year*Est | Year*Bay | Est*Bay |
|---|---|---|---|---|---|---|---|---|
| Mud | 475 | 56.8% | 0.0512 | 0.0001 | 0.0001 | 0.9980 | 0.9992 | 0.0001 |
| Sand | 475 | 37.4% | 0.1246 | 0.0001 | 0.0001 | 0.9997 | 0.9999 | 0.0001 |
| PW | 373 | 53.3% | 0.9727 | 0.0001 | 0.0218 | 0.9999 | 0.9977 | 0.0001 |
| Total C | 338 | 2.02% | 0.2705 | 0.0001 | 0.4513 | 0.1052 | 0.7531 | 0.0001 |
| TOC | 329 | 0.78% | 0.0001 | 0.0001 | 0.2680 | 0.0747 | 0.5791 | 0.0001 |
| Total N | 338 | 0.09% | 0.2198 | 0.0001 | 0.5751 | 0.2042 | 0.3836 | 0.0310 |
| $\delta^{13}$C Total C | 213 | -9.99 | 0.7503 | 0.0001 | 0.0001 | 0.3484 | 0.9493 | 0.0001 |
| $\delta^{13}$C TOC | 204 | -20.07 | 0.0001 | 0.0001 | 0.6078 | 0.0001 | 0.3071 | 0.1100 |
| $\delta^{15}$N | 204 | 6.57 | 0.0001 | 0.0001 | 0.0009 | 0.0009 | 0.2057 | 0.0001 |

**Fig. 8** Average and standard error of total organic carbon (TOC) and TOC stable isotope ($\delta^{13}$C) values over time for all estuaries and bays. HRI data surface (0–3 cm) sediments

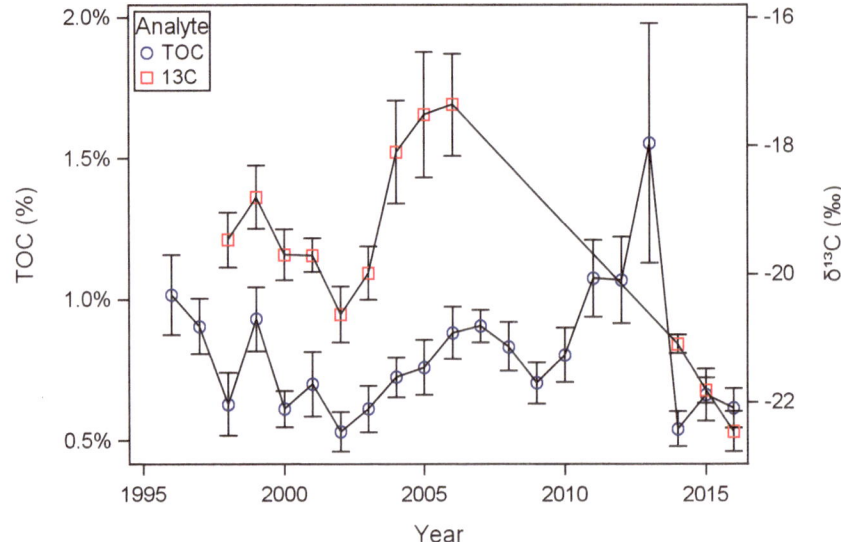

sent primarily benthic microalgae sources. The simple explanation is that when there are more benthic microalgae, the TOC content of sediments is higher.

The one source of variability that was always highly different for every measurement variable was among estuaries (Table 3). So, a subset of the data where all variables were measured ($n = 143$), was analyzed by PCA. The first two components of the PCA for sediments explained 58% of the variance in the dataset (Fig. 9a). Mud versus sand content drove the response, explaining 39% of the variance, and porewater content (PW) was strongly correlated to mud content. Percent nitrogen and TOC content also correlated with mud content. The total carbon (TC) content variable was the main one loading on PC2, which explained 19% of the variance. The estuaries grouped in different quadrants (Fig. 9b). The Lavaca-Colorado (LC) samples grouped in the upper left corner, meaning sediments are the muddiest and lowest in carbon. The Mission-Aransas (MA) was muddy but average in carbon. The Laguna Madre (LM) estuary was muddy but with high carbon. The Nueces estuary (NC) had negative PC1 values indicating the sandiest sediments, but the samples spread across the entire PC2 axis. The Guadalupe Estuary (GE) had a span of muddy and sandy sediments and the highest PC2 values indicating they had the highest carbon content. But there were differences with distance from

the Guadalupe River where upper San Antonio Bay had positive PC2 values and lower San Antonio Bay had negative PC2 values.

## Impacts of Extreme Events

Extreme events, like tropical cyclones and flooding, create major sediment transport events that may redistribute estuarine sediments and reshape shorelines over a period of hours (peak storm intensity) to weeks (residual storm flooding). Landward storm surge is a crucial factor in the evolution of barrier island systems, eroding the shoreface and depositing sediment landward as bayside overwash deposits (Goff et al. 2019). This "rollover" process allows barrier islands to move landward during periods of rising sea level. However, cyclonic storms may also generate seaward-directed flow and sediment transport, which is not typically accounted for in sediment transport or barrier island modeling.

Hurricane Harvey (Harvey) made landfall on the mid-Texas coast as a category 4, major hurricane and proceeded to deliver substantial amounts of precipitation, ~125 billion m$^3$ of rain, over the US (Fritz and Samenow 2017). After landfall, the storm slowly moved east, dropping more than 1.5 m (4.9 ft) of rainfall over Southeastern Texas and causing a massive pulse of floodwater down local canals and rivers that inundated coastal habitats (Williams and Liu 2019).

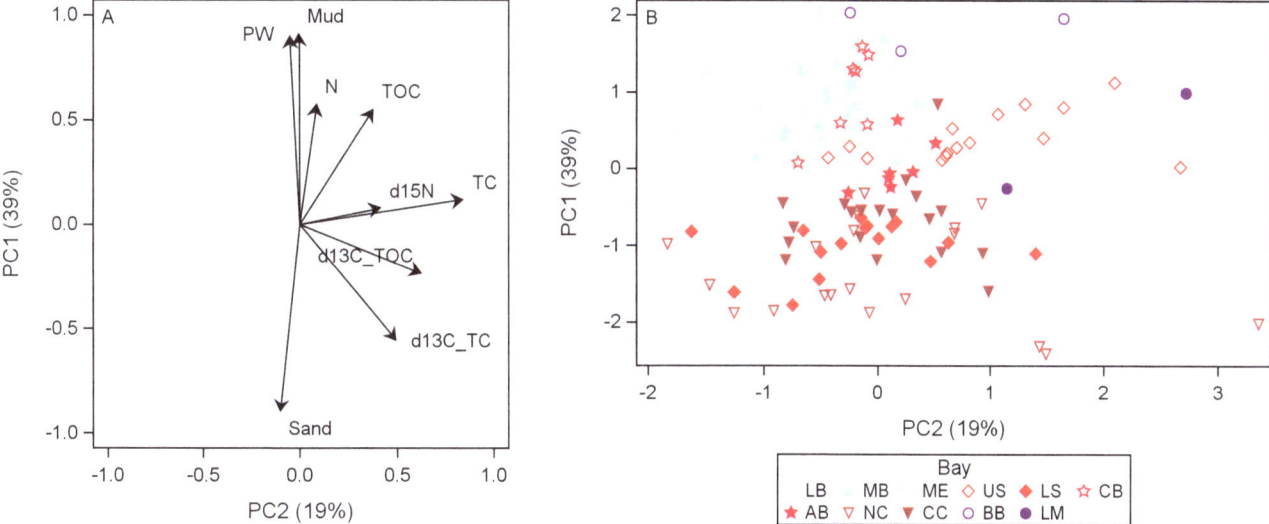

**Fig. 9** PCA on surface (0–3 cm) sediment variables for HRI sediment data. (**a**) Vector loads for variables. (**b**) Sample scores for bays. Abbreviations: d13C_TC = Total Carbon $\delta^{13}C$ (‰), d13C_TOC = Total Organic Carbon $\delta^{13}C$ (‰), d15N = $\delta^{15}$Nitrogen (‰), N = Nitrogen content (%), PW = porewater content (%), and TOC = total organic carbon (%). Bays: MB = Matagorda Bay, ME = Matagorda Bay eastern arm near Colorado River, LB = Lavaca Bay, LS = Lower San Antonio Bay, US = Upper San Antonio Bay, AB = Aransas Bay, CB = Copano Bay, CC = Corpus Christi Bay, NB = Nueces Bay, LM = Upper Laguna Madre, BB = Baffin Bay

Over the course of 53 days, Harvey's floodwaters delivered 14 x 10⁹ m³ of freshwater to Galveston Bay resulting in record flooding of Houston bayous and waterways, all of which drain into the San Jacinto Estuary (SJE) (Dellapenna et al. 2020). Stearns et al. (2023) estimated that Harvey delivered $2.723 \times 10^7$ m³ of sediment to the tributaries of Galveston Bay, which is 40% of the average sediment load of the Mississippi. In addition, some areas, like SJE, experienced massive erosion with at least 48 cm (~1.6 ft) of the sediment column eroded away. This resulted in the deposition of an estimated $131 \times 10^6$ tons of sediment in Galveston Bay, leaving a deposit with an average thickness of 14 cm (5.5 inches) across the entire bay (Dellapenna et al. 2022). Five years later, on average, 80% of the deposit remained (Dellapenna et al. 2023). Transported along with the sediments was an estimated 5 tons of Hg plus from the urbanized SJE and Buffalo Bayou (Dellapenna et al. 2022). In addition, Harvey delivered 28.6 tons of PAH's, 2.48 tons of pyrenes, as well as a variety of other surface-bound and porewater contaminants (Camargo et al. 2021 and personal communication). The floodwaters left muddy flood deposits over much of the coastal bayous and marshes and the eroded sediment in the SJE was replaced by a Harvey storm deposit. Other areas like the McFaddin National Wildlife Refuge, located along the Gulf of Mexico between Sabine Lake and Galveston Bay, also received substantial muddy flood deposits, on average 2.8 cm (1.1 in) thick along a north-south transect, from Harvey's floodwaters that were equivalent to 7 years of "normal" sedimentation in the marsh (Williams and Liu 2019).

In contrast, field observations, satellite imagery, and tide data to the west of Harvey's landfall show that offshore winds on the left side of the storm created seaward-directed surge that pushed resuspended inshore sediments offshore. This seaward surge resulted in a net loss of fine sediments in the Mission-Aransas Estuary (Zhanfei Liu, personal communication, August 2022), substantial erosion around Nueces Estuary including over 3050 m (10,000 feet) of the southern Nueces Bay shoreline eroded and destabilized such that it is threatening to collapse a bike path into the bay (Mott MacDonald 2018), and erosion of the landward shores of the barrier islands and back bays (Goff et al. 2019). Goff et al. (2019) found that the strong outflow at Port Aransas during Harvey was accompanied by significant erosion at the edges of the Corpus Christi Ship Channel and Lydia Ann Channel. Furthermore, large quantities of sand, up to 6.5 m (21.3 ft) thick, accumulated in the depths of Aransas Pass and Lydia Ann Channel, likely as the outflow event waned in strength. Thus, extreme wind and precipitation events may rapidly alter the sediment distribution within estuaries.

## Sedimentation Rates

Sediment accumulation rates vary along the Texas coast but are generally highest in the north and decreases toward the south. Shepard (1953) compared soundings of bay bottom depths that were taken about 65 years apart and estimated the sedimentation rates of most Texas bays (Table 4). Overall, the average sedimentation rate estimated was 0.38 cm year⁻¹.

**Table 4** Sedimentation rates in Texas estuaries based on comparing bathymetry over long time periods (Shepard 1953). Locations listed from northeast to southwest

| Bay/location | Period | Rate (ft 100 year$^{-1}$) | Rate (cm year$^{-1}$) |
|---|---|---|---|
| Galveston Bay | 1854–1933 | 1.44 | 0.44 |
| West Galveston Bay | 1857–1934 | −0.56 | −0.17 |
| East Matagorda Bay | 1859–1934 | 1.56 | 0.48 |
| Eastern Arm Matagorda Bay | 1836–1933 | 3.50 | 1.07 |
| Matagorda Bay | 1858–1934 | −0.23 | −0.07 |
| Lavaca Bay | 1870–1934 | 0.46 | 0.14 |
| Espiritu Santo Bay | 1873–1954 | 0.41 | 0.12 |
| San Antonio Bay | 1875–1955 | 1.23 | 0.37 |
| Mesquite Bay | 1874–1935 | 0.97 | 0.30 |
| St. Charles Bay | 1875–1954 | 0.05 | 0.02 |
| Aransas Bay | 1875–1941 | 1.42 | 0.43 |
| Copano Bay | 1875–1935 | 0.87 | 0.27 |
| Corpus Christi Bay | 1868–1934 | 1.56 | 0.48 |
| Average coast-wide | 1870–1935 | 1.26 | 0.38 |

**Table 5** Sedimentation rates in Texas estuaries based on radioisotope dating profiles

| Location | Method | Sedimentation rate | Source |
|---|---|---|---|
| Sabine-Neches Estuary | $^{239,240}$Pu | 0.4–0.5 cm year$^{-1}$ | Ravichandran et al. (1995) |
| Galveston Bay | $^{210}$Pb, Hg | 0.41 cm year$^{-1}$ | Pekowski (2017) |
| Galveston Bay | $^{137}$Cs, $^{210}$Pb | 0.27 cm year$^{-1}$ | Al Mukaimi et al. (2018) |
| Trinity | $^{210}$Pb | 0.514 cm year$^{-1}$ | White et al. (2002) |
| San Bernard NWR | $^{137}$Cs, $^{210}$Pb | 0.62 (0.15) cm year$^{-1}$ | Callaway (1994) |
| Lavaca-Navidad | $^{210}$Pb | 0.328 cm year$^{-1}$ | White et al. (2002) |
| Lavaca Bay | $^{137}$Cs, $^{210}$Pb | 0.84–1.22 cm year$^{-1}$ | Bronikowski (2004) |
| Lavaca Bay | $^{137}$Cs, $^{239,240}$Pu, Hg | 1.6–2.2 cm year$^{-1}$ | Santschi et al. (1999) |
| Aransas NWR | $^{137}$Cs, $^{210}$Pb | 0.44 (0.16) cm year$^{-1}$ | Callaway (1994) |
| Nueces | $^{210}$Pb | 0.262 cm year$^{-1}$ | White et al. (2002) |
| Corpus Christi Estuary | $^{137}$Cs, $^{239,240}$Pu, Hg | 0.26–0.42 g cm$^{-2}$ year$^{-1}$ | Santschi and Yeager (2004) |
| Nueces Bay | Zn | 0.6 cm year$^{-1}$ | Hill et al. (2014) |
| Nueces Delta | $^{226,228}$Ra, $^{228,230,232}$Th, $^{137}$Cs, $^{210}$Pb | 0.09–0.53 g cm$^{-2}$ year$^{-1}$ | Yeager et al. (2006) |
| Laguna Madre tidal flat | $^{210}$Pb | 0.07–0.83, mean = 0.29 cm year$^{-1}$ | Morton and Holmes (2009) |

Modern approaches to measuring sedimentation rates are based on accretion rates of radioisotopes, such as $^{137}$Cs or $^{210}$Pb, or a known contaminant of local significance, such as Hg or Zn. Studies have been performed in six estuary systems (Table 5) and the overall average is 0.63 cm year$^{-1}$, which is about 40% larger than the rates originally estimated by Shepard (1953). Estuaries tend to be very dynamic systems, with both natural processes, such as both flood erosion as well as deposition (e.g., Dellapenna et al. 2020, 2022), wave generated sediment resuspension and tidal dispersal (e.g., Carlin et al. 2016); Schmidt et al. (2021), as well as hurricane driven storm surge flood and ebb (e.g., Yao et al. 2020). In addition, numerous anthropogenic processes and influences, including maintenance dredging, ship/vessel wakes (e.g., Figlus et al. 2023), shrimp trawling (Dellapenna et al. 2006), oyster dredging and the installation of jetties, pipelines, hydrocarbon production platforms, bridge construction, mitigation of wetlands and oyster reef construction, and anthropogenically driven land subsidence are all factors which can influence the "average" sedimentation rates. Short-lived radio-isotope accumulation rates (e.g., $^{210}$Pb$_{xs}$, $^{137}$Cs, $^{239+240}$Pu) estimate rates based on decadal time-scales and assume steady-state conditions; however, the myriad of previously listed as well as other influences create non-steady state conditions, resulting in an average decadal accumulation rate that does not reflect shorter-term factors which may not be fully preserved in the sedimentary record (e.g., Dellapenna et al. 1998, 2003). Increases in coastal sedimentation rates (cm year$^{-1}$) and mass accumulation rates (g cm$^{-2}$ year$^{-1}$) have doubled across North America in the latter half of the twentieth century despite increased damming of rivers demonstrating that downstream sediment sources may compensate for the river-sediment lost to impoundments (Rodriguez et al. 2020). However, while sedimentation rates currently match or exceed relative sea-level in many U.S. estuaries, the sedimentation rates in Texas and Louisiana coast have generally not kept pace with relative sea level rise (e.g., Al Mukaimi et al. 2018) contributing to the loss of intertidal areas.

## Summary and Recommendations

Freshwater inflows and sediments are intimately linked in several ways:

1. Sediment Transport, Deposition, and Dynamics: Sediments are transported into coastal water bodies through streams and runoff. During high inflow/high energy conditions, the increased water velocity may erode and resuspend previously deposited sediments resulting in a significant amount of sediment being transported and redistributed downstream or within the estua-

rine system. However, during low inflow/low energy conditions or as the water velocity slows, the sediments settle and accumulate in the bays. Alternating periods of high and low inflow/energy and sea level result in the formation of sediment layers which over geologic time scales formed the coastal aquifer systems and the geomorphology of the estuaries (i.e., barrier islands, lagoons, primary and secondary bays, etc.). Freshwater inflows influence sediment dynamics by altering the flow velocity, turbulence, sediment load, and sediment resuspension.

2. Nutrient, Organic Matter, and Contaminant Inputs: Nutrients, organic matter, and various contaminants may bind to sediment particles and be transported and deposited long distances from their origins. Thus, sediments may serve as an important reservoir of nutrients, organic matter, and contaminants in coastal systems with implications for nutrient cycling, productivity, and ecosystem health.

3. Habitat Formation: Deposited sediments create substrate for benthic organisms (e.g., benthic invertebrates) and contribute to the formation of various aquatic habitats (e.g., sand beaches, mid-bay muds, seagrass beds, oyster reefs). The sediment characteristics (i.e., percent sand, porosity, etc.) influence habitat formation, and the species composition and diversity in the estuaries.

The primary data gaps identified here are:

1. Monitoring/measurements of suspended sediment from the inflowing tributaries. Although the major tributaries are gauged to measure freshwater inflow, suspended sediment is neither monitored nor measured.

2. Estimations of subsidence rates. While subsidence rates have been measured around the perimeter of Galveston Bay and estimated across the bay (Al Mukaimi et al. 2018), comparable measurements/estimations for other bays are currently unavailable.

3. Measurements of sedimentation rates. Except for Galveston Bay (Al Mukaimi et al. 2018), recent measurements of sedimentation rates across estuaries are also a key dataset currently unavailable.

Thus, while sediments are one of the three main estuarine responses to freshwater inflow, there are still knowledge gaps that need to be filled through monitoring and measurements to gain a complete picture of estuarine sediment response to freshwater inflow along the Texas coast.

**Acknowledgements** This publication was funded in part through Contract No. 21-155-007-C879 from the Texas General Land Office (GLO) with Gulf of Mexico Energy Security Act of 2006 funding made available to the State of Texas and awarded under the Texas Coastal Management Program. The views contained herein are those of the authors and should not be interpreted as representing the views of the GLO or the State of Texas. Partial funding was also provided by the National Oceanic and Atmospheric Administration (NOAA), Office of Education, Educational Partnership Program (EPP) award (NA21SEC4810004). Its contents are solely the responsibility of the award recipient and do not necessarily represent the official views of the U.S. Department of Commerce, National Oceanic and Atmospheric Administration.

**Data Sources** All data used and presented are publicly available or available by request to the authors. The following datasets are available on GRIIDC: BEG Submerged Lands (https://doi.org/10.7266/D1PNX3HD), Formosa Plastics Corporation (https://doi.org/10.7266/DCNHQD59), and HRI sediment (https://doi.org/10.7266/aa36pm7h).

# References

Alber M (2002) A conceptual model of estuarine freshwater inflow management. Estuaries 25(6):1246–1261. https://doi.org/10.1007/BF02692222

Al Mukaimi ME, Dellapenna TM, Williams JR (2018) Enhanced land subsidence in Galveston Bay, Texas: interaction between sediment accumulation rates and relative sea level rise. Estuar Coastal Shelf Sci 207:183–193. https://doi.org/10.1016/j.ecss.2018.03.023

Anderson JB, Wallace DJ, Simms AR, Rodriguez AB, Milliken KT (2014) Variable response of coastal environments of the northwestern Gulf of Mexico to sea-level rise and climate change: implications for future change. Mar Geol 352:348–366. https://doi.org/10.1016/j.margeo.2013.12.008

Armstrong NE (1982) Responses of Texas estuaries to freshwater inflows. In: Kennedy VS (ed) Estuarine comparisons. Academic Press, pp 103–120. https://doi.org/10.1016/B978-0-12-404070-0.50013-2

Besonen M, Hill EM, Tissot P (2016) Baffin Bay sediment Core profiling for historical water quality. Publication 109 of the coastal bend bays and estuaries program, p 76

Bianchi TS (2006) Biogeochemistry of Estuaries. Oxford University Press

Bianchi TS, Pennock JR, Twilley RR (1999) Biogeochemistry of Gulf of Mexico estuaries. Wiley, New York

Bouillon S, Connolly RM, Gillikin DP (2011) Use of stable isotopes to understand food webs and ecosystem functioning in estuaries. In: Wolanski E, McLusky DS (eds) Treatise on estuarine and coastal science, vol 7. Academic Press, Waltham, pp 143–173

Brock DA (2001) Nitrogen budget for low and high freshwater inflows, Nueces Estuary, Texas. Estuaries 24:509–521

Bronikowski JL (2004) Sedimentary environments and processes in a shallow, Gulf coast estuary-Lavaca Bay, Texas. Master of Science Thesis, Texas A&M University

Callaway JC (1994) Sedimentation processes in selected coastal wetlands from the Gulf of Mexico and northern Europe. Doctoral Dissertation, Louisiana State University

Camargo K, Sericano JL, Bhandari S, Hoelscher C, McDonald TJ, Chiu WA, Wade TL, Dellapenna TM, Liu Y, Knap AH (2021) Polycyclic aromatic hydrocarbon status in post-hurricane Harvey sediments: considerations for environmental sampling in the Galveston Bay/Houston Ship Channel region. Mar Pollut Bull 162:111872. https://doi.org/10.1016/j.marpolbul.2020.111872

Carlin J, Lee G-H, Dellapenna TM, Laverty P (2016) Sediment resuspension by wind, wave, and currents during meteorological frontal passage in a micro-tidal lagoon. Estuar Coast Shelf Sci 172:24–33. https://doi.org/10.1016/j.ecss.2016.01.029

Christopherson RW (2004) Elemental geosystems, 4th edn. Pearson Education, Inc, Upper Saddle River, NJ

Dalrymple RW, Zaitlin BA, Boyd R (1992) Estuarine facies models: conceptual basis and stratigraphic implications. J Sediment Petrol 62(6):1130–1146. https://doi.org/10.1306/D4267A69-2B26-11D7-8648000102C1865D

Davies JL (1973) Geographical variation in coastal development. Hafner, New York

Day J, Hall M, Kemp W, Yanez-Arancibia A (1989) Estuarine ecology. Wiley, New York

Dellapenna TM, Allison MA, Gill GA, Lehman RD, Warnken KW (2006) The impact of shrimp trawling and associated sediment resuspension in mud dominated, shallow estuaries. Estuar Coast Shelf Sci 69:519–530. https://doi.org/10.1016/j.ecss.2006.04.024

Dellapenna TM, Hoelscher C, Hill L, Al Mukaimi ME, Knap A (2020) How tropical cyclone flooding caused erosion and dispersal of mercury-contaminated sediment in an urban estuary: the impact of hurricane Harvey on Buffalo Bayou and the San Jacinto Estuary, Galveston Bay, USA. Sci Total Environ 748:141226. https://doi.org/10.1016/j.scitotenv.2020.141226

Dellapenna TM, Hoelscher CE, Hill L, Critides L, Bartlett V, Bell M, Al Mukaimi DJ, Park KM, Knap A (2022) Hurricane Harvey delivered a massive load of mercury rich sediment to Galveston Bay, Texas, USA. Estuar Coasts 45:428–444. https://doi.org/10.1007/s12237-021-00990-7

Dellapenna TM, Jung N, Schenk R, Sudduth S, Lin P, Figlus J (2023) How subsidence and cyclone driven sediment flux within Galveston Bay has caused elevated siltation within the Bayport Channel and Flare. Conference proceeding: coastal sediments 2023

Dellapenna TM, Kuehl SA, Schaffner LC (2003) Ephemeral deposition, seabed mixing and fine-scale strata formation in the York River estuary, Chesapeake Bay. Estuar Coast Shelf Sci 58:621–643. https://doi.org/10.1016/S0272-7714(03)00174-4

Dellapenna TM, Kuehl SA, Schaffner LC (1998) Sea-bed mixing and particle residence times in biologically and physically dominated estuarine systems: a comparison of lower Chesapeake Bay and the York River subestuary. Estuar Coast Shelf Sci 46:777–795. https://doi.org/10.2307/1352946

Fairbridge R (1980) The estuary: its definition and geodynamic cycle. In: Olausson E, Cato I (eds) Chemistry and biochemistry of estuaries. Wiley, New York, pp 1–35

Fetter CW (2001) Applied hydrogeology, 4th edn. Prentice-Hall, Inc

Figlus J, Joubert JJ, Dellapenna TM (2023) Field investigation of enhanced ship channel shoaling in a shallow bay system. In coastal sediments. World Scientific, pp 2731–2741. https://doi.org/10.1142/9789811275135_0250

Flemming B (2020) Beach sand and its origin. In: DWT J, Short AD (eds) Sandy beach morphodynamics. Elsevier, pp 15–37. https://doi.org/10.1016/B978-0-08-102927-5.00002-3

Folk RL (1980) Petrology of sedimentary rocks. Hemphill Publishing Company, Austin, TX

Fritz A, Samenow J (2017) Harvey unloaded 33 trillion gallons of water in the US. The Washington Post 2

Fry B (2006) Stable isotope ecology. Springer, New York, NY

Goff JA, Swartz JM, Gulick SPS, Dawson CN, de Alegria-Arzaburu AR (2019) An outflow event on the left side of Hurricane Harvey: erosion of barrier sand and seaward transport through Aransas Pass, Texas. Geomorphology 334:44–57. https://doi.org/10.1016/j.geomorph.2019.02.038

Gu X, Evans LJ (2007) Modelling the adsorption of Cd(II), Cu(II), Ni(II), Pb(II), and Zn(II) onto Fithian illite. J Colloid Interface Sci 307:317–325. https://doi.org/10.1016/j.jcis.2006.11.022

Hao W, Flynn SL, Kashiwabara T, Alam MS, Bandara S, Swaren L, Robbins LJ, Alessi DS, Konhauser KO (2019) The impact of ionic strength on the proton reactivity of clay minerals. Chem Geol 529:119294. https://doi.org/10.1016/j.chemgeo.2019.119294

Harper M et al (2015) Python-ternary: ternary plots in python. Zenodo. https://doi.org/10.5281/zenodo.2628066

Hill EM, Besonen M, Nicolau B (2014) Nueces Bay zinc in sediment profiling assessment. Report to coastal bend bays & estuaries program. https://www.cbbep.org/publications/publication1313b.pdf

Hizal J, Apak R (2006) Modeling of copper (II) and lead (II) adsorption on kaolinite-based clay minerals individually and in the presence of humic acid. J Colloid Interface Sci 295(1):1–13. https://doi.org/10.1016/j.jcis.2005.08.005

Hoyt JH (1967) Barrier Island formation. GSA Bulletin 78(9):1125–1136. https://doi.org/10.1130/0016-7606(1967)78[1125:BIF]2.0.CO;2

Kasmarek MC, Robinson JL (2004) Hydrogeology and simulation of ground-water flow and land-surface subsidence in the northern part of the Gulf coast aquifer system, Texas [Report](2004-5102). (Scientific Investigations Report, Issue. U. S. G. Survey. http://pubs.er.usgs.gov/publication/sir20045102

Lee CJ, Glysson GD (2013) Compilation, quality control, analysis, and summary of discrete suspended-sediment and ancillary data in the United States, 1901–2010. U.S. Geological Survey, Reston, VA., 46 p. https://doi.org/10.3133/ds776

Lewis D (1984) Practical sedimentology. Hutchinson Ross, Stroudsburg, PA, pp 80–100

Liu Y, Alessi DS, Flynn SL, Alam MS, Hao W, Gingras M, Zhao H, Konhauser KO (2018) Acid-base properties of kaolinite, montmorillonite and illite at marine ionic strength. Chem Geol 483:191–200. https://doi.org/10.1016/j.chemgeo.2018.01.018

Longley WL (1994) Freshwater inflows to Texas bays and estuaries: ecological relationships and methods for determination of needs. Texas Water Development Board and Texas Parks and Wildlife Department

McGowen JH, Morton RA (1979) Sediment distribution, bathymetry, faults, salt diapirs, and submerged lands of Texas. Univ. Texas at Austin, Bureau of Economic Geology

McKee BA, Baskaran M (1999) Sedimentary processes of Gulf of Mexico estuaries. In: Bianchi T, Pennock J, Twilley R (eds) Biogeochemistry of Gulf of Mexico estuaries. Wiley, New York, pp 63–85

Meade RH, Yuzyk TR, Day J (1990) Movement and storage of sediment in rivers of the United States and Canada. In: Wolman MG, Riggs HC (eds) Surface water hydrology, vol O-1. Geological Society of America

Montagna PA, Kalke RD (1992) The effect of freshwater inflow on meiofaunal and macrofaunal populations in the Guadalupe and Nueces Estuaries, Texas. Estuaries 15:307–326. https://doi.org/10.2307/1352779

Morton RA, Holmes CW (2009) Geological processes and sedimentation rates of wind-tidal flats, Laguna Madre, Texas. Gulf Coast Assoc Geol Soc Trans 59:519–538. https://archives.datapages.com/data/gcags_pdf/2009/MortHolm.pdf

MacDonald M (2018) PCCA hurricane Harvey damage shoreline assessment. Port of Corpus Christi Authority, Corpus Christi, TX. 44 p

Murgulet D, Trevino M, Douglas A, Spalt N, Hu X, Murgulet V (2018) Temporal and spatial fluctuations of groundwater-derived alkalinity fluxes to a semiarid coastal embayment. Sci Total Environ 630:1343–1359. https://doi.org/10.1016/j.scitotenv.2018.02.333

Nichols MM, Biggs RB (1985) Estuaries. In: Davis RA Jr (ed) Coastal sedimentary environments, 2nd Revised, Expanded edn. Springer, New York, pp 77–186

Olsen SB, Padma TV, Richter BD (2006) Managing freshwater inflows to estuaries: a methods guide. United States Agency for International Development (USAID); the nature conservancy; Coastal Resources Center, University of Rhode Island, Washington, DC, p 44

Onuf CP (2007) Laguna Madre. In: Handley L, Altsman D, DeMay R (eds) Seagrass status and trends in the northern Gulf of Mexico: 1940–2002: U.S. Geological Survey Scientific Investigations Report 2006–5287, pp 29–40. https://pubs.usgs.gov/sir/2006/5287/

Opsahl S, Benner R (1997) Distribution and cycling of terrigenous dissolved organic matter in the ocean. Nature 386:480–482. https://doi.org/10.1038/386480a0

Palmer TA, Montagna PA, Pollack JB, Kalke RD, DeYoe HR (2011) The role of freshwater inflow in lagoons, rivers, and bays. Hydrobiologia 667:49–67. https://doi.org/10.1007/s10750-011-0637-0

Pekowski AD (2017) Elevated modern sedimentation rates over the buried Trinity River incised valley suggests elevated, localized subsidence rates, Galveston Bay, TX, USA. Master of Science Thesis, Texas A&M University

Perillo GME (1995) In: Perillo GME (ed) Developments in sedimentology. Elsevier, pp 17–47. https://doi.org/10.1016/S0070-4571(05)80022-6

Pettijohn FJ (1975) Sedimentary rocks, vol 3. Harper & Row, New York

Plumb R (1981) Procedures for handling and chemical analysis of sediment and water samples. Technical report EPA/CE-81-1. US Army Engineers Waterways Experimental Station Vicksburg, MS

Pritchard DW (1952) In: Landsberg HE (ed) Advances in geophysics. Elsevier, pp 243–280. https://doi.org/10.1016/S0065-2687(08)60208-3

Pritchard DW (1967) What is an estuary: physical viewpoint. In: Lauff GH (ed) Estuaries. American Association for the Advancement of Science, Washington, pp 52–63

Ravichandran M, Baskaran M, Santschi PH, Bianchi TS (1995) Geochronology of sediments in the Sabine-Neches estuary, Texas, USA. Chem Geol 125:291–306. https://doi.org/10.1016/0009-2541(95)00082-W

Rodriguez AB, McKee BA, Miller CB, Bost MC, Atencio AN (2020) Coastal sedimentation across North America doubled in the 20th century despite river dams. Nat Commun 11:3249. https://doi.org/10.1038/s41467-020-16994-z

Santschi PH, Allison MA, Asbill S, Perlet B, Cappellino S, Dobbs C, McShea L (1999) Sediment transport and Hg recovery in Lavaca Bay, as evaluated from radionuclide and Hg distributions. Environ Sci Technol 33:378–391. https://doi.org/10.1021/es980378l

Santschi PH, Yeager KM (2004) Quantification of terrestrial and marine sediment sources to a managed fluvial, deltaic and estuarine system: the Nueces-Corpus Christi Estuary, Texas. Report to Texas Water Development Board. https://www.twdb.texas.gov/publications/reports/contracted_reports/doc/2003483001.pdf

Schmidt N, Dellapenna TM, Lin P (2021) Cold front sediment resuspension, age, and residence times of suspended sediment using $^7Be/^{210}Pb_{xs}$ ratio in Galveston Bay. Front Mar Sci 8:703945. https://doi.org/10.3389/fmars.2021.703945

Shepard FP (1953) Sedimentation rates in Texas estuaries and lagoons. Bull Am Assoc Petroleum Geol 37:1919–1934

Shepard FP, Rusnak GA (1957) Texas bay sediments. Publ Inst Mar Sci 4(2):5–13. https://repositories.lib.utexas.edu/handle/2152/19164

Stearns AI, Wellner JS, Kendall JJ, Khan SD (2023) Sediment routing in an incised valley during Hurricane Harvey (2017) in Houston, Texas, USA: Implications for modern sedimentation. Geology 51:995–1000. https://doi.org/10.1130/G51312.1

Syvitski JP, Vörösmarty CJ, Kettner AJ, Green P (2005) Impact of humans on the flux of terrestrial sediment to the global coastal ocean. Science 308(5720):376–380

Vörösmarty CJ, Meybeck M, Fekete B, Sharma K, Green P, Syvitski JP (2003) Anthropogenic sediment retention: major global impact from registered river impoundments. Glob Planetary Chan 39(1–2):169–190

Wentworth CK (1922) A scale of grade and class terms for clastic sediments. J Geol 30(5):377–392. https://doi.org/10.1086/622910

Wermund EG, White WA, Calnan TR, Morton RA (1989) Mapping off-shore coastal areas: 'The Submerged Lands of Texas Atlas'—a review. Oce Shore Man 12:411–426

White WA, Morton RA, Holmes CW (2002) A comparison of factors controlling sedimentation rates and wetland loss in fluvial–deltaic systems, Texas Gulf coast. Geomorphology 44:47–66

Williams H, Liu K-B (2019) Contrasting Hurricane Ike washover sedimentation and Hurricane Harvey flood sedimentation in a southeastern Texas coastal marsh. Mar Geol 417:106011. https://doi.org/10.1016/j.margeo.2019.106011

Yao Q, Liu KB, Williams H, Joshi S, Bianchette TA, Ryu J, Dietz M (2020) Hurricane Harvey storm sedimentation in the San Bernard national wildlife refuge, Texas: Fluvial versus storm surge deposition. Estuar Coasts 43:971–983. https://doi.org/10.1007/s12237-019-00639-6

Yeager KM, Santschi PH, Schindler KJ, Adres MJ, Weaver EA (2006) The relative importance of terrestrial versus marine sediment sources to the Nueces-Corpus Christi estuary, Texas: an isotopic approach. Estuaries Coast 29(3):443–454

# Nutrient-Phytoplankton Dynamics in Texas Estuaries

Michael S. Wetz ⓘ, Laura Beecraft ⓘ, Molly McBride ⓘ, Jamie L. Steichen ⓘ, and Antonietta Quigg ⓘ

## Abstract

Phytoplankton form the base of the food web and are often the main primary producer in estuaries. Freshwater inflow variability will modulate phytoplankton biomass and production in estuaries through effects on nutrient delivery and cycling in estuaries, as well as on light availability and flushing rates. The goal of this chapter is to build upon the conceptual framework of inflow-nutrient-phytoplankton dynamics proposed by Longley (1994) by highlighting advances that have been made since then. Specific objectives are to: quantify coast-wide relationships between inflow, nutrients, and phytoplankton biomass (chlorophyll *a*; hereafter "chlorophyll") in Texas estuaries; quantify variability in phytoplankton biomass in relation to inflow variability; identify key drivers of this variability in individual systems where sufficient data exists (Galveston Bay, Matagorda Bay, Baffin Bay); and highlight deficiencies and future data needs. The overarching conclusion is that freshwater inflow plays a major role in external nutrient loadings to Texas estuaries, but its influence on phytoplankton dynamics is complicated because of effects on not only nutrient availability but also flushing rates and the light environment within estuaries. One recommendation based on the synthesis here is that there is a strong need for more system-specific studies to elucidate mechanistic linkages between inflow, nutrients, and phytoplankton, because this information will be vital for developing management plans in areas where inflow rates are changing and/or where nutrient pressures may be a growing concern.

M. S. Wetz (✉) · L. Beecraft · M. McBride
Texas A&M University-Corpus Christi, Harte Research Institute for Gulf of Mexico Studies, Corpus Christi, TX, USA
e-mail: Michael.wetz@tamucc.edu; laura.beecraft@tamucc.edu

J. L. Steichen · A. Quigg
Texas A&M University at Galveston, Department of Marine Biology, Galveston, TX, USA
e-mail: jamie.steichen@tamu.edu; quigga@tamug.edu

## Keywords

Chlorophyll · Dissolved inorganic nutrients · Dissolved organic nutrients · Nitrogen · Phosphate · Phytoplankton · Silicate

## Introduction

Freshwater inflow plays a vital role in fueling the rich production and diversity within estuaries through delivery of nutrients that support primary producers. Early freshwater inflow studies on Texas estuaries set the stage for concepts of bay and estuary function (reviewed by Longley 1994) and the links between freshwater inflows, nutrients, and productivity. Longley (1994) provided an excellent conceptual diagram of inflow-nutrient dynamics and nutrient processing that largely holds true today (Fig. 1). For example, they demonstrated that high inflow conditions often lead to higher dissolved inorganic nitrogen (DIN) and total phosphorus (TP) concentrations over low inflow conditions, especially in the upper estuary and near river mouths. In Texas estuaries, the upper estuary often serves as a site of intense processing of river-derived nutrients and organic matter by phytoplankton and bacteria, whereas the middle and lower estuary run on recycled nutrients from the water column and benthos (Fig. 1). Depending on the magnitude of an inflow event and estuarine geomorphology, effects can sometimes manifest further downstream in the estuary as higher nutrient concentrations and productivity (Longley 1994; Örnólfsdóttir et al. 2004; Dorado et al. 2015; Roelke et al. 2017; Walker et al. 2020). In addition to the short-term (days-weeks) fertilization effect of inflow events (Mooney and McClelland 2012; Walker et al. 2020), longer-term (months) stimulation of primary production can still occur when nutrients are retained and recycled in the estuary, a common phenomenon in the shallow lagoonal estuaries of the central and south Texas coast (e.g., Mooney and McClelland 2012; Bruesewitz et al. 2013).

P. A. Montagna, A. R. Douglas (eds.), *Freshwater Inflows to Texas Bays and Estuaries*, Estuaries of the World, https://doi.org/10.1007/978-3-031-70882-4_7

**Fig. 1** Conceptual diagram of nutrient processing in an estuary. From Longley (1994)

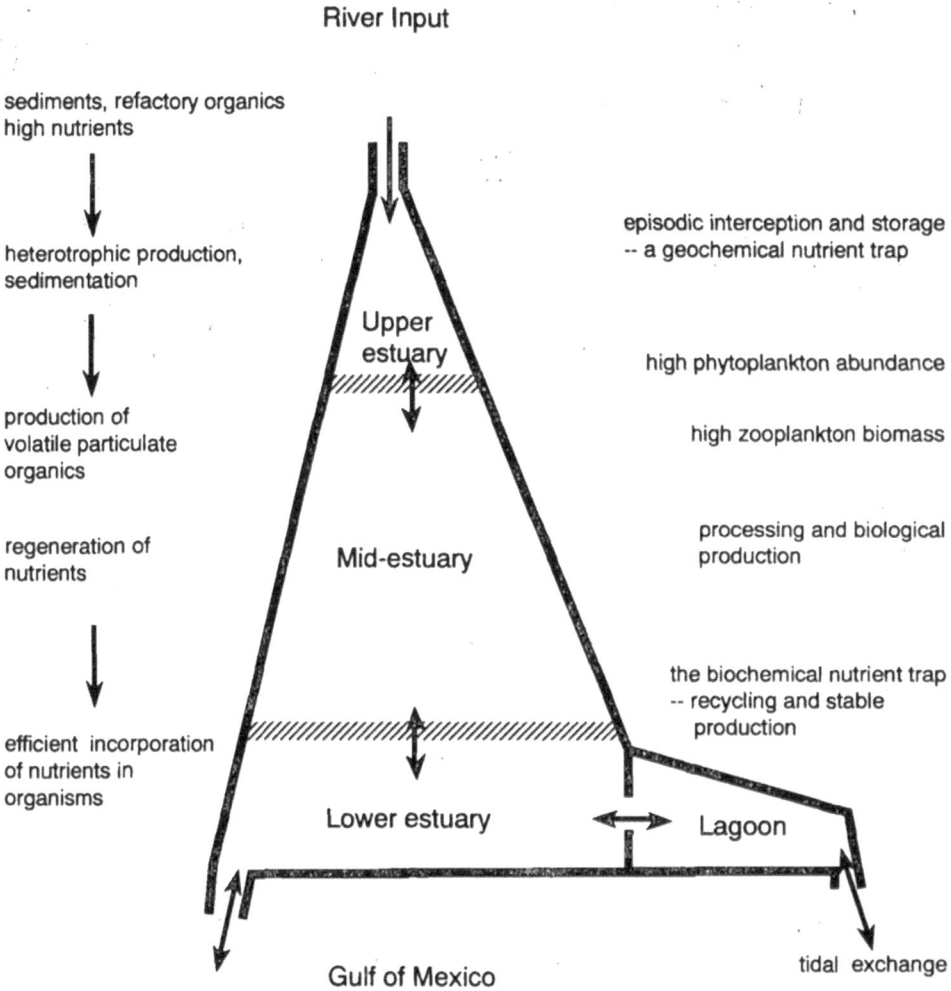

Phytoplankton are sensitive indicators of environmental change because of their ability to respond rapidly to perturbations (Paerl et al. 2003, 2010). Freshwater inflow variability influences estuarine phytoplankton by dictating nutrient, sediment, flushing, and light regimes. While nutrient loadings often scale to the level of freshwater inflow and can stimulate phytoplankton growth (Mallin et al. 1993), high magnitude inflows may also limit biomass accumulation when flushing times exceed phytoplankton growth rates (Paerl et al. 2001; Roelke et al. 2013; Azevedo et al. 2014; Dorado et al. 2015). Higher inflows may also increase sediment loading, resulting in decreased light availability and introducing the potential for light limitation (Lancelot and Muylaert 2011; Geyer et al. 2018). On the other hand, phytoplankton growth can become nutrient limited under prolonged low-flow conditions in river-dominated estuaries (Rask et al. 1999; Örnólfsdóttir et al. 2004; Abreu et al. 2010; Phlips et al. 2010; Wetz et al. 2011), while studies have shown that phytoplankton growth can still be high in lagoonal systems that lack extensive river networks, owing to potential for high rates of internal nutrient loadings and regeneration of nutrients (Gilbert et al. 2010; Wetz et al. 2017).

Over the past century, natural cycles of freshwater delivery to the coastal zone have been altered due to human activity and climate change (Alber 2002; Milliman et al. 2008). For several central-south Texas estuaries, long-term reductions in freshwater inflow have been observed in response to damming, drought, and increasing water withdrawals for human needs (Montagna et al. 2013). A prime case study is the Nueces Estuary, which has experienced a sharp reduction in inflows following the construction of several reservoirs. A recent synthesis of long-term water quality monitoring data showed that chlorophyll has decreased over time in the system, presumably due to the decreasing inflows and river-derived nutrient loadings (Bugica et al. 2020). In contrast, evidence has emerged that rainfall has become "flashier" on the upper Texas coast (Nielsen-Gammon et al. 2020), resulting in intense but short-lived freshwater pulses that deliver high nutrient and sediment loads while also increasing flushing of the estuary. The long-term consequences for the ecosystem are unclear, although work conducted in estuaries outside of Texas has shown that high magnitude flood pulses can lead to dramatic changes in phytoplankton biomass over multiple timescales, often following a pattern of a short-term reduction due to light limitation and increased flushing and a longer-term stimulation over a period of months due to increased nutrient availability (Paerl et al. 2001; Wetz and Yoskowitz 2013; Geyer et al. 2018). In addition to these alterations in the magnitude of freshwater inflow, human population growth and widespread land use change in coastal watersheds have increased nutrient loads to freshwater and estuarine systems worldwide (Handler et al. 2006; Kaushal et al. 2008; Rothenberger et al. 2009; Freeman et al. 2019), resulting in growing prevalence of eutrophication and associated symptoms (Bricker et al. 2008). In Texas, eutrophication is not a widespread phenomenon yet, although clear signatures of nutrient enrichment have been documented in Baffin Bay-Upper Laguna Madre, Oso Bay, and parts of the Galveston Bay complex (Bugica et al. 2020).

There is increasing concern that the combined effects of human alteration of hydrologic regimes, anthropogenic nutrient enrichment, and climate change will lead to less desirable estuarine water quality and habitat conditions in Texas and elsewhere (Flemer and Champ 2006; Freeman et al. 2019). Among the critical needs in projecting estuarine ecosystem response to these environmental changes is the development of mechanistic linkages between specific key features of climatic and anthropogenic drivers (e.g., nutrient loading and variable freshwater delivery) and relevant components of the ecosystem. In this chapter, we provide an updated review of broad-scale patterns in nutrients and phytoplankton biomass (using chlorophyll as a proxy) on the Texas coast in relation to a natural freshwater inflow gradient, assess mechanistic linkages that have been established in select estuaries through investigator-led long-term sampling programs, and provide recommendations for management actions and future research needs.

## Broadscale Salinity, Nutrient, and Chlorophyll Patterns along the Texas Coast

To quantify spatial patterns in various nutrient indicators along the Texas coast, data was obtained from the Texas Commission on Environmental Quality's Surface Water Quality Monitoring program (https://www.tceq.texas.gov/waterquality/monitoring). The 10-year average (from 2010 to 2019) of relevant variables was calculated for up to 130 stations. The sharp salinity gradient, as defined by well-documented precipitation and freshwater inflow gradients (Montagna et al. 2013), was clearly seen in the TCEQ data, with lower salinities in the upper estuaries and much higher salinities to the south (Fig. 2a).

On average, nitrate plus nitrite (N + N) concentrations were ≤ 0.07–0.14 mg/L along much of the coast (Fig. 2b). Exceptions were primarily in Galveston Bay and its tributar-

ies, where concentrations of up to 1.01 mg/L were observed, in the Inner Harbor of Corpus Christi Bay (0.36–0.50 mg/L), and in the lower Laguna Madre adjacent to the Arroyo Colorado outflow (0.45 mg/L) (Fig. 2b). Ammonium concentrations were ≤ 0.08 mg/L along the entire coast except in tributaries of Galveston Bay where slightly higher concentrations (0.14–0.18 mg/L) were observed (Fig. 2c). Total Kjeldahl Nitrogen (TKN) represents a combination of organic nitrogen and ammonium. Average TKN concentrations were 0.31 mg/L at all sites along the Texas coast (Fig. 2d), with higher concentrations noted in parts of Galveston Bay and its tributaries (up to 1.72 mg/L), Port Bay (1.81 mg/L), Redfish Bay (1.71 mg/L), Oso Bay (2.34 mg/L), Baffin Bay (1.18–1.30 mg/L), and an adjacent site in the upper Laguna Madre (1.05 mg/L). Given the low ammonium concentrations that were present, it appears that the majority of the TKN was present as organic nitrogen, making it the dominant form of nitrogen in Texas estuaries. This is consistent with data from more intensive sampling in Baffin Bay (Wetz et al. 2017; Cira and Wetz 2019; Beecraft and Wetz 2023), which has shown only ephemeral increases in inorganic nutrients, typically as a result of high rainfall periods or storm events.

Total phosphorus (TP) concentrations were generally ≤0.15 mg/L, with particularly low (≤ 0.06 mg/L) concentrations observed throughout much of the Laguna Madre and Corpus Christi Bay (Fig. 2e). TP concentrations up to 0.46 mg/L were observed in Galveston Bay and its tributaries. If we assume that the sum of TKN and N + N approximates the total nitrogen (TN) at a site, then we can estimate the ratio of TN:TP for Texas estuaries. Only seven out of the 130 stations displayed evidence of phosphorus limitation, as noted by TN:TP ratios >16. These stations were in Port Bay, Redfish Bay, and in the Upper Laguna Madre-Baffin Bay complex (Fig. 2f). All other sites had a TN:TP <16, indicative of nitrogen limitation. Nutrient addition bioassays conducted in Galveston Bay and Baffin Bay confirmed that phytoplankton growth in those systems was largely nitrogen limited (Örnólfsdóttir et al. 2004; Wetz et al. 2017). This was the case even in Baffin Bay where periods of high (>16 up to several hundred) nitrogen-to-phosphorus ratios are common and can be attributed in part to the ecology of the dominant bloom-forming phytoplankton taxa, the harmful "brown tide" phytoplankton species, *Aureoumbra lagunensis*. This species has been shown to thrive under otherwise phosphorus-limiting conditions (Liu et al. 2001; Cotner et al. 2004), which has been attributed to its ability to utilize organic phosphorus (Sun et al. 2012).

Chlorophyll concentrations averaged ≤10 μg/l at many sites along the Texas coast, but high concentrations were observed in Galveston Bay and tributaries (up to 50 μg/l on average), Oso Bay (16 μg/l), a site adjacent to the Arroyo Colorado outflow in the Lower Laguna Madre (15 μg/l),

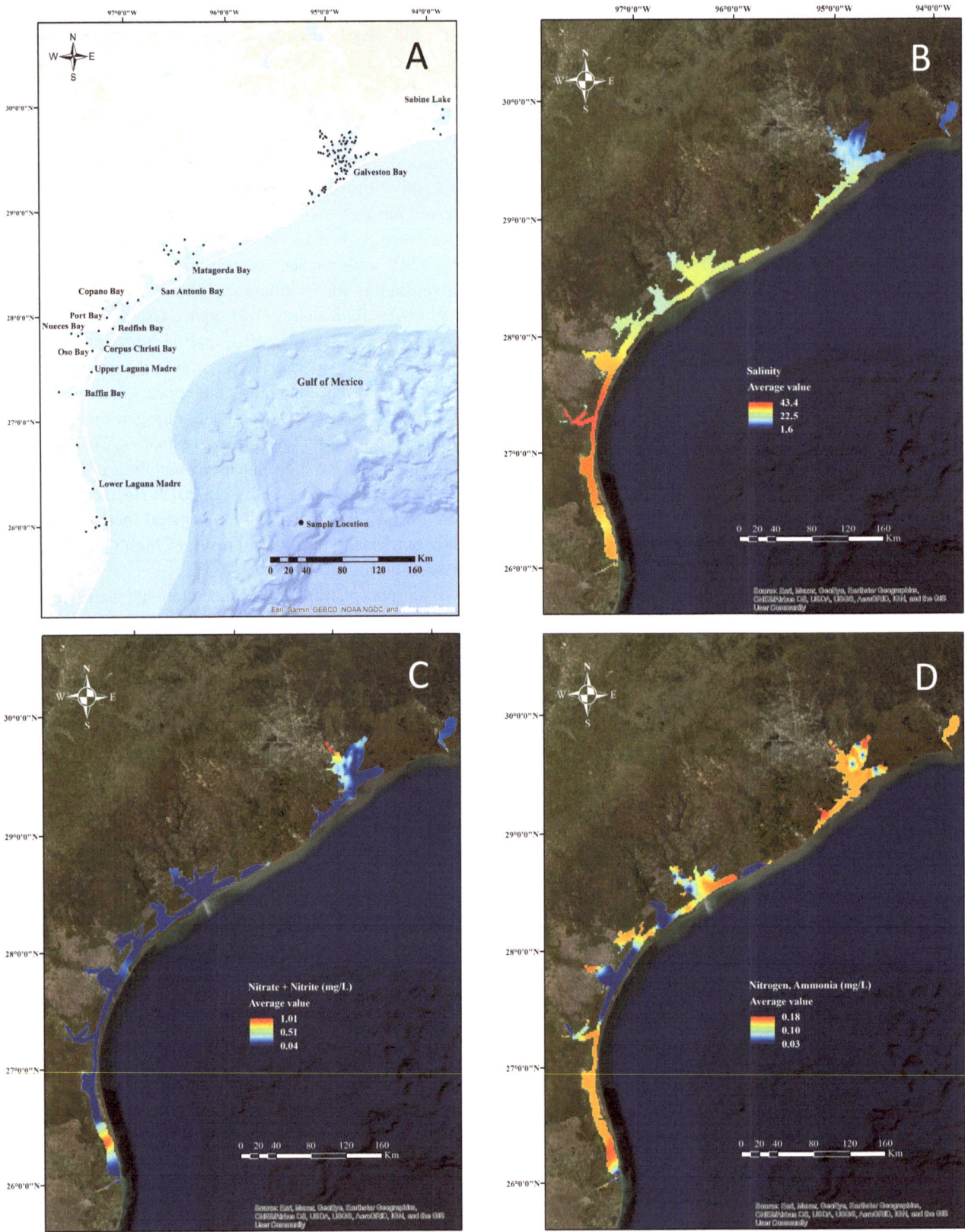

**Fig. 2** (**a**) TCEQ water quality monitoring stations. (**b**) salinity, (**c**) nitrate + nitrite, (**d**) ammonium, (**e**) total Kjeldahl nitrogen, (**f**) total phosphorus, (**g**) TN:TP, and (**h**) chlorophyll

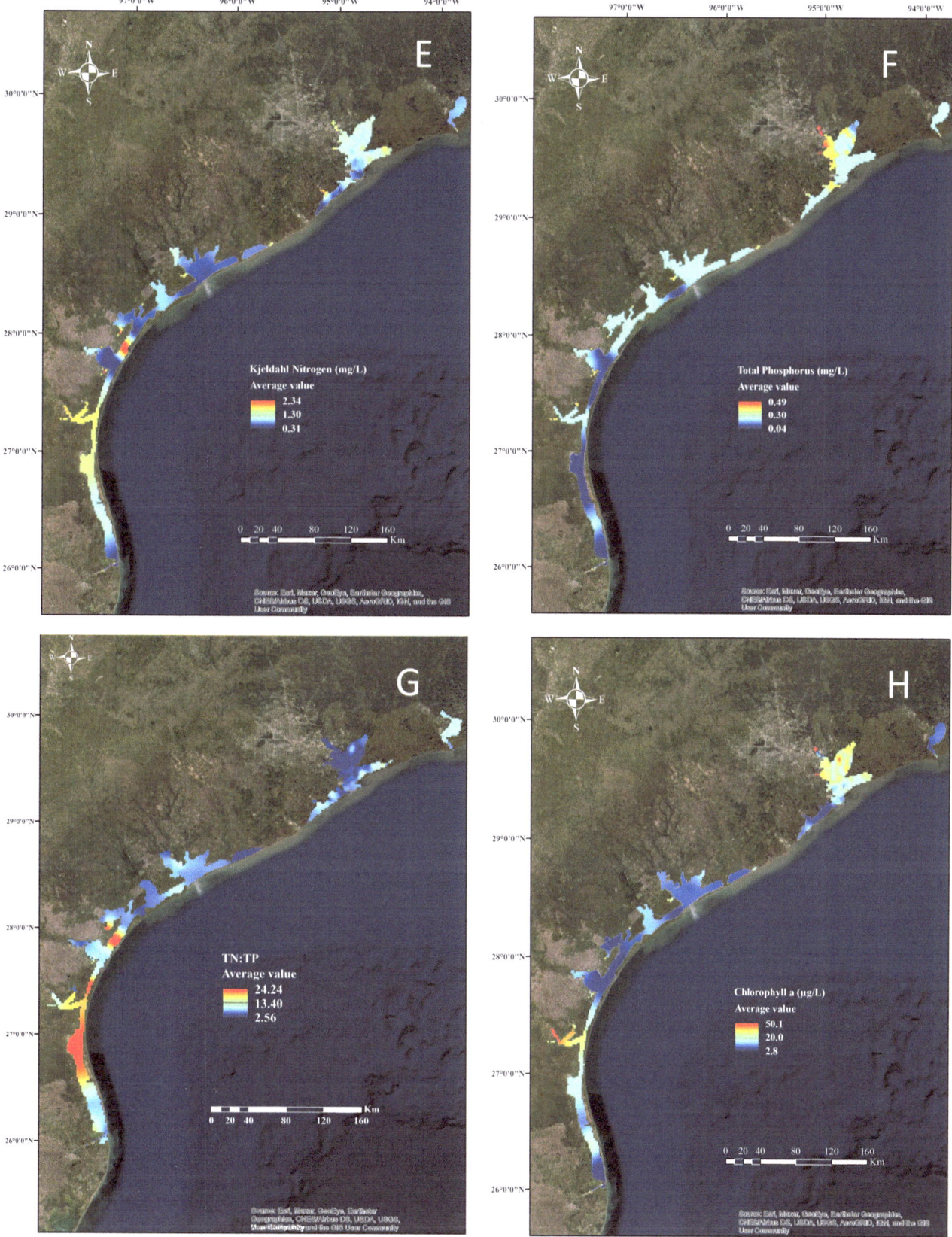

**Fig. 2** (continued)

Baffin Bay (19–29 µg/l) and adjacent sites in the Laguna Madre (11–13 µg/l), and Tres Palacios Bay (11 µg/l) (Fig. 2g). Over the course of the 10-year period, chlorophyll displayed a statistically significant inverse relationship with salinity in the Lavaca-Colorado (slope = −0.27, $r^2$ = 0.28, $p < 0.01$) and Mission-Aransas (slope = −0.16, $r^2$ = 0.14, $p = 0.05$) estuaries, a borderline significant inverse relationship with salinity in Guadalupe Estuary (slope = −0.42, $r^2$ = 0.14, $p = 0.06$), and a significant positive relationship with salinity in the Upper Laguna Madre (slope = 0.55, $r^2$ = 0.22, $p < 0.01$) (Table 1). Chlorophyll showed a significant positive relationship with freshwater discharge in the Lavaca-Colorado Estuary (slope = $4.4 \times 10^{-6}$, $r^2$ = 0.18, $p < 0.01$) and a borderline significant positive relationship with freshwater discharge in Mission-Aransas Estuary (slope = $4.4 \times 10^{-5}$), $r^2$ = 0.11, $p = 0.08$) (Table 1). Although causal relationships are difficult to establish with the quarterly sampling schedule of the TCEQ collections, some inferences can be drawn that will be further strengthened in ensuing sections that focus on estuary-specific sampling programs. For example, these results show that in central coast estuaries such as Lavaca-Colorado, Guadalupe, and Mission-Aransas, there appears to be a stimulatory effect of freshwater inflow on phytoplankton growth and biomass production. In contrast, the positive relationship between salinity and chlorophyll in the Upper Laguna Madre implies that inflow events may actually flush out resident phytoplankton populations or reduce residence times such that growth cannot keep up.

The patterns in the historical data observed here generally conform to a recent, more in-depth assessment of water quality patterns and trends on the Texas coast (Bugica et al. 2020). For example, Bugica et al. (2020) found multiple indicators suggesting that Baffin Bay and adjacent Upper Laguna Madre are undergoing eutrophication, attributed to a combination of point and nonpoint sources as well as sensitivity to nutrient inputs caused by a general lack of flushing. They also found evidence of nutrient enrichment in the wastewater-influenced Oso Bay and the heavily urbanized

Galveston Bay. Longley (1994) speculated that high turbidity might prevent symptoms of eutrophication from manifesting in Texas estuaries, but these findings suggest that to not be the case unfortunately. What was particularly interesting is where Bugica et al. (2020) did not find evidence of nutrient enrichment, specifically the Nueces-Corpus Christi Bay system, which is surrounded by a population of over 300,000 people. Their results showed evidence of a long-term decrease in chlorophyll in Nueces Bay that corresponded to increasing salinity levels, pointing to a long-term decrease in freshwater inflow and associated nutrient loadings as potentially buffering the system from the effects of urbanization.

## Bay-Specific Relationships between Freshwater Inflow Variability, Nutrients, and Chlorophyll

### Galveston Bay

#### Site Description

The Galveston Bay watershed (72,000 km$^2$) includes Houston and Dallas, ranked as the fourth and ninth largest cities and fastest growing cities in the United States, respectively (Census 2024, https://www.census.gov). Galveston Bay (1554 km$^2$) is the second largest estuary in the Gulf of Mexico, located in a region where freshwater from rain, land runoff, large rivers, and local bayous and saltwater from the Gulf of Mexico meet. Gulf waters flow into the bay during tidal fluxes through Bolivar Roads (inlet between Galveston Island and the Bolivar Peninsula), San Luis Pass (western end of Galveston Island), and to a lesser degree through Rollover Pass (east end of the Bolivar Peninsula). Usually, most of the freshwater entering the Galveston Bay system flows down two large rivers in its upper reaches: the Trinity and San Jacinto rivers. The Trinity River originates in the Dallas-Fort Worth region and contributes to the majority of freshwater inflows (55%) into Galveston Bay (Guthrie et al. 2012). The San Jacinto River flows into Lake Houston and then into the Houston Ship Channel and contributes ~26% (along with Buffalo Bayou) of the freshwater inflows (Guthrie et al. 2012). Runoff from the coastal urban watershed and returned flows are the biggest contributors of the remaining ~19% of the freshwater inflows to the upper reaches of the bay.

Human alterations have and continue to exert profound impacts on bay circulation (Wilber and Clarke 1998; Lester and Gonzalez 2011; Freeman et al. 2019). The Houston Ship Channel and the Texas City Dike have caused the most pervasive changes to circulation by greatly increasing the flow of Gulf water into the bay and essentially cutting off West

**Table 1** Regression output from a comparison of chlorophyll versus salinity in Texas estuaries from 2010 to 2019. Bold indicates values where $p < 0.05$

| Estuary | Slope | R$^2$ | $p$-value | $n$ |
|---|---|---|---|---|
| Sabine | 0.11 | 0.04 | 0.25 | 37 |
| Trinity-San Jacinto | −0.22 | 0.04 | 0.19 | 41 |
| Lavaca-Colorado | **−0.27** | **0.28** | **<0.01** | **60** |
| Guadalupe | −0.42 | 0.14 | 0.06 | 26 |
| Mission-Aransas | **−0.16** | **0.14** | **0.05** | **28** |
| Nueces | −0.07 | 0.01 | 0.65 | 37 |
| Upper Laguna Madre | **0.55** | **0.22** | **<0.01** | **36** |
| Lower Laguna Madre | −0.15 | 0.01 | 0.53 | 32 |

Bay waters, respectively. Changes in land use and land cover since the early 1970s were synthesized in Quigg et al. (2017) and H-GAC (2017) and summarized in Freeman et al. (2019). It was found that there was an ~11% increase in in area used for development in Harris County (where Houston is located) from 1996 to 2011 (H-GAC 2017). An examination of the counties immediately surrounding the bay revealed that land cover as forest has experienced the greatest loss, primarily due to development (urbanization). Forest area was also lost to grasslands during this period; other significant losses were associated with agricultural (cultivated) lands and wetlands. In most cases, wetlands were converted into developed lands, to shrubs and grasslands associated with urban community centers, particularly those connected to waterways.

## Inflow-Nutrient-Phytoplankton Dynamics in Galveston Bay

From 2010 to 2019, water sampling was conducted by TCEQ stations in Galveston Bay. The annual median Trinity and San Jacinto River discharge was 594 m³ s⁻¹ for the sampling period and was highly variable, ranging from 102 to 1120 m³ s⁻¹ (see Chapter "Plankton Dynamics in Texas Estuaries"; Fig. 1). This is not surprising given that during this period, there was a drought (2010–2014) and three exceptional high

discharge events in May 2015 (Houston Memorial Day Flood), April 2016 (Houston Tax Day Flood,) and August/September 2017 (Hurricane Harvey).

Mean nitrate + nitrite (N + N) concentration was variable and always less than 0.80 mg/L during this sampling period (Fig. 3a). Higher N + N were observed at lower salinities, which are those typically measured nearest to the river mouth. The N + N relationship with salinity however was not significantly negative (Fig. 3a). Mean ammonium concentrations ranged from 0.03 to 0.20 mg/L, and like N + N, ammonium did not a significant negative relationship with salinity in Galveston Bay (Fig. 3b). Total Kjeldahl nitrogen (TKN) varied from 0.27 to 2.20 mg/L with generally higher values measured at lower salinities and vice versa, but the relationship was not significant (Fig. 3c). Total phosphorus was highly variable in Galveston Bay with a concentration ranging from <0.1 to 0.6 mg/L, with the maximum mean concentrations occurring in the middle salinity ranges that typically occur in Galveston Bay (Fig. 3d). Orthophosphate was not measured by TCEQ, and so this data is not available. TKN:TP was consistently <16, indicating that the system is generally N-limited (Fig. 3e). Chlorophyll concentrations ranged from 8 to 32 ug/L (Fig. 3f). A variety of studies conducted in Galveston Bay have consistently reported that phytoplankton

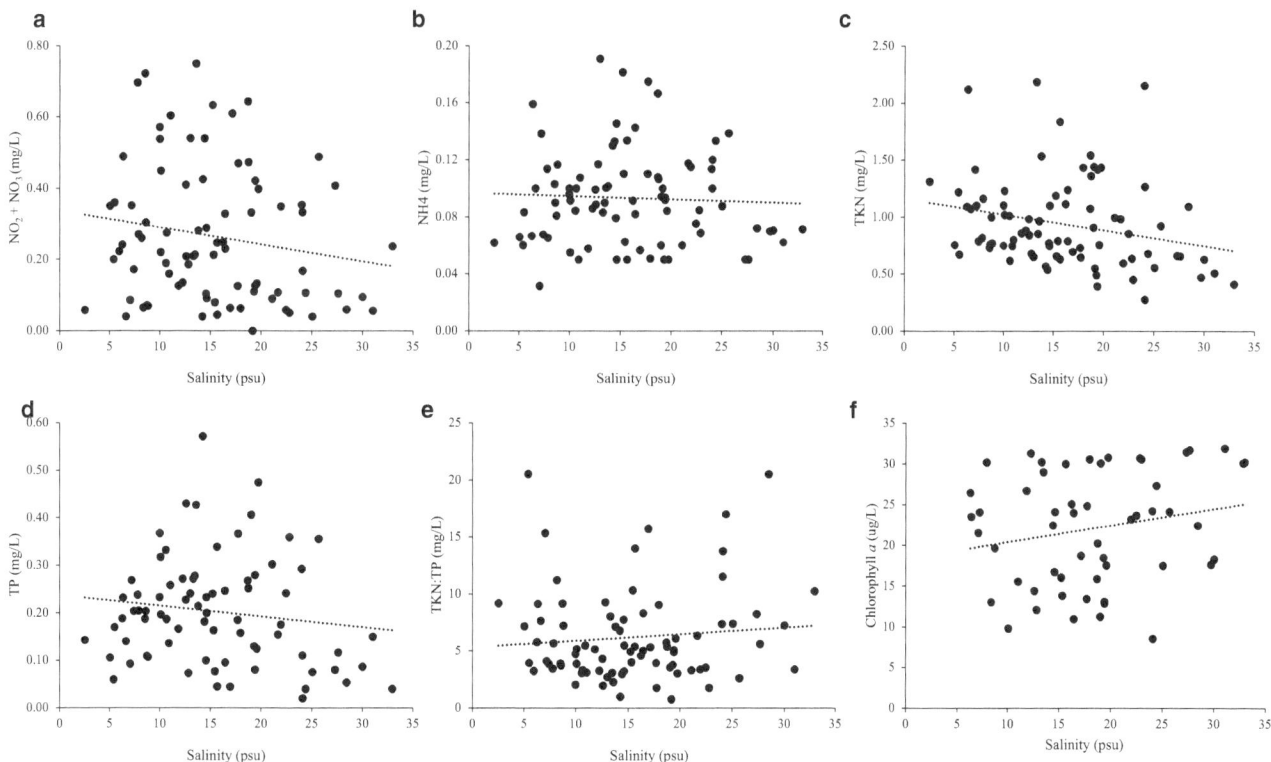

**Fig. 3** Property-property plots for Galveston Bay. Salinity was used a proxy for freshwater inflows and plotted versus (**a**) N + N, (**b**) ammonium, (**c**) TKN, (**d**) total phosphorus (TP), (**e**) TKN:TP, and (**f**) chlorophyll

communities are typically N-limited (see, e.g.: Örnólfsdóttir et al. 2004; Pinckney et al. 2001; Pinckney 2006; Quigg et al. 2009; Quigg 2011; Roelke et al. 2013; Dorado et al. 2015; Williams and Quigg 2019).

## Matagorda Bay

### Site Description

The Matagorda Bay system is located on the central Texas coast and has the second largest surface area of Texas estuaries. The system is lagoonal, mostly isolated from the Gulf of Mexico, and comprised of a main bay (Matagorda Bay) and several subsystems including Lavaca Bay and East Matagorda Bay (Ward and Armstrong 1980). The mean depth of Matagorda Bay is 2.8 meters, and the residence time is about 2.5 months (Ward and Armstrong 1980; Armstrong 1982; Palmer et al. 2011). The tidal range can be semidiurnal at about 0.2 meters or diurnal at about 0.8 meters (Ward and Armstrong 1980). Circulation in the bay is also influenced by wind-driven wave action, mainly from south to southeastern winds of the Gulf of Mexico that keeps the bay well-mixed (Ward and Armstrong 1980). Land use in the Lower Colorado River watershed is heavily influenced by agricul-

ture, whereas the Upper Colorado River watershed is dominated by scrub lands, although urbanization is becoming more prevalent, especially in the vicinity of the rapidly growing city of Austin, Texas (NOAA Coastal Change Analysis Program; Breyer et al. 2018).

## Inflow-Nutrient-Phytoplankton Dynamics in Matagorda Bay

Between November 2019 and October 2021, monthly water sampling was conducted at 6 stations in the eastern arm of Matagorda Bay that receives freshwater input from the Colorado River. Detailed results can be found in McBride (2022) and are synthesized here. The mean Colorado River discharge was 276 m$^3$ s$^{-1}$ for the sampling period and was relatively invariant except for two high discharge periods in May 2021 (~tenfold higher than the average) and September 2021 (~fivefold higher than the average).

Mean monthly nitrate + nitrite (N + N) concentration ranged from <0.01 to 0.22 mg/L for the duration of the study, although concentrations >0.2 mg/L were frequently observed at the site closest to the river mouth (Fig. 4a). N + N had a significant negative relationship with distance from the river mouth (R$^2$ = 0.16, p < 0.01) and decreased sharply in the first ~15 km downstream of the river mouth (Fig. 4a). Mean

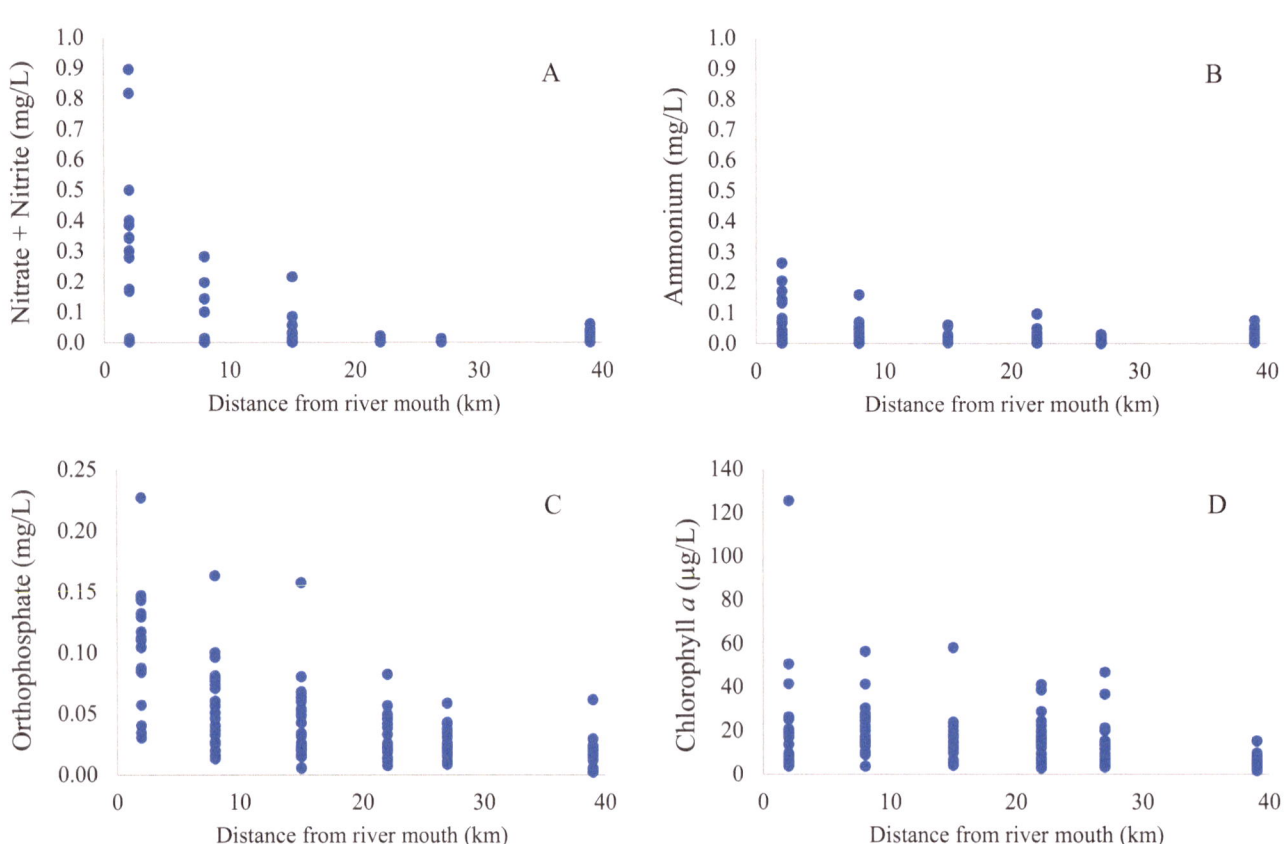

**Fig. 4** Plots of water quality variables versus distance from the Colorado River mouth in Matagorda Bay. (**a**) N + N, (**b**) ammonium, (**c**) orthophosphate (TP), and (**d**) chlorophyll

monthly ammonium concentration ranged from <0.01 to 0.10 mg/L for the duration of the study, and like N + N, ammonium had a significant negative relationship with increasing distance from the river ($R^2$ = 0.09, p < 0.01) (Fig. 4b). The mean monthly orthophosphate concentration ranged from <0.03 to 0.11 mg/L, with the maximum mean concentration occurring coincident with the peak discharge in May 2021 (Fig. 4c). Orthophosphate had a significant negative relationship with increasing distance from the river mouth ($R^2$ = 0.40, p < 0.01) (Fig. 4c). Based on a $DIN:PO_4$ that was consistently <16, this would indicate that the system is generally N-limited. Mean monthly silicate concentration ranged from 0.17 to 3.76 mg/L and had a significant negative relationship with increasing distance from the river mouth. Potential silicate limitation (DIN:Si > 1) only occurred in January (1.39) and February (1.45) of 2020 (data not shown).

The mean monthly chlorophyll concentration was 16.0 µg/L, with a minimum of 5.7 µg/L and maximum of 38.4 µg/L that was observed in June 2021 following the high discharge event in May 2021. Chlorophyll had a significant negative relationship with distance from river mouth ($R^2$ = 0.13, p < 0.01), although concentrations >20 µg/L were commonly observed at stations encompassing a broad region from the river mouth to ~30 km downstream (Fig. 4d). Chlorophyll was also negatively correlated with salinity (p < 0.01) and positively correlated with water temperature (p = 0.02). The inverse relationship between chlorophyll and salinity observed over the course of this two-year study is consistent with findings from the aforementioned analysis of TCEQ water quality data for Matagorda Bay, which also showed both an inverse relationship with salinity and a positive relationship with discharge. These relationships, along with the decreasing chlorophyll and inorganic nutrients moving away from the river mouth, highlight the importance of the Colorado River in terms of fueling phytoplankton growth in this region of Matagorda Bay.

The apparent importance of the Colorado River for primary production suggests that long-term decreases in Colorado River discharge that are expected with increased human populations in its watershed as well as from future drought and increased evaporation may have important implications for the ecosystem. For example, the most obvious potential impact would be decreased phytoplankton productivity and biomass due to reduced riverine nutrient inputs. However, it is unclear what the net effect will be on non-phytoplankton primary productivity. For example, McBride (2022) found that chlorophyll was inversely related to light penetration in the system, and thus, decreased discharge and phytoplankton biomass may actually increase light availability to benthic producers (Rask et al. 1999; Murrell et al. 2009), leading to a shift that favors benthic production over pelagic production. This would obviously have food web implications, but more work is clearly needed to assess future scenarios of discharge and impacts on productivity, as well as potential implications of shifts to a benthic-dominated producer base.

## Baffin Bay

### Site Description

Baffin Bay is a shallow (mean depth of 2.3 m) tributary of the Laguna Madre, connecting at its mouth with the Upper Laguna Madre to the east. Watershed land use coverage for Baffin Bay is dominated by agriculture (primarily cotton farming; 44%), followed by scrub (35%) and grassland (8%) (NOAA Coastal Change Assessment Program, 2016). In comparison, human development is relatively limited within the watershed, with the city of Kingsville being the largest populated area (pop. 25,315 as of 2019 U.S. Census). Winds can be relatively strong in the region and are thought to play a dominant role in hydrography of the system, with the water column generally being well mixed throughout much of the year (Tunnell 2002). The nearest inlets that allow for exchange between the Laguna Madre and Gulf of Mexico are Packery Channel (~41 km north of Baffin Bay), Aransas Pass (~70 km north of Baffin Bay), and Port Mansfield (~80 km south of Baffin Bay). These distances, along with diurnal tidal ranges of only ~2–3 cm, result in minimal overall tidal influence on the system.

### Inflow-Nutrient-Phytoplankton Dynamics in Baffin Bay

Evidence is emerging that lagonal, low-inflow estuaries such as Baffin Bay are particularly vulnerable to eutrophication (Knoppers et al. 1991; Scavia and Liu 2009; Lemley et al. 2017, 2021). The high TKN, high and increasing chlorophyll, and recent tracer studies documenting a significant signature of human-derived nutrients (D. Felix, unpubl. data) provide solid evidence that eutrophication is well underway in Baffin Bay (Wetz et al. 2017; Bugica et al. 2020). In addition, Baffin Bay and the adjacent Laguna Madre have experienced multiple near mono-specific blooms of the harmful "brown tide," *Aureoumbra lagunensis*, since 1990 (e.g., Onuf 1996, 2000; Buskey et al. 2001; Cira and Wetz 2019; Chapter "Plankton Dynamics in Texas Estuaries"). Because of this, there has been a need to understand phytoplankton dynamics and the conditions that trigger phytoplankton bloom development in the system.

Although Baffin Bay receives low inflow on average, sporadic high-inflow events bring increased land-based nutrient and sediment loads into the system (Wetz et al. 2017; Wetz unpubl. data). Cira and Wetz (2019) and Cira et al. (2021) quantified nutrient and phytoplankton dynamics over a three-year period in Baffin Bay (2013–2016), with the first two years of sampling characterized by drought and the third year marked by high rainfall. During drought conditions from May 2013 to March 2015, the estuary was hypersaline (salinity >40), and residence time was estimated to be >1 year. In contrast, salinity reached <10 in parts of the bay and residence time decreased to <1 month during a high rainfall period in spring-summer 2015. Thereafter, salinity increased to >20, and residence time was again on the order of months to >1 year. Average $NH_4^+$ and N + N concentrations were generally higher under lower salinity conditions ($NH_4^+$: 0.08 ± 0.10 mg/L; N + N: 0.04 ± 0.10 mg/L) compared to prior hypersaline conditions ($NH_4^+$: 0.04 ± 0.05 mg/L; N + N: 0.02 ± 0.06 mg/L), indicative of watershed sources of those nutrients. Dissolved organic nitrogen (DON) was the largest dissolved pool of N during the three-year study, averaging 0.92 ± 0.20 mg/L. Ongoing source-tracking studies indicate that a notable fraction of the DON pool is human-derived (i.e., sewage and agricultural sources; Felix and Campbell 2019). Orthophosphate concentrations were higher under low salinity conditions (low salinity: 0.04 ± 0.07 mg/L; high salinity: 0.02 ± 0.05 mg/L) and had distinctive peaks during the spring of 2015, also indicative of a watershed source. Silicate concentrations were lower under low salinity conditions (1.53 ± 1.35 mg/L) compared to high salinity conditions (2.20 ± 0.47 mg/L), which was likely due to diatom uptake during low salinities. Chlorophyll concentrations were lower on average under the lower salinity conditions (16.2 ± 14.3 µg L$^{-1}$; April 2015—April 2016) compared to the hypersaline conditions (25.7 ± 11.7 µg L$^{-1}$; May 2013 to March 2015), which Cira et al. (2021) attributed to higher rates of flushing and a shorter residence time that was unfavorable to the *A. lagunensis* bloom that was in place when the rainy period began.

A recent analysis of an 8-year dataset (2013–2021) of monthly sampling found relationships between inflow/salinity and chlorophyll that were similar to those found by Cira and Wetz (2019) and Cira et al. (2021), with a higher proportion of "bloom" chlorophyll concentrations observed at high salinity (24% of samples at high salinity had bloom levels) compared to intermediate (2% of samples) and low salinity conditions (12% of samples) (Beecraft and Wetz 2023). Those authors also found an interesting relationship between chlorophyll and DON concentration, where DON varied inversely with chlorophyll (Fig. 5). For example, DON peaks in both mid-late 2016 and late 2019 both corresponded with relatively low chlorophyll levels, while DON "troughs" in late 2013 to mid-2015 and early 2018 to early 2019 corresponded with higher chlorophyll levels, suggesting that phytoplankton may have been utilizing the fraction of the DON pool that increased in their absence or decreased in their presence. At one site in Baffin Bay where *A. lagunensis* blooms are particularly common, the DON pool decreased by ~60 ± 6% from non-bloom to bloom periods. The C:N of the dissolved organic matter (DOM) that was utilized (5.0 ± 0.6) was lower than that of the bulk DOM pool

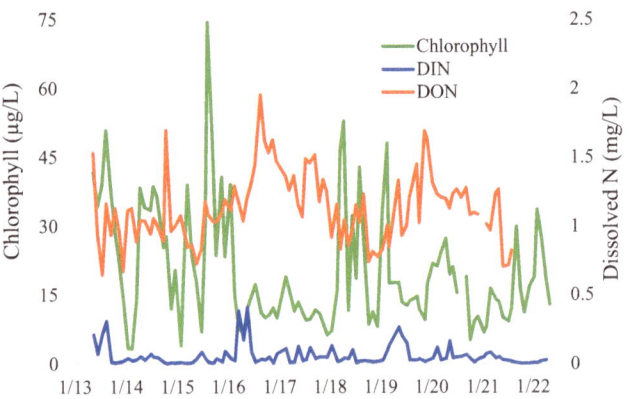

**Fig. 5** Time-series plots of chlorophyll, dissolved inorganic nitrogen (ammonium + nitrate + nitrite) and dissolved organic nitrogen in Baffin Bay

(13.5 ± 3.0), suggesting selective utilization of DON derived from sources that produce labile organic matter (such as e.g., wastewater). Previous studies have documented the ability of phytoplankton to utilize certain DON compounds in support of growth (reviewed Bronk et al. 2007). Most germane to Baffin Bay are the studies of Muhlstein and Villareal (2007) and Agostoni and Erdner (2011), both of which confirmed the mixotrophic capabilities of *A. lagunensis* and its ability to utilize organic nitrogen compounds for growth. Baffin Bay is also subjected to very high light levels for a large part of the year that could promote photoammonification of the DON (Felix and Campbell 2019).

In river-influenced estuaries such as the Chesapeake Bay or Galveston Bay/Matagorda Bay here in Texas, increased nutrient availability brought about by freshwater inflow may stimulate phytoplankton growth, while under low inflow conditions, nutrient limitation is possible (e.g., this chapter; see also Wetz et al. 2011; Phlips et al. 2020). In Baffin Bay, internal nutrient sources such as sediment porewaters and organic nutrients appear to be able to sustain phytoplankton growth and allow for biomass accumulation during drought. For example, Lopez et al. (2018) found very high $NH_4^+$ concentrations in sediment porewater and argued that fluxes of sediment-derived nutrients can be significant in this shallow waterbody. In addition, the aforementioned studies on DON availability via either uptake of organic compounds or photoammonification imply that the DON is a ready source of nitrogen for phytoplankton during low inflow conditions. Consequently, whereas chlorophyll concentrations in some estuaries of the central Texas coast (e.g., Matagorda Bay, Guadalupe, Mission-Aransas) appear to correlate with inflow (or inversely with salinity), the opposite pattern is often observed in Baffin Bay, especially when the resident phytoplankton community is dominated by slow-growing taxa such as *A. lagunensis* that cannot maintain positive biomass accumulation under high inflow conditions. One can speculate that these stark differences in inflow-phytoplankton relationships between Baffin Bay and the more river-influenced systems of the central and upper Texas coast may have arisen because of nutrient pollution to Baffin Bay that has allowed the internal reservoir of nutrients to accumulate in the system over the years, consistent with work from other systems showing challenges of addressing so-called "legacy nutrients" that accumulate as a result of eutrophication (Pinckney et al. 2001).

## Synthesis and Recommendations

Results presented here highlight the important influence that freshwater inflow has on estuarine nutrient and phytoplankton dynamics. The earlier review by Longley (1994) noted that nutrient loadings to and concentrations along the river-estuary-ocean continuum will be influenced by freshwater inflow magnitude. Newer findings generally continue to agree with this, as exemplified by the higher TP concentrations observed in estuaries of the upper Texas coast where freshwater inflow is higher compared to lower coast estuaries, by the general trend of decreasing nutrient concentrations from the upper estuary to the lower estuary as seen in both the TCEQ data analysis section or in the Matagorda Bay study, or by the higher nutrient concentrations that are frequently observed following inflow pulses in many Texas estuaries. Yet as can be seen in the Baffin Bay example, which is influenced by nutrient sources (wastewater, septic) that don't necessarily scale to the magnitude of freshwater inflow and by nutrients that are derived from sediments, there is probably a need for localized nutrient loading/source models if the goal is to assess nutrient budgets or issues related to excessive nutrients. Aside from affecting nutrient loadings and concentrations, there have been several studies conducted since Longley (1994) that have linked freshwater inflow variability to various nutrient processes. For example, Gardner et al. (2006) showed that the relative importance of denitrification (a nitrogen removal process) and dissimilatory nitrate reduction to ammonium (a nitrogen retention process) is strongly influenced by inflow and ultimately salinity conditions in the estuary, with higher salinities favoring the latter. Thus, assuming other factors are favorable, long-term declines in freshwater inflows may increase nitrogen retention in Texas' estuaries and cause them to become more sensitive to external loads. Further work is clearly needed to elucidate the implications of long-term reductions in inflows on nitrogen processing, especially considering the predominant role that nitrogen plays in phytoplankton growth.

In terms of the phytoplankton response to inflows, results presented here highlight commonalities as well as advances in our knowledge since the earlier review by Longley (1994).

First, this synthesis continues to show that inflow modulates several factors that are important in terms of regulating phytoplankton growth, namely, nutrient availability, light, and flushing. However, it also provides a more nuanced, process-based view of the phytoplankton response to inflow conditions owing to newer data that has become available. As seen from the studies highlighted in this chapter, inflow events appear to have three possible outcomes in Texas estuaries over the timescales of days to weeks post-event:

1. A reduction in phytoplankton growth and biomass in upper regions of an estuary if an inflow event is of high magnitude, primarily due to flushing rates that exceed growth rates.
2. Movement of the zone of high phytoplankton growth and biomass downstream in the estuary.
3. Increased phytoplankton growth and biomass accumulation, either in the upper estuary or throughout the estuary depending on inflow magnitude, if the flushing rate is less than the phytoplankton growth rate.

A second key advance that this chapter highlights is our understanding of the overall sensitivity of Texas estuaries to nutrient enrichment. At the time of Longley (1994), symptoms of eutrophication were not apparent in Texas estuaries and the author surmised that this was due to the high turbidity of the estuaries. However, the more recent observations reported here suggest this may not be the case, with those symptoms now becoming apparent in a large, urbanized estuarine complex influenced by freshwater inflow, Galveston Bay, but also in a system that only sees ephemeral high inflow conditions, the Baffin Bay-Upper Laguna Madre complex. Interestingly, several studies have now shown that the shallow lagoonal estuaries of the central and south Texas coast are efficient at retaining nutrients that can support phytoplankton growth even under low inflow conditions (Mooney and McClelland 2012; Bruesewitz et al. 2013; Beecraft and Wetz 2023). This argues for the need for more system-specific studies to elucidate mechanistic linkages between inflow and phytoplankton, information that is vital for developing management plans in areas where nutrient pressures may be a growing concern.

**Acknowledgments** The Baffin Bay and Matagorda Bay data were collected with the help of dedicated citizen scientist volunteers.

Funding for this synthesis chapter was provided by Contract No. 21-155-007-C879 from the Texas General Land Office (GLO) with Gulf of Mexico Energy Security Act of 2006 funding made available to the State of Texas and awarded under the Texas Coastal Management Program (CMP). The views contained herein are those of the authors and should not be interpreted as representing the views of the GLO or the State of Texas.

Funding for the TCEQ coastwide data analysis was provided from the following sources to MSW: a grant from the GLO/CMP (award no.

NA16NOS4190174); Texas OneGulf Center for Excellence; Coastal Bend Bays & Estuaries Program.

Funding for the individual bay data collections was provided from the following sources to MSW: a grant (award no. NA14OAR4170102) from the Texas Sea Grant College Program, grants from the GLO/CMP (award numbers NA14NOS4190139, NA17NOS4190139), a contract (21-060-017-C677) from the GLO with Gulf of Mexico Energy Security Act of 2006 funding, Celanese Corporation, Kleberg County, Coastal Conservation Association, Saltwater Fisheries Enhancement Association, Coastal Bend Bays & Estuaries Program, and a grant from the Texas Comptroller's Office (contract 19-6799CS). MSW and MM were also supported by awards from the National Oceanic and Atmospheric Administration, Office of Education Educational Partnership Program (NA21SEC4810004, NA16SEC4810009). Its contents are solely the responsibility of the award recipients and do not necessarily represent the official views of the U.S. Department of Commerce, National Oceanic and Atmospheric Administration.

# References

Abreu PC, Bergesch M, Proenca LA, Garcia CAE, Odebrecht C (2010) Short- and long-term chlorophyll a variability in the shallow microtidal Patos Lagoon estuary, southern Brazil. Estuar Coasts 33:554–569

Agostoni M, Erdner DL (2011) Analysis of ammonium transporter and urease gene expression in *Aureoumbra lagunensis*. Harmful Algae 10:549–556

Alber M (2002) A conceptual model of estuarine freshwater inflow management. Estuaries 25:1246–1261

Armstrong NE (1982) Responses of Texas estuaries to freshwater inflows. In: Kennedy VS (ed) Estuarine comparisons. Academic Press

Azevedo IC, Bordalo AA, Duarte P (2014) Influence of freshwater inflow variability on the Douro estuary primary productivity: a modelling study. Ecol Model 272:1–15

Beecraft L, Wetz MS (2023) Temporal variability in water quality and phytoplankton biomass in a low inflow estuary (Baffin Bay, Texas). Estuar Coasts 46:2064–2075

Breyer B, Zipper SC, Qiu J (2018) Sociohydrological impacts of water conservation under anthropogenic drought in Austin, TX (USA). Water Resour Res 54:3062–3080

Bricker SB, Longstaff B, Dennison W, Jones A, Boicourt K, Wicks C, Woerner J (2008) Effects of nutrient enrichment in the nation's estuaries: a decade of change. Harmful Algae 8:21–32

Bronk DA, See JH, Bradley P, Killberg L (2007) DON as a source of bioavailable nitrogen for phytoplankton. Biogeosciences 4:283–296

Bruesewitz DA, Gardner WS, Mooney RF, Pollard L, Buskey EJ (2013) Estuarine ecosystem function response to flood and drought in a shallow, semiarid estuary: nitrogen cycling and ecosystem metabolism. Limnol Oceanogr 58:2293–2309

Bugica K, Sterba-Boatwright B, Wetz MS (2020) Water quality trends in Texas estuaries. Mar Pollut Bull 152:110903

Buskey EJ, Liu H, Collumb C, Bersano JGF (2001) The decline and recovery of a persistent Texas brown tide algal bloom in the Laguna Madre (Texas, USA). Estuaries 24:337–346. https://doi.org/10.2307/1353236

Cira EK, Wetz MS (2019) Spatial-temporal distribution of *Aureoumbra lagunensis* ("brown tide") in Baffin Bay, Texas. Harmful Algae 89:101669

Cira EK, Palmer TA, Wetz MS (2021) Phytoplankton dynamics in a low-inflow estuary (Baffin Bay, TX) during drought and high-rainfall conditions associated with an El Niño event. Estuar Coasts 44:1752–1764

Cotner JB, Suplee MW, Chen NW, Shormann DE (2004) Nutrient, sulfur and carbon dynamics in a hypersaline lagoon. Estuar Coast Shelf Sci 59:639–652

Dorado S, Booe T, Steichen J, McInnes AS, Windham R, Shepard A, Lucchese AEB, Preischel H, Pinckney JL et al (2015) Towards an understanding of the interactions between freshwater inflows and phytoplankton communities in a subtropical estuary in the Gulf of Mexico. PLoS One 10:e0130931

Felix JD, Campbell J (2019) Investigating reactive nitrogen sources that stimulate algal blooms in Baffin Bay. Report 129 of the Coastal Bend Bays & Estuaries Program. 30 pp.

Flemer DA, Champ MA (2006) What is the future fate of estuaries given nutrient over-enrichment, freshwater diversion and low flows? Mar Poll Bull 52:247–258

Freeman LA, Corbett DR, Fitzgerald AM, Lemley DA, Quigg A, Steppe CN (2019) Impacts of urbanization and development on estuarine ecosystems and water quality. Estuar Coasts 42:1821–1838

Gardner WS, McCarthy MJ, An S, Sobolev D, Sell KS, Brock D (2006) Nitrogen fixation and dissimilatory nitrate reduction to ammonium (DNRA) support nitrogen dynamics in Texas estuaries. Limnol Oceanogr 51:558–568

Geyer N, Huettel M, Wetz MS (2018) Biogeochemistry of a river-dominated estuary (Apalachicola Bay, Florida) influenced by drought and storms. Estuar Coasts 41:2009–2023

Gilbert PM, Boyer JN, Heil CA, Madden CJ, Sturgis B, Wazniak CS (2010) In: Kennish MJ, Paerl HW (eds) Blooms in lagoons: different from those of river-dominated estuaries. CRC Press

Guthrie CG, Solis RS, Matsumoto J (2012) Analysis of the influence of water plan strategies on inflows and salinity in Galveston Bay. Report submitted to the United States Army Corps of Engineers Texas Water Allocation Assistance Program Contract #R0100010015. 70 pp.

Handler NB, Paytan A, Higgins CP, Luthy RG, Boehm AB (2006) Human development is linked to multiple water body impairments along the California coast. Estuar Coasts 29:860–870

Houston-Galveston Area Council (HGAC) (2017). http://www.h-gac.com/community/socioeconomic/land-use-data/default.aspx (Last Accessed June 2017)

Kaushal SS, Groffman PM, Band LE, Shields CA, Morgan RP, Palmer MA, Belt KT, Swan CM, Findlay SEG, Fisher GT (2008) Interaction between urbanization and climate variability amplifies watershed nitrate export in Maryland. Environ Sci Technol 42:5872–5878

Knoppers B, Kjerfve B, Carmouze J (1991) Trophic state and water turnover time in six choked coastal lagoons in Brazil. Biogeochemistry 14:149–166

Lancelot C, Muylaert K (2011) Trends in estuarine phytoplankton ecology. In: Treatise on estuarine and coastal science. Elsevier Inc., pp 5–15

Lemley DA, Adams JB, Strydom NA (2017) Testing the efficacy of an estuarine eutrophic condition index: does it account for shifts in flow conditions? Ecol Indic 74:357–370

Lemley DA, Lamberth SJ, Manuel W, Nunes M, Rishworth GM, van Niekerk L, Adams JB (2021) Effective management of closed hypereutrophic estuaries requires catchment-scale interventions. Front Mar Sci 8:1–17

Lester LJ, Gonzalez LA (eds) (2011) The state of the bay: a characterization of the Galveston Bay ecosystem, 3rd edn. Texas Commission on Environmental Quality, Galveston Bay Estuary Program, Houston, TX, p 356

Liu H, Laws EA, Villareal TA, Buskey EJ (2001) Nutrient-limited growth of *Aureoumbra lagunensis* (Pelagophyceae), with implications for its capability to outgrow other phytoplankton species in phosphate-limited environments. J Phycol 37:500–508

Longley WL (ed) (1994) Freshwater inflows to Texas bays and estuaries: ecological relationships and methods for determination of needs. Texas Water Development Board and Texas Parks and Wildlife Department, Austin, TX, p 386

Lopez CV, Murgulet D, Douglas A, Murgulet V (2018) Impacts of temporal and spatial variation of submarine groundwater discharge on nutrient fluxes to Texas coastal embayments, Phase III (Baffin Bay). Final report to the Texas General Land Office

Mallin MA, Paerl HW, Rudek J, Bates PW (1993) Regulation of estuarine primary production by watershed rainfall and river flow. Mar Ecol Prog Ser 93:199–203

McBride MR (2022) Influence of the Colorado River discharge variability on phytoplankton communities in Matagorda Bay, Texas. M.S. thesis. Texas A&M University-Corpus Christi

Milliman JD, Farnsworth KL, Jones PD, Xu KH, Smith LC (2008) Climatic and anthropogenic factors affecting river discharge to the global ocean, 1951-2000. Glob Planet Change 62:187–194

Montagna PA, Palmer TA, Beseres Pollack J (2013) Hydrological changes and estuarine dynamics. Springer, New York

Mooney RF, McClelland JW (2012) Watershed export events and ecosystem responses in the Mission-Aransas National Estuarine Research Reserve, South Texas. Estuar Coasts 35:1468–1485

Muhlstein HI, Villareal TA (2007) Organic and inorganic nutrient effects on growth rate-irradiance relationships in the Texas brown-tide alga *Aureoumbra lagunensis* (Pelagophyceae). J Phycol 43:1223–1226

Murrell MC, Campbell JG, Hagy JD, Caffrey JM (2009) Effects of irradiance on benthic and water column processes in a Gulf of Mexico estuary: Pensacola Bay, Florida, USA. Estuar Coastal Shelf Sci 81:501–512

Nielsen-Gammon JW, Banner JL, Cook BI et al (2020) Unprecedented drought challenges for Texas water resources in a changing climate: what do researchers and stakeholders need to know? Earth's Future 8. https://doi.org/10.1029/2020EF001552

Onuf CP (1996) Seagrass responses to long-term light reduction by brown tide in upper Laguna Madre, Texas: distribution and biomass patterns. Mar Ecol Prog Ser 138:219–231

Onuf CP (2000) Seagrass responses to and recovery from seven years of brown tide. Pacific Conserv Biol 5:306–313

Örnólfsdóttir EB, Lumsden SE, Pinckney JL (2004) Nutrient pulsing as a regulator of phytoplankton abundance and community composition in Galveston Bay, Texas. J Exp Mar Biol Ecol 303:197–220

Paerl HW, Bales JD, Ausley LW, Buzzelli CP, Crowder LB, Eby LA, Fear JM, Go M, Peierls BL, Richardson TL, Ramus JS (2001) Ecosystem impacts of three sequential hurricanes (Dennis, Floyd, and Irene) on the United States' largest lagoonal estuary, Pamlico Sound, NC. Proc Natl Acad Sci 98:5655–5660

Paerl HW, Dyble J, Moisander PH, Noble RT, Piehler MF, Pinckney JL, Steppe TF, Twomey L, Valdes LM (2003) Microbial indicators of aquatic ecosystem change: current applications to eutrophication studies. FEMS Microbiol Ecol 46:233–246

Paerl HW, Rossignol KL, Hall NS, Peierls BL, Wetz MS (2010) Phytoplankton community indicators of short- and long-term ecological change in the anthropogenically and climatically impacted Neuse River Estuary, North Carolina, USA. Estuar Coasts 33:485–497

Palmer TA, Montagna PA, Beseres Pollack J, Kalke RD, DeYoe HR (2011) The role of freshwater inflow in lagoons, rivers, and bays. Hydrobiologia 667:49–67

Phlips EJ, Badylak S, Christman MC, Lasi MA (2010) Climatic trends and temporal patterns of phytoplankton composition, abundance, and succession in the Indian River Lagoon, Florida, USA. Estuar Coasts 33:498–512

Phlips EJ, Badylak S, Nelson NG, Havens KE (2020) Hurricanes, El Niño and harmful algal blooms in two sub-tropical Florida estuaries: direct and indirect impacts. Sci Rep 10:1910

Pinckney J (2006) System-scale nutrient fluctuations in Galveston Bay, Texas (USA). In: Kromkamp JC, de Brouwer JFC, Blanchard GF, Forster RM, Creách V (eds) Functioning of microphytobenthos in estuaries. Royal Netherlands Academy of Arts and Sciences, Amsterdam, pp 141–164

Pinckney JL, Paerl HW, Tester P, Richardson TL (2001) The role of nutrient loading and eutrophication in estuarine ecology. Environ Health Perspect 109:699–706

Quigg A (2011) Understanding the role of nutrients in defining phytoplankton responses in the Trinity-San Jacinto Estuary. Final report for the Texas Water Development Board. Interagency Cooperative Contract TWDB Contract No. 1104831134.pp 48

Quigg A, Roelke DF, Davis SE (2009) Freshwater inflows and the health of Galveston bay: Influence of nutrient and sediment load on the base of the food web. Final report of the Coastal Coordination Council pursuant to National Oceanic and Atmospheric Administration Award No. NA07NOS4190144. pp 54

Quigg A, Steichen J, Windham R (2017) Galveston Bay: changing land use patterns and nutrient loading. Causal or casual relationship with water quality, quantity, and patterns? Prepared in cooperation with the Galveston Bay Estuary Program (GBEP) a program of the Texas Commission on Environmental Quality (TCEQ) and the U.S. Environmental Protection Agency (EPA) TCEQ Contract # 582-15-53393, Federal ID #CE-00655005. pp 55

Rask N, Pederson SE, Jensen MH (1999) Response to lowered nutrient discharges in the coastal waters around the Island of Funen, Denmark. Hydrobiologia 393:69–81

Roelke DL, Li HP, Hayden NJ, Miller CJ, Davis SE, Quigg A, Buyukates Y (2013) Co-occurring and opposing freshwater inflow effects on phytoplankton biomass, productivity and community composition of Galveston Bay, USA. Mar Ecol Prog Ser 477:61–76

Roelke DL, Li H-P, Miller-DeBoer CJ, Gable GM, Davis SE (2017) Regional shifts in phytoplankton succession and primary productivity in the San Antonio Bay System (USA) in response to diminished freshwater inflows. Mar Freshw Res 68:131–145

Rothenberger MB, Burkholder JM, Brownie C (2009) Long-term effects of changing land use practices on surface water quality in a coastal river and lagoonal estuary. Environ Manag 44:505–523

Scavia D, Liu Y (2009) Exploring estuarine nutrient susceptibility. Environ Sci Technol 43:3474–3479

Sun MM, Sun J, Qiu JW, Jing HM, Liu HB (2012) Characterization of the proteomic profiles of the brown tide alga Aureoumbra lagunensis under phosphate- and nitrogen-limiting conditions and of its phosphate limitation-specific protein with Alkaline Phosphatase activity. Appl Environ Microbiol 78:2025–2033

Tunnell JW Jr (2002) Geography, climate and hydrography. In: Tunnell JW Jr, Judd FW (eds) The Laguna Madre of Texas and Tamaulipas. Texas A&M University Press, College Station, pp 7–27

Walker LM, Montagna PA, Hu X, Wetz MS (2020) Timescales and magnitude of water quality change in three Texas estuaries induced by passage of Hurricane Harvey. Estuar Coasts 44:960–971

Ward G, Armstrong NE (1980) Matagorda Bay, Texas: its hydrography, ecology, and fishery resources. No. FWS/OBS-81/52. Fish and Wildlife Service, U.S. Department of the Interior, Washington, D.C.

Wetz MS, Yoskowitz DA (2013) An extreme future for estuaries? Effects of extreme climatic events on estuarine water quality and ecosystem dynamics. Mar Pollut Bull 69:7–18

Wetz MS, Hutchinson E, Lunetta R, Paerl H, Taylor JC (2011) Severe droughts reduce planktonic production in estuaries with cascading effects on higher trophic levels. Limnol Oceanogr 56:627–638

Wetz MS, Cira EK, Sterba-Boatwright B, Montagna PA, Palmer TA, Hayes KC (2017) Exceptionally high organic nitrogen concentrations in a semi-arid South Texas estuary susceptible to brown tide blooms. Estuar Coast Shelf Sci 188:27–37

Wilber DH, Clarke DG (1998) Estimating secondary production and benthic consumption in monitoring studies: a case study of the impacts of dredged material disposal in Galveston Bay, Texas. Estuaries 21:230–245

Williams AK, Quigg A (2019) Spatiotemporal variability in autotrophic and heterotrophic microbial plankton abundance in a subtropical estuary (Galveston Bay, Texas). J Coast Res 35:434–444

# Physical and Biogeochemical Conditions and Trends in Texas Estuaries

Xinping Hu and Hang Yin

## Abstract

To explore long-term changes in Texas estuaries, both physical (temperature and salinity) and biogeochemical parameters (dissolved oxygen, total organic carbon, total titration alkalinity, and pH) were examined for their long-term trends. Warming is found in all estuaries, consistent with observed warming in the Gulf of Mexico, and long-term salinity increase is also found in most of the examined stations, indicating a reduction in freshwater inflow, except in the northernmost estuary (i.e., Sabine Lake) and few secondary bays to the south, and the latter are subject to inflow management. Dissolved oxygen concentration decrease predominantly in mid-coast estuaries exceeds the extent that can be explained by warming and salinity increase, and this decrease appears to be consistent with the decrease in total organic carbon concentration. Freshwater inflow reduction may be responsible for the decrease in allochthonous organic carbon input, and both inflow-related nutrient reduction and nutrient pollution management also lead to reduced autochthonous organic carbon production. Freshwater inflow decline, the resulted increase in estuarine residence time, and associated more pronounced biogeochemical reactions may all contribute to total alkalinity consumption and increase in estuarine acidity. These changes highlight the importance of freshwater inflow on estuarine biogeochemistry in this climate transition zone, which requires sustained monitoring and continued investigations.

## Keywords

Estuary · Freshwater inflow · Total organic carbon · Dissolved oxygen · Alkalinity · pH

## Abbreviations

| | |
|---|---|
| CTBS | Center for Texas Beaches and Shores |
| DIC | Total dissolved inorganic carbon |
| DO | Dissolved oxygen |
| DOC | Dissolved organic carbon |
| ENSO | El Niño-southern oscillation |
| GAM | Generalized additive model |
| GRIIDC | Gulf of Mexico Research Initiative information and data cooperative |
| OC | Organic carbon |
| SWQM | Surface water quality monitoring |
| TA | Total titration alkalinity |
| TCEQ | Texas commission on environmental quality |
| TKN | Kjeldahl nitrogen |
| TOC | Total organic carbon |
| TPWD | Texas parks and wildlife department |

X. Hu (✉)
Harte Research Institute for Gulf of Mexico Studies, Texas A&M University-Corpus Christi, Corpus Christi, TX, USA

Present Address: Marine Science Institute, The University of Texas at Austin, Port Aransas, TX, USA
e-mail: xinping.hu@austin.utexas.edu

H. Yin
Harte Research Institute for Gulf of Mexico Studies, Texas A&M University-Corpus Christi, Corpus Christi, TX, USA
e-mail: Hang.Yin@tamucc.edu

## Introduction

Estuaries are coastal water bodies, either open (an extension to continental shelves) or semi-closed, where ocean water meets with freshwater draining from river watersheds (Potter et al. 2010), and mixing between the freshwater and the salty ocean water is usually the dominant process occurring in river-dominated estuaries (Dürr et al. 2011; Regnier et al. 2013). Meanwhile, estuaries also act as filters that not only remove particulate matter (via sedimentation and flocculation) from river input but also alter dissolved components

© The Author(s) 2025
P. A. Montagna, A. R. Douglas (eds.), *Freshwater Inflows to Texas Bays and Estuaries*, Estuaries of the World,
https://doi.org/10.1007/978-3-031-70882-4_8

through a host of biogeochemical processes. To investigate these various processes, both dissolved and particulate species are often analyzed in conjunction with hydrographic measurements from headwaters to the receiving water bodies, including estuaries and the coastal ocean. Many of these parameters reflect changing source water conditions that are caused by land use changes, population growth, and the resultant river water diversion (Hopkinson and Vallino 1995; Liu et al. 2013; Raimonet and Cloern 2017; Russell et al. 2006; Smith and Hollibaugh 1997). Because changes in upper reaches of headwaters have cascading effects on how estuaries respond and process materials transported by rivers, large-scale climate factors, that is, long-term warming and climate variabilities, which either increase or decrease precipitation hence streamflow, will ultimately affect estuaries downstream (Ni et al. 2019).

Warming has been widely reported both on land and in the ocean, with different rates across the globe (Ranasinghe et al. 2021). On the regional scale, the Gulf of Mexico large marine ecosystem has warmed 0.31 °C (or ~ 0.13 °C per decade) between 1982 and 2006 (Belkin 2009), and this rate is projected to increase to 0.37 °C per decade toward the last three decades of the twenty-first century (Alexander et al. 2018). Texas alone was estimated to warm by ~2.2 °C by the middle of the twenty-first century compared to the last few decades in the century prior (Nielsen-Gammon 2009). Both long-term warming and climate variability can lead to changes in hydrological cycle (Held and Soden 2006; Joyce et al. 1999; Kundzewicz 2008; Tolan 2007), by affecting precipitation patterns and the subsequent freshwater delivery from land to the coast. Warming and hydrological condition changes will have large impacts on estuaries, especially those in climate transition zones such as the Texas coast along the northwestern Gulf of Mexico (Montagna et al. 2011b; Montagna et al. 2007). Changes in hydrological cycle as well as freshwater inflow will affect estuarine salinity, and long-term decrease in freshwater inflow has led to salinity increases in Texas estuaries (Bugica et al. 2020; Marshall et al. 2021), despite episodic freshwater inflow pulses caused by large-scale climate variability (Tolan 2007).

Changes in temperature and salinity typically have the first-order effect on the solubility of dissolved gases as warmer and saltier water dissolves less gas. On land, warming caused increase in net primary production may be cancelled by larger increase in respiration (Vukićević et al. 2001). In estuaries, temperature increase is associated with increasing heterotrophy (Kenney et al. 1988), hence decreasing dissolved oxygen (DO) concentration (Tassone et al. 2022), and warming causes further decline in estuarine DO under eutrophic conditions (Jutras et al. 2020; Ni et al. 2019). While DO concentration is controlled by both physical conditions (salinity and temperature as well as air-water exchange) and the balance between production and con-

sumption, which indicates trophic state of an estuary (Odum and Hoskin 1956), the trophic state is also reflected by organic carbon dynamics. Estuaries are known as "hotspots" for organic carbon (OC) processing and transformation, including allochthonous (or terrestrial) input, in situ production, remineralization ($CO_2$), burial, and export to the coastal ocean. Land use changes can alter the inputs of both allochthonous OC and nutrients. Allochthonous OC stimulates estuarine respiration and nutrient regeneration. Both land-derived and regenerated nutrients enhance estuarine primary production of autochthonous OC.

Among the materials mobilized by rivers across watersheds, dissolved inorganic carbon (DIC) is a product of continental weathering (Eq. 1) and is present predominantly in the form of bicarbonate ion ($HCO_3^-$), which is also the major component of titration alkalinity (TA) in most river waters, and the latter is defined as acid neutralizing capacity and also includes carbonate ($CO_3^{2-}$), borate ($B(OH)_4^-$), and organic components that contribute to proton binding during the titration process.

$$CaCO_3 + CO_2 + H_2O \rightarrow Ca^{2+} + 2HCO_3^-$$

$$CaSiO_3 + 2CO_2 + 2H_2O \rightarrow Ca^{2+} + 2HCO_3^- + H_2SO_3 \quad (1)$$

TA determines the buffering capacity of the aquatic system, and greater TA level (or concentration) in the aqueous system represents greater buffer against acidity change (as measured by pH, or proton concentration). Along the same vein, higher TA in natural waters generally corresponds to greater carbonate saturation state. There are two common carbonate minerals produced by marine and estuarine organisms, calcite (commonly found in adult oyster shells) and aragonite (found in corals and oysters at the larval stage) (Gattuso et al. 1998; Miller et al. 2009). Therefore, a decrease in carbonate saturation state affects the aragonite producing organisms and life stages the most.

A host of biogeochemical processes alters TA, including photosynthesis (consumption of inorganic nutrient such as nitrate increases TA), aerobic remineralization (production of mineral acids decrease TA), carbonate precipitation (removes $HCO_3^-$ hence decreases TA), and dissolution (opposite of carbonate precipitation), to name a few (Wolf-Gladrow et al. 2007). In addition to these commonly understood reactions in the water column and surface sediments, benthic processes such as anaerobic organic carbon remineralization (including denitrification, metal, and sulfate reduction) also produce TA (Chen 2002; Hu and Cai 2011). Conversely, exposure of buried reduced compounds to oxygen, especially in the coastal and estuarine environment (e.g., draining coastal wetland), produces acid, which titrates down TA, and sometimes, the excess acid can cause extremely low pH conditions (de Weys et al. 2011; Sammut et al. 1996).

Along the Texas coast, there are seven major estuarine systems, all sharing similar physical characteristics (i.e., geomorphology and temperature), along a < 4° narrow latitudinal range that lies in a climatic gradient of decreasing rainfall and freshwater inflow from northeast to southwest (Montagna et al. 2013). Along this gradient, rainfall decreases by a factor of two, but freshwater inflow balance (inflow minus evaporation and diversion) decreases by nearly two orders of magnitude. Water resource development and severe drought have led to long-term reduction in flows to the southernmost estuaries (Montagna et al. 2011a). The net effect is a sharp increase in average salinity levels of estuaries moving southwest along the coastline. In addition to this spatial gradient, large-scale climate variability imparts a temporal signature in the salinity levels of these estuaries. For example, the Southern Oscillation Index is inversely correlated with total inflow to the Texas coast (Tolan 2007) and primary production (Kim et al. 2014). These latitudinal and year-to-year variations make the Texas coast an ideal area to study effects of altered inflows into estuaries and dynamics of particulate and dissolved species. Another feature of this coastal region is that estuaries are submerged river valleys from the pre-Holocene era, and the estuaries are all separated from the Gulf of Mexico by sandbars; hence, water exchange between estuaries and the coastal ocean is always somewhat restricted, and the amount of freshwater inflow controls the residence time of estuarine waters (Solis and Powell 1999). This type of restricted water exchange in theory would allow sufficient time for biogeochemical processes to alter solute concentrations, especially during the low inflow periods.

There have been several studies examining long-term trends of both temperature and water quality parameters in Texas estuaries. For example, Montagna et al. (2011b) revealed coastwide warming (0.0428 °C yr.$^{-1}$) and decrease in DO concentration (−0.0532 mg L$^{-1}$ yr.$^{-1}$), but they did not find trends in salinity. Bugica et al. (2020) analyze long-term trends of salinity, pH, nutrients (phosphorus and nitrogen), chlorophyll, and DO at 36 stations along the Texas coast using data up until 2016. They reported largely no significant changes in temperature on the annual scale but increasing salinity at 17 stations. In addition, these authors found decreasing pH trends from Matagorda Bay down the coast at 16 stations (−0.004 ~ −0.007 yr.$^{-1}$) but variable DO trends at 11 stations, ranging from increase in DO at a Lavaca-Colorado Estuary and a Nueces Estuary station (0.03 ~ 0.04 mg L$^{-1}$ yr.$^{-1}$) and decreasing DO trends (−0.02 ~ −0.11 mg L$^{-1}$ yr.$^{-1}$) at other locations. Hu et al. (2015) analyzed temporal trends in both TA and pH using data up until 2010 and found widespread TA and pH decline in most estuaries.

While warming can be attributed to the climate change (Montagna et al. 2011a; b), both salinity increase (Bugica

et al. 2020) and decrease in TA (Hu et al. 2015) are thought to be caused by declining freshwater inflow into these estuaries. As rivers in this region typically have moderate to high levels of TA (except the northernmost rivers), which generally increases toward the south, and are often greater than the oceanic TA values, decreasing river input; hence, TA delivery is responsible to the estuarine TA decrease, and the changes are largely a chronic shortage of fresh water in this climate transition zone. Changing freshwater inflow not only alters estuarine salinity and TA levels, total organic carbon (TOC) and DO that are significant to estuarine biogeochemistry may have also been impacted due to the changing nutrient conditions. The past studies (Montagna et al. 2011a; b; Hu et al. 2015; and Bugica et al. 2020) analyzed data primarily collected from late 1970s to early to mid-2010s. As Texas Commission on Environmental Quality's (TCEQ) Surface Water Quality Monitoring (SWQM) program maintains water quality survey, it is desirable to examine these interrelated parameters in a more holistic fashion using the most up-to-date data.

This chapter examines both spatial distribution and multidecadal trends of several important estuarine parameters, including physical conditions (temperature and salinity) and biogeochemical parameters, DO, TOC, TA, and pH. The latter two parameters reflect estuarine carbonate chemistry state, acid buffering capacity, and carbonate saturation state.

## Data Sources and Statistical Methods

All estuarine and river water quality data (temperature, salinity, DO, TOC, TA, and pH) are from the SWQP website hosted by TCEQ (https://www80.tceq.texas.gov/SwqmisPublic/index.htm). All data collection methods are documented in TCEQ's analytical procedures handbook (TCEQ 2012).

Raw water quality data files in text format were parsed using a MatLab® script and sorted by time and station. These data were further processed using R Studio (Ver. 2022.02.2). Both salinity (every 5 salinity units in salinity 0–35 and all >35) and time intervals (5-years) were used, and measurements that were three times the interquartile range (IQR) above quartile three (Q3) or three times the IQR below Q1 within bracketed subsets of the data were considered as outliers (Hu et al. 2015; McCutcheon and Hu 2022).

Similar to the earlier studies (Hu et al. 2015; McCutcheon and Hu 2022), we conducted Theil-Sen (TS) regression between water quality parameters and time (decimal year) except for temperature. Water temperature trends were calculated after deseasonalization using generalized additive model (GAM) and then simple least square regression. The

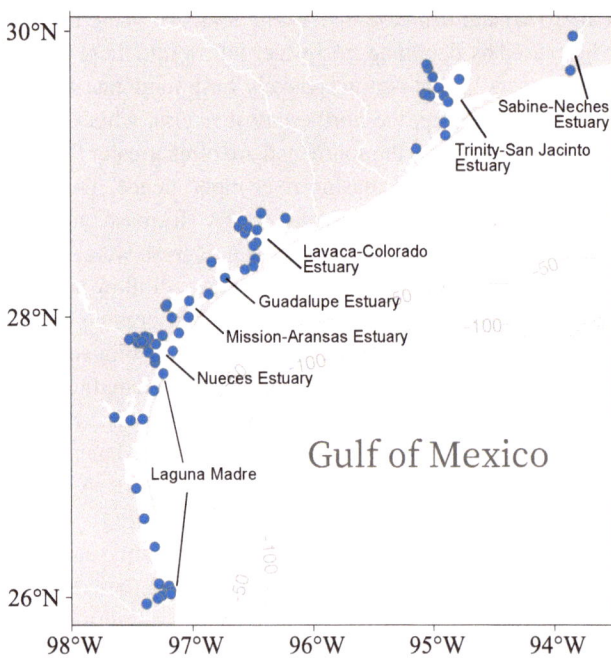

**Fig. 1** TCEQ's SWQM stations with at least one parameter (other than temperature) measured twice a year and at least a dataset length of 20 years. See Table 1 for station ID and their coordinates

TS approach is nonparametric method and robust to outliers, and this method has been widely used in the literature to explore long-term trends of water quality parameters (Kaushal et al. 2013; Stets et al. 2014). The change rates were reported for the statistically significant regression values ($p \leq 0.05$). Instead of combining all sampling stations together as in Hu et al. (2015), we chose to analyze individual sampling stations and reveal spatial difference within studied estuaries. The criteria for the selection of the monitoring stations are that at least 20 years of data are present, and each year there are at least two measurements. Note TCEQ conducts seasonal sampling in most cases. The selection criteria narrowed down to 67 stations for salinity; 71 stations for temperature; 62 stations for both TA and DO although they do not completely overlap; 68 stations for pH; and 30 stations for TOC, out of a total of 381 stations monitored by the SWQM program (Fig. 1). The number of stations that shows significant trends in one or more parameters is 68 (Table 1 and Fig. 1). The number of stations analyzed here is greater than those in McCutcheon and Hu (2022) as they used more stringent requirements to limit the data records that have simultaneous measurements of pH, TA, temperature, and salinity, which are necessary for carbonate speciation calculations. In addition, many of these long-term stations do not have up-to-date monitoring data as the sample collection has stopped more than two decades before the end of the entire time series.

**Table 1** TCEQ water quality stations examined in this study. "x" indicates parameters with significant temporal trends ($p \leq 0.05$)

| Station ID | Longitude | Latitude | Temp | Sal | DO | TOC | TA | pH |
|---|---|---|---|---|---|---|---|---|
| 12945 | −97.2209 | 28.0740 | | | x | x | | |
| 13285 | −97.2138 | 26.0729 | | | | | x | |
| 13298 | −93.8681 | 29.7292 | x | | | x | | |
| 13302 | −93.8472 | 29.9667 | x | x | | x | x | x |
| 13303 | −94.9078 | 29.5492 | | | | x | x | x |
| 13315 | −94.7867 | 29.6650 | | | | x | x | |
| 13332 | −95.0229 | 29.5495 | x | x | | x | x | |
| 13334 | −95.0711 | 29.5619 | | | | x | x | |
| 13335 | −95.0481 | 29.5556 | | | x | x | x | |
| 13342 | −95.0400 | 29.7433 | | | | x | | x |
| 13344 | −95.0514 | 29.7678 | x | | | x | | x |
| 13346 | −95.1417 | 29.1819 | | | | x | | x |
| 13355 | −94.9995 | 29.6819 | | | | x | | x |
| 13361 | −94.9045 | 29.3629 | x | x | | | | |
| 13364 | −94.8751 | 29.5087 | x | x | | x | | |
| 13378 | −96.4667 | 28.5256 | x | x | | | | |
| 13382 | −96.2260 | 28.6959 | | x | | | x | |
| 13383 | −96.6092 | 28.6389 | | x | x | | | x |
| 13384 | −96.5625 | 28.5958 | | x | | | | x |
| 13386 | −96.5374 | 28.6329 | | x | | | | x |
| 13387 | −96.4608 | 28.6150 | | | | | | x |
| 13388 | −96.4287 | 28.7327 | | x | | x | | x |
| 13396 | −96.4792 | 28.4097 | | x | x | | x | x |
| 13397 | −96.7233 | 28.2767 | x | | | | x | x |
| 13400 | −96.8617 | 28.1633 | x | x | | | x | |
| 13402 | −97.0278 | 28.0014 | | x | | | | |
| 13404 | −97.0259 | 28.1137 | | | x | x | x | x |
| 13405 | −97.1684 | 27.9959 | | | x | x | | |
| 13407 | −97.3015 | 27.8114 | x | | | | x | x |
| 13409 | −97.2482 | 27.8689 | x | | | | x | |
| 13410 | −97.3883 | 27.8100 | | | x | | x | x |
| 13411 | −97.3650 | 27.7514 | x | x | | | x | |
| 13420 | −97.3603 | 27.8528 | x | x | | | | x |
| 13421 | −97.3767 | 27.8397 | | x | | | | x |
| 13422 | −97.4103 | 27.8425 | | | x | | | |
| 13423 | −97.3908 | 27.8608 | x | | | x | | |
| 13425 | −97.4745 | 27.8564 | | x | | x | | |
| 13426 | −97.1097 | 27.8889 | x | | x | x | x | |
| 13430 | −97.4262 | 27.8183 | | x | | | x | x |
| 13432 | −97.4497 | 27.8200 | | x | | | x | x |
| 13439 | −97.5200 | 27.8433 | x | x | | | x | x |
| 13440 | −97.3101 | 27.6787 | x | | | x | | |
| 13442 | −97.3075 | 27.7092 | x | | x | x | | x |
| 13443 | −97.2400 | 27.6000 | | | x | | x | |
| 13444 | −97.4103 | 27.2764 | x | | | x | | |
| 13445 | −97.3208 | 27.4792 | x | x | | | | |
| 13446 | −97.2000 | 26.0833 | | x | | | | |
| 13447 | −97.3167 | 26.3667 | | x | | | x | |
| 13448 | −97.4000 | 26.5667 | | | | | | x |
| 13449 | −97.4667 | 26.7833 | x | | | x | | |
| 13450 | −97.5100 | 27.2683 | x | x | | | x | |
| 13452 | −97.6433 | 27.2900 | x | | | x | x | |

(continued)

**Table 1** (continued)

| Station ID | Longitude | Latitude | Temp | Sal | DO | TOC | TA | pH |
|---|---|---|---|---|---|---|---|---|
| 13459 | −97.1833 | 26.0500 | | x | | | | |
| 13460 | −97.2575 | 26.0147 | x | x | | | | |
| 13563 | −96.5822 | 28.6797 | | | | x | | |
| 14355 | −97.1625 | 27.7611 | x | | | x | | x |
| 14560 | −94.9523 | 29.6067 | | x | | | | |
| 14622 | −94.8992 | 29.2778 | | x | | | | |
| 14726 | −96.4889 | 28.5065 | x | x | | x | x | x |
| 14732 | −96.4919 | 28.3619 | | x | | | | |
| 14733 | −96.5627 | 28.3371 | | x | | | | |
| 14783 | −97.2092 | 28.0875 | | | | x | | |
| 14833 | −97.4167 | 27.8275 | | x | | | | |
| 14865 | −97.1858 | 26.0264 | x | x | | | | |
| 14870 | −97.2806 | 26.1000 | | x | | | | |
| 14871 | −97.3850 | 25.9556 | | x | | | | |
| 14875 | −97.2926 | 25.9964 | | x | | x | | |
| 14956 | −96.8382 | 28.3918 | | x | | x | | |

## Spatial Distributions of Physical and Chemical Parameters

Average temperature, salinity, TA, and DO all exhibit clear latitudinal trends from northeast to southwest. There are increasing trends down the coast for temperature, salinity, and TA (Fig. 2a, b, e), while DO shows a decreasing trend (Fig. 2c). The transition zone of these parameters appears to coincide with the few estuaries in the mid coast region, from Lavaca-Colorado Estuary to Nueces Estuary, which have been examined in the past (Montagna et al. 2018). In comparison, neither TOC nor pH shows any clear latitudinal trend. Rather, TOC concentration is generally higher in primary bays (Fig. 2d), where pH is usually lower (Fig. 2f).

In the aquatic environment, DO concentration decreases with increasing salinity and increasing temperature. The observed DO concentrations appear to be overwhelmingly greater than their corresponding equilibrium values at the same temperature and salinity (open circles in Fig. 3a) at stations with higher average salinity and temperature (the southern estuaries), suggesting greater autotrophy at these stations, although stations with lower average salinity can be either autotrophic or heterotrophic (Fig. 3a). TA (but not pH) shows a generally increasing trend with salinity (Fig. 3b). In addition, estuarine TA is mostly higher than the surface GOM level (salinity 36, TA = 2450 µM, the horizontal line in Fig. 3b). TOC concentration shows a U-shaped distribution along the salinity gradient, higher at both low salinity (greater freshwater influence) and hypersaline conditions. While oxygen saturation is more variable at low salinity waters, indicating varying metabolic states likely corresponding to the presence of both autochthonous and allochthonous OC, high TOC concentration is associated with supersaturated

DO at hypersaline conditions, suggesting mostly autochthonous OC production (Fig. 3c).

## Long-Term Trends of Physical and Biogeochemical Parameters

### Physical Condition Changes

There are 30 stations that exhibit long-term warming trends at a rate of 0.02–0.08 °C yr.$^{-1}$ (Fig. 4a). However, two stations in a secondary bay of Nueces Estuary (Oso Bay, 13,440 and 13,442), one station in the primary bay of Nueces Estuary (Corpus Christi Bay, 14,355), and one station in Laguna Madre (South Bay, 14,865) show negative temperature trends. The negative temperature trends at stations 14,355 and 14,865 may be caused by the substantially shorter data records (1998–2020 for Station 14,355 and 1996–2019 for Station 14,865, respectively), compared with all other stations that have data extending to the early 1970s. The possible reason that the two stations in Oso Bay exhibit negative temperature trends is more frequent (weekly) measurements in several years after 2000, which might have caused bias in the trend analysis. Nevertheless, the increases in temperature in all other stations are consistent with the coastwide warming rate of ~0.04–0.05 °C yr.$^{-1}$ (mid-1970s to 2008) from a previous study (Montagna et al. 2011b).

Out of the 38 stations where long-term salinity trends are present (Fig. 4b), most (32) of the stations exhibit long-term salinity increase, and six stations that are all located in upper estuaries or secondary bays of mid latitude estuaries showed long-term decrease. The six stations that showed negative trends are 13,302 (−0.12 yr.$^{-1}$, Sabine-Neches Estuary), 13,332 and 13,364 (−0.29 and −0.17 yr.$^{-1}$, both in Trinity-San Jacinto Estuary), 13,388 (−0.94 yr.$^{-1}$, Lavaca-Colorado Estuary), and 13,420 and 14,833 (−0.49 and −0.40 yr.$^{-1}$, respectively, both in Nueces Bay). The largest increase rate (0.68 yr.$^{-1}$) is observed at station 14,733 in Guadalupe Estuary.

### DO and TOC

Ten stations show significant long-term DO trends (Fig. 5a). The only station that shows a positive rate, 0.04 mg L$^{-1}$ yr.$^{-1}$, is at 13335 (Clear Lake in Trinity-San Jacinto Estuary). This station is adjacent to the other Clear Lake station (13332) that shows a decreasing salinity trend. The other stations show decreasing DO trends −0.02 ~ −0.05 mg L$^{-1}$ yr.$^{-1}$, mostly in mid coast estuaries, that is, Lavaca-Colorado, Mission-Aransas, and Nueces estuaries (Fig. 5a).

**Fig. 2** Spatial distribution of average temperature (**a**), salinity (**b**), DO (**c**), TOC (**d**), TA (**e**), and pH (**f**) in Texas estuaries

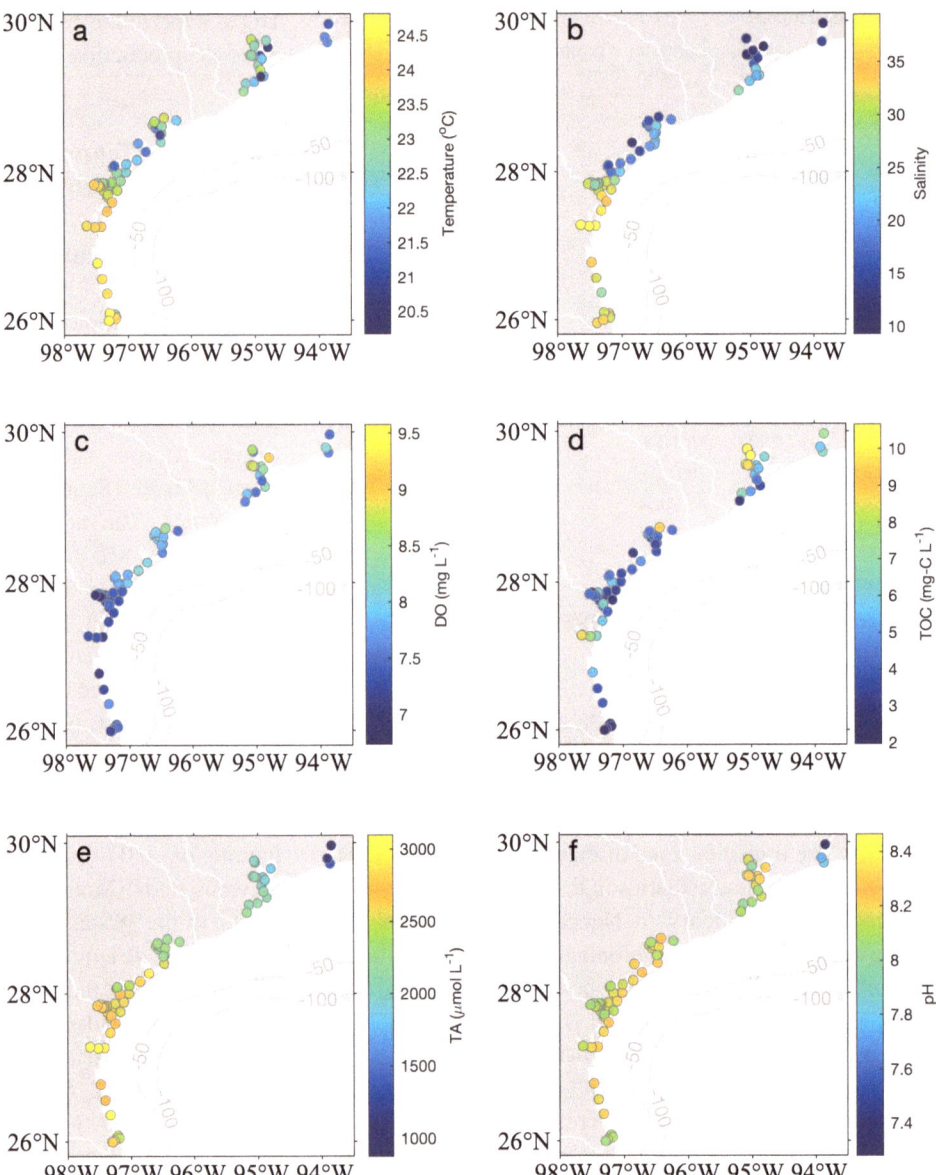

Most (24 out 30) stations show declining TOC concentration over time with rates as much as −0.47 mg-C L$^{-1}$ yr.$^{-1}$ (13,346 in Chocolate Bay in lower Trinity-San Jacinto Estuary, Fig. 5b). A total of six stations show long-term increase in TOC concentration, including two stations in Lavaca-Colorado Estuary (13,563 and 14,726, 0.24 and 0.14 mg-C L$^{-1}$ yr.$^{-1}$, respectively), one station in Mission-Aransas Estuary (14,783, 0.20 mg-C L$^{-1}$ yr.$^{-1}$), two stations in Nueces Estuary (13,425 and 14,355, 0.12 and 0.13 mg-C L$^{-1}$ yr.$^{-1}$, respectively), and one station in Lower Laguna Madre (14,875, 0.09 mg-C L$^{-1}$ yr.$^{-1}$). It is noted that the upper coast estuaries show exclusively negative trends, but those in the lower coast are variable.

## Carbonate Chemistry (TA and pH)

Most of the 25 stations that show significant trends have declining TA (Fig. 6a). Six stations, including two stations in the upper coast (13,302 in Sabine-Neches Estuary and 13,315 in Trinity-San Jacinto Estuary), one station in the mid coast (14,726 in Lavaca-Colorado Estuary), and three stations in the lower coast (13,447 and 13,452 in Baffin Bay, and 13,285 in Lower Laguna Madre) have long-term TA increase.

The same number of stations (25) exhibit significant pH trends as those for TA although the stations do not entirely overlap (Fig. 6b). Two stations in the upper coast (13,302 in

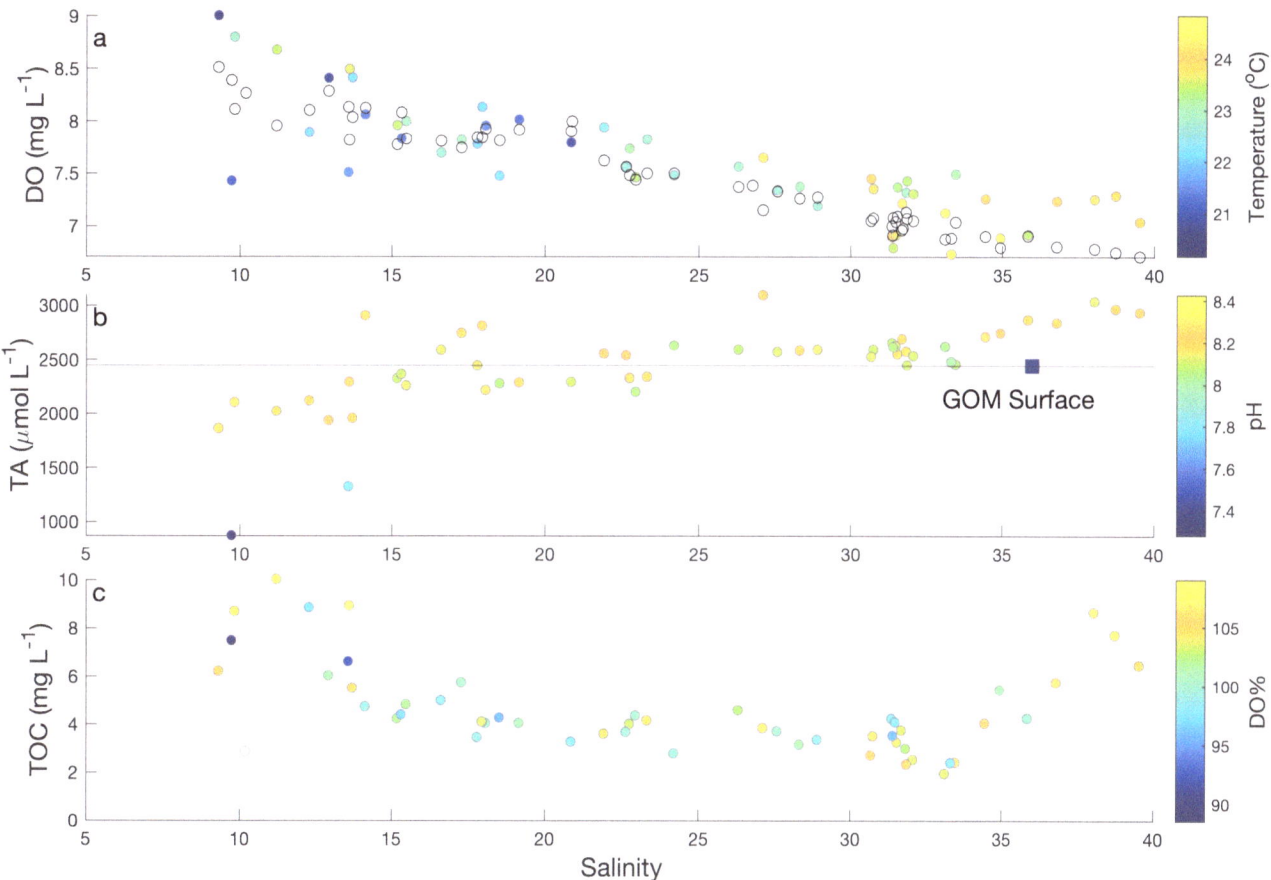

**Fig. 3** Average DO (**a**), TA (**b**) and TOC (**c**) concentrations versus salinity across Texas estuaries. Color bars represent temperature (**a**), pH (**b**), and percent DO saturation (**c**). The open circles in (**a**) represent equilibrium concentrations at the corresponding average temperature and salinity of examined stations. In (**b**), the blue square and the horizontal line represent surface GOM TA (2450 μmol L$^{-1}$ at salinity 36)

Sabine-Neches Estuary and 13,344 in Trinity-San Jacinto Estuary) and two stations in the mid coast (14,726 in Lavaca-Colorado Estuary and 14,355 in Nueces Estuary) show increasing pH trends and the rest of the stations (21) have declining pH.

## Discussion

Freshwater inflow is the primary driver for productivity in Texas estuaries (Buzan et al. 2009; Kim and Montagna 2012; Kim et al. 2014; Montagna and Kalke 1992), and the amount of freshwater delivered into the estuaries and subsequent salinity variation control the distribution of calcifiers including the economically important eastern oyster (Dekshenieks et al. 2000; Pollack et al. 2011; Powell et al. 2003). However, both the changing climate (warming) and the resultant hydrological condition changes as well as increasing demand for freshwater in the upstream of river watersheds have long put a strain on freshwater resources in Texas, which directly affect the amount of freshwater inflow reaching the estuaries (Montagna et al. 2011a), hence their productivity. As a result, Texas Senate Bill 3 (SB3) was passed in 2007 in an effort to

establish the freshwater needs in inflow management. Nevertheless, long-term decline in freshwater availability has been an issue that may have significant impact on estuarine biogeochemistry in the context of climate change.

## Changes in Climate and Hydrological Conditions

Using the same SWQM dataset (till 2016), Bugica et al. (2020) analyzed estuarine temperature trends in Texas estuaries. They did not find significant temperature changes based on annual data although summer temperature shows an appreciable increase. Using weekly data collected by Texas Parks and Wildlife Department (TPWD) from 1976–2007, Montagna et al. (2011b) found an overall warming rate of 0.0428 °C yr.$^{-1}$, and based on a longer time-series TPWD dataset from 1976 to 2020, the same method generated a coastwide warming rate of 0.0325 °C yr.$^{-1}$ (Montagna, pers. comm.). The latter rate is consistent with the average rate increase calculated using the GAM method, 0.0353 °C yr.$^{-1}$ (with a range of 0.0114–0.0586 °C yr.$^{-1}$) based on the same TPWD dataset, and these results also agree with the average

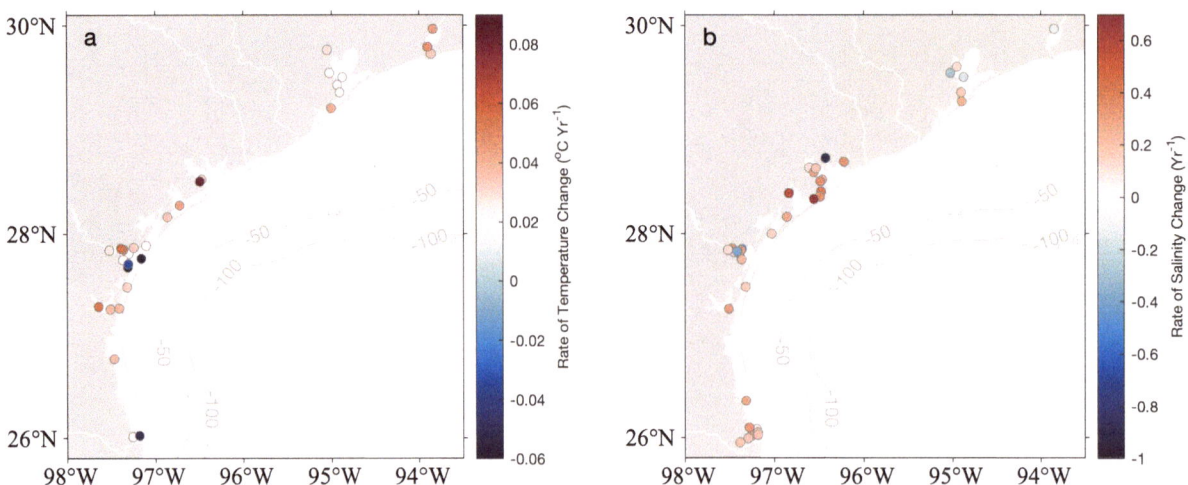

**Fig. 4** Long-term temperature (**a**) and salinity (**b**) trends at SWQM stations in Texas estuaries

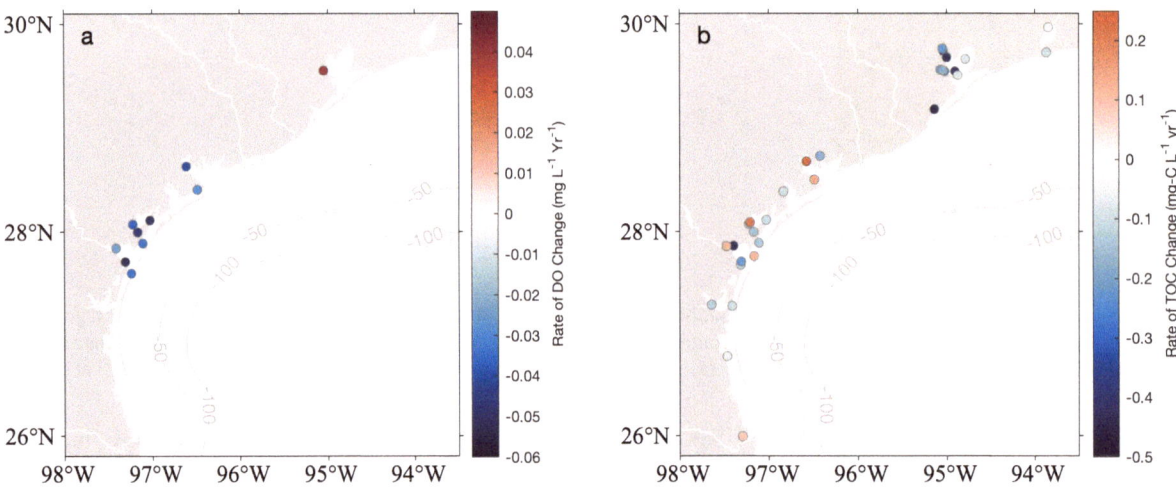

**Fig. 5** Long-term DO (**a**) and TOC (**b**) trends at SWQM stations in Texas estuaries

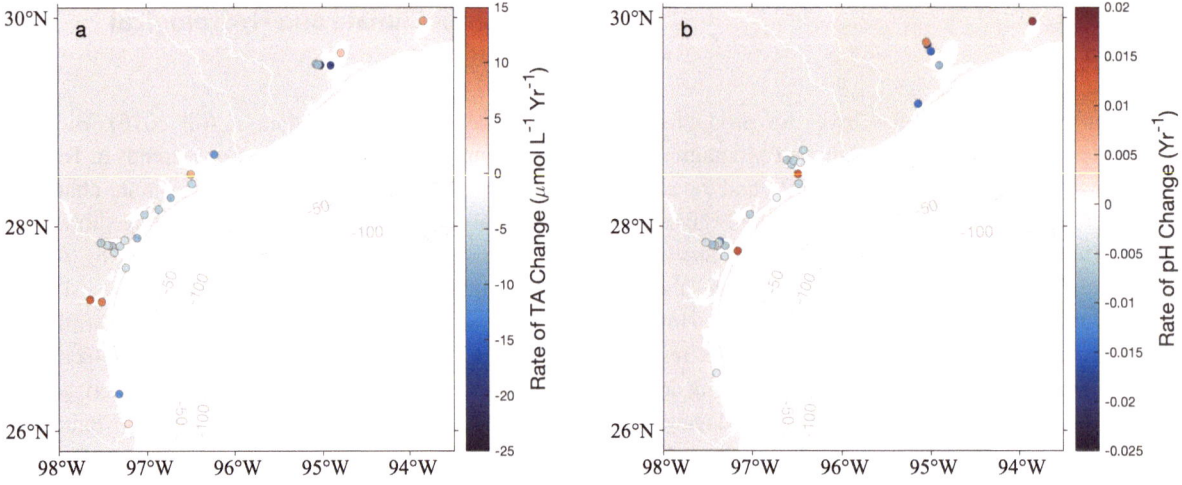

**Fig. 6** Long-term (**a**) total TA and (**b**) pH trends at SWQM stations in Texas estuaries

warming rate across all the stations analyzed here using the SWQM data (0.0366 °C yr.$^{-1}$) after excluding the four stations with negative rates (due to either shorter temperature record or uneven measurement frequency, Fig. 4a). The warming rates in estuaries are consistent with that in offshore waters. For example, a long-term monitoring in Flower Garden Banks National Marine Sanctuary at the northwestern Gulf of Mexico found a warming rate of 0.35 °C per decade (1985–2020) (Dias et al. 2023).

During the past nearly four decades (salinity data were collected starting from the 1980s), there is overwhelming evidence that Texas estuaries have kept becoming saltier (Fig. 4b). Bugica et al. (2020) attributed the salinity increase to long-term freshwater inflow decline and suggested that climate variability such as the El Niño-Southern Oscillation (ENSO) has little to do with the observed trends, in contrast to what was suggested in Tolan (2007). Previous studies have examined multi-decadal trends in river discharge in this region using different approaches (modeling and observation), though due to the differences in both study period and methodology, the results are not entirely consistent. However, the general findings point to stable or perhaps slight increase discharge in upper coast rivers but declining freshwater delivery in the lower coast. For example, Liu et al. (2013), using a process-based land ecosystem model, suggested that the Texas Basin as a whole experiences an increase in river discharge (by 0.86% per year) and the Rio Grande Basin to the south has a decrease of −0.01% per year during the period from 1979 to 2008. Milliman et al. (2008) found a 30% increase in precipitation in the Colorado River watershed but no appreciable increase in runoff, while the border river with Mexico—Rio Grande—experienced a 38% decrease in runoff even though precipitation increased by 22%, both during the period from 1950 to 2000. The somewhat contradiction between the increase in precipitation in river basins versus the invariable or even declining river runoff is likely a result of freshwater diversion to meet the demand of increasing population. Nevertheless, this latitudinal pattern in long-term river runoff trend appears to be consistent with the salinity change rate (Fig. 4b), that is, slightly negative to slightly positive trends in the upper estuaries (Sabine-Neches and Trinity-San Jacinto estuaries) but greater salinity increases down the coast. Note the three negative trends found at mid to lower coast stations (13,388, 13,420, 14,833) are all located in secondary bays that have managed freshwater inflow (Lavaca-Colorado Estuary and Nueces Estuary) (Montagna et al. 2009; Olson 2017).

## Potential Changes in Estuarine Metabolism

Other than the only station that shows an increasing DO trend in Clear Lake within Trinity-San Jacinto Estuary (Station 13,355), which also experiences a decrease in TOC

concentration (Fig. 5b), all other stations with significant regression values all show deceasing DO trends (Fig. 5a), mostly in the mid coast region (Lavaca-Colorado, Mission-Aransas, and Nueces estuaries, as well as Upper Laguna Madre). These trends are largely consistent with those found in Bugica et al. (2020) across the entire Texas coast and that in Nueces Estuary (~ −0.06 mg L$^{-1}$ yr.$^{-1}$, 1982–2002) as reported in Applebaum et al. (2005). In the latter study, the trend in summer months (from July to August) is as much as ~ −0.9 mg L$^{-1}$ yr.$^{-1}$. Using the calculated equilibrium DO concentrations at the mid-coast stations that show long-term temperature and salinity trends, if assuming an average warming rate of 0.0366 °C yr.$^{-1}$ and an average salinity increase rate of ~0.4 yr.$^{-1}$, DO decrease rate is slightly higher than −0.02 mg L$^{-1}$ yr.$^{-1}$, with warming contributing about 20% and salinity increase accounting for the rest of the decline. Nevertheless, the DO decline rates in the mid-coast estuaries are all greater than the estimated value due to the combination of warming and salinity increase effect. On the other hand, for the Clear Lake station (13355) that does not show significant temperature and salinity trends, if we use the adjacent Station 13,332 (salinity trend −0.29 yr.$^{-1}$ and warming rate 0.022 °C yr.$^{-1}$), DO increase rate is around 0.01 mg L$^{-1}$ yr.$^{-1}$, which is also lower than the actual trend. Based on these simple comparisons, other factors should contribute to the DO trends in addition to the physical condition (temperature and salinity) induced changes.

Using data collected in four mid-coast estuaries, Montagna et al. (2018) suggested that particulate organic carbon (POC) concentration mirrors that of chlorophyll, which shows a U-shaped pattern along the salinity gradient. In addition, dissolved organic carbon (DOC) concentration is also higher at both low and high salinity but the lowest at intermediate salinities (Montagna et al. 2018). The above observations using the smaller number of estuaries are consistent with data collected along the entire coast (Fig. 3c). TOC concentration in upper estuaries (Trinity-San Jacinto and Sabine-Neches) uniformly decrease over time, and the declining trend in Trinity-San Jacinto Estuary is consistent with a prior report by Center for Texas Beaches and Shores (CTBS 2017) using a slightly shorter time series data (from the early 1970s to mid-2010s), while the mid to lower coast estuaries exhibit variable rates, with positive values mostly appearing in secondary bays. The declining TOC in Trinity-San Jacinto Estuary is co-occurring with the declining nutrient level (CTBS 2017), suggesting improvement from earlier nutrient pollution condition (Pinckney 2006). Similarly, Bugica et al. (2020) found decreasing trends of total phosphorus and orthophosphate at both a Galveston Bay (13364) and a Trinity Bay (13315) station although the two stations show divergent total Kjeldahl nitrogen (TKN) trends, decreasing at the Galveston Bay station but increasing at the Trinity Bay station. However, Thronson and Quigg (2008) suggested that Galveston Bay exhibits long-term eutrophication symptoms (i.e., low oxygen) based on fish kill observations spanning from 1951 to 2006, which seems contradictory to the improving nutrient conditions based on TOC and nutrient trends, hence further study is needed to elucidate the mechanisms. In addition, it is worth examining the relative changes in the proportions of marine (autochthonous) versus terrestrial (allochthonous) organic matter in these estuaries. Cifuentes et al. (1999) suggested that most Texas estuaries (other than Sabine-Neches Estuary and Laguna Madre) have lower levels of allochthonous organic matter input than that from autochthonous production due to relatively small river input and long estuarine residence time, and the ratio between the two organic fractions decreases from north to south. This latitudinal difference appears to be consistent with the temporal freshwater decline; hence, the estuaries become relatively more autotrophic (or less heterotrophic) under increasing freshwater deficient conditions even though the overall TOC concentration decreases from north to south (except the hypersaline Baffin Bay, Fig. 5b), indicating a likely decrease in both terrestrial export and in situ production. Moreover, from the carbon cycle perspective, the fact that the largest estuary in the Texas coast (Trinity-San Jacinto) has declining TOC level over time suggests that estuary-coastal exchange in this area will likely also see a decline in TOC export. Temporal changes in organic carbon sources and its subsequent consequences on coastal oceans will require more detailed examination in the context of freshwater inflow changes.

## The Changing Estuarine Inorganic Carbon System

In typical river-dominated estuaries as well as river-influenced continental shelves, TA often exhibits linear relationships with salinity, mostly reflecting conservative mixing between river and ocean waters, as biogeochemical reactions often cannot cause appreciable TA changes that deviate from such linear relationship within the timescale of estuarine mixing. The slope and intercept of the mixing line are primarily controlled by river endmember values (Cai et al. 2010). Therefore, both freshwater endmember change and salinity fluctuation have the first-order influence on estuarine TA. It is known that river TA is controlled by precipitation, catchment mineralogy, soil weathering status, and land use type (Lauerwald et al. 2013). Similar to this study on the continental scale, river TA shows a clear latitudinal gradient in Texas, lower in the north and higher in the south (Fig. 7a, also see Hu et al. 2015), consistent with the precipitation gradient. The river TA gradient, along with the river inflow balance, shapes latitudinal distribution of river TA (Fig. 2e).

Clearly, rivers that empty into Sabine-Neches Estuary and those that empty into Trinity-San Jacinto Estuary have much lower TA, while most mid to lower coast rivers have higher TA than the ocean water value. For the above two estuaries that receive low TA river water, slightly declining salinity at the station in Sabine-Neches Estuary (13,302, Fig. 4b) and no change in salinity in Trinity Bay but increasing TA in both

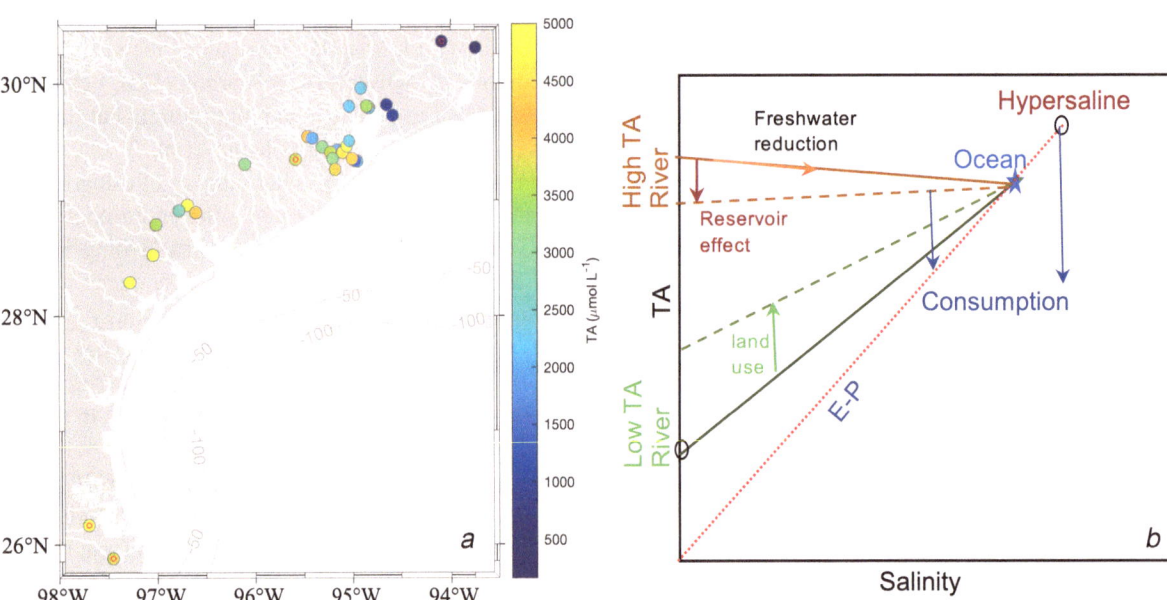

**Fig. 7** (**a**) Latitudinal distribution of river and stream TA along the Texas coast. The data were extracted from the National Water Information System (https://nwis.waterdata.usgs.gov). Note most stations only have a few years of sample collection from 1990 to 2010. The few stations with longer than 10 years of data are marked in red circle.

Color infill represents mean of multiple measurements at each station. As the reference, the ocean water at salinity 36.3 is 2460 µM (Hu et al. 2018). (**b**) A schematic figure for different scenarios of long-term TA change (see text for details)

estuaries (Fig. 6a) indicate that freshwater endmember TA may have been increasing (Fig. 7b). Indeed, TA record from USGS station #8041000 on Neches River suggests an increasing rate of $49 \pm 10$ $\mu$M yr.$^{-1}$, even though the record period (1989–2007) is much shorter than the estuarine station (13,302, 1973–2020). Lauerwald et al. (2013) found positive correlation between agricultural land and river TA and a large watershed where liming is used and contributes to increasing river TA (Oh and Raymond 2006). As a part of the Texas Basin (Liu et al. 2013), the watersheds of Sabine and Neches rivers as well as that for Trinity River have experienced significant land use changes, that is, from forests to cropland and from grassland to cropland, respectively. These changes may have contributed to increase in soil weathering in these areas.

For estuaries that receive river input with higher TA than the ocean water, the mechanisms for causing TA decrease can be more complex. From the physical mixing perspective alone, the reduction in freshwater input leads to higher salinity along with decreasing TA because the mixing line has a negative slope (Fig. 7b), which is the case for most mid to lower coast estuaries. In addition, most medium and large rivers in Texas are heavily managed, and multiple reservoirs are often present along their flow paths (Montagna et al. 2011a). Because of the relatively high TA levels and carbonate saturation states (Stets et al. 2014), precipitation of carbonate minerals may occur due to $CO_2$ degassing, as the $CO_2$ in reservoirs with moderate to high TA is supersaturated (Marce et al. 2015; McDonald et al. 2013). Therefore, carbonate precipitation in reservoirs reduces river TA export (Fig. 7b). Dias (2022) used coupled USGS river flow and TCEQ TA data and found that decreases in freshwater discharge (Aransas River), area-weighted TA yield (Colorado and Aransas rivers), and flow-weighted TA concentration (Trinity River) may all contributed to long-term TA reduction in their respective estuaries.

Other than the changes in river flow and chemistry, biogeochemical processes occurring in estuaries can consume alkalinity. Intuitively, the effect of these processes should be more pronounced in estuaries with long water residence time, that is, toward the lower coast (Solis and Powell 1999). Based on the TA-S diagram (Fig. 7b), in the case of river-ocean mixing, TA/S ratio (or specific TA, Hu et al. 2015) of the ocean endmember should theoretically be the lowest value. However, this ratio often dips below the theoretical minimum under hypersaline conditions from Mission-Aransas Estuary down south (Hu et al. 2015; Yao and Hu 2017). Moreover, even under non-hypersaline conditions TA consumption may still occur, although the TA/S ratio alone cannot be used to detect this process (Fig. 7b).

Hu et al. (2015) attributed TA decrease, especially in the mid-coast estuaries to the presence of calcifying organisms such as the eastern oysters, and its commercial harvesting

does not go south of Mission-Aransas Estuary. The abundance of calcifying organisms appears to be consistent with the observed TA trends, especially in the mid-coast estuaries. In comparison, at the two stations in Baffin Bay (13,452 and 13,450), TA shows increasing trends of $13 \pm 6$ and $10 \pm 5$ $\mu$M yr.$^{-1}$, respectively (Fig. 6a). While only the outer station (13450) shows detectable salinity increasing trend of 0.31 yr.$^{-1}$ (Fig. 4b), these increasing TA trends suggest that TA consumption because of long water residence time under hypersaline conditions is unlikely as large as in other estuaries due to less abundant calcifying organisms in Baffin Bay. Certainly, one cannot rule out the possibility that human caused eutrophication in this system may contribute to additional alkalinity upon the reduction of nitrate (Wolf-Gladrow et al. 2007) delivered by wastewater treatment plants, the only major freshwater source to Baffin Bay (Bugica et al. 2020).

In addition to calcification, redox processes either in the water column or at the sediment-water interface can also affect estuarine TA balance. The most extreme cases can be found in eastern Australian estuaries, where hydrological changes due to human disturbance (dredging and draining coastal wetland for constructions) expose reduced sulfur buried in sediments, and upon oxidation, the produced strong acid (sulfuric) can seriously acidify the water (de Weys et al. 2011; Sammut et al. 1996). In theory, this added sulfate in estuarine waters will cause "excess" sulfate compared with the conservative mixing (similar to mixing for TA, Fig. 7b). Other than these extreme cases, excessive sulfate is also reported in a North Carolina estuary (Matson and Brinson 1985), which is caused by the oxidation of sedimentary reduced sulfur under drought condition, the reason being the presence of chemoautotrophic bacteria that outcompete heterotrophs when nutrient-driven primary production thus organic matter supply is low. No concomitant carbonate chemistry data were collected in this study though. Preliminary examination of estuarine sulfate data suggests that significant sulfate enrichment also occurs during drought periods that correspond with hypersalinity along with low TA/S ratios than the ocean endmember (Yin et al. 2023). Nonetheless, estuarine redox reaction in response to freshwater reduction (hence decrease in estuarine production) may play some role in regulating carbonate chemistry and is worth further examination. The fact that most estuaries with decreasing TA levels suggest that these estuaries have been exporting less inorganic carbon to the coastal ocean.

Unlike the open ocean where pH closely follows the increasing $CO_2$ levels in the atmosphere (Bates et al. 2013), both freshwater end-member and metabolic effect are important drivers for pH variation in coastal and estuarine environments (Hu and Cai 2013), and large range of long-term pH trends often deviate from the open ocean trends (both positive and negative). Carstensen and Duarte (2019) conducted

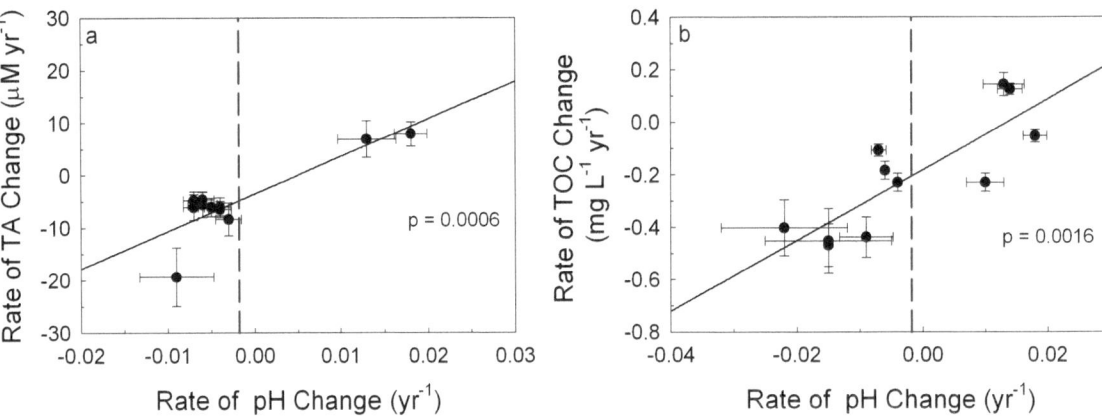

**Fig. 8** Relationship between the rates of TA and pH changes (**a**) and relationship between the rates of TOC and pH changes (**b**). The vertical dashed lines represent global ocean pH decrease rate − 0.0018 yr.⁻¹. Note the two sets of stations do not entirely overlap

a global synthesis and found that pH trends of −0.023–0.023 pH units yr.⁻¹, in contrast to relatively consistent open ocean rates of −0.0018 yr.⁻¹ (Carstensen and Duarte 2019). Similarly, trends at the 25 stations examined here range from −0.022 to 0.018 yr.⁻¹, a range that is wider than that in Hu et al. (2015) (−0.0079–0.0062 yr.⁻¹), which are based on individual water bodies though, not individual stations. Similar to the findings in Hu et al. (2015), TA and pH trends at 11 stations show high degree of correlation ($r = 0.86$, $p < 0.001$, Fig. 8a). The two stations (13,302 in Sabine-Neches Estuary and 14,726 in Lavaca-Colorado Estuary) have TA increase rate of 7.9 ± 2.3 and 6.9 ± 3.5 μM yr.⁻¹, corresponding to pH increase rates of 0.018 ± 0.002 and 0.013 ± 0.003 yr.⁻¹, respectively. Similarly, estuarine water TA increase as a result of freshwater TA change has been associated with pH increase elsewhere, even canceling out the effect of ocean acidification (e.g., Gomez et al. 2021).

In addition to TA change, it has long been recognized that nutrient input modulates pH variations in the aquatic system (e.g., Borges and Gypens 2010), and the metabolic pH responses to nutrient addition in coastal waters affect the autotrophic period more than the heterotrophic period (e.g., Nixon et al. 2015). The changing nutrient-driven production appears to apply to Texas estuaries as well. This can be seen from the highly correlated TOC and pH trends ($r = 0.83$, $p < 0.002$, Fig. 8b). The four stations that have the highest TOC decline (13,342, 13,355, 13,303, and 13,346, note 13,303 also has declining TA) are all located in Trinity-San Jacinto Estuary, which has experienced decreasing nutrient levels in the past a few decades (CTBS 2017), suggesting decreasing autochthonous OC production. On the other hand, station 14,726 in Lavaca-Colorado Estuary and station 14,355 in lower Nueces Estuary both have increasing TOC levels along pH increase, suggesting divergent changes in nutrient condition and associated production in estuaries with lower freshwater input.

Although not calculated here, both TA and pH trends found in this study with extended data records are consistent with those reported in McCutcheon and Hu (2022). Therefore, both Ω and buffer capacity against further pH changes in these estuaries have continued to decrease. Even though these changes appear to be predominantly affected by the changes on the terrestrial side (land use change, hydrological cycle), the increasing acidification in the coastal ocean may ultimately have an appreciable effect on estuarine water chemistry. Hence, despite the favorable carbonate saturation, continued monitoring is needed especially in areas with abundant shellfish production.

## Summary

Hydrological condition changes in watershed and human intervention including freshwater diversion and land use changes have profound impact on estuaries, ranging from water quality to productivity, due to the altered material (both carbon and nutrients) and water fluxes. TCEQ's SWQP monitoring program provides an excellent historical context and up-to-date dataset based on which multidecadal trends of both physical and biogeochemical parameters are examined. The results provide a holistic understanding of changing estuarine conditions along the Texas coast on a climatic gradient.

Similar to the open Gulf of Mexico waters, Texas estuaries have also experienced multidecadal warming, irrespective of data sources. Meanwhile, even though there has been no systematic examination of freshwater inflow rates from individual rivers except a few large-scale studies, salinity can be used as a sensitive indicator for the reduced inflow conditions in most estuaries.

Freshwater inflow reduction has important implications to estuarine biogeochemistry as both allochthonous OC and nutrient export may decrease. In addition, the effect of active nutrient management is evident in an estuary close to a large metropolitan area (Trinity-San Jacinto Estuary). Because Texas estuaries have overall greater input of autochthonous than allochthonous OC, freshwater inflow reduction not only leads to a decline of TOC levels but also changes their relative proportions, with likely further decreases in allochthonous to autochthonous ratio. Even though both warming and salinity increase reduce DO level, the extent of multidecadal DO decrease in mid-coast estuaries is greater than predicted by physical condition changes alone. This excess DO decrease can be likely attributed to declining productivity. Since the allochthonous OC input may have greater decline than in situ production, these estuaries still maintain relative autotrophy.

Changing hydrological conditions and land use changes in watersheds affect the production and transport of weathering products (as measured by TA) through the river systems. Land use changes including conversion to croplands in upper Texas coast lead to increase in TA levels in rivers, hence in the downstream estuary (Sabine-Neches Estuary and upper Trinity-San Jacinto Estuary). However, freshwater retention along the river flow paths leads to a reduction in river TA export to mid and lower coast estuaries. In addition to changes in the watershed, inflow reduction increases estuarine residence time. Because of longer residence time, calcification may result in more extensive TA removal, and such removal is especially excessive under drought conditions. Although not completely understood in Texas estuaries, benthic redox reactions such as oxidation of reduced sulfur may also consume TA under drought conditions (Dias et al. 2022). Whether this oxidation process is caused by hydrological means (physical exposure of reduced sulfur followed by oxidation) or chemosynthetic pathway needs further study. From the carbon cycle perspective, the declining TOC and TA levels in most estuaries suggest that carbon export to the coastal ocean has been decreasing in the past few decades.

TA change rate is positively correlated with pH change, suggesting TA's dominant control on acid-base equilibria in estuarine waters; pH changing rate also shows a positive correlation with that of TOC, with the lowering TOC predominantly in Trinity-San Jacinto Estuary that has active nutrient control, corresponding to a decrease in nutrient-driven production. The decreases in both TA and pH have resulted in a decrease in carbonate saturation state and buffer capacity in most Texas estuaries. Even though carbonate saturation states in most estuaries are within the optimal range, ocean acidification may affect these estuaries in the long term.

## Future Research Directions

Climate change, human alteration to river watersheds, and land use changes will continue affecting both freshwater inflow and material flux to Texas estuaries and subsequent

coastal ocean. However, currently, there have been few systematic examination of material fluxes in most Texas rivers other than large-scale studies that do not differentiate individual systems (Preston et al. 2011; Shih et al. 2010). Given the importance of freshwater inflow on estuarine productivity, it is desirable to quantify the amounts of both OC and nutrient inputs based on long-term water quality and flow records maintained by both TCEQ and United States Geological Survey and using this information to construct more quantitative relationship between freshwater inflow and estuarine productivity.

While this study uses TOC and DO dynamics to infer riverine allochthonous OC export and autochthonous OC production, direct and sustained net ecosystem metabolism (NEM) measurements are scarce and mostly in mid-coast estuaries (Bruesewitz et al. 2013; Russell and Montagna 2007; Russell et al. 2006). Moreover, the only estuarine system on the Texas coast that is being continuously monitored is Mission-Aransas Estuary since 2007 (Evans et al. 2012), where NEM can be estimated using high resolution DO data, although long-term trend probably will not emerge soon given its large temporal variability. Nevertheless, to examine temporal changes or use the salinity gradient as a proxy for increasing drought conditions, more extensive NEM measurements and analyses are needed.

Estuarine carbonate chemistry is strongly modulated by freshwater inflow, river endmember changes, and within estuarine processes. Given the importance of carbonate saturation state on calcifying organisms, there is a need to include high precision measurements using state-of-the-art chemical sensors with additional parameters (e.g., total dissolved inorganic carbon) for more precise characterization of the carbonate system. Such work will be especially important in upper coast estuaries given the lower river TA levels as these estuaries will likely experience greater pH decrease from both ocean acidification and respiration (Hu and Cai 2013).

This study has benefited from the long-term water quality monitoring conducted by TCEQ, which is among the few such programs both nationally and internationally (e.g., Carstensen and Duarte 2019). These data, despite somewhat crude precision on some field-collected data (for example, pH) compared to those collected in academic laboratories but due to constant methodology (Hu et al. 2015), offer valuable sights in changing environmental conditions. Therefore, continued monitoring and analysis of temporal trends and interpretation of the driving factors are needed especially under the current climate conditions.

**Acknowledgments** This chapter was funded in part through Contract No. 21-155-007-C879 from the Texas General Land Office (GLO) with Gulf of Mexico Energy Security Act of 2006 funding made available to the State of Texas and awarded under the Texas Coastal Management Program. The views contained herein are those of the authors and should not be interpreted as representing the views of the GLO or the State of Texas. Hu also acknowledges partial support from the National Science Foundation (OCE #1654232) during the development of this chapter.

# References

Alexander MA, Scott JD, Friedland KD, Mills KE, Nye JA, Pershing AJ, Thomas AC (2018) Projected sea surface temperatures over the 21st century: changes in the mean, variability and extremes for large marine ecosystem regions of northern oceans. Elem Sci Anth 6

Applebaum S, Montagna PA, Ritter C (2005) Status and trends of dissolved oxygen in Corpus Christi Bay, Texas, U.S.A. Environ Monit Assess 107:297–311

Bates NR et al (2013) A time-series view of changing surface ocean chemistry due to ocean uptake of anthropogenic CO2 and ocean acidification. Oceanography 27:126–141

Belkin IM (2009) Rapid warming of large marine ecosystems. Prog Oceanogr 81:207–213

Borges AV, Gypens N (2010) Carbonate chemistry in the coastal zone responds more strongly to eutrophication than to ocean acidification. Limnol Oceanogr 55:346–353

Bruesewitz DA, Gardner WS, Mooney RF, Pollard L, Buskey EJ (2013) Estuarine ecosystem function response to flood and drought in a shallow, semiarid estuary: nitrogen cycling and ecosystem metabolism. Limnol Oceanogr 58:2293–2309

Bugica K, Sterba-Boatwright B, Wetz MS (2020) Water quality trends in Texas estuaries. Mar Pollut Bull 152:110903

Buzan D, Lee W, Culbertson J, Kuhn N, Robinson L (2009) Positive relationship between freshwater inflow and oyster abundance in Galveston Bay, Texas. Estuar Coast 32:206–212

Cai W-J, Hu X, Huang W-J, Jiang L-Q, Wang Y, Peng T-H, Zhang X (2010) Surface ocean alkalinity distribution in the western North Atlantic Ocean margins. J Geophys Res 115:C08014

Carstensen J, Duarte CM (2019) Drivers of pH variability in coastal ecosystems. Environ Sci Technol 53:4020

Chen C-TA (2002) Shelf-vs. dissolution-generated alkalinity above the chemical lysocline. Deep-Sea Res II Top Stud Oceanogr 49:5365–5375

Cifuentes LA, Coffin RB, Morin J, Bianchi TS, Peter ME (1999) Particulate organic matter in Gulf of Mexico estuaries - implications for net heterotrophy. In: Bianchi TS, Pennock JR, Twilley RR (eds) Biogeochemistry of gulf of Mexico estuaries. Wiley, pp 239–267

CTBS (2017) Galveston Bay status and trends final report. Texas A&M University, Galveston, TX

de Weys J, Santos IR, Eyre BD (2011) Linking groundwater discharge to severe estuarine acidification during a flood in a modified wetland. Environ Sci Technol 45:3310–3316

Dekshenieks MM, Hofmann EE, Klinck JM, Powell EN (2000) Quantifying the effects of environmental change on an oyster population: a modeling study. Estuaries 23:593

Dias LM (2022) Hydrological and Biogeochemical Controls on Estuarine Carbonate Chemistry along a Climate Gradient. Texas A&M University-Corpus Christi, Ph.D. Dissertation

Dias LM, Yin H, Hu X (2022) A biogeochemical alkalinity sink in a shallow, semiarid estuary of the northwestern Gulf of Mexico. Aquat Geochem:49–71

Dias LM et al (2023) Rapid Climate Vulnerability Assessment for Flower Garden Banks National Marine Sanctuary. National Oceanographic and Atmospheric Administration's Flower Garden Banks National Marine Sanctuary. https://doi.org/10.25923/jdpb-zw04

Dürr HH, Laruelle GG, van Kempen CM, Slomp CP, Meybeck M, Middelkoop H (2011) Worldwide typology of nearshore coastal systems: defining the estuarine filter of river inputs to the oceans. Estuar Coast 34:441–458

Evans A, Madden K, Palmer SM (eds) (2012) The ecology and sociology of the Mission-Aransas estuary - an estuarine and watershed profile. University of Texas Marine Science Institute, Port Aransas, TX

Gattuso JP, Frankignoulle M, Bourge I, Romaine S, Buddemeier RW (1998) Effect of calcium carbonate saturation of seawater on coral calcification. Glob Planet Chang 18:37–46

Gomez FA, Wanninkhof R, Barbero L, Lee S-K (2021) Increasing river alkalinity slows ocean acidification in the northern Gulf of Mexico. Geophys Res Lett 48:e2021GL096521

Held IM, Soden BJ (2006) Robust responses of the hydrological cycle to global warming. J Clim 19:5686–5699

Hopkinson CS, Vallino JJ (1995) The relationships among man's activities in watersheds and estuaries: a model of runoff effects on patterns of estuarine community metabolism. Estuaries 18:598–621

Hu X, Cai W-J (2011) An assessment of ocean margin anaerobic processes on oceanic alkalinity budget. Global Biogeochem Cy 25:GB3003

Hu X, Cai W-J (2013) Estuarine acidification and minimum buffer zone—A conceptual study. Geophys Res Lett 40:5176–5181

Hu X, Beseres Pollack J, McCutcheon MR, Montagna PA, Ouyang Z (2015) Long-term alkalinity decrease and acidification of estuaries in northwestern Gulf of Mexico. Environ Sci Technol 49:3401–3409

Hu X et al (2018) Seasonal variability of carbonate chemistry and decadal changes in waters of a marine sanctuary in the northwestern Gulf of Mexico. Mar Chem 205:16–28

Joyce TM, Pickart RS, Millard RC (1999) Long-term hydrographic changes at 52 and 66°W in the North Atlantic Subtropical Gyre & Caribbean. Deep-Sea Res II Top Stud Oceanogr 46:245–278

Jutras M, Dufour CO, Mucci A, Cyr F, Gilbert D (2020) Temporal changes in the causes of the observed oxygen decline in the St. Lawrence Estuary. J Geophys Res-Oceans 125:e2020JC016577

Kaushal SS, Likens GE, Utz RM, Pace ML, Grese M, Yepsen M (2013) Increased river alkalinization in the eastern U.S. Environ Sci Technol 47:10302–10311

Kenney BE, Litaker W, Duke CS, Ramus J (1988) Community oxygen metabolism in a shallow tidal estuary. Estuar Coast Shelf S 27:33–43

Kim H-C, Montagna PA (2012) Effects of climate-driven freshwater inflow variability on macrobenthic secondary production in Texas lagoonal estuaries: a modeling study. Ecol Model 235–236:67–80

Kim H-C, Son S, Montagna P, Spiering B, Nam J (2014) Linkage between freshwater inflow and primary productivity in Texas estuaries: downscaling effects of climate variability. J Coast Res 68:65–73

Kundzewicz ZW (2008) Climate change impacts on the hydrological cycle. Ecohydrol Hydrobiol 8:195–203

Lauerwald R, Hartmann J, Moosdorf N, Kempe S, Raymond PA (2013) What controls the spatial patterns of the riverine carbonate system? — a case study for North America. Chem Geol 337-338:114–127

Liu M, Tian H, Yang Q, Yang J, Song X, Lohrenz SE, Cai W-J (2013) Long-term trends in evapotranspiration and runoff over the drainage basins of the Gulf of Mexico during 1901–2008. Water Resour Res 49:1988–2012

Marce R, Obrador B, Morgui J-A, Lluis Riera J, Lopez P, Armengol J (2015) Carbonate weathering as a driver of CO2 supersaturation in lakes. Nat Geosci 8:107–111

Marshall DA, La Peyre MK, Palmer TA, Guillou G, Sterba-Boatwright BD, Beseres Pollack J, Lebreton B (2021) Freshwater inflow and responses from estuaries across a climatic gradient: an assessment of northwestern Gulf of Mexico estuaries based on stable isotopes. Limnol Oceanogr n/a:3568–3581

Matson EA, Brinson MM (1985) Sulfate enrichments in estuarine waters of North Carolina. Estuaries 8:279–289

McCutcheon MR, Hu X (2022) Long-term trends in estuarine carbonate chemistry in the northwestern Gulf of Mexico. Front Mar Sci 9:793065

McDonald CP, Stets EG, Striegl RG, Butman D (2013) Inorganic carbon loading as a primary driver of dissolved carbon dioxide concentrations in the lakes and reservoirs of the contiguous United States. Global Biogeochem Cy 27:285–295

Miller AW, Reynolds AC, Sobrino C, Riedel GF (2009) Shellfish face uncertain future in high CO2 world: influence of acidification on oyster larvae calcification and growth in estuaries. PLoS One 4:e5661

Milliman JD, Farnsworth KL, Jones PD, Xu KH, Smith LC (2008) Climatic and anthropogenic factors affecting river discharge to the global ocean, 1951–2000. Glob Planet Chang 62:187–194

Montagna P, Kalke R (1992) The effect of freshwater inflow on meiofaunal and macrofaunal populations in the Guadalupe and Nueces Estuaries, Texas. Estuaries 15:307–326

Montagna PA, Gibeaut JC, Tunnell JW (2007) South Texas climate 2100: coastal impacts. In: Norwine J, John K (eds) The changing climate of South Texas 1900–2100: problems and prospects, impacts and implications. Texas A&M University, Kingsville, pp 57–77

Montagna IA, Hill EM, Moulton B (2009) Role of science-based and adaptive management in allocating environmental flows to the Nueces Estuary, Texas, USA. In: Brebbia CA, Tiezzi E (eds) Ecosystems and sustainable development VII. WIT Press, Southampton, pp 559–570

Montagna P, Vaughan B, Ward G (2011a) The importance of freshwater inflows to Texas estuaries. In: Griffin RC (ed) Water policy in Texas: responding to the rise of scarcity. The RFF Press, Washington DC, pp 107–127

Montagna PA, Brenner J, Gibeaut J, Morehead S (2011b) Coastal impacts. In: Schmandt J, Clarkson J, North GR (eds) The impact of global warming on Texas, 2nd edn. University of Texas Press, Austin, pp 96–123

Montagna P, Palmer TA, Beseres Pollack J (2013) Hydrological changes and estuarine dynamics, Springer briefs in environmental sciences, vol 8. Springer, New York

Montagna PA, Hu X, Palmer TA, Wetz MS (2018) Effect of hydrological variability on the biogeochemistry of estuaries across a regional climatic gradient. Limnol Oceanogr 63:2465–2478

Ni W, Li M, Ross AC, Najjar RG (2019) Large projected decline in dissolved oxygen in a eutrophic estuary due to climate change. J Geophys Res-Oceans 124:8271–8289

Nielsen-Gammon J (2009) The impact of global warming on Texas. In: Schmandt J, Clarkson J, North GR (eds) The impact of global warming on Texas, 2nd edn. University of Texas Press, Austin

Nixon S, Oczkowski A, Pilson MQ, Fields L, Oviatt C, Hunt C (2015) On the response of pH to inorganic nutrient enrichment in well-mixed coastal marine waters. Estuar Coast 38:232–241

Odum HT, Hoskin CM (1956) Primary production in flowing waters. Limnol Oceanogr 1:102–117

Oh N-H, Raymond PA (2006) Contribution of agricultural liming to riverine bicarbonate export and $CO_2$ sequestration in the Ohio River basin. Global Biogeochem Cy 20:GB3012/3011–GB3012/3017

Olson CM (2017) Will small diversions of freshwater inflow affect water quality?. MS Thesis. Texas A&M University-Corpus Christi

Pinckney JL (2006) System-scale nutrient fluctuations in Galveston Bay, Texas (USA). In: Kromkamp J, Brouwer J, Blanchard G, Forster R, Creach V (eds) Functioning of microphytobenthos in estuaries: proceedings of the colloquium, vol 103. Amsterdam, pp 141-164

Pollack J, Kim H-C, Morgan E, Montagna P (2011) Role of flood disturbance in natural oyster (*Crassostrea virginica*) population maintenance in an estuary in South Texas, USA. Estuar Coast 34:187–197

Potter IC, Chuwen BM, Hoeksema SD, Elliott M (2010) The concept of an estuary: a definition that incorporates systems which can become closed to the ocean and hypersaline. Estuar Coast Shelf S 87:497–500

Powell EN, Klinck JM, Hofmann EE, McManus MA (2003) Influence of water allocation and freshwater inflow on oyster production: a hydrodynamic–oyster population model for Galveston Bay, Texas, USA. Environ Manag 31:0100–0121

Preston SD, Alexander RB, Schwarz GE, Crawford CG (2011) Factors affecting stream nutrient loads: a synthesis of regional SPARROW model results for the continental united States1. JAWRA J Am Water Resour Assoc 47:891–915

Raimonet M, Cloern JE (2017) Estuary–ocean connectivity: fast physics, slow biology. Glob Chang Biol 23:2345–2357

Ranasinghe R et al (2021) Climate change information for regional impact and for risk assessment. In: Masson-Delmotte V et al (eds) Climate change 2021: the physical science basis. Contribution of working group I to the sixth assessment report of the Intergovernmental Panel on Climate Change. Cambridge University Press, Cambridge/New York, pp 1767–1926. https://doi.org/10.1017/9781009157896.014

Regnier P, Arndt S, Goossens N, Volta C, Laruelle GG, Lauerwald R, Hartmann J (2013) Modelling estuarine biogeochemical dynamics: from the local to the global scale. Aquat Geochem 19:591–626

Russell M, Montagna P (2007) Spatial and temporal variability and drivers of net ecosystem metabolism in western Gulf of Mexico estuaries. Estuar Coast 30:137–153

Russell MJ, Montagna PA, Kalke RD (2006) The effect of freshwater inflow on net ecosystem metabolism in Lavaca Bay, Texas. Estuar Coast Shelf S 68:231–244

Sammut J, White I, Melville M (1996) Acidification of an estuarine tributary in eastern Australia due to drainage of acid sulfate soils. Mar Freshw Res 47:669–684

Shih JS, Alexander RB, Smith RA, Boyer EW, Schwarz GE, Chung S (2010) An initial SPARROW model of land use and in-stream controls on total organic carbon in streams of the conterminous United States. United States Geological Survey Open File Rep, Reston, VA. https://doi.org/10.3133/ofr20101276

Smith SV, Hollibaugh JT (1997) Annual cycle and interannual variability of ecosystem metabolism in a temperate climate embayment. Ecol Monogr 67:509–533

Solis RS, Powell GL (1999) Hydrography, mixing characteristics, and residence times of Gulf of Mexico estuaries. In: Bianchi TS, Pennock JR, Twilley RR (eds) Biogeochemistry of Gulf of Mexico estuaries. Wiley, pp 29–61

Stets EG, Kelly VJ, Crawford CG (2014) Long-term trends in alkalinity in large rivers of the conterminous US in relation to acidification, agriculture, and hydrologic modification. Sci Total Environ 488–489:280–289

Tassone SJ, Besterman AF, Buelo CD, Walter JA, Pace ML (2022) Co-occurrence of aquatic heatwaves with atmospheric heatwaves, low dissolved oxygen, and low pH events in estuarine ecosystems. Estuar Coast 45:707–720

TCEQ (2012) Surface water quality monitoring procedures, Volume 1: Physical and chemical monitoring methods. TCEQ, Austin, TX

Thronson A, Quigg A (2008) Fifty-five years of fish kills in coastal Texas. Estuar Coast 31:802–813

Tolan JM (2007) El Niño-southern oscillation impacts translated to the watershed scale: estuarine salinity patterns along the Texas Gulf Coast, 1982 to 2004. Estuar Coast Shelf S 72:247–260

Vukićević T, Braswell BH, Schimel D (2001) A diagnostic study of temperature controls on global terrestrial carbon exchange. Tellus B 53:150–170

Wolf-Gladrow DA, Zeebe RE, Klaas C, Kortzinger A, Dickson AG (2007) Total alkalinity: the explicit conservative expression and its application to biogeochemical processes. Mar Chem 106:287–300

Yao H, Hu X (2017) Responses of carbonate system and $CO_2$ flux to extended drought and intense flooding in a semiarid subtropical estuary. Limnol Oceanogr 62:S112–S130

Yin H, Hu X, Dias LM (2023) Sulfate enrichment in estuaries of the northwestern Gulf of Mexico: The potential effect of sulfide oxidation on carbonate chemistry under a changing climate. Limnol Oceanogr Lett 8:742–750

# Coastal Wetland Habitats in Texas

James C. Gibeaut ⓘ, Jessica Magolan, Pu Huang ⓘ,
and Paul A. Montagna ⓘ

## Abstract

Coastal wetlands rely on freshwater inflow for two reasons: the water itself that is needed by plant foundation species and the sediment that helps to build wetland habitats. There is a climatic gradient along the Texas coast with decreasing rain and inflow from the northeast to the southwest. Along this gradient, the percentage contribution of freshwater requiring habitats, for example, saltwater marsh, decreases. In contrast, freshwater-sensitive habitats, for example, seagrass and tidal flats, increase along the climatic gradient with drier conditions.

## Keywords

Estuary · Freshwater inflow · Freshwater marsh · Mangrove · Oyster reef · Saltwater marsh · Seagrass · Swamp · Tidal flat

## Abbreviations

| | |
|---|---|
| ac-ft | acre-feet (1233.48 m³) |
| ac-ft/y | acre-feet per year |
| BEG | Bureau of Economic Geology |
| ENSO | El Niño Southern Oscillation |
| GLO | Texas General Land Office |
| NOAA | National Oceanic and Atmospheric Administration |
| NWI | National Wetlands Inventory |
| USFWS | US Fish and Wildlife Service |

J. C. Gibeaut (✉) · J. Magolan · P. Huang · P. A. Montagna
Texas A&M University-Corpus Christi, Corpus Christi, TX, USA
e-mail: James.Gibeaut@tamucc.edu; Jessica.Magolan@tamucc.edu; Pu.Huang@tamucc.edu; Paul.Montagna@taumcc.edu

## Introduction

The Texas coast is a flat, coastal plain, with eight coastal river basins (Wurbs and Zhang 2014). The river basins culminate in wetland habitats, such as marshes and deltas, that depend on freshwater inflow for maintenance. Thus, tidal wetland habitats are transition zones between the land and the sea. The climatic gradient across the Texas coast, which ranges from the wetter northeast to drier southwest, has a large influence on the magnitude, extent, and persistence of the wetland habitats (Longley 1995; Montagna et al. 2007).

The freshwater inflow climatic gradient is regional and spans the entire northwestern Gulf of Mexico (Osland et al. 2014). The global aridity index ranges from about 1.2 in Mississippi and Alabama and decreases to about 1.0 westward near the Texas border with Louisiana and further decreases to about 0.4 southwestward toward the Texas border with Mexico. Freshwater availability drives community dynamics of the wetland foundation species. There are nonlinear relationships between freshwater inflow and relative abundance of wetland plant species. Small changes in precipitation in the drier areas leads to larger, landscape-level, changes in wetland plant species across the northern Gulf of Mexico.

There are at least two major hydrological threats to wetlands: reduced inflow as water diversions or water uses increase and climate change that can result in less rainfall or increased evaporation. For example, water resource development led to a 45% decrease in freshwater flow into the Nueces Delta marsh from 1983 to 2000, and this led to marsh degradation (Alexander and Dunton 2002). Because of reduced inflows, marsh plant coverage can only increase in the Nueces Delta Marsh during wet climatic periods (Montagna et al. 2017). These hydrological pressures on marsh habitats have led to a call for engineering solutions to reduce water overuse and increase hydrological restoration (Middleton and Montagna 2018).

© The Author(s) 2025

P. A. Montagna, A. R. Douglas (eds.), *Freshwater Inflows to Texas Bays and Estuaries*, Estuaries of the World,
https://doi.org/10.1007/978-3-031-70882-4_9

Two previous studies have examined the relationship between freshwater inflow and wetland habitats in Texas. Generally, fresh marshes, salt and brackish marshes, and swamps increase in area with increasing freshwater inflow volumes (Longley 1995). In contrast, seagrass and wind-tidal flat habits decrease with freshwater inflow volumes. Areal extent of oyster reefs also increases as freshwater inflow increases among estuaries (Montagna et al. 2007). Habitat change over time was estimated for three periods from the 1950s to 2000s (Montagna et al. 2007). There was a severe drought in the 1950s, and it was wetter in the latter period. Estuarine marsh doubled in area over the period as predicted. Tidal flats and seagrass habitat decreased by half as predicted. The purpose of the current study is to reexamine trends in habitat change over space and time.

## Data and Methods

The same approach used by Longley (1995) to characterize freshwater inflow and estuarine habitat distribution in Texas estuaries is used here but with updated inflow, habitat, topographic, and bathymetric data and processing techniques. Furthermore, groupings of HUC10 watersheds, as mapped by the US Geological Survey, were used to define the estuarine extents and bordering areas for habitat analysis. Eight primary estuarine areas were defined (Fig. 1).

An estuary's freshwater inflow regime is expressed as the annual volume of freshwater flowing into an estuary (ac-ft/y) divided by the estuary volume (ac-ft). This value is the number of times during a year freshwater is replaced in an estuary. The Texas Water Development Board (TWDB) recently computed annual inflow volumes for years 1977 to 2020 for Upper and Lower Laguana Madre and 1941 to 2020 for the other six major estuaries (dataset version TWDB202101). Total inflow is the sum of gauged, modeled, and return flow minus diversions for human use. This study assesses estuarine inflow for the period from 1977 to 2020. Using this later period allows treating Upper and Lower Laguna Madre as separate features and also corresponds better with available habitat data.

The volume of each estuary was calculated using a merged bathymetric and topographic (topo/bathy) model, which was developed by modifying a computational mesh for the hydrodynamic model, ADCIRC. The mesh was originally constructed for the Coastal Texas Flood Insurance Study conducted by the US Army Corps of Engineers and the Federal Emergency Management Agency (USACE 2011). The mesh, designated as TX2008R35H, consists of unstructured triangular elements with increasing resolution from 100 s of meters offshore to 20 m in the estuaries. The original

bathymetric data came from a variety of sources including NOAA and the USACE. For this study, the existing bathymetry was not altered, but the topographic data along the Texas coast was updated using a 2-m resolution lidar Digital Elevation Model (DEM) of the Texas coastal plain (Su et al. 2021a; Su et al. 2021b). The updated topographic elevation nodes and the bathymetry nodes of the mesh were gridded as a seamless 115-m DEM relative to the NAVD88 vertical datum. The area of each estuary was clipped from the DEM and vertically shifted by the average offset between Mean Sea Level (MSL) and NAVD88 at tide gauges to yield elevations relative to local MSL. MSL contours of the DEM defined the water surface of each estuary below which the volume was calculated.

Geospatial habitat data was obtained from the US Fish and Wildlife Service, National Wetlands Inventory (NWI), which is available at https://www.fws.gov/program/national-wetlands-inventory/wetlands-data (USFWS 2017). Data covering each of the estuarine watersheds (Fig. 1) was extracted from the Texas and Louisiana state-wide data sets. The NWI uses the Cowardin System to classify wetlands and deepwater habitats (Cowardin et al. 1979). The Cowardin System is a hierarchical system that describes the aquatic system, type of substrate, and water regime of wetlands and became a National Standard in 1996 (FGDC-STD-004). The maps are derived from manual digitization and interpretation of high-resolution aerial imagery of the National Agriculture Imagery Program plus integration with other data such as topography and field verification. Habitat classes for analysis were created from the NWI data by integrating across the Cowardin System hierarchy to obtain seven tidally influenced habitats of interest: saltwater marsh, mangrove, tidal freshwater marsh, tidal swamp, oyster, seagrass, and tidal flats (Table 1). Oyster reefs are not well represented in the NWI data; therefore, a layer, which is a compilation of oyster reef data, was used (Anson et al. 2011).

NWI and oyster reef data cover the estuarine watershed study areas with data spanning from the 1990s to approximately 2010, which aligns well for comparing with average freshwater inflow data from 1977 to 2020. To gain further insight into habitat dynamics, additional map data was obtained from the University of Texas, Bureau of Economic Geology (BEG) (BEG 2021). The BEG has conducted a coastal mapping program of select areas using the same Cowardin System and techniques as the NWI. Figure 2 shows where the BEG mapped habitats in the Corpus Christi, Matagorda, and Sabine Lake estuarine study areas for three time periods: 1950s, 1979/1983, and 2001/02/04/05. The BEG data was summarized into the same seven classes as was done for the NWI data and examined for changes from the deep drought period of the 1950s to more recent periods.

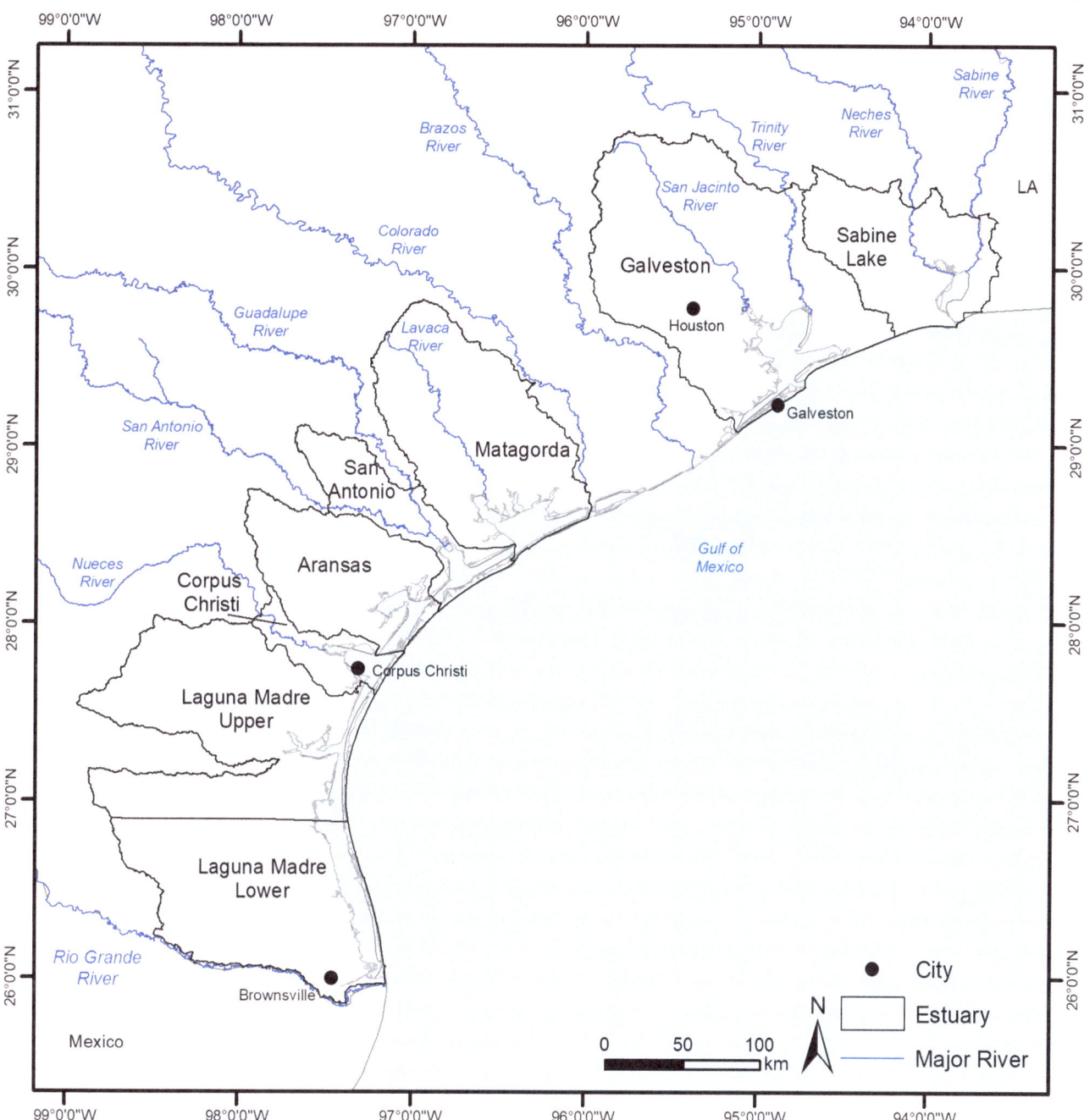

**Fig. 1** Estuarine watershed boundaries along the Texas coast

**Table 1** Crosswalk of NWI classification codes to those habitats used in this study

| Habitat (this study) | USFWS NWI | BEG NWI | | |
| --- | --- | --- | --- | --- |
| | | 1950s | 1970s | 2000s |
| Saltwater marsh | E2EM1/SS1P, E2EM1/SS3N, E2EM1/SS3P, E2EM1/USN, E2EM1/USP, E2EM1N, E2EM1N6, E2EM1Nh, E2EM1Ns, E2EM1Nx, E2EM1P, E2EM1P5, E2EM1P6, E2EM1Pd, E2EM1Ph, E2EM1Ps, E2EM1Px, E2EM5P, E2EM5Ps | E2EM, E2EMFL, E2EMP, E2EMPU | E2EM, E2EM1F, E2EM1N, E2EM1N/ E2FLN, E2EM1NX, E2EM1P, E2EM1P/E2FLP, E2EM1PH, E2EM1PH/ E2OWPH, E2EM1PX | E2EM1N, E2EM1Nd, E2EM1Nh, E2EM1Ns, E2EM1Nx, E2EM1P, E2EM1Pd, E2EM1Ph, E2EM1Ps |
| Mangrove | E2FO3P6, E2SS3/EM1P, E2SS3N, E2SS3Ns, E2SS3P, E2SS3P6, E2SS3Ps | E2SS | E2SS, E2SS1P | E2SS, E2SS3 |
| Freshwater marsh | PEM1/SS1R, PEM1/SS1T, PEM1R, PEM1Rh, PEM1Rx, PEM1S, PEM1Sh, PEM1T, PEM1Ts, PEM1Tx | | PEM1R, PEM1S, PEM1SH, PEM1T, PEM1TH, PEM1Y | PEM1R, PEM1S, PEM1T, PEM1Th, PEM1Tx, PEM1V |
| Swamp | PFO1/2R, PFO1/2 T, PFO1/EM1R, PFO1/EM1T, PFO1/SS1R, PFO1/SS1T, PFO1R, PFO1Rs, PFO1S, PFO1Ss, PFO1Sx, PFO1T, PFO2R, PFO2T, PSS1/2R, PSS1/2 T, PSS1/3R, PSS1R, PSS1Rh, PSS1Rs, PSS1S, PSS1Ss, PSS1T, PSS3R, PSS3S, PUSR | | PFO1R, PFO1S, PFO1T, PFO4S, PFO6R, PSS1R, PUSR | PFO1R, PFO1S, PFO1T, PFO2R, PSS1R, PSS1S |
| Oyster | E1RF2L, E2RF2M, E2RF2Nr | E1RF2M, E2RF | E2RF2M, E2RF2M | E1RF2L, E2RF2M |
| Seagrass | E1AB3L, E1AB3L5, E1AB3L6, E1AB3Lx | E1AB, E1ABFL, E1ABOW | E1AB, E1AB2L | E1AB, E1AB1, E1AB3, E1AB3x, E1AB4, E1AB5, E1AB5x, E1AB6x |
| Tidal flat | E2AB1N, E2AB3M, E2AB3Ms, E2AB3Mx, E2AB4M, E2ABM, E2ABMs, E2ABN, E2ABNh, E2ABNs | E2BB, E2FL | E2FL, E2FL6N, E2FL6Y, E2FLM, E2FLMH, E2FLN, E2FLP, E2FLPH, E2FLUH | E2AB1P, E2AB1Ps |

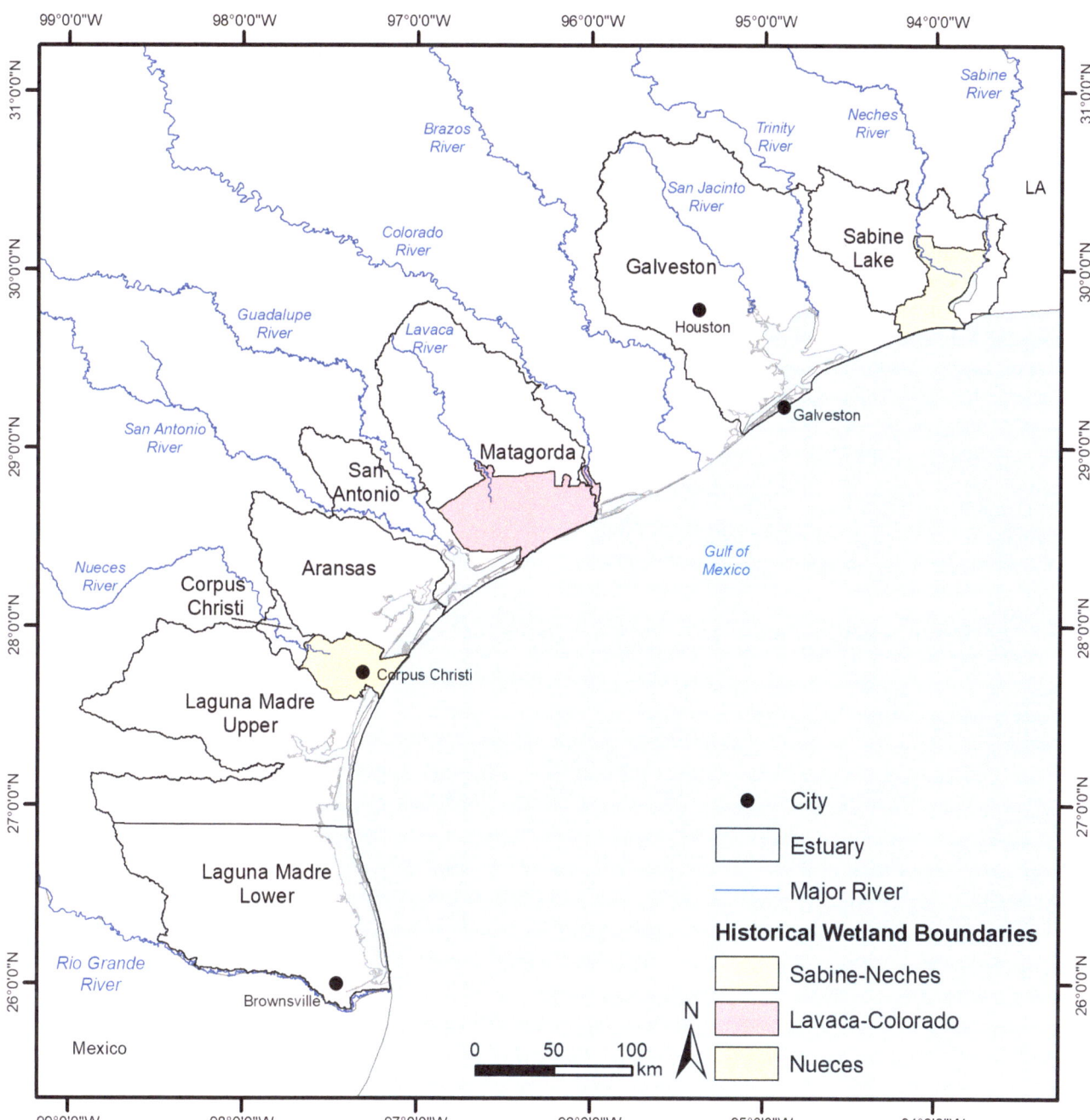

**Fig. 2** NWI study areas compared to BEG study areas

## Distribution and Dynamics of Principal Coastal Habitats

The extracted and integrated NWI data quantifies coverage for seven tidally influenced habitat types occurring within eight estuarine study areas (Table 2, Fig. 3). Not included is subtidal bay bottom habitat not covered by seagrass or oyster reefs. The estuaries in Table 2 are listed from northeast to southwest and align with the climatic gradient from wet to dry. The climatic gradient along the Texas coast is correlated to a salinity gradient as well (Montagna et al. 2013). The "bookends," Sabine Lake with the lowest average salinities and Lower Laguna Madre with high average salinities, have the largest total tidally influenced habitat areas. But they are different in that Sabine Lake is primarily saltwater marsh habitat and Lower Laguna Madre is primarily seagrass and tidal flat habitat (Fig. 3). These differences exhibit the role of freshwater inflow and influences of salinity.

Variation in total amount of tidally influenced habitat is largely dependent on geomorphology with the "bookends" accommodating more tidal habitat but in different ways. The Sabine Lake area includes a broad, low, and flat chenier plain, which accommodates salt and brackish marsh, and the lower Laguna Madre area is shallow and clear and receives wash over sediments, which promotes large areas of flats and seagrass. Estuaries from Galveston Bay to Corpus Christi Bay are developed from the erosion of incised river valleys and have steep, elevated margins restricting the reach of tidal influence.

Figure 4 shows more detail of how the dominant habitats shift along the coast. The percentage of saltwater marsh within estuaries decreases from northeast to southwest along the climatic gradient. In contrast, the percentage of seagrass and tidal flats increases along the gradient. Freshwater marshes and wooded swamps are limited to the upper estuaries, Sabine Lake, Galveston Bay System, and the Matagorda Bay System. Notably, oyster habitat is absent in the northeastern most estuary, Sabine Lake, and in the most southwestern areas of upper and lower Laguna Madre but prevalent in the estuaries along the central coast.

**Table 2** Distribution and area (km²) of habitat types in Texas estuaries from the NWI database

| Habitat | Sabine Lake | Galveston | Matagorda | San Antonio | Aransas | Corpus Christi | Laguna Madre—Upper | Laguna Madre—Lower |
|---|---|---|---|---|---|---|---|---|
| Saltwater marsh | 648.6 | 321.8 | 137.4 | 99.8 | 104.7 | 33.6 | 12.7 | 66.0 |
| Mangrove | 3.3 | 0.3 | 0.5 | 1.3 | 6.8 | 0.0 | 0.0 | 3.2 |
| Oyster | 0.0 | 61.8 | 97.4 | 14.5 | 31.7 | 7.2 | 0.0 | 0.0 |
| Seagrass | 10.9 | 2.5 | 9.4 | 39.5 | 83.8 | 42.3 | 223.2 | 347.6 |
| Tidal flat | 7.0 | 0.1 | 2.2 | 4.0 | 2.6 | 1.0 | 70.0 | 160.2 |
| Freshwater marsh | 23.3 | 19.8 | 16.2 | 0.1 | 0.7 | 0.0 | 1.4 | 0.0 |
| Swamp | 4.3 | 23.9 | 1.3 | 0.0 | 0.0 | 0.4 | 0.0 | 0.4 |
| Total critical habitats | 697.4 | 430.2 | 264.4 | 159.2 | 230.3 | 84.5 | 307.3 | 577.4 |

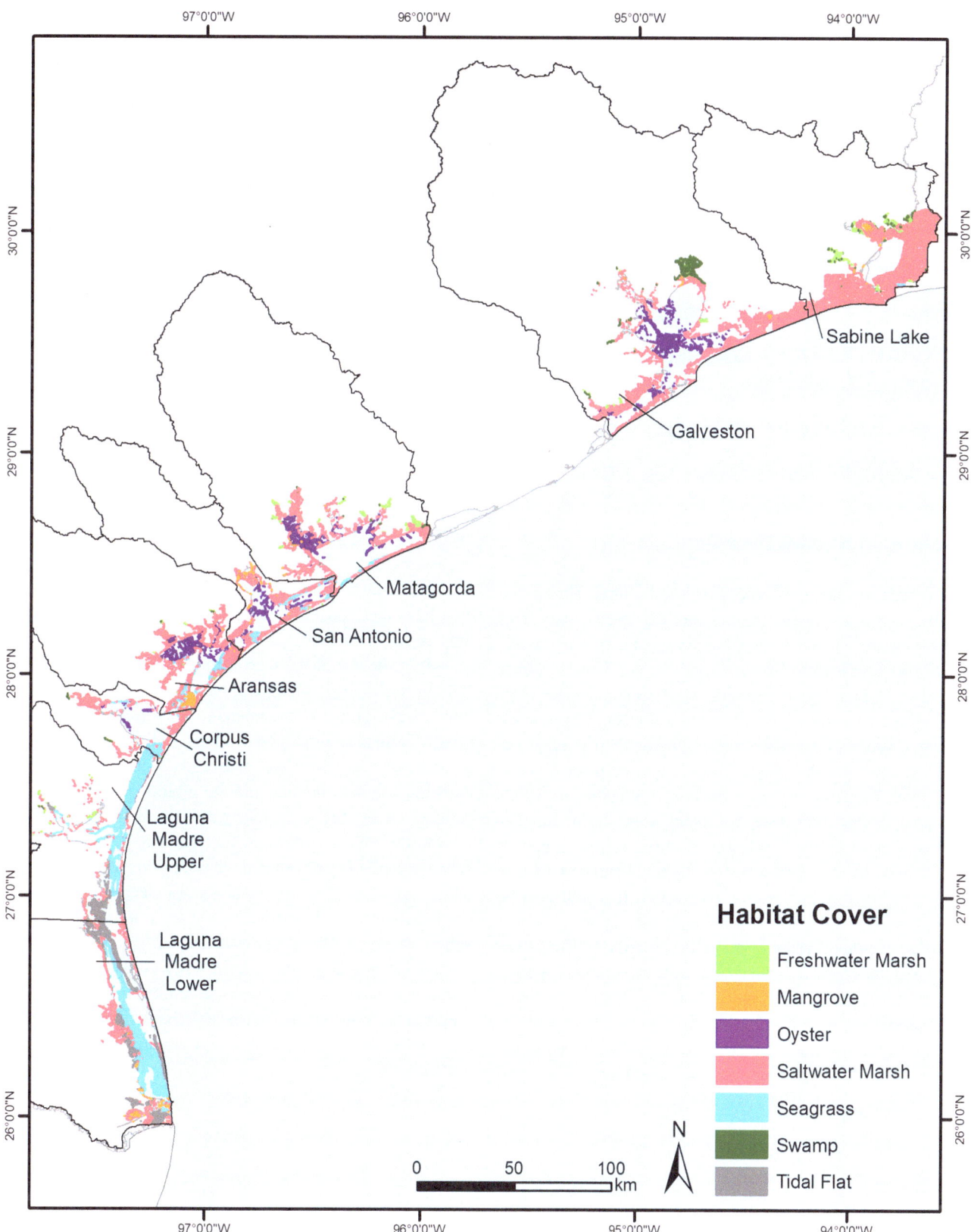

**Fig. 3** Tidally influenced wetlands in estuarine study areas

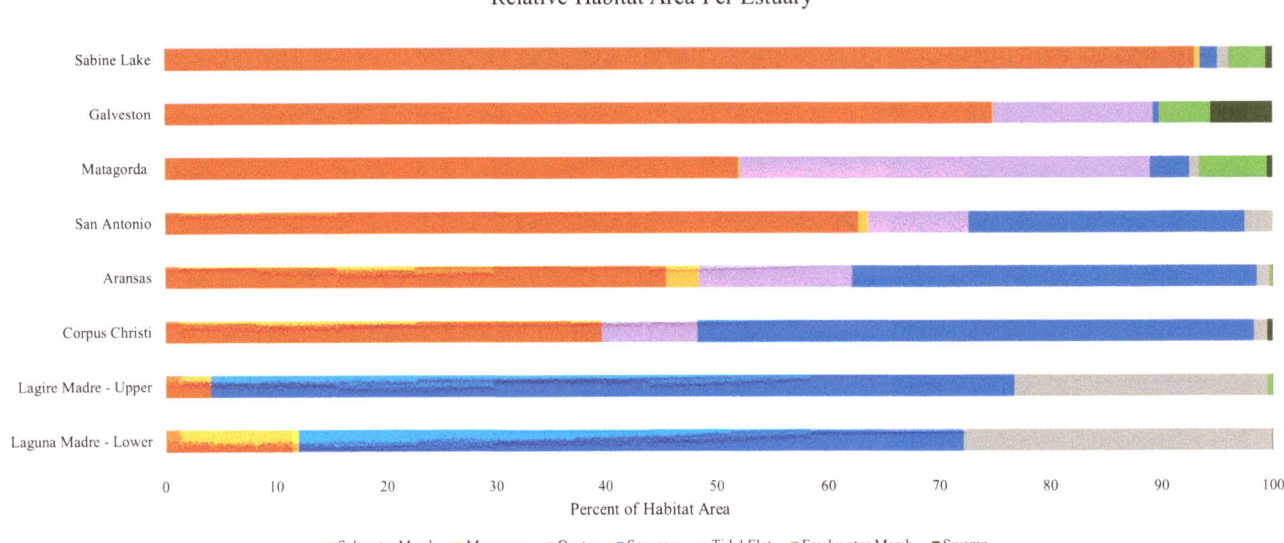

**Fig. 4** Relative proportion (%) of tidally influenced habitat areas within Texas estuaries

## Changes in Habitat Over Time

There was an extensive and extreme drought during the 1950s, so it is interesting to compare that period with later periods because it is expected that those habitats that are sensitive to freshwater inputs would respond with either increases or decreases in habitat area. However, the BEG historical data does not show expected patterns of change when only considering freshwater inputs (Table 3). It is expected that saltwater marsh habitats would increase from 1956 to 1979 with increasingly wet conditions following the 1950s drought, but that did not happen in either the Sabine Lake or Corpus Christi Estuaries. However, saltwater marsh habitat did increase by 19% in the Matagorda Bay Estuary. Changes from 1979 to the 2000s were very small. It would also be expected that salinity-tolerant habitats, such as seagrass and tidal flats would decrease from the 1950s to the 1970s/2000s, but overall seagrass increased or at least began to rebound in the 2000s (Matagorda Bay and Sabine Lake) while tidal flats decreased.

The historical data time series is imperfect in depicting trends. The 1950s photography was black and white and lower quality than later photography, making some earlier environments more difficult to discern than in later years. Also, short-term conditions during or prior to aerial surveys may have an undesired influence on land cover trends in a time series. However, the consistent mapping techniques, including the use of the Cowardin System (Cowardin et al. 1979) for classifying habitats, helps when comparing data. Importantly, grouping the mapped units into seven broader, tidally influenced categories improves the reliability of comparisons through time. It is clear from the alongshore mapping of environments and the relationship of dominant habitats with the climatic gradient that salinity and freshwater inflow are important drivers of habitat type. When considering decadal-scale temporal trends, however, other factors such as sea level rise or decadal-scale water level variation combined with geomorphology are also key. White et al. (2006) suggested the barrier islands in the Corpus Christi Bay and Aransas Bay study areas experienced decreases in tidal flat concomitant with increases in seagrass as water levels rose across flats from the 1950s to the 2000s, and this process could explain the trends presented here.

**Table 3** Areal change of habitats for portions of three estuaries covered by the BEG database

| Habitat | 1956 | | 1979 | | 2000s | |
|---|---|---|---|---|---|---|
| | km² | acres | km² | acres | km² | acres |
| **Sabine Lake (Sabine-Neches Estuary)** | | | | | | |
| Saltwater marsh | 298.7 | 73,811 | 281.3 | 69,513 | 272.2 | 67,272 |
| Mangrove | | | 0.0 | 11 | 0.6 | 137 |
| Oyster | | | | | | |
| Seagrass | | | | | 1.5 | 374 |
| Tidal flat | 0.7 | 171 | 0.1 | 21 | | |
| Freshwater marsh | | | 33.4 | 8248 | 10.0 | 2464 |
| Swamp | | | 3.3 | 820 | 2.4 | 586 |
| **Matagorda Bay (Lavaca-Colorado Estuary)** | | | | | | |
| Saltwater marsh | 147.8 | 36,519 | 176.2 | 43,537 | 141.1 | 34,864 |
| Mangrove | 0.0 | 12 | 0.0 | 7 | 0.0 | 10 |
| Oyster | 2.5 | 615 | 1.0 | 246 | 0.3 | 73 |
| Seagrass | 16.8 | 4161 | 4.2 | 1036 | 11.1 | 2754 |
| Tidal flat | 56.0 | 13,846 | 18.8 | 4634 | 0.8 | 202 |
| Freshwater marsh | | | 3.3 | 816 | 6.2 | 1535 |
| Swamp | | | | | 0.0 | 9 |
| **Corpus Christi Bay (Nueces Estuary)** | | | | | | |
| Saltwater marsh | 47.0 | 11,624 | 46.3 | 11,429 | 48.6 | 12,022 |
| Mangrove | | | | | 2.0 | 482 |
| Oyster | | | | | 2.3 | 560 |
| Seagrass | 23.1 | 5697 | 48.3 | 11,943 | 50.3 | 12,437 |
| Tidal flat | 80.2 | 19,810 | 37.7 | 9320 | 7.8 | 1931 |
| Freshwater marsh | | | 3.5 | 871 | 0.6 | 152 |
| Swamp | | | 0.1 | 22 | | |

## Future Concerns Regarding Sea Level Rise and Freshwater Inflow Level

The long-term year-to-year variation in hydrology along the Texas coast is controlled by the long-term El Niño Southern Oscillation (ENSO) cycle (Tolan 2007). The ENSO signals are correlated to salinity structure within Texas estuaries within four to six months. During El Niño events, salinities in Texas estuaries decrease because of increased freshwater flows to the coasts. During La Niña periods, salinities increase because of the drier climatic conditions. These cycles occur with a periodicity of 3.55, 5.33, and 10.67 years. The ENSO is dominated by the 3.55- and 5.33-year periods, and the 10.67-year period is defined by the Pacific Decadal Oscillation.

As the freshwater inflow regime increases, four habitats (freshwater marsh, swamp, saltwater marsh, and oyster reefs) increase exponentially (Montagna et al. 2007). However, if bays become too fresh because of very high inflow rates (as in Sabine Lake), then oyster populations plummet. Thus, increasing freshwater can be bad for some organisms. Seagrass and wind tidal flats decrease exponentially with increasing freshwater inflow regime (Montagna et al. 2007).

Sea level rise and the interplay of water level and geomorphology may mask, reinforce, or counter effects caused by freshwater inflow.

## Recommendations for Future Research and Management

Better mapping, particularly in terms of higher resolution maps with 2-m ground pixel size or smaller, is needed to track change more reliably. More rapid and frequent mapping is also important to decipher causes of change over time.

**Acknowledgments** Work on this manuscript was funded in part by the Harte Research Institute for Gulf of Mexico Studies; Contract No. 21-155-007-C879 from the Texas General Land Office (GLO) with Gulf of Mexico Energy Security Act of 2006 funding made available to the State of Texas and awarded under the Texas Coastal Management Program and the National Oceanic and Atmospheric Administration (NOAA), Office of Education Educational Partnership Program award number NA21SEC4810004. Additionally, the views and contents are solely the responsibility of the award recipient and do not necessarily represent the official views of the US Department of Commerce, NOAA, the GLO, or the State of Texas.

# References

Alexander HD, Dunton KH (2002) Freshwater inundation effects on emergent vegetation of a hypersaline salt marsh. Estuaries 25:1424–1435

Anson K, Arnold W, Banks P, Berrigan M, Pollack J, Randall B, Reed (2011) Eastern oyster in Gulf of Mexico data atlas. Stennis Space Center (MS): National Centers for Environmental Information. Available from https://gulfatlas.noaa.gov/

Bureau of Economic Geology (BEG) (2021) Status and trends of Texas wetlands, 1950's to 2000's. Retrieved from https://www.beg.utexas.edu/research/programs/coastal/wetlands/trends

Cowardin LM, Carter V, Golet FC, LaRoe ET (1979) Classification of wetlands and Deepwater habitats of the United States. United States Department of Interior, Fish and Wildlife Service, Washington DC, p 131

Longley WL (1995) Estuaries. In: North GR, Schmandt J, Clarkson J (eds) The impact of global warming on Texas. University of Texas Press, Austin, pp 88–118

Middleton BA, Montagna PA (2018) Turning on the faucet to a healthy coast. Solutions J 9(3):14. https://www.thesolutionsjournal.com/article/turning-faucet-healthy-coast/

Montagna PA, Gibeaut JC, Tunnell JW Jr (2007) South Texas climate 2100: coastal impacts. In: Norwine J, John K (eds) South Texas climate 2100: problems and prospects, impacts and implications. Texas A&M University-Kingsville, Kingsville, TX, pp 57–77

Montagna PA, Palmer TA, Pollack JB (2013) Hydrological changes and estuarine dynamics. SpringerBriefs in Environmental Sciences, New York, p 94

Montagna PA, Sadovski AL, King SA, Nelson KK, Palmer TA, Dunton KH (2017) Modeling the effect of water level on the Nueces Delta marsh community. Wetlands Ecol Man 25:731–742. https://doi.org/10.1007/s11273-017-9547-x

Osland MJ, Enwright N, Stagg CL (2014) Freshwater availability and coastal wetland foundation species: ecological transitions along a rainfall gradient. Ecology 95:2789–2802

Su L, Subedee M, Dotson M, Gibeaut J, Lupher B, Reisinger A, Bezore R (2021a) 2-meter topographic Lidar digital elevation model (DEM) of the lower Texas coast. Distributed by: Gulf of Mexico Research Initiative information and data cooperative (GRIIDC). Harte Research Institute, Texas A&M University–Corpus Christi. https://doi.org/10.7266/Z7WG9EGN

Su L, Subedee M, Dotson M, Gibeaut J, Lupher B, Reisinger A, Bezore R (2021b) 2-meter topographic Lidar digital elevation model (DEM) of the upper Texas coast. Distributed by: Gulf of Mexico Research Initiative information and data cooperative (GRIIDC). Harte Research Institute, Texas A&M University–Corpus Christi. https://doi.org/10.7266/2MYPTJ7Y

Tolan JM (2007) El Niño-southern oscillation impacts translated to the watershed scale: estuarine salinity patterns along the Texas Gulf Coast, 1982 to 2004. Estuar Coast Shelf Sci 72:247–260

U.S. Army Corps of Engineers (USACE) (2011) Flood insurance study, coastal counties, Texas intermediate submission 2: offshore water levels and waves. USACE, Vicksburg, MS, p 150

U.S. Fish and Wildlife Service (USFWS) (2017) National wetlands inventory. Retrieved from http://www.fws.gov/wetlands/

White WA, Tremblay TA, Waldinger RL, Calnan TR (2006) Status and trends of wetland and aquatic habitats on Texas barrier islands, coastal bend. Coastal Coordination Division, Texas General Land Office. Final Report No. NA04NOS4190058

Wurbs R, Zhang Y (2014) River system hydrology in Texas. Technical Report No. 461. Texas Water Resources Institute, College Station, TX. https://hdl.handle.net/1969.1/152426

# Submerged Aquatic Vegetation, Marshes, and Mangroves

Kyle A. Capistrant-Fossa [ID], Berit E. Batterton [ID],
and Kenneth H. Dunton [ID]

## Abstract

The intertidal and subtidal wetlands of vascular vegetation on the Texas coast are among the most diverse flora of any coastal state in the United States. Marsh, mangrove, and seagrass distributional patterns reflect the unique latitudinal gradient in rainfall from the wet Sabine-Neches estuary to the arid Lower Laguna Madre, in which precipitation decreases by over 50% over Texas' 5400 km coastline. The estuarine vegetation changes predictably in response to increasing salinity, from brackish emergent marsh systems in the north, to mixed mangrove-marsh assemblages on the central coast, to hypersaline systems dominated by submerged seagrasses and wind-tidal flats in the south. These foundation species are largely responsible for the enormous secondary productivity of the Texas coastal system as reflected in strong fisheries that include many estuarine dependent species, from oysters to redfish. These vegetated habitats are also critical to the amazing resilience of the Texas coastal zone to storms and other natural disturbances. In this chapter, we describe vegetative spatial distributions in relation to freshwater inflow and the role of nutrients, light, and soils on plant productivity and carbon sequestration. Special emphasis is placed on the value of wetlands as long-term integrators of regional climate, sea level rise, nutrient loading, and salinity.

## Keywords

Marshes · Mangroves · Seagrasses · Wetlands · Emergent vegetation · Estuaries · Salinity · Precipitation

## Abbreviations

| | |
|---|---|
| rSLR | Relative sea level rise |
| TDWR | Texas department of water resources |
| TPWD | Texas parks and wildlife department |
| SAV | Submerged aquatic vegetation |
| USGS | United States geological survey |

## Introduction

Freshwater availability plays a pivotal role in shaping the abundance and distribution of foundational wetland plant species. The Texas coast is especially unique because it has a natural latitudinal precipitation gradient (Nielsen-Gammon et al. 2020), a phenomenon where rainfall decreases more than 50% from the northeast at the Louisiana border (140 cm yr.$^{-1}$) south to the Rio Grande at our border with Mexico (63 cm yr.$^{-1}$). The northern coast receives sufficient fresh water to support highly productive coastal forests and herbaceous salt marshes. Along the southern coast, the climate is hot and dry, leading to elevated levels of evaporation that form hypersaline tidal flats. These systems are often covered by highly productive cyanobacterial mats that can produce a substantial nitrogen and carbon subsidy for the entire ecosystem (Osland et al. 2014; Cuddy and Dunton 2023).

Rainfall and aridity (the ratio of mean annual precipitation to mean annual potential evapotranspiration; Zomer et al. 2006) have previously been identified as the most important climatic predictors of foundation plant species coverage in tidal wetlands. The Laguna Madre and Nueces estuaries have Global Aridity Index values <0.64. The Mission-Aransas, Guadalupe, and Lavaca-Colorado estuaries fall around ~0.75, while the remaining estuaries fall around ~0.9. Only Sabine-Neches is greater than 1 (Osland et al. 2014). The freshwater inflow balance (net water gain or loss) ranges from −677 to 17,073 million m$^3$ yr.$^{-1}$ from

K. A. Capistrant-Fossa · B. E. Batterton · K. H. Dunton (✉)
University of Texas at Austin, Marine Science Institute,
Port Aransas, TX, USA
e-mail: kyle.capistrantfossa@utexas.edu; b.batterton@utexas.edu;
ken.dunton@utexas.edu

P. A. Montagna, A. R. Douglas (eds.), *Freshwater Inflows to Texas Bays and Estuaries*, Estuaries of the World,
https://doi.org/10.1007/978-3-031-70882-4_10

Laguna Madre in the south to Sabine Lake in the north. The relative abundance of vegetated wetlands (i.e., marshes) within estuaries increases from 7% in the Lower Laguna Madre to nearly 100% at Sabine Lake. The mean annual precipitation, aridity index, estuarine freshwater inflow balance, estuarine freshwater replacement time, estuarine salinity, and relative abundance of tidal wetland plants are highly correlated (Osland et al. 2014) along 5400 km of Texas coastline.

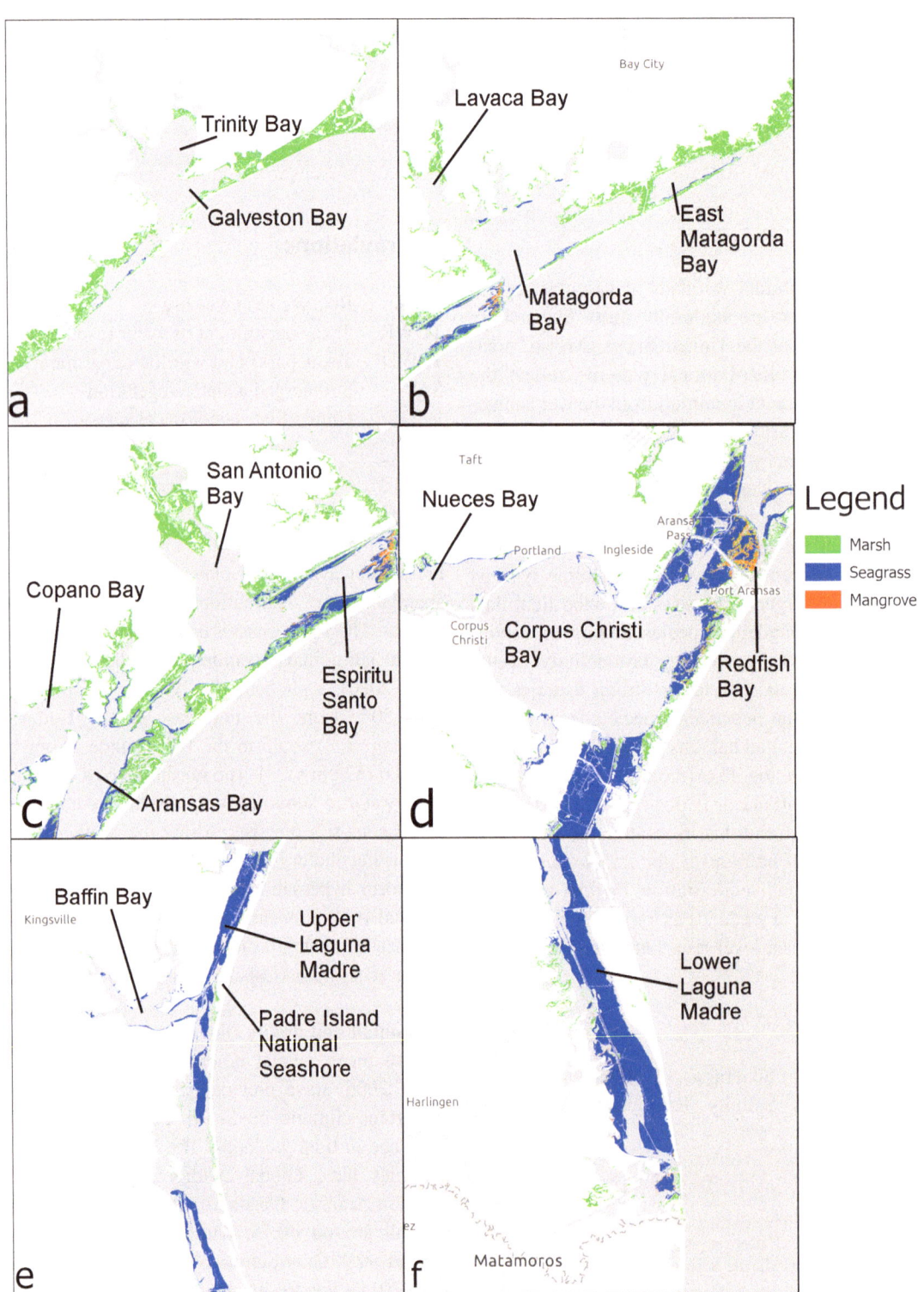

**Fig. 1** The distribution of Texas' estuarine vegetation within the (**a**) Sabine-Neches and Trinity-San Jacinto, (**b**) Lavaca-Colorado, (**c**) Guadalupe and Mission-Aransas, (**d**) Nueces, (**e**) Upper Laguna Madre, and (**f**) Lower Laguna Madre estuaries. Data provided by the Coastal Change Analysis Program (NOAA, 2016)

The distribution, abundance, and relative frequency of wetland vegetation (Fig. 1) reflects the long-term hydrological gradients reflected in precipitation and aridity (threshold for vegetated vs unvegetated wetlands is 765 mm yr.$^{-1}$ and 0.61, respectively). In general, submerged seagrasses and emergent mangrove communities are prevalent on the southern coast with marshes replacing mangroves north of the Nueces estuary at 27° N latitude, which reflect the strong deceasing salinity gradient from south to north.

## Submerged Aquatic Vegetation

### Coastwide Distribution

#### Controls on Community Composition

Submerged aquatic vegetation (SAV) refers to the numerous species of fully aquatic vascular plants that each have different tolerances to environmental conditions (e.g., salinity, light, and nutrients). Consequently, SAV distributions within Texas' estuaries reflect the local environmental conditions that favor or prohibit the growth of different species. Typically, estuaries have a strong salinity gradient resulting from freshwater inputs (e.g., rivers, groundwater, and surface runoff) mixing with saline ocean water. However, some of Texas' estuaries (e.g., Laguna Madre) are largely disconnected from both freshwater and seawater, allowing evaporation to create hypersaline conditions. Furthermore, salinities at a fixed location vary temporally from hourly to decadal time scales because of numerous processes (e.g., tides, storms, weather, climate, El Niño-Southern Oscillation, or Lunar Nodal Cycles). Aquatic plants are especially sensitive to changes in salinity compared to many other organisms because they lack behavioral responses to avoid suboptimal conditions and instead must alter their physiology to match their environment. These plants are either euryhaline (able to survive a wide range of salinities) or stenohaline (have a limited salinity tolerance range). Therefore, the preferred salinities of SAV species and their ability to tolerate changes in salinity are among the most crucial factors in determining estuarine community composition.

Texas has two distinct groups of SAV based on their preferred salinity range: seagrasses and freshwater macrophytes. Seagrasses refer to one of ~65 plant species whose ancestors recolonized the world's oceans ~120 million years ago (Larkum et al. 2018). Texas has five seagrass species that are fairly euryhaline across their distribution (Fig. 2; Table 1). Seagrasses found near the mouth (ocean endmember) of Texas' estuaries are often one of three seagrass species (*Thalassia testudinum*, *Syringodium filiforme*, and *Halophila engelmannii*) because they are sensitive to freshwater inflows. For example, greater freshwater inflows in regions of Biscayne Bay (FL) decreased salinities and limited

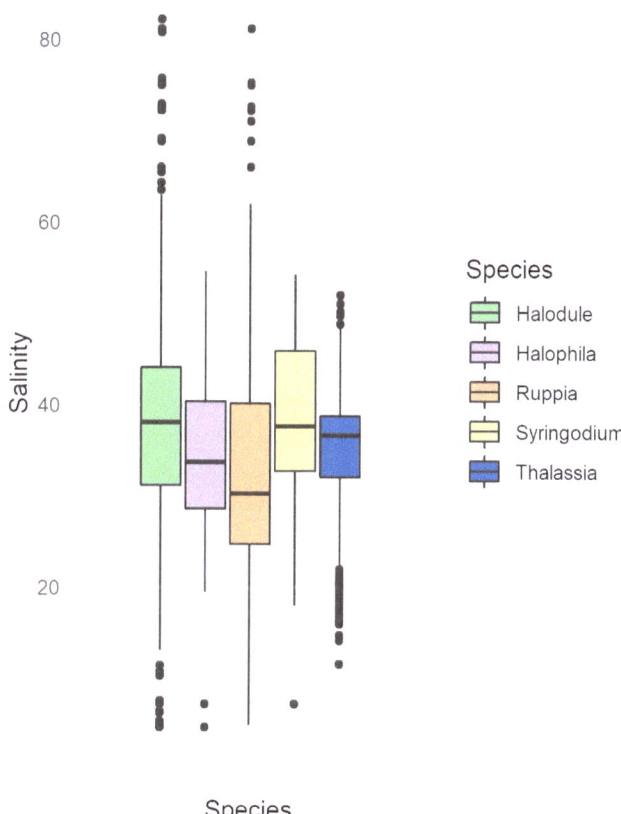

**Fig. 2** Boxplots show the salinity distribution for each species of seagrass during rapid annual surveys along the Texas Coast from 2011 to 2021. The boxes for each species represent the 75th, 50th (median), and 25th quartiles of observations, while the ends of the vertical lines represent the interquartile range (IQR = 75th percentile - 25th percentile). Individual dots represent extreme values. Data are provided by the Texas Seagrass Monitoring Program (Dunton et al. 2018)

*Thalassia testudinum* productivity (Irlandi et al. 2002). Likewise, decreased salinity (< 20) from hurricanes in Florida caused widespread losses of *Syringodium* (Buzzelli et al. 2012). On the other hand, *Ruppia maritima* and *Halodule wrightii* can be located near the head (river endmember) of estuaries because they are highly tolerant to freshwater. *Halodule* in the Mississippi River Delta can live at salinities <5 for 12 months (Biber 2022). Likewise, *Ruppia* is found in a variety of purely freshwater habitats, survives salinities of 70+, and has a nearly global distribution (Kantrud 1991). High salinity also limits the distribution of seagrasses during periods of decreased freshwater inflow (i.e., droughts). For example, drought-induced high salinities (50–70) caused a large-scale die-off of *Syringodium* from Laguna Madre in 2012–2013 (Wilson and Dunton 2018).

The heads of Texas' estuaries often contain halotolerant freshwater macrophytes because they only tolerate moderate salinities (<15–20; Table 1). Therefore, freshwater availability restricts these organisms to hyposaline regions of estuaries. Common species include *Heteranthera dubia*, *Hydrilla*

**Table 1** Extreme salinity limits and optimum ranges for selected Texas estuarine-dependent plants

| Species | Minimum salinity | Maximum salinity | Optimal salinity | References |
|---|---|---|---|---|
| *Thalassia testudinum* | ~5 | 50–60 | 30–40 | Tomasko and Hall (1999); Chollett et al. (2007); Lirman and Cropper (2003); McMillan (1974); Kahn and Durako (2006); McMillan and Moseley (1967) |
| *Syringodium filiforme* | 10 | 40–50 | | Wilson and Dunton (2018); McMahan (1968); McMillan (1974) |
| *Halophila engelmannii* | 5–10 | 35–75 | 25 | Dawes et al. (1987); McMillan (1976); McMillan and Moseley (1967) |
| *Halodule wrightii* | 5–25 | 50–80 | 25–45 | Doering et al. (2002); McMahan (1968); McMillan (1974); Biber (2022); Pulich (1985) |
| *Ruppia maritima* | 0–20 | 30–200 | 10–30 | Kantrud (1991) |
| *Hydrilla verticillata* | 0 | ~20 | ~7 | Steward and Van (1987) |
| *Myriophyllum spicatum* | 0 | ~15 | | Haller et al. (1974) |
| *Heteranthera dubia* | 0 | 10 | 5 | Shields and Moore (2016) |
| *Stuckenia pectinata* | 0 | 22 | 5–10 | Kantrud (1990) |
| *Vallisneria americana* | 0 | 15 | 0 | Doering et al. (2002) |
| *Najas guadalupensis* | 0 | 10 | ~0 | Haller et al. (1974) |
| *Spartina alterniflora* | 0 | >50 | ~25 | Subudhi and Baisakh (2011); Hester et al. (1998); Tang et al. (2014); Stachelek and Dunton (2013); El-Haddad and Noaman (2001) |
| *Salicornia virginica* | 0 | >100 | ~35 | He et al. (2017); Callaway and Zedler (1997) |
| *Salicornia bigelovii* | 0 | >100 | ~35 | Flowers et al. (1977); Rivers and Weber (1971) |
| *Avicennia germinans* | 0 | 50 | ~25 | Suárez and Medina (2005); Madrid et al. (2014) |
| *Scirpus maritimus* | 0 | 25 | ~15 | Lillebø et al. (2003); Hroudová et al. (2014) |
| *Suaeda linearis* | 0 | 40 | ~15 | Flowers et al. (1977); Alhdad and Flowers (2021); Yeo and Flowers (1980) |
| *Distichlis spicata* | 0 | ~50 | ~15 | Flowers et al. (1977); Aschenbach (2006) |
| *Juncus roemerianus* | 0 | 30–40 | ~10 | El-Haddad and Noaman (2001) |
| *Phragmites australis* | 0 | 15–35 | ~10 | Vasquez et al. (2006); Chambers et al. (1999); Tang et al. (2014) |
| *Batis maritima* | | ~60 | ~20 | El-Haddad and Noaman (2001); Debez et al. (2010); Pennings and Richards (1998) |
| *Spartina patens* | 0 | 50 | ~10 | Hester et al. (2001); Merino et al. (2010) |
| *Spartina spartinae* | 0 | 18 | ~2 | Scifres et al. (1980); Riche (2020) |
| *Spartina cynosuroides* | | 12 | ~5 | Constantin et al. (2019); Odum et al. (1984); Failon et al. (2020) |

*verticillata, Myriophyllum spicatum, Najas guadalupensis, Stuckenia pectinata,* and *Vallisneria americana* (Fig. 3). These species have differing salinity tolerances, which help determine their community structure. *S. pectinata* and *V. americana* had no change in biomass when grown at salinities 0–10, but *H. dubia* biomass significantly decreased at 10 salinity (Shields and Moore 2016). However, it is important to consider ecotypes (regional variants) because they may allow for more stress-tolerant varieties. The lethal salinity for *Vallisneria* varied for different ecotypes between 0.5 and 9 (Tootoonchi et al. 2020). Likewise, two variants of *Hydrilla verticillata* could survive maximum salinities of 19.5 or only 14.5, respectively (Steward and Van 1987).

Beyond salinity, numerous other biotic and abiotic factors shape SAV community composition and distribution. Therefore, the presence or absence of species may provide evidence of the environmental conditions that formed the community. *Myriophyllum* and *Ruppia* form mixed stands in Louisiana, but *Ruppia* dominates under high light/high salinity conditions, whereas *Myriophyllum* dominates during low

salinity (Hillmann and La Peyre 2019). Therefore, the presence of *Myriophyllum* instead of *Ruppia* can indicate periods of high freshwater inflows. Additionally, *Myriophyllum* often encroaches into *Vallisneria* meadows. However, diebacks of *M. spicatum* can indicate degradation of light conditions because of its higher light requirements than *V. americana* (Titus and Adams 1979). Seagrass succession experiments by Williams (1990) and others have shown a typical pattern of early colonization by *Halodule*, replacement by *Syringodium*, and finally climax succession by *Thalassia*. Therefore, the presence of a *Thalassia* meadow indicates stabler environmental conditions than a *Halodule* meadow. However, these climax communities are more susceptible to long-term damage from sporadic extreme weather, such as the huge freshwater input during hurricanes (Congdon et al. 2019).

## Distribution

SAV is a common sight within Texas' estuaries, yet the community composition varies significantly among systems due

**Fig. 3** Freshwater macrophytes of Texas: (**a**) *Heteranthera dubia*, (**b**) *Hydrilla verticillata*, (**c**) *Myriophyllum spicatum*, (**d**) *Najas guadalupensis*, (**e**) *Stuckenia pectinata*, and (**f**) *Vallisneria americana*. All images obtained under a creative commons license

**Table 2** Spatial distribution (km²) of aquatic vegetation among Texas' major estuaries

| Estuary | Seagrass extent | Marsh extent | Mangrove extent |
|---|---|---|---|
| Sabine-Neches | 0[a] | 253.96 | 0 |
| Trinity-San-Jacinto | 13.87 | 664.51 | 0 |
| Lavaca-Colorado | 22.01 | 521.66 | 0.004 |
| Guadalupe | 72.14 | 189.64 | 17.56 |
| Mission Aransas | 111.2 | 208.47 | 8.17 |
| Nueces | 75.54 | 54.37 | 1.57 |
| Laguna Madre | 744.2 | 67.32 | 2.83 |
| Total | 1039.8 | 1959.32 | 30.13 |

[a]*Ruppia* is present within this system (e.g., Griffith and Bechler, 1995), but was not estimated for this dataset because of the relatively low coverage

to their distinct environmental characteristics and differences in freshwater inflows. In total, there is an estimated 1040 km² of seagrass habitat (Table 2; Fig. 1), but the total spatial distribution of freshwater macrophytes is still unknown due to limited data. Seagrasses are the second most expansive habitat type in Texas (Marsh > Seagrass > Mangrove) and generally follows a north-south trend coinciding with the precipitation gradient. The northernmost estuary, the Sabine-Neches, has the lowest seagrass coverage, while the greatest is within the Laguna Madre. Likewise, the distribution of Texas' 11 SAV species can vary among estuaries because of their individual environmental tolerances and biotic interactions. *Thalassia testudinum,* a late successional species, dominates the Redfish Bay region of the Mission-Aransas/Nueces estuaries and south of the Arroyo Colorado River in Lower Laguna Madre (Fig. 4; Table 3). *Halodule wrightii,* a pioneer species, is present in most estuaries (except the Sabine-Neches) but is outcompeted by *Thalassia.* *Syringodium filiforme,* an intermediate succession species, is often intermixed within the *Halodule* beds of northern Laguna Madre and *Thalassia* beds of Redfish Bay (Fig. 4; Table 3). *Halophila engelmannii* forms an understory canopy within seagrass meadows making it difficult to map but is present in all estuaries except the Sabine-Neches. *Ruppia maritima* is found in all estuarine systems (although limited in Laguna Madre) but is often confined to isolated dense pockets near freshwater, or it can intermix with *Halodule* depending on the salinity.

Many of the freshwater macrophyte communities found in Texas' estuaries are ephemeral because of their restricted salinity tolerance. However, populations from deeper in the watershed may spread down during favorable conditions (e.g., high freshwater inflows). The distributions of these species are relatively unknown because they have rapidly changing biomass, cover a broad geographic range, and are difficult to study. *Vallisneria* is common only within the upper reaches of the Trinity-San Jacinto system (Fig. 5; Table 3; Adair et al. 1994). *Najas* is documented in all estua-

rine systems except for the Nueces and Laguna Madre. Similarly, *Stuckenia* is found everywhere except the Mission-Aransas and Trinity-San Jacinto estuaries. *Myriophyllum* and *Hydrilla* are both reported north of the Guadalupe estuary. Conversely, *Heteranthera* is limited to the south of the Lavaca-Colorado estuary.

## Restoration and Conservation Efforts

Efforts to support healthy, productive ecosystems vary significantly between seagrasses and freshwater SAV. Seagrass meadows provide numerous ecosystem services including fisheries habitat, nutrient remediation, sediment stabilization, storm attenuation, and carbon sequestration, but they have declined at a rate of ~7% per year because of a combination of anthropogenic and environmental factors (Waycott et al. 2009). Common drivers of loss include heat waves, eutrophication, algal blooms, climate change, sea level rise, and habitat destruction (Orth et al. 2006; Capistrant-Fossa and Dunton 2024). Seagrass decline is most pronounced in the Trinity-San Jacinto and Lower Laguna Madre estuaries where coverage dropped, respectively, from 5.9 and 590 km² in the 1950s/1960s to 2.1 and 460 by 1998 (Pulich and Onuf, 2007). In response to statewide concern, Texas Parks and Wildlife (1999) coordinated creation of the "Seagrass Conservation Plan for Texas" as a framework for seagrass conservation, regulation, and monitoring. This led to the adoption of the Texas Seagrass Monitoring Plan to provide routine meadow assessments and investigate drivers of habitat change (Dunton et al. 2011). The current coverage of seagrass distributions across Texas' estuaries requires frequent reassessments (every 3 years) to provide accurate distributional change data. Each hectare of seagrass supplies ecosystem services valued at least $19,000 (Costanza et al. 1997) providing a large economic incentive for keeping seagrass meadows thriving.

Seagrass restoration is an attractive option for seagrass conservation, but long-term success relies on remediation of the original drivers of plant loss (van Katwijk et al. 2016). Furthermore, only ~37% of seagrass restoration projects across the globe have been successful because the planting areas are typically undersized (van Katwijk et al. 2016). In some cases, the seagrass species lost is replaced with another species, which may not be ecologically equivalent. Large-scale projects are expensive and difficult to implement, but smaller-scale projects have been successful in Texas. Programs in the Laguna Madre, Trinity-San Jacinto, and Mission-Aransas estuaries have been successful in restoring the carbon sequestration within seagrass meadows (Thorhaug et al. 2017). Success of these projects requires planting in suitable environmental conditions (e.g., adequate freshwater inflow, light, and nutrient availability). However, no legislation in Texas exists to protect water quality with reference to seagrass health.

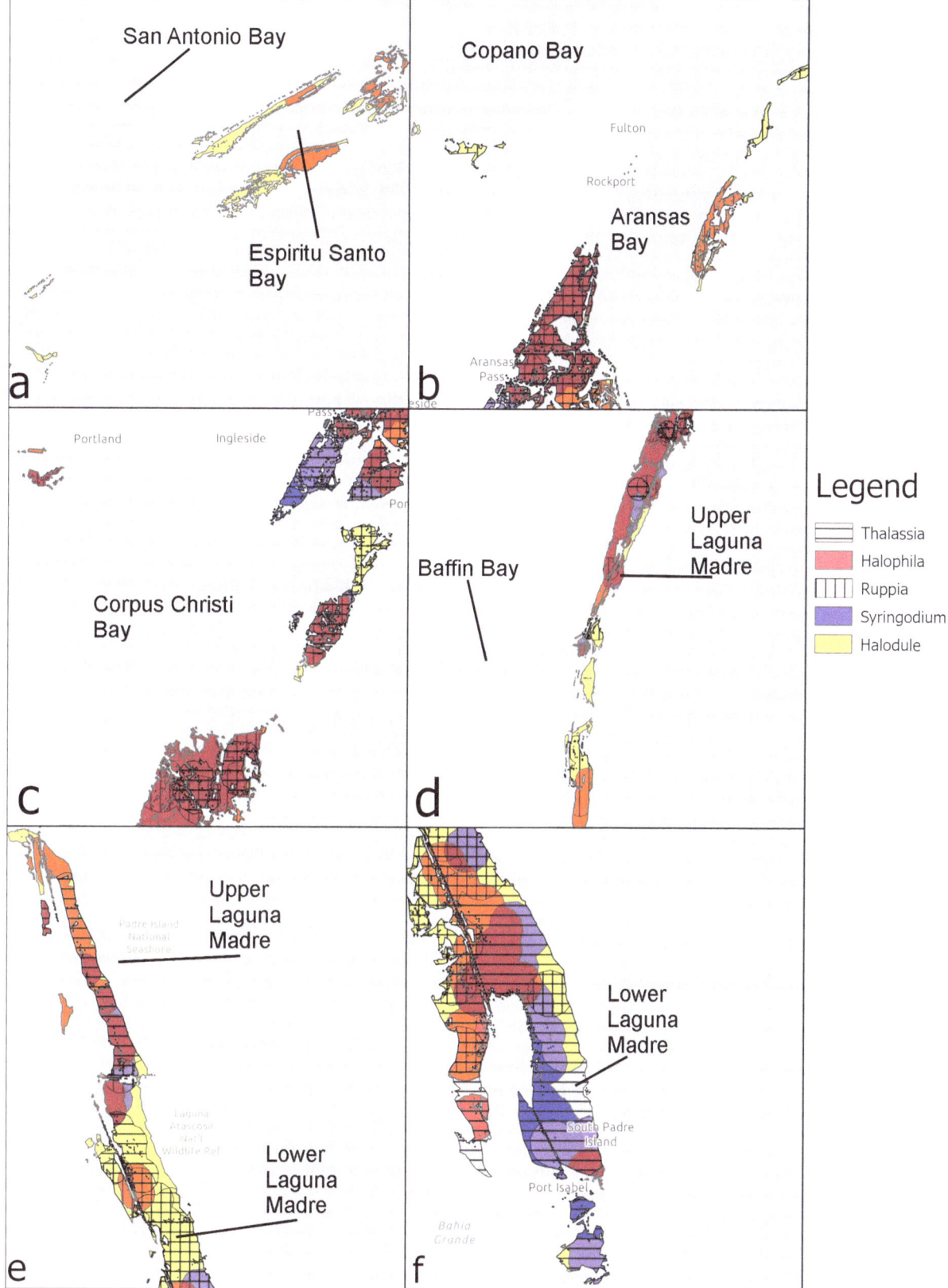

**Fig. 4** Known habitats for each of Texas' seagrass species from 2011 to 2018 within the (**a**) Guadalupe, (**b**) Mission-Aransas, (**c**) Nueces, (**d**) upper Laguna Madre, (**e**) upper half of Lower Laguna Madre, and (**f**) lower half of Lower Laguna Madre estuaries. Data provided by the Texas Seagrass Monitoring Program (Dunton et al. 2018)

Texas state legislation serves to protect seagrass meadows by limiting direct human removal of seagrasses. For example, in 2000, the Redfish Bay State Scientific Area was created in the Mission-Aransas estuary to allow for the "education, scientific research, and preservation of flora and fauna of scientific or educational value" (Texas Administrative Code §57.921). Furthermore, in 2013, Texas made it a Class C misdemeanor to damage seagrass meadows using a boat propeller because of issues with "prop scarring" (Texas Parks and Wildlife Code §66.024). Before this law passed, over a third of the 57 km², of seagrass meadows in the Nueces Estuary surveyed by Dunton and Schonberg (2002) had significant prop damage. *Halodule* meadows take around 3 years to recover from prop damage (Martin et al. 2008) but *Thalassia* can take up to 7.5 years (Dawes et al. 1997). The combination of regulations, conservation, and restoration has led to a substantial increase (876–1040 km²) in statewide seagrass coverage since the mid-1990s (Pulich and Onuf 2007). For example, seagrasses in Galveston Bay significantly declined between the 1950s and 1970s due to human impacts, but coverage has since increased over tenfold, indicating a reduction in anthropogenic pressure (Pulich 2007). However, further analyses are needed to tease apart the contributions of human intervention and environmental changes in this seagrass expansion.

Conversely, the goals of freshwater conservation are the eradication and prevention of non-native species. Texas Parks and Wildlife classifies *Hydrilla* and *Myriophyllum* as "potentially harmful exotic species." This prohibits possession of these species by individuals within the state and forces boaters to immediately remove the plants if found on boats or equipment (Texas Administrative Code §57.112; Texas Parks and Wildlife Code §66.0071). These laws help to protect SAV native to Texas (*Heteranthera, Najas, Stuckenia*, and *Vallisneria*).

## Additional Impacts of Freshwater Inflow

### Sediment Conditions

SAV retains many features from their ancestral land plants such as root/rhizome systems. These structures are important for anchoring, nutrient acquisition, and vegetative growth of the meadow. Consequently, SAV require sediments for growth and maintenance, similar to how many terrestrial plants require soil. Sediments consist of inorganic and organic particles of varying sources and sizes. The primary source of sediment nourishment for SAV meadows is from the riverine input to the estuary, which results from substantial weathering and erosion across the watershed. Therefore, times of increased freshwater inflows can increase sediment influx to seagrass meadows. SAV meadows act as particle traps by slowing water velocities, increasing water retention

times, and decreasing wave kinetic energy (de Boer 2007). Consequently, SAV meadows continue to accumulate new sediments throughout their existence. An important factor in determining substrate suitability is the formation of an anoxic reduction zone; therefore sediments with lower water column exchange (e.g., fine clays and muds) create better seagrass habitats than those with higher exchange rates (e.g., sands and gravels; Patriquin 1972). Maintenance of this reduction zone is critical because it promotes the recycling and fixation of nutrients by bacteria (particularly N and P) critical for plant health. Oligotrophic tropical waters are typically too nutrient depleted to promote *Thalassia* growth; instead, they receive nearly all their nitrogen from the sediments (Patriquin 1972). The availability of nutrients produced in these sediments (viz., ammonium or phosphate) is one of the primary drivers for differences in seagrass biomass between estuarine regions. *Thalassia* growing in the Nueces estuary has significantly higher sediment ammonium than Lower Laguna Madre (100 μM vs 30 μM) resulting in higher above ground biomass in Nueces (Lee and Dunton 1999a, b).

## Light Availability and Inflow Turbidity

Light availability is the ultimate factor that controls the productivity and survivability for SAV because plants require adequate quantities and qualities of light to perform photosynthesis, thereby meeting their metabolic demands. While emergent vegetation is often light saturated, submerged organisms have the added challenge of light attenuation by the overlying water column and its suspended/dissolved constituents. Compounding this issue, seagrasses (Gattuso et al. 2006) and other aquatic vegetation (Middelboe and Markager 1997) require >10% of surface irradiance, whereas seaweeds and phytoplankton only require 0.1–1% (Gattuso et al. 2006). Reduction from 50% surface irradiance (ambient conditions) to 13–16% (*Halodule*) and 10% (*Thalassia*) surface irradiance caused complete die-offs of seagrasses in Corpus Christi Bay (Czerny and Dunton 1995). Furthermore, the light reduction caused by sea level rise in Upper Laguna Madre (>14 mm yr.⁻¹) has caused seagrasses to vanish from long term monitoring stations (Capistrant-Fossa and Dunton, 2024). This significant difference in light requirements between groups is due to respiration by the complex root/rhizome structures of plants compared to algae (Hemminga 1998). In situ light reductions can harm plants by reducing their metabolic potential for growth and reproduction by allowing toxic sulfides to build up in the plant's rhizosphere (Lee and Dunton 2000a, b; Czerny and Dunton 1995). Additionally, plants must utilize their long-term carbon reserves in rhizomes for energy utilization, which causes a significant decrease in their belowground biomass (Lee and Dunton 1997). Therefore, high light requirements make submerged aquatic vegetation particularly sensitive to reduc-

**Table 3** Dominant plants of Texas estuaries and respective communities (from Pulich 1990 and others)

| Scientific name | Common name | Community Type Submerged | Salt marsh | Brackish marsh | Fresh marsh | Estuary where commonly found |
|---|---|---|---|---|---|---|
| *Thalassia testudinum* | Turtle grass | S[a] | | | | M-A[b], LM[c] |
| *Syringodium filiforme* | Manatee grass | S | | | | M-A, N[d], LM |
| *Halophila engelmannii* | Star grass | S | | | | M-A, N, LM |
| *Halodule wrightii* | Shoal grass | S, B[e] | | | | All, except S-N[f] |
| *Ruppia maritima* | Widgeon grass | S, B, F[g] | | | | All |
| *Vallisneria americana* | Wild celery | B, F | | | | T-SJ[h] |
| *Najas guadalupensis* | Water nymph | F | | | | All, except N, L-C[i] |
| *Stuckenia pectinata* | Sago pondweed | F | | | | All, except M-A T-SJ |
| *Myriophyllum spicatum* | Eurasian milfoil | F | | | | North of G[j] |
| *Hydrilla verticillate* | Water thyme | F | | | | North of G |
| *Heteranthera dubia* | Water star grass | F | | | | South of L-C |
| *Heliotropium curassavicum* | Beach heliotrope | | X | | | All |
| *Salicornia bigelovii* | Annual glasswort | | X | | | All |
| *Sesuvium portulacustrum* | Sea purslane | | X | | | M-A, N, LM |
| *Suaeda linearis* | Sea blite | | X | | | All |
| *Aster tenuifolius* | Salt marsh aster | | X | X | | All |
| *Avicennia germinans* | Black mangrove | | X | X | | G, M-A, N, L-C, LM |
| *Batis maritima* | Saltwort | | X | X | | All |
| *Borrichia frutescens* | Sea oxeye daisy | | X | X | | All |
| *Distichlis spicata* | Salt grass | | X | X | | All |
| *Iva frutescens* | Marsh elder | | X | X | | All |
| *Limonium nashii* | Sea lavender | | X | X | | All |
| *Lycium carolinianum* | Carolina wolfberry | | X | X | | All |
| *Monanthochloe littoralis* | Shore grass | | X | X | | All |
| *Salicornia virginica* | Perennial glasswort | | X | X | | All |
| *Spartina alterniflora* | Smooth cordgrass | | X | X | | All (except LM) |
| *Spartina patens* | Marsh hay cordgrass | | X | X | X | All |
| *Spartina spartinae* | Gulf cordgrass | | X | X | X | All |
| *Juncus roemerianus* | Black needlerush | | | X | | S-N, T-SJ, L-C |
| *Scirpus maritimus* | Bulrush | | | X | | All except LM |
| *Spartina cynosuroides* | Big cordgrass | | | X | | S-N, T-SJ |
| *Bacopa monnieri* | Water hyssop | | | X | X | All |
| *Hydrocotyle* spp. | Pennywort | | | X | X | All |
| *Paspalum* spp. | Paspalum | | | X | X | S-N, T-SJ |
| *Phragmites australis* | Common reed | | | X | X | All |
| *Typha* spp. | Cattail | | | X | X | All |
| *Alternanthera philoxeroides* | Alligator weed | | | | X | S-N, T-SJ |
| *Eichhornia crassipes* | Water hyacinth | | | | X | T-SJ, L-C, G |
| *Sagittaria* spp. | Arrowhead | | | | X | S-N, T-SJ, L-C, G |

[a]S: Submerged saltwater habitat
[b]M-A: Mission-Aransas estuary
[c]LM: Laguna Madre estuary
[d]N: Nueces estuary
[e]B: Submerged brackish water habitat
[f]S-N: Sabine-Neches estuary
[g]F: Submerged freshwater habitat
[h]T-SJ: Trinity-San Jacinto estuary
[i]L-C: Lavaca-Colorado estuary
[j]G: Guadalupe estuary

tions in light intensity. Texas estuaries have two major mechanisms of light degradation: particulate resuspension and influx.

Particulate resuspension, defined as the remixing of previously settled sediments, can occur due to natural or anthropogenic drivers. Common natural mechanisms include tectonic action, rapid freshwater inflow, winds, tides, and

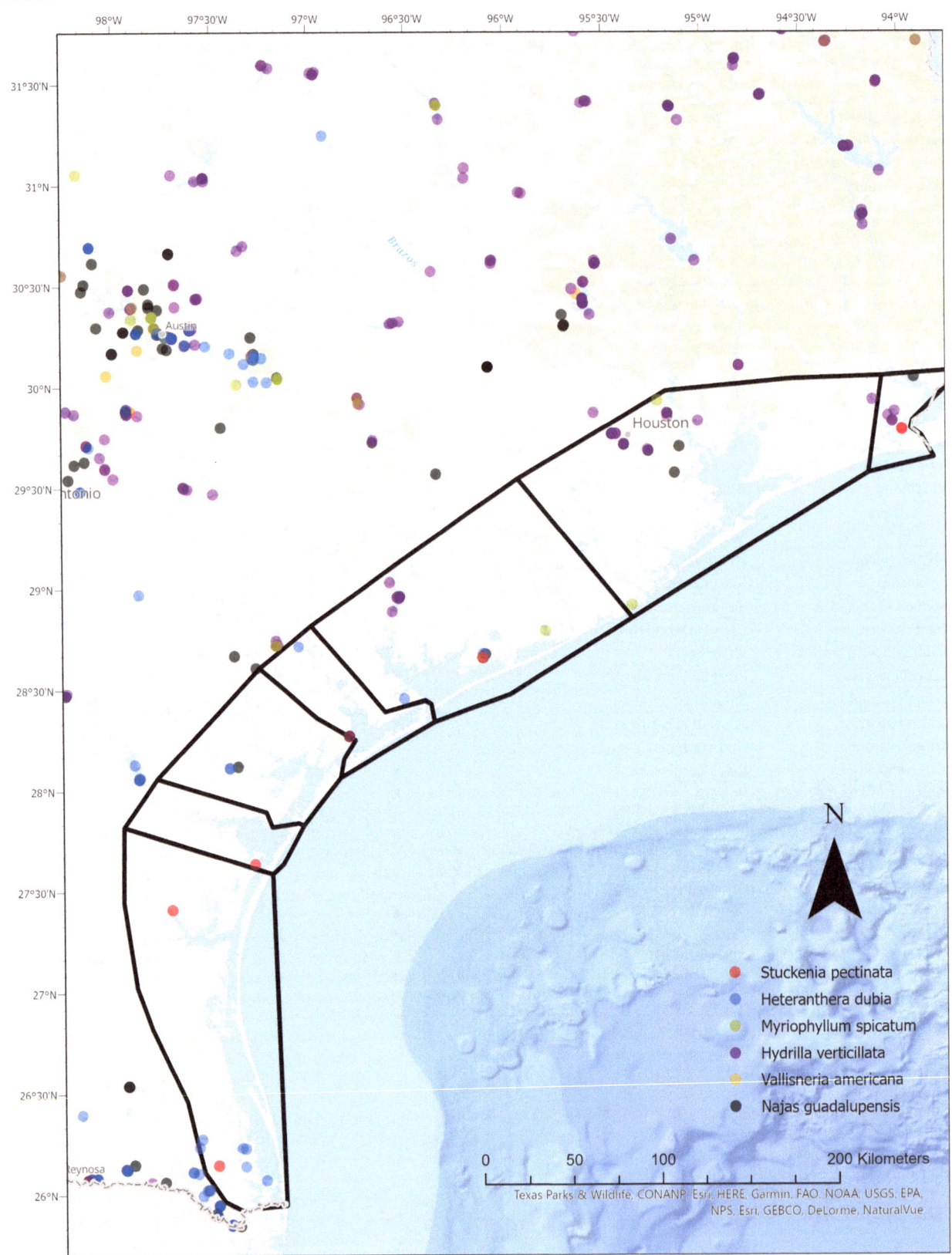

**Fig. 5** The distribution of freshwater SAV across Texas estuaries. The black boundaries depict each of the seven major estuarine systems identified by the TWDB on the Texas coast (see https://www.twdb.texas.gov/surfacewater/bays/major_estuaries/inde.g.,asp). The occurrence data for each species provided by the Global Biodiversity Information Facility (GBIF) combines research-grade citizen science observations and herbarium records (DOIs: 10.15468/dl.v4369n, 10.15468/dl.835ty3, 10.15468/dl.7ymbxt, 10.15468/dl.pm7zn6, 10.15468/dl.nd3erg, and 10.15468/dl.qecnez)

currents, but each estuarine system is locally controlled. For example, Corpus Christi Bay is a wind-forced system, and water column turbidity is significantly correlated with the preceding 30-hours' wind conditions (Shideler 1984). In contrast, waves and currents generated during cold front passages were more correlated to sediment resuspension than winds or currents alone in West Galveston Bay (Carlin et al. 2016). Dredging is the most common anthropogenic mechanism of particulate resuspension, but its spatiotemporal impacts vary depending on the exact source. For example, shrimp fisheries in Galveston can resuspend sediments on the same magnitude as a 9–10 m/s wind, but these effects last only ~15 min (Dellapenna et al. 2006). However, a large-scale dieback of seagrasses in Laguna Madre caused by dredging in 1988 affected meadows up to 2 km away over a year later (Onuf 1994).

Particle influx refers to the advection of new particles into a system, usually from rivers through freshwater inflows. Sources of particulate matter include weathered products from rocks, humic matter from decaying organic material, leached ions, and anthropogenically introduced compounds (e.g., plastics). Typically, the ocean-end of estuaries are much clearer because of limited inputs into the system (e.g., phytoplankton production and waste products from animals). Despite particles falling out along the way, typically there is an estuarine turbidity maximum somewhere along the estuary because of a combination of riverine input, tidal resuspension, and flocculation of particulate matter when freshwater meets saltwater. These regions have significantly higher light attenuation than river or ocean and make it difficult for SAV to persist there.

## Nutrient Loading to Estuaries

Submerged aquatic vegetation, like terrestrial plants, requires exogenous inputs of numerous nutrients (e.g., Fe, N, and P) to sustain normal physiological function. However, rapid and/or excessive input of nutrients (i.e., eutrophication) can lead to degradation of the ecosystem and prove harmful to SAV. The primary mechanism for eutrophication is runoff carried by freshwater inflows into the estuarine basin. For example, decomposition products from leftover agriculture (e.g., husks, roots, and seeds) or high algal biomasses in Petronilla creek can be flushed into Baffin Bay during periods of strong freshwater inflows (Wetz et al. 2017). Integrated over the entire basin, this can elevate the organic nitrogen concentration of the water (~60 µM) to two times that of other estuaries across Texas (Wetz et al. 2017). Similarly, Oso and Galveston Bays are both considered eutrophic systems because of nearby urbanization (Bugica et al. 2020) and contain degraded seagrass meadows. Some Texas estuaries (e.g., Laguna Madre) are among the most nitrogen loaded estuaries in the United States, but the average regional load-

ing (Western Gulf Coast) is still lesser than most other regions of the country (Castro et al. 2003).

Eutrophication through nutrient loaded freshwater inflows poses both direct and indirect effects to SAV meadows. Directly, ammonium, the preferred nitrogen source for seagrasses, is toxic in high concentrations (~600 µM), thereby making it difficult for seagrasses to colonize ammonium-rich substrates such dredged sediments (Kaldy et al. 2004). Nutrient enrichment can also cause deplete plant belowground biomass (roots/rhizomes) as they rapidly produce leafy material and increase their carbon requirements (Lee and Dunton 1999a, b). This causes plants to have less stored carbon and increases their vulnerability during times of decreased productivity (e.g., winter, storms, and dredging). However, not all plants are susceptible to this effect because some are already living in nutrient saturated conditions (e.g., Corpus Christi Bay). Experimental nutrient additions have negligible effect on seagrass growth characteristics in Corpus Christi Bay, but plants in Lower Laguna Madre are more nutrient limited and have greater responses (Lee and Dunton 2000a, b).

In contrast, phytoplankton and macroalgae are more effective at rapidly utilizing nutrients than submerged plants (Burkholder et al. 2007). These organisms grow rapidly, deplete nutrients available to seagrasses, and shade out canopies reducing light available for photosynthesis (Lamote and Dunton 2006). Baffin Bay currently maintains standing populations of *Aureoumbra lagunensis*, a pelagophyte that formed a multiyear algal bloom in Laguna Madre during the 1990s following large nutrient inputs from a fish kill (DeYoe and Suttle 1994; Wetz et al. 2017). This harmful algal bloom led to a reduction in seagrasses in Laguna Madre (Onuf 1996) and caused *Halodule* to lose around 50% of its belowground biomass between 1989 and 1992 alone (Dunton 1994). Additionally, the abundance of drift macroalgae (e.g., *Gracilaria* spp., *Laurencia* spp., *Dictyota* spp.) is related to nutrients added to systems through freshwater inflows (Kopecky and Dunton 2006). Once these seaweeds are advected onto SAV meadows, they shade out the plants, reduce their photosynthetic potential, and promote the buildup of toxic hydrogen sulfide in the environment (Lamote and Dunton 2006). Similarly, the abundant epiphytes living on seagrass blades tend to increase under nutrient loading and can significantly reduce the amount of light available for seagrasses (e.g., Frankovich and Fourqurean 1997).

## Productivity and Sequestration

### Seagrasses

Estuaries act as funnels and can accumulate high concentrations of organic matter from both rivers and the open ocean. Periods of increased freshwater inflow can further raise organic matter concentrations through increased weathering and erosion. This allochthonous production not only fuels estuaries to be a highly productive ecosystems for numerous animals but also causes many systems to be net heterotrophic (i.e., they release more carbon through respiration than they take up; Smith and Hollibaugh 1993). However, a global synthesis of community metabolism found that the average photosynthetic rates in seagrass meadows (225 mmol $O_2$ m$^{-2}$ d$^{-1}$) were significantly greater than respiration rates (188 mmol $O_2$ m$^{-2}$ d$^{-1}$), indicating they are net autotrophic (Duarte et al. 2010). Laguna Madre is highly productive, but ecosystem-level photosynthesis (45–304 mmol $O_2$ m$^{-2}$ d$^{-1}$) and respiration (102–381 mmol $O_2$ m$^{-2}$ d$^{-1}$) are nearly balanced (Ziegler and Benner 1998). Seagrasses in Laguna Madre only contribute ~33% of the system's primary production, while the remaining production (67%) is from seaweeds and microalgae (Kaldy et al. 2002). However, seagrasses ecosystem also host epiphytic algae attached to plants (Morgan and Kitting 1984) that can rival seagrass biomass (~20%–90%) and production (max *Halodule* = ~2.5 mg C g$^{-1}$ dry wt h$^{-1}$; max epiphyte = ~4.0 mg C g$^{-1}$ dry wt h$^{-1}$). Epiphyte abundances vary significantly through space and time (e.g., depths, sites, and seasons) and are affected by the same factors that control seagrass productivity (e.g., light, nutrients, temperature, and salinity; Dunton 1990). Epiphyte-free *Halodule wrightii* in Texas has a lower maximum photosynthetic efficiency (34.5 µmol $O_2$ h$^{-1}$ mg chl$^{-1}$; Dunton and Tomasko 1994) than *Thalassia* (67.5 µmol $O_2$ h$^{-1}$ mg chl$^{-1}$; Dawes and Tomasko 1988) but greater than *Myriophyllum* (28.1 µmol $O_2$ h$^{-1}$ mg chl$^{-1}$; Nielsen and Sand-Jensen 1989).

In Texas estuaries, seagrass productivity can vary significantly between species, environmental conditions, and seasons. *Halodule* meadows can quickly establish themselves and thrive under a wide variety of environmental conditions. Despite significant differences in environmental conditions between the upper Laguna Madre, Nueces, and Guadalupe estuaries, there was no significant difference in *Halodule* photosynthetic capability or seasonal signal in biomass between sites (Dunton 1996). In contrast, *Halodule* from Lower Laguna Madre showed strong seasonality with peak biomass between August and September (Kowalski et al. 2009). *Thalassia*'s peak biomass in Lower Laguna Madre is earlier (July to August; Kaldy and Dunton 2000), whereas the Nueces estuary is later (October to November; Lee and Dunton 1996).

Carbon fixed by seagrasses during photosynthesis is allocated disproportionately between a seagrass plant's belowground (roots, rhizomes, and short shoots; about 85% of fixed biomass) and aboveground (leaves) biomass (Fourqurean and Zieman 1991). Belowground biomass is a carbon reserve and a site for intense plant respiration (Fourqurean and Zieman 1991). Global estimates of the sea-

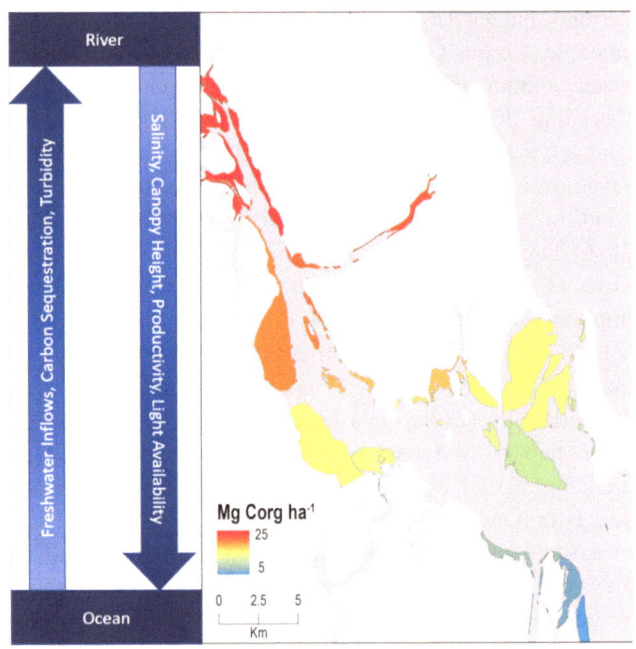

**Fig. 6** An example of the seagrass carbon sequestration along an estuarine gradient (modified from Ricart et al. 2020). Seagrass meadows near the head of an estuary are smaller and less dense but sequester more carbon than those near the mouth

grass carbon stock range between 4.2 and 19.9 Pg C, nearly the same amount of $CO_2$ released by human activities in a year (Fourqurean et al. 2012; Friedlingstein et al. 2020). The carbon fixed within seagrass meadows has different fates including export (14% of carbon fixed), burial (25%), and respiration (17%; Mateo and Romero 1997). The portion allocated to sequestration/burial coined "blue carbon" and represents long-term removal of carbon from the system into the lithosphere. Currently, there is an estimated 1.5 million Mg of carbon stored in Texas' seagrass meadows (Thorhaug et al. 2017). Ricart et al. (2020) demonstrated that seagrass canopy density was less important than particulate input for establishing carbon sequestration potential within an estuarine system (Fig. 6). They found a 50-fold difference in organic carbon burial from the upper to lower estuary driven by allochthonous input. This indicates that regions of the Texas coast experiencing larger rates of freshwater inflows will likely have more carbon sequestration regardless of SAV meadow condition.

## Freshwater Macrophytes

The suite of halotolerant freshwater SAV species often maintain high biomasses because they are highly productive and rapid colonizers. *Hydrilla* and *Myriophyllum* are invasive species that outcompete native SAV in many parts of the world. To this effect, *Hydrilla*, "the perfect aquatic weed," can grow up to 2 cm d$^{-1}$ and accumulate 8.5 g m$^{-2}$ d$^{-1}$ of new biomass (deBusk et al. 1977; Langeland 1996). Similarly,

*Myriophyllum* is known to produce up to $8.1 \text{ g m}^{-2} \text{ d}^{-1}$ (Grace and Wetzel 1978). In contrast, *Heteranthera* and *Stuckenia* produce significantly less biomass ($1.2 \text{ g m}^{-2} \text{ d}^{-1}$ and $1.4 \text{ g m}^{-2} \text{ d}^{-1}$, respectively) than these invasive species, which puts them at risk for being outcompeted (Wilkinson 1961; Kantrud 1990). In general, productivity rates are underreported for freshwater macrophytes, therefore, more investigation into these systems is needed (e.g., seasonal changes, missing species, and environmental drivers).

Analogous to blue carbon, the term teal carbon has recently emerged in literature for carbon dynamics in non-tidal freshwater wetlands. However, research that falls under this category primarily focuses on emergent marsh species (e.g., *Juncus*, *Phragmites*, *Pontederia*, *Scirpus*, and *Typha*) and does not include submerged halotolerant species. This creates a literature gap because blue carbon research and independent lines of investigation often ignore the carbon sequestration potential of these freshwater SAV species. Therefore, little is known about the carbon sequestration potential of individual species. However, at the community level, long-term burial does not seem to vary significantly (Hillmann et al. 2020). When comparing between fresh (*Hydrilla* and *Najas*), intermediate (*Heteranthera, Myriophyllum, Najas,* and *Ruppia*), brackish (*Myriophyllum, Najas, Ruppia,* and *Stuckenia*), and saline (*Myriophyllum, Ruppia,* and *Stuckenia*) communities, there was no significant difference in organic carbon concentration over the top 100 cm of sediments (Hillmann et al. 2020). This highlights that the carbon sequestration potential of freshwater SAV is on the same order of magnitude as seagrasses (although limited to a much smaller spatial coverage), but more research is needed to understand the long-term fate and controls on storage of this carbon.

## Estuarine Marshes

### Coastwide Distribution

#### Controls on Distribution

Estuarine marshes are wetlands composed of a diverse community of emergent herbaceous vegetation. Found along temperate and subtropical coasts worldwide, their distribution is controlled by large-scale factors such as temperature, precipitation, and shoreline relief (Fig. 1). Specifically, marsh development is promoted in areas of low relief, abundant rainfall, and moderate climate (e.g., Atlantic and Gulf coasts of the United States; Crump et al. 2022). For example, Texas marshes are responsive to interannual variation in precipitation (Dunton et al. 2001), freshwater inundation (Alexander and Dunton 2002), pulses of freshwater inflow (Batterton and Dunton 2022), and local climate change (Forbes and Dunton 2006).

Marsh distributions are influenced both by tides and freshwater inflows. Texas coasts are microtidal (tidal range typically less than 1 m), largely influenced by meteorological forces (i.e., winds, storm, and tidal surges), and driven by secular tides, where water levels are higher in the spring and fall and low in the summer and winter (Smith 1979). Geographically, tidal influence is strongest in the lagoons and bays located close to the Gulf of Mexico and is attenuated along the estuarine gradient. Salinity decreases with tidal influence due to the supply of freshwater from precipitation, runoff, and riverine flow. Marsh assemblages track this gradient: salt marshes along the coast to brackish and freshwater marshes upstream. Notable exceptions to these classic salinity dynamics include the hypersaline Laguna Madre and reverse estuaries, such as the Nueces (Palmer et al. 2002).

Texas' small and inconsistent tidal range causes reduced tidal flushing of marsh soil pore water. This leads to hypersalinity in soils, which, in turn, causes predictable structuring within the marsh vegetation community (i.e., zonation). The degrees of soil hypersalinity and waterlogging vary strongly and rapidly with elevation, causing distinct and regular bands of vegetation within the small intertidal zone (Pennings and Callaway 1992; Pennings and Bertness 2001; Pennings and Moore 2001). The only exceptions to this distribution are the extremely salt-tolerant halophytes that may grow at the margins of barren flats and emergent marsh vegetation (Alexander and Dunton 2002). Broadly, marsh plant zonation in Texas is mainly driven by maximum summer salinity, which occurs when low water level and rainfall create highly stressful abiotic conditions (Forbes and Dunton 2006).

### Species Composition of Texas Estuarine Marshes

Texas' estuaries contain ~1960 $km^2$ of marsh habitats that encompass the northern half of the Texas coastline from the Nueces estuary northward to the Sabine-Neches estuary (Fig. 1; Tables 2 and 3). As noted below, the Nueces estuary marks the transition zone where salt marshes compete with mangroves that are migrating northward as winters become milder on the central coast. Mangroves are most common along the southern coast of Texas, where they live on fringing barrier islands and primary bays. Brackish and freshwater marshes become more prevalent as rainfall increases northward and with proximity to freshwater inflow sources (Table 3). As Texas' climate continues to change (e.g., warmer temperature and decreased precipitation), the distribution of marshes will shift, and community diversity will decrease (Feher et al. 2017).

### Sabine-Neches

The brackish marshes of the Sabine-Neches estuary are dominated by *Spartina* species (*alterniflora, patens, spartinae,* and *cynosuroides*), *Juncus roemerianus* (black needle rush),

*Scirpus maritimus* (bulrush), *and Phragmites australis* (common reed), an invasive reed (TDWR 1981; White et al. 1987).

## Trinity-San Jacinto

Emergent vegetation in the brackish Trinity-San Jacinto estuary is mostly composed of *S. alterniflora* (smooth cordgrass), *J. roemerianus, S. maritimus, S. patens* (gulf cordgrass), *Aster tenuifolius* (salt marsh aster), *Echinochloa muricata* (rough barnyard grass), *Alternanthera philoxeroides* (alligator weed), *Paspalum lividum* (longtom), *P. australis, Polygonum punctatum* (water smartweed), and *Sagittaria graminea* (grassy arrowhead) (TDWR 1982; White et al. 1985, 1988).

## Lavaca-Colorado

Along the central coast, the Lavaca-Colorado estuary exhibits a mixture of salt marsh and brackish marsh vegetation. The Lavaca River Delta supports a highly diverse marsh plant community (TDWR 1980b). Recent work by Batterton and Dunton (2022) found that the dominant species include typical salt marsh plants as well as the brackish *J. roemerianus, S. maritimus, A. tenuifolius,* and *S. spartinae* (gulf cordgrass). Upstream, dense stands of *Typha* sp. (cattail), *P. australis,* and *Iva frutescens* (marsh elder) were found. There are also small regions of *Avicennia germinans* (black mangrove) along Matagorda Bay (White et al. 1989a).

## Guadalupe

The Guadalupe estuary is dominated by *S. spartinae, S. patens* (marsh hay cordgrass), *S. maritimus, Distichlis spicata* (salt grass), *Monanthochloe littoralis* (shore grass), *Borrichia frutescens* (sea oxeye daisy), and the invasive *P. australis* (TWDR 1980a; White et al. 1989a).

## Mission-Aransas

The Mission-Aransas National Estuarine Research Reserve conducts annual vegetation surveys. Their data reveal high abundances of *A. germinans, S. alterniflora, B. frutescens, Batis maritima* (saltwort), and several other salt-tolerant marsh species. *Lycium carolinianum* (Carolina wolfberry) is often found in low abundances in high marsh zones (Madrid 2021). Mangrove coverage has expanded since the 1980s in the Mission-Aransas estuary as the marsh-mangrove transition zone moves northward (White et al. 1983; Osland et al. 2020).

## Nueces

The Nueces estuary marshes are composed of hypersaline salt marsh plant communities, with abundant *Salicornia virginica* (glasswort), *B. frutescens,* and *B. maritima* and sparse salt meadows (*D. spicata, M. littoralis, Limonium nashii* (sea lavender), and *S. spartinae*; Rasser 2009, Rasser et al. 2013, White et al. 1983). At upstream locations, however, brackish marsh communities (*I. frutescens, A. tenuifolius, S. maritimus,* and *Typha* sp.) are found (<1% of relative cover over 10 years; Forbes and Dunton 2006, Rasser et al. 2013). Approximately 40% of Nueces estuary marshes are unvegetated, due to erosion-driven land loss (Batterton et al. 2023) or semiarid climatic conditions; however, bare patches may be seasonally colonized by the highly halophytic *Salicornia bigelovii* (annual glasswort; Rasser et al. 2013; Dunton et al. 2019).

## Laguna Madre

Marshes in the Laguna Madre are mostly composed of bare tidal or wind flats. Due to the semiarid climate of the south Texas coast, vegetation community diversity is limited to highly salt-tolerant halophytes such as *A. germinans* and the succulents *S. virginica, S. bigelovii, B. maritima, Suaeda linearis* (sea blite), and *Sesuvium portulacastrum* (sea purslane; White et al. 1986, White et al. 1989b).

## Controls on Community Composition

Marsh vegetation often exhibits clear zonation patterns along environmental gradients due to differences in reproduction and growth, abiotic stress tolerance, and competition (Fig. 7). Marsh species have typical elevational ranges, and plant communities are distinct based on elevation (Rasser et al. 2013). Very small differences in inundation frequency, duration, and timing along an elevational gradient can account for large variations in plant community distributions. For example, as little as a 7% variation in tidal inundation period can result in different plant assemblages (Sadro et al. 2007).

Gradients in tidal inundation frequency and salinity, the main abiotic stressors, are mainly driven by elevation change from the water's edge to the upland boundary (Bertness et al. 1992; Pennings and Callaway 1992). Low marshes are flooded more frequently, with salinity regimes similar to seawater. Intertidal, low marsh vegetation (e.g., *S. alterniflora,* found ~20 cm above MSL) possess mechanisms for tolerance of extended submergence and hypoxia, as oxygen levels are typically lower in waterlogged soils (Adam 1990; Gleason and Zieman 1981). Abundance of *S. alterniflora* abundance is positively correlated with tidal amplitude (McKee and Patrick 1988). However, at consistently flooded elevations (i.e., below MLW), no marsh plants are found due to the physiological stress of frequent inundation (Dunton et al. 2001). High marshes are flooded irregularly and are thus exposed to desiccation, leading to hypersaline soils. High marsh species are typically halophytes (e.g., *S. virginica*). Non-halophytic species are excluded from the high marsh due to competition (Pennings and Moore 2001). Mid-elevation bare patches are typically caused by hypersaline soils left produced by the evaporation of infrequent tidal inundation and low freshwater flushing (Pennings and Bertness 2001; Rasser et al. 2013). These small bare spaces

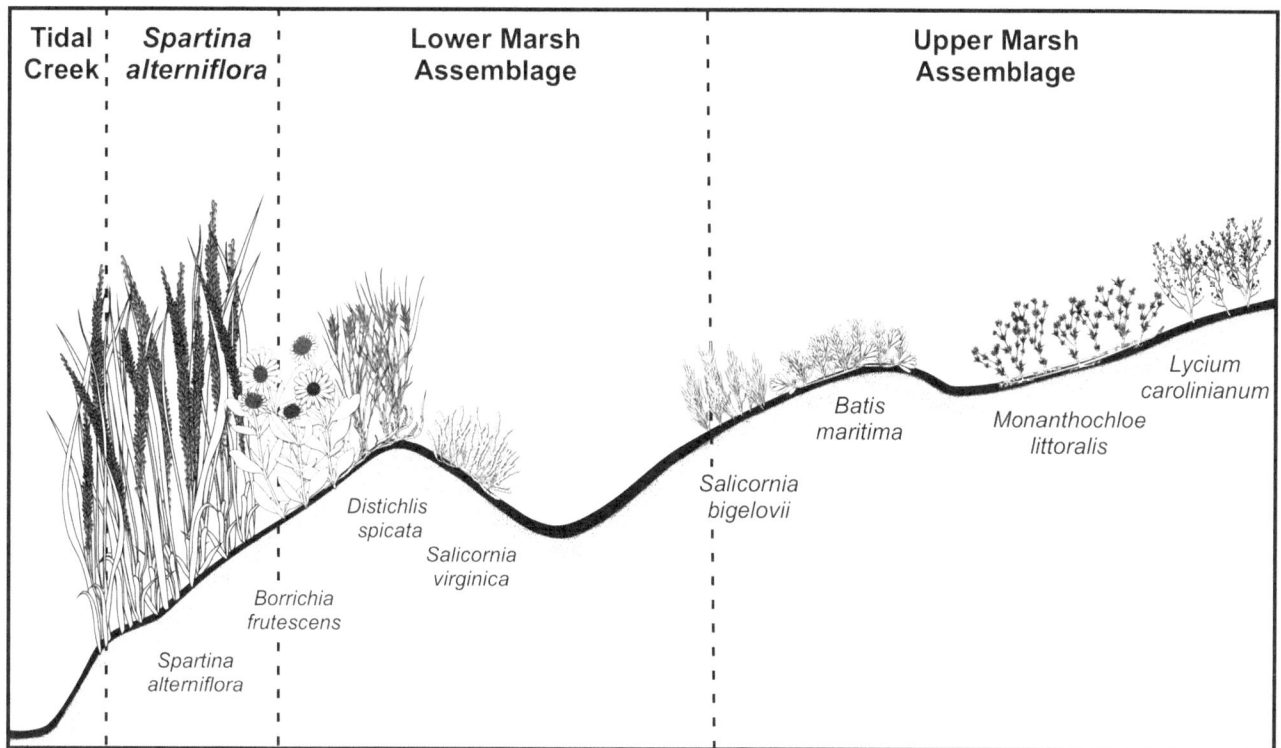

**Fig. 7** Zonation of salt marsh vegetation in the Nueces River Delta. Modified from Rasser et al. (2013) by H.S. Rempel

are different from the large salt flats characteristic of high marshes (Fig. 7; Forbes and Dunton 2006). Because Texas marshes are microtidal, low, and high marsh zones often occur in narrow bands depending on the elevational gradient. Competitively subordinate plants dominate the most stressful habitats, while competitively superior plants dominate habitats where abiotic stress is mildest (Pennings and Callaway 1992). However, the Texas Coast has unpredictable water levels and high abiotic stresses, such as hypersalinity, which may make competitive exclusion less common or important (Kunza and Pennings 2008; Forbes and Dunton 2006).

Marsh plant species respond differently to changes in freshwater availability. For example, some marsh species, such as *B. frutescens*, expanded coverage following large precipitation events (Weilhoefer 1998) and treated wastewater inflow (Alexander and Dunton 2006). Conversely, *S. virginica* abundance increased during periods of reduced precipitation (drought), indicating high salinity tolerance. Furthermore, Rasser (2009) found that *B. frutescens'* growth was impaired at salinity of 30, whereas *S. virginica* isn't. Conversely, Batterton and Dunton (2022) found that most marsh species have a significant interaction with inflows across time scales ranging from 3 to 24 months. This indicates that, while there are species-level differences in responses to inflows, most marsh plants will shift their distributions or abundances due to changes in freshwater.

In conjunction with physiological gradients and competition, plant distribution can also be impacted by physical disturbance. Functional classifications, groupings based on traits related to resource acquisition, reproduction, and competitive strategies, allow for analysis of general patterns of plant distribution and prediction of changes in community structure along environmental gradients in response to disturbance (Grime 1974). Examples of functional groups of marsh plants include clonal stress-tolerators (small plants with slow growth rates that spread laterally and occupy marginal habitats; *B. maritima, D. spicata, M. littoralis,* and *S. virginica*), clonal dominants (elevated leaf canopy, high relative growth rates, and extensive lateral spread; *S. alterniflora* and *B. frutescens*), facultative annuals (no clonal spreading, exist in mixed cultures; *L. carolinianum* and *S. linearis*), and obligate annuals (rapid seedling establishment and growth and high seed production; *S. bigelovii*). Extreme climate variability and periodic disturbance favor rapid germination of weedy species (i.e., obligate annuals) in newly created bare areas. For example, pioneer annuals, such as *S. bigelovii*, are abundant near large bare areas indicating they can quickly colonize these spaces during a freshwater inflow event (Fig. 8; Alexander and Dunton 2002). This may be because *S. bigelovii* produces a viable latent seed bank that persists through periods of hypersalinity and can quickly germinate once hypersalinity is ameliorated (Forbes and Dunton 2006; Dunton et al. 2001; Alexander and Dunton

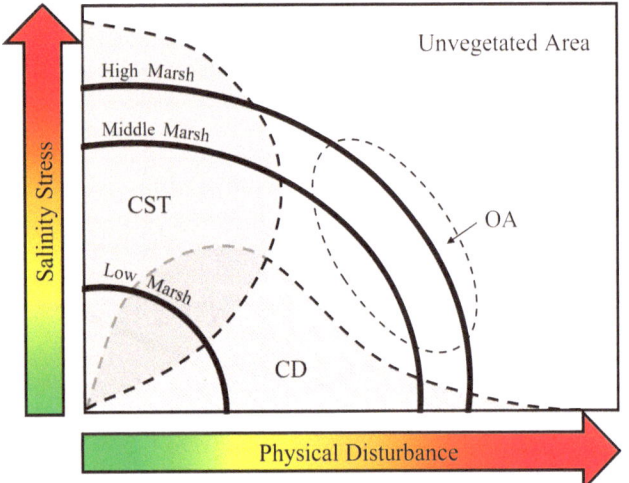

**Fig. 8** Conceptual model of the response of marsh vegetation functional groups to physical, flood disturbance and salinity stress. CST stands for clonal stress-tolerators; CD stands for clonal dominants; OA stands for obligate annuals. Modified from Forbes and Dunton (2006)

2002). *Distichlis spicata* is also common in disturbed marsh habitats (Levine et al. 1998) but is susceptible to competitive displacement by clonal plants such as *B. frutescens* and *S. virginica* (Pennings and Bertness 2001).

Additional important factors can co-vary with inundation and salinity. For example, soil ammonium and sulfide concentrations are typically correlated with salinity (Valiela et al. 1985). Porewater inorganic N (ammonium) shows similar spatial patterns to those of porewater salinity, with the highest concentrations observed in the high marsh. However, unlike porewater salinity, porewater N is not correlated with rainfall or freshwater inflows (Forbes and Dunton 2006). Increased ammonium concentrations at inland sites have been attributed to reduced assimilation of soil nitrogen by plants caused by salinity and other stressors (Buresh et al. 1980; Mitsch and Gosselink 2000). Thus, high ammonium at sites with low vegetation cover is indicative of abiotic, rather than interspecific competition, control of plant distribution (Forbes and Dunton 2006). Sulfide concentrations are inversely correlated with tidal water movement and drainage; thus, soils that are irregularly flooded and are prone to waterlogging tend to accumulate toxic sulfides. This in turn reduces plant nutrient uptake ability and stunts production and growth (King et al. 1982; Mendelssohn and Morris 2000).

## Indicator Species

Tiner (1999) classified marsh species into different groups because they often produce clear bands caused by varying abiotic (mostly salinity) regimes. *Batis maritima, L. carolinianum*, and *M. littoralis* are indicators of hypersaline salt marsh (>40), while *S. virginica* is euhaline (30–40), and *S.*

*alterniflora, B. frutescens,* and *D. spicata* are euhaline/polyhaline (18–40; Table 1). In fact, freshwater inflow management targets for estuaries are often determined by the physiological requirements of indicator species, particularly those species that are commercially important or display a high sensitivity to environmental conditions (Montagna et al. 2020). Once critical salinity thresholds are identified, corresponding freshwater inflows are determined and used to estimate specific inflow requirements (Montagna et al. 2020). Vascular marsh plants are excellent condition indicators because they are immobile and will integrate environmental factors over longer time scales (Kutcher et al. 2022).

Stachelek and Dunton (2013) expanded upon the idea of using marsh plants as key indicators of salinity and overall estuarine hydrological condition over longer time scales. The abundance of *S. alterniflora*, the target indicator species, was significantly correlated with porewater salinity and freshwater inflow. Porewater salinities exceeding 25 resulted in dramatic declines in cover of *S. alterniflora*. Achieving this porewater salinity value required $2.87 \times 10^8$ m$^3$ yr.$^{-1}$ of discharge from the Nueces River (Stachelek and Dunton 2013).

## Impacts of Freshwater Inflows

The frequency, duration, and seasonal distribution of freshwater inflows regulate environmental characteristics such as nutrient concentrations and salinity. The relative impact of inflows is dependent on tidal regime, precipitation frequency, geomorphology, and water residence time (Solis and Powell 1999; Brock 2001). Estuaries with small volumes and large tidal ranges will flush with minimal inflows, whereas microtidal semiarid estuaries with long residence times require large inflow events to remove accumulated salts and nutrients (Solis and Powell 1999). Thus, an estuary-specific approach to studying freshwater inflows is required. In Texas, there is a natural latitudinal gradient of estuary types, from north to south, with the southernmost estuaries being reflective of the semiarid, hypersaline type.

### Inundation

Salt marsh plant communities are substantially impacted both by a lack of freshwater (drought) and excess freshwater (flooding). Drought induces gradual reductions in cover, while flooding causes rapid and severe plant die back, followed by germination of annuals. Drought can reduce primary productivity (Zedler and Onuf 1984), diminish carbon fixation in soils (Heinsch et al. 2004), decrease vegetative cover (Keer and Zedler 2002), and further increase soil salinities and desiccation (Bertness and Callaway 1994). Forbes and Dunton (2006) found that drought is better tolerated by clonal stress-tolerators than clonal dominants, but clonal dominants recovered rapidly following flooding and subsequent wet periods. Within the clonal stress-tolerators group, succulents fare better than grasses during drought, due to water retention strategies characteristic of these plants (Forbes and Dunton 2006). In addition, Stachelek and Dunton (2013) found that vegetation assemblages in the Nueces Delta were distinct between dry and wet hydroclimatic periods.

Moderate flooding has been shown to alleviate salt stress, increase germination, and accelerate clonal expansion (Zedler and Onuf 1984; Alexander and Dunton 2002; Zedler et al. 2003). High flooding can cause widespread die-off, reductions in productivity, and changes in community composition (Zedler et al. 1986; Zedler and Beare 1986; Allison 1996). Resulting debris and wrack from large-scale disturbance-induced die-off represent a large export of organic carbon for both terrestrial and aquatic consumers (Forbes and Dunton 2006).

## Salinity

Estuarine marsh vegetation requires a specific range of salinities for seed germination, seedling growth, and reproduction. Annual plant species located at the colonizing margin are replaced by germination and establishment of seeds each year. This process is facilitated by a reduction in salinity (< 20; Alexander and Dunton 2002, Woodell 1985). In spring, soil temperatures increase, and secular (semi-annual) tides provide inundation of the entire intertidal zone, enhancing seed germination by providing moisture and dissolving accumulated salts in the soil. Seedling growth is also facilitated by extended immersions in normal seawater, increasing spring temperatures, longer day lengths, and increased sun angles. Freshwater inflows due to spring rains also promote further reduction of salinity levels (< 10; Longley 1994, Zimmerman et al. 1990). In contrast, perennial species are not as sensitive to fluctuations in salinity. Many marsh species can grow in both saline and nonsaline soils. However, at lower salinities (< 10), obligate halophytes lose their competitive advantage with more aggressive and fast-growing non-salt-tolerant species (Longley 1994; Forbes and Dunton 2006). This leads to species-specific optimum salinity values for marsh vegetation (Table 1; Pearcy and Ustin 1984).

Vegetation is dependent on rainfall for the replenishment of saline groundwater with freshwater to decrease watertable depth and salinity (Allen et al. 2018; Boorman 2019). In deltaic marshes, the upstream distance of salinity influence on marsh vegetation is a function of the dilution of tidal seawater with freshwater runoff. Because Texas is microtidal, the intertidal zone influenced by regular seawater inundation is restricted. However, with rising sea level, the area inundated with seawater is increasing, which will subject the existing vegetation to greater salinity stress. Further upstream in tidal brackish and freshwater marshes, lower salinities (< 5) and

broad expanses of low elevation landscapes contribute to high diversity and abundance of vascular plants (TDWR 1980b; Zimmerman et al. 1990).

Porewater salinities vary in accordance with bay and tidal creek salinities but are generally higher (up to 150), especially in high marsh areas that flood irregularly (Dunton et al. 2019). Forbes and Dunton (2006) found that porewater salinities were significantly related to rainfall, which accounted for 35.7% of its variability over time. Porewater salinities can decrease rapidly after flooding or large freshwater inflow events, but without sustained freshwater, soil salinities quickly rebound. In addition, the range of porewater salinities in the high marsh is higher than in the low marsh (Forbes and Dunton 2006). This reflects the relative independence of low marsh porewater to climatic influence due to the regularity of tidal flooding.

## Sediment Nourishment

Sea level rise is a growing threat to coastal regions globally, and marshes are no exception. As relative sea level rise (rSLR) increases to an approximate rate of >10 mm yr.$^{-1}$ on the Texas coastline, microtidal marshes are at high risk of being converted to open water (Dangendorf et al. 2023; Paine et al. 2012; Kearney and Turner 2016). Much smaller changes in sea level or flooding frequency may have a greater impact on the microtidal marshes of Texas (Crump et al. 2022). For example, the transition from 100% mud flat at elevations <10 cm above MSL to less than 5% mud at elevations from 18 to 25 cm reflects a rapid shift across a small difference in elevation. In addition, the high rate of rSLR on the Texas coast (regional subsidence, due to local fluid extraction, occurring at 0.5–1.2 cm yr.$^{-1}$) has the potential to exacerbate marsh elevation loss (White and Calnan 1990). Coastal regions are prone to high development for industrial, residential, and recreational purposes. Agriculture, cattle-ranching, and oil production result in alterations to the natural hydrologic regimes through construction of artificial levees, canals, and reservoirs. The sediment trapping ability of these structures reduces sediment loads to the estuaries, and marshes can fall behind sea level rise promoting extensive submergence. The western Gulf of Mexico is one of the few regions with watersheds that would be able to provide enough sediment to keep up with sea level rise otherwise (Ensign et al. 2023). Reduced freshwater inflow from damming may also lead to a reversal of riverine flow and saltwater intrusion into wetlands. Increase in soil salinity due to saltwater intrusion will allow colonization by salt marsh vegetation or upland semiterrestrial species such as *Baccharis* spp. or *Tamarix* spp. into previously brackish or fresh wetlands (TDWR 1980b). The overall consequence of the combination of sea level rise, subsidence, and hydrologic regime alterations is an increased frequency of exposure of naturally brackish marshes to higher salinities, which can have significant consequences on marsh vegetation diversity, productivity, and community composition. In the last century, salt marshes along the Texas Coast have expanded and migrated upland at the expense of other crucial ecosystems such as brackish and freshwater marshes and coastal prairie grasslands (White et al. 1985).

## Nutrients

Nutrient transport is a critical process in estuarine marshes. Freshwater runoff from land often includes high concentrations of inorganic-N (nitrate and ammonium), which is a major source of primary production for phytoplankton and wetland marsh vegetation. Fluxes of inorganic nutrients between the marsh and estuary depend on external nutrient loading from upstream, the degree of eutrophication, and the release of nutrients through the decay of organic matter. Marshes are normally net exporters of dissolved inorganic N, though this can vary, depending on uptake rate and maturity of the marsh (Boorman 2019; Longley 1994). Nutrient addition due to freshwater inflow can increase vegetation growth but will affect species composition by enhancing the abundance of vigorous, N-demanding species and competitively excluding less vigorous species s (Mendelssohn and Morris 2000; Pennings and Bertness 2001; Boorman 2019). In addition to inorganic nutrients, marshes can transport significant fluxes of dissolved inorganic carbon (DIC), dissolved organic matter (DOM), and particulate organic matter (POM) (Santos et al. 2021).

## Estuarine Mangrove Communities

### Coastwide Distribution

#### Controls on Community Composition

Mangroves are a group of euryhaline trees (~60 species) that colonize the land-sea interface of tropical estuarine systems worldwide. However, the diversity of mangroves in Texas is low with only two species: black (*Avicennia germinans*) and red mangroves (*Rhizophora mangle*). Mangroves have a variety of anatomical, physiological, and meta-organismal adaptations that enhance their salinity tolerance, enabling them to thrive in both hyposaline and hypersaline environments (Parida and Jha 2010; Subedi et al. 2022; Table 1). Furthermore, their lifestyles allow them to survive in both submerged and fully exposed environments. For example, seedlings formed successful root systems when submerged under 0–2.5 cm of water at salinities between 0 and 57 (McMillan 1971). Therefore, salinity is not as important in driving the distribution of mangroves across Texas' estuaries as other environmental factors. Instead, the transition zone between the mangrove-dominated and salt marsh-dominated plant assemblages is related to winter climatic conditions that determine the northern limit of mangrove encroachment into salt marshes.

Typically, the most limiting factor in Texas has been the presence of freezing temperatures, but prior to February 2021, the last major freeze in south Texas occurred in late December 1989. Consequently, the 20-year period from 1990 to 2021 allowed for new mangrove germination, growth, and expansion (Osland et al. 2020). In February 2021, another freeze event (Winter Storm Uri) effectively eliminated over 99% of the mangrove populations in south Texas (Martinez et al. 2024). Seedlings of *Avicennia* suffer greater than 50% mortality when exposed to 5.7 to −6.5 °C temperatures for as little as 24 h (McMillan 1971). Furthermore, periods of greater freshwater inflows increase the plants' susceptibility to freezing damage because they are less cold tolerant under low salinity conditions (Coldren and Proffitt 2017).

Beyond environmental conditions, biotic factors also shape mangrove communities. Mangroves are viviparous and produce positively buoyant propagules that can immediately begin growing as seedlings (Parida and Jha 2010; Lonard et al. 2017). Successful recruitment of propagules requires retention in favorable environmental conditions because they have no dormancy period. Additionally, turbulence prevents propagule recruitment; therefore, shorelines must stabilize propagules to enhance recruitment (McMillan 1971). Unvegetated shorelines retain less than 10% of propagules after 4 months, whereas shores vegetated with cooccurring marsh species have a 75% retention rate (Donnelly and Walters 2014). Generally, marshes in northern Texas facilitate mangrove establishment, whereas those in southern regions actively compete for resources (Guo et al. 2013). These negative effects disappear once trees have grown into juveniles because mangroves are often indicative of climax communities and are resilient to competition effects (Berger et al. 2006; Guo et al. 2013). Therefore, the presence of other vegetation is beneficial to the establishment of mangrove encroachment, but once established, the mangroves outcompete and displace the original vegetation. Mangrove encroachment caused only about 6% of the 77.8 km² marsh losses between 1990 and 2010 in Texas (Armitage et al. 2015). Following plant establishment, the major mechanisms of mangrove displacement include degraded environmental conditions, extreme weather conditions (e.g., Winter Storm Uri in 2021), or introduction of a highly competitive invasive species such as *Schinus terebinthifolius* (Donnelly et al. 2008).

### Coastwide Distribution

Texas' estuaries contain 30 km² of mangrove forests that extend southward from the Lavaca-Colorado estuary (Table 2; Fig. 1). This marks the transition zone where mangrove freeze intolerance and established marsh habitat prevent further northward expansion. Over half (17.56 km²) of Texas' mangroves lie within the Guadalupe estuary, and coverage generally decreases along a southward gradient

(Table 2). The Lower Laguna Madre contains the highest concentration of mangroves within the estuary. *Avicennia* dominates most habitats, but the distribution of *Rhizophora* is difficult to ascertain because they are rarer and often heavily intermixed.

## Additional Impacts of Freshwater Inflow

### Sediment Deposition and Erosion

Rivers provide high concentrations of sediments and organic matter produced through weathering and erosion that mangrove forests trap in their pneumatophore/prop root systems. Therefore, mangrove sediment accumulation rates are sensitive to changes in freshwater inflows. Mangrove stands are more efficient than surrounding marshes at capturing sediments because of their extensive prop root and/or pneumatophore systems; therefore, they typically lie at higher elevations. Mature mangrove forests within the Nueces estuary have accumulated an extra 4 cm of sediment compared to surrounding marshes because the mangrove sediment accumulation rate (0.7 cm yr.$^{-1}$) was four times higher than the marsh (Comeaux et al. 2012). Mangroves are also more effective at stabilizing sediments and preventing erosion than marsh systems. Salt marsh plots lost ~20% of their shoreline during the passage of Hurricane Harvey, whereas mangroves only lost ~5% (Pennings et al. 2021). With increasing rates of relative sea level rise along the Texas coast, sufficient sediment nourishment to coastal wetlands is crucial for the survival of these ecosystems.

Interestingly, mangroves also directly change the chemical structure of their sediments by removing water through transpiration. This can increase the salinity of the soils by up to 20 (Comeaux et al. 2012). This creates hypersaline sediments that prevent colonization by marsh plants (Comeaux et al. 2012) and would require large freshwater inflows to reset (e.g., hurricane and floods). Mangrove removal by freeze events may not allow marsh recolonization.

### Inundation

Living on the land-sea interface, mangroves can experience a variety of inundation regimes: fully inundated, frequently inundated, or rarely inundated conditions. Many of Texas' mangroves are either classified as scrub (interior, dwarf) or fringe (exterior, water edge) communities (Lugo and Snedaker 1974). Furthermore, the relative abundance of these physiological states may change dynamically depending on freshwater inflows where times of decreased inflow may lower water levels and shift the community upward (i.e., fringe becoming scrub) and vice versa. These regimes drastically change the mangrove sediments and affect the physiology of the plants. Scrub communities have restricted access to water and must utilize either rainfall or tidal water depending on the season, whereas fringe communities

always use seawater (Lin and Sternberg 1992). Consequently, scrub communities have greater water use efficiency but have a decreased photosynthetic potential compared to fringe communities (Lin and Sternberg 1992). Furthermore, scrub communities are less resilient to environmental impacts. Nitrogen-enriched scrub communities took twice as long to recover from hurricane damage than enriched transition or fringe communities and controls (Feller et al. 2015). Furthermore, nitrogen-enriched scrub trees grow taller, denser stands that are more resilient to freeze susceptibility, aiding in encroachment (Weaver and Armitage 2018).

Beyond direct physiological effects, inundation also affects the chemical properties of mangrove sediments. Sediments with >1 mg $O_2$ $L^{-1}$ have a positive (~400 mV) redox potential (strength of a molecule to lose or gain electrons), but as sediments become waterlogged and anoxic, values become significantly negative (~ − 300 mV; Søndergaard 2009). Often, scrub mangroves have intermediate values (69.8 ± 25.5 in Honduras; Castañeda-Moya et al. 2006), whereas inundated ones are usually negative (−90 mV ± 17 in Belize; McKee et al. 1988). To cope with anaerobic soils, inundated mangroves actively maintain an oxygenated rhizosphere around their prop roots and pneumatophores to raise redox potential values compared to nearby unvegetated areas (~ − 170 mV; McKee et al. 1988). When grown in soils of varying redox potential (400 mV to −180 mV), the net photosynthetic output of seedlings of both *Avicennia* and *Rhizophora* reduced by ~70% at −180 mV compared to 400 mV soils, highlighting the susceptibility of young stages to soil conditions (Pezeshki et al. 1997). However, adult *Avicennia* and *Rhizophora* systems oxidize soils similarly despite different anatomies to reduce the concentration of toxic sulfides in pore waters by three to five times compared to unvegetated sediment (McKee et al. 1988). Furthermore, waterlogging of an *Avicennia* scrub community caused porewater sulfides to double and cause damage to an otherwise healthy stand (Pérez-Ceballos et al. 2022). These sulfides can reduce nutrient uptake ability and reduce growth. However, freshwater inundation is also important for mangroves because the soil inundation during the rainy season in Mexico controls the reproductive capacity and timing of *Rhizophora* (Agraz Hernández et al. 2015). Mangrove propagules are strictly dispersed via water; thus, inundation is crucial for distribution and reproduction.

## Productivity and Sequestration

Mangroves are one of the most productive ecosystems worldwide, rivaling more well-known terrestrial biomes such as tropical rainforests (Crump et al. 2022). The combination of high photosynthetic activity and slow decomposition rates allows mangrove habitats to sequester carbon in sediments over extended periods of time (Crump et al. 2022). Across the Gulf of Mexico, mangroves have less spatial coverage than seagrasses but represent a larger carbon stock (Thorhaug et al. 2019). Likewise, they have a carbon stock nearly twice the size of marshes despite having similar coverage (Thorhaug et al. 2019). Under normal circumstances, microbial respiration and decomposition are normally greater in mangrove forests than in salt marshes. However, labile organic matter broke down faster in marsh systems than mangroves after Hurricane Harvey (Kominoski et al. 2022). This highlights the balance of mangrove systems and their potential to transform their carbon dynamics under rapidly changing conditions.

Under ambient environmental conditions, there is not a significant difference in the carbon fixation rates between *Avicennia* (0.015 g $CO_2$ $m^{-2}$ $min^{-1}$) and *Rhizophora* (0.012 g $CO_2$ $m^{-2}$ $min^{-1}$) growing in Florida (Snedaker and Araújo 1998). When atmospheric concentrations of carbon increased from 346–360 ppm to 361–485 ppm productivity suffered in both plants (Snedaker and Araújo 1998). Currently, the atmospheric concentration of $CO_2$ has exceeded 400 ppm and is continuing to rise, suggesting that mangrove net productivity has decreased since the 1990s. However, freshwater inflows can have a significant effect on the productivity of mangroves under rising $CO_2$ concentrations. For example, *Avicennia* grown in lower salinity levels at 280 ppm were significantly more productive (0.021 g $CO_2$ $m^{-2}$ $min^{-1}$) than those grown at higher salinity, but by 400+ ppm, there was no difference in productivity rates (Fig. 9; Reef et al. 2015).

This highlights that increases in freshwater inflows could potentially compensate for increasing atmospheric carbon concentrations; conversely, decreased freshwater inflows during drought conditions could compound decreased productivity.

**Acknowledgments** This publication was funded in part through Contract No. 21-155-007-C879 from the Texas General Land Office (GLO) with Gulf of Mexico Security Act of 2006 funding made available to the State of Texas and awarded under the Texas Coastal Management Program. The views contained herein are those of the authors and should not be interpreted as representing the views of the GLO or the State of Texas. The authors would like to thank Hannah S. Rempel for her artistic contributions towards figure modifications.

## References

Adair SE, Moore JL, Onuf CP (1994) Distribution and status of submerged vegetation in estuaries of the upper Texas coast. Wetlands 14(2):110–121. https://doi.org/10.1007/BF03160627

Adam P (1990) Saltmarsh ecology. Cambridge studies in ecology. Cambridge University Press, Cambridge, 461 pp. https://doi.org/10.1017/CBO9780511565328

Agraz Hernández CM, Chan Keb CA, Iriarte-Vivar S, Posada Venegas G, Vega Serratos B, Osti Sáenz J (2015) Phenological variation of *Rhizophora mangle* and ground water chemistry associated to changes of the precipitation. Hidrobiológica 25(1):49–61

Alexander HD, Dunton KH (2002) Freshwater inundation effects on emergent vegetation of a hypersaline salt marsh. Estuaries 25:1426–1435

Alexander HD, Dunton KH (2006) Treated wastewater effluent as an alternative freshwater source in a hypersaline salt marsh: impacts on salinity, inorganic nitrogen, and emergent vegetation. J Coast Res 22:377–392

Alhdad GM, Flowers TJ (2021) Salt tolerance in the halophyte *Suaeda maritima* L. Dum. —the effect of oxygen supply and culture medium on growth. J Soil Sci Plant Nutr 21(1):578–586. https://doi.org/10.1007/s42729-020-00384-x

Allen ST, Stagg CL, Brenner J, Goodin KL, Faber-Langendoen D, Gabler CA, Ames KW (2018) Ecological resilience indicators for salt marsh ecosystems. USGS Publications Warehouse. Wetlands and Aquatic Research Center, Lafayette, Louisiana, p 53. http://pubs.er.usgs.gov/publication/70197950

Allison SK (1996) Recruitment and establishment of salt marsh plants following disturbance by flooding. Am Midl Nat 136:232–247

Armitage AR, Highfield WE, Brody SD, Louchouarn P (2015) The contribution of mangrove expansion to salt marsh loss on the Texas Gulf Coast. PLoS One 10(5):e0125404. https://doi.org/10.1371/journal.pone.0125404

Aschenbach TA (2006) Variation in growth rates under saline conditions of *Pascopyrum smithii* (Western wheatgrass) and *Distichlis spicata* (Inland Saltgrass) from different source populations in Kansas and Nebraska: implications for the restoration of salt-affected plant communities. Restor Ecol 14(1):21–27. https://doi.org/10.1111/j.1526-100X.2006.00101.x

Batterton BE, Dunton KH (2022) Lavaca River Delta marsh assessment. Final report to the Texas Water Development Board for contract #2000012439. The University of Texas at Austin, Port Aransas, TX, p 95

Batterton BE, Swanson KM, Dunton KH (2023) Relative Sea Level Rise and Habitat Assessment in the Nueces Delta. Final Report for Project 2321. Coastal Bend Bays and Estuaries Program Publication 163. Corpus Christi, TX, p 23

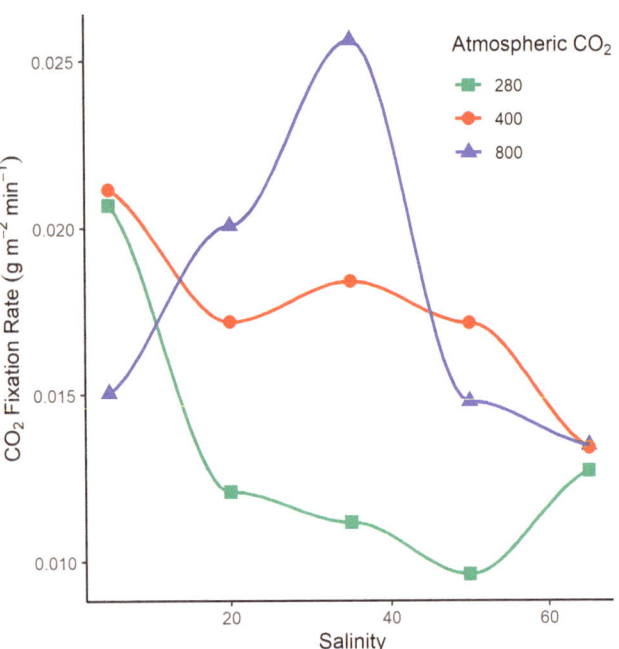

**Fig. 9** The relationship between $CO_2$ uptake rate by black mangrove (*Avicennia germinans*) and salinity varies under different atmospheric concentrations of $CO_2$ (adapted from Reef et al. 2015)

Batterton BE, Person MA, Dunton KH (2024) Use of Historical Data to Assess Climate Change Effects: Nueces Delta. Final Report for Project 2425. Coastal Bend Bays and Estuaries Program Publication 184. Corpus Christi, TX, p 43

Berger U, Adams M, Grimm V, Hildenbrandt H (2006) Modelling secondary succession of neotropical mangroves: causes and consequences of growth reduction in pioneer species. Perspect Plant Ecol 7(4):243–252. https://doi.org/10.1016/j.ppees.2005.08.001

Bertness MD, Callaway R (1994) Positive interactions in communities. Trends Ecol Evol 9:191–193

Bertness MD, Gough L, Shumway SW (1992) Salt tolerances and the distribution of fugitive salt marsh plants. Ecology 73(5):1842–1851

Biber P (2022) Prolonged low salinity tolerance in *Halodule wrightii* Asch. Aquat Bot 178:103498. https://doi.org/10.1016/j.aquabot.2022.103498

Boorman LA (2019) The role of freshwater flows on salt marsh growth and development. In: Perillo GME, Wolanski E, Cahoon DR, Hopkinson CS (eds) Coastal wetlands: an integrated ecosystem approach. Elsevier, Amsterdam, pp 597–618

Brock DA (2001) Nitrogen budget for low and high freshwater inflows, Nueces estuary, Texas. Estuaries 24:509–521

Bugica K, Sterba-Boatwright B, Wetz MS (2020) Water quality trends in Texas estuaries. Mar Pollut Bull 152:110903. https://doi.org/10.1016/j.marpolbul.2020.110903

Buresh RJ, DeLaune RD, Patrick WH (1980) Nitrogen and phosphorus distribution and utilization by *Spartina alterniflora* in a Louisiana Gulf Coast marsh. Estuaries 3(2):111–121

Burkholder JM, Tomasko DA, Touchette BW (2007) Seagrasses and eutrophication. J Exp Mar Biol Ecol 350(1–2):46–72. https://doi.org/10.1016/j.jembe.2007.06.024

Buzzelli C, Robbins R, Doering P, Chen Z, Sun D, Wan Y, Welch B, Schwarzschild A (2012) Monitoring and modeling of *Syringodium filiforme* (manatee grass) in southern Indian River lagoon. Estuar Coast 35(6):1401–1415. https://doi.org/10.1007/s12237-012-9533-8

Callaway JC, Zedler JB (1997) Interactions between a salt marsh native perennial (*Salicornia virginica*) and an exotic annual (*Polypogon monspeliensis*) under varied salinity and hydroperiod. Wetl Ecol Manag 5(3):179–194. https://doi.org/10.1023/A:1008224204102

Capistrant-Fossa KA, Dunton KH (2024) Rapid sea level rise causes loss of seagrass meadows. Communications Earth & Environment 5(87). https://doi.org/10.1038/s43247-024-01236-7

Carlin JA, Lee G, Dellapenna TM, Laverty P (2016) Sediment resuspension by wind, waves, and currents during meteorological frontal passages in a micro-tidal lagoon. Estuar Coast Shelf S 172:24–33. https://doi.org/10.1016/j.ecss.2016.01.029

Castañeda-Moya E, Rivera-Monroy VH, Twilley RR (2006) Mangrove zonation in the dry life zone of the Gulf of Fonseca, Honduras. Estuar Coast 29(5):751–764. https://doi.org/10.1007/BF02786526

Castro MS, Driscoll CT, Jordan TE, Reay WG, Boynton WR (2003) Sources of nitrogen to estuaries in the United States. Estuaries 26(3):803–814. https://doi.org/10.1007/BF02711991

Chambers RM, Meyerson LA, Saltonstall K (1999) Expansion of *Phragmites australis* into tidal wetlands of North America. Aquat Bot 64(3):261–273. https://doi.org/10.1016/S0304-3770(99)00055-8

Chollett I, Bone D, Pérez D (2007) Effects of heavy rainfall on *Thalassia testudinum* beds. Aquat Bot 87(3):189–195. https://doi.org/10.1016/j.aquabot.2007.05.003

Coldren GA, Proffitt CE (2017) Mangrove seedling freeze tolerance depends on salt marsh presence, species, salinity, and age. Hydrobiologia 803(1):159–171. https://doi.org/10.1007/s10750-017-3175-6

Comeaux RS, Allison MA, Bianchi TS (2012) Mangrove expansion in the Gulf of Mexico with climate change: implications for wet-

land health and resistance to rising sea levels. Estuar Coast Shelf S 96:81–95. https://doi.org/10.1016/j.ecss.2011.10.003

Congdon VM, Bonsell C, Cuddy MR, Dunton KH (2019) In the wake of a major hurricane: differential effects on early vs. late successional seagrass species. Limnol Oceanogr Lett 4(5):155–163. https://doi.org/10.1002/lol2.10112

Constantin AJ, Broussard WP, Cherry JA (2019) Environmental gradients and overlapping ranges of dominant coastal wetland plants in Weeks Bay, AL. Southeast Nat 18(2):224–239. https://doi.org/10.1656/058.018.0202

Costanza R, D'Arge R, de Groot R, Farber S, Grasso M, Hannon B, Limburg K, Naeem S, O'Neill RV, Paruelo J, Raskin RG, Sutton P, van den Belt M (1997) The value of the world's ecosystem services and natural capital. Nature 387(6630):253–260. https://doi.org/10.1038/387253a0

Crump BC, Testa JM, Dunton KH (2022) Estuarine Ecology, 3rd edn. John Wiley and Sons, Hoboken, NJ, p 467

Cuddy MR, Dunton KH (2023) Seagrass isoscapes and stochioscapes reveal linkages to inorganic nitrogen sources in the lower Laguna Madre, Western Gulf of Mexico. Estuaries Coast 46:2115. https://doi.org/10.1007/s12237-023-01206-w

Czerny AB, Dunton KH (1995) The effects of in situ light reduction on the growth of two subtropical seagrasses, *Thalassia testudinum* and *Halodule wrightii*. Estuaries 18(2):418. https://doi.org/10.2307/1352324

Dangendorf, Sönke, et al. "Acceleration of US Southeast and Gulf coast sea-level rise amplified by internal climate variability." Nature Communications 14.1 (2023): 1–11.

Dawes CJ, Tomasko DA (1988) Depth distribution of *Thalassia testudinum* in two meadows on the west coast of Florida. Mar Ecol 9:123–130

Dawes CJ, Chan M, Chinn R, Koch EW, Lazar A, Tomasko D (1987) Proximate composition, photosynthetic and respiratory responses of the seagrass *Halophila engelmannii* from Florida. Aquat Bot 27(2):195–201. https://doi.org/10.1016/0304-3770(87)90067-2

Dawes CJ, Andorfer J, Rose C, Uranowski C, Ehringer N (1997) Regrowth of the seagrass *Thalassia testudinum* into propeller scars. Aquat Bot 59(1–2):139–155. https://doi.org/10.1016/S0304-3770(97)00021-1

de Boer WF (2007) Seagrass–sediment interactions, positive feedbacks and critical thresholds for occurrence: a review. Hydrobiologia 591(1):5–24. https://doi.org/10.1007/s10750-007-0780-9

Debez A, Saadaoui D, Slama I, Huchzermeyer B, Abdelly C (2010) Responses of *Batis maritima* plants challenged with up to two-fold seawater NaCl salinity. J Plant Nutr Soil Sc 173(2):291–299. https://doi.org/10.1002/jpln.200900222

deBusk TA, Williams LD, Ryther JH (1977) Growth and yields of freshwater weeds, *Eichhornia crassipes* (water hyacinth), *Lenna minor* (duckweed) and *Hydrilla verticilata*. In: Ryther JH (ed) Cultivation of macroscopic marine algae and freshwater aquatic weeds, Progress report. US Department of Energy, pp 275–295

Dellapenna TM, Allison MA, Gill GA, Lehman RD, Warnken KW (2006) The impact of shrimp trawling and associated sediment resuspension in mud dominated, shallow estuaries. Estuar Coast Shelf S 69(3–4):519–530. https://doi.org/10.1016/j.ecss.2006.04.024

DeYoe HR, Suttle CA (1994) The inability of the Texas "brown tide" alga to use nitrate and the role of nitrogen in the initiation of a persistent bloom of this organism. J Phycol 30(5):800–806. https://doi.org/10.1111/j.0022-3646.1994.00800.x

Doering PH, Chamberlain RH, Haunert DE (2002) Using submerged aquatic vegetation to establish minimum and maximum freshwater inflows to the Caloosahatchee estuary, Florida. Estuaries 25(6):1343–1354. https://doi.org/10.1007/BF02692229

Donnelly MJ, Walters LJ (2014) Trapping of *Rhizophora mangle* propagules by coexisting early successional species. Estuar Coast 37(6):1562–1571. https://doi.org/10.1007/s12237-014-9789-2

Donnelly MJ, Green DM, Walters LJ (2008) Allelopathic effects of fruits of the Brazilian pepper *Schinus terebinthifolius* on growth, leaf production and biomass of seedlings of the red mangrove *Rhizophora mangle* and the black mangrove *Avicennia germinans*. J Exp Mar Biol Ecol 357(2):149–156. https://doi.org/10.1016/j.jembe.2008.01.009

Duarte CM, Marbà N, Gacia E, Fourqurean JW, Beggins J, Barrón C, Apostolaki ET (2010) Seagrass community metabolism: assessing the carbon sink capacity of seagrass meadows. Global Biogeochem Cy 24(4):1–8. https://doi.org/10.1029/2010GB003793

Dunton KH (1990) Production ecology of *Ruppia maritima* L. s.l. and *Halodule wrightii* Aschers, in two subtropical estuaries. J Exp Mar Biol Ecol 143(3):147–164. https://doi.org/10.1016/0022-0981(90)90067-M

Dunton KH (1994) Seasonal growth and biomass of the subtropical seagrass in relation to continuous measurements of underwater irradiance. Mar Biol 120(3):479–489. https://doi.org/10.1007/BF00680223

Dunton KH (1996) Photosynthetic production and biomass of the subtropical seagrass *Halodule wrightii* along an estuarine gradient. Estuaries 19(2):436–447. https://doi.org/10.2307/1352461

Dunton KH, Schonberg SV (2002) Assessment of propeller scarring in seagrass beds of the South Texas coast. J Coast Res 37:100–110. http://www.jstor.org/stable/25736346

Dunton KH, Tomasko DA (1994) In situ photosynthesis in the seagrass *Halodule wrightii* in a hypersaline subtropical lagoon. Mar Ecol Prog Ser 107(3):281–293. http://www.jstor.org/stable/24842684

Dunton KH, Hardegree B, Whiteledge TE (2001) Response of estuarine marsh vegetation to interannual variations in precipitation. Estuaries 24:851–861

Dunton KH, Pulich W Jr, Mutchler T (2011) A seagrass monitoring program for Texas coastal waters. Final report to the coastal bend and bays estuary program for contract #0627. The University of Texas at Austin, Port Aransas, TX, p 39

Dunton KH, Jackson K, Wilson S, Congdon V, Cuddy M, Hall W, Becker M, Meiman J, Whiteaker T, Bohannon P, Grubbs F, Hobson (2018) Seagrass canopy height, water depth, chlorophyll-a concentration and other plant and water quality indicators in coastal waters of Texas. NOAA National Centers for Environmental Information. https://www.ncei.noaa.gov/archive/accession/0181898

Dunton KH, Whiteaker T, Rasser MK (2019) Patterns in the emergent vegetation of the Rincon Bayou Delta, 2005–2016. Final report to the Texas Water Development Board report for contract #1600011971. The University of Texas at Austin, Port Aransas, TX, p 44

El-Haddad E-SH, Noaman MM (2001) Leaching requirement and salinity threshold for the yield and agronomic characteristics of halophytes under salt stress. J Arid Environ 49:865–874

Ensign, Scott H., Joanne N. Halls, and Erin K. Peck. "Watershed sediment cannot offset sea level rise in most US tidal wetlands." Science 382.6675 (2023): 1191–1195.

Failon CM, Wittyngham SS, Johnson DS (2020) Ecological associations of *Littoraria irrorata* with *Spartina cynosuroides* and *Spartina alterniflora*. Wetlands 40(5):1317–1325. https://doi.org/10.1007/s13157-020-01306-4

Feher LC, Osland MJ, Griffith KT, Grace JB, Howard RJ, Stagg CL, Enwright NM, Krauss KW, Gabler CA, Day RH, Rogers K (2017) Linear and nonlinear effects of temperature and precipitation on ecosystem properties in tidal saline wetlands. Ecosphere 8(10). https://doi.org/10.1002/ecs2.2017.8.issue-1010.1002/ecs2.1956

Feller IC, Dangremond EM, Devlin DJ, Lovelock CE, Proffitt CE, Rodriguez W (2015) Nutrient enrichment intensifies hurricane impact in scrub mangrove ecosystems in the Indian River Lagoon, Florida, USA. Ecology 96(11):2960–2972. https://doi.org/10.1890/14-1853.1

Flowers TJ, Troke PF, Yeo AR (1977) The mechanism of salt tolerance in halophytes. Ann Rev Plant Physio 28:89–121

Forbes MG, Dunton KH (2006) Response of a subtropical estuarine marsh to local climatic change in the southwestern Gulf of Mexico. Estuar Coast 29(68):1242–1254

Fourqurean JW, Zieman JC (1991) Photosynthesis, respiration and whole plant carbon budget of the seagrass *Thalassia testudinum*. Mar Ecol Prog Ser 69(1/2):161–170. http://www.jstor.org/stable/44634775

Fourqurean JW, Duarte CM, Kennedy H, Marbà N, Holmer M, Mateo MA, Apostolaki ET, Kendrick GA, Krause-Jensen D, McGlathery KJ, Serrano O (2012) Seagrass ecosystems as a globally significant carbon stock. Nat Geosci 5(7):505–509. https://doi.org/10.1038/ngeo1477

Frankovich TA, Fourqurean JW (1997) Seagrass epiphyte loads along a nutrient availability gradient, Florida Bay, USA. Mar Ecol Prog Ser 159:37–50

Friedlingstein P, O'Sullivan M, Jones MW, Andrew RM, Hauck J, Olsen A, Peters GP, Peters W, Pongratz J, Sitch S, Le Quéré C, Canadell JG, Ciais P, Jackson RB, Alin S, Aragão LEOC, Arneth A, Arora V, Bates NR, Becker M, Benoit-Cattin A, Bittig H, Bopp L, Bultan S, Chandra N, Chevallier F, Chini LP, Evans W, Florentie L, Forster PM, Gasser T, Gehlen M, Gilfillan D, Gkritzalis T, Gregor L, Gruber N, Harris I, Hartung K, Haverd V, Houghton RA, Ilyina T, Jain AK, Joetzjer E, Kadono K, Kato E, Kitidis V, Korsbakken JI, Landschutzer P, Lefevre N, Lenton A, Lienert S, Liu Z, Lombardozzi D, Marland G, Metzl N, Munro D, Nabel JEMS, Nakaoka S-I, Niwa Y, O'Brien K, Ono T, Palmer P, Pierrot D, Poulter B, Resplandy L, Robertson E, Rodenbeck C, Schwinger J, Seferian R, Skjelvan I, Smith AJP, Sutton AJ, Tanhua T, Tans PP, Tian H, Tilbrook B, van der Werf G, Vuichard N, Walker AP, Wanninkhof R, Watson AJ, Willis D, Wiltshire AJ, Wenping Y, Yue X, Zaehle S (2020) Global carbon budget 2020. Earth Syst Sci Data 12(4):3269–3340. https://doi.org/10.5194/essd-12-3269-2020

Gattuso JP, Gentili B, Duarte CM, Kleypas JA, Middelburg JJ, Antoine D (2006) Light availability in the coastal ocean: impact on the distribution of benthic photosynthetic organisms and their contribution to primary production. Biogeosciences 3(4):489–513. https://doi.org/10.5194/bg-3-489-2006

Gleason ML, Zieman JC (1981) Influence of tidal inundation on internal oxygen supply of *Spartina alterniflora* and *Spartina patens*. Estuar Coast Shelf S 13(1):47–57

Grace JB, Wetzel RG (1978) The production biology of eurasian watermilfoil. J Aquat Plant Manage 16(1):1–11

Griffith SA, Bechler DL (1995) The distribution and abundance of the bay Anchovy, Anchoa mitchilli, in a Southeast Texas Marsh Lake System. Gulf Res Rep 9(2):117–122. https://doi.org/10.18785/grr.0902.06

Grime J (1974) Vegetation classification by reference to strategies. Nature 250:26–31

Guo H, Zhang Y, Lan Z, Pennings SC (2013) Biotic interactions mediate the expansion of black mangrove (*Avicennia germinans*) into salt marshes under climate change. Glob Chang Biol 19(9):2765–2774. https://doi.org/10.1111/gcb.12221

Haller WT, Sutton DL, Barlowe WC (1974) Effects of salinity on growth of several aquatic macrophytes. Ecology 55(4):891–894

He Q, Silliman BR, Cui B (2017) Incorporating thresholds into understanding salinity tolerance: a study using salt-tolerant plants in salt marshes. Ecol Evol 7(16):6326–6333. https://doi.org/10.1002/ece3.3209

Heinsch FA, Heilman JL, McInnes KJ, Cobos DR, Zuberer DA, Roelke DL (2004) Carbon dioxide exchange in a high marsh on the Texas Gulf Coast: effects of freshwater availability. Agric For Meteorol 125(1–2):159–172

Hemminga M (1998) The root/rhizome system of seagrasses: an asset and a burden. J Sea Res 39(3–4):183–196. https://doi.org/10.1016/S1385-1101(98)00004-5

Hester MW, Mendelssohn IA, McKee KL (1998) Intraspecific variation in salt tolerance and morphology in *Panicum hemitomon* and *Spartina alterniflora* (Poaceae). Int J Plant Sci 159(1):127–138. https://doi.org/10.1086/297530

Hester MW, Mendelssohn IA, McKee KL (2001) Species and population variation to salinity stress in *Panicum hemitomon*, *Spartina patens*, and *Spartina alterniflora*: morphological and physiological constraints. Environ Ex Bot 46(3):277–297. https://doi.org/10.1016/S0098-8472(01)00100-9

Hillmann ER, La Peyre MK (2019) Effects of salinity and light on growth and interspecific interactions between *Myriophyllum spicatum* L. and *Ruppia maritima* L. Aquat Bot 155:25–31. https://doi.org/10.1016/j.aquabot.2019.02.007

Hillmann ER, Rivera-Monroy VH, Nyman JA, La Peyre M (2020) Estuarine submerged aquatic vegetation habitat provides organic carbon storage across a shifting landscape. Sci Total Environ 717(137217). https://doi.org/10.1016/j.scitotenv.2020.137217

Hroudová Z, Zákravský P, Flegrová M (2014) The tolerance to salinity and nutrient supply in four European Bolboschoenus species (*B. maritimus*, *B. laticarpus*, *B. planiculmis* and *B. yagara*) affects their vulnerability or expansiveness. Aquat Bot 112:66–75. https://doi.org/10.1016/j.aquabot.2013.07.012

Irlandi E, Orlando B, Maciá S, Biber P, Jones T, Kaufman L, Lirman D, Patterson ET (2002) The influence of freshwater runoff on biomass, morphometrics, and production of *Thalassia testudinum*. Aquat Bot 72(1):67–78. https://doi.org/10.1016/S0304-3770(01)00217-0

Kahn AE, Durako MJ (2006) *Thalassia testudinum* seedling responses to changes in salinity and nitrogen levels. J Exp Mar Biol Ecol 335(1):1–12. https://doi.org/10.1016/j.jembe.2006.02.011

Kaldy JE, Dunton KH (2000) Above- and below-ground production, biomass and reproductive ecology of *Thalassia testudinum* (turtle grass) in a subtropical coastal lagoon. Mar Ecol Prog Ser 193:271–283. https://doi.org/10.3354/meps193271

Kaldy JE, Onuf CP, Eldridge PM, Cifuentes LA (2002) Carbon budget for a subtropical seagrass dominated coastal lagoon: how important are seagrasses to total ecosystem net primary production? Estuaries 25(4):528–539. https://doi.org/10.1007/BF02804888

Kaldy JE, Dunton KH, Kowalski JL, Lee K-S (2004) Factors controlling seagrass revegetation onto dredged material deposits: a case study in lower Laguna Madre, Texas. J Coastal Res 20(1):292–300. https://doi.org/10.2112/1551-5036(2004)20[292:fcsrod]2.0.co;2

Kantrud HA (1990) Sago pondweed (*Potamogeton pectinatus* L.): a literature review. Resource publication 176. US Fish and Wildlife Service, Washington, DC, p 88

Kantrud HA (1991) Widgeongrass (*Ruppia maritima* L.): a literature review, Fish and wildlife research, vol 10. US Fish and Wildlife Service, Washington, DC, p 64

van Katwijk MM, Thorhaug A, Marbà N, Orth RJ, Duarte CM, Kendrick GA, Althuizen IHJ, Balestri E, Bernard G, Cambridge ML, Cunha A, Durance C, Giesen W, Han Q, Hosokawa S, Kiswara W, Komatsu T, Lardicci C, Lee K-S, Meinesz A, Nakaoka M, O'Brien KR, Paling EI, Pickerell C, Ransijn AMA, Verduin JJ (2016) Global analysis of seagrass restoration: the importance of large-scale planting. J Appl Ecol 53(2):567–578. https://doi.org/10.1111/1365-2664.12562

Kearney MS, Turner RE (2016) Microtidal marshes: can these widespread and fragile marshes survive increasing climate–sea level variability and human action? J Coast Res 32(3):686–699

Keer GH, Zedler JB (2002) Salt marsh canopy architecture differs with the number and composition of species. Ecol Appl 12(2):456–473

King GM, Klug MJ, Wiegert RG, Chalmers AG (1982) Relation of soil water movement and sulfide concentration to *Spartina alterniflora* production in a Georgia salt marsh. Science 218(4567):61–63

Kominoski JS, Weaver CA, Armitage AR, Pennings SC (2022) Coastal carbon processing rates increase with mangrove cover following a hurricane in Texas, USA. Ecosphere 13:e4007. https://doi.org/10.1002/ecs2.4007

Kopecky AL, Dunton KH (2006) Variability in drift macroalgal abundance in relation to biotic and abiotic factors in two seagrass dominated estuaries in the Western Gulf of Mexico. Estuar Coast 29(4):617–629. https://doi.org/10.1007/BF02784286

Kowalski JL, DeYoe HR, Allison TC (2009) Seasonal production and biomass of the seagrass, *Halodule wrightii* Aschers (shoal grass), in a subtropical Texas lagoon. Estuar Coast 32(3):467–482. https://doi.org/10.1007/s12237-009-9146-z

Kunza AE, Pennings SC (2008) Patterns of plant diversity in Georgia and Texas salt marshes. Estuar Coast 31:673–681

Kutcher TE, Raposa KB, Roman CT (2022) A rapid method to assess salt marsh condition and guide management decisions. Ecol Indic 138:108841. https://doi.org/10.1016/j.ecolind.2022.108841

Lamote, Morgane, and Kenneth H. Dunton. "Effects of drift macroalgae and light attenuation on chlorophyll fluorescence and sediment sulfides in the seagrass Thalassia testudinum." Journal of Experimental Marine Biology and Ecology 334.2 (2006): 174-186.

Langeland KA (1996) *Hydrilla verticillata* (L.F.) Royle (Hydrocharitaceae), "the perfect aquatic weed". Castanea 61(3):293–304. http://www.jstor.org/stable/4033682

Larkum AWD, Waycott M, Conran JG (2018) Evolution and biogeography of seagrasses. In: Larkum AWD, Kendrick GA, Ralph PJ (eds) Seagrasses of Australia. Springer International Publishing, Manhattan, NY, pp 3–29. https://doi.org/10.1007/978-3-319-71354-0_1

Lee K-S, Dunton KH (1996) Production and carbon reserve dynamics of the seagrass *Thalassia testudinum* in Corpus Christi Bay, Texas, USA. Mar Ecol Prog Ser 143:201–210. https://doi.org/10.3354/meps143201

Lee K-S, Dunton KH (1997) Effect of in situ light reduction on the maintenance, growth and partitioning of carbon resources in *Thalassia testudinum* banks ex König. J Exp Mar Biol Ecol 210(1):53–73. https://doi.org/10.1016/S0022-0981(96)02720-7

Lee K-S, Dunton KH (1999a) Influence of sediment nitrogen-availability on carbon and nitrogen dynamics in the seagrass *Thalassia testudinum*. Mar Biol 134(2):217–226. https://doi.org/10.1007/s002270050540

Lee K-S, Dunton KH (1999b) Inorganic nitrogen acquisition in the seagrass *Thalassia testudinum*: development of a whole-plant nitrogen budget. Limnol Oceanogr 44(5):1204–1215. https://doi.org/10.4319/lo.1999.44.5.1204

Lee K-S, Dunton KH (2000a) Effects of nitrogen enrichment on biomass allocation, growth, and leaf morphology of the seagrass *Thalassia testudinum*. Mar Ecol Prog Ser 196:39–48. https://doi.org/10.3354/meps196039

Lee K-S, Dunton KH (2000b) Diurnal changes in pore water sulfide concentrations in the seagrass *Thalassia testudinum* beds: the effects of seagrasses on sulfide dynamics. J Exp Mar Biol Ecol 255(2):201–214. https://doi.org/10.1016/S0022-0981(00)00300-2

Levine JM, Brewer JS, Bertness MD (1998) Nutrients, competition and plant zonation in a New England salt marsh. J Ecol 86(2):285–292

Lillebø AI, Pardal MA, Neto JM, Marques JC (2003) Salinity as the major factor affecting *Scirpus maritimus* annual dynamics: evidence from field data and greenhouse experiment. Aquat Bot 77(2):111–120. https://doi.org/10.1016/S0304-3770(03)00088-3

Lin G, Sternberg LDSL (1992) Comparative study of water uptake and photosynthetic gas exchange between scrub and fringe red mangroves, *Rhizophora mangle* L. Oecologia 90(3):399–403. https://doi.org/10.1007/BF00317697

Lirman D, Cropper WP (2003) The influence of salinity on seagrass growth, survivorship, and distribution within Biscayne Bay, Florida: field, experimental, and modeling studies. Estuaries 26(1):131–141. https://doi.org/10.1007/BF02691700

Longley WL (ed) (1994) Freshwater inflows to Texas bays and estuaries: ecological relationships and methods for determination of needs. Texas Water Development Board and Texas Parks and Wildlife Department, Austin, TX, p 386

Lonard RI, Judd FW, Summy KR, DeYoe H, Salter R (2017) The biological flora of coastal dunes and wetlands: Avicennia germinans (L.). L. J Coastal Res 33(1):191–207. https://doi.org/10.2112/JCOASTRES-D-16-00013.1

Lugo AE, Snedaker SC (1974) The ecology of mangroves. Annu Rev Ecol Syst 5(1):39–64. https://doi.org/10.1146/annurev.es.05.110174.000351

Madrid M (2021) Monitoring for resilience: detecting and responding to coastal wetland change at the Mission-Aransas National Estuarine Research Reserve. MS Thesis. University of Texas at Austin, Austin, TX, p 106

Madrid EN, Armitage AR, López-Portillo J (2014) Avicennia germinans (black mangrove) vessel architecture is linked to chilling and salinity tolerance in the Gulf of Mexico. Front Plant Sci 5:503. https://doi.org/10.3389/fpls.2014.00503

Martin SR, Onuf JP, Dunton KH (2008) Assessment of propeller and off-road vehicle scarring in seagrass bed ad wind-tidal flats of the southwestern Gulf of Mexico. Bot Mar 51:79–91

Martinez M, Osland MJ, Grace JB, Enwright NM, Stagg CL, Kaalstad S, Anderson GH, Armitage AR, Cebrian J, Cummins KL, Day RH, Devlin DJ, Dunton KH, Feher LC, Fierro-Cabo A, Flores EA, From AS, Hughes AR, Kaplan DA, Langston AK, Miller C, Proffitt CE, Reaver NGF, Sanspree CR, Snyder CM, Stetter AP, Swanson KM, Thompson JE, Zamora-Tovar C (2024) Integrating Remote Sensing with Ground-based Observations to Quantify the Effects of an Extreme Freeze Event on Black Mangroves (Avicennia germinans) at the Landscape Scale. Ecosystems 27(1):45–60. https://doi.org/10.1007/s10021-023-00871-z

Mateo M, Romero J (1997) Detritus dynamics in the seagrass Posidonia oceanica: elements for an ecosystem carbon and nutrient budget. Mar Ecol Prog Ser 151:43–53. https://doi.org/10.3354/meps151043

McKee KL, Patrick WH (1988) The relationship of smooth cordgrass (Spartina alterniflora) to tidal datums: a review. Estuaries 11(3):143–151

McKee KL, Mendelssohn IA, Hester MW (1988) Reexamination of pore water sulfide concentrations and redox potentials near the aerial roots of Rhizophora mangle and Avicennia germinans. Am J Bot 75(9):1352–1359. https://doi.org/10.1002/j.1537-2197.1988.tb14196.x

McMahan CA (1968) Biomass and salinity tolerance of shoalgrass and manateegrass in lower Laguna Madre, Texas. J Wildlife Manage 32(3):501–506. https://doi.org/10.2307/3798928

McMillan C (1971) Environmental factors affecting seedling establishment of the black mangrove on the central Texas coast. Ecology 52(5):927–930. https://doi.org/10.2307/1936046

McMillan C (1974) Salt tolerance of mangroves and submerged aquatic plants. In: Reimold RJ, Queen WH (eds) Ecology of halophytes. Academic Press Inc, New York, pp 379–390. https://doi.org/10.1016/b978-0-12-586450-3.50014-7

McMillan C (1976) Experimental studies on flowering and reproduction in seagrasses. Aquat Bot 2:87–92. https://doi.org/10.1016/0304-3770(76)90011-5

Mendelssohn IA, Morris JT (2000) Eco-physiological controls on the productivity of Spartina Alterniflora Loisel. In: Weinstein MP, Kreeger DA (eds) Concepts and controversies in tidal marsh ecology. Springer Netherlands, Dordrecht, pp 59–80

Merino JH, Huval D, Nyman AJ (2010) Implication of nutrient and salinity interaction on the productivity of Spartina patens. Wetl Ecol Manag 18(2):111–117. https://doi.org/10.1007/s11273-008-9124-4

Middelboe AL, Markager S (1997) Depth limits and minimum light requirements of freshwater macrophytes. Freshw Biol 37(3):553–568. https://doi.org/10.1046/j.1365-2427.1997.00183.x

Mitsch WJ, Gosselink JG (2000) Wetlands. Wiley, Hoboken, NJ, p 752

Montagna PA, Cockett PM, Kurr EM, Trungale J (2020) Assessment of the relationship between freshwater inflow and biological indicators in Lavaca Bay. Final report to the Texas Water Development Board for contract # 1800012268. Texas A&M University-Corpus Christi, Corpus Christi, TX, p 115

Morgan MD, Kitting CL (1984) Productivity and utilization of the seagrass Halodule wrightii and its attached epiphytes. Limnol Oceanogr 29(5):1066–1076. https://doi.org/10.4319/lo.1984.29.5.1066

Morgane L, Dunton KH (2006) Effects of drift macroalgae and light attenuation on chlorophyll fluorescence and sediment sulfides in the seagrass Thalassia testudinum. J Exp Mar Biol Ecol 334(2):174–186. https://doi.org/10.1016/j.jembe.2006.01.024

National Oceanic and Atmospheric Administration (2016) Office for Coastal Management. Coastal Change Analysis Program (C-CAP) Regional Land Cover, NOAA Of ice for Coastal Management, conus_2016_ccap_landcover_20200311.tif.

Nielsen SL, Sand-Jensen K (1989) Regulation of photosynthetic rates of submerged rooted macrophytes. Oecologia 81:364–368

Nielsen-Gammon J, Escobedo J, Ott C, Dedrick J, van Fleet A (2020) Assessment of Historic and Future Trends of Extreme Weather in Texas, 1900–2036. Report OSC-202001. Texas A&M University, College Station, TX, p 40

Odum WE, Smith TJ III, Hoover JK, McIvor CC (1984) The ecology of tidal freshwater marshes of the United States East Coast: a community profile. Report FWS/OBS-83/17. National Coastal Ecosystems Team, U.S. Fish and Wildlife Service, Slidell, Louisiana, p 177

Onuf CP (1994) Seagrasses, dredging and light in Laguna Madre, Texas, USA. Estuar Coast Shelf S 39(1):75–91. https://doi.org/10.1006/ecss.1994.1050

Onuf C (1996) Seagrass responses to long-term light reduction by brown tide in upper Laguna Madre, Texas: distribution and biomass patterns. Mar Ecol Prog Ser 138:219–231. https://doi.org/10.3354/meps138219

Orth RJ, Carruthers TJB, Dennison WC, Duarte CM, Fourqurean JW, Heck KL, Hughes AR, Kendrick GA, Kenworthy WJ, Olyarnik S, Short FT, Waycott M, Williams SL (2006) A global crisis for seagrass ecosystems. Bioscience 56(12):987–996. https://doi.org/10.1641/0006-3568(2006)56[987:AGCFSE]2.0.CO;2

Osland NJ, Enwright N, Stagg CL (2014) Freshwater availability and coastal wetland foundation species: ecological transitions along a rainfall gradient. Ecology 95(10):2789–2802

Osland MJ, Day RH, Hall CT, Feher LC, Armitage AR, Cebrian J, Dunton KH, Hughes AR, Kaplan DA, Langston AK, Macy A, Weaver CA, Anderson GH, Cummins K, Feller IC, Snyder CM (2020) Temperature thresholds for black mangrove (Avicennia germinans) freeze damage, mortality and recovery in North America: refining tipping points for range expansion in a warming climate. J Ecol 108(2):654–665. https://doi.org/10.1111/1365-2745.13285

Paine JG, Mathew S, Caudle T (2012) Historical shoreline change through 2007, Texas Gulf Coast: rates, contributing causes, and Holocene context. Gulf Coast Assoc of Geol Soc J 1:13–26

Palmer TA, Montagna PA, Kalke RD (2002) Downstream effects of restored freshwater inflow to Rincon Bayou, Nueces Delta, Texas, USA. Estuaries 25(6):1448–1456. https://doi.org/10.1007/BF02692238

Parida AK, Jha B (2010) Salt tolerance mechanisms in mangroves: a review. Trees 24(2):199–217. https://doi.org/10.1007/s00468-010-0417-x

Patriquin DG (1972) The origin of nitrogen and phosphorus for growth of the marine angiosperm *Thalassia testudinum*. Mar Biol 15(1):35–46. https://doi.org/10.1007/BF00347435

Pearcy RW, Ustin SL (1984) Effects of salinity on growth and photosynthesis of three California tidal marsh species. Oecologia 62(1):68–73. https://doi.org/10.1007/BF00377375

Pennings SC, Bertness MD (2001) Salt marsh communities. In: Bertness MD, Gaines SD, Hay ME (eds) Marine community ecology. Sinauer Associates, Sunderland, MA, pp 289–316

Pennings SC, Callaway RM (1992) Salt marsh plant zonation: the relative importance of competition and physical factors. Ecology 73(2):681–690

Pennings SC, Moore DJ (2001) Zonation of shrubs in western Atlantic salt marshes. Oecologia 126(4):587–594

Pennings SC, Richards CL (1998) Effects of wrack burial in salt-stressed habitats: *Batis Maritima* in a Southwest Atlantic salt marsh. Ecography 21(6):630–638

Pérez-Ceballos R, Zaldívar-Jiménez A, Melgarejo-Salas S, Canales-Delgadillo J, López-Portillo J, Merino-Ibarra M, Celis-Hernandez O, Lara-Domínguez AL, Ochoa-Gómez J (2022) Porewater sulfide: the most critical regulator in the degradation of mangroves dominated by tides. Forests 13(8):1–15. https://doi.org/10.3390/f13081307

Pezeshki SR, DeLaune RD, Meeder JF (1997) Carbon assimilation and biomass partitioning in *Avicennia germinans* and *Rhizophora mangle* seedlings in response to soil redox conditions. Environ Exp Bot 37(2–3):161–171. https://doi.org/10.1016/S0098-8472(96)01051-9

Pulich WM (1985) Seasonal growth dynamics of *Ruppia maritima* L. s.l. and *Halodule wrightii* Aschers. In southern Texas and evaluation of sediment fertility status. Aquat Bot 23(1):53–66. https://doi.org/10.1016/0304-3770(85)90020-8

Pulich WM (1990). Effects of freshwater inflows on estuarine vascular plants of Texas bay systems. I. San Antonio Bay. Report to Texas Water Development Board, by Resource Protection Division, Texas Parks and Wildlife Department, Austin, TX.

Pulich W Jr (2007) Galveston Bay System. In: Handley L, Altsman D, DeMay R (eds) Seagrass status and trends in the Northern Gulf of Mexico: 1940–2002. US Geological Survey Scientific Investigations Report 2006–5287 and US Environmental Protection Agency Report 855-R-04-003, pp 7–16

Pulich W Jr, Onuf C (2007) Statewide Summary for Texas. In: Handley L, Altsman D, DeMay R (eds) Seagrass status and trends in the Northern Gulf of Mexico: 1940–2002. US Geological Survey Scientific Investigations Report 2006–5287 and US Environmental Protection Agency Report 855-R-04-003, pp 7–16

McMillan C, Moseley FN (1967) Salinity Tolerances of Five Marine Spermatophytes of Redfish Bay. Texas. Ecology 48(3):503–506. https://doi.org/10.2307/1932688

Pennings SC, Glazner RM, Hughes ZJ, Kominoski JS, Armitage AR (2021) Effects of mangrove cover on coastal erosion during a hurricane in Texas, USA. Ecology 102(4). https://doi.org/10.1002/ecy.v102.410.1002/ecy.3309

Rasser MK (2009) The role of biotic and abiotic processes in the zonation of salt marsh plants in the Nueces River Delta, Texas. Ph.D. Dissertation. The University of Texas at Austin, Austin, TX, p 166

Rasser MK, Fowler NL, Dunton KH (2013) Elevation and plant community distribution in a microtidal salt marsh of the western Gulf of Mexico. Wetlands 33(4):575–583

Reef R, Winter K, Morales J, Adame MF, Reef DL, Lovelock CE (2015) The effect of atmospheric carbon dioxide concentrations on the performance of the mangrove *Avicenia germinans* over a range of salinities. Physiol Plantarum 154(3):358–368. https://doi.org/10.1111/ppl.12289

Ricart AM, York PH, Bryant CV, Rasheed MA, Ierodiaconou D, Macreadie PI (2020) High variability of blue carbon storage in seagrass meadows at the estuary scale. Sci Rep 10(1):5865. https://doi.org/10.1038/s41598-020-62639-y

Riche, C (2020) Plant guide for gulf cordgrass [*Spartina spartinae* (Trin.) Merr. ex A.S. Hitchc.]. U.S.D.A. Natural Resources Conservation Service, Golden Meadow Plant Materials Center, Galliano, Louisiana, 2 pp.

Rivers WG, Weber DJ (1971) The influence of salinity and temperature on seed germination in *Salicornia bigelovii*. Physiol Plantarum 24(1):73–75. https://doi.org/10.1111/j.1399-3054.1971.tb06719.x

Sadro S, Gastil-Buhl M, Melack J (2007) Characterizing patterns of plant distribution in a southern California salt marsh using remotely sensed topographic and hyperspectral data and local tidal fluctuations. Remote Sens Environ 110(2):226–239

Santos IR, Burdige DJ, Jennerjahn TC, Bouillon S, Cabral A, Serrano O, Wernberg T, Filbee-Dexter K, Guimond JA, Tamborski JJ (2021) The renaissance of Odum's outwelling hypothesis in "blue carbon" science. Estuar Coast Shelf S 255(107361):1–11

Scifres CJ, McAtee JW, Drawe DL (1980) Botanical, edaphic, and water relationships of gulf cordgrass (*Spartina spartinae* [Trin.] Hitchc.) and associated communities. Southwest Nat 25(3):397–409. https://doi.org/10.2307/3670697

Shideler G (1984) Suspended sediment responses in a wind-dominated estuary of the Texas Gulf Coast. J Sediment Res 54(3):731–745. https://doi.org/10.1306/212F84E5-2B24-11D7-8648000102C1865D

Shields EC, Moore KA (2016) Effects of sediment and salinity on the growth and competitive abilities of three submersed macrophytes. Aquat Bot 132:24–29. https://doi.org/10.1016/j.aquabot.2016.03.005

Smith NP (1979) Tidal dynamics and low-frequency exchanges in the Aransas Pass, Texas. Estuaries 2(4):218–227. https://doi.org/10.2307/1351568

Smith SV, Hollibaugh JT (1993) Coastal metabolism and the oceanic organic carbon balance. Rev Geophys 31(1):75–89. https://doi.org/10.1029/92RG02584

Snedaker SC, Araújo RJ (1998) Stomatal conductance and gas exchange in four species of Caribbean mangroves exposed to ambient and increased $CO_2$. Mar Freshw Res 49(4):325–327. https://doi.org/10.1071/MF98001

Solis RS, Powell G (1999) Hydrography, mixing characteristics, and residence times of Gulf of Mexico estuaries. In: Bianchi TS, Pennock JR, Twilley RR (eds) Biogeochemistry of Gulf of Mexico estuaries. Wiley, Hoboken, NJ, pp 29–61

Søndergaard M (2009) Redox Potential. In: Likens G (ed) Encyclopedia of inland waters. Pergamon Press, Oxford, pp 852–859. https://doi.org/10.1016/B978-012370626-3.00115-0

Stachelek J, Dunton KH (2013) Freshwater inflow requirements for the Nueces Delta, Texas: *Spartina alterniflora* as an indicator of ecosystem condition. Texas Water J 4(2):62–73

Steward KK, Van TK (1987) Comparative studies of monoecious and dioecious *Hydrilla* (*Hydrilla verticillata*) biotypes. Weed Sci 35(2):204–210. https://doi.org/10.1017/S0043174500079066

Suárez N, Medina E (2005) Salinity effect on plant growth and leaf demography of the mangrove, *Avicennia germinans* L. Trees 19(6):722–728. https://doi.org/10.1007/s00468-005-0001-y

Subedi SC, Allen P, Vidales R, Sternberg L, Ross M, Afkhami ME (2022) Salinity legacy: foliar microbiome's history affects mutualist-conferred salinity tolerance. Ecology 103(6):1–14. https://doi.org/10.1002/ecy.3679

Subudhi PK, Baisakh N (2011) *Spartina alterniflora* Loisel., a halophyte grass model to dissect salt stress tolerance. In Vitro Cell Dev – Pl 47(4):441–457. https://doi.org/10.1007/S11627-011-9361-8

Tang L, Gao Y, Li BO, Wang Q, Wang C-H, Zhao AB, Tang L, Gao Y, Li B, Wang Q, Wang C-H, Zhao B (2014) *Spartina alterniflora* with high tolerance to salt stress changes vegetation pattern by outcompeting native species. Ecosphere 5(9):1–18. https://doi.org/10.1890/ES14-00166.1

Texas Department of Water Resources (1980a) Guadalupe estuary: a study of the influence of freshwater inflows. Report LP-107. Texas Department of Water Resources, Austin, TX, p 339

Texas Department of Water Resources (1980b) Lavaca-Tres Palacios estuary: a study of the influence of freshwater inflows. Report LP-106. Texas Department of Water Resources, Austin, TX, p 349

Texas Department of Water Resources (1981) Sabine-Neches estuary: a study of the influence of freshwater inflows. LP-116. Texas Department of Water Resources, Austin, TX, p 322

Texas Department of Water Resources (1982) Trinity-San Jacinto estuary: an analysis of bay segment boundaries, physical characteristics, and nutrient processes. Report LP-086. Texas Department of Water Resources, Austin, TX, p 83

Texas Parks and Wildlife Department (1999) Seagrass conversation plan for Texas. Texas Parks and Wildlife, Austin, TX, p 84

Thorhaug AL, Poulos HM, López-Portillo J, Ku TC, Berlyn GP (2017) Seagrass blue carbon dynamics in the Gulf of Mexico: stocks, losses from anthropogenic disturbance, and gains through seagrass restoration. Sci Total Environ 605–606:626–636. https://doi.org/10.1016/j.scitotenv.2017.06.189

Thorhaug AL, Poulos HM, López-Portillo J, Barr J, Lara-Domínguez AL, Ku TC, Berlyn GP (2019) Gulf of Mexico estuarine blue carbon stock, extent and flux: mangroves, marshes, and seagrasses: a north American hotspot. Sci Total Environ 653:1253–1261. https://doi.org/10.1016/j.scitotenv.2018.10.011

Tiner R (1999) Wetland indicators: A guide to Wetland identification, delineation, classification, and mapping, 1st edn. CRC Press LLC, Boca Raton, FL, p 418

Titus JE, Adams MS (1979) Coexistence and the comparative light relations of the submersed macrophytes *Myriophyllum spicatum* L. and *Vallisneria americana* Michx. Oecologia 40(3):273–286. https://doi.org/10.1007/BF00345324

Tomasko DA, Hall MO (1999) Productivity and biomass of the seagrass *Thalassia testudinum* along a gradient of freshwater influence in Charlotte Harbor, Florida. Estuaries 22(3):592–602. https://doi.org/10.2307/1353047

Tootoonchi M, Gettys LA, Thayer KL, Markovich IJ, Sigmon JW, Sadeghibaniani S (2020) Ecotypes of aquatic plant *Vallisneria americana* tolerate different salinity concentrations. Diversity 12(2):1–16. https://doi.org/10.3390/d12020065

Valiela I, Teal JM, Allen SD, Van Etten R, Goehringer D, Volkmann S (1985) Decomposition in salt marsh ecosystems: the phases and major factors affecting disappearance of above-ground organic matter. J Exp Mar Biol Ecol 89(1):29–54

Vasquez EA, Glenn EP, Guntenspergen GR, Brown JJ, Nelson SG (2006) Salt tolerance and osmotic adjustment of *Spartina alterniflora* (Poaceae) and the invasive M haplotype of *Phragmites australis* (Poaceae) along a salinity gradient. Am J Bot 93(12):1784–1790. https://doi.org/10.3732/AJB.93.12.1784

Waycott M, Duarte CM, Carruthers TJB, Orth RJ, Dennison WC, Olyarnik S, Calladine A, Fourqurean JW, Heck KL, Hughes AR, Kendrick GA, Kenworthy WJ, Short FT, Williams SL (2009) Accelerating loss of seagrasses across the globe threatens coastal ecosystems. P Natl Acad Sci USA 106(30):12377–12381. https://doi.org/10.1073/pnas.0905620106

Weaver CA, Armitage AR (2018) Nutrient enrichment shifts mangrove height distribution: implications for coastal woody encroachment. PLoS One 13(3):e0193617. https://doi.org/10.1371/journal.pone.0193617

Weilhoefer CL (1998) Effects of freshwater inflow, salinity and nutrients on salt marsh vegetation in South Texas. MS Thesis. University of Texas at Austin, Port Aransas, TX

Wetz MS, Cira EK, Sterba-Boatwright B, Montagna PA, Palmer TA, Hayes KC (2017) Exceptionally high organic nitrogen concentrations in a semi-arid South Texas estuary susceptible to brown tide blooms. Estuar Coast Shelf S 188:27–37. https://doi.org/10.1016/j.ecss.2017.02.001

White WA, Calnan TR (1990) Sedimentation and historical changes in fluvial-deltaic wetlands along the Texas Gulf Coast with emphasis on the Colorado and Trinity River deltas. Interagency contract (88–89) 1423. Texas Parks and Wildlife Department, Austin, TX, p 124

White WA, Calnan TR, Morton RA, Kimble RS, Littleton TG, McGowen JH, Nance HS, Schmedes KE (1983) Submerged lands of Texas, Corpus Christi area: sediments, geochemistry, benthic macroinvertebrates, and associated wetlands. University of Texas at Austin, Bureau of Economic Geology, Austin, TX, p 154

White WA, Calnan TR, Morton RA, Kimble RS, Littleton TG, McGowen JH, Nance HS, Schmedes KE (1985) Submerged lands of Texas, Galveston - Houston area: sediments, geochemistry, benthic macroinvertebrates, and associated wetlands. University of Texas at Austin, Bureau of Economic Geology, Austin, TX, p 145

White WA, Calnan TR, Morton RA, Kimble RS, Littleton TG, McGowen JH, Nance HS, Schmedes KE (1986) Submerged lands of Texas, Brownsville-Harlingen area: sediments, geochemistry, benthic macroinvertebrates, and associated wetlands. University of Texas at Austin, Bureau of Economic Geology, Austin, TX, p 138

White WA, Calnan TR, Morton RA, Kimble RS, Littleton TG, McGowen JH, Nance HS, Schmedes KE (1987) Submerged lands of Texas, Beaumont – Port Arthur area: sediments, geochemistry, benthic macroinvertebrates, and associated wetlands. University of Texas at Austin, Bureau of Economic Geology, Austin, Texas, p 110

White WA, Calnan TR, Morton RA, Kimble RS, Littleton TG, McGowen JH, Nance HS, Schmedes KE (1988) Submerged lands of Texas, Bay City - Freeport area: sediments, geochemistry, benthic macroinvertebrates, and associated wetlands. University of Texas at Austin, Bureau of Economic Geology, Austin, TX, p 130

White WA, Calnan TR, Morton RA, Kimble RS, Littleton TG, McGowen JH, Nance HS, Schmedes KE (1989a) Submerged lands of Texas, Kingsville area: sediments, geochemistry, benthic macroinvertebrates, and associated wetlands. University of Texas at Austin, Bureau of Economic Geology, Austin, TX, p 137

White WA, Calnan TR, Morton RA, Kimble RS, Littleton TG, McGowen JH, Nance HS, Schmedes KE (1989b) Submerged lands of Texas, Port Lavaca area: sediments, geochemistry, benthic macroinvertebrates, and associated wetlands. University of Texas at Austin, Bureau of Economic Geology, Austin, TX, p 165

Wilkinson RE (1961) Effects of reduced sunlight on water stargrass (*Heteranthera dubia*). Weeds 9(3):457–462. https://doi.org/10.2307/4040864

Williams SL (1990) Experimental studies of Caribbean seagrass bed development. Ecol Monogr 60(4):449–469. https://doi.org/10.2307/1943015

Wilson SS, Dunton KH (2018) Hypersalinity during regional drought drives mass mortality of the seagrass *Syringodium filiforme* in a subtropical lagoon. Estuar Coast 41(3):855–865. https://doi.org/10.1007/s12237-017-0319-x

Woodell SRJ (1985) Salinity and seed germination in coastal plants. Plant Ecol 61:223–229

Yeo AR, Flowers TJ (1980) Salt tolerance in the halophyte *Suaeda maritima* L. Dum.: evaluation of the effect of salinity upon growth. J Exp Bot 31(4):1171–1183. https://doi.org/10.1093/jxb/31.4.1171

Zedler JB, Beare PA (1986) Temporal variability of salt marsh vegetation: the role of low-salinity gaps and environmental stress. In: Wolfe D (ed) Estuarine Variability. Academic Press, New York, NY, pp 295–306

Zedler JB, Onuf CP (1984) Biological and physical filtering in arid-region estuaries: seasonality, extreme events, and effects of watershed modification. In: Kennedy VS (ed) The estuary as a filter. Academic Press, New York, NY, pp 415–432

Zedler JB, Covin J, Nordby C, Williams P, Boland J (1986) Catastrophic events reveal the dynamic nature of salt-marsh vegetation in southern California. Estuaries 9(1):75–80

Zedler JB, Morzaria-Luna H, Ward K (2003) The challenge of restoring vegetation on tidal, hypersaline substrates. Plant Soil 253:259–273

Ziegler S, Benner R (1998) Ecosystem metabolism in a subtropical, seagrass-dominated lagoon. Mar Ecol Prog Ser 173:1–12. https://doi.org/10.3354/meps173001

Zimmerman RJ, Minello TJ, Smith DL, Kostera J (1990) The use of *Juncus* and *Spartina* marshes by fisheries species in Lavaca Bay, Texas, with reference to effects of floods. NOAA technical memorandum NMFS-SEFC-251. National Marine Fisheries Service, Galveston, TX, p 40

Zomer RJ, Trabucco A, van Straaten O, Bossio DA (2006) Carbon, land and water: a global analysis of the hydrologic dimensions of climate change mitigation through afforestation/reforestation. IWMI research report 101. International Water Management Institute, Colombo, p 44

# Effect of Freshwater Inflow on Benthic Infauna

Paul A. Montagna ⓘ, Richard D. Kalke, and Larry J. Hyde ⓘ

## Abstract

Benthic organisms are ideal bioindicators of freshwater inflow effects in bays and estuaries because they are fixed in space and integrate ephemeral processes in the overlying water column over long periods of time. Freshwater inflow regulates water quality, which drives benthic abundance, productivity, diversity, and community structure. Texas estuaries have different long-term characteristic fauna that reflect the long-term average salinity and sediment conditions in each bay system. Within estuary systems, the secondary bays have distinct communities compared to the primary bays because secondary bays are closer to freshwater inflow sources and are more oligohaline and/or brackish in nature than primary bays that are more marine influenced. Similar responses occur within bay types over time when conditions change with droughts, floods, freezes, and major events, such as hurricanes. Bioindicators of freshwater inflow effects include four dominant species: the polychaetes, *Mediomastus ambiseta*, and *Streblospio benedicti*; the bivalve *Mulinia lateralis*; and the amphipod *Ampelisca abdita*. Each of these species is primarily found in secondary bays with similar salinities where abundances are higher than in primary bays. Because of the relationship between prevailing salinity conditions and benthos community structure, assessment of benthic conditions can be used to determine if a "sound ecological environment" exists in a given bay system.

## Keywords

Estuary · Infauna · Macrofauna · Sediments · Freshwater inflow

## List of Abbreviations

| | |
|---|---|
| AB | Aransas Bay |
| BB | Baffin Bay |
| CB | Copano Bay |
| CC | Corpus Christi Bay |
| EM | Eastern Arm of Matagorda Bay |
| gCm2 | Dry weight biomass converted to carbon, extrapolated to g per m$^2$ |
| GE | Guadalupe Estuary = San Antonio Bay System |
| gm2 | Dry weight biomass extrapolated to g per m$^2$ |
| H | Shannon Diversity index, *H'* |
| J | Pielou's Evenness index, *J'* |
| LB | Lavaca Bay |
| LC | Lavaca-Colorado Estuary = Matagorda Bay System |
| LM | Laguna Madre Estuary = Laguna Madre System, or Laguna Madre as a bay |
| Ln_gCm2 | Natural logarithm biomass converted to carbon, extrapolated to g C per m$^2$ |
| Ln_gm2 | Natural logarithm biomass extrapolated to g per m$^2$ |
| Ln_nm2 | Natural logarithm number extrapolated to n per m$^2$ |
| LS | Lower San Antonio Bay |
| MA | Mission-Aransas Estuary = Aransas Bay System |
| MB | Matagorda Bay |
| N1 | Hill's *N*1, number of dominant species |
| NB | Nueces Bay |
| NC | Nueces Estuary = Corpus Christi Bay System |

**Supplementary Information** The online version contains supplementary material available at https://doi.org/10.1007/978-3-031-70882-4_11.

P. A. Montagna (✉) · R. D. Kalke · L. J. Hyde
Texas A&M University-Corpus Christi, Harte Research Institute for Gulf of Mexico Studies, Corpus Christi, TX, USA
e-mail: Paul.Montagna@taumcc.edu; Rick.Kalke@tamucc.edu; Larry.Hyde@tamucc.edu

| nm2 | Number of individuals extrapolated to number per $m^2$ |
| R | Richness, the number of species (S) |
| RB | Rincon Bayou |
| SN | Sabine-Neches Estuary = Sabine Lake System |
| TCEQ | Texas Commission of Environmental Quality |
| TPWD | Texas Parks & Wildlife Division |
| TS | Trinity-San Jacinto Estuary = Galveston Bay System |
| TWDB | Texas Water Development Board |
| US | Upper San Antonio Bay |
| USGS | United States Geological Survey |

## Introduction

The term benthos refers to freshwater or marine bottom sediments and associated biological communities typically including a diverse range of both plants and animals. There are five categories of benthic animals defined by their relative size and the microhabitats they occupy. These classifications are based on size and habitat (Fig. 1). The microfauna are < 0.063 mm. These unicellular organisms, also known as protists, are ubiquitous throughout in the upper few centimeters of sediment but are most abundant in the more highly oxygenated sediments near the surface. Meiofauna range in size between 0.063 and 0.5 mm and typically burrow in sediment or live on the surface. Macrofauna are >0.5 mm in size and burrow, dwell in tubes, or crawl on the surface of sediments. Macrofauna that live on the surface are called epifauna and macrofauna that live in the sediment are called infauna. Megafauna are the largest with a length of 1 cm or greater. They occur both above and below the sediment surface.

The size ranges relate to trophic dynamics (Montagna et al. 1996). Microfauna are unicellular protists and are mostly decomposers. The meiofauna feed on microfauna as well as each other. Macrofauna feed on microfauna, phytoplankton, microphytobenthos, detritus, and meiofauna and can be cannibalistic. Larger mobile epifauna (such as shrimp and crabs) and demersal fish feed on macrofauna, as do birds.

In this chapter, the focus is on macrofauna. There are three dominant taxa: Polychaeta (which are annelid worms), Crustacea, and Mollusca. Typically, the polychaetes are dominant, with mollusks, and crustaceans following.

## Why Benthos?

Analysis of benthic invertebrate communities has been widely used as bioindicators in assessment and monitoring studies worldwide. These organisms serve as proverbial "canaries in the coal mine." The pattern of their presence or health is an indicator of environmental health or stress. At a minimum, indicator organisms should have (at least) five characteristics that make them useful to detect change (Soule 1988): (1) They should direct our attention to qualities of the environment. (2) They should provide a sign that some environmental characteristic is present. (3) They should express a generalization about the environment. (4) They should suggest a cause, outcome, or remedy. (5) Finally, they should show a need for action.

Benthic organisms have been especially useful as bioindicators broadly in environmental research (Dauer 1993; Borja et al. 2019). Specifically, benthic community indices have been widely recognized as one of the best indicators of ecosystem health, as illustrated by the adoption of estuarine biotic integrity indices by the US Environmental Protection Agency (Engle and Summers 1999; USEPA 1999). To assess environmental stress in general, biotic indices have been calibrated to adjust for salinity effects (Gillett et al. 2015). There are many reasons benthic organisms are good indicators of environmental health. They are relatively long-lived, fixed in place, integrate variations in the overlying water column over time, and are forage for commercial and recreational fish species. Benthos are usually the first organisms affected by environmental disturbances. Gravity coupled with other natural forces ensure that diverse biotic and abiotic materials from atmospheric, terrestrial, and aquatic sources are ultimately deposited and concentrated in the sediments of low energy coastal ecosystems. The organic material from terrestrial, freshwater, and marine sources ends up in the detrital food chain, which is utilized by the benthos. Pollutants are usually tightly coupled to organic matrices; therefore, benthos have great exposure through their niche (food) and habitat (living spaces) to pollutants. Benthos are also key components of the food web and are often forage for commercial and recreational fish species. Because benthos are relatively long-lived and sessile, so they integrate pollutants effects of over long temporal and spatial scales. Bioturbation and irrigation of sediments by benthos affect the mobilization and burial of xenobiotic materials. Benthic invertebrates are also sensitive to change in environmental conditions, such as salinity, dissolved oxygen, and pollutants. Thus, benthic biodiversity loss associated with disturbance or environmental change is an excellent indicator of environmental stress.

## Concept of Sediment Quality

Sediments act as the memory of the ecosystem. Because of gravity, there is a record of past events and ecosystem change over time that accumulates and preserves evidence of ecological events in bottom sediments. These sediments become

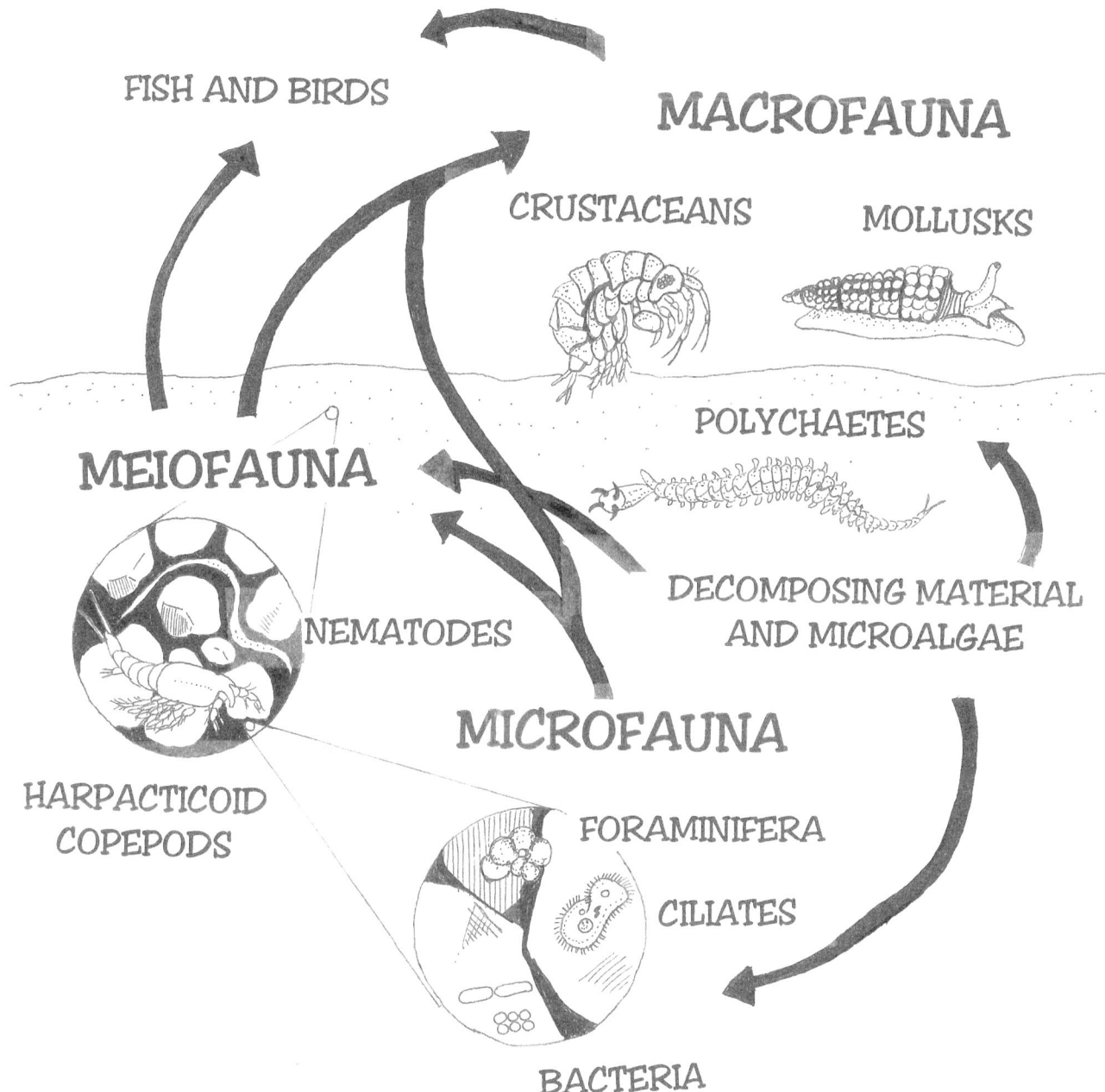

FISH AND BIRDS

MACROFAUNA

CRUSTACEANS          MOLLUSKS

POLYCHAETES

MEIOFAUNA

NEMATODES

DECOMPOSING MATERIAL
AND MICROALGAE

MICROFAUNA

HARPACTICOID
COPEPODS

FORAMINIFERA

CILIATES

BACTERIA

**Fig. 1** Size classes of sediment-bound benthic invertebrates and tropic relationships. From smallest to largest: one-celled microfauna, meiofauna, and macrofauna (Montagna et al. 1996)

intimately associated with exogenous natural and anthropogenic materials that enter the system. The analysis of biodiversity and community structure of benthos provide powerful metrics to detect changes among sensitive species, which decrease in number or die out, versus tolerant species, which survive or thrive, during prolonged unfavorable conditions. Thus, analysis of estuarine benthic abundance, biomass, and diversity data has been used to evaluate spatial and temporal change in environmental health.

The original paradigm for freshwater inflow effects was based on a simple conceptual model of "*grow = flow*" where

inflow was expected to have a *direct* impact on community structure and population size. It is now recognized that freshwater inflow has very important *indirect* effects where inflow drives water quality conditions and water quality drives habitat quality, which is sediment quality for benthic organisms (Montagna 2021). This idea was first formalized into a management strategy by Alber (2002). The conceptual model used by Alber (2002) was based on a previous quantitative model using the cumulative impacts on ecosystem processes as a function of changes in freshwater, sediment, and nutrient inflows created by Sklar and Browder (1998). This paradigm of indirect effects was adopted by the statewide Science Advisory Committee (SAC 2009). The SAC was created by Texas 2007 Senate Bill 3 to provide guidance to all environmental flow science and stakeholder teams responsible for making inflow recommendations to the Texas Commission on Environmental Quality (TCEQ). The conceptual model developed by these earlier efforts was refined based on benthic studies by Palmer et al. (2011) and Montagna et al. (2013) and named the Domino Theory. Benthic studies conducted in Texas estuaries have demonstrated that long-term hydrological cycles affecting freshwater inflow also drive water quality (Montagna et al. 2013; Palmer et al. 2011; Paudel and Montagna 2014; Palmer and Montagna 2015) and benthic abundance (Pollack et al. 2011), productivity (Montagna and Li 2010; Kim and Montagna 2012), diversity (Montagna et al. 2002; Van Diggelen and Montagna 2016), and community structure (Montagna and Kalke 1992, 1995; Ritter et al. 2005).

## Bioindicator Approach

There are ecological models that provide a scientific basis for interpreting the effects of ecological disturbances regardless of natural or anthropogenic origin. These models are based on the use of many individual species, community studies, and statistical models. Among these models, one of the most important is the succession model proposed by Rhoads et al. (1978) and Pearson and Rosenberg (1978). They applied theories of ecological succession and its relation to productivity and community structure to suggest ways to assess risk due to dredge-spoil disposal and organic waste enrichment. The underlying concept in both papers is that distance from a source is analogous with time since disturbance. The idea is that succession after a natural disturbance proceeds in a predictable process over a given time period; thus, succession will proceed in an analogous way with distance from a pollution source. In both cases, disturbed communities have pioneer species (*r*-selected life history strategies, small, surface-dwelling infauna, numerous, and low diversity), and undisturbed communities have climax species (*k*-selected, large, deeper dwelling infauna, low abundances, and more diverse). One important prediction of this theory is that normal sediments will have a diverse assemblage of deeper dwelling organisms compared to a polluted or disturbed environment. Thus, we have a scientific explanation and justification for using community structure and biological diversity studies as an endpoint.

Application of the succession model by many studies after 1978 has demonstrated that benthic community structure and biodiversity are an excellent indicator of environmental health. The endpoints for benthic bioindicator

projects can be univariate (abundance, biomass, and/or diversity) or multivariate (community structure). Early studies relied almost exclusively on species-independent, univariate metrics of diversity. Diversity indices are influenced by sample size, and two environments can have different species, or distributions of species, and have the same diversity index (Ludwig and Reynolds 1988). Multivariate analysis of species-dependent data was successfully employed to remedy this problem and create a more statistically robust and sensitive method to detect environmental differences (Green 1979). Multivariate methods have been extended to synthesize interdisciplinary data including environmental factors (Green and Montagna 1996; Long et al. 2013). The later approach is a statistically robust way to simplify to reduce and synthesize large complex data sets.

## Methods

Many long-term studies were initiated and completed in Texas estuaries between 1984 (Kalke and Montagna 1991) and 2019 (Montagna 2021, 2023) to identify freshwater inflow effects on benthos. Sediment samples were collected at stations in the Lavaca-Colorado Estuary (LC), Guadalupe Estuary (GE), Mission-Aransas (MA), Nueces Estuary (NC), and Laguna Madre Estuary (LM). Water column samples were always processed within 30 days after collection, but the benthic samples were processed much later. Most of these projects were supported by the Texas Water Development Board.

Field sampling locations were selected along freshwater to saltwater gradients, and a common naming system was adopted. Stations A and B were located in secondary bays closer to sources of freshwater inflow. Following this gradient rationale, stations C and D were located in primary bays closer to exchange points with the Gulf of Mexico (Table 1, Fig. 2).

To identify effects of the Colorado River, stations E and F were added in January 1993 to the eastern arm of Matagorda Bay (EM). The eastern arm of Matagorda Bay is not East Matagorda Bay, which is northeast of the Colorado River and is a minor bay that receives no freshwater inflow (Palmer et al. 2011). To increase resolution of Colorado River influence, stations 8 and 15 were added in 2002 in EM. To examine potential influences of the Formosa Plastics industrial discharge into Lavaca Bay (Harris et al. 2023), station FD was added in 2007.

The Laguna Madre Estuary was unique compared to other Texas estuaries because it has no major river flowing into tertiary bays or Baffin Bay, which is the secondary bay. There were two stations (24 and 6, analogous to A and B, respectively) in Baffin Bay and two stations (155 and 189) in seagrass habitat in Laguna Madre (which is analogous to a

**Table 1** Locations of bays and stations (Sta) within the Guadalupe (GE), Lavaca-Colorado (LC), Nueces (NC), and Laguna Madre (LM) estuaries (Est) with start and end dates, and number of benthic samples

| Est | Bay (abbreviation) | Sta | Latitude | Longitude | Start | End | Samples |
|-----|--------------------|-----|----------|-----------|-------|-----|---------|
| LC | Lavaca (LB) | A | 28.67467 | –96.5827 | Nov-1984 | Jul-2019 | 411 |
| LC | Lavaca (LB) | FD | 28.68096 | –96.5822 | Apr-2007 | Jul-2019 | 147 |
| LC | Lavaca (LB) | B | 28.63868 | –96.5844 | Apr-1988 | Jul-2019 | 369 |
| LC | Matagorda (MB) | C | 28.54672 | –96.4689 | Apr-1988 | Jul-2019 | 369 |
| LC | Matagorda (MB) | D | 28.48502 | –96.2897 | Apr-1988 | Jul-2019 | 372 |
| LC | Matagorda (MB) | E | 28.55450 | –96.2155 | Jan-1993 | Jul-2019 | 276 |
| LC | Matagorda (MB) | 8 | 28.57639 | –96.1192 | Jul-2008 | Jul-2019 | 135 |
| LC | Eastern Matagorda (EM) | 15 | 28.61493 | –96.0236 | Jul-2008 | Jul-2019 | 135 |
| LC | Eastern Matagorda (EM) | F | 28.60463 | –96.0460 | Jan-1993 | Jul-2019 | 276 |
| GE | Upper San Antonio (US) | A | 28.39352 | –96.7724 | Jan-1987 | Jul-2019 | 348 |
| GE | Upper San Antonio (US) | B | 28.34672 | –96.7442 | Jan-1987 | Jul-2019 | 348 |
| GE | Lower San Antonio (LS) | C | 28.24934 | –96.7584 | Jan-1987 | Jul-2019 | 348 |
| GE | Lower San Antonio (LS) | D | 28.30210 | –96.6844 | Jan-1987 | Jul-2019 | 348 |
| MA | Copano (CB) | A | 28.07460 | –97.2191 | Apr-1988 | Apr-2003 | 36 |
| MA | Copano (CB) | B | 28.13228 | –97.0344 | Apr-1988 | Apr-2003 | 36 |
| MA | Aransas (AB) | C | 28.08882 | –96.9625 | Apr-1988 | Apr-2003 | 36 |
| MA | Aransas (AB) | D | 27.97975 | –97.0287 | Apr-1988 | Apr-2003 | 36 |
| NC | Nueces (NB) | A | 27.86069 | –97.4736 | Oct-1987 | Jul-2019 | 369 |
| NC | Nueces (NB) | B | 27.85708 | –97.4103 | Oct-1987 | Jul-2019 | 369 |
| NC | Corpus Christi (CC) | C | 27.82533 | –97.3521 | Oct-1987 | Jul-2019 | 375 |
| NC | Corpus Christi (CC) | D | 27.71280 | –97.1787 | Oct-1987 | Jul-2019 | 369 |
| NC | Corpus Christi (CC) | E | 27.79722 | –97.1508 | Oct-1990 | Jul-2019 | 341 |
| LM | Baffin (BB) | 6 | 27.27697 | –97.4269 | Nov-1988 | Jul-2000 | 171 |
| LM | Baffin (BB) | 24 | 27.26388 | –97.5514 | Apr-1988 | Jul-2000 | 156 |
| LM | Laguna Madre (LM) | 155G | 27.42413 | –97.3413 | Mar-1989 | Oct-1990 | 12 |
| LM | Laguna Madre (LM) | 155S | 27.42413 | –97.3413 | Mar-1989 | Jul-1989 | 9 |
| LM | Laguna Madre (LM) | 189G | 27.3499 | –97.3924 | Mar-1989 | Jul-2000 | 165 |
| LM | Laguna Madre (LM) | 189S | 27.3499 | –97.3924 | Apr-1988 | Jul-2000 | 174 |

primary bay). Sampling locations were located in and outside of seagrass beds denoted with a suffix G for seagrass and S for Sand.

## Field and Laboratory Methods

The benthic collection and analysis methods are described in detail in previous studies (Kalke and Montagna 1991; Montagna and Kalke 1992). Briefly, sediment samples were collected using a 6.7 cm diameter core tube (35.4 cm$^2$ area). Three replicates were taken per station at each sample date. Cores were sectioned at 0–3 and 3–10 cm to determine macrofauna distribution by depth, but depth differences are not reported here. Organisms were extracted on a 0.5 mm sieve and enumerated to the lowest taxonomic level possible, usually the species level. Biomass was determined for higher taxonomic groupings by drying at 55 °C for 24 h. Calcium carbonate mollusk shells were dissolved by acid fumigation and not included in the biomass measurements. Biomass was converted from dry weight (DW) to carbon (C) using conversion factors found in Brey (2001) as described in Brey et al. (2010).

Several diversity indices were calculated (Ludwig and Reynolds 1988). Species richness ($R$) is the number of species in a sample regardless of abundance. The Shannon index ($H'$) accounts for species number by estimating the average uncertainty per species with proportional abundances. The Shannon index is calculated by: $H' = -\sum[(n_i/n)\ \ln(n_i/n)]$, where $n_i$ is the number of individuals belonging to the $i$th of S species in the sample and $n$ is the total number of individuals in the sample. Dominance diversity is calculated using Hill's diversity number one ($N1$). It is a measure of the effective number of species in a sample and indicates the number of abundant species. It is calculated as the exponentiated form of the Shannon diversity index: $N1 = e^{H'}$. As diversity decreases, $N1$ will tend toward 1. Evenness is an index to identify the extent to which all species in a sample are equally abundant. The most common form is Pielou's $J'$, and it expresses $H'$ relative to the maximum value of $H'$: $J' = \ln(N1) / \ln(R)$.

Water quality measurements were made just beneath the surface (i.e., within the top 10 cm) and at the bottom of the water column (i.e., within 10 cm of the bottom) at all stations on every sampling date. Analytical procedures are described in detail in previous studies (Paudel and Montagna 2014;

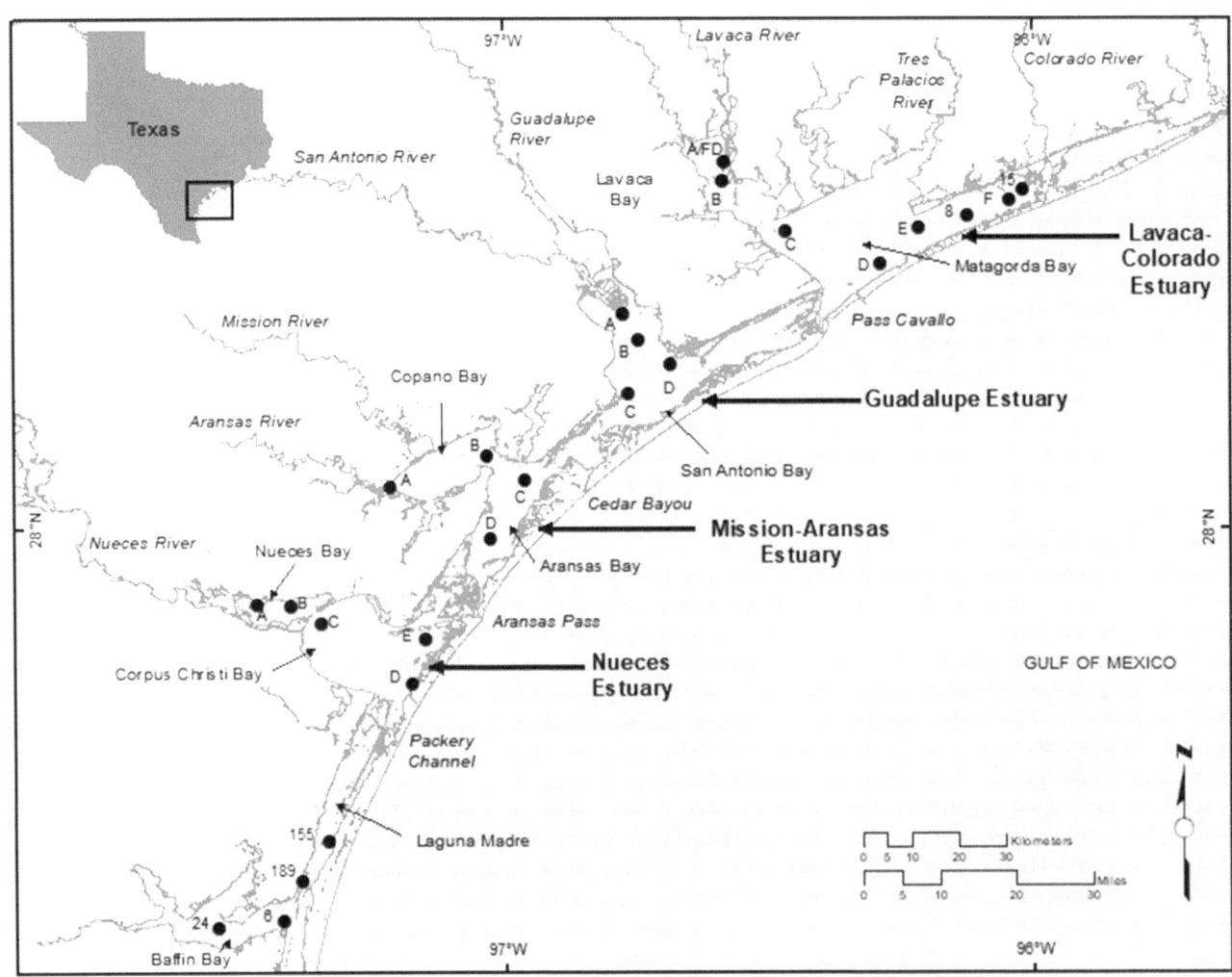

**Fig. 2** Station locations in the five Coastal Bend estuaries

Montagna et al. 2018b). Briefly, salinity, temperature, pH, and dissolved oxygen (DO) were measured with multiparameter sondes. Water was filtered, chlorophyll was measured fluorometrically, and nutrients were measured with an autoanalyzer.

## Analytical Approach

For univariate and multivariate statistical analyses, abundance was transformed by the square root of abundance or count per sample. Transformations improve the performance of analyses by making the distributions more nearly normal and decreasing the weight of the dominant species. Diversity and water quality metrics were not transformed because the distributions were more approximately normal. The analytical methods below were developed during previous studies (Montagna et al. 2002; Pollack et al. 2009; Montagna 2022, 2023).

## Estuary and Bay Differences

Univariate analysis of variance (ANOVA) was used to determine if there were differences among estuaries, bays, and sampling dates. A partially hierarchical analysis design was used because bays are unique to estuaries, that is, bays are nested within estuaries. Also, each station is unique to each bay within an estuary. Thus, estuaries, bays, and stations are completely nested, random effects. In addition, stations can be thought of as a form of replication for bay differences. Sampling dates, however, are a fixed effect variable. Thus, the ANOVA model is a two-way, partially hierarchical design that can be described by the following formula: $Y_{ijkl} = \mu + \alpha_j + \beta_k + \beta\gamma_{k(l)} + \beta\gamma\delta_{k(lm)} + \alpha\beta\gamma\delta_{jk(l)} + e_{(i)jklm}$ where $Y_{ijklm}$ is the dependent response variable; $\mu$ is the overall sample mean; $\alpha_j$ is the main fixed effect for sampling dates where $j = 1, 2, 3, \ldots, 139$ for each quarter; $\beta_k$ is the main fixed effect for estuary where $k = 1, 2, \ldots, 5$ for Lavaca-Colorado Estuary, Guadalupe Estuary, Mission-Aransas Estuary, Nueces Estuary, and Laguna Madre Estuary; $\beta\gamma_{k(l)}$ is the main effect for bays that are nested (or unique) within each estuary and are thus a random effect as denoted by the parentheses around the subscript $l$ that represents the 11 bays (Lavaca Bay, Matagorda Bay, eastern arm of Matagorda Bay, Upper San Antonio Bay, Lower San Antonio Bay, Copano Bay, Aransas Bay, Nueces Bay, Corpus Christi Bay, Baffin Bay, and Laguna Madre); $\beta\gamma\delta_{k(lm)}$ is the main effect for stations that are nested with bays; $\alpha\beta\gamma\delta_{jk(lm)}$ is the interaction term for date, estuary, bay, and station; and $e_{(i)jkl}$ is the random error term for each of the $i$ replicate measurements. The expected mean squares for the $F$-tests were calculated for each source of variation and the interaction term. For water quality, there were no replicates per stations, so the interaction term is deleted, so that the model is not over-specified.

## Community Structure

Multivariate community structure of macrofauna species was analyzed by nonmetric multidimensional scaling (nMDS) and cluster analysis using a Bray-Curtis similarity matrix (Clarke and Gorley 2015; Clarke et al. 2014). The nMDS was used to compare numbers of individuals of each species for each station-date combination. The distance between station-date combinations can be related to community similarities or differences between different stations. Cluster analysis determines how much each station-date combination resembles each other based on species abundances. The percent resemblance can then be displayed on the nMDS plot to elucidate grouping of station-date combinations. The group average cluster mode was used for the cluster analysis to identify different groups.

## Linking Sediment Communities to Inflow and Water Quality

Multivariate analyses were used to analyze how the physical-chemical environment changes over time. The physical-chemical water column characteristics were analyzed using Principal Component Analysis (PCA). PCA reduces multiple environmental variables into a smaller set of uncorrelated variables that explain much of the variation in the original data. The new variables are component loads, which describe the variance of the underlying structure in a data set. Components are extracted in order of importance based on the eigenvalue, or weight, of each factor on the overall model, and factors with eigenvalues greater than 1 are usually considered. Data were normalized to a mean of zero and standard deviation of one prior to PCA. A nMDS was also performed on water quality variables, using a normalization transformation and Euclidean distances to create the resemblance matrix.

Community structure is linked with environmental variables using the nonmetric multivariate BIO-ENV and RELATE procedures calculated with PRIMER software (Clarke and Gorley 2015). The BIO-ENV procedure calculates weighted Spearman rank correlations ($\rho$w) between sample ordinations from all environmental variables and an ordination of biotic variables. Correlations are then compared to determine the best match. The null hypothesis of no agreement in the multivariate patterns was tested using the rho ($\rho$) statistic.

Linkage between biotic response and water column conditions was also examined with correlation analysis using the Spearman rank correlation method.

Many benthic organisms have preferred salinity ranges. The max bin regression technique was used to identify the optimal salinity range for total community metrics. The approach is based on fitting maximum benthic response values in salinity bins using a log-normal, three-parameter model (Montagna et al. 2002; Turner and Montagna 2016):

$$Y = a \times \exp\left(-0.5 \times \left(\ln\frac{\left(\dfrac{X}{c}\right)}{b}\right)^2\right)$$

where $a$ is the peak of the dependent value ($Y$-axis benthic metric), $b$ is the skewness or rate of change of the response as a function of $X$ (salinity), and $c$ the optimal value of the independent variable on the $X$-axis (salinity). Maximum values in bins are used because responses can be low, even near zero, even during good environmental conditions because of seasonality, prior disturbance, cyclical patterns, or a variety of other reasons. Fitting the maximum values from bins fits the response curve through theoretical maximum responses. Regressions were based on the average bay-quarter time series for muddy stations only, that is, seagrass stations in Laguna Madre were deleted because benthic metrics are influenced by habitat and salinity.

## Time Series Analysis

The fundamental assumption when using long-term data is that changes over time in the drivers (which is freshwater inflow rates here) are affecting the response variables (which are the biological indicators here). However, there are several aspects of time series data that must be addressed because the change of the response variables from one time step to the next is dependent on the preceding environmental conditions and community state. Thus, autocorrelation is a key factor in time series data. Additionally, biological responses are not necessarily instantaneous, and there are usually lags in response to change because of the life cycles and growth rates of the organisms effected. The software package PROC ARIMA (SAS 2017) was used to calculate the Ljung-Box Chi-Square Test for white noise to determine if the series was random with no pattern. The software package PROC TIMESERIES (SAS 2017) was used to estimate the trend components.

For time series analysis, benthic abundance and biomass were averaged for all replicates in primary and secondary bays (Table 1) for each station-quarter combination, and the stations were averaged to create values by bay for each quarter. Two bay values were necessary because the secondary bay has more river influence than the primary bay with more marine influence. For diversity, the species counts in all three replicates were summed per station-quarter combination, and the stations were average by bay-quarter combination. Thus, diversity is reported for 105 cm$^2$ (i.e., 0.01 m$^2$). There is a data gap from 2000 to 2004 for the Guadalupe Estuary; therefore, the data from 2000 to 2004 is estimated using PROC EXPAND.

## Results

### Water Column Effects

An ANOVA of hydrographic metrics of dissolved oxygen, salinity, temperature, pH, ammonium (NH4), nitrite + nitrate (NO23), phosphate (PO4), silicate (SiO4), and chlorophyll (Chl) were different by date ($P$-Value = <0.0001) and bay (Table 2). There was no difference across estuaries.

Dissolved oxygen concentration is lowest at station D in the Nueces estuary (Table S1). The lowest salinity value was found in Upper San Antonio Bay, which also had the second highest value for dissolved oxygen and pH (Table 3). Salinity increases from stations (A and B) located closest to freshwater inflow to stations (C and D) located closest to exchange

with the Gulf of Mexico in all estuaries. In Matagorda Bay, station F is closest to the Colorado River, and E is in between D and F.

All nutrient concentrations decrease from near rivers to the sea (Table S2). Upper San Antonio Bay is different from the other secondary bays in that it has nitrate + nitrite (NO23) concentrations that are 17 times higher than Lavaca Bay, 10 times higher than Copano Bay and Nueces Bay, and 12 times higher than Baffin Bay (Table 4). Baffing Bay had the highest NH4, and the Eastern arm of Matagorda Bay had the second highest NH4.

Estuary conditions are defined by the relationship between freshwater inflow and water quality variables. This relationship is understood through the application of multivariate analysis and other statistical tools to classify stations. Principal Components (PC) analysis was performed on the water quality data. The first axis (PC1) explained 28% of variance in the data set represented by high nutrient and chlorophyll concentrations correlated with low salinities (Fig. 3a). Thus, PC1 is the new variable representing freshwater inflow and estuary condition effects. The second axis (PC2) explained 21% of the variability and is represented by low values of dissolved oxygen (DO) correlated with high temperatures. Thus, PC2 is the new variable related to seasonal effects. The third axis (PC3) explained 13% of the variability and is represented by high chlorophyll and pH values correlated with low NH4 values. The PC3 variable represents metabolic processes demonstrating that as chlorophyll and photosynthesis increase, carbon dioxide is consumed through a reduction process leading to higher pH. In contrast, ammonium levels increase as organic

**Table 2** ANOVA results for water column metrics. (**a**) Salinity (Sal), temperature (Temp), dissolved oxygen (DO), hydrogen ion concentration (pH). (**b**) Ammonium (NH4), nitrite + nitrate (NO23), phosphate (PO4), silicate (SiO4), chlorophyll a (Chl)

**(a)**

| N | $F$-Test | Source | DF | $P$-Value | | | |
| --- | --- | --- | --- | --- | --- | --- | --- |
| | | | | Sal | Temp | DO | pH |
| 1 | 1/5 | Date | 139 | <0.0001 | <0.0001 | <0.0001 | <0.0001 |
| 2 | 2/(3 + 4) | Est | 4 | 0.0155 | 0.0207 | 0.1286 | 0.1659 |
| 3 | 3/4 | Bay(Est) | 6 | <0.0001 | 0.2742 | <0.0001 | 0.0103 |
| 4 | 4/5 | Sta(Est*Bay) | 16 | <0.0001 | <0.0001 | <0.0001 | <0.0001 |
| 5 | | Error DF | | 4879 | 4878 | 4757 | 4645 |
| | | $R^2$ | | 82% | 90% | 57% | 41% |
| | | CV | | 21.52 | 8.86 | 16.86 | 3.33 |

**(b)**

| N | $F$-Test | Source | DF | NH4 | NO23 | PO4 | SiO4 | Chl |
| --- | --- | --- | --- | --- | --- | --- | --- | --- |
| 1 | 1/5 | Date | 127 | <0.0001 | <0.0001 | <0.0001 | <0.0001 | <0.0001 |
| 2 | 2/(3 + 4) | Est | 4 | 0.5026 | 0.2515 | 0.5675 | 0.4993 | 0.7020 |
| 3 | 3/4 | Bay(Est) | 6 | 0.0006 | 0.0003 | <0.0001 | <0.0001 | 0.0002 |
| 4 | 4/5 | Sta(Est*Bay) | 16 | <0.0001 | <0.0001 | <0.0001 | <0.0001 | <0.0001 |
| 5 | | Error DF | | 3732 | 3750 | 3726 | 3714 | 2871 |
| | | $R^2$ | | 47% | 51% | 63% | 66% | 54% |
| | | CV | | 58.69 | 84.67 | 42.47 | 15.42 | 24.13 |

Abbreviations: N, row number; DF, degrees of freedom

**Table 3** Overall sample means (and standard error in parentheses) for water column variables measured during benthic sampling

| Est | Bay | Salinity (PSU) | | Temperature (°C) | | DO (mg/L) | | pH | |
|-----|-----|-----|-----|-----|-----|-----|-----|-----|-----|
| LC | Lavaca (LB) | 16.69 | (0.38) | 22.19 | (0.25) | 8.12 | (0.06) | 8.10 | (0.02) |
| LC | Matagorda (MB) | 25.41 | (0.23) | 22.17 | (0.22) | 7.50 | (0.06) | 8.12 | (0.01) |
| LC | East Matagorda (EM) | 19.69 | (0.49) | 22.28 | (0.34) | 8.60 | (0.15) | 8.21 | (0.02) |
| GE | Upper San Antonio (US) | 10.70 | (0.34) | 23.56 | (0.25) | 8.24 | (0.11) | 8.27 | (0.02) |
| GE | Lower San Antonio (LS) | 18.35 | (0.38) | 23.34 | (0.27) | 7.88 | (0.08) | 8.19 | (0.01) |
| MA | Copano (CB) | 14.26 | (0.76) | 23.35 | (0.45) | 7.61 | (0.13) | 8.06 | (0.03) |
| MA | Aransas (AB) | 19.68 | (0.63) | 23.11 | (0.40) | 7.62 | (0.12) | 8.15 | (0.03) |
| NC | Nueces (NB) | 27.35 | (0.40) | 22.79 | (0.28) | 7.52 | (0.06) | 8.13 | (0.01) |
| NC | Corpus Christ (CC)i | 31.92 | (0.17) | 22.57 | (0.22) | 6.93 | (0.06) | 8.12 | (0.01) |
| LM | Baffin (BB) | 37.76 | (0.69) | 22.97 | (0.39) | 6.74 | (0.13) | 8.16 | (0.03) |
| LM | Laguna Madre (LM) | 37.87 | (0.63) | 23.72 | (0.46) | 8.04 | (0.15) | 8.37 | (0.03) |

**Table 4** Overall sample means (and standard error in parentheses) for water column nutrients measured during benthic sampling

| Est | Bay | NH4 | | NO23 | | PO4 | | SiO4 | | Chl | |
|-----|-----|-----|-----|-----|-----|-----|-----|-----|-----|-----|-----|
| LC | Lavaca | 2.16 | (0.25) | 1.24 | (0.54) | 0.94 | (0.09) | 78.21 | (5.59) | 15.37 | (3.16) |
| LC | Matagorda | 1.51 | (0.08) | 1.46 | (0.15) | 0.83 | (0.03) | 45.28 | (1.48) | 7.91 | (0.26) |
| LC | Eastern arm Matagorda | 3.22 | (0.31) | 9.50 | (1.11) | 1.92 | (0.13) | 59.70 | (2.78) | 14.05 | (0.67) |
| GE | Upper San Antonio | 2.44 | (0.16) | 21.29 | (1.64) | 2.91 | (0.14) | 139.73 | (5.90) | 17.03 | (0.85) |
| GE | Lower San Antonio | 1.57 | (0.11) | 4.33 | (0.49) | 1.72 | (0.09) | 98.94 | (3.85) | 10.08 | (0.50) |
| MA | Copano | 1.59 | (0.17) | 2.22 | (0.45) | 1.88 | (0.14) | 174.84 | (7.28) | 6.85 | (0.53) |
| MA | Aransas | 1.39 | (0.18) | 1.91 | (0.38) | 1.02 | (0.05) | 129.96 | (5.94) | 7.04 | (0.51) |
| NC | Nueces | 2.05 | (0.14) | 2.03 | (0.16) | 1.70 | (0.06) | 104.34 | (3.30) | 7.53 | (0.33) |
| NC | Corpus Christi | 1.30 | (0.09) | 0.72 | (0.06) | 0.55 | (0.03) | 46.61 | (1.43) | 5.44 | (0.17) |
| LM | Baffin | 4.72 | (0.59) | 1.80 | (0.23) | 1.10 | (0.06) | 96.49 | (3.98) | 22.64 | (3.99) |
| LM | Laguna Madre | 2.50 | (0.27) | 5.43 | (0.48) | 2.02 | (0.28) | 95.01 | (2.79) | 8.20 | (0.29) |

Abbreviations: NH4, ammonium ($\mu$mol/L); NO23, nitrite + nitrate ($\mu$mol/L); PO4, orthophosphate ($\mu$mol/L); SiO4, silicate ($\mu$mol/L); Chl, chlorophyll $a$ (ug/L)

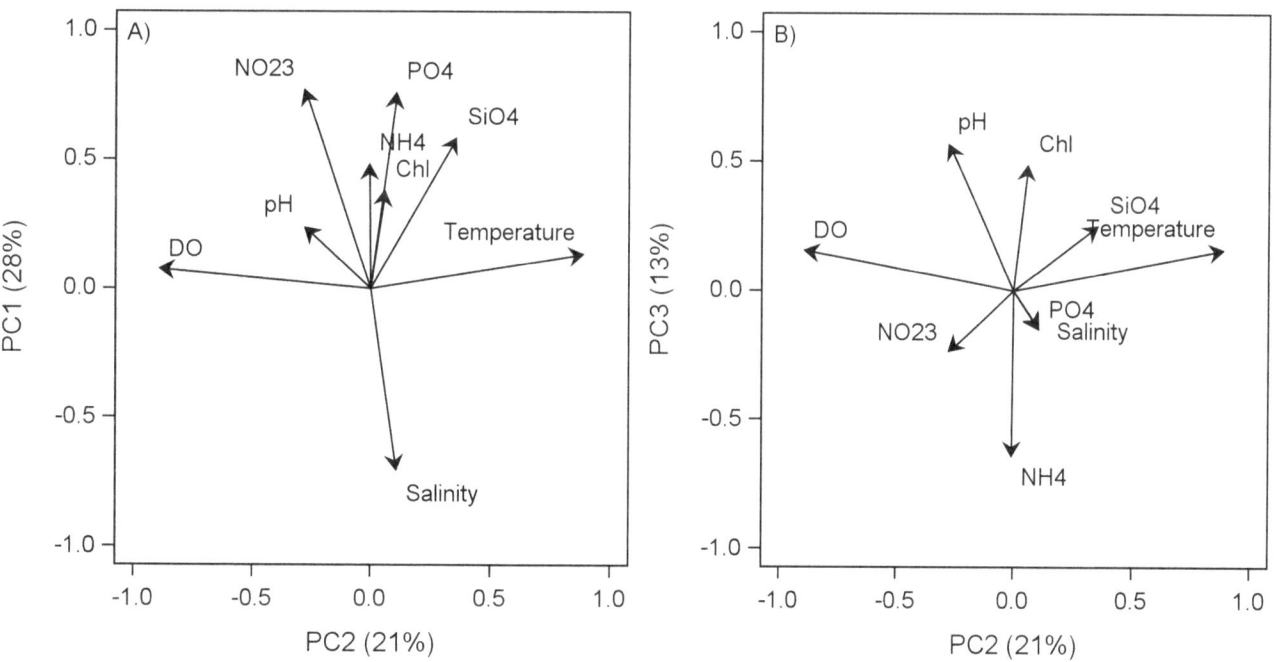

**Fig. 3** Principal components variable loads (PC) for estuary water column condition indicators during benthic sampling. Based on 971 combinations of values averaged by bay for each sampling date. (**a**) Variable loads for PC1 versus PC2. (**b**) Variable loads for PC3 versus PC2

material decomposes. Elevated ammonium and the co-occurrence of increased nitrogenous waste products contribute to anaerobic conditions.

When PC sample scores are plotted with different symbols, it is possible to classify the samples by the new interpretable PC axes. When estuaries are used as symbols for samples, the samples from the Guadalupe Estuary (GE) cluster on the top of PC1, and Nueces Estuary (NC) cluster on the bottom of the axis because inflow has greater effects (i.e., larger volumes of freshwater) in GE than in NC (Fig. 4a). Primary and secondary bays have different water qualities (Fig. 4b). The primary bays (symbol = 1) always have lower PC1 scores than the secondary bays (symbol = 2) indicating higher salinities and lower nutrients in the primary bays. The

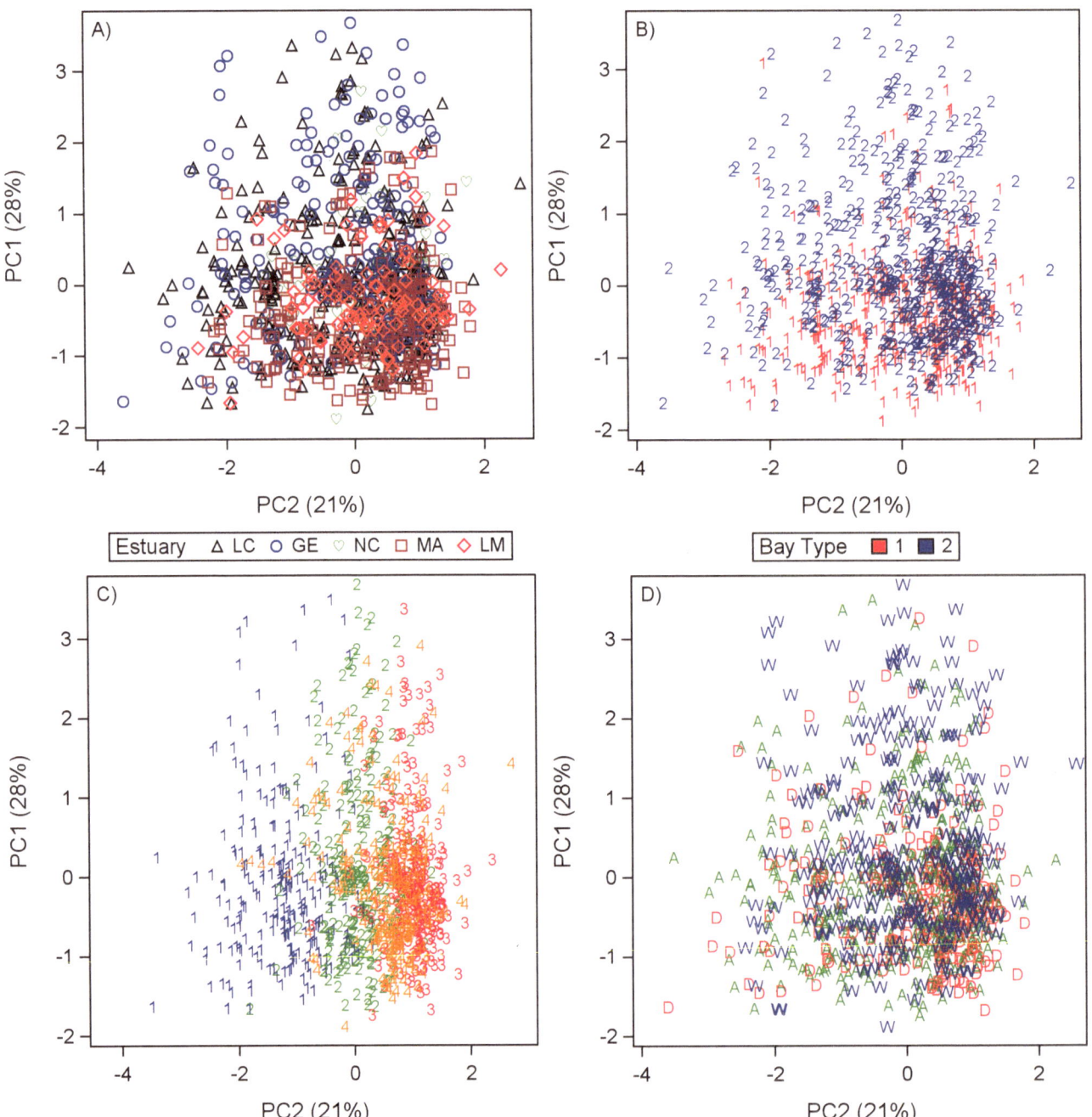

**Fig. 4** Principal component (PC) sample scores for estuary water column conditions during benthic sampling. (**a**) Estuaries as markers where labels abbreviated as in Table 1. (**b**) Bay types as markers where (1) primary bays and (2) secondary bays. (**c**) Season as markers where (1) winter, (2) spring, (3) summer, and (4) fall. (**d**) Climatic periods based on TPWD salinity data as markers where D: dry (< 25%), A: average (25–75%), and W: wet (>75%)

Guadalupe estuary is different because its values are mostly higher on PC1 and lower on PC2 compared to the other estuaries indicating a more prominent role of freshwater inflow into that system. Nueces and Laguna Madre estuaries are uniquely lower in PC1 and PC2 indicating generally higher water temperatures and salinities. Water column data was most divergent for the Lavaca-Colorado Estuary.

When samples are plotted by season, there is scatter along the entire freshwater inflow axis (PC1), meaning a variety of inflow scenarios are possible at any time during the year (Fig. 4c). However, winter samples cluster on the left of the seasonal axis (PC2) and summer and fall samples cluster on the right of the axis because negative PC2 values represent cold temperatures and positive PC2 values represent warm temperatures. Climatic periods were classified using the TPWD Coastal Fisheries salinity data aggregated by quarter (Fig. 4d). The TPWD data is used because it provides a continuous record for all bays. Periods with average and above average rainfall (wet) are spread along the PC1 axis. In contrast, periods with below average rainfall primarily (dry) cluster in the negative PC1 and positive PC2 quadrant. This pattern indicates that low flow-low nutrient-high salinity conditions predominate during dry periods, summer, and fall.

## Sediment Effects

Sediments samples were collected in all bays in October each year to measure sediment texture and biogeochemical characteristics as described in chapter "Influence of Inflows on Estuary Sediments" (Douglas and Montagna 2024). Sediment measurements included: grain size (i.e., rubble, sand, silt, and clay), total carbon $\delta^{13}C$ (‰), total organic carbon $\delta^{13}C$ (‰), total nitrogen $\delta^{15}Nitrogen$ (‰), nitrogen content (%), PW = porewater content (%), carbon content (%), and total organic carbon (TOC, %). Multivariate analysis indicates the Lavaca-Colorado estuary (LC) had mainly muddy sediments, the Guadalupe estuary (GE) had a span of muddy and sandy sediments, and Nueces Estuary had sandier sediments. Muddier sediments usually had higher porewater, nitrogen, and TOC content and more negative carbon stable isotope values.

## Benthic Effects

Benthic metrics across estuaries cluster in similar ranges (Fig. S1). When averages across bays were compared (Fig. S2), the highest abundances and biomass were found in Laguna Madre (the only station with seagrass habitats), and the lowest values were found in Copano Bay (Table 5). The low values in Copano Bay could be attributed to lower sampling effort and data available for Copano Bay. The highest richness ($R$) was found in Laguna Madre, but the highest diversity ($N1$) was found in Corpus Christi Bay. The lowest richness ($R$) and diversity ($H'$ and $N1$) were found in Baffin Bay. The highest evenness ($J'$) was found in Matagorda Bay, and the lowest value was found in Baffin Bay.

To test for estuary and bay differences, macrofauna data were grouped by quarter; thus, 492 separate sampling dates reduced to 139 quarters. In this analysis, date was treated as a categorical variable. An analysis of variance (ANOVA) for differences in macrofauna metrics (abundance, biomass, diversity, and evenness) was different by date ($P < 0.0001$), station ($P < 0.0001$), and the interaction terms ($P < 0.0001$) (Table 6). The differences by date are expected and dealt with in the temporal analysis section below. While the models fit well with high $R^2$, there was little or no evidence for differences among estuaries and bays. The main exception was for log-transformed abundance among estuaries and biomass among bays. There also appears to be differences in evenness among bays. Box plots for benthic metrics by estuary (Table S1) and bay (Table S2) are found in the supplement.

## Temporal Dynamics

In temporal analyses, date was treated as a continuous variable. There appears to be a declining trend over time for benthic abundance in some bays (Figs. S3, S4, S5, and S6). Benthic infauna abundance and biomass declined in Lavaca and Matagorda Bays of the Lavaca-Colorado Estuary, but there was a decline in diversity only in Matagorda Bay (Table 7). There was a decline in abundance only in upper and lower San Antonio Bay within the Guadalupe Estuary. In contrast, diversity increased over time in Nueces Bay, Corpus Christi Bay, and Baffin Bay.

Time series analysis was performed by bay on benthic metrics. Because Copano and Aransas Bays were mostly measured only once per year, the time series for these two bays were based on annual averages. Very few of the series were white noise except Lower San Antonio, Copano, Aransas, and Baffin Bays (Table 8). A white noise series is best explained by the overall temporal means. Biomass was a white noise series in Lower San Antonio, Aransas, and Baffin Bays, which was the most of all the benthic metrics. Diversity metrics were white noise in Copano Bay.

There are three types of signal components in a time series: a trend over time, seasonal changes, or irregular component cycles. In general, temporal variability in abundance, biomass, and diversity was high in all bays (Figs. 5, 6, 7, 8, 9, 10, 11, 12 and 13). The Mission-Aransas Estuary (Copano and Aransas Bays) were sampled only once per year, so their seasonal decomposition could not be estimated. Within estuaries, the primary bays connected to the Gulf of Mexico had higher benthic abundance, biomass, and diversity compared

**Table 5** Averages for all macroinfauna variables sampled quarterly in each estuary from 1984 to 2019. Primary bays include Matagorda Bay (MB), Lower San Antonio Bay (LS), and Corpus Christi Bay (CC). Secondary bays include Lavaca Bay (LB), Eastern arm of Matagorda Bay (EM), Upper San Antonio (US), and Nueces Bay (NB)

| Metrics | Estuary and bay | | | | | | | | | | |
| | LC | | | GE | | MA | | NC | | LM | |
| | LB | MB | EM | US | LS | CB | AB | NB | CC | BB | LM |
|---|---|---|---|---|---|---|---|---|---|---|---|
| N | 927 | 1152 | 411 | 696 | 696 | 72 | 72 | 742 | 1109 | 327 | 360 |
| Abundance | 5432 | 9091 | 10,385 | 19,119 | 9585 | 4522 | 9297 | 11,402 | 15,691 | 13,149 | 28,099 |
| Abundance SD | 4881 | 10,988 | 8746 | 20,409 | 9768 | 4582 | 8583 | 9926 | 13,637 | 20,167 | 24,716 |
| Ln abundance | 8.056 | 8.589 | 8.932 | 9.335 | 8.766 | 7.827 | 8.796 | 9.012 | 9.255 | 8.606 | 9.880 |
| Ln abundance SD | 1.547 | 1.187 | 0.831 | 1.239 | 1.095 | 1.403 | 0.825 | 0.901 | 1.239 | 1.863 | 0.918 |
| Biomass DW | 3.028 | 4.264 | 3.811 | 11.924 | 5.960 | 0.672 | 2.868 | 8.381 | 9.447 | 1.797 | 19.900 |
| Biomass DW SD | 19.336 | 8.993 | 5.953 | 30.343 | 33.200 | 1.017 | 3.381 | 11.591 | 11.436 | 3.394 | 68.473 |
| Ln biomass DW | 0.684 | 1.145 | 1.158 | 1.609 | 1.065 | 0.400 | 1.060 | 1.718 | 1.888 | 0.696 | 1.857 |
| Ln biomass DW SD | 0.765 | 0.891 | 0.823 | 1.176 | 0.924 | 0.432 | 0.736 | 1.026 | 0.998 | 0.696 | 1.168 |
| Biomass C | 1.095 | 1.263 | 1.112 | 4.263 | 1.848 | 0.247 | 0.848 | 2.890 | 2.986 | 0.645 | 7.731 |
| Biomass C SD | 7.043 | 2.721 | 1.562 | 10.895 | 10.182 | 0.371 | 0.853 | 3.861 | 3.824 | 1.203 | 27.763 |
| Ln biomass C | 0.362 | 0.596 | 0.597 | 1.000 | 0.588 | 0.189 | 0.527 | 1.049 | 1.108 | 0.365 | 1.177 |
| Ln biomass C SD | 0.532 | 0.559 | 0.495 | 0.927 | 0.637 | 0.235 | 0.405 | 0.750 | 0.713 | 0.443 | 0.992 |
| Richness $R$ | 4.497 | 8.518 | 7.401 | 5.658 | 6.602 | 3.500 | 6.639 | 9.705 | 14.807 | 3.232 | 15.064 |
| Richness SD | 2.862 | 5.015 | 4.059 | 2.705 | 4.579 | 1.808 | 2.569 | 6.463 | 7.440 | 2.311 | 7.496 |
| Diversity $H'$ | 0.964 | 1.542 | 1.296 | 1.024 | 1.150 | 0.818 | 1.277 | 1.593 | 2.018 | 0.577 | 1.937 |
| Diversity $H'$ SD | 0.515 | 0.529 | 0.545 | 0.386 | 0.574 | 0.443 | 0.364 | 0.666 | 0.670 | 0.538 | 0.594 |
| Diversity $N1$ | 2.977 | 5.322 | 4.234 | 2.985 | 3.760 | 2.463 | 3.828 | 6.074 | 8.996 | 2.053 | 8.060 |
| Diversity $N1$ STD | 1.711 | 2.754 | 2.399 | 1.140 | 2.557 | 1.012 | 1.413 | 3.890 | 4.686 | 1.330 | 3.985 |
| Evenness $J'$ | 0.667 | 0.783 | 0.695 | 0.632 | 0.658 | 0.653 | 0.715 | 0.761 | 0.782 | 0.426 | 0.750 |
| Evenness $J'$ SD | 0.263 | 0.179 | 0.188 | 0.199 | 0.207 | 0.322 | 0.161 | 0.161 | 0.181 | 0.348 | 0.147 |

Units: Abundance ($n/m^2$); biomass ($g/m^2$); richness (S/35 $cm^2$); diversity $H'$ ($H'$/35 $cm^2$); diversity $N1$ ($N1$/35 $cm^2$); evenness $J'$ ($J'$/35 $cm^2$)
Abbreviations: N, number of individuals/sample; SD, standard deviation; DW; dry weight; C, carbon

**Table 6** Results for ANOVA test for date, estuary, bay, and station differences among macrofauna metrics. (A) Stocks: total abundance ($n/m^2$) and biomass ($g/m^2$) for dry weight (DW) and carbon (C). (B) Diversity: Richness (S/35 $cm^2$), diversity ($H'$/35 $cm^2$), diversity ($N1$/35 $cm^2$), and evenness ($J'$/35 $cm^2$)

| Source | 1-Date | 2-Estuary | 3-Bay(Est) | 4-Sta(Est*Bay) | 5-Date*Sta(Est Bay) | $R^2$ | |
|---|---|---|---|---|---|---|---|
| DF | 138 | 4 | 6 | 17 | 1958 | | |
| $F$-Test | 1/5 | 2/3 | 3/4 | 4/5 | 5/6† | | CV |
| Abun. | <0.0001 | 0.4540 | 0.0989 | <0.0001 | <0.0001 | 84% | 57.53 |
| Ln Abun. | <0.0001 | 0.3997 | 0.0310 | <0.0001 | <0.0001 | 84% | 7.34 |
| Biom. DW | <0.0001 | 0.1136 | 0.8690 | <0.0001 | <0.0001 | 70% | 231.45 |
| Ln Biom. DW | <0.0001 | 0.0467 | 0.4339 | <0.0001 | <0.0001 | 83% | 40.01 |
| Biom. C | <0.0001 | 0.0467 | 0.4339 | <0.0001 | <0.0001 | 74% | 234.84 |
| Ln Biom. C | <0.0001 | 0.0300 | 0.6193 | <0.0001 | <0.0001 | 82% | 50.37 |
| Richness $R$ | <0.0001 | 0.1053 | 0.1761 | <0.0001 | <0.0001 | 90% | 28.33 |
| Diversity $H'$ | <0.0001 | 0.1097 | 0.0669 | <0.0001 | <0.0001 | 88% | 21.29 |
| Diversity $N1$ | <0.0001 | 0.0546 | 0.2272 | <0.0001 | <0.0001 | 88% | 30.71 |
| Evenness $J'$ | <0.0001 | 0.2968 | 0.0186 | <0.0001 | <0.0001 | 69% | 21.63 |

Abbreviations: Abun, abundance; Biom., biomass; $N1$, row number; $F$-Test, row number as numerator and denominator for $F$-Test and probability ($P$-value) for rejecting null hypothesis; DF, degrees of freedom; Est, estuary; 6†, mean square error = 4440; $R^2$, goodness of fit; CV, coefficient of variation; total observations = 6564

to their respective secondary bays in all estuaries, except for the Guadalupe Estuary, which is a closed bay system without direct access to the Gulf.

Primary bays had higher diversity than secondary bays in all estuaries except one. This anomaly was found in the Lavaca-Colorado Estuary where the eastern arm of Matagorda Bay (which is influenced by the Colorado River) responded more like Matagorda Bay than Lavaca Bay (Fig. 7). This was especially true for abundance. Biomass and diversity in the eastern arm of Matagorda Bay had more of an average response and appears to fall between Lavaca Bay (Fig. 5) and Matagorda Bay (Fig. 6). A decreasing trend

**Table 7** Linear regression for abundance, biomass, and diversity over time by bay within estuary

| Metric | Bay | Regression model | $R^2$ (%) | $P$ |
|---|---|---|---|---|
| (A) Abundance ($\ln (n + 1)/\text{m}^2$) | LB | $A = 4.547 - 0.00006 * \text{Date}$ | 29.77 | <0.0001 |
| | MB | $A = 4.989 - 0.00007 * \text{Date}$ | 37.47 | <0.0001 |
| | EM | $A = 3.957 - 0.000005 * \text{Date}$ | 0.27 | 0.6234 |
| | US | $A = 4.839 - 0.00005 * \text{Date}$ | 16.90 | <0.0001 |
| | LS | $A = 4.244 - 0.00003 * \text{Date}$ | 8.53 | 0.0015 |
| | CB | $A = 5.172 - 0.00013 * \text{Date}$ | 54.07 | 0.0096 |
| | AB | $A = 4.767 - 0.00007 * \text{Date}$ | 25.18 | 0.0965 |
| | NB | $A = 3.849 + 0.000004 * \text{Date}$ | 0.25 | 0.5711 |
| | CC | $A = 4.141 - 0.000007 * \text{Date}$ | 0.58 | 0.3800 |
| | BB | $A = 4.582 - 0.00007 * \text{Date}$ | 1.86 | 0.2982 |
| | LM | $A = 4.037 + 0.00002 * \text{Date}$ | 1.05 | 0.4403 |
| (B) Biomass ($\ln (\text{g C} + 1)/\text{m}^2$) | LB | $B = 0.679 - 0.00003 * \text{Date}$ | 18.36 | <0.0001 |
| | MB | $B = 0.773 - 0.00003 * \text{Date}$ | 27.38 | <0.0001 |
| | EM | $B = 0.29 - 0.000002 * \text{Date}$ | 0.18 | 0.6866 |
| | US | $B = 0.539 - 0.000007 * \text{Date}$ | 1.03 | 0.2774 |
| | LS | $B = 0.271 - 0.000001 * \text{Date}$ | 0.04 | 0.8226 |
| | CB | $B = 0.301 - 0.00002 * \text{Date}$ | 18.25 | 0.1661 |
| | AB | $B = 0.399 - 0.00001 * \text{Date}$ | 4.15 | 0.5255 |
| | NB | $B = 0.097 + 0.000022 * \text{Date}$ | 12.18 | 0.0001 |
| | CC | $B = 0.314 + 0.00001 * \text{Date}$ | 5.98 | 0.0044 |
| | BB | $B = 0.362 - 0.00002 * \text{Date}$ | 1.33 | 0.3806 |
| | LM | $B = 1.193 - 0.00006 * \text{Date}$ | 6.69 | 0.0480 |
| (D) Diversity ($N1/35 \text{ cm}^2$) | LB | $D = 3.195 - 0.00002 * \text{Date}$ | 0.26 | 0.5513 |
| | MB | $D = 11.121 - 0.00035 * \text{Date}$ | 33.14 | <0.0001 |
| | EM | $D = 3.804 + 0.000017 * \text{Date}$ | 0.07 | 0.8023 |
| | US | $D = 3.514 - 0.00003 * \text{Date}$ | 2.54 | 0.0875 |
| | LS | $D = 3.387 + 0.000023 * \text{Date}$ | 0.23 | 0.6072 |
| | CB | $D = 2.919 - 0.00003 * \text{Date}$ | 1.22 | 0.7323 |
| | AB | $D = 2.354 + 0.00011 * \text{Date}$ | 6.61 | 0.4197 |
| | NB | $D = 1.375 + 0.000293 * \text{Date}$ | 16.54 | <0.0001 |
| | CC | $D = 4.327 + 0.00029 * \text{Date}$ | 19.11 | <0.0001 |
| | BB | $D = -1.035 + 0.000248 * \text{Date}$ | 9.32 | 0.0177 |
| | LM | $D = 8.433 - 0.00004 * \text{Date}$ | 0.04 | 0.8881 |

Abbreviations: Bay abbreviations as in Table 1, $R^2$ = model goodness of fit, $P$ = probability level that slope = 0. Data and regression in Figs. S3, S4, S5, and S6

**Table 8** Chi-square ($\chi^2$) test for white noise in a time series

| Metric | Lag to | Bay LB | MB | EM | US | LS | CB | AB | NB | CC | BB | LM |
|---|---|---|---|---|---|---|---|---|---|---|---|---|
| nm2 | 6 | <0.0001 | <0.0001 | <0.0001 | <0.0001 | <0.0001 | <0.0001 | 0.0151 | 0.0074 | 0.0003 | 0.0900 | 0.0022 |
| | 12 | <0.0001 | <0.0001 | <0.0001 | <0.0001 | <0.0001 | <0.0001 | 0.0002 | 0.0120 | 0.0008 | 0.1157 | 0.0043 |
| lnm2 | 6 | <0.0001 | <0.0001 | <0.0001 | <0.0001 | <0.0001 | <0.0001 | 0.0201 | 0.0098 | <.0001 | 0.0423 | 0.0095 |
| | 12 | <0.0001 | <0.0001 | <0.0001 | <0.0001 | <0.0001 | <0.0001 | 0.0004 | 0.0212 | 0.0003 | 0.0317 | 0.0200 |
| gCm2 | 6 | <0.0001 | <0.0001 | <0.0001 | <0.0001 | 0.2021 | 0.0159 | 0.2964 | <0.0001 | <0.0001 | 0.9577 | <0.0001 |
| | 12 | 0.0002 | <0.0001 | <0.0001 | <0.0001 | 0.6801 | 0.0059 | 0.2371 | 0.0010 | <0.0001 | 0.9507 | 0.0009 |
| lgCm2 | 6 | <0.0001 | <0.0001 | <0.0001 | <0.0001 | <0.0001 | 0.0104 | 0.2248 | <0.0001 | <0.0001 | 0.7517 | 0.0002 |
| | 12 | <0.0001 | <0.0001 | <0.0001 | <0.0001 | <0.0001 | 0.0016 | 0.3782 | <0.0001 | <0.0001 | 0.3883 | 0.0097 |
| R | 6 | <0.0001 | <0.0001 | <0.0001 | <0.0001 | <0.0001 | 0.1556 | 0.1465 | <0.0001 | <0.0001 | <0.0001 | 0.0001 |
| | 12 | <0.0001 | <0.0001 | <0.0001 | <0.0001 | <0.0001 | 0.2544 | 0.1367 | <0.0001 | <0.0001 | <0.0001 | 0.0013 |
| H' | 6 | <0.0001 | <0.0001 | <0.0001 | <0.0001 | <0.0001 | 0.0136 | 0.0339 | <0.0001 | <0.0001 | <0.0001 | 0.0002 |
| | 12 | <0.0001 | <0.0001 | <0.0001 | <0.0001 | <0.0001 | 0.0775 | <0.0001 | <0.0001 | <0.0001 | <0.0001 | 0.0004 |
| N1 | 6 | <0.0001 | <0.0001 | <0.0001 | 0.0006 | <0.0001 | 0.0042 | 0.0329 | <0.0001 | <0.0001 | <0.0001 | 0.0021 |
| | 12 | <0.0001 | <0.0001 | <0.0001 | 0.0002 | <0.0001 | 0.0708 | <0.0001 | <0.0001 | <0.0001 | <0.0001 | 0.0176 |
| J' | 6 | <0.0001 | 0.0002 | <0.0001 | <0.0001 | <0.0001 | 0.0003 | 0.0129 | <0.0001 | <0.0001 | 0.0001 | 0.0200 |
| | 12 | <0.0001 | <0.0001 | <0.0001 | <0.0001 | <0.0001 | <0.0001 | <0.0001 | <0.0001 | <0.0001 | 0.0018 | 0.0316 |

Abbreviations: nm2, abundance ($n/\text{m}^2$); lnm2, abundance ($\ln (n + 1)/\text{m}^2$); gCm2, biomass ($\text{g C}/\text{m}^2$); lgCm2, biomass ($\ln (\text{g C} + 1)/\text{m}^2$); $R$, richness ($S/105 \text{ cm}^2$); $H'$, diversity ($H'/105 \text{ cm}^2$); $N1$, diversity ($N1/105 \text{ cm}^2$); $J'$, evenness ($J'/105 \text{ cm}^2$). Bay abbreviations as in Table 1

**Fig. 5** Trend decomposition analysis for Lavaca Bay (LB). Symbols are average over all stations for each quarter (blue: observed, red: imputed by join interpolation) with overlying black line as fitted trend. (**a**) Abundance ($\ln(n + 1)/m^2$). (**b**) Biomass ($\ln(g\,C + 1)/m^2$). (**c**) Diversity ($H'/0.01\,m^2$)

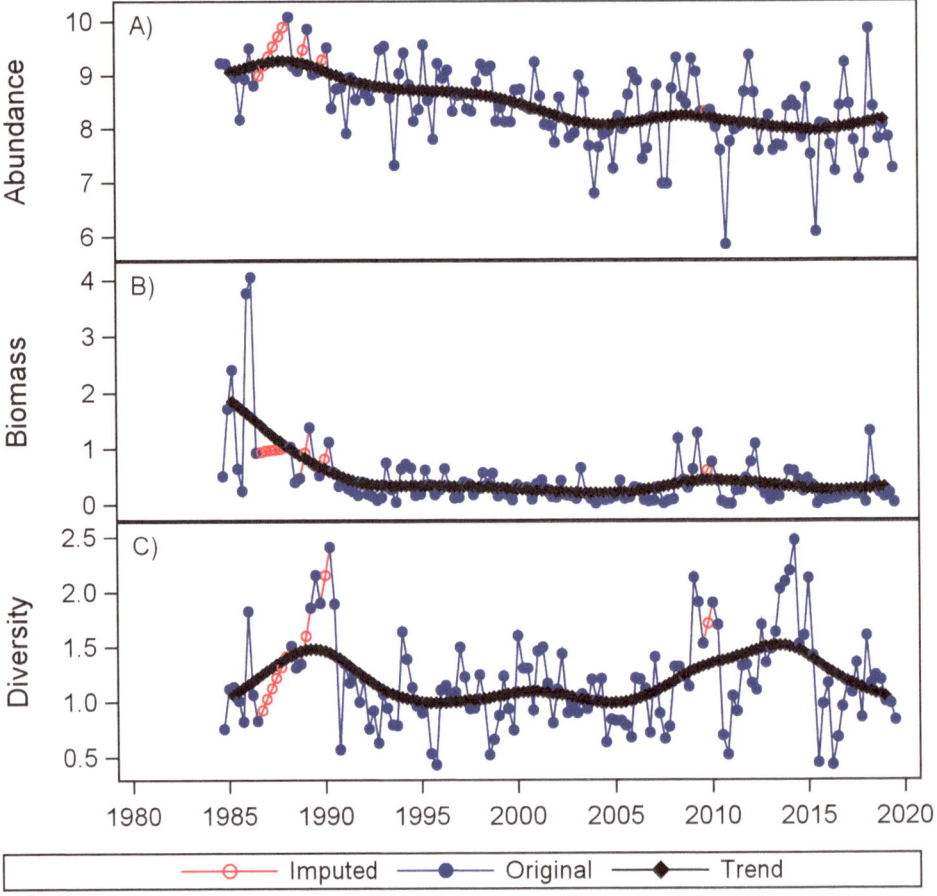

**Fig. 6** Trend decomposition analysis for Matagorda Bay (MB). Symbols are average over all stations for each quarter (blue: observed, red: imputed by join interpolation) with overlying black line as the fitted trend. (**a**) Abundance ($\ln(n + 1)/m^2$). (**b**) Biomass ($\ln(g\,C + 1)/m^2$). (**c**) Diversity ($H'/0.01\,m^2$)

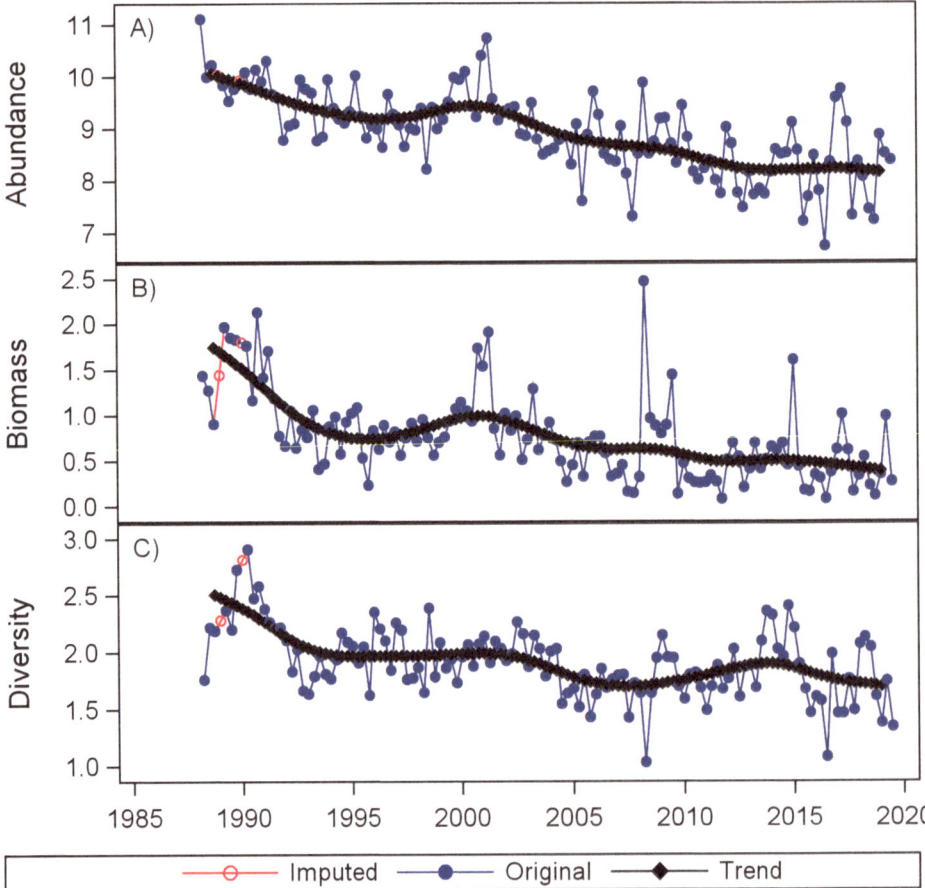

**Fig. 7** Trend decomposition analysis for eastern Matagorda Bay (EM). Symbols are average over all stations for each quarter (blue: observed, red: imputed by join interpolation) with overlying black line as the fitted trend. (**a**) Abundance ($\ln(n + 1)/m^2$). (**b**) Biomass ($\ln(g\,C + 1)/m^2$). (**c**) Diversity ($H'/0.01\ m^2$)

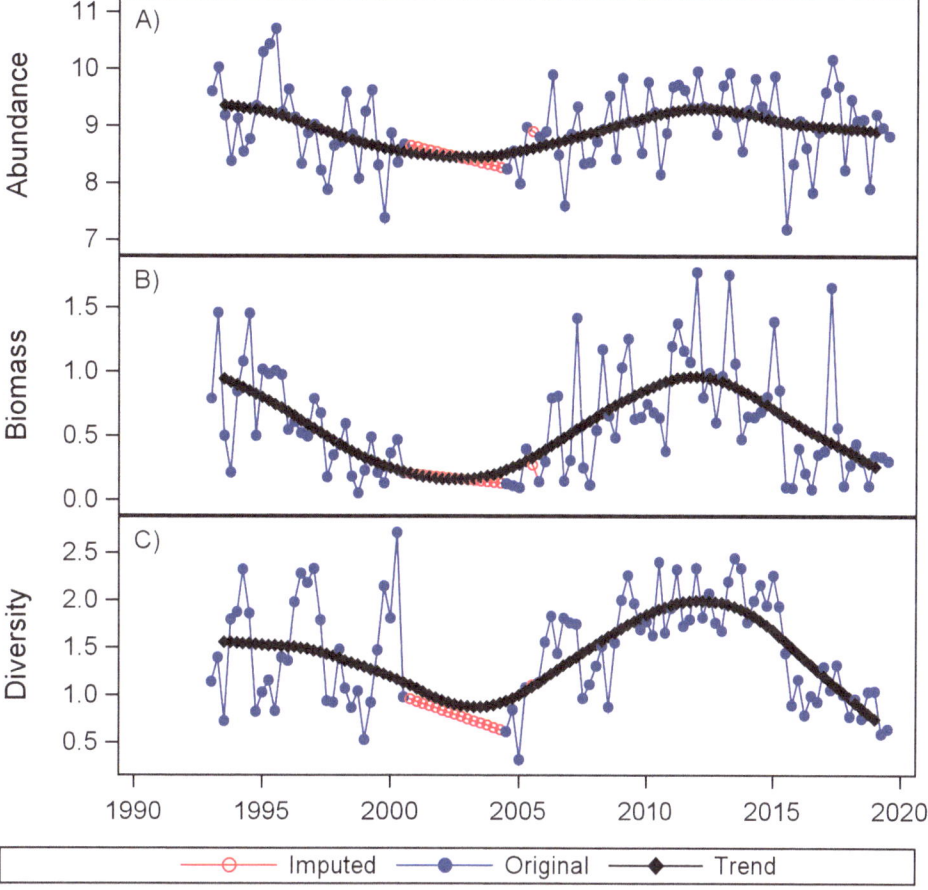

**Fig. 8** Trend decomposition analysis for Upper San Antonio Bay (US). Symbols are average over all stations for each quarter (blue: observed, red: imputed by join interpolation) with overlying black line as the fitted trend. (**a**) Abundance ($\ln(n + 1)/m^2$). (**b**) Biomass ($\ln(g\,C + 1)/m^2$). (**c**) Diversity ($H'/0.01\ m^2$)

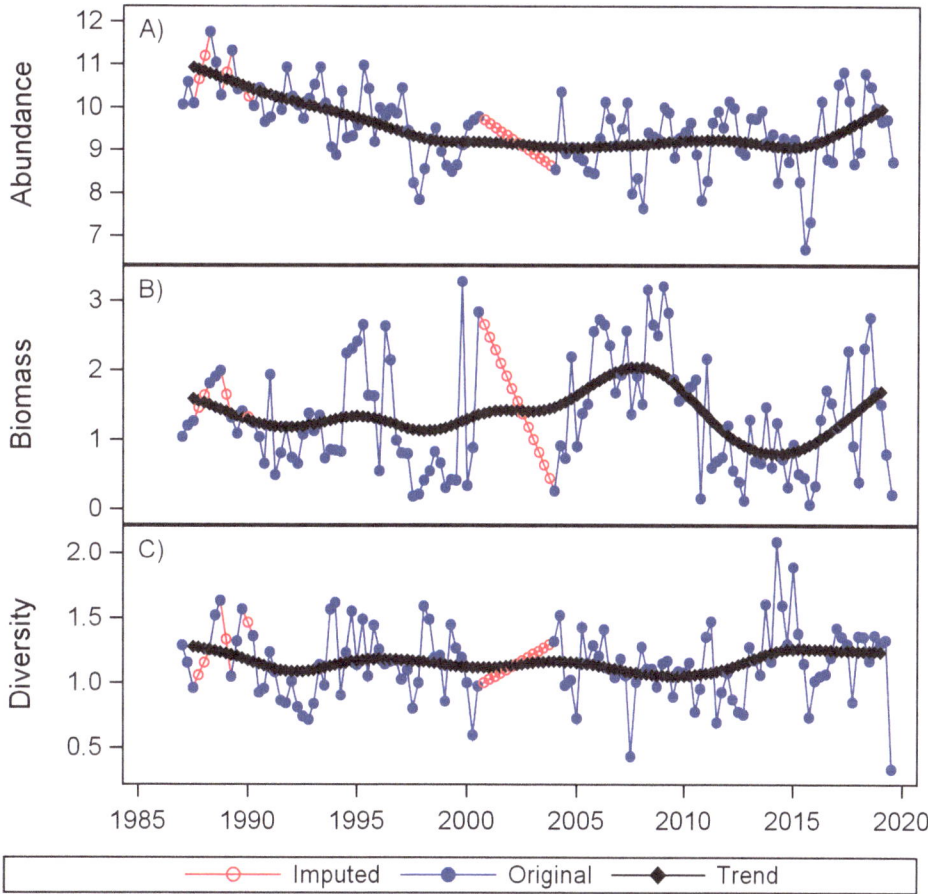

**Fig. 9** Trend decomposition analysis for Lower San Antonio Bay (LS). Symbols are average over all stations for each quarter (blue: observed, red: imputed by join interpolation) with overlying black line as the fitted trend. (**a**) Abundance ($\ln(n + 1)/m^2$). (**b**) Biomass ($\ln(g\,C + 1)/m^2$). (**c**) Diversity ($H'/0.01\ m^2$)

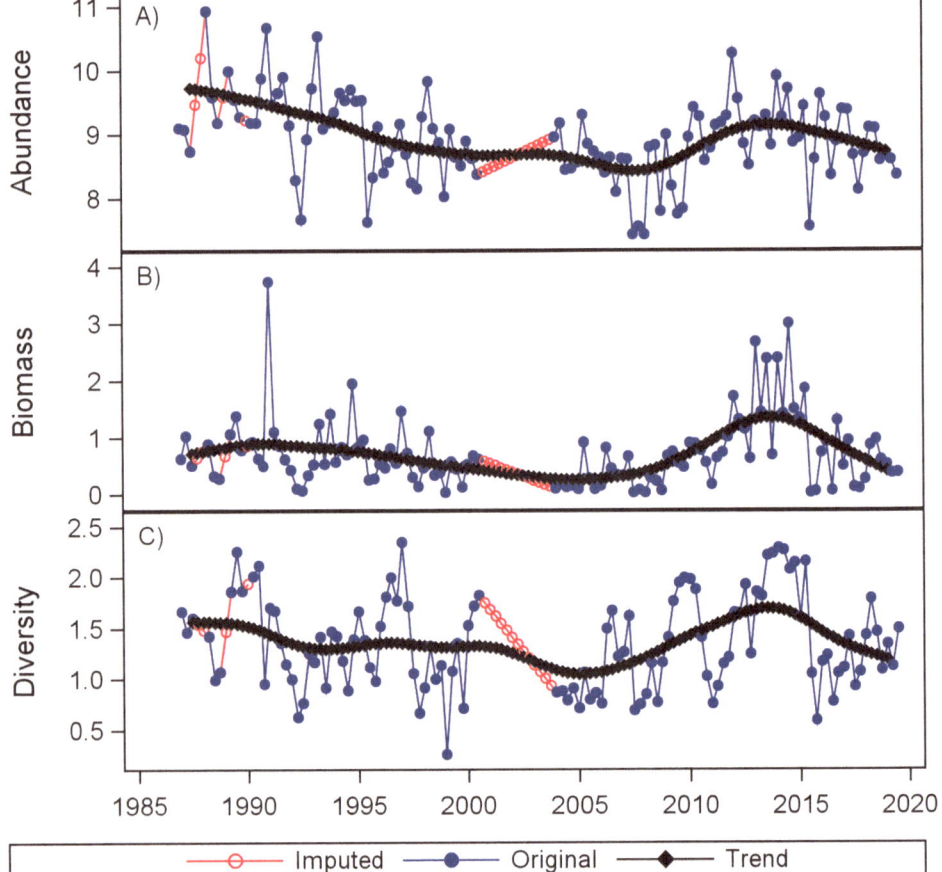

**Fig. 10** Trend decomposition analysis for Nueces Bay (NB). Symbols are average over all stations for each quarter (blue: observed, red: imputed by join interpolation) with overlying black line as the fitted trend. (**a**) Abundance ($\ln(n + 1)/m^2$). (**b**) Biomass ($\ln(g\,C + 1)/m^2$). (**c**) Diversity ($H'/0.01\ m^2$)

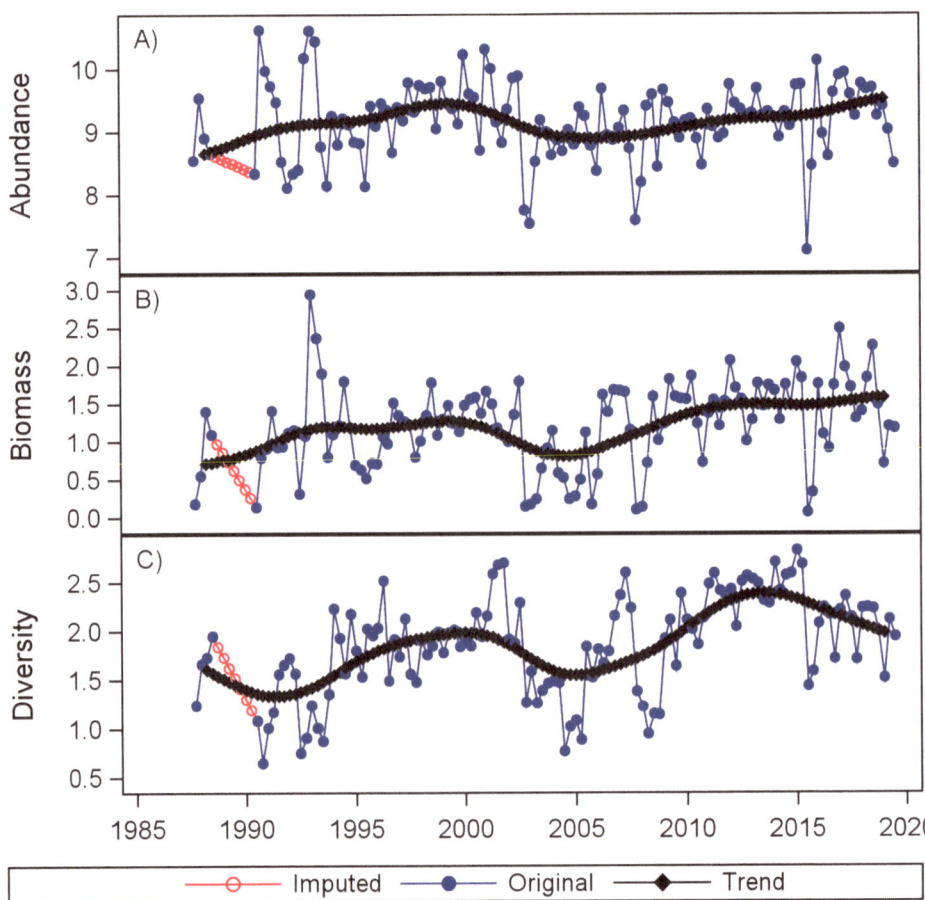

**Fig. 11** Trend decomposition analysis for Corpus Christi Bay (CC). Symbols are average over all stations for each quarter (blue: observed, red: imputed by join interpolation) with overlying black line as the fitted trend. (**a**) Abundance ($\ln(n + 1)/m^2$). (**b**) Biomass ($\ln(g\,C + 1)/m^2$). (**c**) Diversity ($H'/0.01\,m^2$)

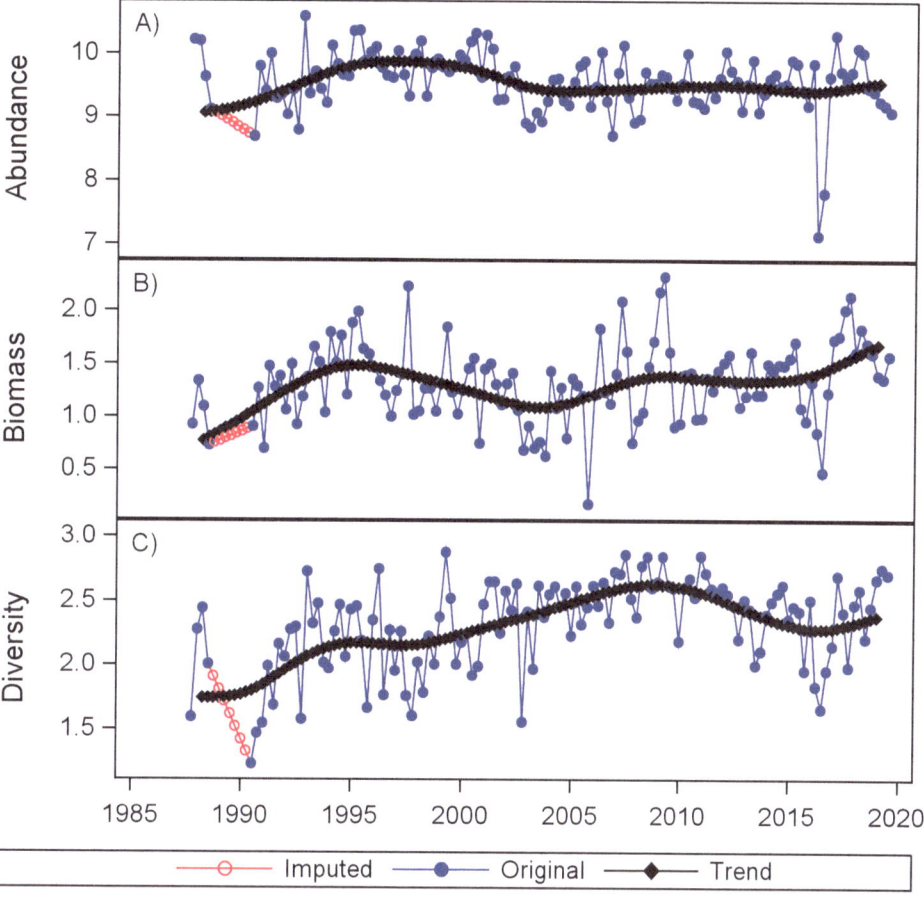

**Fig. 12** Trend decomposition analysis for Baffin Bay (BB). Symbols are average over all stations for each quarter (blue: observed) with black line as the fitted trend. (**a**) Abundance ($\ln(n + 1)/m^2$). (**b**) Biomass ($\ln(g\,C + 1)/m^2$). (**c**) Diversity ($H'/0.01\,m^2$)

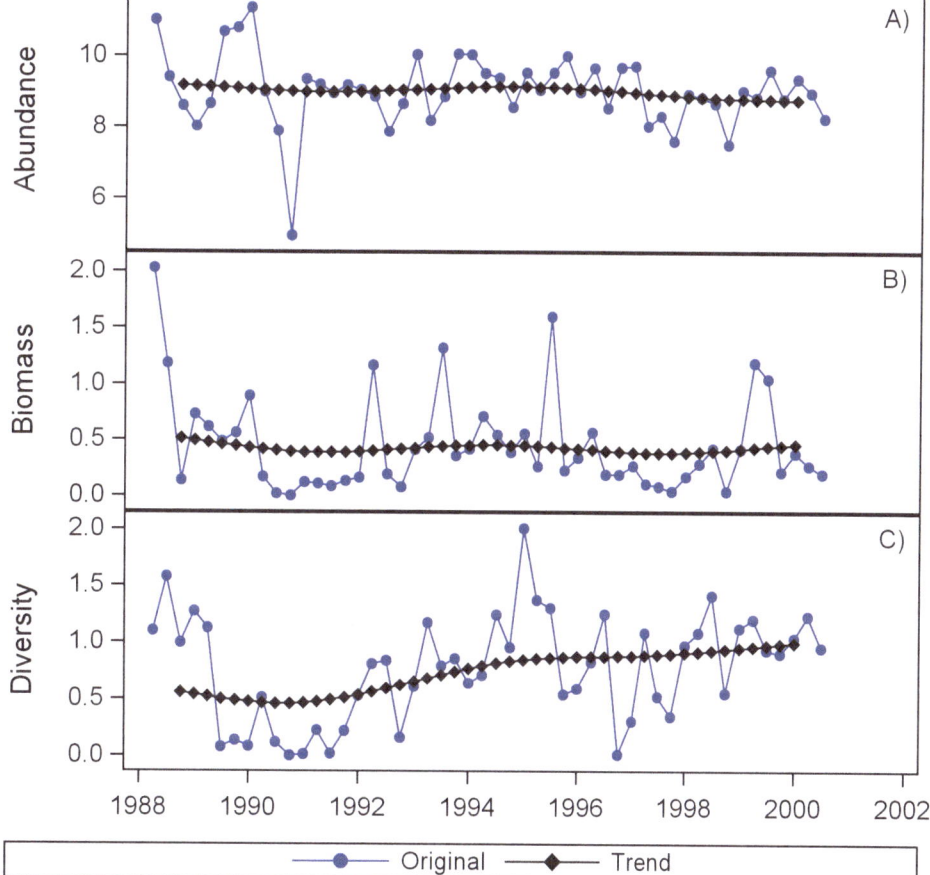

**Fig. 13** Trend decomposition analysis for Laguna Madre (LM). Symbols are average over all stations for each quarter (blue: observed) with black overlying line as the fitted trend. (**a**) Abundance $(\ln(n + 1)/m^2)$. (**b**) Biomass $(\ln(g\,C + 1)/m^2)$. (**c**) Diversity $(H'/0.01\,m^2)$

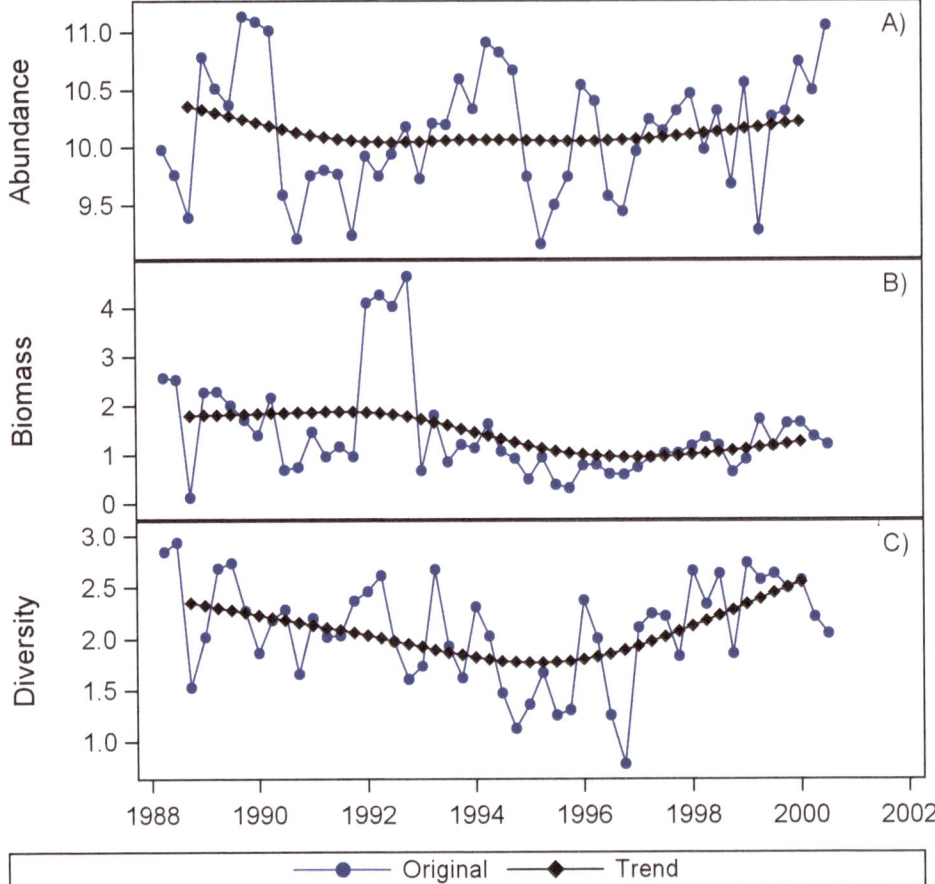

in abundance and biomass was seen in Lavaca and Matagorda Bays. A decrease in diversity, however, appears to be isolated to Matagorda Bay. Long-term changes in Lavaca Bay could be related to wastewater discharge from the Formosa Plastics Corp in Port Lavaca. Eastern Matagorda Bay, fed by fresh water from the Colorado River and tidal exchange from the Gulf of Mexico, appears to be stable (Fig. 7).

In the Guadalupe Estuary, benthic abundance and biomass were higher in the secondary bay closer to the river source (Fig. 8) than in the primary bay closer to Gulf of Mexico connections (Fig. 9). The higher abundance and biomass in the upper bay are related to the long-term lower salinity stability from the Guadalupe River inflow. Abundance appears to be declining in both bays, while biomass and diversity appear to remain stable.

Unlike trends seen for estuaries on the upper Texas coast, trends in abundance, biomass, and diversity all appear to be increasing in the Nueces estuary. This is true for both Nueces Bay (Fig. 10) and Corpus Christi Bay (Fig. 11).

The time series in the Laguna Madre estuary is only 12 years long. During this time frame, little change is observed for abundance, biomass, and diversity. This appears to be the case for both Baffin Bay (Fig. 12) and Laguna Madre (Fig. 13).

**Table 9** Analysis of covariance, test for homogeneity of slopes among bays within estuaries. Estuary abbreviations as in Table 1

| Estuary | nm2 | Ln_nm2 | gm2c | Ln_gm2c | R | N1 |
|---|---|---|---|---|---|---|
| LC | <0.0001 | <0.0001 | 0.0143 | 0.0011 | <0.0001 | <0.0001 |
| GE | 0.0063 | 0.1003 | 0.6850 | 0.4590 | 0.6453 | 0.2521 |
| MA | 0.8824 | 0.2893 | 0.9291 | 0.8701 | 0.5039 | 0.3864 |
| NC | 0.0751 | 0.3016 | 0.4606 | 0.0658 | 0.0123 | 0.9710 |
| LM | 0.0272 | 0.2113 | 0.2325 | 0.2297 | 0.7147 | 0.3493 |

## Linking Annual Change in Infauna to Climate Change

The primary and secondary bays had similar trends over time within each estuary because the test for homogeneity of slopes (i.e., a parallel response) was accepted in all cases except for the Lavaca-Colorado Estuary (Table 9). Thus, estuary-wide averages are another way to examine spatial and temporal differences over time.

Climate change effects likely occur at regionals scales. For example, in the 32 years between 1976 and 2007, seawater temperatures increased on average 1.37 °C (2.47 °F) overall the Texas estuaries (Montagna et al. 2011). So, it is useful to determine if there are long-term annual changes that are linked to known climate change. The weighted average annual benthic metrics were calculated for each estuary by averaging stations within estuaries by year to remove seasonality and bay differences (Fig. 14). Abundance decreased in all estuaries through 2000, remained level between 2002 and 2016, and increased in 2017 and 2018. Biomass was highest in 1985 and 1986 in the Lavaca-Colorado Estuary (LC) and Laguna Madre (LM) in 1992, and these high points act as levers. Biomass in the LC was always the lowest after 2004. Species richness was always highest in Nueces Estuary (NC) after 1992. The LC and Guadalupe Estuary (GE) have similar richness responses over time.

An analysis of covariance tests for similar response and change among estuaries over time, and it occurs in three steps (Table 10). The first step is to test for the homogeneity of slopes, to determine if the estuaries responding in a similar or parallel fashion over time, which is represented by the interaction term between year*estuary in Table 10A. The test is nonsignificant (if barely) for both abundance and biomass, meaning all estuaries are behaving in a similar way over time at a regional scale. This is not true for diversity measured as species richness. The second step is a test for elevation, or differences among the parallel slopes (Table 10B). Only abundance is significant, indicating there are estuary differ-

ences in abundances over time but not biomass or richness. The third step is to determine the overall slope or change over time (Table 10C). Only abundance is changing over time, and it is declining. The linear regression equation is abundance = 621652 − 304.196*Year. This means coast-wide, about 300 individuals per meter square are being lost every year between 1984 and 2019 (35 years).

When the long-term decline in benthos was first noticed, it was suggested that it was related to long-term variability in climate factors (Pollack et al. 2011). The TPWD Coastal Fisheries program measured water quality data from 1977

**Table 10** Analysis of covariance of the annual average temporal benthic metrics (Fig. 14). (A) Test for homogeneity of slopes. (B) Test for elevation of main effects and slope. (C) Test for overall slope effect

|     | Source | Abundance $(n/m^2)$ | Biomass $(g\ C/m^2)$ | Richness $(S/35\ cm^2)$ |
|-----|--------|------------|---------|----------|
| (A) | Year | 0.0003 | 0.1348 | 0.8463 |
|     | Estuary | 0.0496 | 0.0987 | 0.0005 |
|     | Year*Estuary | 0.0503 | 0.0982 | 0.0004 |
| (B) | Year | <0.0001 | 0.1452 | 0.7570 |
|     | Estuary | <0.0001 | 0.3683 | <0.0001 |
| (C) | Year | <0.0001 | 0.1394 | 0.7722 |

**Fig. 14** Annual average change in benthic metrics among five estuaries. (**a**) Abundance $(n/m^2)$. (**b**) Biomass $(g\ C/m^2)$. (**c**) Species richness $(S/0.1\ m^2)$. Overlying black solid line is the linear regression over all estuaries. Estuary abbreviations as in Table 1

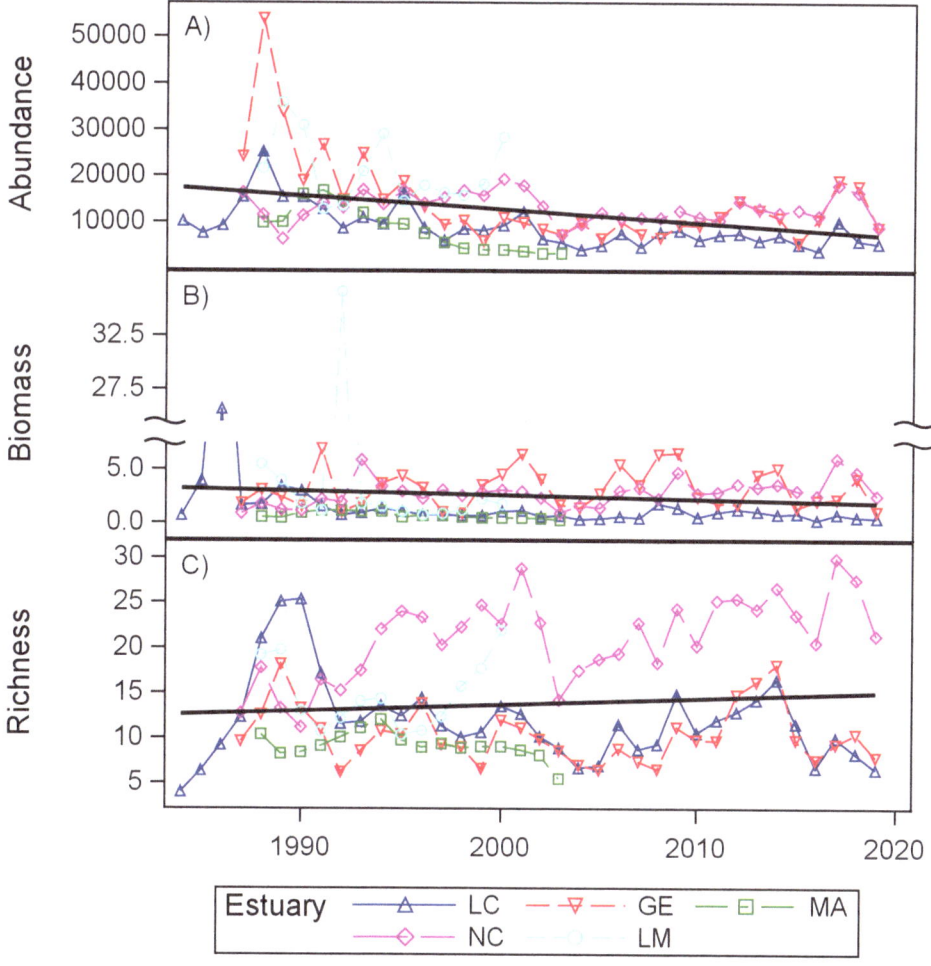

across all Texas estuaries. Both sea-surface temperature and dissolved oxygen are highly seasonal (Fig. S7). The TPWD Coastal Fisheries data was averaged over years and indicates there has been a continual rise in seawater temperatures in all five estuaries on the central and southern coast of Texas (Fig. 15). Temperature and salinity change in the same way over time, as indicated by the analysis of covariance, that is, there is homogeneity of slopes among estuaries over time (Table 11A). Conversely, dissolved oxygen (DO) behaves differently over time among estuaries. It appears this is due to the lowest DO values being present in the Guadalupe Estuary in the early part of the record (Fig. 15b). There are estuary differences for temperature, DO, and salinity (Table 11B). Temperature is increasing over time, but DO is decreasing (Table 11C). The overall composite regressions for the five estuaries yield equations of: Temperature = –41.23 + 0.03233*Year; and DO = 59.70 – 0.02599*Year. Meaning temperature is increasing at a rate of 0.032 °C/year and DO is decreasing at a rate of 0.026 mg/L/year.

The solubility of gas in seawater is known to decrease with increasing seawater temperature. Thus, DO is expected to decrease as the temperature of estuaries continue to increase. This has obvious implications for future habitat stability. While there is an inverse correlation between DO and other benthic metrics, only diversity and richness are significant. Also, there is a positive correlation between all benthic metrics and salinity (Table 12).

## Community Structure Differences

The nonmetric multidimensional scaling (nMDS) analysis of macrofauna community structure based on long-term aver-

**Table 11** Analysis of covariance of the annual average physical conditions in Coastal Bend estuaries from TPWD data (Fig. 12). (A) Test for homogeneity of slopes. (B) Test for elevation of main effects and slope. (C) Test for overall slope effect

| | Source | Temperature (°C) | Dissolved oxygen (mg/L) | Salinity (psu) |
|---|---|---|---|---|
| (A) | Year | 0.3513 | 0.0040 | 0.9634 |
| | Estuary | <0.0001 | <0.0001 | 0.0006 |
| | Year*estuary | 0.3146 | 0.0040 | 0.9718 |
| (B) | Year | <0.0001 | <0.0001 | <0.0001 |
| | Estuary | <0.0001 | <0.0001 | 0.0006 |
| (C) | Year | <0.0001 | <0.0001 | 0.7722 |

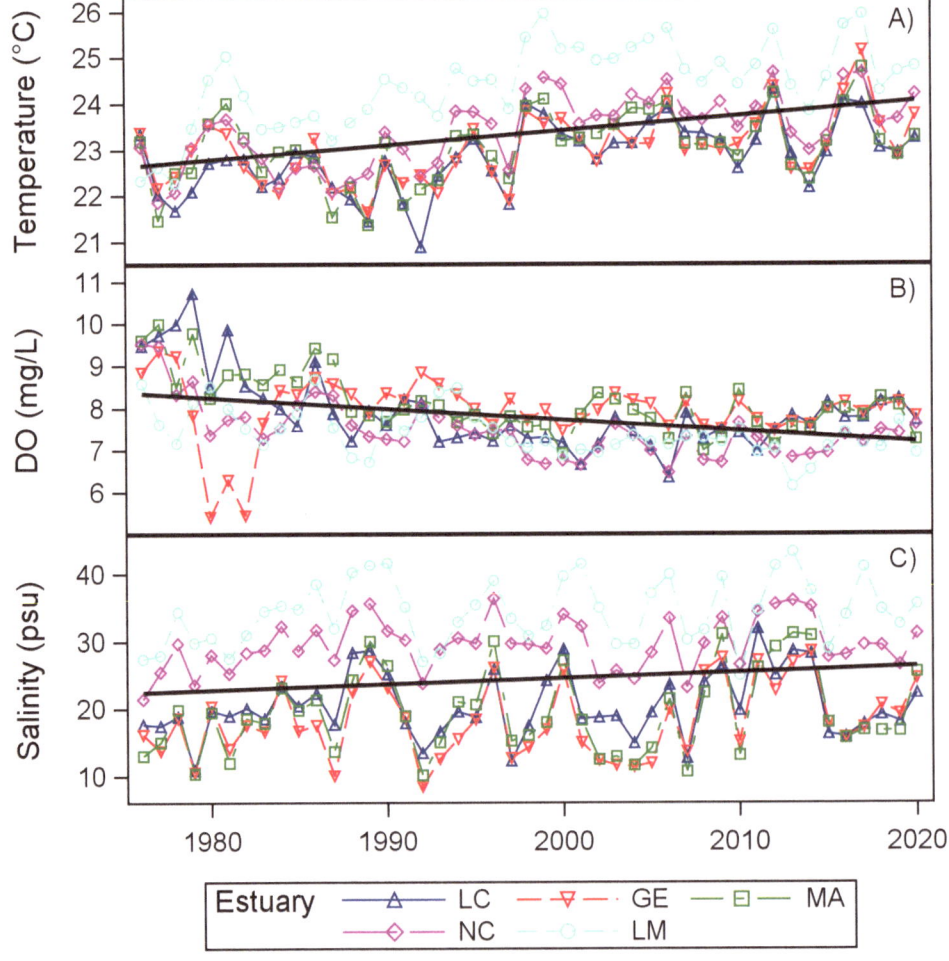

**Fig. 15** Annual Coastal Bend estuaries physical characteristics based on TPWD data. Solid black line is linear regression over all estuaries. Abbreviations as in Table 1. (**a**) Sea surface temperature. (**b**) Dissolved oxygen (DO). (**c**) Surface salinity

age abundance found three different groupings of bays (Analysis of Similarities, ANOSIM, $P = 0.007$; Fig. 16). The first group contains Laguna Madre, which shared only 24% of species in common with other bays. Laguna Madre was also the only bay where sampling was confined to seagrass habitats, so it is not surprising that it is unique with regard to species abundance. The second group is the largest containing both Upper and Lower San Antonio Bays (US, LS), Matagorda Bay (MB), eastern arm of Matagorda Bay (EM), Lavaca Bay (LB), Aransas Bay (AB), Corpus Christi Bay (CC), and Nueces Bay (NB). Within this group, CC and NB, in the Nueces Estuary, clustered at the 60% similarity level.

**Table 12** Spearman correlation ($r$) and significance level ($P$) of physical attributes with benthic characteristics on an annual basis

| Benthic metric | Temperature | | DO | | Salinity | |
|---|---|---|---|---|---|---|
| | $r$ | $P$ | $r$ | $P$ | $r$ | $P$ |
| Abundance (Ln($n + 1$)/m²) | 0.07 | 0.4467 | –0.10 | 0.2566 | 0.40 | <0.0001 |
| Abundance (n/m²) | 0.14 | 0.1186 | –0.10 | 0.2888 | 0.48 | <0.0001 |
| Biomass (g C/m²) | 0.09 | 0.3387 | –0.14 | 0.1403 | 0.37 | <0.0001 |
| Biomass (Ln(g C + 1)/m²) | 0.11 | 0.2144 | –0.25 | 0.0070 | 0.50 | <0.0001 |
| Diversity (N1/35 cm²) | 0.21 | 0.0195 | –0.57 | <0.0001 | 0.67 | <0.0001 |
| Richness (S/35 cm²) | 0.20 | 0.0249 | –0.54 | <0.0001 | 0.70 | <0.0001 |

Also clustering at the 60% level are LS, EM, and LB. Within the second group of bays, the two clusters with 60% similarity share a coincident 50% similarity with US, MB, and AB. The third group of bays includes Baffin Bay (BB) and Copano Bay (BB), which have different salinities, but both have fewer samples. In contrast, LM with a sand/seagrass bottom and high salinity and BB with a muddy bottom and higher salinity had very different community structures. Interestingly, LB influenced by the Lavaca River and EM influenced by the Colorado River were similar at the 60% level but different from Matagorda Bay, which is located between them.

Analysis of similarities for macrofauna by bay (Fig. 16) is correlated with salinity (Table 3). This correlation is revealed in the left to right pattern depicted for bay similarity where salinity (psu) follows decreasing values as CC = 32, NB = 27, MB = 25, EM = 20, LS = 18, LB = 17, and US =11. Bays with similar salinity group most closely together indicating salinity is driving community structure.

A shade plot overlaid with the cluster analysis illustrates which species were responsible for the trends (Fig. 17). Two species (*Streblospio benedicti* and *Mediomastus ambiseta*) were dominant and found everywhere (Table 13). *Streblospio* and *Mediomastus* are estuarine species and are not typical freshwater indicators. They cluster based on long-term average bay similarities and habitat preferences. *Streblospio* is more of an opportunistic species because of its ability to colonize adverse habi-

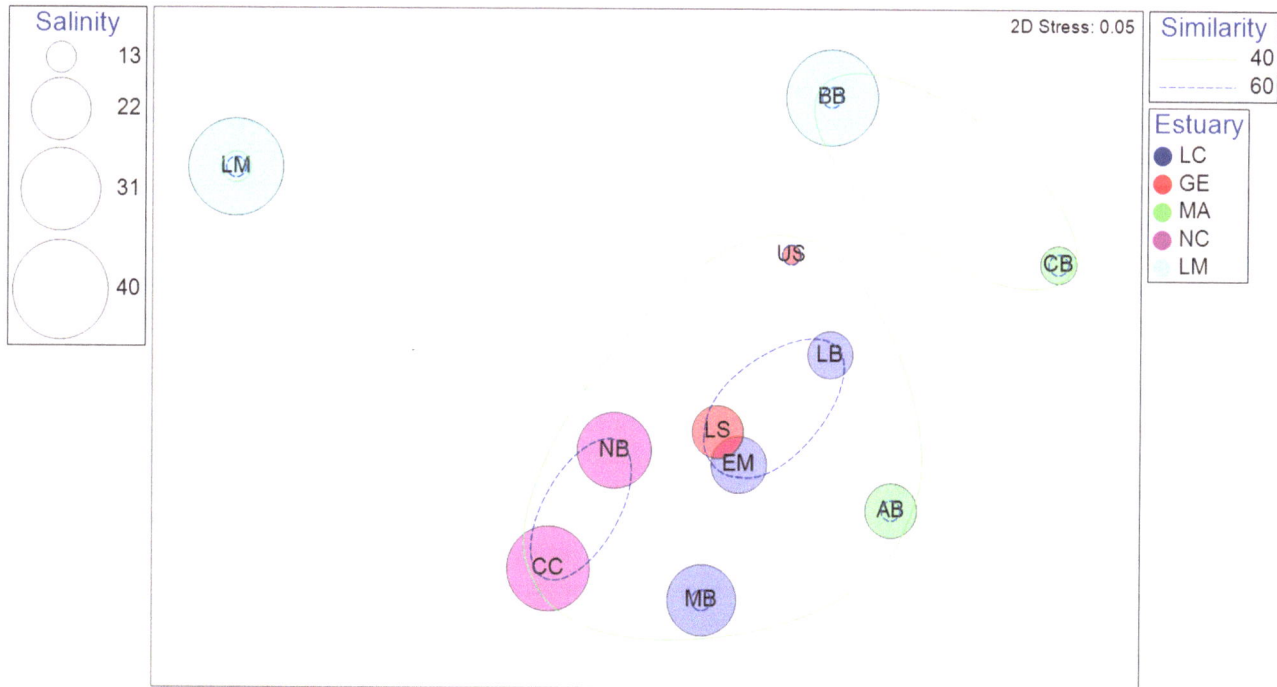

**Fig. 16** Analysis of similarities by multidimensional scaling for macrofauna community structure based on long-term average abundance within bays. Label abbreviations for estuary and bay as in Table 1

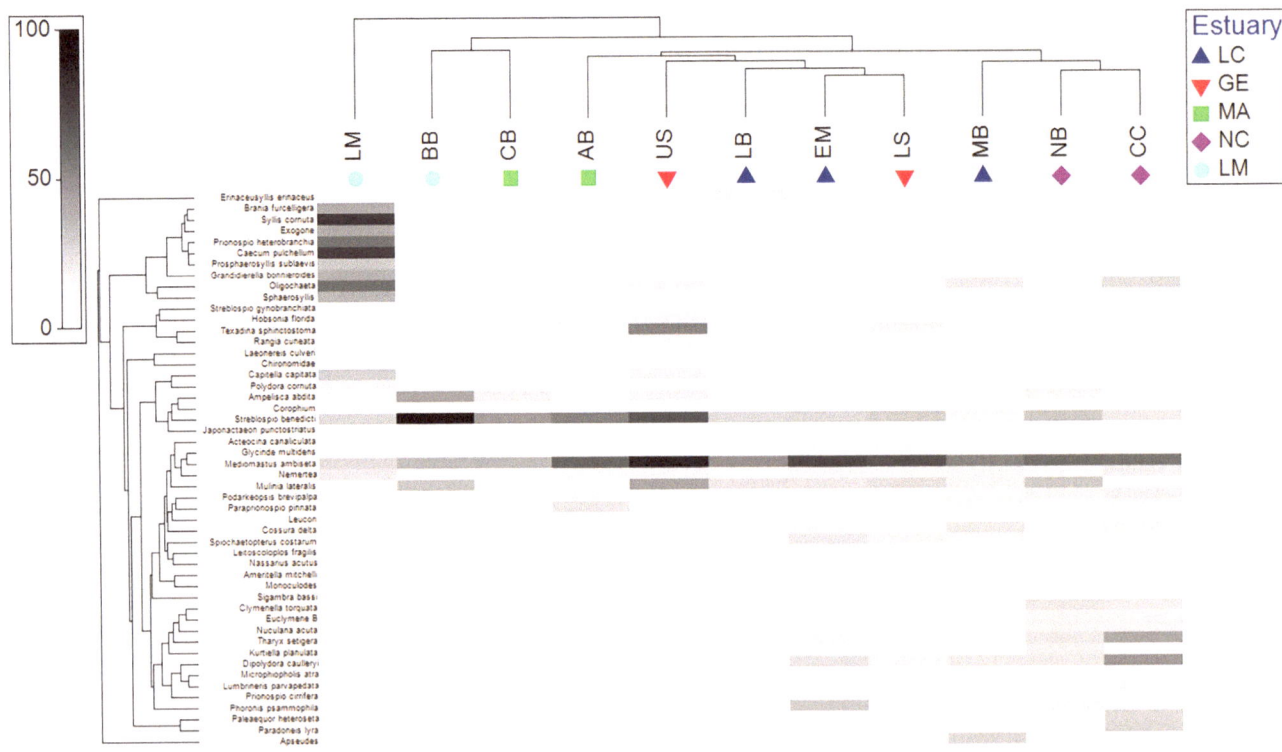

**Fig. 17** Relative abundance and similarity of species among bays. Label abbreviations for estuary and bay as in Table 1

tats, for example, higher salinity Baffin Bay, Rincon Bayou lakes, and lower salinity San Antonio and Lavaca Bays. LM had eight dominant species (*Brania furcelligera*, *Syllis cornuta*, *Exogene* sp., *Prionospio heterobranchia*, *Caecum pulchellum*, *Prosphaerosyllis sublaevis*, *Grandidierella bonnieroides*, and *Sphaerosyllis* sp.). These are typical of high salinity, seagrass, and habitats but were uncommon in the other bays. The Upper San Antonio Bay was unique in that it was dominated by *Texadina sphinctostoma*, a riverine upper bay gastropod associated with freshwater inflow events. This species also occurred in lower abundance in upper Lavaca Bay. A comprehensive list of species for all bays is presented in Table S1.

There were important differences among stations (Table S2). For example, the fifth dominant species *Caecum pulchellum* was found only in the seagrass stations of Laguna Madre (155 and 189), and if only muddy stations were considered, it would be one of the rarest species found. The dominant polychaetes, *Mediomastus ambiseta* and *Streblospio benedicti*, were found everywhere.

A total of 429 species were found in all bays and all samples, but only 22 comprised 80% of all species (Table 13). Of these 22 species, 14 were polychaetes. Dominant polychaetes included three species: *Mediomastus ambiseta*, *Streblospio benedicti*, and *Syllis cornuta*. The fourth and fifth most dominant among the 22 species were both mollusks and included the clam *Mulinia lateralis* and the snail *Caecum pulchellum*. Two ecological characteristics of the above spe-

cies are worth noting. First, *Syllis cornuta* and *Caecum pulchellum* were both associated with grass beds and the higher salinities of Laguna Madre. Second, *Streblospio benedicti* and *Mulinia lateralis* are opportunistic colonizers of defaunated areas that can additionally tolerate euryhaline conditions, for example, 0–86 ppt (Flint and Younk 1983). A list ranking all species is given in Table S3.

## Linking Benthos to Environmental Factors

### Univariate Metrics

Recruitment and/or growth of some benthic species can be dependent on freshwater inflow events. Reduction of Salinity following an inflow event can, independent of other environmental changes, induce spawning among some species, for example, *Crassostrea virginica* and *Mulinia lateralis*. Also associated with freshwater inflow is increased nutrient loading, which can fuel the benthic food web. This plays an important role in restoring benthic communities defaunated by the shock of abrupt salinity changes or physical alterations caused by storms. A delayed response, or lag, in estuarine salinities can be expected following large freshwater inflow events. To account for this time delay, benthic metrics were compared to monthly lags in salinity change following inflow events to determine possible correlations with a benthic response (Fig. 18). Wet and dry periods were identified as salinity quartiles from the TPWD Coastal Fisheries water quality data (Fig. S9).

**Table 13** Species rank dominance and percentage contributions averaged over all bays. Average abundance over all bays in n/m$^2$

| Rank | Phylum | Class | Order | Family | Species | Abundance | Percent | Cum% |
|------|--------|-------|-------|--------|---------|-----------|---------|------|
| 1 | Annelida | Polychaeta | Sedentaria | Capitellidae | *Mediomastus ambiseta* | 3388 | 26.1 | 26.1 |
| 2 | Annelida | Polychaeta | Spionida | Spionidae | *Streblospio benedicti* | 2099 | 16.2 | 42.2 |
| 3 | Annelida | Polychaeta | Phyllodocida | Syllidae | *Syllis cornuta* | 583 | 4.5 | 46.7 |
| 4 | Mollusca | Bivalvia | Venerida | Mactridae | *Mulinia lateralis* | 560 | 4.3 | 51.0 |
| 5 | Mollusca | Gastropoda | Littorinimorpha | Caecidae | *Caecum pulchellum* | 530 | 4.1 | 55.1 |
| 6 | Annelida | Clitellata | Clitellata | Clitellata | Oligochaeta | 445 | 3.4 | 58.5 |
| 7 | Annelida | Polychaeta | Spionida | Spionidae | *Dipolydora caulleryi* | 351 | 2.7 | 61.2 |
| 8 | Annelida | Polychaeta | Spionida | Spionidae | *Prionospio heterobranchia* | 288 | 2.2 | 63.4 |
| 9 | Mollusca | Gastropoda | Littorinimorpha | Cochliopidae | *Texadina sphinctostoma* | 287 | 2.2 | 65.6 |
| 10 | Arthropoda | Malacostraca | Amphipoda | Ampeliscidae | *Ampelisca abdita* | 279 | 2.1 | 67.8 |
| 11 | Annelida | Polychaeta | Terebellida | Cirratulidae | *Tharyx setigera* | 226 | 1.7 | 69.5 |
| 12 | Nemertea | Nemertea | Nemertea | Nemertea | Nemertea | 181 | 1.4 | 70.9 |
| 13 | Annelida | Polychaeta | Phyllodocida | Syllidae | *Sphaerosyllis* | 138 | 1.1 | 72.0 |
| 14 | Annelida | Polychaeta | Phyllodocida | Syllidae | *Brania furcelligera* | 138 | 1.1 | 73.1 |
| 15 | Annelida | Polychaeta | Phyllodocida | Syllidae | *Exogone* | 136 | 1.0 | 74.1 |
| 16 | Arthropoda | Malacostraca | Amphipoda | Aoridae | *Grandidierella bonnieroides* | 132 | 1.0 | 75.1 |
| 17 | Phoronida | Phoronida | Phoronida | Phoronidae | *Phoronis psammophila* | 115 | 0.9 | 76.0 |
| 18 | Annelida | Polychaeta | Sedentaria | Capitellidae | *Capitella capitata* | 115 | 0.9 | 76.9 |
| 19 | Annelida | Polychaeta | Sedentaria | Cossuridae | *Cossura delta* | 114 | 0.9 | 77.8 |
| 20 | Annelida | Polychaeta | Spionida | Spionidae | *Paraprionospio pinnata* | 110 | 0.8 | 78.6 |
| 21 | Annelida | Polychaeta | Sedentaria | Chaetopteridae | *Spiochaetopterus costarum* | 109 | 0.8 | 79.4 |
| 22 | Annelida | Polychaeta | Phyllodocida | Hesionidae | *Podarkeopsis brevipalpa* | 99 | 0.8 | 80.2 |
| Subtotal dominant species | | | | | 22 species | 10,425 | 80.2 | |
| Subtotal rare species | | | | | 407 species | 2573 | 19.8 | |
| Total | | | | | 429 species | 12,998 | 100.0 | |

The response of benthic metrics to freshwater inflow lags varies by bay (Figs. S1 and S2). Bays with high average salinity respond to reduced salinity with a corresponding increase in diversity. Conversely, low salinity bays have higher diversity during droughts because increased salinity provides colonization opportunities for marine species with a greater tolerance for higher salinities, for example, *Thais haemastoma*. Only species richness (*R*) has a consistent positive response to increasing salinity. Interesting characteristics are also seen among stations with long-term stable salinities. For example, station E in CC maintains a highly diverse benthic community over time because salinity does not vary much over time. In BB, stations 6 and 24 occupy a stable environment characterized by high salinity and a muddy bottom. These two stations are dominated by the polychaete *Streblospio benedicti* and bivalve *Mulinia lateralis*, which comprise the base of the food web supporting the commercial and recreational Black Drum fishery.

Benthic metrics had optimal salinity ranges from 8 to 25 psu (Table 14). Log transformed abundance and diversity had the lowest optimal salinity values, and diversity metrics had the highest optimal salinity ranges. Peak abundance and biomass typically occur in secondary bays, such as Upper San Antonio Bay (US) and Lavaca Bay (LB) with the lowest salinity values (Fig. 19. Highest diversity occurs in seawater salinity ranges in Corpus Christi Bay. For example, NC-E is located in the northeast corner of Corpus Christi Bay near the ship channel, which results in long term daily flushing of stable salinity Aransas Pass water resulting in a high diversity community. NC-D is located on the backside of Musang Island, closer to the Laguna Madre entrance, and is subject to periodic cells of hypersaline Laguna Madre waters, which stratify the water column and cause hypoxia (Ritter and Montagna 1999). This can cause low abundance and low diversity at NC-D. Baffin Bay, with the highest salinity ranges, has the lowest abundance, biomass, and diversity.

**Fig. 18** Pearson correlation coefficients between benthic metrics and water quality for each bay over time. Benthic metrics averaged bay-wide by quarter. Salinity averaged bay-wide by month from TPWD sampling, thus salinity = .0 time lag and Salinitylagx = $x$ months lag. Abbreviations: Bays as in Table 1, nm², abundance (n/m²); gCm², biomass (g C/m²); $R$, richness (S/sample), $N1$, diversity ($N1$/sample); $H$, diversity ($H'$/sample); $J$, evenness ($J'$/sample). Colors: red = positive, white = neutral, blue = negative. Sample size (months or quarters): LB = 137, MB = 135, US = 116, LS = 116, CB = 12, AB = 12, NB = 123, CC = 124, BB = 59, LM = 59

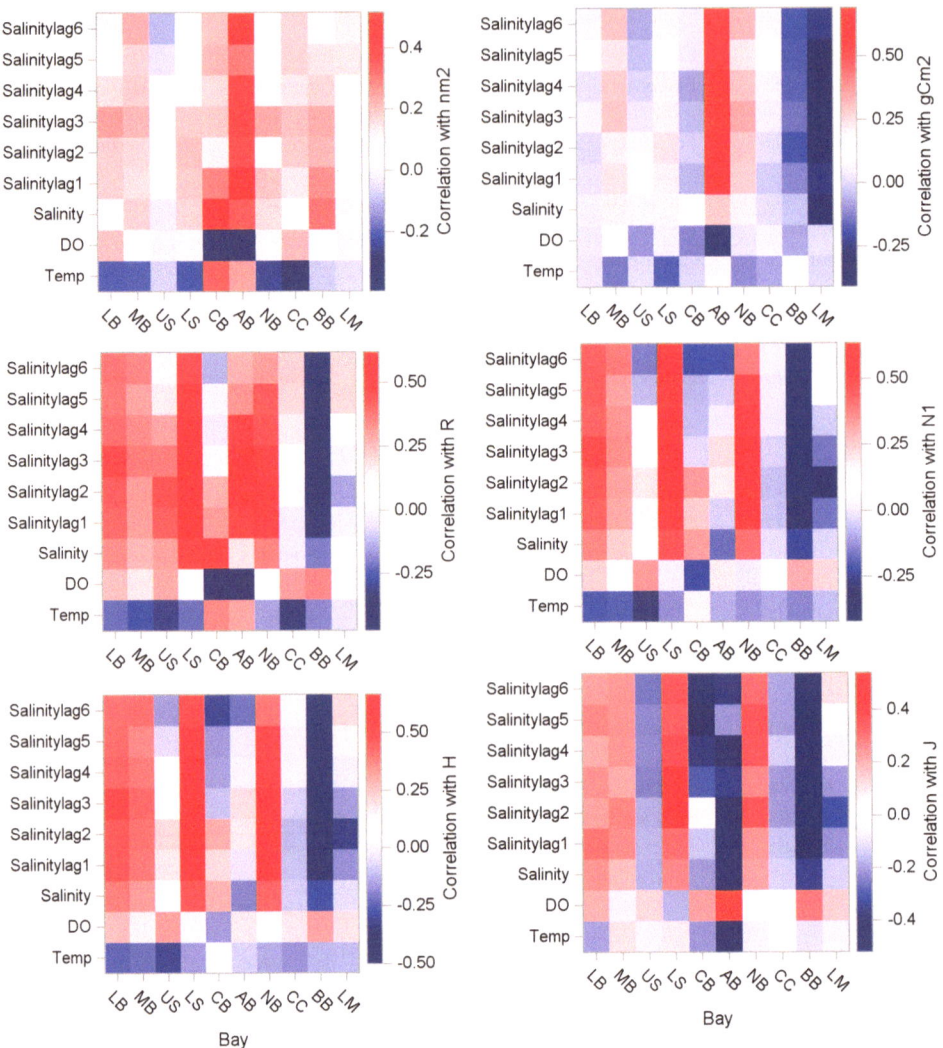

**Table 14** Parameters estimated by log-normal regressions to predict macrofauna metrics from salinity. Mean and standard error in parentheses for (a) maximum benthic metric values, (b) rate of change, (c) maximum salinity value, and ($R^2$) coefficient of determination

| Benthic metrics | a | | b | | c | | $R^2$ (%) |
|---|---|---|---|---|---|---|---|
| Abundance (n/m²) | 68,843 | (11,682) | 1.41 | (0.50) | 10.16 | (4.18) | 18 |
| Abundance (Ln n/m²) | 11.03 | (0.21) | 5.39 | (2.48) | 8.17 | (5.77) | 0 |
| Biomass (g C/m²) | 33.66 | (6.16) | 0.53 | (0.12) | 16.13 | (2.03) | 26 |
| Biomass (Ln g C/m²) | 3.41 | (0.30) | 1.12 | (0.17) | 11.59 | (1.66) | 5 |
| Richness (S/0.01 m²) | 39.37 | (3.52) | 0.56 | (0.07) | 23.42 | (1.48) | 7 |
| Diversity ($N1$/0.01 m²) | 18.62 | (1.99) | 0.52 | (0.08) | 25.10 | (1.73) | 10 |
| Diversity ($H'$/0.01 m²) | 2.83 | (0.32) | 1.02 | (0.20) | 16.46 | (2.66) | 10 |
| Evenness ($J'$/0.01 m²) | 1.00 | (0.10) | 1.45 | (0.32) | 10.94 | (2.68) | 7 |

*Streblospio benedicti* and *Mulinia lateralis* have adapted to the long-term high salinity environment of Baffin Bay.

## Multivariate Metrics

The multivariate analysis of water column conditions was described in section "Water Column Effects." Recall that the principal component 1 (PC1) axis is summarizing inflow effects where positive values are high inflow, PC2 is summarizing seasonal effects where positive values are summer and fall, and PC3 is summarizing trophic state because positive values represent high chlorophyll values (an indicator of autotrophic conditions) and negative values represent high

**Fig. 19** Log-normal regression on maximum benthic metric values for 20 salinity bins (filled circles). Bins were constructed for average bay-quarter times series data. Bay abbreviations as in Table 1

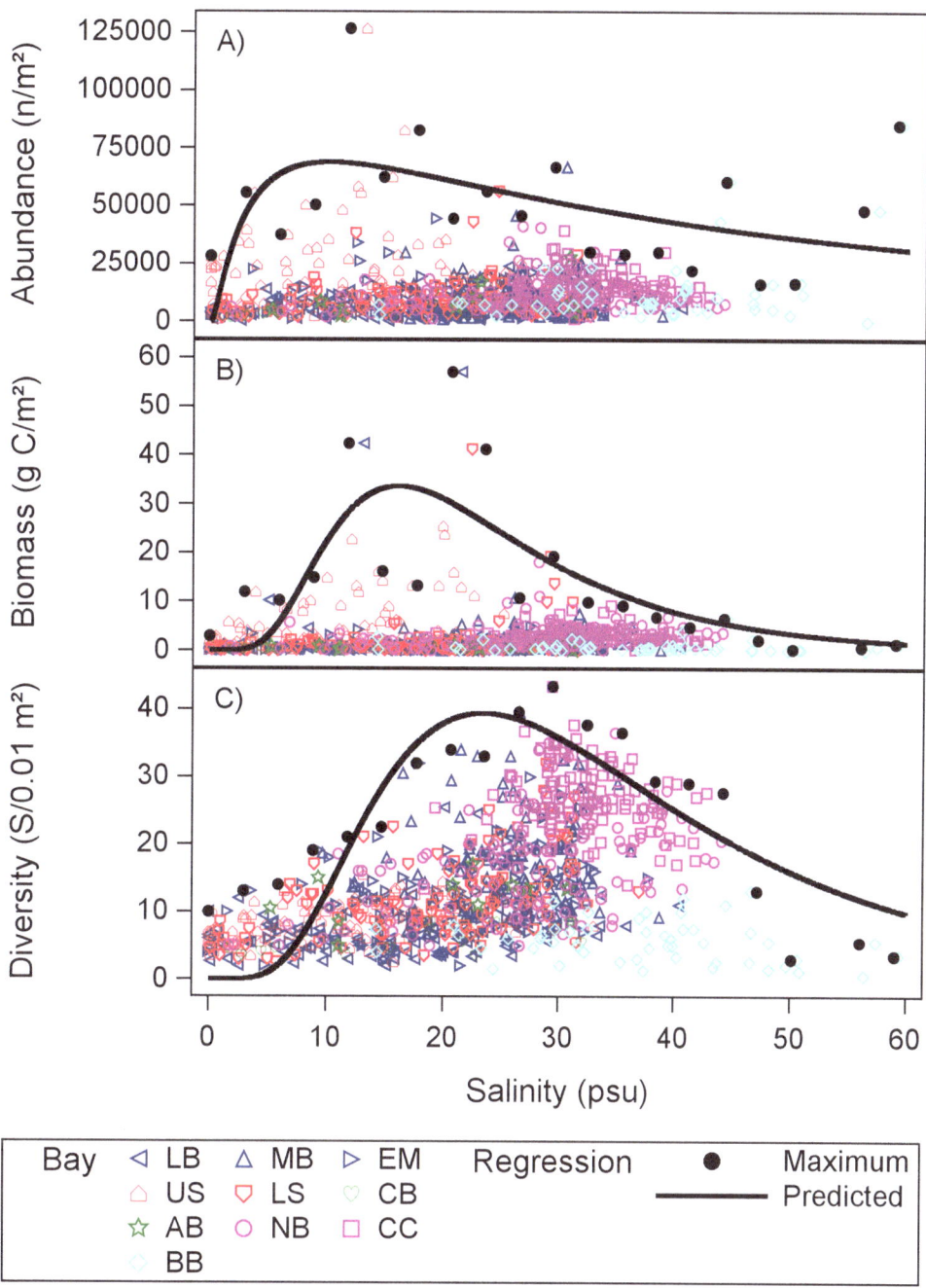

ammonium values (an indicator of heterotrophic conditions). All diversity and biomass metrics were inversely correlated with the inflow axis (PC1, Table 15), meaning diversity biomass decreases during floods. Abundance and diversity metrics were inversely correlated with the seasonal axis; meaning they decreased in summer. Evenness and biomass did not correlate with season. All benthic metrics were inversely correlated with the trophic index, indicating that benthic metrics increased with increased heterotrophy.

**Table 15** Correlation between and water column PCA sample scores (Figs. 3 and 4) and benthic metrics by 885 bay and quarterly sampling date combinations

| Metrics | PC1 (Inflow) | | PC2 (Season) | | PC3 (Trophic) | |
|---|---|---|---|---|---|---|
| | r | P | r | P | r | P |
| Abundance (n/m²) | −0.08 | 0.0124 | −0.17 | <0.0001 | −0.17 | <0.0001 |
| Abundance (Ln(n + 1)/m²) | −0.06 | 0.0852 | −0.22 | <0.0001 | −0.17 | <0.0001 |
| Richness (S/35 cm²) | −0.31 | <0.0001 | −0.13 | 0.0001 | −0.34 | <0.0001 |
| Diversity (H′/35 cm²) | −0.32 | <0.0001 | −0.10 | 0.0034 | −0.35 | <0.0001 |
| Diversity (N1/35 cm²) | −0.34 | <0.0001 | −0.07 | 0.0344 | −0.35 | <0.0001 |
| Evenness (J′/35 cm²) | −0.22 | <0.0001 | 0.02 | 0.6485 | −0.19 | <0.0001 |
| Biomass (g DW/m²) | −0.10 | 0.0018 | −0.06 | 0.0959 | −0.22 | <0.0001 |
| Biomass (g C/m²) | −0.09 | 0.0054 | −0.06 | 0.0746 | −0.21 | <0.0001 |
| Biomass (Ln(g C + 1)/m²) | −0.10 | 0.0023 | −0.08 | 0.0245 | −0.23 | <0.0001 |

Abbreviations: $r$, Spearman correlation coefficient; $P$, probability ($P$) that $r$ equals zero

## Discussion

While the information about long-term benthic dynamics as it relates to changes in salinity is useful for developing and evaluating environmental freshwater inflow standards, it is also important to compare effects among bays and estuaries to identify the general ecological principles that drive organismal response to inflow in estuaries. This is because the domino theory predicts community structure and function is controlled by long-term water quality dynamics (Alber 2002; Montagna et al. 2013). This leads to spatial gradients within estuaries from river to sea and among estuaries from northeast to southwest and temporal variation related to inflow events.

An important temporal factor is tides. The tidal cycles in Texas estuaries typically vary by annual quarters where January and July have extreme low tides and April and October have extreme high tides (Opdyke et al., chapter "Hydrology, Circulation, and Salinity"). Texas bays and estuaries are the nursery areas for the commercial and recreational fish, shellfish, and many forage species (Montagna et al. 2021). These species have estuarine-dependent life cycles with spawning in the Gulf of Mexico or in the passes and growth in nursery habitat areas. The species with offshore spawning key on annual tidal movement cycles for their larval dispersal into the estuary nursery areas. Their life cycles are staggered throughout the year to decrease the competition for niche preference between competing species. The seasonal tidal cycles lead to seasonal dynamics in Texas benthos.

Many species have multiple food resource requirements (i.e., plankton, benthos, epifauna, and nekton) during different stages of their development (Riera et al. 2000). Species have different habitat preferences, for example, seagrass beds, open bay bottoms, oyster reefs, and mud bottoms associated with upper deltas and marshes, which leads to spatial variability.

While the focus here is on the synoptic, long-term studies in the middle Texas coast, there have been benthic studies in other bays. The most extensive spatial studies in Galveston Bay (the Trinity-San Jacinto Estuary) were conducted by the USEPA Environmental Monitoring and Assessment program (EMAP) (USEPA 1999). Other individual studies were performed in Galveston Bay (e.g., Copeland and Bechtel 1971; Holland et al. 1973; Nance 1991; Harper 1992; Carr 1993). Additional studies have been performed in rivers and minor bays (Palmer et al. 2011). An attempt was made to incorporate these studies, but they could not because of the lack of standardized monitoring along a freshwater gradient and the limited time frame of the studies.

### Spatial Considerations

Because the Texas coast lies in a climatic gradient, different bay systems have different long-term water quality dynamics (Montagna et al. 2018b). So, it is expected that the different bay systems have different mollusk communities (Montagna and Kalke 1995), different diversity patterns (Van Diggelen and Montagna 2016), and different secondary productivity patterns (Montagna and Li 2010; Kim and Montagna 2012). All of these previous findings are confirmed here for water quality differences (Fig. 3) and for benthic community differences (Figs. 16 and 17).

Most authors have organized Texas estuaries into zoogeographic zones, which we have outlined in Table 2 (Ladd 1951; Parker 1959; Mackin 1971; Blanton et al. 1971; Harper 1973; Matthews et al. 1974; Gilmore et al. 1976; White et al. 1983, 1985; Jones et al. 1986). Typically, these zones ranged from the freshwater influenced, upper or secondary bays, along a gradient to marine influence in the lower or primary bays. The authors have either defined their own, or used different terms that describe the zones and their associated fauna from the upper to lower bay. For example, *Rangia cuneata* can occupy a low-salinity zone by slightly overlapping the range of typically freshwater clams such as mussels (Unionidae) on its inland border and slightly overlapping the oyster zone on its seaward border (Hopkins et al. 1973). Thus, *Rangia* is associated with freshwater animals on one end of its range and euryhaline members of the oyster community, estuarine fishes, and crustaceans on the other end of its range.

There are three generic salinity zones in Texas estuaries with characteristic benthic communities along this gradient. These are a freshwater zone, an estuarine zone, and a marine zone. The estuarine zone is where fresh and saltwater are mixed, and salinities are intermediate. The boundaries of the estuarine zone are the most susceptible to intra- and interannual climatic variations.

Community differences were found between the Lavaca-Colorado Estuary and Nueces Estuary (Table 5, Figs. 16 and 17), which were open to the Gulf of Mexico, and the Guadalupe Estuary, which does not have a continuously open and direct connection to the Gulf of Mexico (Table 3). Although separated geographically, the Lavaca-Colorado Estuary and the Nueces Estuary are more similar to each other than each is to the Guadalupe Estuary (Fig. 16). Both the Lavaca and Nueces estuaries have an upper secondary bay and a large open primary bay directly connected to the Gulf of Mexico via passes. The Guadalupe Estuary is very different. San Antonio Bay is divided into upper and lower San Antonio Bay and does not have direct access to the Gulf. Additionally, Cedar Bayou, the pass to the Gulf of Mexico is intermittently open and closed (Ward 2010).

The total number of species in both open and closed systems increases along a salinity gradient from the freshwater influenced upper bay to the marine influenced lower bay. In the open estuarine system, species common to the upper secondary bay are usually replaced by more marine tolerant species in the lower bay. Total abundance and biomass also typically increase from upper to the lower bay (Table 6).

Dominant species in the upper bay of the closed estuarine system (San Antonio Bay) are typically part of the dominant fauna in the lower bay during flood years but can be replaced by marine fauna in the lower bay during drought years. Infaunal abundance and biomass are usually higher in the upper San Antonio Bay and decrease in lower San Antonio Bay. This response is most likely due to nutrient input and sediment loading during periodic flooding and to a stable long-term low salinity environment.

## Temporal Considerations

The long-term data set is important because ecological relationships can be obscured in short-term studies by common features such as time lags, natural variability, nonlinear relationships, interactive drivers, or relatively slow processes (Hampton et al. 2019). Thus, long-term research provides a unique perspective on environmental processes, dynamics of populations, and communities of organisms and has led to major scientific discoveries.

Over time, climatic events can drive benthic community structure, and these include random storm events, floods, freezes, tropical storms and hurricanes, and droughts. Harmful algal blooms can also effect benthos. There was large defaunation benthos due to the brown tide, *Aureoumbra lagunensis*, in Baffin Bay and Laguna Madre during the 1990s (Montagna et al. 2010). Seasonal and year-to-year variability also effects community structure and secondary production. The Texas coast does not have very cold temperatures during winter, yet the biological responses are still what one would expect of a temperate estuary, that is, fall and winter die-offs and spring and summer blooms (Montagna 2023). The cycle of floods and droughts moderate or exacerbate the natural seasonal cycles.

We jump to the wrong conclusions by looking at the noise (over short-term periods) rather than the signal (over long-term periods). This is especially evident when we examine the benthic response after hurricanes (Montagna 2023). There was near defaunation in San Antonio Bay from Hurricane Harvey but recovery 6 months later because a recruitment by the opportunistic mollusk *Mulinia lateralis*. The short-term view leads to the conclusion that benthos are vulnerable but resilient, meaning they die off but recover after a period of time. On the other hand, the long-term view is very different and leads to the conclusion that the benthos responses to the storm are a bit larger than expected but still within the range of error. Thus, the long-term view is that benthos are actually both resistant (meaning they bend without breaking) because of seasonal dynamics and resilient (meaning they can recover when knocked down) because of responses many months after a storm.

## Bioindicators of Salinity Zone Habitats

The complex structure of estuarine habitats in Texas has long been recognized. In early descriptions, habitats were referred to as biotopes, that is, a region with uniform environmental conditions and populations of animals and plants for which it is the habitat (Oppenheimer and Gordan 1972). Oppenheimer and Gordan (1972) listed 18 biotypes in Texas estuaries. Even though these habitats fall in distinct zones with respect to distance from river sources, salinity was not considered the driver of these habitat associations. However, the importance of salinity in controlling organisms in Texas bays was well established in the 1950s and 1960s (Galtsoff 1964; Copeland 1966). More recent descriptions of Texas habitats have specifically focused on the role of freshwater inflow and salinity in organizing communities (Blackburn 2004). Inflow and thus salinity gradients with secondary and primary bays are key drivers of the spatial distribution of habitats in the Texas Coastal Bend (Fig. 1, Kalke and Montagna 1991; Montagna et al. 1996).

Infaunal benthic organisms (e.g., polychaetes, amphipod crustaceans, and mollusks) are good indicators of salinity effects because they are relatively immobile compared to epibenthos (e.g., large mobile crustacean like shrimp and crabs and demersal fish), plankton, and nekton (e.g., fish). Their immobility means that benthos must adapt or survive to changing conditions in their habitat because they cannot move. Variability of benthos communities is shaped by morphology of habitats, but the "hydrological seascape," that is, the interaction between habitats and salinity, is the major driver that explains the role benthos play in estuaries (Tenore et al. 2006). The importance of benthic indicators was recognized by the BBESTs. Five of the seven BBESTs used oysters, *Rangia*, or benthos to guide derivation of inflow standards. Benthos are also at the base of food webs and are thus important forage for higher trophic levels, that is, crab, shrimp, and fish. There are three main types of feeding strategies among benthic organisms: predation/omnivory, deposit feeding, and filter/suspension/epistrate feeding (Montagna and Li 2010; Kim and Montagna 2010, 2012). The filter/suspension/epistrate feeders are selecting food from the water column or the surface sediments (Tenore et al. 2006) and are thus most directly affected by inflow dynamics, which can influence nutrient loading and primary productivity.

Benthic community structure in the mid-Texas bays is linked to salinity (Fig. 15). There is extreme dominance in Texas estuaries, and two polychaete species, *Mediomastus ambiseta* and *Streblospio benedicti*, represent 42% of all individuals found. Both species are more abundant in the secondary intermediate salinity bays than primary bays, and their distribution follows salinity distributions. For example, where the long-term average salinities for bays are similar, the abundance of these two species is similar (Fig. 17). The fourth dominant species at 4% of total is the bivalve *Mulinia lateralis*, but it is particularly dominant in upper San Antonio Bay (US) but less dominant in the other low salinity bays. However, the response to Hurricane Harvey indicates *Mulinia* recruitment is very dependent on the large salinity changes and defaunation events brought by large floods and devastating storms. (Montagna 2023). The fifth dominant species at 3.8% of total is the gastropod *Texadina sphinctostoma*. *Texadina* occurs primarily in upper San Antonio Bay and Lavaca Bay and is one of the most dominant gastropod inhabitants of the river influenced upper bays of the Texas coast. Abundance increases are typically associated with freshwater inflow events. The only infaunal crustacean that is an inflow indicator is the amphipod *Ampelisca abdita*, which

made up 1.2% of total organisms, and occurs primarily in the secondary bays.

There are also marine indicators. The polychaete *Dipolydora caulleryi* is the fourth dominant species at 4.6% of all species and is found primarily in the primary bays. *Dipolydora caulleryi* has been found to colonize deeper sediments in association with larger deep burrowing organisms, that is, the hemichordate *Schizocardium* sp. in Corpus Christi Bay (Flint and Kalke 1986). The predatory worm Nemertea at 1.6% is also found in primary bays. The tenth dominant species at 1.5% is the polychaete *Cossura delta* and is found primarily in primary bays.

## Linking Inflow, Salinity, and Ecological Response

We have learned that salinity is an important driver of estuarine benthic community structure. This is especially true within estuaries along the salinity gradient and among estuaries along the coastal climatic gradient. However, we have also learned that climate variability is an important driver of salinity in Texas estuaries (Kim et al. 2014; Pollack et al. 2011; Tolan 2007). Instead of starting with inflow, the conceptual model in Fig. 2 should start with climate because climate drives the hydrologic cycle and thus the amount of freshwater inflow delivered to the coast. Texas estuaries are a suitable location to study the effects of climate variation because they are physically similar, each estuary drains one or two watersheds, and they lie in a climatic gradient with decreasing precipitation from the northeast to southwest. The local climatic gradient and ENSO are influencing hydrological (Tolan 2007) and ecological (Kim et al. 2014) dynamics in Texas estuaries.

Estuarine organisms exhibit optimal salinity tolerances for growth, development, and reproduction (Patillo et al. 1997). Droughts, floods, tropical storms (i.e., hurricanes), and occasional freezes cause salinity changes and are the main factors controlling distribution and diversity of macroinfaunal communities. This is because benthic organisms are especially sensitive to changes in salinity because they typically are fixed in place and can't move if conditions are unfavorable. Sometimes, events cause defaunation. Changes in salinity alter macroinfauna diversity (Van Diggelen and Montagna 2016) and biomass (Palmer et al. 2011) in Texas. Similar results were found in the Gulf of Riga in the North Sea (Kotta et al. 2009), in estuaries in India (Mulik et al. 2020), in the Yangtze Estuary in China (Wu et al. 2019), and many other places.

In Texas, the primary bays are different from secondary bays. The similarities in macroinfauna communities within bays were likely driven by similarities in salinity within bays. In Tees Bay, United Kingdom, long-term changes in macrobenthos abundance, diversity, and community structure changed differently near the river mouth compared to far from it (Warwick et al. 2002). Functional infauna diversity will decrease with changes in freshwater inflow, and benthic infauna communities will acclimate to the changes in salinity, and more (or less) salt-tolerant species will dominate the communities depending on the long-term salinity averages (Kim and Montagna 2009, 2012; Montagna et al. 2002; Palmer et al. 2002).

## Comparisons with Historical Studies

Ecological studies over the years have demonstrated the importance of salinity as a factor in affecting the distribution of marine and estuarine organisms. The number of species, but not necessarily the observed total biomass increases as one proceeds along a salinity gradient from the freshwater side of a large estuary to the open sea (Springer and Woodburn 1960; Gunter 1961). Sessile and nonmotile organisms have optimum salinities at which they grow best and variations from the optimum inhibit growth (Gunter 1961). The life histories of many motile species, that is, commercially important penaeid shrimps and menhaden, are migratory. Spawning occurs offshore, followed by the migration of larvae and juveniles to low salinity nursery areas at the heads of estuaries (Gunter 1957; Baldauf 1970). Individuals of marine species of fish that utilize freshwater nursery areas are predominantly juveniles (Gunter 1957).

Some of these trends change when the estuary becomes hypersaline. For example, in the Laguna Madre when the salinity increases there are fewer species, the number of individuals of each species increases (e.g., *Streblospio benedicti* and *Mulinia lateralis*), the average individual of each vertebrate species is larger, and the average individual of many invertebrate species, that is, blue crabs and barnacles, decreases (Simmons 1957). When salinity and temperature are variable and extreme in Laguna Madre, only a few species of marine invertebrates and individuals of each species may survive, but when hydrographic conditions are stable and salinities and temperature are in the extreme range, a few tolerant species may become extremely abundant. When physical conditions are stable and within the normal range for marine environments, there will be many species but fewer individuals of each species (Parker 1959).

Variations in salinity may be adverse for some organisms and beneficial for others. In general, when salinity changes within limits, the biomass remains fairly constant while the diversity or number of species changes (Hopkins et al. 1973). Salinity changes may add or eliminate species from an environment. For example, because of the worst drought in Texas history (1948–1956), Mesquite Bay, a low to medium salinity bay, had salinity over 50 psu in August 1956 (Hoese

1960). Low salinity species, *Rangia cuneata*, *Texadina sphinctostoma*, and *Brachidontes recurvus*, apparently suffered complete mortality over much of their original range. The fauna of Mesquite Bay in early 1957 was predominantly marine, but by May 1957, salinities in Mesquite Bay dropped from over 35 to 3 psu due to flooding from the Guadalupe River system. The marine community suffered complete morality and overtime was replaced with typical estuarine species.

## Biology of Dominant Species

The dominant benthic macrofauna species collected from San Antonio Bay in quantitative studies were the gastropod, *Texadina sphinctostoma*, the bivalves, *Rangia cuneata* and *Mulinia lateralis*, and the polychaetes, *Mediomastus ambiseta*, *Streblospio benedicti*, and *Hypaniola gunneri floridus* (Harper 1973; Matthews et al. 1974). The distribution of these species is strongly linked to long-term lower salinity environmental conditions, although responses to flood conditions may result in rapid population changes.

*Texadina sphinctostoma*, a gastropod, populations increase following peaks in freshwater inflow (Harper 1973; Matthews et al. 1974). This is apparently a recruitment response caused by a salinity decline (Harper 1973). *Texadina* carries its eggs on the shell and undergoes direct development with the young ready to assume adult existence upon emerging from the egg. *Texadina sphinctostoma* is commonly reported as one of the most dominant gastropod inhabitants of the river influenced upper bays of the Texas coast (Ladd 1951; Ladd et al. 1957; Parker 1955, 1959; Harper 1973; Matthews et al. 1974; Gilmore et al. 1976; White et al. 1983, 1985; Staff et al. 1985; Cummins et al. 1986; Montagna and Kalke 1995).

*Rangia cuneata*, a brackish water clam in the family Mactridae, is an excellent indicator of ecological effects of salinity changes in coastal waters and has been comprehensively studied by Hopkins et al. (1973). It is commonly the dominant species in the 0–15 psu salinity zone, and since 1955, its range has extended from along the Gulf of Mexico coastal estuaries to along the Atlantic coast from Georgia to Maryland. The well-being of the species is not dependent on the physiology of the adult, because the adults can tolerate salinities from 0 to 38 psu and temperatures from 10 to 35 °C. Spawning is induced by a change in salinity either up from near 0 psu or down from 15 ppt (Hopkins et al. 1973). The embryos and early larvae can survive only in salinities between 2 and 15 psu. *Rangia* is another species for which low salinity, in the range of 1–15 psu, is optimal, and it evidently cannot maintain a population outside this range (Hopkins and Andrews 1970). Adult *Rangia* are abundant far up tidal rivers where salinity may stay below 1 psu continuously for months or even years. No living *Rangia* were found in San Antonio Bay in the drought summer of 1951 and spring 1952 although extensive collections were made throughout the bay (Parker 1955). Individuals of *Rangia* are progressively larger in size from the center of San Antonio Bay to the Guadalupe River delta, into the mouth of the river and Mission Lake (Ladd 1951). *Rangia cuneata* in the 90 cm size range occur in the Nueces River near Labonte Park. The salinity in this area can be in the range for *Rangia* larval development, but no sampling was done in this area to evaluate their presence. The size ranges from juvenile to long-lived adult *Rangia* are so large that different gear types, that is, trawls, oyster dredges, benthic grabs, cores, rakes, and quadrats, have been used to collect *Rangia*. Some studies target *Rangia*, while some *Rangia* data is a byproduct of other studies, that is, trawl data, and is size selective. *Rangia* shell middens are common throughout the Nueces Bay delta at historic Indian campsites. This indicates that approximately 100 years ago freshwater inflow into the Nueces Delta was greater and supported large populations of *Rangia cuneata* used as a food source by the local Indians.

*Mulinia lateralis* is widely reported from other bays around the globe. Spawning was observed in the Tred Avon River, Maryland, and Chesapeake Bay where it was observed to have a continuous period of setting from a single spawning cycle from May through November (Shaw 1965; Holland et al. 1977). In Alazan Bay, Texas Cornelius (1984) observed juveniles in all months except December, and Poff (1973) observed year-round spawning in Trinity Bay, Texas. *Mulinia lateralis* has a very short generation time and is capable of successfully spawning at 3 mm in length, which is approximately 60 days old (Calabrese 1969a). Embryo survival and development for *Mulinia* as it is with *Rangia cuneata* are dependent on certain salinity and temperature ranges. *Mulinia lateralis* developed into normal larvae throughout the salinity range of 15–35 psu and the temperature range of 10–30 °C (Calabrese 1969b). This clam is an important food item to bottom feeding organisms, that is, the black drum (Pearson 1929; Breuer 1957; Simmons and Breuer 1962; Martin 1979) and to the greater and lesser scaup ducks (Cronan 1957). Large rafts of scaup ducks were observed in upper San Antonio Bay in November 1988 corresponding to high densities of *Mulinia lateralis*.

The polychaete *Mediomastus ambiseta* is a euryhaline species reaching peak abundance in San Antonio Bay at 12.5 psu and gradually declining at higher and lower salinities. Population densities were not affected by flood conditions (Harper 1973). Matthews et al. (1974) collected *M. ambiseta* in brackish to higher salinity waters of 6–16 psu.

*Streblospio benedicti*, a polychaete, preferred the salinity range of 10–12 psu according to Harper (1973). It was restricted by higher salinities and virtually disappeared from upper San Antonio Bay following the flood. It was described as a brackish water species by Matthews et al. (1974), being associated only with the mid-bay zone. In other high salinity bays, that is, Baffin Bay and Rincon Bayou, high densities of *Streblospio benedicti* occur in salinities of 30–55 psu and 1.0–86 psu, respectively. *Streblospio benedicti* is an important food source to shrimp and fish, has the ability to colonize adverse high salinities, and can produce a community with high worm abundance but low diversity during peak larval shrimp and fish migrations (Riera et al. 2000; Montagna et al. 2018a).

Populations of the polychaete, *Hypaniola gunneri floridus*, were highest in the upper bay from June to August 1972 when the salinity was lowest. This species was not common above 10 ppt (Harper 1973). Increased abundance of *H. gunneri floridus* was attributed to freshwater inflow by Matthews et al. (1974). Highest abundance of *Hobsonia florida* occurred with high freshwater inflow in Lavaca Bay from 1984 to 1986 (Kalke and Montagna 1991).

## Using Benthos to Evaluate Inflow Standards

While it is desirable to use benthic responses to evaluate current freshwater inflow standards adopted for Texas estuaries, it is nearly impossible. This was because there was essentially one number for whole bay systems, and you could calculate how that number, and its year-to-year variability would affect salinity and thus biological responses. A good example is the application of the domino theory to the Caloosahatchee River in Florida (Palmer et al. 2016). Biological resources in estuaries are affected by salinity more than inflow by itself, so the links between flow, salinity, and biology will determine the relationship between inflow and living resources. The first step is to identify the resource to be protected. The second step is to identify the salinity range or requirements of the resource in both space and time. The third step is to calculate the flow regime needed to support the required distribution of salinity.

In contrast, Texas rules are quite complex and describe different attainment frequencies over 4–6-year periods of time, in different seasons, and under different conditions. The adopted Texas environmental flow standards can be found in Chapter 298 of TCEQ's rules (https://www.tceq. texas.gov/permitting/water_rights/wr_technical-resources/ eflows/rulemaking). The standards are complex, consisting of multiple tables that describe flow regimes that vary in three dimensions: over components or climatic periods (such as wet and dry years), over seasons, and spatially within rivers, streams, and bay systems. Additionally, terminology in the rules varies. For example, the component climatic periods for inflow regimes are called "wet, average, dry, and subsistence" for Lavaca Bay, "level" for Matagorda Bay and Nueces Bay, and by season name (i.e., spring, summer, summer, and fall) for San Antonio Bay. The environmental flow standards were based on statistical evaluations of historical occurrence frequencies of different flow level hydrological categories, which also vary (Opdyke et al. 2014; Anchor 2021).

There is little in common in the structure of the standards among or within the basins. For example, in Matagorda Bay, the standards are based on "monthly" thresholds annually and four "seasonal" thresholds at different "levels" where levels are defined as different inflow regimes. For Lavaca Bay, there are annual attainment frequencies for fall and spring only and for defined "regimes" (subsistence, base dry, base average, and base wet). In San Antonio Bay, there are separate tables for spring and summer attainment frequencies based on six consecutive years. For Nueces Bay and Delta, there are attainment frequencies for three time periods (November to February, March to June, and July to October) at three different "levels" where levels are defined as three different inflow regimes (wet, average, and dry). Thus, it is impossible to link specific benthic responses in any 1 year to a specific standard; instead, the response must be linked to the attainment frequency of the standards.

## Conclusions and Recommendations

In general, diversity is most influenced by Gulf exchange and productivity (as indicated by biomass) is most influenced by latitude and interannual variability in freshwater inflow. Species diversity increases with increasing salinity at the marine parts of the estuaries. Biomass and abundance in an open system, like the Lavaca-Colorado and Nueces Estuaries, increase near the opening to the Gulf but in a closed system, like the Guadalupe Estuary, decrease toward the marine influenced zone. All Texas estuaries can be divided into marine, estuarine, and freshwater zoogeographic zones, but the boundaries of the zones are regulated by three factors. First is the geographic location of the estuary. Second is whether the bay has a direct opening to the Gulf of Mexico. Third is the interannual variability in rainfall and freshwater inflow. These generalizations indicate that the estuaries must be treated independently when making management decisions but that the goal to develop a generic model, which describes the effect of freshwater inflow on estuarine benthic dynamics, can be achieved.

Benthic monitoring is useful for a variety of environmental studies and assessments. This is because bottom-dwelling organisms are relatively immobile (and can't move when exposed to stressors), effected by deposition of material from the water column (because of gravity), and highly diverse (so that during disturbances, tolerant species increase or stay the same, and sensitive species decrease or die out). Thus, benthos are sensitive bioindicators of environmental change. However, using any bioindicator to evaluate current Texas inflow standards is problematic because of the complexity of the standards. Simplifying standards is necessary.

The environment associated with the Texas estuaries is subject to hurricanes, inland flooding, droughts, and temperature extremes, which result in an estuarine environment, which is variable. However, the extremes are cyclical in a chaotic fashion, that is, storms occur at predictable intervals over the long term. The most important effect of these events is on the variability of freshwater inflow, which in turn affects the salinity, nutrients, and sediment-load input to the estuary. This controls the ultimate effect on the benthic communities. The variability in freshwater inflow cycle results in predictable changes in the estuary (Fig. 20).

Flood conditions introduce nutrient-rich waters into the estuary, which result in lower salinity. This usually happens very rapidly. During these periods, the spatial extent of the freshwater fauna is increased. The estuarine fauna may even replace the marine fauna. The high level of nutrients can stimulate a burst of benthic productivity of predominantly freshwater and estuarine communities. This is followed by a transition to low inflow resulting in higher salinities, lower nutrient, marine fauna, and drought conditions. At first, the marine fauna may respond with a burst of productivity as the remaining nutrients are utilized, but eventually, nutrients are depleted. The cycle is repeated with flooding and high freshwater inflow.

This model is supported by data in the Guadalupe Estuary. During successive wet years, densities decrease (stages 1, and 4–1). When a dry year follows a wet year, the densities increase (stages 1–2). The same pattern also occurs in the Lavaca-Colorado Estuary and the Nueces Estuary. Other aspects of the model are supported by the Nueces Estuary data. Although there was intervening wet years, densities decreased during successive dry years (stages 3–4).

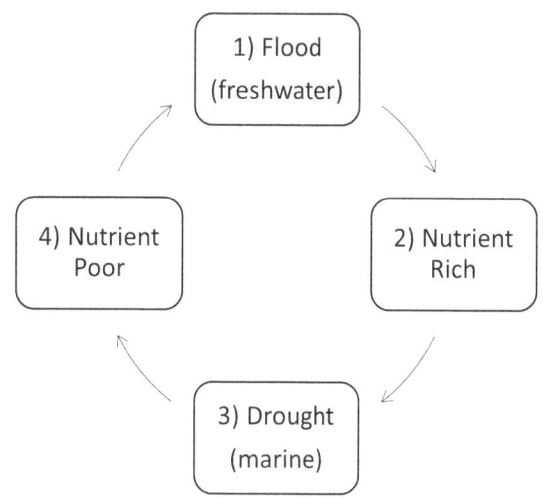

**Fig. 20** Effects of the hydrological cycle on water column conditions that drive benthic communities

The results of benthic sampling depend on what state this cycle is in during the study. For example, benthic studies in Texas estuaries have often reported a response following a flood period which results in higher abundance and biomass of particular estuarine species (Mackin 1971; Harper 1973; Matthews et al. 1974; Gilmore et al. 1976; Kalke and Montagna 1991; Flint et al. 1981; Flint and Rabalais 1981). The boundaries that authors draw on the various zones will also be a function on the state of the cycle that the estuary is in. The length of time that the estuaries are maintained in any given state will be a function of the periodicity of storms, floods, and droughts.

**Acknowledgments** Work on this manuscript was funded in part by the Harte Research Institute for Gulf of Mexico Studies; Contract No. 21-155-007-C879 from the Texas General Land Office (GLO) with Gulf of Mexico Energy Security Act of 2006 funding made available to the State of Texas and awarded under the Texas Coastal Management Program; the National Oceanic and Atmospheric Administration (NOAA), Office of Education Educational Partnership Program award number NA21SEC4810004; Texas Sea Grant Program award number NA18-2020-R/HCE-WQ-4 via prime award from NOAA NA18OAR4170088. Additionally, the Texas Water Development Board provided funding from 1986 to 2020 for field, laboratory, and synthesis studies on Texas benthos. The views and contents are solely the responsibility of the award recipient and do not necessarily represent the official views of the US Department of Commerce, NOAA, the GLO, or the State of Texas.

# References

Alber M (2002) A conceptual model of estuarine freshwater inflow management. Estuaries 25:1246–1261

Anchor QEA (2021) Evaluating the attainment of environmental flow standards. Final report to the Texas Water Development Board. Anchor QEA, LLC, Austin, TX, USA, 1239 pp

Baldauf RJ (1970) A study of selected chemical and biological conditions of the lower Trinity River and the upper Trinity Bay. Technical report, to Water Resources Institute. Texas A&M University, College Station, TX, 168 pp

Blackburn JB (2004) The book of Texas bays. Texas A & M University Press, College Station, Texas, USA, p 200

Blanton WG, Culpepper TJ, Bischoff HW, Smith AL, Blanton CJ (1971) A study of the total ecology of a secondary bay (Lavaca Bay). Final report to Aluminum Co. of America. Aluminum Co. of America, Point Comfort, TX, 306 pp

Borja A, Chust G, Muxika I (2019) Forever young: the successful story of a marine biotic index. Adv Mar Biol 82:93–127

Breuer JP (1957) An ecological survey of Baffin and Alazan Bay, TX. Publ Inst Mar Sci 4:134–155

Brey T (2001) Conversion factors for aquatic organisms, version 4. Population dynamics in benthic invertebrates. A virtual handbook. Version 01.2. http://www.thomas-brey.de/science/

Brey T, Müller-Wiegmann C, Zittier Z, Hagen W (2010) Body composition in aquatic organisms—a global data bank of relationships between mass, element composition and energy content. J Sea Res 64:334–340. https://doi.org/10.1016/j.seares.2010.05.002

Calabrese A (1969a) Individual and combined effects of salinity and temperature on embryos and larvae of the coot clam, *Mulinia lateralis*. Biol Bull 137:417–428

Calabrese A (1969b) Reproductive cycle of the coot clam, *Mulinia lateralis*, in Long Island Sound. Veliger 12:265–269

Carr RS (1993) Sediment quality assessment survey of the Galveston Bay System. The Galveston Bay National Estuary Program, Publication GBNEP-30

Clarke KR, Gorley RN (2015) PIMER v7: user manual/tutorial. Primer-E, Plymouth, UK

Clarke KR, Gorley RN, Somerfield PJ, Warwick RM (2014) Change in marine communities: an approach to statistical analysis and interpretation, 3rd edn. Primer-E, Plymouth, UK

Copeland BJ (1966) Effects of decreased river flow on estuarine ecology. J Water Pollut Control Fed 38:1831–1839

Copeland BJ, Bechtel TJ (1971) Species diversity and water quality in Galveston Bay, Texas. Water Air Soil Pollut 1:89–105

Cornelius SE (1984) An ecological survey of Alazan Bay, Texas. Caesar Kleberg Wildlife Research Found, Kingsville, TX. 87 pp

Cronan JM Jr (1957) Food and Feeding habits of the scaups in Connecticut waters. Auk 74:459–468

Cummins H, Powell EN, Newton HJ, Stanton RJ Jr (1986) Assessing transportation by the covariance of species with comments on contagious and random distributions. Lethaia 19:22

Dauer DM (1993) Biological criteria, environmental health and estuarine macrobenthic community structure. Mar Pollut Bull 26:249–257

Douglas AR, Montagna PA (2024) Influence of inflows on estuary sediments. In: Montagna PA, Douglas AR (eds) Freshwater inflows to Texas bays and estuaries: A regional-scale review, synthesis, and recommendations. Springer, Switzerland

Engle VD, Summers JK (1999) Refinement, validation, and application of a benthic condition index for northern Gulf of Mexico estuaries. Estuaries 22:624–635

Flint RW, Kalke RD (1986) Biological enhancement of estuarine benthic community structure. Mar Ecol Prog Ser 31:23–33

Flint RW, Rabalais SC (1981) Estuarine benthic community dynamics related to freshwater inflow to Corpus Christi Bay Estuary. In: Cross RD, Williams DL (eds) Proceedings of the international symposium of freshwater inflow to estuaries. US Dept. Int. Fish & Wildlife, Washington, DC, pp 489–508

Flint RW, Younk JA (1983) Estuarine benthos: Long-term variations, Corpus Christi, Texas. Estuaries 6:126–141

Flint RW, Kalke RD, Rabalais SC (1981) Quantification of extensive freshwater input to estuarine benthos. Report to the Texas Water Development Board. University of Texas, Marine Science Institute, Port Aransas, TX, 55 pp & Appendices

Galtsoff PS (1964) The American oyster, Crassostrea virginica Gmelin. Fishery Bull USFWS 64:1–477

Gillett DJ, Weisberg SB, Grayson T, Hamilton A, Hansen V et al (2015) Effect of ecological group classification schemes on performance of the AMBI benthic index in US coastal waters. Ecol Indic 50:99–107

Gilmore G, Dailey J, Garcia M, Hannebaum N, Means J (1976) IV Benthos. Technical report to the Texas Water Development Board. A study of the effects of fresh water on the plankton benthos, and nekton assemblages of the Lavaca Bay system, Texas. Texas Parks & Wildlife Dept., Austin, TX, 113 pp

Green RH (1979) Sampling design and statistical methods for environmental biologists. Wiley, New York

Green RH, Montagna P (1996) Implications for monitoring: study designs and interpretation of results. Can J Fish Aquat Sci 53:2629–2636. https://doi.org/10.1139/cjfas-53-11-2629

Gunter G (1957) Predominance of the young among marine fishes found in fresh water. Copeia 1:13–16

Gunter G (1961) Some relations of estuarine organisms to salinity. Limnol Oceanogr 6:182–190

Hampton SE, Scheuerell MD, Church MJ, Melack JM (2019) Long-term perspectives in aquatic research. Limnol Oceanogr 64:S2–S10

Harper DE Jr (1973) The distribution of benthic and nektonic organisms in undredged control areas of San Antonio Bay. Environmental impact assessment of shell dredging in San Antonio Bay, TX. Report to U.S. Army Engineer District. Texas A&M Res. Found., College Station, TX, 157 pp

Harper DE Jr (1992) Characterization of open bay benthic assemblages of the Galveston Estuary and adjacent estuaries from the Sabine River to San Antonio Bay. In: Status and trends of selected living resources in the Galveston Bay System, publication GBNEP-19 June 1992. https://wayback.archive-it.org/414/20210309074213/https://www.tceq.texas.gov/assets/public/comm_exec/pubs/gbnep/gbnep-19/index.html. Accessed 2 Dec 2022

Harris EK, Montagna PA, Douglas AR, Vitale L, Buzan D (2023) Influence of an industrial discharge on long-term dynamics of abiotic and biotic resources in Lavaca Bay, Texas, USA. Environ Monit Assess 195:40. https://doi.org/10.1007/s10661-022-10665-w

Hoese HD (1960) Biotic changes in a bay associated with the end of a drought. Limnol Oceanogr 5(3):326–336

Holland JS, Maciolek NJ, Oppenheimer CH (1973) Galveston Bay benthic community structure as an indicator of water quality. Contrib Mar Sci 17:169–188

Holland FA, Mountford NK, Mihursky JA (1977) Temporal variation in upper bay mesohaline benthic communities the 9-m mud habitat. Chesap Sci 18:370–378

Hopkins SH, Andrews JD (1970) *Rangia cuneata* on the East Coast: Thousand mile range extension, or resurgence. Science 167:868–869

Hopkins SH, Anderson JW, Horvath K (1973) The brackish water clam *Rangia cuneata* as indicator of ecological effects of salinity changes in coastal waters. U.S. Army Engineer Waterways Experiment Station. Texas A&M Research Foundation, College Station, TX, p 257

Jones RS, Cullen JJ, Lane RG, Yoon W, Rosson RA, Kalke RD, Holt SA, Arnold CR (1986) Studies of freshwater inflow effects on the Lavaca River Delta and Lavaca Bay, TX. Report to the Texas Water Development Board. The Univ of Texas Mar Sci Inst, Port Aransas, TX, 423 pp

Kalke R, Montagna PA (1991) The effect on freshwater inflow on macrobenthos in the Lavaca River delta and upper Lavaca Bay, Texas. Contrib Mar Sci 32:49–77. https://repositories.lib.utexas.edu/handle/2152/18043

Kim H-C, Montagna PA (2009) Implications of Colorado River freshwater inflow to benthic ecosystem dynamics: a modeling study. Estuar Coast Shelf Sci 83:491–504

Kim H-C, Montagna PA (2010) Effect of climatic variability on freshwater inflow, benthic communities and secondary production in Texas lagoonal estuaries. Final report to the Texas Water Development Board, contract number 08 483 0791. Harte Research Institute for Gulf of Mexico Studies, Texas A&M University Corpus Christi, Corpus Christi, Texas. 119 pages.

Kim H-C, Montagna PA (2012) Effects of climate-driven freshwater inflow variability on macrobenthic secondary production in Texas lagoonal estuaries: A modeling study. Ecol Model 235–236:67–80. https://doi.org/10.1016/j.ecolmodel.2012.03.022

Kim H-C, Son S, Montagna P, Spiering B, Nam J (2014) Linkage between freshwater inflow and primary productivity in Texas estuaries: Downscaling effects of climate variability. J Coastal Res 68:65–73

Kotta J, Kotta I, Simm M, Põllupüü M (2009) Separate and interactive effects of eutrophication and climate variables on the ecosystem elements of the Gulf of Riga. Estuar Coastal Shelf Sci 84:509–518

Ladd HS (1951) Brackish water and marine assemblages of the Texas coast, with special reference to mollusks. Publ Inst Mar Sci 2:125–164

Ladd HS, Hedgepeth JW, Post R (1957) Environments and facies of existing bays on the central Texas coast. In Treatise on marine—ecology and paleoecology. Geol Soc Am Mem 67:599–640

Long ER, Dutch M, Partridge V, Weakland S, Welch K (2013) Revision of sediment quality triad indicators in Puget Sound (Washington, USA): I. a Sediment Chemistry Index and targets for mixtures of toxicants. Integr Environ Assess Manag 9:31–49

Ludwig JA, Reynolds JF (1988) Statistical ecology. Wiley, New York

Mackin JG (1971) A study of the effect of oilfield brine effluents of biotic communities in Texas estuaries. Report to Humble Oil & Refining Co. & Amoco Production Co. Texaco Inc. Texas A&M Research Foundation, College Station, TX, 73 pp

Martin JH (1979) A study of the feeding habits of the black drum (*Pogonius cromis* Linnaeus) in Alazan Bay and the Laguna Salada, TX. MS thesis. Texas A & M Univ., Kingsville, TX, 103 pp

Matthews GA, Marcin CA, Clements GL (1974) A plankton and benthos survey of the San Antonio Bay system March 1972-July 1974. Technical report to the Texas Water Development Board. Texas Parks and Wildlife Dept., Austin, TX, 75 pp

Montagna PA (2021) How a simple question about freshwater inflow to estuaries shaped a career. Gulf Carib Res 32:ii–xiv. https://doi.org/10.18785/gcr.3201.04

Montagna PA (2022) Long-term benthic data: adaptive management of three basins. Final Report to the Texas Water Development Board, Contract # 2000012436. Available via https://tamucc-ir.tdl.org/handle/1969.6/93830

Montagna PA (2023) Benthic infauna are resistant and resilient to hurricane disturbance. Mar Ecol Prog Ser 707:1–13. https://doi.org/10.3354/meps14263

Montagna PA, Kalke RD (1992) The effect of freshwater inflow on meiofaunal and macrofaunal populations in the Guadalupe and Nueces Estuaries, Texas. Estuaries 15:266–285. https://doi.org/10.2307/1352779

Montagna PA, Kalke RD (1995) Ecology of infaunal Mollusca in south Texas estuaries. Am Malacol Bull 11:163–175. https://www.biodiversitylibrary.org/page/45940964

Montagna PA, Li J (2010) Effect of freshwater inflow on nutrient loading and macrobenthos secondary production in Texas lagoons. In: Kennish MJ, Paerl HW (eds) Coastal lagoons: critical habitats of environmental change. CRC, Taylor & Francis, Boca Raton, FL, pp 513–539

Montagna PA, Li J, Street GT (1996) A conceptual ecosystem model of the Corpus Christi Bay National Estuary Program Study Area. Corpus Christi Bay National Estuary Program, Publication CCBNEP-08. Texas Natural Resource Conservation Commission, Austin, TX, USA, 114 pp. http://hdl.handle.net/2152/46975

Montagna PA, Kalke RD, Ritter C (2002) Effect of restored freshwater inflow on macrofauna and meiofauna in upper Rincon Bayou, Texas, USA. Estuaries 25:1436–1447. https://doi.org/10.1007/BF02692237

Montagna PA, Kalke RD, Conley MF, Stockwell DA (2010) Relationship between macroinfaunal diversity and community stability, and a disturbance caused by a persistent brown tide bloom in Laguna Madre, Texas. In: Kennish MJ, Paerl HW (eds) Coastal lagoons: critical habitats of environmental change. CRC, Taylor & Francis, Boca Raton, FL, pp 117–136

Montagna PA, Brenner J, Gibeaut J, Morehead S (2011) Coastal impacts. In: Schmidt J, North GR, Clarkson J (eds) The impact of global warming on Texas, 2nd edn. University of Texas Press, Austin, TX, pp 96–123. http://www.jstor.org/stable/10.7560/723306.8

Montagna PA, Palmer TA, Beseres Pollack J (2013) Hydrological changes and estuarine dynamics. SpringerBriefs in environmental sciences. Springer, New York. 94 pp. https://doi.org/10.1007/978-1-4614-5833-3

Montagna PA, Chaloupka C, DelRosario EA, Gordon AM, Kalke RD, Palmer TA, Turner EL (2018a) Managing environmental flows and water resources. WIT Trans Ecol Environ 215:177–188. https://doi.org/10.2495/EID180161

Montagna PA, Hu X, Palmer TA, Wetz M (2018b) Effect of hydrological variability on the biogeochemistry of estuaries across a regional climatic gradient. Limnol Oceanogr 63:2465–2478. https://doi.org/10.1002/lno.10953

Montagna PA, McKinney L, Yoskowitz D (2021) Focused flows to maintain natural nursery habitats. Texas Water J 12(1):129–139. https://doi.org/10.21423/twj.v12i1.7123

Mulik J, Sukumaran S, Srinivas T (2020) Factors structuring spatio-temporal dynamics of macrobenthic communities of three differently modified tropical estuaries. Mar Pollut Bull 150:110767

Nance JM (1991) Effects of oil/gas field produced water on the macrobenthic community in a small gradient estuary. Hydrobiologia 220:189–204

Opdyke DR, Oborny EL, Vaugh SK, Mayes KB (2014) Texas environmental flow standards and the hydrology-based environmental flow regime methodology. Hydrol Sci J 59:820–830

Oppenheimer CH, Gordan KG (1972) Texas coastal zone biotopes: an ecography. The University of Texas Marine Science Institute, Port Aransas, Texas, Interim Report for the Bay and Estuary Management Program (CRMP). http://hdl.handle.net/2152/43939

Palmer TA, Montagna PA, Kalke RD (2002) Downstream effects of restored freshwater inflow to Rincon Bayou, Nueces Delta, Texas, USA. Estuaries 25:1448–1456

Palmer TA, Montagna PA (2015) Impacts of droughts and low flows on estuarine water quality and benthic fauna. Hydrobiologia 753:111–129. https://doi.org/10.1007/s10750-015-2200-x

Palmer TA, Montagna PA, Pollack JB, Kalke RD, DeYoe HR (2011) The role of freshwater inflow in lagoons, rivers and bays. Hydrobiologia 667:49–67. https://doi.org/10.1007/s10750-011-0637-0

Palmer TA, Montagna PA, Chamberlain RH, Doering PH, Wan Y, Haunert K, Crean DJ (2016) Determining the effects of freshwater inflow on benthic macrofauna in the Caloosahatchee Estuary, Florida. Integr Environ Assess Manag 12:529–539

Parker RH (1955) Changes in the invertebrate fauna, apparently attributable to salinity changes, in the bays of central Texas. J Paleontol 29:193–211

Parker RH (1959) Macro-invertebrate assemblages of central Texas coastal bays and Lagune Madre. Bull Am Assoc Pet Geol 43:2100–2166

Patillo ME, Czapla TE, Nelson DM, Monaco ME (1997) Distribution and abundance of fishes and invertebrates in Gulf of Mexico estuaries. Species life history summaries. ELMR Report No. 11, vol. II. NOAA/NOS Strategic Environmental Assessment Division, Silver Spring, MD, p 377

Paudel B, Montagna PA (2014) Modeling inorganic nutrient distributions among hydrologic gradients using multivariate approaches. Ecol Inform 24:35–46. https://doi.org/10.1016/j.ecoinf.2014.06.003

Pearson JC (1929) Natural history and conservation of redfish and other commercial sciaenids of the Texas coast. Bull Bur Fish 44:129–144

Pearson TH, Rosenberg R (1978) Macrobenthic succession in relation to organic enrichment and pollution of the marine environment. Ocean Mar Biol Annu Rev 16:229–311

Poff MJ (1973) Species composition, distribution and abundance of macrobenthic organisms in the intake and discharge areas after construction and operation of the Cedar Bayou Electric Power Station. M.S. thesis, Texas A & M Univ., College Station, TX, 348 pp

Pollack JB, Kinsey JW, Montagna PA (2009) Freshwater Inflow Biotic Index (FIBI) for the Lavaca-Colorado Estuary, Texas. Environ Bioindic 4:153–169. https://doi.org/10.1080/15555270902986831

Pollack JB, Palmer TA, Montagna PA (2011) Long-term trends in the response of benthic macrofauna to climate variability in the Lavaca-Colorado Estuary, Texas. Mar Ecol Prog Ser 436:67–80. https://doi.org/10.3354/meps09267

Rhoads DC, McCall PL, Yingst JY (1978) Disturbance and production on the estuarine seafloor. Am Sci 66:577–586

Riera P, Montagna PA, Kalke RD, Richard P (2000) Utilization of estuarine organic matter during growth and migration by juvenile brown shrimp Penaeus aztecus in a South Texas estuary. Mar Ecol Prog Ser 199:205–216. http://www.int-res.com/abstracts/meps/v199/p205-216/

Ritter C, Montagna PA (1999) Seasonal hypoxia and models of benthic response in a Texas bay. Estuaries 22:7–20. https://doi.org/10.2307/1352922

Ritter C, Montagna PA, Applebaum S (2005) Short-term succession dynamics of macrobenthos in a salinity-stressed estuary. J Exp Mar Biol Ecol 323:57–69. https://doi.org/10.1016/j.jembe.2005.02.018

SAS Institute Inc (2017) SAS/ETS® 14.3 User's Guide. SAS Institute Inc.,Cary, NC

Science Advisory Committee (2009) Methodologies for establishing a freshwater inflow regime for Texas estuaries within the context of the Senate Bill 3 Environmental Flows Process. Report # SAC 2009 03 Rev1., June 5, 2009. Available at https://tamucc-ir.tdl.org/handle/1969.6/94344

Shaw WN (1965) Seasonal setting patterns of five species of bivalves in the Tred Avon River, Maryland. Chesap Sci 6:33–37

Simmons EG (1957) An ecological survey of the upper Laguna Madre of Texas. Publ Inst Mar Sci 4:156–200

Simmons EG, Breuer JP (1962) A study of redfish Sciaenops ocellata Linnaeus and black drum Pogonias cromis Linnaeus. Publ Inst Mar Sci 8:184–211

Sklar FH, Browder JA (1998) Coastal environmental impacts brought about by alterations to freshwater flow in the Gulf of Mexico. Environ Manag 22:547–562

Soule DF (1988) Marine organisms as indicators: reality or wishful thinking? In: Soule DF, Kleppel GS (eds) Marine organisms as indicators. Springer, New York, pp 1–11

Springer VG, Woodburn KD (1960) An ecological study of the fishes of the Tampa Bay area. Prof. Papers No 1, Florida State Board Conservation Marine Lab 1:104

Staff G, Powell EN, Stanton RJ Jr, Cummins H (1985) Biomass: is it a useful tool in paleocommunity reconstruction? Lethaia 18:209–232

Tenore KR, Zajac RN, Terwin J, Andrade F, Blanton J, Boynton W, Carey D, Diaz R, Holland AF, Lopez-Jamar E, Montagna P, Nichols F, Rosenberg R, Queiroga H, Sprung M, Whitlatch RB (2006) Characterizing the role benthos plays in large coastal seas and estuaries: A modular approach. J Exp Mar Biol Ecol 330:392–402

Tolan JM (2007) El Niño-Southern Oscillation impacts translated to the watershed scale: estuarine salinity patterns along the Texas Gulf Coast, 1982 to 2004. Estuar Coastal Shelf Sci 72:247–260

Turner EL, Montagna PA (2016) The max bin regression method to identify maximum bioindicators responses to ecological drivers. Ecol Inform 36:118–125. https://doi.org/10.1016/j.ecoinf.2016.10.007

USEPA (1999) Ecological condition of estuaries in the Gulf of Mexico. EPA 620-R-98-004. U.S. Environmental Protection Agency, Gulf Breeze, FL

Van Diggelen AD, Montagna PA (2016) Is salinity variability a benthic disturbance in estuaries? Estuaries Coasts 39:967–980

Ward GH (2010) A time line of Cedar Bayou. Report to the Texas Water Development Board, The University of Texas at Austin, Austin, TX. https://repositories.lib.utexas.edu/bitstream/handle/2152/69939/0900010973_CedarBayou.pdf. Accessed 6 Sept 2022

Warwick RM, Ashman CM, Brown AR, Clarke KR, Dowell B, Hart B, Lewis RE, Shillabeer N, Somerfield PJ, Tapp JF (2002) Inter-annual changes in the biodiversity and community structure of the macrobenthos in Tees Bay and the Tees estuary, UK, associated with local and regional environmental events. Mar Ecol Prog Ser 234:1–13

White WA, Calnan TR, Morton RA, Kimble RS, Littleton TG, McGowen JH, Nance HS, Schmedes KE (1983) Submerged lands of Texas Corpus Christi, TX: Bureau of Economic Geology. University of Texas at Austin Bureau of Economic Geology, Austin, TX. 101 pp

White WA, Calman TR, Morton RA, Kimble RS, Littleton TG, McGowen JH, Nance HS (1985) Submerged lands of Texas, Port Lavaca Area: sediments, geochemistry, benthic macroinvertebrates and associated wetlands. Report: Bureau of Economic Geology. University of Texas Bureau of Economic Geology, Austin, TX, 309 pp

Wu F, Tong C, Feng H, Gu J, Song G (2019) Effects of short-term hydrological processes on benthic macroinvertebrates in salt marshes: A case study in Yangtze Estuary, China. Estuar Coastal Shelf Sci 218:48–58

# Effects of Climate-Driven Salinity Regimes on Disease Dynamics of the Eastern Oyster, a Key Estuarine Resource and Bioindicator

Kelley B. Savage ⓘ, Terence A. Palmer ⓘ,
Paul A. Montagna ⓘ, and Jennifer Beseres Pollack ⓘ

## Abstract

This chapter addresses the connections between long-term trends in climate variability, freshwater inflow dynamics, and salinity patterns and their effects on *Perkinsus marinus* infection of eastern oysters (*Crassostrea virginica*) at local and regional scales in Texas estuaries. Salinities were highest during droughts compared to normal and wet climatic conditions. At the local, within-estuary scale, salinities increased longitudinally from reefs closest to the freshwater inflow source to reefs closest to the Gulf of Mexico inlet. At the regional scale, salinities increased latitudinally moving from estuaries in the northeast to the southwest. Relationships between salinity and *P. marinus* infection levels were strongest for market size (≥76 mm) oysters. At the local scale, mean salinity had positive relationships with infection prevalence and severity of infection (= weighted prevalence) for market-sized oysters; salinity explained 94% of the variance in infection prevalence and 82% of the variance in severity of infection. Relationships at the regional scale were less strong; salinity was positively correlated with infection severity but not infection prevalence in market-sized oysters; salinity explained 64% of the variation in infection prevalence and 71% of variation in infection severity. Results were used to develop salinity recommendations of between 21.0 and 24.9 for maintaining low (≤50%) prevalence of *P. marinus*-infected oysters and low severity of infection (≤1.0) in order to support development of water resource management plans that account for variability in climate patterns.

## Keywords

*Crassostrea virginica* · Dermo · Drought · Freshwater inflow · *Perkinsus marinus* · Texas · USA

## Introduction

Freshwater inflow is fundamental to the functioning of estuarine systems. Delivery of freshwater to estuaries increases productivity, maintains biodiversity, delivers sediments and nutrients, and moderates marine salinities (Benson 1981; Tolan 2007; Beseres Pollack et al. 2011; Montagna et al. 2013). Inflow hydrology (e.g., frequency, timing, velocity, and quantity) drives estuarine condition (e.g., salinity, dissolved material), which in turn influences estuarine resources (e.g., benthic abundance, biomass, and diversity) (Palmer et al. 2011; Montagna et al. 2013). Thus, estuarine resources can be used as indicators of change in estuarine conditions resulting from changes in freshwater inflow.

Eastern oysters *Crassostrea virginica* are important indicators of changing estuarine conditions related to varying freshwater inflow in estuaries along the US Atlantic and Gulf of Mexico (Volety et al. 2009; Beseres Pollack et al. 2011). Oyster populations depend on freshwater delivery to estuaries to maintain suitable salinities and reduce predators and parasites (Andrews and Ray 1988; Soniat and Brody 1988; Soniat and Gauthier 1989; Buzan et al. 2009; Beseres Pollack et al. 2011). Salinity, in turn, influences oyster reproduction, growth, and development (Butler 1949; Shumway 1996). Freshwater inflow and salinity have been linked to indicators of oyster health, including density, recruitment, and parasite infections, which can therefore be linked to water management (Livingston et al. 2000; La Peyre et al. 2003; Volety et al. 2009).

The protozoan oyster parasite, *Perkinsus marinus*, the causative agent of Dermo disease, causes severe mortalities in oyster populations in the Gulf of Mexico (Ray 1966). Transmission of *P. marinus* occurs via suspension feeding

K. B. Savage · T. A. Palmer · P. A. Montagna · J. Beseres Pollack (✉)
Harte Research Institute, Texas A&M University-Corpus Christi, Corpus Christi, TX, USA
e-mail: ksavage1@islander.tamucc.edu; terry.palmer@tamucc.edu; paul.montagna@tamucc.edu; jennifer.pollack@tamucc.edu

P. A. Montagna, A. R. Douglas (eds.), *Freshwater Inflows to Texas Bays and Estuaries*, Estuaries of the World, https://doi.org/10.1007/978-3-031-70882-4_12

when oysters are exposed to infective stages released from dead or dying oysters, or via feces or pseudofeces of live infected oysters (Mackin 1962; Bushek et al. 2002). *Perkinsus marinus* causes reduction in growth and meat condition followed by mortality of susceptible infected oysters, although oyster mortalities do not generally occur until the second year of exposure (Ford and Tripp 1996). The parasite can accumulate in oyster tissues over time, with older (larger) oysters tending to have higher levels of *P. marinus* infection than younger (smaller) oysters (Ray 1954; Andrews and Ray 1988). *Perkinsus marinus* can control oyster population dynamics and threaten economic viability of the oyster fishery as well as dramatically affect reef conservation efforts (Soniat et al. 2005). The parasite increases in prevalence in oysters at higher salinities and is intolerant of very low salinities (Chu and Greene 1989; Ragone and Burreson 1993; Hofmann et al. 1995; La Peyre et al. 2003; Powell et al. 2003). In areas with salinity consistently <9, *P. marinus* infections are less severe and do not result in oyster mortality (Burreson and Ragone Calvo 1996). Salinities <5 can decrease *P. marinus* infection intensities in individual oysters (La Peyre et al. 2009). Increasing freshwater inflow can regulate *P. marinus* prevalence and severity by reducing salinities and facilitating oyster population recovery (Beseres Pollack et al. 2011). Prevalence and severity of *P. marinus* infection within oyster populations can thus serve as a practical indicator of salinity regimes in estuaries.

This chapter addresses the connections between long-term trends in climate variability, freshwater inflow dynamics, and salinity patterns, and their effects on *P. marinus* infection at local and regional scales. *P. marinus* infection prevalence and severity in oyster populations were utilized as indicators to determine how changes in salinity (i.e., estuarine condition) affect estuarine resources (i.e., *C. virginica*). The goal is to provide optimal salinity recommendations at local and regional scales that are adequate to control *P. marinus* infection, an indicator of estuarine health. Because environmental factors that influence oysters likely influence other estuarine species in similar ways (Shumway 1996), gaining insight into optimal salinity conditions for controlling *P. marinus* infection supports the development of water

resource management approaches that account for variability in climate patterns.

## Methods

### Study Area

Texas estuaries are hydrologically diverse despite having similar morphology (Table 1, Fig. 1; Montagna et al. 2013). All estuaries (except for Sabine-Neches) are separated from the Gulf of Mexico by barrier islands and typically consist of a primary bay, secondary bay, and connection to the Gulf of Mexico (Perillo 1995). Freshwater inflow occurs via rivers flowing into the secondary bays, which connect to primary bays, which then connect to the Gulf of Mexico. A latitudinal climatic gradient exists such that northeastern estuaries have an inflow balance—sum of freshwater inputs (gaged, modeled runoff, direct precipitation, and return flows) minus outputs (diversions and evaporation)—that is 50 times greater than southwestern most estuaries.

The effects of climate and water quality on *P. marinus* were evaluated on two different spatial scales: a local (within estuary) scale and a regional (among estuaries) scale. The within-estuary study focuses on the Mission-Aransas Estuary (Fig. 2), which has moderate salinities (mean of ~15) relative to the other major Texas estuaries (means of ~8–36). The among-estuaries study focuses on all six estuaries in Texas where harvestable *C. virginica* reefs occur: Nueces, Mission-Aransas, Guadalupe, Lavaca-Colorado, Trinity-San Jacinto, and Sabine-Neches estuaries.

### Determining Climatic Conditions

The salinity of an estuary is influenced by the volume of freshwater inflow that it receives. Biweekly-collected salinity and temperature data from all estuaries were obtained from Texas Parks and Wildlife Department-Coastal Fisheries (TPWD-CF) independent monitoring program (Martinez-Andrade et al. 2005) from 1986 to 2015. Salinity data were

**Table 1** Listed from north to south: area at mean low tide, mean annual precipitation (1951–1980), mean annual freshwater inflow balance (1941–1994; Texas Water Development Board), and mean annual bottom salinity (Orlando et al. 1993). Inflow balance is the sum of freshwater inputs (gaged, modeled runoff, direct precipitation, and plus return flows) minus the outputs (diversions and evaporation)

| Estuary | Estuary abbreviation | Area (km$^2$) | Rainfall (cm year$^{-1}$) | Inflow (10$^6$ m$^3$ year$^{-1}$) | Salinity |
|---|---|---|---|---|---|
| Sabine-Neches | SN | 183 | 142 | 16,894 | 7 |
| Trinity-San Jacinto | TS | 1416 | 112 | 13,495 | 16 |
| Lavaca-Colorado | LC | 1158 | 102 | 3679 | 20 |
| Guadalupe | GE | 551 | 91 | 2677 | 11 |
| Mission-Aransas | MA | 453 | 81 | 278 | 17 |
| Nueces | NU | 433 | 76 | 346 | 26 |

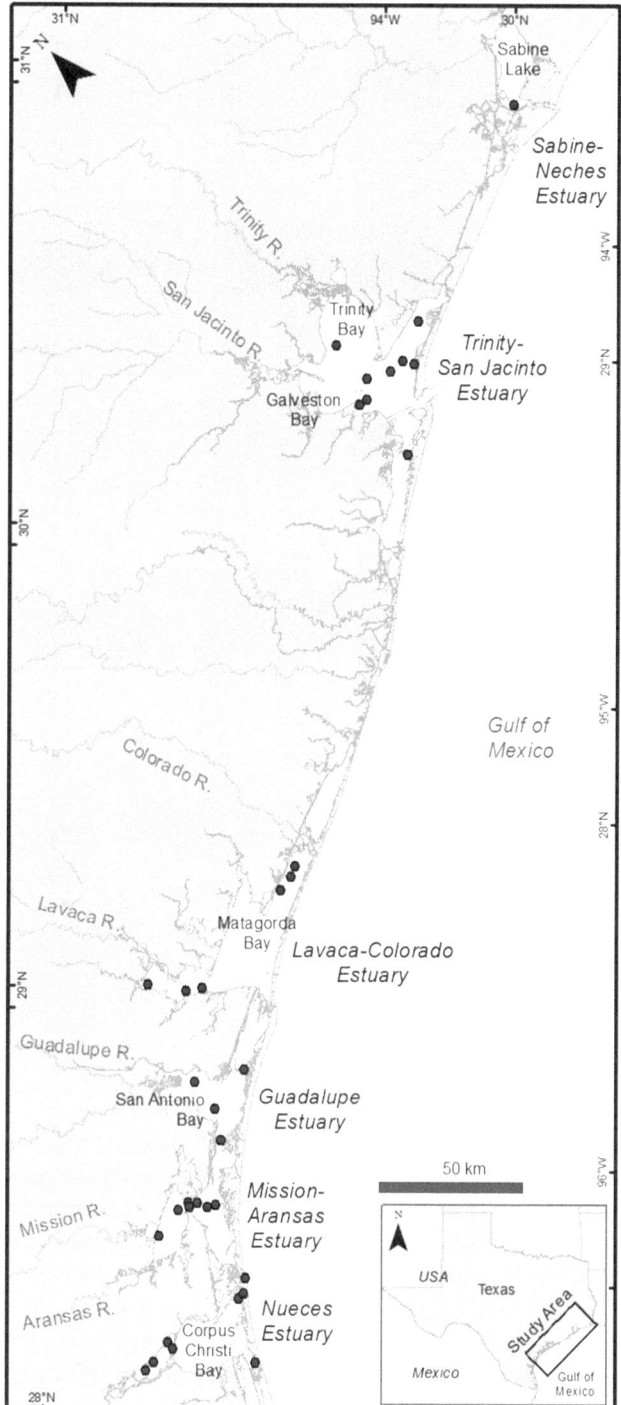

**Fig. 1** Texas estuaries sampled for oysters (black dots) and water quality data for 18 years (1998–2015)

used to hydrologically characterize each estuary under three inflow/salinity regimes: wet/low salinity, drought/high salinity, and normal (Palmer and Montagna 2015). Within Texas estuaries, the salinity of secondary bays is greatly affected by changes in freshwater inflow because they are closest to river inputs (Palmer and Montagna 2015).

Therefore, periods of drought, normal, and wet conditions for each estuary were determined using mean monthly salinities of the largest secondary bay (e.g., Sabine, Trinity, Lavaca, Upper San Antonio, Copano, and Nueces Bays). Inflow/salinity regimes for each estuary were determined to be in drought or wet conditions if mean monthly salinities of the secondary bay were within the upper or lower quartile of monthly salinities from 1986 to 2015. Normal conditions were defined by salinities that were in the interquartile range of historical salinities for the secondary bay. This drought classification method has been verified along the Texas coast by comparing the mean Palmer Drought Severity Index (PDSI) for each drought condition (wet, normal, and drought) for each estuary (Palmer and Montagna 2015). The PDSI is one of the most widely used regional indices of drought and is particularly useful in determining medium-term (several months) dry and wet periods (Alley 1984; Heddinghaus and Sabol 1991).

## Oyster Disease Assessment

For the local scale approach, seven stations were sampled quarterly along a salinity gradient in the Mission-Aransas Estuary from December 2004 to October 2015 to measure oyster characteristics and water quality (Beseres Pollack et al. 2022). At each station, $\geq 20$ oysters (~10 market size ($\geq 76$ mm shell height (SH)) and ~10 submarket size (26–75 mm SH)) were collected using an oyster dredge. At each station, temperature, dissolved oxygen, salinity, and turbidity were measured with a Hydrolab Surveyor II or YSI Pro DSS at 10 cm from the water surface and approximately 10 cm above the bay bottom. In the laboratory, *P. marinus* infection was assessed for each oyster using Ray's Fluid Thioglycollate Method (Ray 1966). A $5 \times 5$ mm section of mantle tissue was removed and incubated in Ray's Fluid Thioglycollate Media (RFTM) for 5–14 days (media was refrigerated after 7 days), modified from the culture method of Ray (1966). Tissue cultures were then stained with 75% Lugol's solution and examined microscopically for *P. marinus* hypnospores (blue/black spheres). *P. marinus* intensity was scored using the 6-point Mackin scale (uninfected (0)—heavily infected (5)) adapted from Mackin (1962) by Craig et al. (1989). Oyster infection prevalence—the proportion of oysters infected with *P. marinus*—was calculated by dividing the number of infected oysters by the number of oysters sampled. Infection severity (= weighted prevalence) was calculated as the mean intensity values multiplied by infection prevalence and is thus a measure of infection within a population. Infection severity values $\geq 2.0$ denote many severe infections and potential for high oyster mortality (Mackin 1962; Burreson et al. 1994).

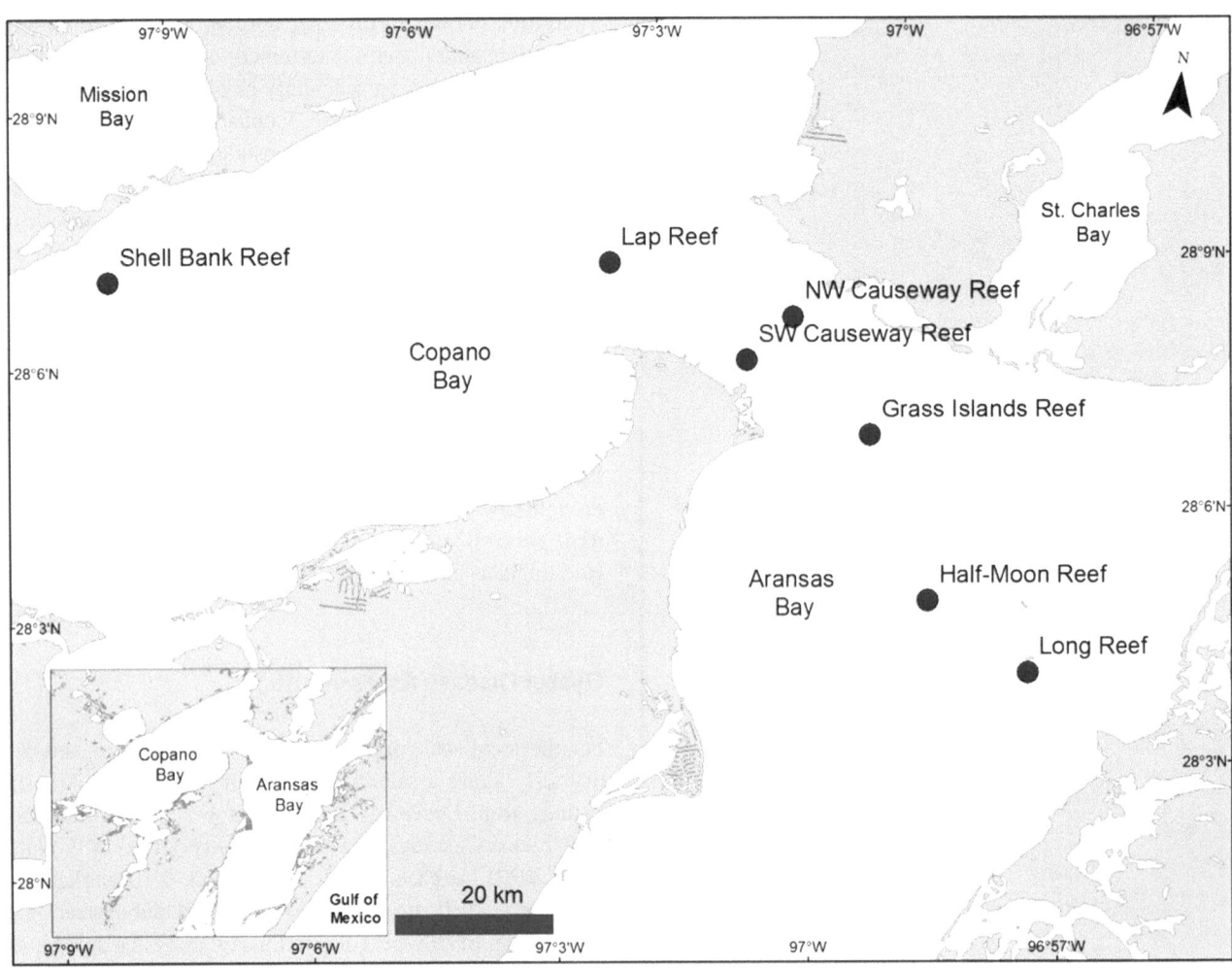

**Fig. 2** Oyster reefs sampled (black dots) along a salinity gradient in the Mission-Aransas Estuary, years 2004–2015

For the regional scale approach, 18 years (1998–2015) of discrete samples of *P. marinus* oyster infection and simultaneously sampled salinity and temperature data were compiled from Oyster Sentinel (https://oystersentinel.cs.uno.edu/), a database maintained by Dr. Tom Soniat at the University of New Orleans (obtained August 2015). Oyster Sentinel manages data on *P. marinus* infection in oysters as part of efforts to monitor the health of estuaries along the northern Gulf of Mexico. Salinity and water temperature are measured simultaneously with the collection of oyster samples.

## Statistical Analyses

Two-way analysis of variance (ANOVA) tests were used to test for differences in salinity, temperature, *P. marinus* prevalence and infection severity among climatic conditions (drought, normal, and wet), and locations: either stations (for local, within-estuary comparisons) or estuaries (for regional, among-estuary comparisons). Because *P. marinus* accumulates in oyster tissue over time and infections tend to increase with size (Andrews and Ray 1988), ANOVA tests were run separately for submarket and market size classes. Dates were nested within climatic conditions for both local- and regional-scale ANOVA tests of *P. marinus* prevalence and severity. Differences among climatic conditions and locations were highlighted using Tukey's Studentized Range Test ($p < 0.05$). Two-way uncrossed ANOVAs were used to determine the variation in *P. marinus* infection explained by salinity and temperature.

Broad-scale effects of freshwater inflow on water quality were determined by calculating Pearson correlations between water quality variables and salinity. Pearson correlations were also used to determine relationships among salinity, temperature, and *P. marinus* infection. Salinity, temperature, and infection prevalence and severity were averaged by station-climate condition combinations for local-scale analyses and by estuary-climate condition combinations for regional-scale analyses of differences in salinity and temperature.

Linear regressions between salinity and *P. marinus* prevalence and severity were used to produce regression parameter estimates both among and within estuaries. These regression parameters were then used to estimate maximum salinities for several *P. marinus* infection thresholds, which indicate low disease effects in oyster populations. An infection prevalence of ≤50% is considered low relative to other populations within the Gulf of Mexico, which generally have mean intensities >50% (Wilson et al. 1990; Powell et al. 1992). An infection severity of 1.5 could be considered a level at which disease-related mortalities are occurring (Soniat 2012), with severe infections and potential for high oyster mortality likely at values ≥2.0 (Mackin 1962; Burreson et al. 1994). Therefore, management targets of ≤50% infection prevalence and ≤1.0 infection severity may be targeted to maintain low infection severities and prevalences in oyster communities.

Statistical analyses were completed using SAS software version 9.4 (SAS Institute Inc. 2014) (ANOVA: PROC MIXED for ANOVAs with random factors, PROC GLM for ANOVAs without random factors, Normality tests: PROC UNIVARIATE, Linear regressions: PROC REG, Pearson correlations: PROC CORR).

## Results

### Local Scale

Water quality variables measured simultaneously with oyster samples varied among climatic conditions (drought, normal, and wet conditions) and among stations within the Mission-Aransas Estuary (Fig. 3). Salinities taken simultaneously with oyster samples were higher in drought conditions than in normal conditions (drought means are 1.4–2.1× higher) and higher in normal conditions than in wet conditions among all stations (normal means are 1.5–2.2× higher, $p < 0.0001$), except Shell Bank Reef where wet and normal conditions were similar. Salinity generally increased moving away from the Mission and Aransas Rivers and toward the Gulf of Mexico. The lowest salinities occurred at Copano Bay oyster reefs (e.g., Shell Bank Reef), and the highest salinities were at Aransas Bay reefs (e.g., Long Reef). There was a 0.1–0.3 year$^{-1}$ increase in salinity from 2004 to 2015 at all stations ($p ≤ 0.0192$), except at NW Causeway Reef (0.1 year$^{-1}$, $p = 0.0517$) and Shell Bank Reef (0.4 year$^{-1}$, $p ≤ 0.0847$). At each reef, there was a positive relationship between salinity and *P. marinus* prevalence and severity in market-sized oysters ($r ≥ 0.43$, $p ≤ 0.0311$) except at Long Reef ($r = 0.35$, $p = 0.0463$ and $r = 0.34$, $p = 0.0564$). In

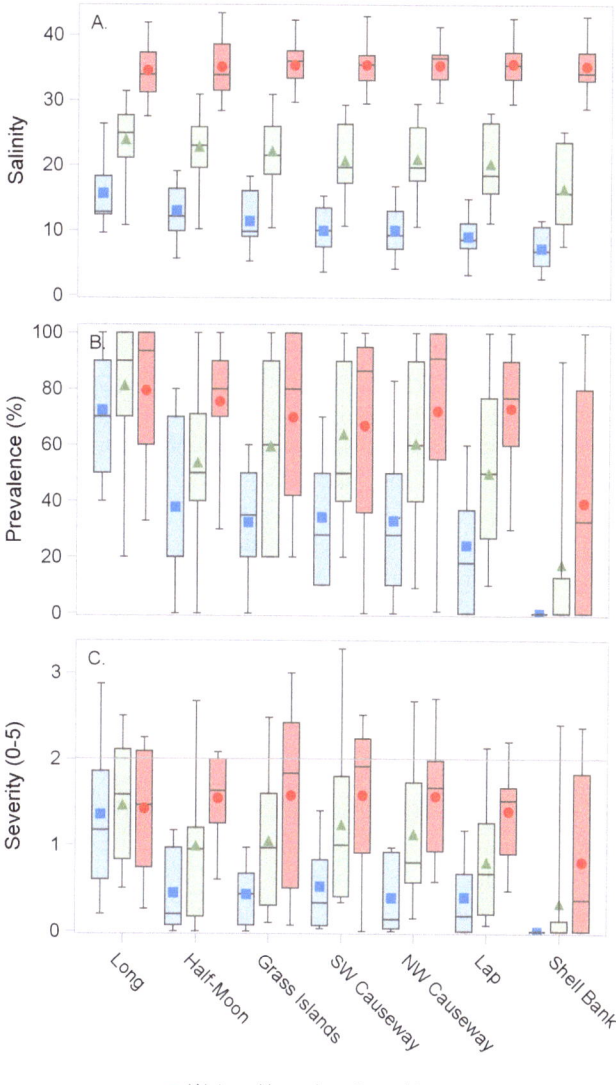

**Fig. 3** Salinity (**a**), prevalence (**b**), and severity (**c**), of *P. marinus* infection in populations of market-sized oysters within the Mission-Aransas Estuary. Stations in order from southeast (Long Reef) to northwest (Shell Bank Reef). Infection severity values ≥2.0 indicates many severe infections and the potential for high oyster mortality (Mackin 1962; Burreson et al. 1994)

submarket-sized oysters, there was a positive relationship between infection prevalence and salinity at Shell Bank Reef ($r = 0.43$, $p < 0.0311$) and a positive relationship between the severity of *P. marinus* infection and salinity at Shell Bank, Lap, NW Causeway, Grass Islands, and Long Reefs ($r ≥ 0.36$, $p ≤ 0.0318$). Salinity explained 94% of the variance in infection prevalence and 82% of the variance in *P. marinus* severity of infection in market-sized oysters. Mean salinity had positive relationships with infection prevalence of market-sized oysters ($r = 0.90$, $p = 0.0062$) and severity of infection ($r = 0.87$, $p = 0.0119$). Salinity variance had negative relationships with proportion of oysters infected

($r = -0.95$, $p = 0.0008$) and severity of infection ($r = -0.93$, $p = 0.0150$).

Temperature had a strong seasonal signal that increased over the study period ($r = 0.19$, $p \leq 0.0033$). Temperature was similar among climatic conditions ($p = 0.3356$) but varied among stations ($p = 0.0004$). Differences in mean temperature were small, ranging from 23.3 °C at Grass Islands to 24.3 °C at NW Causeway Reef. There was a positive relationship between salinity and temperature ($r = 0.55$, $p = 0.0103$) when correlating among station-drought condition means. Temperature explained 5.1% of the variation in *P. marinus* infection prevalence and 11.4% of the variation in *P. marinus* infection severity in market-sized oysters. There were no linear relationships between temperature and *P. marinus* infections for either submarket- or market-sized oyster size classes ($p \geq 0.3469$).

There were no differences in turbidity among stations ($p = 1.0000$) or climatic conditions ($p > 0.3637$). Dissolved oxygen (DO) was higher in wet conditions (8.6 mg L$^{-1}$) than in normal and drought conditions (7.4–7.3 mg L$^{-1}$, $p \leq 0.0330$) but similar among stations ($p = 1.0000$). Turbidity was similar among climatic conditions ($p = 0.3637$) and among stations ($p = 1.0000$). Salinity was negatively correlated with DO ($r = -0.87$, $p < 0.0001$) and turbidity ($r = -0.95$, $p < 0.0001$) when correlating station-drought condition means.

Oyster infection by *P. marinus* varied by climatic conditions and across stations in the Mission-Aransas Estuary (Table 2). Among climatic conditions, there were highly significant positive relationships between salinity and the proportion of market-sized oysters infected with *P. marinus* ($r = 0.735$, $p < 0.0001$) as well as with the severity of infection ($r = 0.827$, $p < 0.0001$, Fig. 4). Among all reefs, *P. marinus* infections in market-sized oysters significantly increased from wet (35.1%, 0.53) to normal (56.8%, 1.03), to drought (68.2%, 1.41) conditions ($p < 0.0001$). In submarket-sized oysters, there was a lower infection prevalence and severity

in wet conditions (28.9%, 0.40) than in normal (43.9%, 0.82) and drought conditions (44.7%, 0.86, $p \leq 0.0461$).

All reefs except Shell Bank Reef and Long Reef had similar prevalence and severity of infection. The prevalence of market-sized oysters infected was lowest at Shell Bank Reef (21.1%) and highest at Long Reef (78.2%) but similar among all other stations (49.1–57.6%, $p < 0.0001$). The relative severity of market-sized oyster infection was lowest at Shell Bank Reef (0.42) but similar among all other stations (0.85–1.42, $p < 0.0001$). The infection prevalence and severity of submarket-sized oysters was also lowest at Shell Bank Reef (19.3%, 0.35) and highest at Long Reef (54.3%, 1.09), although no other reef was distinctly different from another ($p < 0.0001$).

## Regional Scale

Water quality variables measured simultaneously with oyster samples varied among climatic conditions (drought, normal, and wet conditions) and among estuaries. Salinities increased significantly from wet to normal and normal to drought conditions in all estuaries ($p < 0.0001$, Fig. 5). Salinities increased by 0.6–1.4 year$^{-1}$ from 1998 to 2015 in the four southernmost estuaries studied: Nueces ($r = 0.20$, $p \leq 0.0279$), Mission-Aransas ($r = 0.38$, $p \leq 0.0001$), Guadalupe ($r = 0.23$, $p \leq 0.0002$), and Lavaca-Colorado ($r = 0.45$, $p \leq 0.0001$).

There were positive relationships between salinity and *P. marinus* infection prevalence and severity in market- and submarket-sized oysters in the Trinity-San Jacinto and Mission-Aransas Estuaries and in market-sized oysters in the Nueces and Sabine-Neches Estuary ($p < 0.0001$). Salinity explained 63.7% of the variance in infection prevalence and 71.1% of the variance in infection severity in market-sized oysters. Salinity had positive relationships with infection severity ($r = 0.60$, $p = 0.0091$) but not infection prevalence

**Table 2** Results of regression analyses between mean salinity and *P. marinus* infection and severity for oyster populations in the Mission-Aransas Estuary and all Texas estuaries (see Fig. 4). Maximum salinities that are predicted to provide low infection prevalence ($\leq$50%) and severity ($\leq$1.0) were also calculated for the Mission-Aransas Estuary and Among-Estuary data

| Location | Infection variable | n | R | Regression variable | DF | Parameter estimate | Standard error | $t$ Value | Pr > \|t\| | 95% Confidence limits | | Predicted Salinity prevalence 50% | Predicted Salinity severity 1.0 |
|---|---|---|---|---|---|---|---|---|---|---|---|---|---|
| Mission-Aransas | Prevalence | 21 | 0.73 | Intercept | 1 | 16.31 | 8.34 | 1.96 | 0.0654 | −1.15 | 33.76 | 21.0 | |
| | | | | Slope | 1 | 1.60 | 0.34 | 4.72 | 0.0001 | 0.89 | 2.31 | | |
| | Severity | 21 | 0.83 | Intercept | 1 | 0.07 | 0.15 | 0.43 | 0.6717 | −0.26 | 0.39 | | 23.1 |
| | | | | Slope | 1 | 0.04 | 0.01 | 6.41 | <.0001 | 0.03 | 0.05 | | |
| Among estuaries | Prevalence | 18 | 0.34 | Intercept | 1 | 39.11 | 9.80 | 3.99 | 0.0011 | 18.34 | 59.87 | a | |
| | | | | Slope | 1 | 0.64 | 0.43 | 1.47 | 0.1609 | −0.28 | 1.55 | | |
| | Severity | 18 | 0.60 | Intercept | 1 | 0.38 | 0.19 | 1.96 | 0.0681 | −0.03 | 0.78 | | 24.9 |
| | | | | Slope | 1 | 0.03 | 0.01 | 2.97 | 0.0091 | 0.01 | 0.04 | | |

[a]The salinity corresponding to 50% prevalence was not calculated for the regional scale analysis because there was no strong relationship between salinity and prevalence

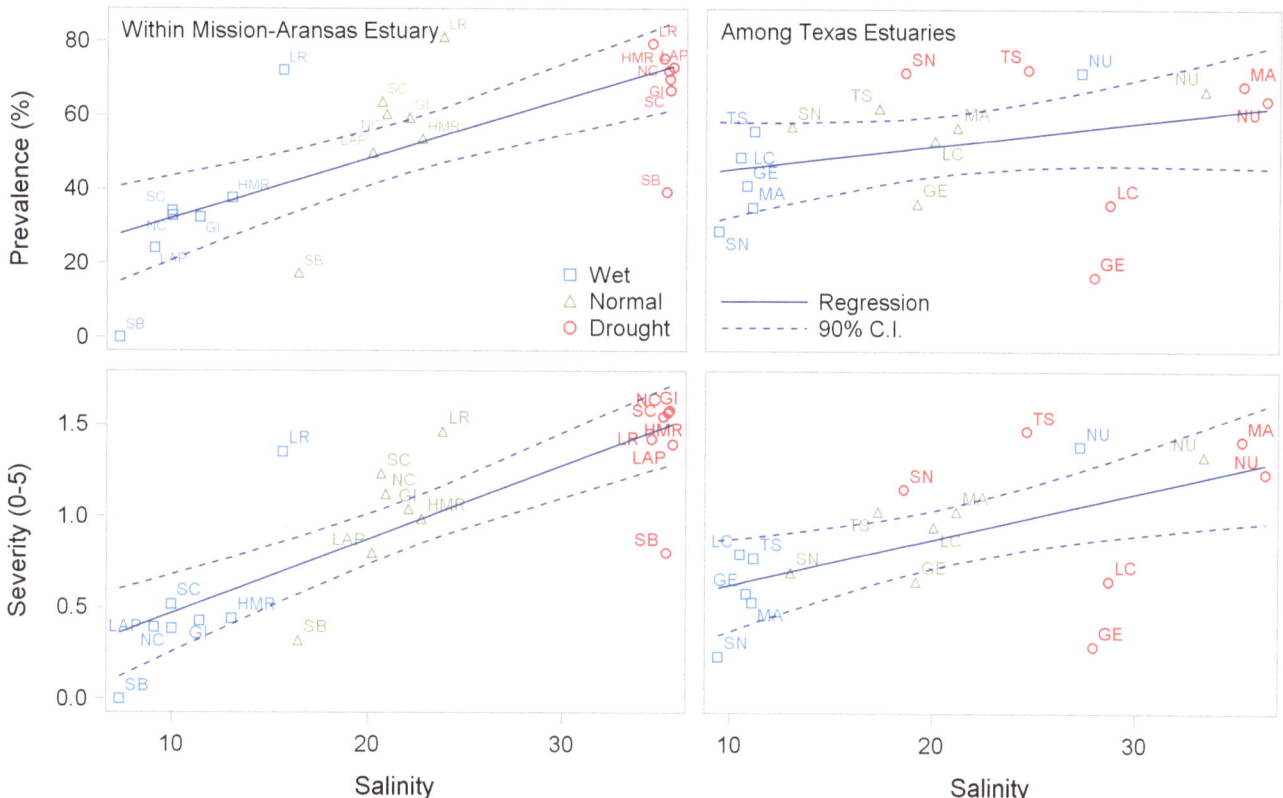

**Fig. 4** Linear regressions for infection prevalence and severity in populations of market-sized oysters among estuaries and climatic conditions among stations within the Mission-Aransas Estuary (left) and among Texas estuaries (right). Reef abbreviations: *LR* long reef, *HMR* half-moon reef, *GI* grass islands reefs, *SC* SW causeway reef *NC* NW causeway reef, *LAP* lap reef, *SB* shell bank reef. Estuary abbreviations in Table 1. *C.I.* confidence interval for the mean

($r = 0.34$, $p = 0.1609$) in market-sized oysters when using estuary-climate condition means (Fig. 4). Salinity variance had negative relationships with infection prevalence ($r = -0.48$, $p = 0.0447$) and infection severity ($r = -0.59$, $p = 0.0107$) in market-sized oysters when using estuaries-condition means.

There was a difference in temperature among climate conditions in the Trinity-San Jacinto Estuary ($p = 0.0023$) but no other estuaries ($p > 0.2365$). Temperature was highest in normal conditions (22.8 °C) than drought (21.1 °C) and wet (20.9 °C) conditions in the Trinity-San Jacinto Estuary. Temperature and salinity were positively correlated in the Trinity-San Jacinto ($r = 0.08$, $p \leq 0.0178$), Mission-Aransas ($r = 0.29$, $p < 0.0001$), and Lavaca-Colorado Estuaries ($r = 0.10$, $p \leq 0.0241$). During the study period, temperature increased in the Mission-Aransas Estuary ($r = 0.19$, $p \leq 0.0034$) and decreased in the Nueces Estuary ($r = -0.24$, $p \leq 0.0086$). Temperature explained 9.4% of the variance in infection prevalence and 25.4% of the variance in infection severity in commercial-sized oysters. Temperature did not have a relationship with market-sized oysters among estuaries. However, temperature was positively correlated with infection prevalence ($r \geq 0.11$, $p \leq 0.0446$) and severity

($r \geq 0.18$, $p \leq 0.0049$) in market-sized oysters within all individual estuaries except the Nueces Estuary (prevalence: $p = 2022$, severity: $p = 0.9294$). There were also positive relationships between submarket-sized oyster infection prevalence and temperature in the Trinity-San Jacinto and Mission-Aransas Estuaries ($r \geq 0.12$, $p \leq 0.0188$) and submarket-sized oyster infection severity and temperature in the Trinity-San Jacinto, Mission-Aransas, and Lavaca-Colorado Estuaries ($r \geq 0.18$, $p \leq 0.0012$).

Oyster infection by *P. marinus* varied among climatic conditions and among estuaries across the Texas coast. Among estuaries and climatic conditions, there was a positive linear trend between salinity and infection severity ($r = 0.60$, $p \leq 0.0091$, Fig. 4) but not salinity and infection prevalence ($r = 0.34$, $p \leq 0.1609$) in populations of market-sized oysters (Table 2).

In individual estuaries, *P. marinus* infection prevalence increased from wet to normal to drought conditions ($p \leq 0.0155$), except in the Guadalupe ($p = 0.4760$) and Lavaca-Colorado ($p = 0.0772$) Estuaries. Infection prevalence increased from wet to normal to drought conditions in all estuaries ($p \leq 0.007$) except for the Nueces ($p = 0.1803$, Guadalupe ($p = 0.1154$) and Lavaca-Colorado ($p = 0.1582$)

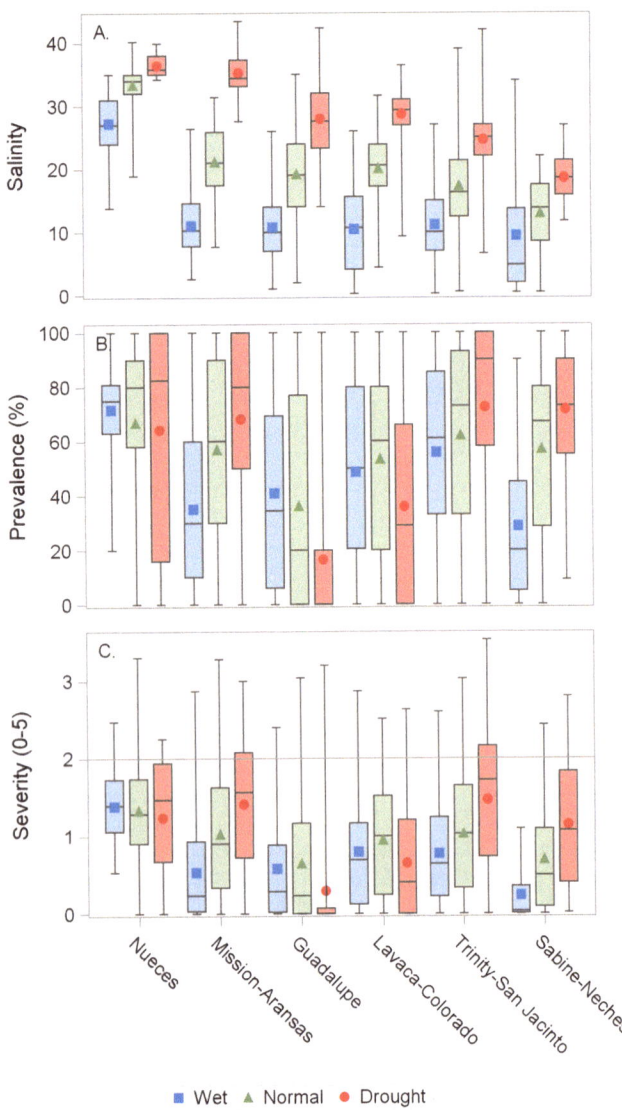

**Fig. 5** Salinity (**a**), prevalence (**b**), and severity (**c**), of *P. marinus* infection in populations of market-sized oysters among Texas Estuaries. Stations in order from southwest (Nueces Estuary) to northeast (Sabine-Neches Estuary). Infection severity values ≥2.0 indicates many severe infections and the potential for high oyster mortality (Mackin 1962; Burreson et al. 1994)

Estuaries, where no differences among climatic conditions occurred. The infection prevalence and severity of market- and submarket-sized oysters were similar among climatic conditions when comparing all estuaries simultaneously ($p \geq 0.0869$).

The proportion of market-sized oysters infected with *P. marinus* was lower in the Guadalupe Estuary (33%), than in all other estuaries (>47%, $p < 0.0001$, Fig. 4). The relative severity of *P. marinus* infection in market-size oysters was lowest in the Guadalupe Estuary (0.55) and highest in the Nueces Estuary (1.33) and similar among all other estuaries (0.82–1.08, $p < 0.0001$). The proportion of submarket-sized oysters infected with *P. marinus* was lowest in the Lavaca-

Colorado (23%) and Guadalupe (28.2%) Estuaries but similar in the other estuaries (40–53%, $p < 0.0001$). The severity of infection in submarket-sized oysters was highest in the Nueces Estuary (1.03) and lowest in the Guadalupe (0.49), Sabine-Neches (0.44), and Lavaca-Colorado (0.37) Estuaries ($p < 0.0001$).

Over the study period, there were significant increases in the proportion of infected oysters and the relative severity of infection in both size classes in the Nueces, Mission-Aransas, and Lavaca-Colorado Estuaries. There were significant decreases in the proportion of infected oysters and the relative severity of infection for market-sized oysters in the Trinity-San Jacinto and Guadalupe Estuaries.

## Salinity Recommendations

At the local scale, in the Mission-Aransas Estuary, the mean salinity recommended to maintain a relatively low (≤ 50%) proportion of *P. marinus*-infected oysters (prevalence) within the population is 21.0 (Table 2). In order to maintain low (≤ 1.0) severity (= weighted prevalence) of *P. marinus* infection in the population, the recommended mean salinities are ≤23.1. At the regional scale, among all six estuaries in Texas where harvestable *C. virginica* reefs occur, recommended mean salinity for maintaining ≤ 50% proportion of *P. marinus* infected oysters could not be calculated because there wasn't a strong relationship between salinity and infection prevalence ($p = 0.1609$). In order to maintain low (≤ 1.0) severity of *P. marinus* infection, recommended mean salinities are ≤24.9.

## Discussion

The positive relationship between freshwater inflow and estuarine health is well established (Copeland 1966; Sklar and Browder 1998; Montagna et al. 2013), as is the suitability of benthic organisms as important indicators of estuarine environmental condition (Montagna and Kalke 1992; Weisberg et al. 1997; Beseres Pollack et al. 2009). Although *P. marinus* is a primary factor limiting oyster growth and survival in the Gulf of Mexico (Cake 1983; Craig et al. 1989; Soniat 1996) and is strongly regulated by salinity and temperature (Soniat 1985; Chu and Greene 1989; Soniat and Gauthier 1989; Chu et al. 1993; Burreson and Ragone Calvo 1996), *P. marinus* infection has not been widely evaluated as an indicator of ecological integrity in estuaries. Combining information on *P. marinus* infection characteristics with data on salinity and temperature during drought, normal, and wet conditions allowed us to investigate the role of climate variability in driving ecosystem dynamics and responses to freshwater inflow. Results can be used to help understand the effects of future climatic conditions on estuarine salinity patterns and *P. marinus* dynamics by indicating that: (1) climatic conditions drive local and regional salinity regimes in Texas estuaries; (2) climate-driven salinity regimes influence the proportion of infected oysters with *P. marinus* and the relative severity of infection, with the strongest relationships occurring at local, within-estuary scales; (3) *P. marinus* infection characteristics increase with increasing salinity among climatic conditions; and (4) temperature does not play a major role in *P. marinus* infection dynamics among Texas estuaries. Relationships between *P. marinus* infection dynamics and salinity were strongest for market-sized oysters, demonstrating their utility as bioindicators of climatic conditions. Utilizing this information can help resource managers determine optimal salinity regimes and make informed decisions about the effects of climate variability on *P. marinus* dynamics as an indicator for ecological integrity.

Climatic conditions had a significant effect on water quality and *P. marinus* infection dynamics at local and regional scales. At the local scale, wet conditions corresponded with lower salinities (mean = 11.1), lesser infection prevalence (mean = 35%), and infection severity (mean = 0.53) in market-sized oysters than during normal and drought periods, corroborating findings from previous studies (La Peyre et al. 2003, 2009; Beseres Pollack et al. 2011). At the regional scale, wet conditions again promoted lower salinities (mean = 11.5) than during normal and drought periods, coinciding with general decreases in infection prevalence (mean = 49%) and severity (mean = 0.73) in market-sized oysters. Low salinities have been shown to result in significant decreases in *P. marinus* infection intensities in both field and laboratory studies, either from a decrease in infection intensities in individual oysters or to the death of the most highly infected individuals (Craig et al. 1989; Soniat 1996; La Peyre et al. 2009). In the current study, it is unknown which mechanism had primacy in driving the observed decreases in *P. marinus* infection characteristics during wet periods.

Drought-related high salinities have been shown to correspond with increases in prevalence and severity of *P. marinus* infections (Andrews 1988; Burreson and Ragone Calvo 1996; Beseres Pollack et al. 2011; Petes et al. 2012). In the current study, droughts were associated at the local scale with higher salinities (mean = 34.5), higher proportions of market-size oysters infected with *P. marinus* (mean = 68.2%), and more severe infections (mean = 1.4) than during wet or normal conditions. At the regional scale, the relationship was generally, but not conclusively, consistent (higher salinity (mean = 27.6), higher prevalence (mean = 56.6%) and severity of infection (mean = 1.11)). Severity (= weighted prevalence) values ≥ 2.0 have been demonstrated to cause high oyster mortality (Mackin 1962; Burreson et al. 1994). Although infection severity in the current study usually

remained <2, even during droughts, previous work has shown that oyster mortalities are likely when average infection severity >1, with oyster deaths potentially preventing infection severity scores from rising substantially above 1.5 (Andrews 1988). Although disease-related oyster mortality was not quantified in this study, it is possible that observed increases in infection severity during droughts may have increased oyster deaths.

For mid-coast estuaries (Guadalupe and Lavaca Colorado), it is remarkable that *P. marinus* infection prevalence (16.5% and 36.0%, respectively) and severity (0.29 and 0.65, respectively) was often low during drought periods. Although the mechanism for this unexpected result is unclear, it may relate to differences in river system-specific water management practices, particularly during droughts (Nielsen-Gammon et al. 2020). Flows of the major rivers in Texas are controlled by over 200 major reservoirs, and water management practices vary greatly with reservoir characteristics as well as water resource allocation and use (Wurbs and Ayala 2014). Rivers supplying the Guadalupe and Lavaca Colorado estuaries also flow through population centers (i.e., San Antonio, Austin) with high municipal demand, challenging water resource managers to provide sufficient water to support the downstream environment (Montagna and Kalke 1992; Miller et al. 2008). Additional research is needed to understand how estuarine conditions and resources are influenced by differences in water allocation and management among river systems during droughts.

In the current study, average salinity was used to identify the effects of freshwater inflow and climate on *P. marinus* infections. However, salinity variability has also been shown to be a useful indicator of benthic community stability (Van Diggelen and Montagna 2016). Both *P. marinus* prevalence and severity had strong relationships with mean salinity and salinity variance within the Mission-Aransas Estuary (mean salinity: $r \geq 0.87$, salinity variance: $r \leq 0.93$) and weaker but obvious relationships among all Texas estuaries (mean salinity: $r \geq 0.34$, salinity variance: $r \leq 0.48$), possibly indicating that *P. marinus* infection dynamics are influenced by both pulse and press inflow events. Acute pulses of freshwater from storms or river diversions can reduce *P. marinus* prevalence and severity by diluting or terminating *P. marinus* stages (La Peyre et al. 2009), limiting the severity of *P. marinus* infection (La Peyre et al. 2003) and/or causing mortality of infected oysters within the populations (La Peyre et al. 2003, 2009; Beseres Pollack et al. 2011). Because Texas estuaries have relatively long residence times (up to 1 year, Solis and Powell 1998), extreme pulses of freshwater can be followed by extended (press) periods of low salinity. Although rapid drops in salinity can increase oyster mortality, the incompatibility of low salinities with disease proliferation may also facilitate oyster population resilience and recovery over time (Beseres Pollack et al. 2011; La Peyre et al. 2003). Results highlight the importance of periodic climate-driven wet conditions (high inflow/low salinity) to promote long-term oyster sustainability in Texas estuaries.

Analogous to the observed increases in *P. marinus* infection prevalence and intensity moving from wet (low salinity) to normal to drought (high salinity) periods, there was also a spatial gradient of increasing *P. marinus* infections moving from lower to higher salinity areas. At the local scale, *P. marinus* infection prevalence and severity increased moving from reefs closest to the freshwater inflow source (Shellbank, 21.1%, 0.42) to reefs closest to the Gulf of Mexico inlet (Long Reef 78.2%, 1.42). At the regional scale, *P. marinus* infection prevalence and severity was greatest in the estuary with highest mean salinity, Nueces Estuary (67.2%, 1.32). However, the lowest infection prevalence and severity occurred in the mid-salinity (mean = 19.6) Guadalupe Estuary (33.0%, 0.54). In Sabine Neches Estuary, the estuary with the lowest mean salinity, infection severity was second lowest (0.82), and prevalence was third highest relative to the other Texas estuaries. Relationships between salinity and infection characteristics were less strong at the regional scale, indicating that variability in the relationship between *P. marinus* infections and salinity is higher among estuaries than within estuaries. Variations in sampling frequency and locations among estuaries and the locations of oyster reefs within different salinity zones within each estuary likely contribute to the variation in the meta-analysis approach in the regional analysis. The sampling regime in the regional-scale approach contrasts with the frequent quarterly sampling of the reefs, which span much of the salinity gradient in the Mission-Aransas Estuary (Beseres Pollack et al. 2012), used in the local-scale analysis. Despite the stronger link between *P. marinus* infection and salinity in the local-scale approaches than the regional-scale approaches, both approaches corroborate with the spatial gradients of increasing *P. marinus* infections and oyster mortality observed in Delaware Bay (Bushek et al. 2012).

Although temperature has a smaller role than salinity in *P. marinus* infection dynamics over long timescales in Texas Estuaries, temperature is known to play a strong role in disease progression over shorter timescales and elsewhere (Chu et al. 1993; Soniat 1996). The relatively warm temperatures and low long-term spatiotemporal variability in temperature in the Mission-Aransas Estuary and other Texas estuaries may be responsible for the weaker relationship between *P. marinus* disease and temperature relative to salinity in the current study. The role of temperature on disease progression is important to consider for short timescales, or in locations with changing temperatures, because high water temperature can be a critical factor in *P. marinus* proliferation (Chu and Greene 1989; Chu et al. 1993; Soniat 1996). Chu et al. (1993) found *P. marinus* infections were higher in temperatures at 20–25 °C and lower in temperatures at 10–15 °C. In the cur-

rent study, temperatures were high (means of 19.5–24.9 °C) relative to other US estuaries and thus may not have had a significant influence on the fluctuation of *P. marinus* infections.

Salinity had a much larger effect on *P. marinus* infection dynamics than temperature in this current study. Salinity explained 64–94% of the variance in *P. marinus* infections, whereas temperature explained 5–25% of the variance in *P. marinus* infections within the Mission-Aransas Estuary and among Texas estuaries. Previous studies have shown that the combination of high salinities and high temperatures cause a proliferation of *P. marinus* infections in oysters (Chu and Greene 1989). Aside from seasonal differences in temperature, salinity variation was much greater than temperature variation in Texas estuaries, which is likely the reason why salinity plays a much larger role in *P. marinus* infections in Texas oysters.

Incorporating optimal salinity regimes to keep *P. marinus* at low levels of infection would be beneficial to sustain oyster populations across periods of known climate variability. In US Gulf of Mexico estuaries, *P. marinus* responds to El Niño-Southern Oscillation (ENSO) via local impacts on salinity. Drought conditions during the La Niña phase of ENSO elevate estuarine salinities, increasing the prevalence and severity of *P. marinus* infections and associated oyster mortalities (Andrews 1988; Powell et al. 1996; Soniat et al. 1998, 2005; Goedken et al. 2005). Previous studies suggest that at least 50% of oysters in Gulf of Mexico estuaries are infected with *P. marinus* (Powell et al. 1992; Wilson et al. 1990), with increases in disease-related mortalities likely occurring at infection severity ≥1.5–2 (Mackin 1962; Andrews 1988). Thus, environmental conditions that limit or maintain the proportions of oysters infected to ≤50% and the relative severity of infection to ≤1.0 would be beneficial in sustaining oyster populations among Texas Estuaries. Results from the current study indicate that the optimal mean salinity to maintain the proportion of *P. marinus* infected oysters at ≤50% is ≤21.0 within the Mission-Aransas Estuary. The optimal mean salinity is 24.9 among estuaries and 23.2 within the Mission-Aransas Estuary to limit *P. marinus* severity of infection to ≤1.0. Salinity recommendations for limiting *P. marinus* infections in Texas estuaries are similar at local and regional scales. These mean salinity recommendations were calculated using three different climate regimes within a 30-year period. However, droughts generally occur over the scale of months. Therefore, water resource managers may want to target salinity ranges to maintain low *P. marinus* infections using salinity averaged over several months, to match the timescale that drought periods occur.

The next steps for a freshwater inflow manager to use a targeted salinity range are to identify a focus area of oyster reefs and determine the relationship between inflow and salinity for that area. The effects of targeted inflows on salinities in other areas of the estuary should also be taken into consideration. The approach taken by Palmer et al. (2015) to relate inflows to salinities in different estuary zones in the Caloosahatchee Estuary, Florida, might be useful in some Texas estuaries. As a more general rule, setting targets for base freshwater flows to maintain salinities ≤25 over targeted oyster reefs could help to sustain oyster populations, especially when the oysters are most susceptible to *P. marinus* infection, such as drier, warmer conditions in summer months. Additional research is warranted to link climatic conditions and salinity regimes to freshwater inflow volumes to inform water resource management decisions in Texas.

## Conclusion

*Perkinsus marinus* infections increase with climate-driven salinity regimes: from wet to normal and from normal to drought. The strong spatiotemporal relationships between salinity and *P. marinus* infection characteristics demonstrated in this chapter may be useful for predicting infection dynamics in estuaries where inflows are increased or decreased. Generation of salinity recommendations was possible using data both from long-term monitoring of a single estuary, and to a lesser degree, from meta-analysis from multiple estuaries. Mean salinities of <23.1 during drought periods will likely ensure that ≤50% of the oyster populations have *P. marinus* infections, whereas mean salinities of 23.1–24.9 during drought periods will likely ensure that infection severities within a population are low (1.0 on a scale of 0–5). Salinity-based predictions such as these are important for understanding the effects of ever-increasing demands on freshwater resources, as well as more frequent and severe drought conditions caused by warming temperatures and rising evapotranspiration (Nielsen-Gammon 2009). Using *P. marinus* as a bioindicator of adequate salinity regimes can provide water resource managers with an effective strategy to recommend base flow regimes to support desired estuarine conditions within drought, normal, or wet climatic periods.

**Acknowledgments** This publication was funded in part through Contract No. 21-155-007-C879 from the Texas General Land Office (GLO) with Gulf of Mexico Energy Security Act of 2006 funding made available to the State of Texas and awarded under the Texas Coastal Management Program. The views contained herein are those of the authors and should not be interpreted as representing the views of the GLO or the State of Texas. Partial funding was also provided by the Texas General Land Office Coastal Management Program, the Rotary Club of Corpus Christi Harvey Weil Grant, the Atlee M. and John W. Cunningham Memorial Research Award, the Hans and Pat Suter Endowment, and the National Oceanic and Atmospheric Administration, Office of Education Educational Partnership Program with Minority Serving Institutions award (NA21SEC4810004). The contents of this publication are solely the responsibility of the award recipient and do not necessarily represent the official views of the U.S. Department of Commerce, National Oceanic and Atmospheric Administration. We would like to acknowledge Dr. Thomas M. Soniat and the late Dr. Sammy M. Ray for collecting and providing data on *Perkinsus marinus* and water quality parameters. Lastly, we thank the Gulf of Mexico Research Initiative Information and Data Cooperative for storing and allowing open access of the data that were generated in this study (Beseres Pollack et al. 2022).

## References

Alley W (1984) The Palmer Drought Severity Index: limitations and assumptions. J Appl Meteorol Climatol 23:1100–1109. https://doi.org/10.1175/1520-0450(1984)023<1100:TPDSIL>2.0.CO;2

Andrews JD (1988) Epizootiology of the disease caused by the oyster pathogen *Perkinsus marinus* and its effects on the oyster industry. Am Fish Soc Spec Publ 18:47–63

Andrews JD, Ray SM (1988) Management strategies to control the disease caused by *Perkinsus marinus*. Am Fish Soc Spec Publ 18:257–264

Benson NG (1981) The freshwater-inflow-to-estuaries issue. Fisheries 6(5):8–10. https://doi.org/10.1577/1548-8446(1981)006<0008:TFI>2.0.CO;2

Beseres Pollack J, Kinsey JW, Montagna PA (2009) Freshwater Inflow Biotic Index (FIBI) for the Lavaca-Colorado Estuary, Texas. Environ Bioindic 4:153–169. https://doi.org/10.1080/15555270902986831

Beseres Pollack J, Kim H-C, Morgan EC, Montagna PA (2011) Role of flood disturbance in natural oyster (*Crassostrea virginica*) population maintenance in an estuary in South Texas, USA. Estuaries Coasts 34:187–197. https://doi.org/10.1007/s12237-010-9338-6

Beseres Pollack J, Cleveland A, Palmer TA, Reisinger AS, Montagna PA (2012) A restoration suitability index model for the eastern oyster (*Crassostrea virginica*) in the Mission-Aransas Estuary, TX, USA. PLoS One 7:e40839. https://doi.org/10.1371/journal.pone.0040839

Beseres Pollack J, Breaux N, Palmer TA, Savage KB, De Santiago K, Downey D (2022) Monitoring of *Perkinsus marinus* (Dermo disease) infection intensity in the *Crassostrea virginica*, and water quality in the Mission-Aransas Estuary, 2014-2021. Distributed by: Gulf of Mexico Research Initiative Information and Data Cooperative (GRIIDC), Harte Research Institute, Texas A&M University–Corpus Christi. https://doi.org/10.7266/JJZYJ9AT

Burreson EM, Ragone Calvo LM (1996) Epizootiology of *Perkinsus marinus* disease of oysters in Chesapeake Bay, with emphasis on data since 1985. J Shellfish Res 15(1):17–34

Burreson EM, Alvarez RS, Martinez W, Macedo LA (1994) *Perkinsus marinus* (Apicomplexa) as a potential source of oyster *Crassostrea virginica* mortality in coastal lagoons of Tabasco, Mexico. Dis Aquat Org 20(1):77–82. https://doi.org/10.3354/dao020077

Bushek D, Ford SE, Chintala MM (2002) Comparison of *in vitro*-cultured and wild type *Perkinsus marinus*. III. Fecal elimination and its role in transmission. Dis Aquat Org 51:217–225. https://doi.org/10.3354/dao051217

Bushek D, Ford SE, Burt I (2012) Long-term patterns of an estuarine pathogen along a salinity gradient. J Mar Res 70(2–3):225–251. https://doi.org/10.1357/002224012802851968

Butler PA (1949) Gametogenesis in the oyster under conditions of depressed salinity. US Fish Wildl Serv Spec Sci Rep 8:1–27

Buzan D, Lee W, Culbertson J, Kuhn N, Robinson L (2009) Positive relationship between freshwater inflow and oyster abundance in Galveston Bay, Texas. Estuaries Coasts 32:206–212. https://doi.org/10.1007/s12237-008-9078-z

Cake EW Jr (1983) Habitat suitability index models: Gulf of Mexico American oyster. FWS/OBS-82/10.57. U.S. Fish and Wildlife Service, Washington, DC, p 37

Chu FE, Greene KH (1989) Effect of temperature and salinity on in vitro culture of the oyster pathogen, *Perkinsus marinus* (Apicomplexa: Perkinsea). J Invertebr Pathol 53:260–269. https://doi.org/10.1016/0022-2011(89)90016-5

Chu FE, La Peyre JF, Burreson CS (1993) *Perkinsus marinus* infection and potential defense-related activities in eastern oysters, *Crassostrea virginica*: salinity effects. J Invertebr Pathol 62:226–232. https://doi.org/10.1006/jipa.1993.1104

Copeland BK (1966) Effects of decreased river flow on estuarine ecology. Estuarine Ecol 38:1831–1839

Craig A, Powell EN, Fay RR, Brooks JM (1989) Distribution of *Perkinsus marinus* in Gulf Coast oyster populations. Estuaries 12:82–91. https://doi.org/10.2307/1351499

Ford SE, Tripp M (1996) Diseases and defense mechanisms. In: Kennedy VS, Newell RIE, Eble AF (eds) The Eastern Oyster: *Crassostrea virginica*. Maryland Sea Grant College, College Park, MD, pp 581–660

Goedken M, Morsey B, Sunila I, Dungan C, de Guise S (2005) The effects of temperature and salinity on apoptosis of *Crassostrea virginica* hemocytes and *Perkinsus marinus*. J Shellfish Res 24(1):177–183. https://doi.org/10.2983/0730-8000(2005)24[177:TEOTAS]2.0.CO;2

Heddinghaus TR, Sabol P (1991) A review of the Palmer Drought Severity Index and where do we go from here? In: Proceeding 7th conference on applied climatology, American Meteorological Society, pp 242–246

Hofmann EE, Powell EN, Klinck JM, Saunders G (1995) Modeling diseased oyster populations. I. Modeling *Perkinsus marinus* infections in oysters. J Shellfish Res 14:121–151

La Peyre MK, Nickens AD, Volety AK, Tolley GS, La Peyre JF (2003) Environmental significance of freshets in reducing *Perkinsus marinus* infection in eastern oysters *Crassostrea virginica*: potential management applications. Mar Ecol Prog Ser 248:165–176. https://doi.org/10.3354/meps248165

La Peyre MK, Gossman B, La Peyre JF (2009) Defining optimal freshwater flow for oyster production: effects of freshet rate and magnitude of change and duration on eastern oysters and *Perkinsus marinus* infection. Estuaries Coasts 32(3):522–534. https://doi.org/10.1007/s12237-009-9149-9

Livingston RJ, Lewis FG, Woodsum GC, Niu X-F, Galperin B, Huang W, Christensen JD, Monacao ME, Battista TA, Klein CJ, Howell RL IV, Ray GL (2000) Modelling oyster population response to variation in freshwater input. Estuar Coast Shelf Sci 50:655–672. https://doi.org/10.1006/ecss.1999.0597

Mackin JG (1962) Oyster diseases caused by *Dermocystidium marinum* and other microorganisms in Louisiana. Publ Inst Mar Sci Univ Tex 7:132–229

Martinez-Andrade F, Campbell P, Fuls B (2005) Trends in relative abundance and size of selected finfishes and shellfishes along the Texas coast: November 1975–December 2003. Management data series number 232. Texas Parks and Wildlife Department, Coastal Fisheries Division. Austin, TX, 136 pp

Miller CJ, Roelke DL, Davis SE, Hsiu-Ping L, Gable G (2008) The role of inflow magnitude and frequency on plankton communities from the Guadalupe Estuary, TX, USA: findings from microcosm experiments. Estuar Coast Shelf Sci 80(1):67–73. https://doi.org/10.1016/j.ecss.2008.07.006

Montagna PA, Kalke RD (1992) The effect of freshwater inflow on meiofaunal and macrofaunal populations in the Guadalupe and Nueces Estuaries, Texas. Estuaries 15:266–285. https://doi.org/10.2307/1352779

Montagna PA, Palmer TA, Beseres Pollack J (2013) Hydrological changes and estuarine dynamics. Springer briefs in environmental science. Springer, New York. https://doi.org/10.1007/978-1-4614-5833-3

Nielsen-Gammon J (2009) The changing climate of Texas. In: Schmandt J, Clarkson J, North GR (eds) The impact of global warming on Texas. University of Texas Press, Austin, TX, pp 39–68

Nielsen-Gammon J, Banner JL, Cook BI, Tremaine DM, Wong CI, Mace RE, Gao H, Yang Z-L, Gonzalez MF, Hoffpauir R, Gooch T, Kloesel K (2020) Unprecedented drought challenges for Texas water resources in a changing climate: What do researchers and stakeholders need to know? Earth's Future 8(8):e2020EF001552. https://doi.org/10.1029/2020EF001552

Orlando SP Jr, Rozas LP, Ward GH, Klein CJ (1993) Salinity characteristics of Gulf of Mexico estuaries. National Oceanic and Atmospheric Administration, Office of Ocean Resources Conservation and Assessment, Silver Spring, MD, p 209

Palmer TA, Montagna PA (2015) Impacts of droughts and low flows on estuarine water quality and benthic fauna. Hydrobiologia 753:111–129. https://doi.org/10.1007/s10750-015-2200-x

Palmer TA, Montagna PA, Beseres Pollack J, Kalke RD, DeYoe H (2011) The role of freshwater inflow in lagoons, rivers, and bays. Hydrobiologia 667:49–67. https://doi.org/10.1007/s10750-011-0637-0

Palmer TA, Montagna PA, Chamberlain RH, Doering PH, Wan Y, Haunert KM, Crean DJ (2015) Determining the effects of freshwater inflow on benthic macrofauna in the Caloosahatchee Estuary, Florida. Integr Environ Assess Manag 12:529–539. https://doi.org/10.1002/ieam.1688

Perillo GME (1995) Definition and geomorphologic classification of estuaries. Dev Sedimentol 53:17–47. https://doi.org/10.1016/S0070-4571(05)80022-6

Petes LE, Brown AJ, Knight CR (2012) Impacts of upstream drought and water withdrawals on the health and survival of downstream estuarine oyster populations. Ecol Evol 2(7):1712–1724. https://doi.org/10.1002/ece3.291

Powell EN, Gauthier JD, Wilson EA, Nelson A, Fay RR, Brooks JM (1992) Oyster disease and climate change. Are yearly changes in *Perkinsus marinus* parasitism in oysters (*Crassostrea virginica*) controlled by climatic cycles in the Gulf of Mexico? Mar Ecol 13:243–270. https://doi.org/10.1111/j.1439-0485.1992.tb00354.x

Powell EN, Klinck JM, Hofmann EE (1996) Modeling diseased oyster populations. II. Triggering mechanisms for *Perkinsus marinus* epizootics. J Shellfish Res 15(1):141–165

Powell EN, Klinck JM, Hofmann EE, McManus MA (2003) Influence of water allocation and freshwater inflow on oyster production: a hydrodynamic-oyster population model for Galveston Bay, Texas, USA. Environ Manag 31:100–121. https://doi.org/10.1007/s00267-002-2695-6

Ragone LM, Burreson EM (1993) Effect of salinity on infection progression and pathogenicity of *Perkinsus marinus* in the eastern oyster, *Crassostrea virginica* (Gmelin). J Shellfish Res 12:1–7

Ray SM (1954) Studies on the occurrence of *Dermocystidium marinum* in young oysters. Proc Natl Shellfish Assoc 44:80–92

Ray SM (1966) A review of the culture method for detecting *Dermocystidium marinum*, with suggested modifications and precautions. Proc Natl Shellfish Assoc 54:55–69

Shumway SE (1996) Natural environmental factors. In: Kennedy VS, Newell RIE, Eble AF (eds) The Eastern oyster *Crassostrea virginica*. Maryland Sea Grant College, College Park, MD, pp 467–513

Sklar FH, Browder JA (1998) Coastal environmental impacts brought about by alterations to freshwater flow in the Gulf of Mexico. Environ Manag 22(4):547–562

Solis RS, Powell GL (1998) Hydrography, mixing characteristics, and residence times of Gulf of Mexico Estuaries. In: Bianchi TS, Pennock JR, Twilley RY (eds) Biogeochemistry of Gulf of Mexico Estuaries. Wiley, New York, pp 29–61

Soniat TM (1985) Changes in levels of infection of oysters by *Perkinsus marinus*, with special reference to the interaction of temperature and salinity upon parasitism. NE Gulf Sci 7(2):171–174

Soniat TM (1996) Epizootiology of *Perkinsus marinus* disease on eastern oysters in the Gulf of Mexico. J Shellfish Res 15:35–43

Soniat TM (2012) Levels of the parasite *Perkinsus marinus* in populations of oysters from the Louisiana Public Seed Grounds: Summer 2012. Louisiana Department of Wildlife and Fisheries, Louisiana

Soniat TM, Brody MS (1988) Field validation of a habitat suitability index model for the American oyster. Estuaries 11:87–95

Soniat TM, Gauthier JD (1989) The prevalence and intensity of *Perkinsus marinus* from the mid northern Gulf of Mexico, with comments on the relationship of the oyster parasite to temperature and salinity. Tulane Stud Zool Bot 27:21–27

Soniat TM, Powell EN, Hofmann EE, Klinck JM (1998) Understanding the success and failure of oyster populations: the importance of sampled variables and sample timing. J Shellfish Res 17:1149–1165

Soniat TM, Klinck JM, Powell EN, Hofmann EE (2005) Understanding the success and failure of oyster populations: climatic cycles and *Perkinsus marinus*. J Shellfish Res 24:83–93. https://doi.org/10.2983/0730-8000(2006)25[83:UTSAFO]2.0.CO;2

Tolan J (2007) El Niño–Southern Oscillation impacts translated to the watershed scale: Estuarine salinity patterns along the Texas Gulf Coast, 1982 to 2004. Estuar Coast Shelf Sci 72:247–260. https://doi.org/10.1016/j.ecss.2006.10.018

Van Diggelen AD, Montagna PA (2016) Is salinity variability a benthic disturbance in estuaries? Estuaries Coasts 39:967–980. https://doi.org/10.1007/s12237-015-0058-9

Volety AK, Savarese M, Tolley SG, Arnold WS, Sime P, Goodman P, Chamberlain RH, Doering PH (2009) Eastern oysters (*Crassostrea virginica*) as an indicator for restoration of Everglades ecosystems. Ecol Indic 9(6):S120–S136. https://doi.org/10.1016/j.ecolind.2008.06.005

Weisberg SB, Ranasinghe JA, Dauer DM, Schaffner LC, Diaz RJ, Frithsen JB (1997) An estuarine benthic index of biotic integrity (B-IBI) for the Chesapeake Bay. Estuaries 20:149–158. https://doi.org/10.2307/1352728

Wilson EA, Powell EN, Craig MA, Wade TL, Brooks JM (1990) The distribution of *Perkinsus marinus* in Gulf Coast oysters: its relationship with temperature, reproduction, and pollutant body burden. Int Rev Ges Hydrobiol Hydrogr 75:533–550. https://doi.org/10.1002/iroh.19900750408

Wurbs RA, Ayala RA (2014) Reservoir evaporation in Texas, USA. J Hydrol 510:1–9. https://doi.org/10.1016/j.jhydrol.2013.12.011

# Plankton Dynamics in Texas Estuaries

Antonietta Quigg ⓘ, Jamie L. Steichen ⓘ, Laura Beecraft ⓘ, and Michael S. Wetz ⓘ

## Abstract

Plankton (phytoplankton, zooplankton) form the base of the food web. Inflow variability modulates plankton concentration, community composition, and productivity in estuaries. This chapter synthesizes data on plankton and their relationship with inflow from individual bay systems along the Texas coast where sufficient data exists. This will be primarily the Trinity-San Jacinto Estuary (also known as Galveston Bay) and Baffin Bay (located in the Upper Laguna Madre complex) with some details for San Antonio Bay, Mission Aransas Estuary, and Nueces-Corpus Christi Bay. In the former two systems, there is a lot of knowledge on key harmful algal bloom-forming taxa. This chapter will also highlight deficiencies in our current knowledge and future data needs. Given the general lack of data on zooplankton in Texas estuaries, this part of chapter is descriptive in nature. We end with providing a synthesis and recommendation for future efforts.

## Keywords

Phytoplankton · Zooplankton · Harmful algal bloom · Freshwater inflow

## Abbreviations

HAB    Harmful algal bloom
HPLC   High-performance liquid chromatography
IFCB   Imaging FlowCytobot
TCEQ   Texas Commission on Environmental Quality
TDSHS  Texas Department of State Health Services
TPWD   Texas Parks and Wildlife Department
TSJE   Trinity-San Jacinto Estuary
TWDB   Texas Water Development Board

## Introduction

Estuaries are dynamic, transitional zones between coastal and freshwaters, and this naturally occurring environmental gradient supports phyto- and zooplankton communities across spatiotemporal scales. Individual estuaries have unique conditions, controlled by geomorphology, watershed characteristics, surface and groundwater inflow, and tidal influence, that lead to differences in plankton community dynamics between them, making estuary-based approaches important for effective management. Of these, the greatest drivers of plankton community composition are climate (chapter "Climate Effects on Inflows"), hydrology and salinity (chapter "Hydrology, Circulation, and Salinity"), nutrients (chapter "Nutrient-Phytoplankton Dynamics in Texas Estuaries"), and physical and biogeochemical conditions (chapter "Physical and Biogeochemical Conditions and Trends in Texas Estuaries"). In this chapter we will focus on the response to freshwater inflows by the plankton commonly found in Texas estuaries between 2010 and 2019. While many studies have been conducted since Longley (1994) was first published, the greatest effort and available data on phytoplankton are available for Trinity-San Jacinto Estuary (TSJE, also known as Galveston Bay) and Baffin Bay (located in the Upper Laguna Madre complex). Similarly, the largest collection of published zooplankton data during this timeframe is available for TSJE. For the other estuaries in Texas, there have been few studies. A detailed examination of the role of nutrients controlling primary production in Texas estuaries, including nutrient concentrations and nutrient loading, is provided in chapter "Nutrient-Phytoplankton Dynamics in Texas Estuaries";

A. Quigg (✉) · J. L. Steichen
Department of Marine Biology, Texas A&M University at Galveston, Galveston, TX, USA
e-mail: quigga@tamug.edu; jamie.steichen@tamug.edu

L. Beecraft · M. S. Wetz
Texas A&M University-Corpus Christi, Harte Research Institute for Gulf of Mexico Studies, Corpus Christi, TX, USA
e-mail: laura.beecraft@tamucc.edu; Michael.wetz@tamucc.edu

© The Author(s) 2025
P. A. Montagna, A. R. Douglas (eds.), *Freshwater Inflows to Texas Bays and Estuaries*, Estuaries of the World, https://doi.org/10.1007/978-3-031-70882-4_13

hence we address it only briefly herein. Further, none of the inflow standards developed for Texas estuaries (chapter "Hydrology, Circulation, and Salinity," www.twdb.texas. gov) make use of plankton to set inflow criteria. Extensive details on plankton, harmful algal blooms (HABs), and their toxins and other details of plankton groups is of little direct use toward evaluating or developing inflow standards, but it can be argued that distinct changes or shifts in these communities is an early warning system on more long-term changes which estuaries will face. Given plankton are at the base of all food webs, respond on fast time-scales to freshwater inflows, and are important indicators of environmental quality, we focus on them in this chapter.

## Primary Producers

Phytoplankton are single-celled, photosynthesizing organisms that make up the base of the food web in all aquatic ecosystems. These primary producers play a crucial role in the energy flow in aquatic ecosystems (Flemer and Champ 2006; Roelke et al. 2013). Phytoplankton biomass and community composition are important indicators of water quality (Reynolds 2006). The dominant types of phytoplankton vary over time and space within estuaries, owing to variability in physical (light, temperature), chemical (salinity, nutrient form, and concentration), and biological (grazing, competition, and infection) factors. Due to their short life cycle, phytoplankton communities can change quickly in response to environmental fluctuations particularly temperature, salinity, and nutrients (Reynolds 2006).

Salinity is an important factor determining the types of phytoplankton that are present in estuaries. Both freshwater inflows and ocean exchange bring different species into and out of an estuary (Cloern and Dufford 2005; Lancelot and Muylaert 2011), with most residential species being euryhaline. Diverse assemblages of phytoplankton species coexist in Texas estuaries, influenced both directly and indirectly by inflow via changes to salinity (chapter "Hydrology, Circulation, and Salinity"), driven in large part by freshwater inflow (see chapter "Nutrient-Phytoplankton Dynamics in Texas Estuaries") as well terrestrial and submarine groundwater (Lecher and Mackey 2018; Chapter 5), direct precipitation and atmospheric deposition, and point and non-point source human inputs (Anderson et al. 2008; Rabalais et al. 2009; McFarlane et al. 2015; Malone and Newton 2020). Freshwater inflows may be so high at times in TSJE that phytoplankton are displaced by flushing (see, e.g., Roelke et al. 2013, Steichen et al. 2020) or so low that multi-year blooms may be able to persist, such as those in Baffin Bay and Laguna Madre (Buskey et al. 2001).

Nutrients are an important driver of phytoplankton growth and community composition, especially the availability of

the macronutrients nitrogen (e.g., nitrate, ammonium, and dissolved organic forms such as urea and amino acids) and phosphorus (e.g., soluble reactive phosphate). The reader is encouraged to review chapter "Nutrient-Phytoplankton Dynamics in Texas Estuaries" for specific details of the relationship between nutrients and phytoplankton in Texas estuaries. Briefly, estuaries receive most of their nutrients via freshwater inflows from rivers and surface runoff from the land. If the macronutrients are present in limiting concentrations, particularly nitrogen, phytoplankton growth can be limited and, in some cases, co-limited by phosphorus (Örnólfsdóttir et al. 2004; Bricker et al. 2008; Quigg et al. 2009; Moore et al. 2013; Ammerman 2018). However, with increasing human populations in coastal zones and changing land uses, particularly the ongoing expansion of hard surfaces (e.g., urban centers, industry, parking areas), there is a co-current increase in the amount of nutrients entering estuaries (Nixon 1995; Chang 2008; Glibert et al. 2010; Chang et al. 2014; Hogan et al. 2014; Freeman et al. 2019; Bugica et al. 2020). Traditionally only inorganic forms (e.g., nitrate or ammonium) of nutrients were considered "usable" by phytoplankton; today it is known that many species can also utilize organic forms of nutrients (e.g., Bronk et al. 2007). The phytoplankton species that can outcompete others in the presence of elevated nutrients are often bloom formers, and indeed, evidence is emerging that links growing organic nutrient loads to increased prevalence of estuarine harmful algal blooms (HABs) (Anderson et al. 2008, 2021; Glibert et al. 2006; Wetz et al. 2017; Glibert 2020; Chapter 7).

Phytoplankton growth and primary production are also directly dependent on light to drive photosynthesis. The amount and composition of phytoplankton is often very heterogeneous or "patchy" in space and time. Light exposure is controlled by water clarity and mixing, and the daily and seasonal changes in light conditions affect both the amount and type of phytoplankton present (Kirk 1994). Periodic pulses of freshwater inflows can have contrasting effects on estuarine phytoplankton: increasing biomass and primary productivity by supplying river-derived nutrients or decreasing the same by decreasing light availability and/or flushing phytoplankton out of the system (Mallin et al. 1993; Harding 1994; Paerl et al. 2013; Roelke et al. 2013). Grazing by zooplankton and other biota (e.g., ciliates, metazoans,) exerts a top-down control on phytoplankton abundance, often in a selective and size-specific manner (Longley 1994; Buskey et al. 1997; Cloern and Dufford 2005).

An algal bloom is a higher-than-normal concentration of phytoplankton or biomass deviating significantly from normal levels for a given time and location (Smayda 1997; Carstensen et al. 2007; Berdalet et al. 2016). HABs can occur when a blooming community is dominated by a species that has negative impacts on the ecosystem, marine life, or humans (Steidinger and Meave del Castillo 2018). There are multiple mechanisms by which blooms can have negative impacts, for example, toxin production by HABs, which can directly harm exposed marine life or cause illness or death in humans from consuming contaminated fish or shellfish. On the other hand, high biomass of some species can cause smothering or shading of benthic organisms or cause anoxic or hypoxic (low oxygen) conditions detrimental to other organisms (Smayda 1997; Berdalet et al. 2016). Over the past few decades, there has been a precipitous increase in the frequency/magnitude of algal blooms in estuarine and coastal waters (Hallegraeff 2003; Glibert et al. 2005; Anderson et al. 2021). This has largely been linked to increasing human-derived nutrient loads which enter estuaries through freshwater inflows combined with a growing influence of climate change (Anderson et al. 2008; Wells et al. 2015). A recent synthesis of water quality data on the Texas coast found "hot spots" for nutrient pressures, including the TSJE (due primarily to urbanization) and the Baffin Bay-Upper Laguna Madre complex (due to a combination of nutrient sources; Bugica et al. 2020). In addition, the lack of tidal flushing and limited connections with the Gulf of Mexico likely makes estuaries of the central and south Texas coast particularly susceptible to the influence of nutrients (Scavia and Liu 2009; Chapter 7; Beecraft and Wetz 2023).

The magnitude and frequency of fish mortalities in Texas estuaries are high relative to those in the other 22 coastal states in the USA. This is of serious concern for scientists, resource managers, and the public alike. An investigation of the major sources and causes of fish kills in coastal Texas from 1951 to 2006 in which more than 383 million fish were killed (72% were Gulf menhaden, *Brevoortia* spp.) revealed that the TSJE and Lavaca-Colorado Estuary have the highest number of fish kill events and total number of fish killed (Thronson and Quigg 2008). The leading cause of fish kills was found to be low dissolved oxygen concentrations caused by both physical and biological factors. For the latter, in most cases, it was after a HAB event in which fish mortalities were observed. However, in most cases, agency personnel were not able to identify the bloom-causing species, either because by the time they arrived at the fish kill event the bloom had declined or because facilities and resources were simply not available to make a timely identification.

Traditional methods of identifying phytoplankton involve microscopic enumeration which requires considerable time and expertise. Over the past few decades, numerous other methods and technologies for quantifying and identifying phytoplankton have been developed, including multi-pigment analysis, fluorometry, cytometry, imaging technology, and molecular techniques. Phytoplankton contain different types of pigments, some of which can be used as markers of specific taxa (Kirk 1994; Pinckney et al. 2001, 2017). These pigments can be quantified using high-performance liquid chromatography (HPLC) and the relative concentrations of different phytoplankton pigment classes estimated by statistical comparisons with known samples (Mackey et al. 1996). However, the output is only as good as the pre-determined defining (of each taxa) pigment ratios and requires verification with local communities/samples via microscopy (Lewitus et al. 2005; Pinckney et al. 2001). Fluorescence techniques use the natural properties of pigments, in particular chlorophyll *a*, to emit fluorescence when exposed to light. A chlorophyll fluorescence technique can be a sensitive way to measure chlorophyll *a* concentration and, when combined with sample excitation at different wavelengths or intensities, can also be used to estimate relative concentration of phytoplankton pigment classes or the photosynthetic condition of cells in the sample (Cosgrove and Borowitzka 2010). Flow cytometry is a technique that produces a very thin flow of liquid past sensors used to detect and measure properties of individual cells. Different instruments combine flow cytometry with fluorometry and microscopic imaging to detect, quantify, and in some cases identify bacteria and phytoplankton. For example, the Imaging FlowCytobot (IFCB) combines these three capabilities with machine learning algorithms to quantify and identify phytoplankton in situ (Sosik and Olson 2007; Olson and Sosik 2007; Campbell et al. 2010). Molecular and genetic techniques include research areas of transcriptomics, proteomics, and metabolomics. A genetic material from natural or laboratory phytoplankton samples is concentrated, extracted, and amplified. This can be followed by identification using molecular markers or sequencing and comparison with the existing sequence databases, allowing for highly specific identification of phytoplankton and even distinguishing between species and strains that cannot be discerned by microscopy (see Steichen et al. 2020). While all methods have pros and cons, the precision and accessibility of these techniques continue to improve.

Given the importance of phytoplankton community composition for trophic transfer and ecosystem health, there is a need for both monitoring and process-based studies to assess composition and its drivers. This is particularly important for efforts to mitigate the negative impacts of HABs given their increasing frequency and magnitude combined with continually increasing human populations living in coastal areas. Prior to the 1990s, there were few studies that evaluated phytoplankton community composition in Texas estuaries. Those that did typically focused on easy-to-identify taxa (i.e., diatoms) and thus provided an incomplete synopsis of community composition dynamics. This chapter provides a brief overview of early information on phytoplankton in Texas estuaries, more recent assessments of phytoplankton community composition, and a general description of the ecology of various HAB taxa that have now been detected in Texas estuaries.

## Effects of Freshwater Inflow on Phytoplankton

Texas estuaries span a gradient in climate, with the highest precipitation in the northeast (upper coast) and decreasing to the southwest (lower coast) (Tolan 2007; Montagna et al. 2013a, b). This in turn influences inflow (magnitude, duration, and frequency) to estuaries, which generally follows the same geographic gradient (see chapters "Climate Effects on Inflows" and "Hydrology, Circulation, and Salinity" for more details) with flows decreasing along a gradient from the upper to lower coast. Freshwater inflow to each estuary is measured at the last non-tidally affected gauge located on each contributing river (Longley 1994; TWDB 2022). Longley (1994) reported that the Sabine Neches Estuary had, on average, the highest annual freshwater inflow rates at $1.61 \times 10^{10}$ m$^3$ year$^{-1}$, while the Mission-Aransas Estuary has the lowest inflow rates at $5.92 \times 10^8$ m$^3$ year$^{-1}$ (based on 1941–1987). Median inflow rates for this same period were $1.00 \times 10^{10}$ m$^3$ year$^{-1}$ for the Sabine-Neches Estuary and $5.92 \times 10^7$ m$^3$ year$^{-1}$ for the Mission-Aransas Estuary. Inflows to the Mission-Aransas Estuary are the lowest of all Texas estuaries as it has the smallest contributing basin. This general pattern still holds true today.

More recently, the Texas Water Development Board (TWDB) estimated freshwater inflows from 2010 to 2019 using gaged flows, modeled ungaged flows, diverted flows, and return flows for each bay system (Fig. 1; TWDB 2022). In the recent decade, the TSJE had an annual higher median flow rate (594 m$^3$ s$^{-1}$) as well as a wider range of flows

(102–1119 m$^3$ s$^{-1}$) than the nearby Sabine-Neches Estuary (median = 518 m$^3$ s$^{-1}$; min–max = 53–1000 m$^3$ s$^{-1}$). In the mid-Texas coast, Lavaca-Colorado Estuary had an annual median flow of 86 m$^3$ s$^{-1}$, which is seven to eight times lower than those in the upper coast (Fig. 1), and a range of flows from 10 to 142 m$^3$ s$^{-1}$ (Fig. 1). The lowest median annual freshwater inflows occur along the lower coast which vary from 14 m$^3$ s$^{-1}$ in Mission-Aransas Estuary to 23 m$^3$ s$^{-1}$ in Upper Laguna Madre Estuary and with much less variability (Fig. 1).

Large-scale disturbance events such as hurricanes impact estuarine phytoplankton communities (Paerl et al. 2001, 2013). Between 2010 and 2019, several tropical storms and hurricanes hit the Texas coast. One of the most impactful was Hurricane Harvey which made landfall in Texas on August 25, 2017, impacting estuaries from the lower coast (Guadalupe, Lavaca-Colorado, and Nueces-Corpus) to the upper coast (TSJE) (Steichen et al. 2020; Walker et al. 2021). The ephemeral increase in salinity observed due to storm surge was followed by a rapid decrease to <1 practical salinity units as floodwaters flushed the estuaries and carried in freshwater phytoplankton, allochthonous organic matter, and nutrients. In all four systems, the increase in inorganic nutrients immediately post-Harvey (days-weeks) returned to pre-storm levels ~6–9 months later. While chlorophyll concentrations (a proxy for phytoplankton biomass) increased in the Lavaca-Colorado due to nutrient-stimulated growth, they decreased in TSJE due to hydraulic displacement of plankton communities immediately following the

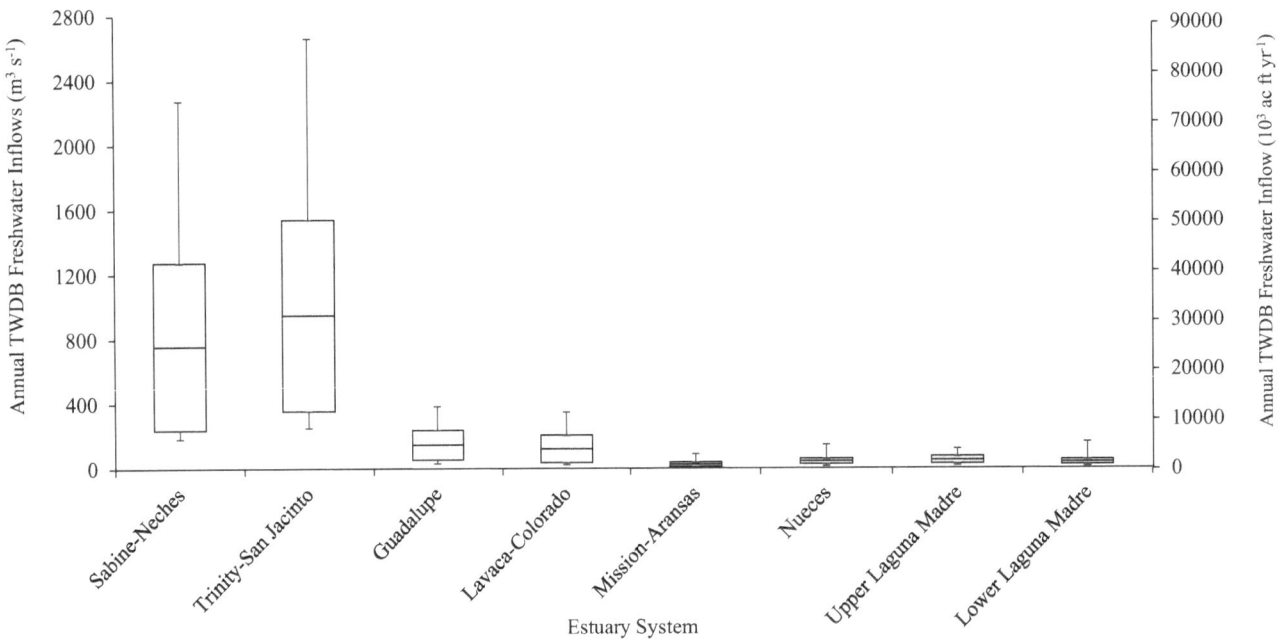

**Fig. 1** Box and whisker plot created using the estimated annual freshwater inflows (m$^3$ s$^{-1}$, left; 10$^3$ ac ft year$^{-1}$ right) for Texas estuaries calculated using TWDB modeled data from 2010 to 2019. The bars in the box and whisker plots display the median values, and the whiskers reflect the standard deviations

flooding event, while no clear impact was observed in the two other estuaries. This may be because the benefits of increased nutrient availability were counterbalanced by opposing effects of increased flushing and/or light limitation. These patterns in chlorophyll have been reported as a response to inflow conditions in many estuaries (Roelke et al. 2013; Azevedo et al. 2014; Dorado et al. 2015). Additionally, it was reported that the phytoplankton community was dominated largely by dinoflagellates in TSJE prior to Harvey making landfall; after the storm, the community shifted to successive blooms of chlorophytes, cyanobacteria, diatoms, and dinoflagellates (Steichen et al. 2020). This rapid change (4–6 weeks) was a result of the high flushing rate associated with flood water after the passage of the storm. The phytoplankton community did not recover to pre-Harvey conditions during the first month after the storm (Steichen et al. 2020) but took substantially longer (Quigg et al. 2019). These measurements were not available for the lower Texas estuaries. These and other studies support the notion that impacts of storms or tropical cyclones on water quality are short-lived, with the magnitude of the disturbance related to its proximity to storm track (Wetz and Paerl 2008).

Reyna et al. (2017) showed that the phytoplankton community changed considerably with an increase in freshwater inflows following a drought period that spanned over 3 years (2012–2015) in the Mission Aransas Estuary system. There was a rapid growth of diatoms in response to an increase in dissolved inorganic nitrogen that was delivered to the estuary with a large freshwater inflow pulse. Throughout the months following the initial pulse of freshwater after the initial increase in diatoms, the smaller-celled cyanobacteria dominated the phytoplankton community in terms of biomass (Reyna et al. 2017). In the nearby Copano Bay, in the spring season before the flooding events in 2015, the phytoplankton community was not only dominated by diatoms but also comprised of cryptophytes followed by dinoflagellates and

prasinophytes, and then during the higher inflow periods, the cyanobacteria dominated the community (Reyna et al. 2017). The shifts in the phytoplankton communities in response to sporadic pulses of freshwater inflows can have major implications for the productivity within an estuarine ecosystem (Cloern and Dufford 2005; Reyna et al. 2017).

## Phytoplankton Biomass and Primary Productivity

Chlorophyll *a* concentration has long been used as a quantitative index of phytoplankton biomass, as it is ubiquitously present in phototrophs. It is the most used measurement by both researchers and resource management agencies, providing our best source of information on estuarine phytoplankton biomass. Comparisons of estuaries based on phytoplankton standing crop require a significant amount of biomass measurements because of the variability of phytoplankton in space and time and the plethora of driving factors which influence this data, as well as caution given changes in protocols over time.

Longley (1994) found the average chlorophyll concentrations among the seven Texas estuaries ranged from 12.9 µg L$^{-1}$ in the TSJE to 3.1 µg L$^{-1}$ in the Mission-Aransas system (Table 1). The average chlorophyll levels in Texas estuaries from the 1960s to the 1980s was in the mid-range of averages for river-dominated estuaries listed by Boynton et al. (1982) and Clement et al. (2001). Chlorophyll concentrations were generally higher in the estuaries located on the upper coast relative to those on the lower coast, noting that data was not included for Baffin Bay or the Lower Laguna Madre (Longley 1994). Examining data collected more recently by Texas Commission on Environmental Quality (TCEQ) for the period 2010–2019 (Table 1), chlorophyll levels among estuaries are generally consistent with those

**Table 1** Phytoplankton biomass and productivity in Texas estuaries

| Estuary | Average annual chlorophyll[a] | Average annual chlorophyll[b] | Average primary production rates[c] | Average primary production rates[d] |
|---|---|---|---|---|
| Sabine-Neches | 6.59 ± 4.78 | 5.5 ± 4.6 | | |
| Trinity-San Jacinto | 18.46 ± 17.60 | 12.9 ± 18.3 | 0.952 ± 0.864 | 2.2 |
| Lavaca-Colorado | 8.24 ± 5.14 | 6.4 ± 7.2 | | 0.5–2.4 |
| Guadalupe | 8.21 ± 7.76 | 9.9 ± 14.4 | | 0.7–1.2 |
| Mission-Aransas | 6.41 ± 4.83 | 3.1 ± 2.9 | | |
| Nueces | 9.24 ± 11.39 | 5.3 ± 5.8 | | 0.5–1.0 |
| Baffin Bay | 28.68 ± 16.99 | | | |
| Upper Laguna Madre | 13.06 ± 9.34 | 9.6 ± 33.4 | | |
| Lower Laguna Madre | 9.53 ± 12.47 | | | |

[a]Data from Texas Commission on Environmental Quality for the period 2010–2020. Units are µg L$^{-1}$
[b]Data from Longley (1994) (Table 5.1.2). Based on "total" data collected from 1968 to 1989 and units of µg L$^{-1}$ (± standard deviation)
[c]Average primary production rates (2010–2019). Units are g C m$^2$ day$^{-1}$. Information available in studies for each estuary
[d]Data from Longley (1994) (Table 5.1.4). Units are g C m$^2$ day$^{-1}$. Sampling period varied (information available in studies for each estuary)

observations reported in the Longley (1994) study. However, we also observed higher values of chlorophyll in the two estuaries at opposite ends of the inflow spectrum, the TSJE and Baffin Bay, respectively. Different processes are likely responsible for the high chlorophyll in these two systems. TSJE phytoplankton are typically supported by higher inflows carrying land-based nutrients (Örnólfsdóttir et al. 2004; Roelke et al. 2013; Dorado et al. 2015; Quigg et al. 2017; Freeman et al. 2019), while Baffin Bay phytoplankton are supported by watershed nutrient loads that are efficiently retained and recycled in the long residence time environment (Wetz et al. 2017; Cira et al. 2021).

Systems undergoing reductions in inflow may have reduced nutrient inputs and in turn decrease in phytoplankton biomass, which has been observed in Nueces-Corpus Christi Bay (Kim et al. 2014; Bugica et al. 2020). Patterns in chlorophyll concentrations in San Antonio and Matagorda Bay are consistent with inflow-derived nutrients supporting phytoplankton abundance (Chin et al. 2022). It was found that chlorophyll concentrations in the Guadalupe Estuary were positively associated with inflow rate and nitrite and negatively associated with salinity and water temperature (Matthews et al. 1975). Whitledge (1989a, b) presented evidence that nitrogen was the chief nutritional limit on phytoplankton growth in the Guadalupe and Nueces estuaries.

Primary productivity of Texas estuaries has been measured by a variety of techniques since the early 1970s (Longley 1994; Quigg et al. 2009). Table 1 presents average rates of production from most major Texas estuaries without respect to those techniques or the caveats; this information is included to provide general trends rather than specific details. In those studies reviewed by Longley (1994), river-dominated estuaries like the TSJE were found to be among the most productive with lower values further south where inflows were also lower (Fig. 1). These high values were like those reported in other estuaries across the USA in the 1970s and 1980s (see, e.g., Boynton et al. 1982). In the work performed in recent decades for TSJE (e.g., Quigg et al. 2009, 2017), the primary productivity values are similar to those reported previously. Future studies are required to measure primary productivity in all estuaries in Texas.

## Common Phytoplankton Functional Groups and Taxa

Phytoplankton encompass remarkable diversity in habitat preferences, motility, nutritional mode, growth form, pigmentation, and production of secondary compounds, among others. Both the abundance and community composition of phytoplankton influence the health and function of aquatic environments. For example, certain species are better food sources to consumers compared to others, some produce harmful toxins, and excessive amounts of phytoplankton can have adverse ecological effects such as smothering or low dissolved oxygen. Phytoplankton can be described by size class, taxonomic group, environment, or common characteristics. There are four size classes of phytoplankton: macro- (>200 nm), micro- (20–200 μm), nano- (2–20 μm), and pico- (0.2–2 μm). Some of the common taxonomic groups of phytoplankton observed in estuaries include cyanobacteria, green algae, cryptophytes, diatoms, and dinoflagellates.

Cyanobacteria, commonly referred to as blue-green algae, occur in both marine and freshwaters, with picoplankton-sized unicellular species common in some estuaries as well as the open ocean (Gaulke et al. 2010; Goleski et al. 2010) and larger colonial and filamentous forms more common in freshwater environments (Paerl 1996). Cyanobacteria do not possess flagella, but some species possess gas vesicles, allowing them to influence vertical positioning in the water column by controlling buoyancy. Numerous members of cyanobacteria are capable of forming blooms and producing a variety of toxins, collectively referred to as cyanotoxins. Cyanobacterial blooms are most common in freshwaters but can also occur in estuarine and marine environments (O'Neil et al. 2012). Various species of cyanobacteria can produce toxins, which can deter feeding or be fatal when consumed by the zooplankton (Burkholder et al. 2018) leading to detrimental cascading effects through the higher trophic levels.

Green algae are species containing the pigments chlorophyll $a$ and $b$ and include chlorophytes and euglenophytes. Chlorophytes include pico- and nanoplankton of unicellular, colonial, and filamentous growth forms, with and without flagella, and tend to be more common in fresh and brackish waters (Chapman 2013; Graham et al. 2016). Euglenophytes are unicellular flagellates with both photosynthetic and heterotrophic species present in fresh, brackish, and coastal waters, in the nanoplankton size class.

Cryptophytes are unicellular flagellates that are photosynthetic and in the nanoplankton size class. They can be found from fresh to coastal waters but are rare in the open ocean (Graham et al. 2016). While this group is often present in phytoplankton assemblages, and may sometimes form blooms, they do not typically form HABs (Pedrós-Alió et al. 1987; Pithart 1997; Hammer et al. 2002; Al-Najjar et al. 2007). Blooms of cryptophytes have been associated with blooms of the ciliate *Mesodinium rubrum* that acquire the chloroplast from their prey via endosymbiosis (Gustafson et al. 2000; Johnson et al. 2016). In the TSJE, cryptophyte blooms (October–January) were found to cause a pinkish coloration of oysters from some commercial reefs in the late 1990s and early 2000s (Paerl et al. 2003). Although the conspicuous color has no obvious effect on oyster condition and it is not known to pose a human health hazard, it adversely affects consumer acceptance and hence oyster marketability.

An "off taste" further detracts from the pink oysters' marketability. The authors concluded that the magnitude and frequency of pink oyster events in the TSJE are a symptom of nutrient-driven eutrophication.

Diatoms (class Bacillariophyceae) are the most diverse groups of phytoplankton with species in all four size classes. They are found across the continuum from fresh to marine waters and are common in both pelagic (open water) and benthic (bottom) environments (Graham et al. 2016). Diatoms produce a silica frustule (shell-like structure) around each cell, with species that are solitary, or forming colonies of varying shapes and sizes. The silica frustule makes diatom cells heavier, and pelagic species rely on water mixing and ornate shapes to slow sinking and remain in the upper part of the water column where light is available. Benthic diatoms live on surfaces (rock, sand, sediment) in water that is shallow enough for light to reach the bottom, and many of these species are capable of gliding motility (Graham et al. 2016). In shallow and well-mixed water bodies, benthic species are regularly mixed into the water column and observed as part of the phytoplankton. All diatoms are autotrophic—relying on photosynthesis to meet their energetic requirements. Most species effectively utilize nitrogen in the form of nitrate, and they have a rapid growth rate under nutrient replete conditions (Cloern and Dufford 2005). Diatoms are generally considered the most important phytoplankton group for supporting higher trophic levels, in part because they produce essential fatty acids required by consumers and benefit the entire food chain (Brett et al. 2009; Parrish 2009). Diatoms are also generally considered a more available food source for zooplankton than cyanobacteria or many green algae (Ryther and Officer 1981).

Dinoflagellates are primarily unicellular flagellates with many being marine and some freshwater species in the nano- and microplankton size classes. Different species are autotrophic, heterotrophic, or both (mixotrophic) and pelagic or benthic (Graham et al. 2016). Many species can form cysts, allowing them to remain dormant under stressful conditions and then re-emerge when environmental conditions become favorable. The ability of many dinoflagellate species to be either mixotrophic or heterotrophic, along with being able to control their vertical position in the water column, allows them to use a variety of nutrient sources. Therefore, they can grow under conditions that may not be favorable to some other groups, such as diatoms, though they typically have a slower growth rate. Most marine toxin-producing phytoplankton are dinoflagellates, and both toxic and non-toxic species can form "red tides" and "mahogany tides"—which are blooms of such high density that the water is colored by their cells (Tango et al. 2005; Steidinger and Meave del Castillo 2018). Those which frequent the Texas coast are summarized in Table 2 along with the toxin they are known to produce.

To assess the general character of estuarine primary production, Longley (1994) grouped phytoplankton by their major taxonomic divisions. The relative importance of these groups was then used to describe the ecological health and functioning of each estuary, how accessible production was to consumers, and whether the ecosystem was dominated by marine or freshwater species. Longley (1994) found diatoms were more prevalent in the bays of the lower coast than in the upper bays and most prevalent in the portions of the bays proximal to the barrier islands. Therefore, it was concluded that phytoplankton productivity in these bays contributed directly to the zooplankton link in the food chain. By contrast, Longley (1994) suggested that because freshwater phytoplankton species are not the preferred food of zooplankton, they are more likely to enter the estuarine food chain through benthic filter feeders (rather than the planktonic food chain). Freshwater phytoplankton species dominate the upper bays particularly during times of high inflow and during floods (Quigg et al. 2019; Steichen et al. 2020). Below we synthesize findings from the last few decades of research on phytoplankton functional groups and taxa within different bay and estuary systems.

## Trinity-San Jacinto Estuary (TSJE)

The TSJE is a subtropical estuarine system located on the upper Texas Gulf coast. It is bordered by low-lying wetlands, a barrier island, and a peninsula. Both the Trinity (55%) and San Jacinto (16%) Rivers contribute significant freshwater inflows into the northern reaches of the bay (Guthrie et al. 2012). TSJE is used extensively for recreational and commercial activities and is one of the largest sources of seafood for Texas, as well as one of the major oyster-producing estuaries in the country (second to Louisiana) (Lester and Gonzalez 2011). The oysters, crabs, shrimp, and finfish harvested from TSJE contribute to one-third of the state's commercial fishing income and more than half of the state's recreational fishing expenditures. Freshwater inflows and return flows, as well as changing land use and land change patterns, are thought to exert the greatest pressure on the flora and fauna in TSJE (Quigg et al. 2009; Dorado et al. 2015; Steichen and Quigg 2018; Windham et al. 2019; Freeman et al. 2019).

Longley (1994) reported that diatoms > greens > cyanobacteria were the most prevalent functional groups in TSJE. An examination of photopigment-based changes in phytoplankton community composition over 10 years (2010–2019) in the TSJE is provided in Fig. 2. The photopigments represent major taxonomic groups from dinoflagellates (peridinin), diatoms (fucoxanthin), alloxanthin (cryptophytes), cyanobacteria (zeaxanthin), and chlorophyll $b$ (a proxy for green algae) (Mackey et al. 1996; Pinckney

**Table 2** Common HAB species observed in Texas estuaries

| Group | Order/family | Genus | Species | Toxin/toxicity | Bay(s) |
|---|---|---|---|---|---|
| Dinoflagellates | Gymnodiniales | Akashiwo | sanguinea | Surfactants, unidentified | TSJE, LC, BB |
| Dinoflagellates | Gymnodiniales | Cochlodinium | sp. | NTK | LC, BB |
| Dinoflagellates | Gymnodiniales | Margalefidinium | sp. (polykrikoides) | Ichthyotoxic | N, LC, BB |
| Dinoflagellates | Gymnodiniales | Karlodinium | sp. | K. veneficum—cytolysins, karlotoxins | N, BB |
| Dinoflagellates | Gymnodiniales | Karenia | spp. | Brevetoxins, ichthyotoxic | N, LC |
| Dinoflagellates | Gymnodiniales | Karenia | brevis | Brevetoxins | N, TSJE |
| Dinoflagellates | Gymnodiniales | Karenia | mikimotoi | Gymnocin-A | BB, N |
| Dinoflagellates | Gymnodiniales | Takayama | sp. | Ichthyotoxic (NTK) | TSJE |
| Dinoflagellates | Peridiniales | Kryptoperidinium | foliaceum | NTK | LC, G, N, BB |
| Dinoflagellates | Peridiniales | Heterocapsa | sp. | Hemolysins (H. circularisquama) | LC, G, N, BB |
| Dinoflagellates | Peridiniales | Peridinium | quadridentatum | NTK | LC, G |
| Dinoflagellates | Gonyaulacales | Gonyaulax | sp. | NTK | G, N, BB |
| Dinoflagellates | Gonyaulacales | Pyrodinium | bahamense var. bahamense | STXs | N, BB |
| Dinoflagellates | Prorocentrales | Prorocentrum | cordatum (= P. minimum) | Hemolytic | LC, G, N, BB |
| Dinoflagellates | Prorocentrales | Prorocentrum | micans | NTK | TSJE, LC, BB |
| Dinoflagellates | Prorocentrales | Prorocentrum | texanum (variants) | OA | TSJE, LC, BB |
| Dinoflagellates | Prorocentrales | Prorocentrum | sp. | Various toxins/NTK varies by species | TSJE, LC, BB |
| Dinoflagellates | Dinophysiales | Dinophysis | ovum | OA | TSJE, LC |
| Dinoflagellates | Dinophysiales | Dinophysis | sp. (acuminata complex) | OA, DTX-1,PTX-2 | TSJE, LC |
| Raphidophytes | Chattonellales | Chattonella | subsalsa (complex) | Ichthyotoxic | LC, G, N, BB |
| Raphidophytes | Chattonellales | Fibrocapsa | japonica | Ichthyotoxic | LC, N |
| Raphidophytes | Chattonellales | Chloromorum | toxicum (= Pseudochattonella verruculosa) | nodularins | BB |
| Pelagophytes | Sarcinochrysidales | Aureoumbra | lagunensis | | BB |
| Diatoms | Bacillariaceae | Pseudo-nitzschia | sp. | Domoic acid | TSJE, LC, G, N, BB |
| Diatoms | Chaetocerotaceae | Chaetoceros | sp. (subgenus Phaeoceros) | Physical irritant | LC, N, BB |

Toxin/toxicity information from Steidinger and Meave del Castillo 2018. Abbreviations: *NTK* no toxin known, but species has been associated with HAB event, for example, by causing low dissolved oxygen, *BB* Baffin Bay, *N* Nueces Estuary, *TSJE* Trinity-San Jacinto Estuary, *LC* Lavaca-Colorado Estuary, *G* Guadalupe Estuary

et al. 2001; Lewitus et al. 2005; Pinckney et al. 2017). Year 2010 is the most unique year in terms of phytoplankton community composition relative to other years, with a dominance of cryptophytes and green algae; in all other years, examined diatoms and dinoflagellates account for the largest fraction of the community (Fig. 2). Diatoms consistently made up ~45% of the community annually, dinoflagellates ~8%, cryptophytes ~11%, cyanobacteria ~23%, and chlorophytes ~13%. While always present, cryptophytes do not exceed more than 10% of the community in terms of seasonal averages, except in 2010 (avg. 42% of the community) and 2018 at 17%. Similarly, chlorophytes account for <14% of the community in terms of annual average, except for 2010 when they made up ~50% of the annual relative abundance. The difference in trends between major taxonomic groups between the work reported in Longley (1994) and this study is that the former was conducted using a microscope, while latter was assessed using HPLC. While HPLC allows greater resolution of groups present compared to using microscopy, there may have been other factors at play in the studies reported in Longley (1994). A review of those documents did not reveal sufficient details.

Recent studies have documented a variety of potential HAB-forming dinoflagellates and diatoms in TSJE (Steichen et al. 2012), in ballast water discharged into the bay (Steichen et al. 2014), and in waters sampled directly from the two major ports (Port of Houston and Port of Galveston) (Steichen et al. 2015). Two genera identified from the ports in TSJE (*Takayama* sp. and *Woloszynskia* sp.) had not been previously reported in published literature in this region (Steichen et al. 2015). It was hypothesized that they may have entered the bay from the discharge of ballast waters which is significant in this estuary. Though the HAB-forming genera *Anabaena* sp., *Karenia brevis*, *Microcystis* sp., and *Pseudonitzschia* sp. are identified, they do not occur in high-enough concentrations to be considered bloomers or harmful (Steichen, pers. obs.). Nonetheless, their appearance (even at one cell per observation) is reported immediately to the Texas Department of State Health Services (TDSHS) which monitors their presence to protect the public through seafood and aquatic life advisories. As a result, from time to time in the last two decades, TSJE hatcheries have been closed due to precautionary concerns associated with HABs, typically those associated with red tide species (www.tdshs.gov).

## Baffin Bay

Prior to the past decade, there were no comprehensive studies on the phytoplankton composition in Baffin Bay. Early

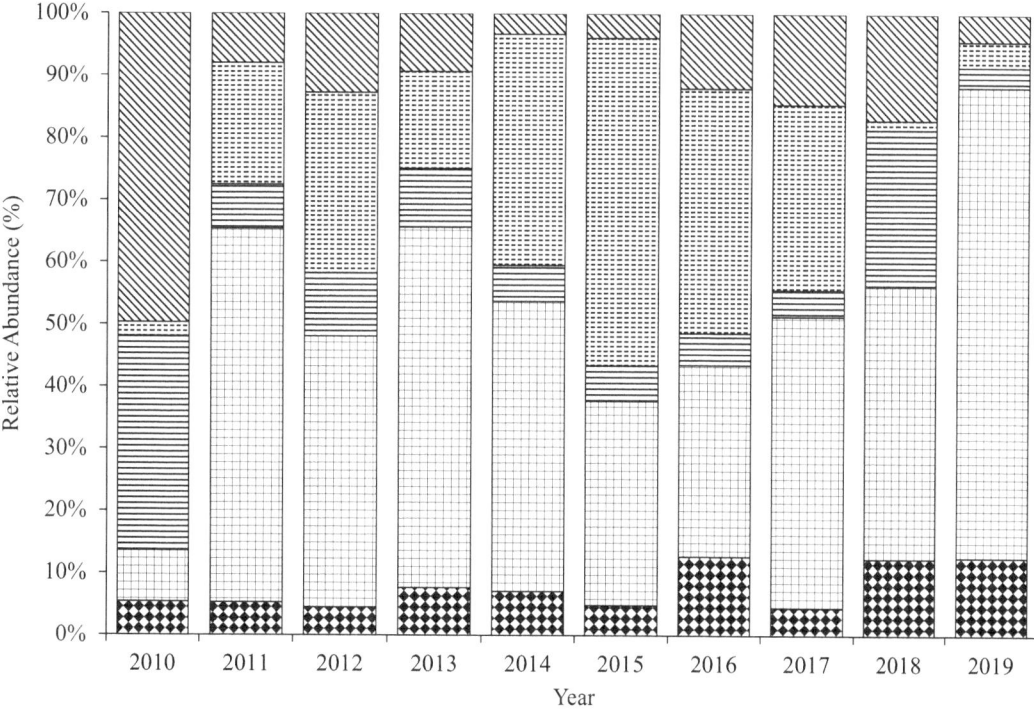

**Fig. 2** Photopigment-based phytoplankton community composition over 10 years (2010–2019) in Trinity-San Jacinto Estuary, Texas. Pigments shown and the taxonomic group of phytoplankton they represent: peridinin (dinoflagellates, black diamonds), fucoxanthin (diatoms, dotted), alloxanthin (cryptophytes, horizontal lines), zeaxanthin (cyanobacteria, horizontal dash lines), and chlorophyll *b* (chlorophytes, diagonal lines)

work centered around documenting and understanding blooms of the harmful brown tide species, *Aureoumbra lagunensis*, which was first detected in 1989, reached bloom levels in early 1990, and persisted in a near-monospecific bloom for more than 7 years in Baffin Bay and the Laguna Madre (Buskey et al. 1997, 2001; see also section "*Aureoumbra lagunensis* DeYoe and Stockwell"). Blooms of *A. lagunensis* form and persist during periods of low inflow in the region. These blooms are disrupted and dispersed by inflow events to the system, after which more typical and mixed estuarine phytoplankton communities are observed. Following high rainfall in 1997 and decreases in *A. lagunensis* concentration, phytoplankton biomass in the system remained high, with blooms of the diatom *Rhizosolenia* spp. followed by cyanobacteria (Buskey et al. 2001). The formation and persistence of *A. lagunensis* blooms may be facilitated in part by low grazing pressure—zooplankton and benthic filter feeder abundances were low during bloom development in 1989–1990 and again in 1998 (Montagna et al. 1993; Buskey et al. 1997, 2001).

Data collected as part of a volunteer water quality monitoring program from 2013 to present has afforded an opportunity to further explore phytoplankton dynamics in Baffin Bay (Fig. 3). Over that timeframe, three low rainfall periods and three wet periods were observed, allowing for evaluation of the influence of extended droughts that are punctuated by episodic high rainfall. Cira et al. (2021) documented

an *A. lagunensis*-dominated phytoplankton community (~90% of phytoplankton community biovolume) during prolonged low rainfall, high salinity, and low inorganic nutrient conditions from May 2013 to March 2015. *A. lagunensis* was particularly dominant during the warmer months, with small contributions of diatoms and dinoflagellates during cooler months. After a period of heavy rainfall in March–May 2015, *A. lagunensis* biovolume dropped sharply throughout much of Baffin Bay. High biovolume of diatoms and other taxa (e.g., the mixotrophic ciliate *Mesodinium* sp.) was observed at different times and locations throughout the fall and winter (Cira et al. 2021). A shorter period of increasing salinity during 2016–2018 was characterized by lower chlorophyll concentrations, with localized high abundances of *A. lagunensis* detected in 2018 at high salinities (Chin et al. 2022). Subsequent rainfall and inflow events in 2018 resulted in increased flushing, dissipation of the high *A. lagunensis* biomass, and a more mixed community. Intermittent blooms dominated by different groups: diatoms, chlorophytes, euglenoids, and mixotrophic ciliates (*Mesodinium*) were observed at different times throughout the year (Chin et al. 2022). Low rainfall and gradually increasing salinities from late 2019 to early 2021 coincided with relatively lower chlorophyll levels, and mixed blooms of diatoms and dinoflagellates formed following unusually high rainfall and inflow in summer 2021 (Beecraft and Wetz 2023).

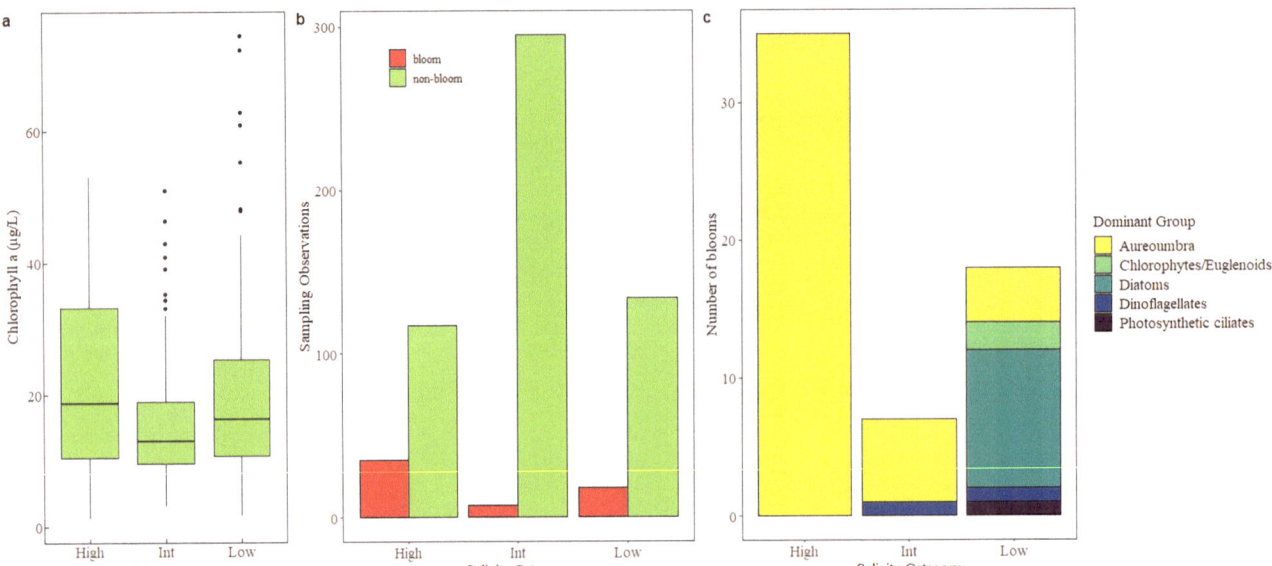

**Fig. 3** Chlorophyll *a* and phytoplankton community composition data from monthly sampling at six locations in Baffin Bay from 2013 to 2021. Observations were categorized as low, intermediate (Int), or high salinity by comparing values to the 25th (31.9) and 75th (49.5) quartiles of observed surface water salinities over the study period: <25th quartile = low, 25th–75th = intermediate, and >75th quartile = high. Chlorophyll *a* observations were categorized as "bloom" when they exceeded the 90th percentile (33.8 µg L$^{-1}$) from the study period and non-bloom when below. (**a**) Box-and-whisker plot of chlorophyll *a* concentration: the box represents the interquartile range (IQR); the lines reach 1.5*IQR past the 25th and 75th quartiles, respectively; and the dots are observations outside this. (**b**) Occurrence of non-bloom and bloom (>33.8 µg L$^{-1}$) level chlorophyll *a* observations. (**c**) Dominant phytoplankton groups comprising bloom observations under high, intermediate (Int), and low salinity conditions. Panel (**c**) as shown in Beecraft and Wetz (2023)

Overall, Baffin Bay can behave similarly to river-dominated estuaries under high rainfall conditions, with ephemeral high inorganic nutrient concentrations and high phytoplankton biomass that is dominated by faster growing taxa but also has characteristics of low inflow estuaries from arid to semiarid regions, namely, high salinity, long residence time conditions, low inorganic nutrient concentrations, and low diversity but high biomass phytoplankton communities that are apparently fueled by regenerated nutrients (Cira and Wetz 2019; Cira et al 2021). Phytoplankton blooms can occur in all seasons and across the continuum of inflow conditions in Baffin Bay, exemplifying the consistent availability of nutrients to support phytoplankton growth in the system. Blooms at different salinities have distinct community composition, and the prevalence of blooms appears to be greater under high salinity conditions (Fig. 3) (Beecraft and Wetz 2023). Furthermore, the high nutrient availability (especially recycled nutrients and dissolved organic nitrogen) as well as anticipated increases in regional temperature and drought may create conditions under which *A. lagunensis* proliferates further in the future. Reducing nutrient loads and maintaining strategic inflow targets to promote phytoplankton diversity in Baffin Bay are likely to be important management strategies to curbing future *A. lagunensis* blooms.

## San Antonio Bay, Nueces-Corpus Christi Bay, and Baffin Bay Comparison

Over an 18-month period during 2018–2019, Chin et al. (2022) compared phytoplankton indicators among San Antonio Bay, which is river-influenced; Nueces-Corpus Christi Bay, which is a neutral estuary based on inflow balance; and Baffin Bay, which has no major inflows and is frequently hypersaline. Their objectives were to assess if the different freshwater inflow regimes led to differences in environmental conditions and phytoplankton communities among the three bays, and if so, how. Chin et al. (2022) found that nitrate and nitrite, phosphate, and silicate concentrations were significantly higher in San Antonio Bay than in the other two bays, while dissolved organic nitrogen concentrations were two- to threefold higher in Baffin Bay than in San Antonio Bay or Nueces-Corpus Christi Bay. On average, phytoplankton biovolume was notably higher in Baffin Bay but lower and roughly equivalent between San Antonio Bay and Nueces-Corpus Christi Bay, whereas chlorophyll was high and equivalent in San Antonio Bay and Baffin Bay but lower in Nueces-Corpus Christi Bay. The observed ratio of chlorophyll/biovolume was highest in San Antonio Bay and lowest in Baffin Bay. The authors argued that despite having higher inorganic nutrient concentrations, the lower phytoplankton biovolume in it was indicative of light limitation, as denoted by its shallower Secchi depth compared to the other two bays. Light limitation can be a common feature in some estuaries, particularly those such as San Antonio Bay that experience both relatively high freshwater inflow and high turbidity due to mixing (Pennock and Sharp 1994). This may also explain the high chlorophyll but low biovolume in San Antonio Bay, as the amount of chlorophyll per cell increases under light-limited conditions (Lewitus et al. 2005; Reynolds 2006).

Despite observing distinct environmental and water chemistry conditions between the three bays, Chin et al. (2022) found some surprising similarities in terms of phytoplankton community composition. In terms of size fractions, the nanoplankton (2–20 μm) and microplankton (>20 μm) were the overall largest contributor to chlorophyll *a* among all three bays, whereas the contribution of picoplankton (<2 μm) was low (<10%) and similar among bays. Likewise, diatoms were a significant contributor to community biovolume in all three bays, which they speculated was a result of high average wind conditions throughout much of the study period (Carlin et al. 2016; Reisinger et al. 2017; see also https://windexchange.energy.gov/maps-data/325). Wind-driven turbulence may competitively favor diatoms by maintaining them in the water column, resuspending benthic taxa, and/or increasing turbidity (Jäger et al. 2008), resulting in reduced light availability and rapidly changing light exposure as cells are transported through the water column—conditions to which many diatoms are specifically well-adapted (Cloern and Dufford 2005; Paerl et al. 2007; Depauw et al. 2012).

Perhaps the most intriguing finding from Chin et al. (2022) was evidence that higher base inflow rates and/or pulsed inflow events may lead to increased diversity in the phytoplankton community (see also Buyukates and Roelke 2005; Beecraft and Wetz 2023). For example, the phytoplankton community was generally more diverse on average in San Antonio Bay than in the other two systems that experience much lower inflows. This was further exemplified by the recurring monospecific blooms of *A. lagunensis* that have been observed in Baffin Bay since 1990, especially during drought conditions (Buskey et al. 2001; Cira et al. 2021). In addition, phytoplankton communities of all three systems tended to see a greater contribution from a larger number of functional groups during and following inflow events, adding further evidence for the role of inflow as a disturbance that increases phytoplankton diversity (Chin et al. 2022; Beecraft and Wetz 2023).

Given the importance of phytoplankton to the base of the estuarine food webs, and specifically the type(s) of phytoplankton that are present, it is tempting to speculate on what the future holds when accounting for impacts of changing freshwater inflow magnitude and frequency of pulses. For example, with the expected long-term reductions in inflow magnitude and/or increasing upstream capture of flood pulses for central-south Texas coast estuaries, can we expect to see less variable and diverse phytoplankton communities in the future? And if so, what other factors will then determine the dominant taxa? For example, the *A. lagunensis* blooms that are common in Baffin Bay during prolonged low inflow conditions are now believed to also be an artifact of the nutrient pressures in that system. In contrast, diatoms dominate in San Antonio Bay under low inflow conditions— is this because San Antonio Bay has not yet felt the nutrient pressures that Baffin Bay has, or is it an artifact of a more turbid environment? Or both factors combined? It is clear that additional work is needed to elucidate the mechanistic linkages between inflow and phytoplankton community composition.

## Harmful Algae Along the Texas Coast

As environmental pressures on ecosystem services along the coast are increasing, there may be a concurrent increase in observations of algal blooms (number, frequency, magnitude) (Anderson et al. 2008, 2021). HABs can be initiated by the parameters that are associated with fluctuating inflows (Roelke and Pierce 2011) including but not limited to floods and droughts. These changes in freshwater inflows can affect HABs in different and complex ways depending on the rate of inflow, bathymetry of the bay system, nutrient limitation before the inflows, etc. These events can cumulatively impact the phytoplankton community and/or blooms that are present at the time of the event. Freshwater inflows bring in nutrients to the estuaries which can fuel HABs (chapter "Hydrology, Circulation, and Salinity" sections) and, depending on their magnitude, may stir up sediments which effectively re-introduces nutrients back into the water column (Roelke and Pierce 2011). And high rates of freshwater inflow can decrease water column stratification bringing nutrients to the surface which can fuel blooms if present. Sediments mixed into the water column can lead to light limitation, subsequently breaking down the bloom (Roelke and Pierce 2011). As freshwater inflows deliver sediments to the estuaries, clay particles can bind to inorganic nutrients such as phosphate and ammonium removing them from the water column and rendering them unusable by the phytoplankton. While the influx of clay particles in the water column may lead to the drawdown of some inorganic nutrients, river inflow rich in nitrate can increase the nutrients available for phytoplankton utilization. Further, during high inflow events, hydraulic flushing physically displaces phytoplankton cells leading to a breakdown of a bloom (Roelke and Pierce 2011). In other areas of a bay system, the presence of "hydraulic storage zones" may exist where residence time is increased compared to the surrounding waters (Roelke and Pierce 2011). In these storage zones, bloom conditions may persist longer, and the blooms may have a shorter recovery time than areas with higher flushing rates.

Nutrient loading and toxin production in HAB species are not straightforward. While the influx of nitrate can fuel diatoms HAB species such as *Pseudo-nitzschia* spp. (Roelke and Pierce 2011), the increase nutrient availability has also shown to be consistent with a decrease in toxin production in phytoplankton species such as *Prymnesium parvum* (Schwierzke-Wade et al. 2011; Roelke and Pierce 2011). Consequences of HABs include but are not limited to low dissolved oxygen zones and fish kills (e.g., Thronson and Quigg 2008), organic biofilm coating and suffocating fish which may or may not lead to fish kills (e.g., McInnes and Quigg 2010), finfish or shellfish hatchery closures (Hallegraeff and Bolch 2016; Roelke et al. 2010), and in some cases illness (respiratory inflammation, gastrointestinal symptoms, nausea, etc.) or even death of humans or marine life (Van Dolah 2000; Fleming et al. 2002; Kite-Powell et al. 2008).

The assessment of HABs in Texas waters typically occurs in response to an event, for example, a fish kill or significant discoloration of a water body. Several monitoring systems are set up in the state which rely on agency personnel (https://tpwd.texas.gov/), funded research programs, and/or citizen science programs. In Texas, programs that measure the changes in phytoplankton species are dependent on intermittent funding, and so comprehensive or even continuous records are not available. In Table 2 we summarized the most commonly observed HAB species and the toxin which they are known to produce and, to the degree that the information is available, the major estuary system in which they have been observed. In this section of the chapter, we consider two types of HABs: those which are known to produce toxins (*Akashiwo sanguinea*, *Dinophysis ovum*, *Karenia brevis*, and *Prorocentrum* spp.) and those which do not produce toxins but are nonetheless harmful because of the damage they cause to ecosystem functions (*A. lagunensis*).

### *Akashiwo sanguinea* (K. Hirasaka) G. Hansen and Moestrup

*A. sanguinea* is a marine mixotrophic, unarmored dinoflagellate formerly known as *Gymnodinium sanguineum*. It is cosmopolitan in its distribution and is typically found in temperate and tropical waters (Hargraves 2011). This species has been found to be both euryhaline and eurythermal as it is able to tolerate salinities ranging from 5 to 40 and grow between 10 and 30 °C (Mended-Deuer and Montalbano 2015). The harmful effects of *A. sanguinea* have been demonstrated toward multiple groups of organisms, from phytoplankton and zooplankton to larger marine animals. *A.*

*sanguinea* blooms have been suspected to be linked to fish deaths, marine mammal standings (Badylak et al. 2014), bird deaths, and coral bleaching (Jessup et al. 2009). Concentrated blooms of *A. sanguinea* can also produce aquatic animal mortalities through oxygen depletion (Hallegraeff 2003). Further, they have been known to form non-toxic blooms that cause red tide (Hargraves 2011), but the blooms are not harmful to humans. The different modes of action have been suggested, including oxygen stress, mucus, and surfactant production, as well as toxic effects—the mechanism of which is still unclear (Wu et al. 2022). The toxic effects of *A. sanguinea* were found by Xu et al. (2017) to be enhanced by increased nutrient supply, suggesting that this species could both directly (via increased growth) and indirectly (e.g., via enhanced toxin production) become more toxic in response to eutrophication.

Locally, *A. sanguinea* was suspected to have caused fish kills and marine mammal strandings in the Gulf of Mexico in the mid-1990s (Harper and Guillen 1989; Robichaux et al. 1998; Steidinger et al. 1998). This includes more than 13 million fish and blue crabs that were killed during a bloom of *A. sanguinea* in the TSJE (Harper and Guillen 1989). It was also found to have caused a large fish kill along Bolivar Peninsula in September 2007 (https://tpwd.texas.gov/). More recently in TSJE, relatively high numbers of this dinoflagellate have been observed from time to time as shown in Fig. 4: August 2017 (8 cells mL⁻¹), March and April 2018 (12 and 24 cells mL⁻¹, respectively), and November 2018 (6–8 cells mL⁻¹) but with no concurrent changes in watercolor or an observable fish kill event (Quigg et al. 2019). *A.*

*sanguinea* is also observed in lower Texas estuaries but not in high abundances (Wetz, pers. obs.).

## *Dinophysis ovum* Schütt (1895)

*Dinophysis* is a marine, planktonic dinoflagellate and an obligate mixotroph. It must consume prey, namely, the photosynthetic ciliate *Mesodinium*, and then uses the prey's ability for photosynthesis in a process called kleptoplastidy for growth and reproduction (Reguera et al. 2012). *Dinophysis* is also highly toxic, producing okadaic acid and its derivatives. *Dinophysis ovum* is known to produce HABs globally (Reguera et al. 2012) and specifically in Texas (Gulf of Mexico). *D. ovum* and *Dinophysis acuminata* form part of the *D. acuminata* complex: a group of *Dinophysis* species currently undistinguishable using light microscopy (Reguera et al. 2012) and even in some cases by molecular methods (see review by Anderson et al. 2021). *D. ovum* at Port Aransas, Texas, was identified by Campbell et al. (2010) using a combination of techniques including IFCB, microscopy, and DNA sequence analysis. It has not been positively identified in this manner in other parts of Texas, with most other recordings relying on microscopy or the IFCB.

Oyster beds (*Crassostrea virginica*) along the Texas coast were closed in 2008 due to the confirmed presence of okadaic acid, the toxin responsible for diarrhetic shellfish poisoning (DSP) (Deeds et al. 2010; Campbell et al. 2010; Swanson et al. 2010). This marked the first shellfish bed closure in the USA due to both elevated concentrations of *D. ovum* and

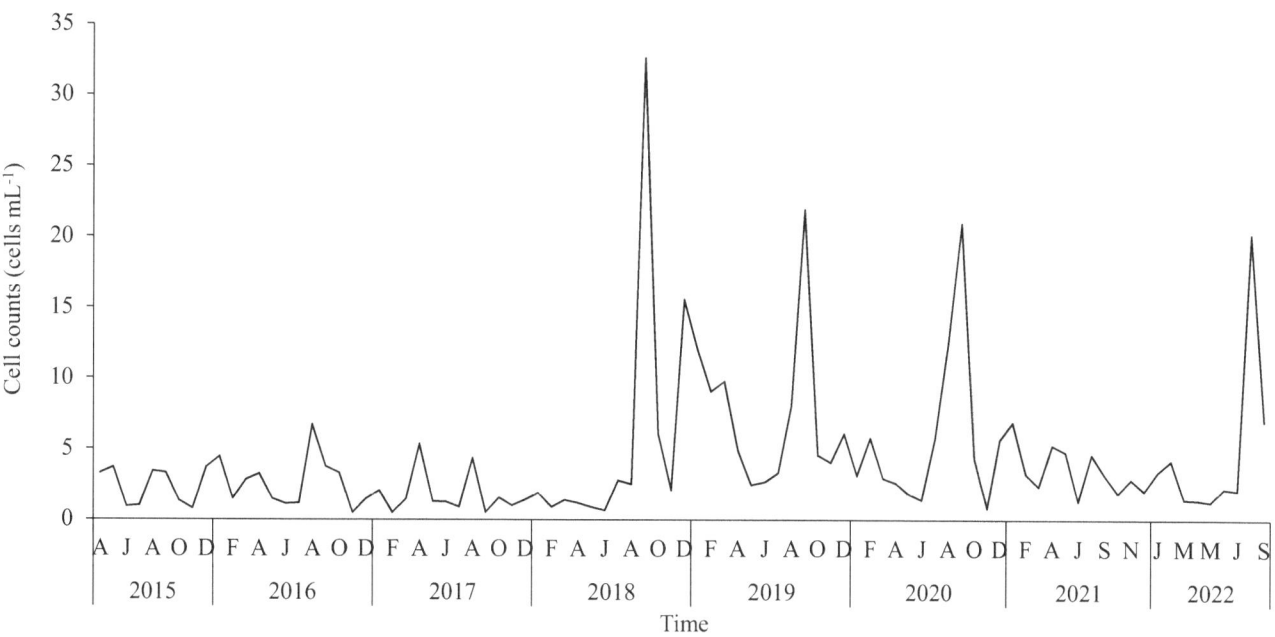

**Fig. 4** Monthly average calculated from the daily cell counts for *Akashiwo sanguinea* in Trinity-San Jacinto Estuary from April 2015 to October 2022 from the TAMUG IFCB (visit https://www.tamug.edu/ phytoplankton/projects/Imaging-FlowCytobot.html for details of how and where these samples are collected and how they are processed)

diarrhetic shellfish poisoning intoxicated shellfish. In an examination of the timing and magnitude of different peak concentrations along the entire Texas coast, Swanson et al. (2010) found that given high offshore abundance, combined with inshore bay distributions with maxima near barrier island passes, an offshore, southern origin for the *Dinophysis* bloom with advection into the bays was the mechanism of distribution during this event. They found no correlation between abundance and environmental parameters, although cell density peaked offshore around 17–18 °C with inshore observations showing a wider temperature range.

Campbell et al. (2010) found the non-toxic ciliate *Mesodinium rubrum* (formerly known as *Myrionecta rubra*, a prey item of *Dinophysis*) was observed before *D. ovum* in 2008, while another potentially toxic dinoflagellate, *Prorocentrum* spp., was observed after the bloom. In a follow-up multi-year study, Harred and Campbell (2014) found that while the *D. ovum* blooms (>200 cells mL$^{-1}$ highest daily count reported; February–March 2008) in the Port Aransas channel had very high cell densities, sampling in the same location in subsequent years revealed *D. ovum* was absent (2009) or present at cell densities which were ten times lower (2010–2012) than the highest recorded during the bloom period. Harred and Campbell (2014) suggested that *M. rubrum* could possibly serve as a predictor of blooms given that the ciliate blooms often precede those of the dinoflagellate. In more recent years, cells of the *D. acuminata* complex have occasionally been observed in Matagorda Bay near channels connecting to the Gulf, but not in significant numbers (Wetz, pers. obs.).

Regional thresholds have recently been defined to trigger enhanced shellfish monitoring or precautionary harvesting closures ranging from 2 to 10 *Dinophysis* cells mL$^{-1}$ (Anderson et al. 2021). While always present in TSJE, the numbers of *D. ovum* are generally very low of <1 cell mL$^{-1}$ (Fig. 5). In general, the highest cell densities were observed late in the fall or early winter in TSJE corresponding to periods of low freshwater inflows (Fig. 5). In the winter of January 2020, the highest number of *D. ovum* in samples was ~20 cells mL$^{-1}$ (Quigg and Steichen 2021), but there were no associated fish kills, and there was no discoloration of the water. An examination of data for TSJE (Steichen and Quigg, pers. obs) revealed no consistent distributions of *D. ovum*, *M. rubrum*, and *Prorocentrum* sp. as recorded in Port Aransas, suggesting that there may be specific conditions required for this trio to co-occur in this sequence. It appears that the *D. acuminata* complex leverages periods of low flows when other phytoplankton are in lower cell densities due to lower nutrient availabilities to bloom; this is consistent also with its mixotrophic lifestyle (Reguera et al. 2012) and their appearance in south Texas (Harred and Campbell 2014) in which freshwater inflows are generally lower (Fig. 1; chapter "Hydrology, Circulation, and Salinity").

### Karenia brevis

*Karenia brevis*, historically known as *Gymnodinium breve*, *Gymnodinium brevis*, and *Ptychodiscus brevis*, is a marine toxic dinoflagellate that can form blooms capable of causing

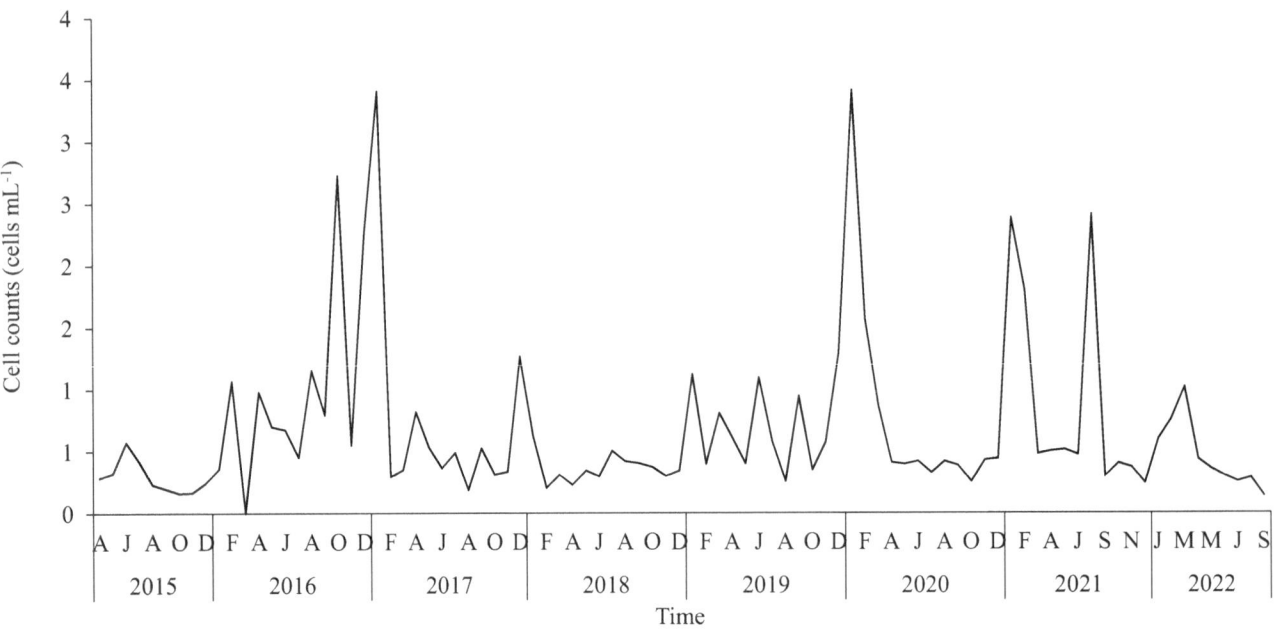

**Fig. 5** Monthly average calculated from the daily cell counts for *Dinophysis ovum* in Trinity-San Jacinto Estuary from April 2015 to October 2022 from the TAMUG IFCB (visit https://www.tamug.edu/ phytoplankton/projects/Imaging-FlowCytobot.html for details of how and where these samples are collect and how they processed)

harmful red tides (Poulson-Ellestad et al. 2014) and/or release neurotoxic shellfish toxins. The neritic *K. brevis* has been known to thrive in salinities ranging from 25 to 40, while other strains have been known to adapt to lower salinities (Hargraves 2011). It can be found to bloom frequently along the lower coast of Texas in the northern Gulf of Mexico (Magaña et al. 2003; Walsh et al. 2006; Henrichs et al. 2013a) as well as almost annually on the southwest Florida coast and occasionally up to the Atlantic off the North Carolina coast (Brand and Compton 2007; Pierce and Henry 2008). Bloom events rarely occur in northern Gulf of Mexico states (Alabama, Mississippi, and Louisiana); when present they have been linked to western advection of northwest Florida blooms (Soto et al. 2018). One hypothesis for the infrequent observation of *K. brevis* along the upper Texas coast is that high freshwater inflows and therefore discharges into the Gulf of Mexico create an environment that is unsuitable for this dinoflagellate. It is only during or after periods of prolonged drought in which *K. brevis* maybe found at the mouth of the TSJE (Quigg, pers. obs.).

Nonetheless, the negative impact of *K. brevis* blooms on natural resources and public health is due to the produced brevetoxins, neurotoxins that interfere with nerve transmission (Pierce and Henry 2008). These toxins have been known to cause mass fish kills as well as marine mammal, sea turtle, sea bird, and benthic community mortalities. Only ~7% of fish kills in coastal Texas have been attributed to red tide from 1951 to 2006 (Thronson and Quigg 2008). In addition, these toxins affect public health through shellfish contamination and exposure to aerosol toxins (Pierce and Henry 2008). The Centers for Disease Control report that the human health effects associated with eating brevetoxin- and other toxin-tainted shellfish are well documented (Van Dolah 2000). Scientists know that people who swim among brevetoxins or inhale brevetoxins dispersed in the air may experience irritation of the eyes, nose, and throat, as well as coughing, wheezing, and shortness of breath (Fleming et al. 2002; Kite-Powell et al. 2008). Additional evidence suggests that people with existing respiratory illnesses, such as asthma, may experience these symptoms more severely.

In Texas, *K. brevis* blooms frequent the area around Port Aransas and neighboring areas but are infrequently reported for TSJE and the upper reaches of the coast. The Texas Parks and Wildlife Department (TPWD) and TDSHS monitor *K. brevis* blooms; citizens including the Red Tide Rangers (https://texasseagrant.org) are also asked to notify agencies immediately. Cell abundances are given the following designations by TPWD: background ($\leq$ 1 cell mL$^{-1}$), very low (1–10 cells mL$^{-1}$), low (10–100 cells mL$^{-1}$), moderate (100–1000 cells mL$^{-1}$), and high ($\geq$1000 cells mL$^{-1}$). From January 2015 to present, *K. brevis* for TSJE were generally in the "background" or "very low" category (Steichen, pers. obs). In 2013, the temporary closure of oyster leases was required to prevent the harvest and sale of shellfish contaminated with brevetoxin in TSJE (www.tdshs.gov). During the period, November 2010 to the end of 2013, the state was experiencing a prolonged and extensive drought (Nielsen-Gammon et al. 2020) such that salinities in the TSJE were frequently >20 psu. This may have created an environment in which *K. brevis* could proliferate (Quigg 2015). By contrast, *K. brevis* blooms occur more frequently around Port Aransas and the nearby Nueces Estuary (Campbell et al. 2013; Tominack et al. 2020). This is thought to be the result of a confluence of physical oceanographic conditions in the Gulf of Mexico (Henrichs et al. 2013a; Thyng et al. 2013), which are summarized in Anderson et al. (2021). Briefly, under strong downwelling conditions, a severe bloom is unlikely to develop along the Texas coast; when alongshore, southwestern transport is weak, blooms will be observed as a convergence zone that concentrates cells near the coast (Thyng et al. 2013). A recent assessment of *K. brevis* trends in the Nueces-Corpus Christi Bay region found an increasing frequency of blooms over the past 30 years, with several factors correlating to bloom frequency including but not limited to increasing salinity levels (Tominack et al. 2020). Interestingly, there was no obvious relationship with nutrients.

## *Prorocentrum* spp.

*Prorocentrum* species are benthic, planktonic, or tychoplanktonic dinoflagellates and can be found worldwide, with some species known to produce fast-acting toxin, okadaic acid, and derivatives, while others can result in high biomass events leading to low dissolved oxygen (Steidinger and Meave del Castillo 2018). This can result in diarrhetic shellfish poisoning (DSP), an alimentary intoxication with non-fatal gastrointestinal symptoms, such as nausea, vomiting, abdominal pain, and most commonly diarrhea. The expression levels of biotoxins in these dinoflagellates are strongly affected by nutritional and environmental factors (see Lee et al. 2016). During February–April 2008, an unusual mortality event occurred in Texas coastal waters that resulted in over 100 bottlenose dolphin (*Tursiops truncatus*) deaths (Fire et al. 2011). This mortality event overlapped spatially and temporally with a HAB composed of *Prorocentrum* spp. and *Dinophysis* spp. and was associated with shellfish bed closures due to elevated toxins.

When present in the highest cell densities in the TSJE, from April to June, high cell densities of *Prorocentrum gracile*, *P. micans*, and *P. texanum* coincide with the end of the spring bloom period and transition to summer and higher temperatures (Fig. 6). It is not clear what if any specific environmental or biological factors may be associated with the very high cell densities, which should be investigated in the future. For *P. gracile*, high cell densities are defined here as

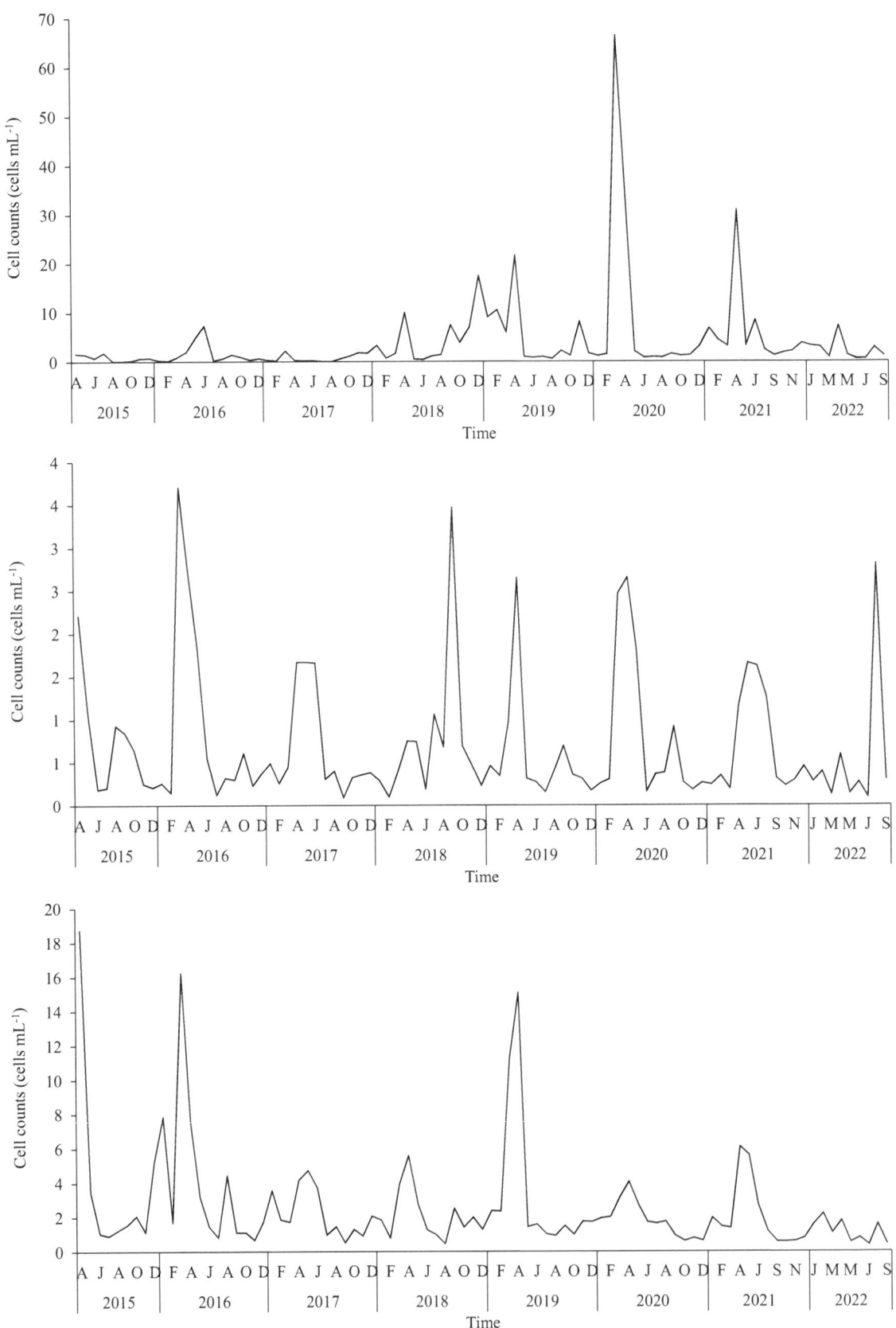

**Fig. 6** Monthly average calculated from the daily cell counts for (**a**) *Prorocentrum gracile*, (**b**) *P. micans*, and (**c**) *P. texanum* in Trinity-San Jacinto Estuary from April 2015 to October 2022 from the TAMUG IFCB (visit https://www.tamug.edu/phytoplankton/projects/Imaging-FlowCytobot.html for details of how and where these samples are collected and how they are processed)

**Fig. 7** Yearly maximum concentrations of *A. lagunensis* in Baffin Bay (Texas). Data not available (na) means no data was collected or no data exists (pers. comm. Wetz)

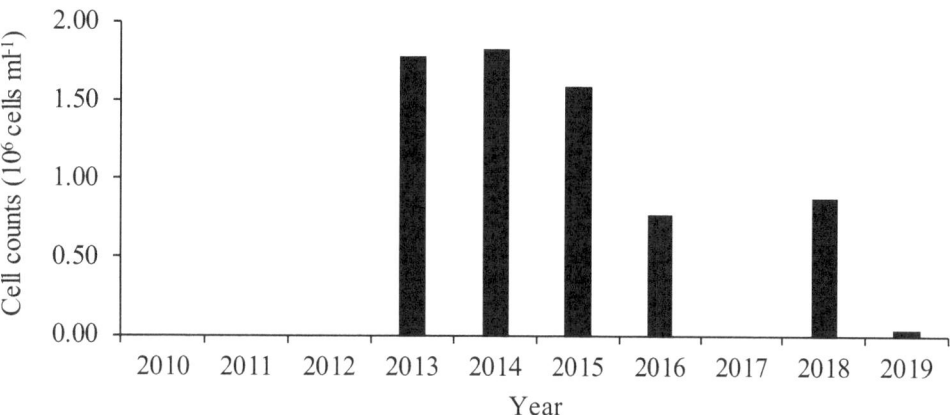

those >20 cells mL$^{-1}$, which was observed in 2019, 2020, and 2021, but not in the other years in the TSJE (Fig. 6a). *P. micans* are frequently measured in TSJE but never at more than 4 cells mL$^{-1}$ (Fig. 6b). *P. texanum* was described in Texas for the first time in Port Aransas as a toxin (okadaic acid) producing dinoflagellate species with two morphologies (Henrichs et al. 2013b). One of the two morphologies (*P. texanum* var. *cuspidatum*) is difficult to distinguish from *P. micans*, particularly when using microscopes or the IFCB. *Prorocentrum* spp. (a complex mixture of *P. cordatum* and *P. texanum*) was observed in TSJE at different times (Fig. 6c), including after the Texas City "Y" oil spill (Lee et al. 2016; Park et al. 2020).

### Pyrodinium bahamense

Another potential HAB species, the dinoflagellate *Pyrodinium bahamense* (unpubl. Texas Parks & Wildlife Department reports), has been observed in Baffin Bay and Corpus Christi Bay. In 2010, a large fish kill was observed that coincided with both hypoxia and a bloom of *P. bahamense* (unpubl. Texas Parks & Wildlife Department reports). *P. bahamense* can produce saxitoxin under certain conditions (Phlips et al. 2015) but is more frequently known for causing low dissolved oxygen events that have led to fish kills elsewhere (e.g., Morrison and Greening 2011). Recent studies suggest that *P. bahamense* is typically most competitive under relatively high nutrient input conditions (Phlips et al. 2015), and bloom formation is favored in the summer (Lopez et al. 2021).

### Aureoumbra lagunensis DeYoe and Stockwell

*Aureoumbra lagunensis* occurs in the Gulf of Mexico from Texas (Buskey et al. 2001; Cira and Wetz 2019) to the east coast of Florida (Lopez et al. 2021). These blooms can occur in such high densities that they shade out sea grass beds

(Onuf 1996) and negatively impact grazers and benthic invertebrates (Anderson et al. 2021). Although these blooms are disruptive to ecosystems, they are harmless to humans. The first *A. lagunensis* brown tide bloom began in shallow lagoons in and around the Laguna Madre and Baffin Bay, Texas, in January 1990 and persisted for almost 8 years, making it the longest continuous HAB event on record (Buskey et al. 2001). Reporting of blooms was sporadic from 1999 to 2009, with at least one bloom occurring during that period (Wetz et al. 2017). The Wetz laboratory has surveyed Baffin Bay annually since 2013 and regularly observed *A. lagunensis* in shallow, protected areas of the bay, as well as higher-density blooms ($10^5$–$10^6$ cells mL$^{-1}$, Fig. 7), most often during the summer of dry years (Cira and Wetz 2019). Over the past decade, the geographic extent of *A. lagunensis* has greatly expanded in South Texas, with high densities occurring in the Indian River Lagoon of Florida and Guantanamo Bay in Cuba (see references in Anderson et al. 2021), but not in TSJE (Steichen, pers. obs). Given *A. lagunensis* blooms have been associated with long residence time conditions and salinities > ~25 (Cira and Wetz 2019; Cira et al. 2021; Lopez et al. 2021), its inability to establish itself in TSJE is thought to be related to the high freshwater inflows (magnitude, frequency) which create a lack of suitable growth conditions.

### Primary Consumers

Zooplankton have relatively short generation times and may respond rapidly to changes in phytoplankton populations associated with inflow and therefore are indicators of water properties and sensitive to climate change and ecosystem shifts (Buskey 1989; Richardson 2008). In the northern Gulf of Mexico, zooplankton comprise a diverse assemblage of copepods, larvaceans, and various meroplanktonic larvae (Dagg 1995). Along with benthic organisms, zooplankton are the primary consumers of phytoplankton produced in estuaries and organic material transported by river flow.

Holoplankton spend their entire life cycle as members of the plankton, while meroplankton spend only part of their life cycle as eggs and larvae in the zooplankton and then either settle as benthic organisms or to develop into fish and shellfish. They are preyed upon by either larger zooplankton or by larval and small fish and invertebrates (Govoni et al. 1983; Minello et al. 1987). In this way, zooplankton provide an essential link in the transfer of energy from organic matter to fish in marine ecosystems. Because of this key role of zooplankton in the early life history of fish, changes in their species composition and abundance have significant implications for the recruitment and dynamics of fish.

In general, zooplankton respond to increased levels of phytoplankton production or organic input resulting from freshwater inflow by increased productivity. The zooplankton populations within Texas estuaries vary temporally and spatially, but there are some predictable changes in zooplankton populations with salinity. Buskey (1989) showed that after times of high freshwater inflows there would be an increase in the zooplankton populations. However, reductions in zooplankton abundance occur after large inflow events resulting from their physical displacement by the large volume of inflowing water (Holland et al. 1975; Jones et al. 1986; Armstrong 1987; Buskey 1989; Liu et al. 2017). If freshwater inflow events are strong enough to mix benthic sediments into the water column, it is speculated that there is a possibility the diapause eggs in the sediments that were previously released by *Acartia tonsa* (a common euryhaline calanoid copepod) may be stimulated by the low salinity water to hatch (Buskey 1989; Marcus 1984). Estuarine zooplankton may be replaced by freshwater species that persist until the salinity increases and freshwater species die from osmotic effects or predation, or they may return in high numbers (Buskey 1989). *A. tonsa* adults have been observed to be the lowest during maximum inflow events, sharply increasing once salinities increase (Kalke 1981). The success of this species after peak inflows is attributed to the influx of nutrients, food, and a decrease in higher salinity predators (Kalke 1981).

Zooplankton vulnerability to disturbance is classified as high, due to overlapping spatial distributions and limited avoidance capacity, but their resilience in these scenarios is also considered high because of their fecundity and connectivity between populations. In the case of the Deepwater Horizon oil spill (2010), zooplankton were found to be at risk because of their consumption of oil droplet-sized particles and laboratory findings which demonstrated lethal/sublethal effects of oil/dispersant, but populations were not observed in the Gulf of Mexico (reviewed in Daly et al. 2021). We were not able to find publications about their responses to natural disturbances such as flooding and tropical storms or hurricanes; we hypothesize it would be similar to that described for phytoplankton above.

## Microzooplankton (0.02–0.2 mm)

Microzooplankton (primarily ciliates and tintinnids) dominate the zooplankton community in the Gulf estuaries (Buskey 1993), and they may provide a large proportion of the diet of mesozooplankton. Between 1972 and 1975, zooplankton populations in the Nueces-Corpus Christi and Copano-Aransas Bay systems were studies by Kalke (1981) in response to freshwater inflows. It was found that estuarine standing groups accounted for a larger fraction of the community after a freshwater inflow event, but this was dependent on the magnitude of the inflow. Buskey (1989) compared the abundance and biomass of microzooplankton in the Nueces and Guadalupe estuaries to find that the average abundance ranged from 28.9 to 60.8 million individuals $m^{-3}$. In the Nueces Estuary in 1987, zooplankton production averaged 1.9 mg C $m^{-3}$ $day^{-1}$ (range 0.05–56 mg C $m^{-3}$ $day^{-1}$) (Buskey 1989). It was also reported that average potential secondary production of *A. tonsa* was similar in San Antonio Bay, the Nueces, and Guadalupe estuaries (Kalke 1981; Longley 1994). No similar such data is available to our knowledge for Trinity-San Jacinto Estuary. What is known from more recent studies is that the total zooplankton abundance averaged $52.67 \pm 9.46$ individuals $m^{-3}$ in 2008–2009 (Liu et al. 2017). Since *A. tonsa* often represents a large fraction of the zooplankton distributed worldwide in nearshore marine environments including those in the Gulf of Mexico and the Texas coast, several studies have attempted to relate its abundance with environmental parameters including season, temperature, salinity, and inflow (Buskey and Schmidt 1992; Escamilla et al. 2011; Liu et al. 2017; Bi and Liu 2017 and references therein).

## Mesozooplankton (0.2–2 mm) and Macrozooplankton (>2 mm)

Mesozooplankton population changes with inflow were reported for the Lavaca-Colorado Estuary in Longley (1994). Their abundance varied from $715 \pm 1051$ individuals $m^{-3}$ in 1984 and 1985 at stations closest to the river mouth to $4128 \pm 3612$ individuals $m^{-3}$ in the middle of the bay on average. Studies in the Lavaca Bay and Guadalupe Estuary reported in Longley (1994) found that lower bay sites had higher macroplankton abundances than locations in or very close to the deltas, while in the Nueces, this pattern was not observed. Season was also found to significantly affect the findings, with the highest abundance of zooplankton generally in the winter. Further, mesozooplankton abundance in Lavaca Bay and the Guadalupe Estuary was found to be significantly lower ($P < 0.01$) during the high-inflow years than during the low-inflow years, but higher abundances were observed in the Nueces Estuary during high flows (Longley

1994). These findings were thought to be related to the relative residence time of these systems. Longley (1994) concluded that freshwater inflows seem to "stimulate" micro- and mesozooplankton population growth, although the point at which displacement overbalances population increase is not clear.

The mesozooplankton population in Texas estuaries is commonly dominated by *A. tonsa* joined by *Oithona* sp., larval polychaetes, *Pseudodiaptomus coronatus*, and *Paracalanus crassirostris* (McAden 1977; Buskey 1989; Buskey and Schmidt 1992). Across TSJE, zooplankton abundance ranged from 3000 to 20,000 individuals m⁻³ (Buskey and Schmidt 1992). This study showed that at the time zooplankton abundances were lower in TSJE than many other Texas estuaries (Buskey and Schmidt 1992). In lab studies, it was shown that *A. tonsa* would die at faster rates when fed only *A. lagunensis* compared to when they were starved showing a toxin effect of this HAB species. Zooplankton play an integral role in the ecology of marine systems by connecting the phytoplankton and higher trophic levels such as nekton and benthos (Buskey 1993). The *A. lagunensis* formed a bloom in Laguna Madre that persisted for over 7 years (Buskey and Hyatt 1995). The persistence of *A. lagunensis* could be detrimental to the mesozooplankton community as they will not utilize that species as a food source. If the bloom is monospecific with *A. lagunensis*, this could lead a sharp decline in the mesozooplankton community in turn leading to less food for larval fishes, etc. (Buskey and Hyatt 1995). This could lead to a negative cascading effect on higher trophic levels.

Historically, there appear to be few studies conducted in TSJE. These include those examining sampling variability of zooplankton at a fixed station in West Bay (Minello and Matthews 1981), community structure of macroplankton in Trinity and Upper Bays (Holt and Strawn 1983), and bay wide species diversity in terms of water quality (Copeland and Bechtel 1971). In a more recent study of TSJE, seasonal variations of mesozooplankton species composition, abundance, and diversity in 2008 and 2009 in relation to environmental factors were examined. Liu et al. (2017) found copepod nauplii, *Acartia tonsa*, *Paracalanus* spp., and *Oithona* spp. were the numerically dominant taxa in samples collected during the day. Zooplankton abundance was significantly and positively correlated to seawater temperature, whereas their diversity was positively related to salinity (Liu et al. 2017). The study also reported that tidal advection of coastal ocean water likely carries diverse copepod species into this estuary.

By contrast, much more work has been done in the lower Texas coast (see Buskey 1989, 1993). Zooplankton samples in the Nueces Estuary collected from September 1987 to October 1988 revealed that microzooplankton were extremely abundant with ciliates (including both aloricate ciliates and tintinnids) ranging in abundance from 5000 to 400,000 individuals m⁻³, with a mean of 38,000 individuals m⁻³. Mesozooplankton abundance averaged 6100 individuals m⁻³ for samples collected during the day and 10,100 individuals m⁻³ for samples collected at night. Mesozooplankton were dominated by *Acartia tonsa* (~50% of total). The copepod *Acartia tonsa* comprised 40–60% of the individual mesozooplankton in the Lavaca-Colorado, Guadalupe, Mission-Aransas, and Nueces estuaries (Holland et al. 1975; Jones et al. 1986; Buskey 1989) and TSJE (Liu et al. 2017). Armstrong (1987) noted that *A. tonsa* contributed to 85% of the standing crop in Sabine Lake, dominated the zooplankton in the upper and lower Laguna Madre, and with barnacle nauplii larvae constituted more than 70% of the standing crop in Trinity Bay. In TSJE, *A. tonsa* blooms in the spring with a second but smaller bloom in the fall (Bi and Liu 2017), closely following phytoplankton bloom dynamics. These temporal dynamics are also driven by changes in temperature. At spatial scales, the high salinity in the lower bays benefits *A. tonsa* populations which has been confirmed by in situ observations that abundance appeared to be higher in the more saline portions of estuaries closer to the Gulf of Mexico than in the upper fresher areas (Longley 1994: Liu et al. 2017).

## Synthesis and Recommendations

A major challenge for determining patterns and trends in plankton communities and bloom events, both through time and in relation to freshwater inflow variability, is that the datasets available at the local, state, and national levels are not in sync in terms of years of coverage, parameters measured, and the spatial-temporal frequency of sampling. In the case of HABs, for example, assessments in Texas estuaries typically occur in response to an event, typically a fish kill or significant discoloration of a water body, with no way of knowing what the antecedent conditions were that contributed to the bloom and little follow-on sampling. Considering the severity of the impacts of HAB events, detection and reporting of HAB species could benefit from more sensitive and sustained measurements than the methods currently employed by state agencies. The use of an Imaging FlowCytobot (IFCB) that can take images of cells allowing us to later identify waterborne cells that pass through it while simultaneously communicating its findings to an electronic database is one option, perhaps coupled with other water quality sensors. Investigator-led programs that measure the changes in other plankton species are dependent on sporadic funding, and so a comprehensive record is not available. This problem is not unique to Texas, but it does curtail the longitudinal assessment of plankton communities. Given the pivotal role plankton play in supporting food webs, a state-wide sampling effort, which uses the same protocols and sampling frequency to examine plankton communities across Texas bays and estuaries is encouraged.

Recent studies of Texas bays and estuaries found that higher base inflow rates and/or pulsed inflow events may lead to increased diversity in the phytoplankton community (Buyukates and Roelke 2005; Roelke et al. 2013, 2017; Chin et al. 2022). While freshwater inflows act as a disturbance that increases phytoplankton diversity (Roelke et al. 2013), floods may initially lead to losses and significant changes in community structure (Steichen et al. 2020; section "Introduction"). In those estuaries which experience much lower inflows and for prolonged periods, diversity declines and blooms may be observed. This was the case with the recurring monospecific blooms of *A. lagunensis* in Baffin Bay since 1990 (Buskey et al. 2001; Cira et al. 2021) and dinoflagellate blooms in TSJE (Quigg et al. 2019; Quigg and Steichen 2021). In the Mission-Aransas Estuary, Reyna et al. (2017) observed a shift in the dominant phytoplankton group from diatoms to cyanobacteria immediately following a freshwater inflow event after prolonged drought. The diatoms took advantage of the increased dissolved inorganic nitrogen available immediately following the freshwater pulse and then months later the community shifted to being dominated by cyanobacteria (Reyna et al. 2017). Reviewing these various case studies shows that the different estuary systems along the coast are unique and highly dynamic. The phytoplankton and subsequent zooplankton response to high and low freshwater inflows exhibit both inter- and intra-variabilities across each estuary making it difficult to determine and describe overall trends along the Texas coast. In terms of management strategies, this needs to be assessed on an individual estuary basis.

While in TSJE, phytoplankton species known to be HABs are frequently observed in daily samples, these rarely if ever were present in numbers high enough to be considered harmful or sufficient to require closures of the fisheries (Quigg and Steichen 2021). In contrast, Port Aransas is the location in which *Dinophysis ovum*, *Karenia brevis*, and other known HABs are not only present in sufficient numbers to cause alerts to communities by agencies (e.g., TDSHS, TPWD), but sometimes blooms occur in such high densities that tourism and the local economy are also impacted. *A. lagunensis* is an example of a species that forms significant blooms which have disrupted ecosystem services in Baffin Bay. We hypothesize that blooms are more frequent and severe in the estuaries located along the central-southern portion of the Texas coast because these estuaries have longer residence times in addition to growing issues with nutrient loading. In contrast, despite similar if not higher nutrient loading in TSJE and nearby, bloom levels of these species are not observed, perhaps because the higher freshwater inflows act as a deterrent to the development of blooms.

Less is known about zooplankton in Texas bays and estuaries despite the important role they play in food webs. Future studies should include an analysis of their community composition as well as the major factors which determine their viability and response to environmental perturbations.

Water resource managers must be able to meet current and future societal demands of freshwater while still meeting inflow criteria to maintain beneficial inflows for bays and estuaries as part of an adaptive management of the environmental flow standards for Texas. Defining "beneficial" flows is a critical step to understanding healthy ecosystems. As a result of the Senate Bill 3 process to determine freshwater inflow needs in Texas bays and estuaries, the Bays and Basins Expert Science Teams (go to www.twdb.texas.gov for full details) developed a list of potential bioindicator species placing an emphasis on sessile species (eastern oyster *Crassostrea virginica*; Atlantic rangia, *Rangia cuneata*), fish (e.g., Blue catfish, *Ictalurus furcatus*; Gulf menhaden, *Brevoortia patronus*; Pinfish, *Lagodon rhomboides*; Atlantic croaker, *Micropogonias undulatus*; Southern flounder, *Paralichthys lethostigma*), and crabs (blue crabs, *Callinectes sapidus*). Freshwater inflows are however also associated with salinity, turbidity, and dissolved nutrients which are important for plankton populations (biomass, diversity, frequency, fecundity). Nonetheless, none of the inflow standards developed made use of plankton to set inflow criteria. Given their ability to respond at the community level on fast time-scales to freshwater inflows, they are important indicators of environmental quality. The appearance of HABs is a signal of ecosystem stress and habitat degradation, particularly when concurrent fish kill events occur and/or persistent multi-year blooms occur. Recommendations for resource managers are the monitoring of plankton populations along with other water quality parameters, particularly nutrients. This will be key to defining "beneficial" flows in Texas and is the next critical step to understanding and preserving healthy ecosystems.

**Acknowledgments** This publication was funded in part through contract no. 21-155-007-C879 from the Texas General Land Office (GLO) with Gulf of Mexico Energy Security Act of 2006 funding made available to the State of Texas and awarded under the Texas Coastal Management Program. The views contained herein are those of the authors and should not be interpreted as representing the views of the GLO or the State of Texas. Funding for the Baffin Bay data collections was provided from the following sources to MSW: a grant (award no. NA14OAR4170102) from the Texas Sea Grant College Program, grants from the GLO/CMP (award numbers NA14NOS4190139, NA17NOS4190139), a contract (21-060-017-C677) from the GLO with Gulf of Mexico Energy Security Act of 2006 funding, Celanese Corporation, Kleberg County, Coastal Conservation Association, Saltwater Fisheries Enhancement Association, and the Coastal Bend Bays & Estuaries Program. Data was collected with the assistance of 18 dedicated "citizen scientists." Funding for the Baffin Bay-Nueces Bay-San Antonio Bay phytoplankton comparison study was provided to MSW by the Texas Water Development Board (contract no.

1800012228). We thank Jessica Hillhouse (Texas A&M University at Galveston) for her help with collating the data and posting it on GRIIDC (doi: 10.7266/CTVAS53B, doi: 10.7266/QOASPQFS) and Ram Neupane (Texas Water Development Board) for sharing the freshwater inflow data. The water quality data from 2010 to 2020 was collected and managed by TCEQ and can be found using their Surface Water Quality data viewer at https://www80.tceq.texas.gov/SwqmisPublic/index.htm. This dataset is publicly available and therefore not entered into GRIIDC.

**Data Availability** The descriptive information can be viewed on the GRIIDC dataset landing page: https://data.gulfresearchinitiative.org/data/HI.x962.000:0001.

Hillhouse, Jessica, Antonietta Quigg, and Jamie Steichen. 2022. Microbial, physical, and chemical data collected from Galveston Bay research cruises from 2010-01-18 to 2019-11-16. Distributed by: Gulf of Mexico Research Initiative Information and Data Cooperative (GRIIDC), Harte Research Institute, Texas A&M University–Corpus Christi. doi: 10.7266/CTVAS53B (https://data.gulfresearchinitiative.org/data/HI.x962.000:0001). Published June 2022.

Hillhouse, Jessica, Antonietta Quigg, and Jamie Steichen. 2022. Phytoplankton pigment and secci data collected from Galveston Bay research cruises from 2010-01-18 to 2019-11-16. Distributed by: Gulf of Mexico Research Initiative Information and Data Cooperative (GRIIDC), Harte Research Institute, Texas A&M University–Corpus Christi. doi: 10.7266/Q0ASPQFS (https://data.gulfresearchinitiative.org/data/HI.x962.000:0002). Published February 2022.

Jamie Steichen. Galveston Bay—IFCB harmful algal bloom species. Distributed by: Gulf of Mexico Research Initiative Information and Data Cooperative (GRIIDC), Harte Research Institute, Texas A&M University–Corpus Christi. doi: 10.7266/emmc0zdt (https://data.gulfresearchinitiative.org/data/HI.x962.000:0008). Published January 2023.

The rest of the data we used is available in the SWQM dataset from TCEQ.

**Coastal Hydrology Data Disclaimer** With respect to data, documents, or information provided, neither the TWDB nor any of their employees, contractors, subcontractors, or their employees makes any warranty, express, or implied, including the warranties of merchantability and fitness for a particular purpose, or assumes any legal liability or responsibility for the accuracy, completeness, or usefulness of any information, product, or process disclosed or represents that its use would not infringe privately owned rights.

# References

Al-Najjar T, Badran MI, Richter C, Meyerhoefer M, Sommer U (2007) Seasonal dynamics of phytoplankton in the Gulf of Aqaba, Red Sea. Hydrobologia 579:69–83

Ammerman J (2018) Phosphorus and estuaries. Long Island Sounds Study. https://longislandsoundstudy.net/2018/10/phosphorus-and-estuaries/

Anderson DM, Burkholder JM, Cochlan WP, Glibert PM, Gobler CJ, Heil CA, Kudela RM, Parsons ML, Rensel JJ, Townsend DW, Trainer VL (2008) Harmful algal blooms and eutrophication: Examining linkages from selected coastal regions of the United States. Harmful Algae 8:39–53. https://doi.org/10.1016/J.HAL.2008.08.017

Anderson DM, Fensin E, Gobler CJ, Hoeglund AE, Hubbard KA, Kulis DM, Landsberg JH, Lefebvre KA, Provoost P, Richlen ML, Smith JL, Solow AR, Trainer VL (2021) Marine harmful algal blooms (HABs) in the United States: History, current status and future trends. Harmful Algae 102:101975. https://doi.org/10.1016/j.hal.2021.101975

Armstrong NE (1987) The ecology of open bay bottoms of Texas: a community profile. US Fish Wildlife Service Biology Rep 85, 104 pp

Azevedo IC, Bordalo AA, Duarte P (2014) Influence of freshwater inflow variability on the Douro estuary primary productivity: A modelling study. Ecol Model 272:1–15. https://doi.org/10.1016/j.ecolmodel.2013.09.010

Badylak S, Philips EJ, Mathews AL, Kelley K (2014) *Akashiwo sanguinea* (Dinophyceae) extruding mucous from pores on the cell surface. Algae 29:197–201

Beecraft L, Wetz MS (2023) Temporal variability in water quality and phytoplankton biomass in a low-inflow estuary (Baffin Bay, Texas). Estuaries Coasts 46:2064–2075. https://doi.org/10.1007/s12237-022-01145-y

Berdalet E, Fleming LE, Gowen R, Davidson K, Hess P, Backer LC, Moore SK, Hoagland P, Enevoldsen H (2016) Marine harmful algal blooms, human health and wellbeing: Challenges and opportunities in the 21st century. J Mar Biol Assoc UK 96:61–91. https://doi.org/10.1017/S0025315415001733

Bi R, Liu H (2017) Effects of variability among individuals on zooplankton population dynamics under environmental conditions. Mar Ecol Progr Ser 564:9–28

Boynton WR, Kemp WM, Keefe CW (1982) A comparative analysis of nutrients and other factors influencing estuarine phytoplankton production. In: Kennedy VS (ed) Estuarine comparisons. Academic, New York, pp 69–91

Brand LE, Compton A (2007) Long-term increase in *Karenia brevis* abundance along the Southwest Florida Coast. Harmful Algae 6:232–252

Brett MT, Müller-Navarra DC, Persson J (2009) Crustacean zooplankton fatty acid composition. In: Arts MT, Brett MT, Kainz M (eds) Lipids in aquatic ecosystems. Springer, New York, pp 115–146

Bricker SB, Longstaff B, Dennison W, Jones A, Bolcourt K, Wick C, Woener J (2008) Effects of nutrient enrichment in the nation's estuaries: A decade of change. Harmful Algae 8:21–32. https://doi.org/10.1016/j.hal.2008.08.028

Bronk DA, See JH, Bradley P, Killberg L (2007) DON as a source of bioavailable nitrogen for phytoplankton. Biogeosciences 4:283–296. https://doi.org/10.5194/bg-4-283-2007

Bugica K, Sterba-Boatwright B, Wetz MS (2020) Water quality trends in Texas estuaries. Mar Pollut Bull 152:110903. https://doi.org/10.1016/j.marpolbul.2020.110903

Burkholder JM, Shumway SE, Glibert PM (2018) Food web and ecosystem impacts of harmful algae. In: Shumway SE, Burkholder JAM, Morton SL (eds) Harmful algal blooms: a compendium desk reference. Wiley Blackwell, Chichester, pp 243–336

Buskey EJ (1989) Effects of freshwater inflow on the zooplankton of Texas coastal bays. Report to Texas Water Development Board, by Marine Science Institute, University of Texas at Austin, Port Aransas, TX, p 212

Buskey EJ (1993) Annual pattern of micro- and mesozooplankton abundance and biomass in a subtropical estuary. J Plankton Res 15:907–924

Buskey EJ, Hyatt CJ (1995) Effects of the Texas (USA) "brown tide" alga on planktonic grazers. Mar Ecol Prog Ser 126:285–292. https://doi.org/10.3354/meps126285

Buskey EJ, Schmidt K (1992) Characterization of plankton from the Galveston estuary. Status and Trends of Selected Living Resources in the Galveston Bay System. Galveston Bay Natl Estuary Prog 19:347–375

Buskey EJ, Montagna PA, Amos AF, Whitledge TE (1997) Disruption of grazer populations as a contributing factor to the initiation of the Texas brown tide algal bloom. Limnol Oceanogr 42:1215–1222. https://doi.org/10.4319/LO.1997.42.5_PART_2.1215

Buskey EJ, Liu H, Collumb C, Bersano JGF (2001) The decline and recovery of a persistent Texas brown tide algal bloom in the

Laguna Madre (Texas, USA). Estuaries 24:337–346. https://doi.org/10.2307/1353236

Buyukates Y, Roelke DL (2005) Influence of pulsed inflows and nutrient loading on zooplankton and phytoplankton community structure and biomass in microcosm experiments using estuarine assemblages. Hydrobiologia 548:233–249

Campbell L, Olson RJ, Sosik HM, Abraham A, Henrichs DW, Hyatt CJ, Buskey EJ (2010) First harmful Dinophysis (Dinophyceae, Dinophysiales) bloom in the US is revealed by automated imaging flow cytometry. J Phycol 46:66–75

Campbell L, Henrichs DW, Olson RJ, Sosik HM (2013) Continuous automated imaging-in-flow cytometry for detection and early warning of Karenia brevis blooms in the Gulf of Mexico. Environ Sci Pollut Res 20:6896–6902

Carlin JA, Lee G-H, Dellapenna TM, Laverty P (2016) Sediment resuspension by wind, waves, and currents during meteorological frontal passages in a micro-tidal lagoon. Estuar Coast Shelf Sci 172:24–33. https://doi.org/10.1016/j.ecss.2016.01.029

Carstensen J, Henriksen P, Heiskanen A-S (2007) summer algal blooms in shallow estuaries: definition, mechanisms, and link to eutrophication. Limnol Oceanogr 52:370–384

Chang H (2008) Spatial analysis of water quality trends in the Han River basin, South Korea. Water Res 42:3285–3304

Chang H, Thiers P, Netusil NR, Yeakley JA, Rollwagen-Bollens G, Bollens SM, Singh S (2014) Relationships between environmental governance and water quality in a growing metropolitan area of the Pacific Northwest, USA. Hydrol Earth Syst Sci 18:1383–1395

Chapman RL (2013) Algae: the world's most important "plants"—an introduction. Mitig Adapt Strat Global Change 18:5–12

Chin T, Beecraft L, Wetz M (2022) Phytoplankton biomass and community composition in three Texas estuaries differing in freshwater inflow regime. Estuar Coast Shelf Sci 277:108059. https://doi.org/10.1016/J.ECSS.2022.108059

Cira EK, Wetz MS (2019) Spatial-temporal distribution of Aureoumbra lagunensis "brown tide" in Baffin Bay, Texas. Harmful Algae 89:101669. https://doi.org/10.1016/j.hal.2019.101669

Cira EK, Palmer TA, Wetz MS (2021) Phytoplankton dynamics in a low-inflow estuary (Baffin Bay, Texas) during drought and high-rainfall conditions associated with an El Niño event. Estuaries Coasts 44:1752–1764. https://doi.org/10.1007/s12237-021-00904-7

Clement C, Bricker SB, Pirhalla DE (2001) Eutrophic conditions in estuarine waters. In: NOAA's State of the Coast Report. National Oceanic and Atmospheric Administration, Silver Spring, MD. http://state-of-coast.noaa.gov/ bulletins/html/eut_18/eut.html

Cloern JE, Dufford R (2005) Phytoplankton community ecology: Principles applied in San Francisco Bay. Mar Ecol Prog Ser 285:11–28. https://doi.org/10.3354/meps285011

Copeland BJ, Bechtel TJ (1971) Species diversity and water quality in Galveston Bay, Texas. Water Air Soil Pollut 1:89–105

Cosgrove J, Borowitzka MA (2010) Chlorophyll fluorescence terminology: an introduction. In: Sugget DJ, Prasil O, Borowitzka MA (eds) Chlorophyll a fluorescence in aquatic sciences: methods and applications. Springer, New York, pp 1–18

Dagg MJ (1995) Ingestion of phytoplankton by the microzooplankton and mesozooplankton communities in a productive subtropical estuary. J Plankton Res 17:845–857

Daly KL, Remsen A, Outram O, Broadbent H, Kramer K, Dubikas K (2021) Resilience of the zooplankton community in the Northeast Gulf of Mexico during and after the Deepwater Horizon oil spill. Mar Pollut Bull 163:111882. https://doi.org/10.1016/j.marpolbul.2020.111882

Deeds JR, Wiles K, Heideman VIGB, White KD, Abraham A (2010) First U.S. report of shellfish harvesting closures due to confirmed okadaic acid in Texas Gulf coast oysters. Toxicon 55:1138–1146

Depauw FA, Rogato A, Ribera d'Alcalá M, Falciatore A (2012) Exploring the molecular basis of responses to light in marine

diatoms. J Exp Bot 63:1575–1591. https://doi.org/10.1093/jxb/ers005

Dorado S, Booe T, Steichen J, McInnes AS, Windham R, Shepard A, Lucchese AEB, Preischel H, Pinckney JL, Davis SE, Roelke DL, Quigg A (2015) Towards an understanding of the interactions between freshwater inflows and phytoplankton communities in a subtropical estuary in the Gulf of Mexico. PLoS One 10:e0130931. https://doi.org/10.1371/journal.pone.0130931

Escamilla BJ, Ordóñez-López U, Suárez-Morales E (2011) Spatial and seasonal variability of Acartia (Copepoda) in a tropical coastal lagoon of the southern Gulf of Mexico. Rev Biol Mar Oceanogr 46:379–390

Fire SE, Wang Z, Byrd M, Whitehead HR, Paternoster J, Morton SL (2011) Co-occurrence of multiple classes of harmful algal toxins in bottlenose dolphins (Tursiops truncatus) stranding during an unusual mortality event in Texas, USA. Harmful Algae 10:330–336. https://doi.org/10.1016/j.hal.2010.12.001

Flemer DA, Champ MA (2006) What is the future fate of estuaries given nutrient over enrichment, freshwater diversion and low flows? Mar Pollut Bull 52:247–258

Fleming LE, Backer L, Rowan A (2002) The epidemiology of human illness associated with HABs. In: Massaro E (ed) Handbook of neurotoxicology. Humana, Clifton, NJ, pp 363–381

Freeman LA, Corbett DR, Fitzgerald AM, Lemley DA, Quigg A, Steppe CN (2019) Impacts of urbanization and development on estuarine ecosystems and water quality. Estuaries Coasts 42:1821–1838

Gaulke AK, Wetz MS, Paerl HW (2010) Picophytoplankton: A major contributor to planktonic biomass and primary production in a eutrophic, river-dominated estuary. Estuar Coast Shelf Sci 90:45–54. https://doi.org/10.1016/j.ecss.2010.08.006

Glibert PM (2020) Harmful algae at the complex nexus of eutrophication and climate change. Harmful Algae 91:101583. https://doi.org/10.1016/j.hal.2019.03.001

Glibert PM, Anderson DM, Gentien P, Granéli E, Sellner KG (2005) The global, complex phenomena of harmful algal blooms. Oceanography 18:136–147. https://doi.org/10.5670/oceanog.2005.49

Glibert PM, Harrison J, Heil C, Seitzinger S (2006) Escalating worldwide use of urea—a global change contributing to coastal eutrophication. Biogeochemistry 77:441–463

Glibert PM, Boyer JN, Heil CA, Madden CJ, Sturgis B, Wazniak CS (2010) Blooms in lagoons: different from those of river-dominated estuaries. In: Kennish MJ, Paerl HW (eds) Coastal lagoons: critical habitats of environmental change. CRC, Boca Raton, FL, pp 91–113

Goleski JA, Koch F, Marcoval MA, Wall CC, Jochem FJ, Peterson BJ, Gobler CJ (2010) The role of zooplankton grazing and nutrient loading in the occurrence of harmful cyanobacterial blooms in Florida Bay, USA. Estuaries Coasts 33:1202–1215. https://doi.org/10.1007/s12237-010-9294-1

Govoni JJ, Hoss DE, Chester AJ (1983) Comparative feeding of three species of larval fishes in the northern Gulf of Mexico: Brevortiapatronus, Leiostomus xanthurus, and Micropogonias undulatus. Mar Ecol Prog Ser 13:189–199

Graham LE, Graham JM, Wilcox LW, Cook ME (2016) Algae, 3rd edn. LJLM, Upper Saddle River, NJ

Gustafson DE, Stoecker DK, Johnson MD, Van Heukelem WF, Sneider K (2000) Cryptophyte algae are robbed of their organelles by the marine ciliate Mesodinium rubrum. Nature 405:1049–1052

Guthrie CG, Schoenbaechler C, Matsumoto J, Lu Q (2012) TxBLEND Model calibration and validation for the Trinity-San Jacinto Estuary. Texas Water Development Board, Austin, TX, p 56

Hallegraeff GM (2003) Harmful algal blooms: a global overview. In: Hallegraeff GM, Anderson DM, Cembella AD (eds) Manual on harmful marine microalgae. UNESCO, Paris, pp 25–49

Hallegraeff G, Bolch C (2016) Unprecedented toxic algal blooms impact on Tasmanian seafood industry. Microbiol Aust 37:143–144

Hammer A, Schumann R, Schubert H (2002) Light and temperature acclimation of *Rhodomonas salina* (Cryptophyceae): photosynthetic performance. Aquat Microb Ecol 29:287–296

Harding LW (1994) Long-term trends in the distribution of phytoplankton in Chesapeake Bay: roles of light, nutrients and streamflow. Mar Ecol Prog Ser 104:267–291. https://doi.org/10.3354/meps104267

Hargraves PE (2011) Indian River Lagoon Species Inventory. https://www.sms.si.edu/irlspec/Kareni_brevis.htm. Retrieved 13 Mar 2019

Harper DE Jr, Guillen G (1989) Occurrence of a dinoflagellate bloom associated with an influx of low salinity water at Galveston, Texas, and coincident mortalities of demersal fish and benthic invertebrates. Contrib Mar Sci 31:147–161

Harred LB, Campbell L (2014) Predicting HABs: a case study with *Dinophysis ovum* in the Gulf of Mexico. J Plankton Res 36:1434–1445. https://doi.org/10.1093/plankt/fbu070

Henrichs DW, Renshaw MA, Gold JR, Campbell L (2013a) Population-genetic structure of the toxic dinoflagellate *Karenia brevis* from the Gulf of Mexico. J Plankton Res 35:427–432

Henrichs DW, Scott PS, Steidinger KA, Errera RM, Abraham A, Campbell L (2013b) Morphology and phylogeny of *Prorocentrum texanum* sp. nov. (Dinophyceae): A new toxic dinoflagellate from the Gulf of Mexico coastal waters exhibiting two distinct morphologies. J Phycol 49:143–155

Hogan DM, Jarnagin ST, Loperfido JV, Van Ness K (2014) Mitigating the effects of landscape development on streams in urbanizing watersheds. J Am Water Resour Assoc 50:163–178

Holland JS, Maciolek NJ, Kalke RD, Mullins L, Oppenheimer CH (1975) A benthos and plankton study of the Corpus Christi, Copano and Aransas bay systems III. Report on data collected during the period July, 1974–May, 1975 and summary of the three-year project. Report to Texas Water Development Board, by Marine Science Institute, University of Texas at Austin, Port Aransas, TX, 174 pp

Holt J, Strawn K (1983) Community structure of macrozooplankton in Trinity and upper Galveston Bays. Estuaries 6:66–75

Jäger CG, Diehl S, Schmidt GM (2008) Influence of water-column depth and mixing on phytoplankton biomass, community composition, and nutrients. Limnol Oceanogr 53:2361–2373. https://doi.org/10.4319/LO.2008.53.6.2361

Jessup DA, Miller MA, Ryan JP, Nevins HP, Kerkering HA, Mekebri A, Crane DB, Johnson TA, Kudela RM (2009) Mass stranding of marine birds caused by a surfactant-producing red tide. PLoS One 4(2):e4550. https://doi.org/10.1371/journal.pone.0004550

Johnson MD, Beaudoin DJ, Laza-Martinez A, Dyhrman ST, Fensin E, Lin S, Merculief A, Nagai S, Pompeu M, Setälä O, Stoecker DK (2016) The genetic diversity of *Mesodinium* and associated cryptophytes. Front Microbiol 7:2017. https://doi.org/10.3389/fmicb.2016.02017

Jones RS, Cullen JJ, Lane R, Yoon W, Rosson RA, Kalke RD, Holt SA, Arnold CR, Parker PL, Pulich WM, Scalan RS (1986) Studies of freshwater inflow effects on the Lavaca River delta and Lavaca Bay, Texas. Report to Texas Water Development Board, by Marine Science Institute, University of Texas at Austin, Port Aransas, TX. Technical Report Number TR/86-006, 423 pp

Kalke RD (1981) The effects of freshwater inflow on salinity and zooplankton populations at four stations in the Nueces Corpus Christi and Copano-Aransas Bay systems, Texas from October 1972-May 1975. In: Cross RD, Williams DL (eds) Proceedings of the national symposium on freshwater inflow to estuaries, vol I, pp 454–471

Kim H-C, Son S, Montagna P, Spiering B, Nam J (2014) Linkage between freshwater inflow and primary productivity in Texas estuaries: Downscaling effects of climate variability. J Coast Res 68:65–73

Kirk JTO (1994) Light and photosynthesis in aquatic ecosystems. Cambridge University Press, Cambridge

Kite-Powell HL, Fleming LE, Backer LC, Faustman EM, Hoagland P, Tsuchiya A, Younglove LR, Wilcox BA, Gast RJ (2008) Linking the oceans to public health: current efforts and future directions. Environ Health 7(Suppl 2):S6. https://doi.org/10.1186/1476-069X-7-82-S6

Lancelot C, Muylaert K (2011) Trends in estuarine phytoplankton ecology. In: Wolanski E, McLusky D (eds) Treatise on estuarine and coastal science. Elsevier, Amsterdam, pp 5–15

Lecher AL, Mackey KRM (2018) Synthesizing the effects of submarine groundwater discharge on marine biota. Hydrology 5:60. https://doi.org/10.3390/hydrology5040060

Lee TC-H, Fong FL-Y, Ho K-C, Lee FW-F (2016) The mechanism of diarrhetic shellfish poisoning toxin production in *Prorocentrum* spp. Physiological and molecular perspectives. Toxins 8:272. https://doi.org/10.3390/toxins8100272

Lester LJ, Gonzalez LA (2011) The state of the bay: a characterization of the Galveston Bay Ecosystem, 3rd edn. Texas Commission on Environmental Quality, Galveston Bay Estuary Program, Houston, TX. 356 pp

Lewitus AL, White DL, Tymowski RG, Geesey ME, Hymel SN, Noble PA (2005) Adapting the CHEMTAX method for assessing phytoplankton taxonomic composition in Southeastern U.S. Estuaries 28:160–172

Liu H, Zhang X, Yang Q, Zui T, Quigg A (2017) Mesozooplankton dynamics in relation to environmental factors and juvenile fish in a subtropical estuary of the Gulf of Mexico. J Coastal Res 33:1038–1050. https://doi.org/10.2112/JCOASTRES-D-16-00155.1

Longley WL (1994) Freshwater inflows to Texas bays and estuaries: ecological relationships and methods for determination of needs. Texas Water Development Board and Texas Parks and Wildlife Department, Austin, TX, 426 pp. https://www.twdb.texas.gov/publications/reports/other_reports/doc/FreshwaterInflowstoTexasBays.pdf

Lopez CB, Tilney CL, Muhlbach E, Bouchard JN, Villac MC, Henschen KL, Markley LR, Abbe SK, Shankar S, Shea CP, Flewelling L, Garrett M, Badylak S, Phlips EJ, Hall LM, Lasi MA, Parks AA, Paperno R, Adams DH, Edwards DD, Schneider JE, Wald KB, Biddle AR, Landers SL, Hubbard KA (2021) High-resolution spatiotemporal dynamics of harmful algae in the Indian River Lagoon (Florida)—a case study of *Aureoumbra lagunensis*, *Pyrodinium bahamense*, and *Pseudo-nitzschia*. Front Mar Sci 8:769877. https://doi.org/10.3389/fmars.2021.769877

Mackey MD, Mackey DJ, Higgins HW, Wright SW (1996) CHEMTAX-a program for estimating class abundances from chemical markers: application to HPLC measurements of phytoplankton. Mar Ecol Prog Ser 114:265–283

Magaña HA, Contreras C, Villareal TA (2003) A historical assessment of *Karenia brevis* in the western Gulf of Mexico. Harmful Algae 2:163–171

Mallin MA, Paerl HW, Rudek J, Bates PW (1993) Regulation of estuarine primary production by watershed rainfall and river flow. Mar Ecol Prog Ser 93:199–203. https://doi.org/10.3354/meps093199

Malone TC, Newton A (2020) The globalization of cultural eutrophication in the coastal ocean: Causes and consequences. Front Mar Sci 7:670

Marcus NM (1984) Recruitment of copepod nauplii into the plankton: importance of diapause eggs and benthic processes. Mar Ecol Prog Ser 15:47–54

Matthews GA, Marcin CA, Welch D (1975) A plankton and benthos survey of the San Antonio Bay system, March 1972-July 1974. Report to Texas Water Development Board, by Coastal Fisheries Branch, Texas Parks and Wildlife Department, Austin, TX, 75 pp

McAden DC (1977) Species composition, distribution and abundance of zooplankton (including icthyoplankton) in the intake and discharge canals of a steam-electric generating station located on Galveston Bay. M. S. thesis, Texas A&M University, College Station, TX

McFarlane R, Leskovskaya A, Lester J, Gonzalez L (2015) The effect of four environmental parameters on the structure of estuarine shoreline communities in Texas, USA. Ecosphere 6:1–9

McInnes A, Quigg A (2010) Near-annual fish kills in small embayments: Casual versus causal factors. J Coast Res 26:957–966. https://doi.org/10.2112/JCOASTRES-D-10-00006.1

Mended-Deuer S, Montalbano AL (2015) Bloom formation potential in the harmful dinoflagellate *Akashiwo sanguinea*: Clues from movement behaviors and growth characteristics. Harmful Algae 47:75–85. https://doi.org/10.1016/j.hal.2015.06.001

Minello TJ, Matthews GA (1981) Variability of zooplankton tows in a shallow estuary. Contrib Mar Sci 24:81–92

Minello TJ, Zimmerman RJ, Maninez EX (1987) Fish predation on juvenile brown shrimp, *Penaeus aztecus:* Effects of turbidity and substratum on predation rates. U.S. Natl Mar Fish Serv Fish Bull 85:59–70

Montagna PA, Stockwell DA, Kalke RD (1993) Dwarf surfclam Mulinia lateralis (say, 1822) populations and feeding during the Texas brown tide event. J Shellfish Res 12:433–442

Montagna PA, Palmer TA, Beseres Pollack J (2013a) Hydrological changes and estuarine dynamics. Springer, New York

Montagna PA, Palmer TA, Pollack JB (2013b) Conceptual model of estuary ecosystems. In: Hydrological changes and estuarine dynamics. Springer, New York, pp 5–22

Moore CM, Mills MM, Arrigo KR, Berman-Frank I, Bopp L, Boyd PW, Galbraith ED, Geider RJ, Guieu C, Jaccard SL, Jickells TD, La Roche J, Lenton TM, Mahowald NM, Marañón E, Marinov I, Moore JK, Nakatsuka T, Oschlies A, Saito MA, Thingstad TF, Tsudaand A, Ulloa O (2013) Processes and patterns of oceanic nutrient limitation. Nat Geosci 6:701–710. https://doi.org/10.1038/NGEO1765

Morrison G, Greening H (2011) Water quality. In: Yates KK, Greening H, Morrison G (eds) Integrating science and resource management in Tampa Bay, Florida. U.S. Geological Survey Circular 1348, 280 pp

Nielsen-Gammon JW, Banner JL, Cook BI, Tremaine DM, Wong CI, Mace RE, Gao H, Yang Z-L, Flores Gonzalez M, Hoffpauir R, Gooch T, Kloesel K (2020) Unprecedented drought challenges for Texas water resources in a changing climate: What do researchers and stakeholders need to know? Earth's Future 8. https://doi.org/10.1029/2020EF001552

Nixon SW (1995) Coastal marine eutrophication: a definition, social causes and future concerns. Ophelia 41:199–219

Olson RJ, Sosik HM (2007) A submersible imaging-in-flow instrument to analyze nano-and microplankton: Imaging FlowCytobot. Limnol Oceanogr Methods 5:195–203. https://doi.org/10.4319/LOM.2007.5.195

O'Neil JM, Davis TW, Burford MA, Gobler CJ (2012) The rise of harmful cyanobacteria blooms: The potential roles of eutrophication and climate change. Harmful Algae 14:313–334. https://doi.org/10.1016/j.hal.2011.10.027

Onuf CP (1996) Seagrass responses to long-term light reduction by brown tide in upper Laguna Madre, Texas: distribution and biomass patterns. Mar Ecol Prog Ser 138:219–231

Örnólfsdóttir EB, Lumsden SE, Pinckney JL (2004) Nutrient pulsing as a regulator of phytoplankton abundance and community composition in Galveston Bay, Texas. J Exp Mar Biol Ecol 303:197–220. https://doi.org/10.1016/j.jembe.2003.11.016

Paerl HW (1996) A comparison of cyanobacterial bloom dynamics in freshwater, estuarine and marine environments. Phycologia 35:25–35

Paerl HW, Bales JD, Ausley LW, Buzzelli CP, Crowder LB, Eby LA, Fear JM, Go M, Peierls BL, Richardson TL, Ramus JS (2001) Ecosystem impacts of three sequential hurricanes (Dennis, Floyd, and Irene) on the United States' largest lagoonal estuary, Pamlico Sound, NC. Proc Natl Acad Sci USA 98:5655–5660

Paerl HW, Valdes LM, Pinckney JL, Piehler MF, Dyble J, Moisander PH (2003) Phytoplankton photopigments as indicators of estuarine and coastal eutrophication. BioScience 53:953–964. https://doi.org/10.1641/0006-3568(2003)053[0953:PPAIOE]2.0.CO;2

Paerl HW, Valdes-Weaver LM, Joyner AR, Winkelmann V (2007) Phytoplankton indicators of ecological change in the eutrophying Pamlico sound system, North Carolina. Ecol Appl 17:S88–S101. https://doi.org/10.1890/05-0840.1

Paerl HW, Hall NS, Peierls BL, Le Rossignol K, Joyner AR (2013) Hydrologic variability and its control of phytoplankton community structure and function in two shallow, coastal, lagoonal ecosystems: The Neuse and New River Estuaries, North Carolina, USA. Estuaries Coasts 37:31–45. https://doi.org/10.1007/s12237-013-9686-0

Park BS, Erdner DL, Bacosa HP, Liu Z, Buskey EJ (2020) Potential effects of bacterial communities on the formation of blooms of the harmful dinoflagellate Prorocentrum after the 2014 Texas City "Y" oil spill (USA). Harmful Algae 95:101802. https://doi.org/10.1016/j.hal.2020.101802

Parrish C (2009) Essential fatty acids in aquatic food webs. In: Arts MT, Brett MT, Kainz MJ (eds) Lipids in aquatic ecosystems. Springer, New York, pp 309–326

Pedrós-Alió C, Gasol JM, Guerrero R (1987) On the ecology of a *Cryptomonas phaseolus* population forming a metalimnetic bloom in Lake Cisó, Spain: Annual distribution and loss factors. Limnol Oceanogr 32:285–298

Pennock JR, Sharp WM (1994) Temporal alternation between light-and nutrient-limitation of phytoplankton in a coastal plain estuary. Mar Ecol Prog Ser 111:275–288

Phlips EJ, Badylak S, Lasi MA, Chamberlain R, Green WC, Hall LM, Hart JA, Lockwood JC, Miller JD, Morris LJ, Steward JS (2015) From red tides to green and brown tides: bloom dynamics in a restricted subtropical lagoon under shifting climatic conditions. Estuaries Coasts 38:886–904

Pierce RH, Henry MS (2008) Harmful algal toxins of the Florida red tide (*Karenia brevis*): natural chemical stressors in South Florida coastal ecosystems. Ecotoxicology 17:623–631

Pinckney JL, Richardson TL, Millie DF, Paerl HW (2001) Application of photopigment biomarkers for quantifying microalgal community composition and in situ growth rates. Org Geochem 32:585–595. https://doi.org/10.1016/S0146-6380(00)00196-0

Pinckney JL, Quigg A, Roelke DL (2017) Interannual and seasonal patterns of estuarine phytoplankton diversity in Galveston Bay, Texas, USA. Estuaries Coasts 40:310–316. https://doi.org/10.1007/s12237-016-0135-8

Pithart D (1997) Diurnal vertical migration study during a winter bloom of cryptophyceae in a floodplain pool. Int Rev Ges Hydrobiol 82:33–46

Poulson-Ellestad K, Mcmillian E, Montoya JP, Kubanek J (2014) Are offshore phytoplankton susceptible to *Karenia brevis* allelopathy? J Plankt Res 36:1344–1356

Quigg A (2015) Sources, fate, transport and effects of nutrients on downstream ecological processes in the Galveston Bay estuary. Final report to US Environmental Protection Agency Cooperative agreement MX954527, p 107

Quigg A, Steichen J (2021) Raising awareness of harmful (and/or toxic) algae blooms (HABs) in Galveston Bay and their impacts. Final report of the Texas Parks and Wildlife Department Contract No. 528672, p 44

Quigg A, Roelke DF, Davis SE (2009) Freshwater inflows and the health of Galveston Bay: influence of nutrient and sediment load on the base of the food web. Final Report of the Coastal Coordination Council pursuant to National Oceanic and Atmospheric Administration Award No. NA07NOS4190144, p 54

Quigg A, Steichen J, Windham R (2017) Galveston Bay: changing land use patterns and nutrient loading. Causal or casual relationship with Water Quality, Quantity, and Patterns? Prepared in cooperation with the Galveston Bay Estuary Program a program of the Texas Commission on Environmental Quality and the U.S. Environmental Protection Agency TCEQ Contract # 582-15-53393, Federal ID #CE-00655005, p 55

Quigg A, Steichen J, Windham R (2019) Freshwater inflows in Galveston Bay: relationship to (harmful) algal blooms (HABs). Prepared in cooperation with the Galveston Bay Estuary Program a program of the Texas Commission on Environmental Quality and the U.S. Environmental Protection Agency, TCEQ Contract # 582-17-70187, Federal ID # CE-00655006, p 53

Rabalais NN, Turner RE, Díaz RJ, Justic D (2009) Global change and eutrophication of coastal waters. ICES J Mar Sci 66:1528–1537. https://doi.org/10.1093/ICESJMS/FSP047

Reguera B, Velo-Suárez L, Raine R, Park MG (2012) Harmful *Dinophysis* species: A review. Harmful Algae 14:87–106. https://doi.org/10.1016/j.hal.2011.10.016

Reisinger A, Gibeaut JC, Tissot PE (2017) Estuarine suspended sediment dynamics: observations derived from over a decade of satellite data. Front Mar Sci 4. https://doi.org/10.3389/fmars.2017.00233

Reyna NE, Hardison AK, Liu Z (2017) Influence of major storm events on the quantity and composition of particulate organic matter and the phytoplankton community in a subtropical estuary, Texas. Front Mar Sci 4:43. https://doi.org/10.3389/fmars.2017.00043

Reynolds CS (2006) Ecology of phytoplankton (ecology, biodiversity and conservation). Cambridge University Press, Cambridge

Richardson AJ (2008) In hot water: Zooplankton and climate change. ICES J Mar Sci 65:279–295

Robichaux RJ, Dortch Q, Wrenn JH (1998) Occurrence of *Gymnodinium sanguineum* in Louisiana and Texas coastal waters, 1989-1994. NOAA Tech Rep NMFS 143:19–25

Roelke DL, Pierce RH (2011) Effects of inflow on harmful algal blooms: some considerations. J Plankton Res 33:205–209. https://doi.org/10.1093/plankt/fbq143

Roelke DL, Gable GM, Valenti TW, Grover JP, Brooks BW, Pinckney JL (2010) Hydraulic flushing as a *Prymnesium parvum* bloom-terminating mechanism in a subtropical lake. Harmful Algae 9:323–332

Roelke DL, Li HP, Hayden NJ, Miller CJ, Davis SE, Quigg A, Buyukates Y (2013) Co-occurring and opposing freshwater inflow effects on phytoplankton biomass, productivity and community composition of Galveston Bay, USA. Mar Ecol Prog Ser 477:61–76. https://doi.org/10.3354/meps10182

Roelke DL, Li H-P, Miller-DeBoer CJ, Gable GM, Davis SE (2017) Regional shifts in phytoplankton succession and primary productivity in the San Antonio Bay System (USA) in response to diminished freshwater inflows. Mar Freshwat Res 68:131–145

Ryther JH, Officer CB (1981) Impact of nutrient enrichment on water uses. In: Neilson BJ, Cronin LE (eds) Estuaries and nutrients. Humana, Clifton, NJ, pp 247–262

Scavia D, Liu Y (2009) Exploring estuarine nutrient susceptibility. Environ Sci Technol 43:3474–3479. https://doi.org/10.1021/es803401y

Schwierzke-Wade L, Roelke DL, Brooks BW, Grover JP (2011) *Prymnesium parvum* bloom decline following an inflow and nutrient-loading event: roles of dilution and grazing. J Plankton Res. 33:309–317. https://doi.org/10.1093/plankt/fbq108

Smayda TJ (1997) What is a bloom? A commentary. Limnol Oceanogr 42:1132–1136

Sosik HM, Olson RJ (2007) Automated taxonomic classification of phytoplankton sampled with imaging-in-flow cytometry. Limnol Oceanogr Methods 5:204–216. https://doi.org/10.4319/lom.2007.5.204

Soto IM, Cambazoglu MK, Boyette AD, Broussard K, Sheehan D, Howden SD, Shiller AM, Dzwonkowski B, Hode L, Fitzpatrick PJ, Arnone RA, Mickle PF, Cressman K (2018) Advection of *Karenia brevis* blooms from the Florida Panhandle towards Mississippi coastal waters. Harmful Algae 72:46–64

Steichen JL, Quigg A (2018) Fish species as indicators of freshwater inflow within a subtropical estuary in the Gulf of Mexico. Ecol Indic 85:180–189

Steichen J, Windham R, Brinkmeyer R, Quigg A (2012) Ecosystem under pressure: Ballast water discharge into Galveston Bay, Texas (USA) from 2005 to 2010. Bull Mar Pollut 64:779–789. https://doi.org/10.1016/j.marpolbul.2012.01.028

Steichen J, Schulze A, Brinkmeyer R, Quigg A (2014) All aboard! A biological survey of ballast water onboard vessels travelling the North Atlantic Ocean. Bull Mar Pollut 87:201–210. https://doi.org/10.1016/j.marpolbul.2014.07.058

Steichen JL, Denby A, Windham R, Brinkmeyer R, Quigg A (2015) A tale of two ports: dinoflagellate and diatom communities found in the high ship traffic region of Galveston Bay, Texas (USA). J Coast Res 31:407–416. https://doi.org/10.2112/JCOASTRES-D-13-00225.1

Steichen JL, Labonté JM, Windham R, Hala D, Kaiser K, Setta S, Faulkner P, Bacosa H, Yan G, Kamalanathan M, Quigg A (2020) Microbial, physical, and chemical changes in Galveston Bay following an extreme flooding event, Hurricane Harvey. Front Mar Sci 7:186. https://doi.org/10.3389/fmars.2020.00186

Steidinger KA, Meave del Castillo ME (eds) (2018) Guide to the identification of harmful microalgae in the Gulf of Mexico, Volume I: Taxonomy. Florida Fish and Wildlife Research Institute, St. Petersburg, FL

Steidinger KA, Stockwell DA, Truby EW, Wardle WJ, Dortch Q, Van Dolah FM (1998) Phytoplankton blooms off Louisiana and Texas, May-June 1994. NOAA Tech Rep Natl Mar Fish Serv 143:13–17

Swanson KM, Flewelling LJ, Byrd M, Nunez A, Villareal TA (2010) The 2008 Texas *Dinophysis ovum* bloom: Distribution and toxicity. Harmful Algae 9:190–199. https://doi.org/10.1016/j.hal.2009.10.001

Tango PJ, Magnien R, Butler W, Luckett C, Luckenbach M, Lacouture R, Poukish C (2005) Impacts and potential effects due to *Prorocentrum minimum* blooms in Chesapeake Bay. Harmful Algae 4:525–531

Texas Water Development Board (TWDB) (2022) Coastal hydrology. Available at: https://www.twdb.texas.gov/surfacewater/bays/coastal_hydrology/index.asp, personal communication with Ram Neupane. Accessed 22 Mar 2022

Thronson A, Quigg A (2008) Fifty five years of fish kills in Coastal Texas. Estuaries Coasts 31:802–813. https://doi.org/10.1007/s12237-008-9056-5

Thyng KM, Hetland RD, Ogle MT, Zhang X, Chen F, Campbell L (2013) Origins of *Karenia brevis* harmful algal blooms along the Texas coast. Limnol Oceanogr Fluids Environ 3:269–278

Tolan JM (2007) El Niño-Southern Oscillation impacts translated to the watershed scale: Estuarine salinity patterns along the Texas Gulf Coast, 1982 to 2004. Estuar Coast Shelf Sci 72:247–260. https://doi.org/10.1016/J.ECSS.2006.10.018

Tominack SA, Coffey KZ, Yoskowitz D, Sutton G, Wetz MS (2020) An assessment of trends in the frequency and duration of *Karenia brevis* red tide blooms on the South Texas coast (western Gulf of Mexico). PLoS One 15. https://doi.org/10.1371/journal.pone.0239309

Van Dolah FM (2000) Marine algal toxins, health effects, and their increasing occurrence. Environ Health Perspect 108:133–141

Walker LM, Montagna PA, Hu X, Wetz MS (2021) Timescales and magnitude of water quality change in three Texas estuaries induced by passage of Hurricane Harvey. Estuaries Coasts 44:960–971. https://doi.org/10.1007/s12237-020-00846-6

Walsh JJ, Jolliff JK, Darrow BP, Lenes JM, Milroy SP, Remsen A, Dieterle DA, Carder KL, Chen FR, Vargo GA, Weisberg RH, Fanning KA, Muller-Karger FE, Shinn E, Steidinger KA, Heil CA, Tomas CR, Prospero JS, Lee TN, Kirkpatrick GJ, Whitledge TE, Stockwell DA, Villareal TA, Jochens AE, Bontempi PS (2006) Red tides in the Gulf of Mexico: where, when, and why? J Geophy Res 111:C11003. https://doi.org/10.1029/2004JC002813

Wells ML, Trainer VL, Smayda TJ, Karlson BSO, Trick CG, Kudela RM, Ishikawa A, Bernard S, Wulff A, Anderson DM, Cochlan WP (2015) Harmful algal blooms and climate change: Learning from the past and present to forecast the future. Harmful Algae 49:68–93. https://doi.org/10.1016/j.hal.2015.07.009

Wetz MS, Paerl HW (2008) Estuarine phytoplankton responses to hurricanes and tropical storms with different characteristics (trajectory, rainfall, winds). Estuaries Coasts 31:419–429

Wetz MS, Cira EK, Sterba-Boatwright B, Montagna PA, Palmer TA, Hayes KC (2017) Exceptionally high organic nitrogen concentrations in a semi-arid South Texas estuary susceptible to brown tide blooms. Estuar Coast Shelf Sci 188:27–37. https://doi.org/10.1016/j.ecss.2017.02.001

Whitledge TE (1989a) Data synthesis and analysis, nitrogen process study (NIPS): nutrient distributions and dynamics in San Antonio Bay in relation to freshwater inflow. Report to Texas Water Development Board, by Marine Science Institute, University of Texas at Austin, Port Aransas, TX

Whitledge TE (1989b) Data synthesis and analysis, nitrogen process study (NIPS): nutrient distributions and dynamics in Nueces-Corpus Christi bays in relation to freshwater inflow. Report to Texas Water Development Board, by Marine Science Institute, University of Texas at Austin, Port Aransas, TX

Windham R, Nunnally A, Quigg A (2019) Investigating *Rangia cuneata* clams as bioindicators of ecological response to three decades of variable inflows. J Coast Res 35:1260–1270. https://doi.org/10.2112/JCOASTRES-D-18-00164.1

Wu X, Yang Y, Yang Y, Zhong P, Xu N (2022) Toxic characteristics and action mode of the mixotrophic dinoflagellate *Akashiwo sanguinea* on co-occurring phytoplankton and zooplankton. Int J Environ Res Public Health 19:404. https://doi.org/10.3390/ijerph19010404

Xu N, Wang M, Tang Y, Zhang Q, Duan S, Gobler CJ (2017) Acute toxicity of the cosmopolitan bloom-forming dinoflagellate *Akashiwo sanguinea* to finfish, shellfish, and zooplankton. Aquat Microb Ecol 80:209–222. https://doi.org/10.3354/ame01846

# Freshwater Inflow and Salinity Shape Nekton Diversity and Community Structure Within Texas Estuaries

Daniel M. Coffey ⓘ, Gregory W. Stunz ⓘ, and Paul A. Montagna ⓘ

## Abstract

The Texas coast climatic gradient of decreasing precipitation and freshwater inflow from northeast to southwest shapes estuarine ecosystems and biodiversity. By employing broad-scale diversity and community structure metrics and habitat suitability models of key indicator species, this study explored how multiple abiotic environmental factors, freshwater inflows, and drought and flood events shape estuarine communities across different spatial and temporal scales. A long-term fishery-independent dataset captured the latitudinal, seasonal, and interannual range of environmental conditions, providing insights into the general patterns and trends across different ecosystems and demonstrating how freshwater inflow regimes affect estuarine nekton communities. Across estuaries, salinity was a primary driver of estuarine diversity and community structure, though nekton diversity and community structure can be highly dynamic. In general, species diversity metrics decreased with increasing salinity; however, community composition shifted across latitudinal hydrological regimes so that a suite of species could inhabit each estuary in response to divergent patterns in freshwater inflows and environmental variables. These spatial patterns indicate that resource managers may be well-served to incorporate community dynamics and hydrological regimes when developing adaptive manage-ment plans to maintain freshwater inflow and target salinities for important fisheries species and estuarine communities to produce a sound ecological environment.

## Keywords

Salinity · Nekton · Diversity · Habitat suitability · Indicator species · *Callinectes sapidus* · *Micropogonias undulatus* · Nueces Estuary

## Introduction

Estuarine species comprise 46% by weight and 68% by value of US commercial fish and shellfish landings and 80% of recreational harvest (by weight) nationwide (Lellis-Dibble et al. 2008). In the Gulf of Mexico, 97% of commercial landings (by weight) and 93% of their economic value are attributed to estuarine fish and shellfish. Thus, the Gulf of Mexico is the largest contributor to the nation's commercial estuarine landings (by weight) by providing 38% of all national commercial estuarine pounds valued at more than $3.5 billion (31% of the national economic revenue gained via commercial estuarine landings). Moreover, 87% of the recreational harvest in the Gulf of Mexico was composed of estuary-using species, contributing to the largest proportion (35% by weight) of the nation's total recreational estuarine harvest. Consequently, the protection and restoration of estuarine habitats that provide the ecological foundation for the majority of US commercial and recreational fisheries, particularly in the Gulf of Mexico, is critical.

Estuaries in the Gulf of Mexico are under threat from both human activities and natural disturbances, posing risks to ecosystem productivity, ecological function, and resilience (Sheridan 2004; Thronson and Quigg 2008; Wetz and Yoskowitz 2013; Patrick et al. 2020; Harris et al. 2023; Martin et al. 2023). The combined effects of shifting rainfall patterns, rising sea levels, and altered salinity, along with local agricultural, municipal, and industrial water uses, are

D. M. Coffey (✉)
Department of Life Sciences, Texas A&M University-Corpus Christi, Corpus Christi, TX, USA
e-mail: Daniel.Coffey@tamucc.edu

G. W. Stunz
Harte Research Institute for Gulf of Mexico Studies, Texas A&M University-Corpus Christi, Corpus Christi, TX, USA
e-mail: Greg.Stunz@tamucc.edu

P. A. Montagna
Harte Research Institute for Gulf of Mexico Studies, Texas A&M University-Corpus Christi, Corpus Christi, TX, USA
e-mail: Paul.Montagna@tamucc.edu

© The Author(s) 2025
P. A. Montagna, A. R. Douglas (eds.), *Freshwater Inflows to Texas Bays and Estuaries*, Estuaries of the World, https://doi.org/10.1007/978-3-031-70882-4_14

expected to disrupt hydrological regimes and impact the diverse communities that rely on estuaries (Sklar and Browder 1998; Montagna et al. 2011a, b, 2013). Consequently, there is a growing need to enhance our understanding of the factors that regulate estuarine communities and predict how species will respond to these disturbances (Mahoney and Bishop 2017).

The Texas coast comprises seven major estuaries along a latitudinal gradient (Fig. 1), resulting in varying environmental conditions that substantially impact estuarine communities. The climatic gradient of decreasing precipitation and freshwater inflow from northeast to southwest is one of the most distinctive features of the Texas coast and plays a significant role in shaping these estuarine ecosystems and overall biodiversity (Tolan 2013). The amount of freshwater inflow is subject to seasonal and interannual variability driven by climate and rainfall patterns and water management practices. This freshwater input influences salinity levels, nutrient availability, and sedimentation rates in estuaries, subsequently impacting the distribution, assemblage, and productivity of estuarine communities (Montagna et al. 2011a, b). For example, higher freshwater inflow can lead to decreased salinity levels and increased nutrient availability, fostering high primary productivity and supporting diverse estuarine food webs (Marshall et al. 2021; Chin et al. 2022; Breaux et al. 2019; Whaley et al. 2023). The latitudinal gradient of salinity and freshwater inflow creates a complex mosaic of habitats for mobile nekton occupying these areas with specific habitat preferences and evolved physiological adaptations to tolerate and thrive under fluctuating salinity conditions to a degree. Understanding the dynamic interplay between estuarine nekton and environmental variables such as salinity is crucial for effective conservation and management strategies to preserve these invaluable coastal ecosystems.

In Texas, early attempts to address the management of freshwater inflows on estuarine diversity and communities focused on identifying the relationship between freshwater inflow volume and commercial fisheries harvest alone (Longley 1994; Powell et al. 2002). However, commercial

**Fig. 1** Major estuaries of Texas named for their primary contributing rivers

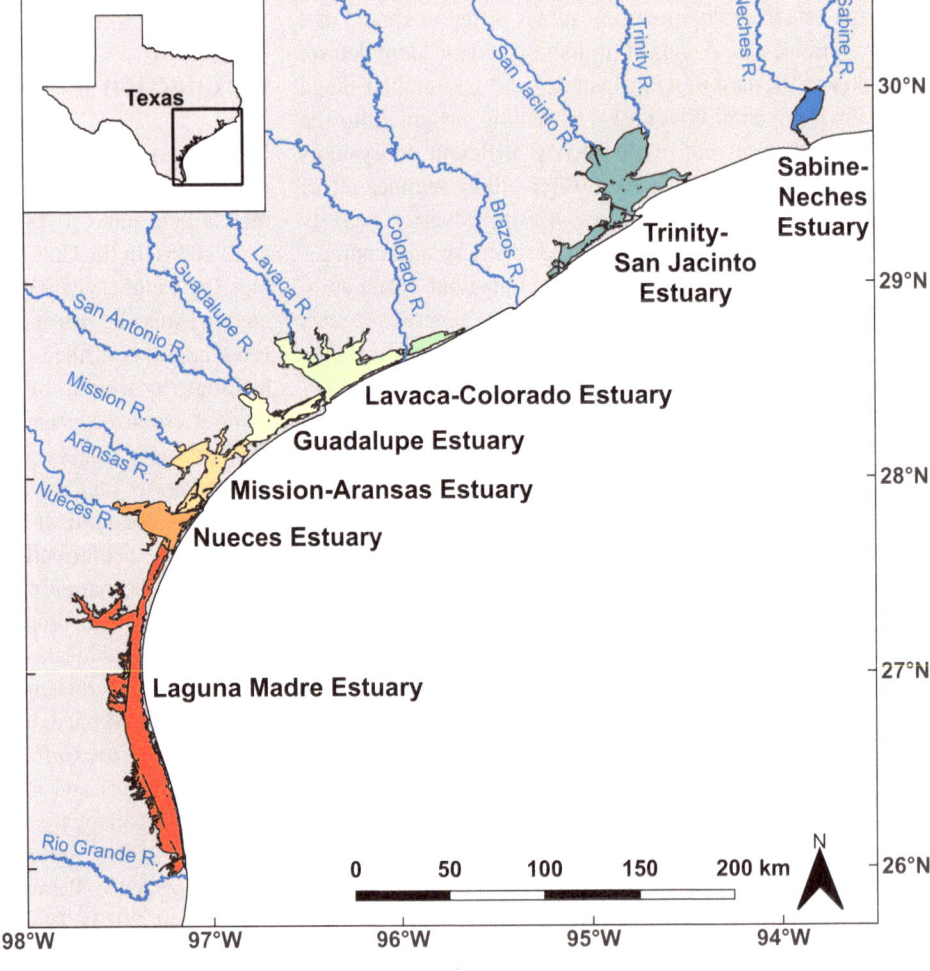

and recreational fisheries harvests are fishery-dependent economic factors driven in part by pricing and fishery regulation and, therefore, may not be a suitable measure of estuary health and function (Montagna et al. 2021b). Furthermore, a single species-based management approach does not consider that large seasonal and interannual variability in climate, which leads to periods of floods and droughts and can dramatically affect the spatial distribution of estuarine nekton with specific environmental requirements differently, especially those that utilize estuaries as critical nursery habitats or serve as hotspots of secondary production that support higher trophic levels. In contrast, long-term abundance data collected independent of fisheries (fishery independent) means the inherent variability in the data does not drive the patterns being investigated and allows for the robust examination of freshwater inflow and other specific environmental parameters on species' abundances more effectively (Steichen and Quigg 2018). With an emerging focus on ecosystem-based management (Imperial and Hennessey 1996), coastal fisheries managers seek to identify the ecological structure and function of estuarine communities to better understand the responses of fisheries to natural and anthropogenic factors, such as the availability of freshwater inflows for environmental uses.

The Texas Parks and Wildlife Department (TPWD) initiated a standardized fishery-independent monitoring program in 1975 to assess changes in the relative abundance and size of fishes and invertebrates in Texas estuaries, their spatial and temporal distributions, species compositions of the community, and selected abiotic environmental parameters known to influence their distribution and abundance (Martinez-Andrade et al. 2005). Fishes and invertebrates are collected using several gear types, including gillnets (1975–present) set during spring (April–June) and fall (September–November) and monthly bag seines (1977–present), bay trawls (1982–present), and oyster dredges (1984–present), within ten bay systems across the seven major estuaries on the Texas coast. Surveys are conducted using a stratified cluster sampling design with each bay system serving as non-overlapping strata with a fixed number of samples per month (or season for gillnets). Sample locations within $1' \times 1'$ grid cells (stations) are drawn independently and without replacement for each combination of gear, stratum, and month (or season). Multiple gear types and standardized protocols reduce sampling biases by mitigating gear selectivity, capturing a broader range of species and life history stages, and ensuring representative coverage of different habitats. Collectively, this program facilitates an ecosystem-based approach that integrates species-specific and community-level responses to changes in salinity within and across Texas estuaries to evaluate potential impacts on local population dynamics and fishery productivity.

## Latitudinal Freshwater Inflow Gradients on Estuarine Nekton Diversity and Communities

Estuarine nekton diversity and communities play a crucial role in connecting primary production with higher trophic levels, including game fishes, dolphins, and wading birds. Consequently, estuarine nekton diversity and community structure can serve as valuable indicators of the overall condition and functioning of the ecosystem (Fausch et al. 1990; Deegan et al. 1997; Whitfield and Elliott 2002). Moreover, estuarine nekton communities often comprise juveniles of commercially and recreationally important fish and shellfish species that have economic importance as adults (Whaley et al. 2007). Hence, employing a community-based methodological approach can be a practical and efficient alternative to individually assessing numerous species (Clarke 1993; Dray et al. 2012). Additionally, community-based approaches provide valuable insights into species co-occurrences, which can aid in establishing relationships between species and their environment for monitoring ecosystem processes or determining priority areas for conservation and management within estuaries (Fujiwara et al. 2016; Gonzalez et al. 2021; Livernois et al. 2021).

Historically, studies focused on the relationship between a single estuarine species and one or a small number of environmental variables, examining them in isolation. For example, the seminal report by Longley (1994) compared the mean relative abundance of several economically and ecologically important fish (southern flounder, *Paralichthys lethostigma*; Atlantic croaker; *Micropogonias undulatus*; black drum, *Pogonias cromis*; Gulf menhaden, *Brevoortia patronus*; pinfish, *Lagodon rhomboides*; red drum, *Sciaenops ocellatus*; spotted seatrout, *Cynoscion nebulosus*; striped mullet, *Mugil cephalus*) and shellfish (blue crab, *Callinectes sapidus*; white shrimp, *Litopenaeus setiferus*; brown shrimp, *Farfantepenaeus aztecus*) species collected using TPWD bay trawls and gillnets among Texas estuaries and different salinity regimes. Significant relationships between mean relative abundance and mean salinities among the estuaries were observed for shellfish and juvenile and smaller fish species captured in bay trawls. In contrast, no relationship was observed for larger fishes captured using gillnet sampling gear. Still, differences in mean catch rates were observed for all of the selected fish species (regardless of sampling gear) among estuaries.

Estuarine ecosystems, however, are complicated by interactions among numerous species and environmental variables, and these characteristics and components of ecosystems cannot easily be isolated, especially across broad spatial scales. While freshwater inflow volume and salinity are dominant drivers of functional shifts in Texas estuarine fish and shellfish communities, the general approach in this study was not to isolate the most important environmental variable influencing community structure but to use spatial structuring common to multiple environmental variables and estuarine species to predict general nekton distribution patterns across Texas estuaries. Combining analyses of estuarine functional group abundances with environmental variables can provide insight into ecosystem-level shifts across time and space to both short- and long-term physical processes such as freshwater inflows and flood and drought events. Based on the findings by Longley (1994), which identified relationships between estuarine salinities and shellfish and juvenile and smaller fish species captured in bay trawls, we re-examined contemporary and historical trends in nekton collected using TPWD bay trawls to investigate the influence of latitudinal freshwater inflow gradients on a broader range of estuarine species through diversity and community metrics.

The set number of monthly trawl samples varies with the aerial coverage of appropriate deep-water areas for a given bay system (here set at ≥1 m for the appropriate deployment of the sampling gear; Martinez-Andrade 2015). The larger bay systems with deep-water surface areas ≥ 20,000 ha (Galveston, Matagorda, San Antonio, Aransas, and Corpus Christi) are stratified into equal-sized upper and lower bay zones with ten monthly trawl samples collected in the upper zone and ten in the lower zone to ensure the suitable spatial distribution of samples. Ten trawl samples are collected monthly within each smaller bay system with a deep-water surface area of <10,000 ha (Sabine Lake, East Matagorda, upper Laguna Madre, and lower Laguna Madre). No trawl

samples are collected from the minor Cedar Lakes estuary. East Matagorda Bay, located just northeast of the Lavaca-Colorado estuary, receives freshwater inflow only from run-off of surrounding coastal watersheds and from direct precipitation with no direct sources of river inflow into this minor estuary; therefore, trawl samples collected from this bay were excluded. Estuarine ecosystems are uniquely affected by marine and freshwater inputs, and thus proximity to tidal passes and/or freshwater sources can considerably influence estuarine habitat quality (Furey and Rooker 2013; Dance and Rooker 2016). Texas estuaries are bay complexes, typically with two geographical components: a primary bay behind the barrier island connected to the Gulf of Mexico characterized by higher salinity and one or more secondary or tertiary bays into which streams and rivers flow characterized by lower salinity. Hence, the bay trawl dataset was further divided into primary and secondary bay systems within each estuary, and the Laguna Madre was subdivided into an upper and lower section (Withers et al. 2023), providing a total of eight estuaries (Tolan 2013; Table 1).

Bay trawl samples ($n$ = 54,615) collected from 1982 to 2020 were analyzed. Otter trawls (6.1 m wide with 38-mm-stretched nylon multifilament mesh and 1.2 × 0.5 m trawl doors) were deployed in open bay habitats (≥1 m depth) and towed at 4.82 km/h for 10 min away from the shoreline in a circular manner (further details of sampling protocols are described in Martinez-Andrade 2015). While otter trawls typically sample the bay bottom, they likely capture a large portion of the water column given the shallow nature of Texas bays (~2 m average depth; Heim-Ballew and Olsen 2019). Hence, captured fish and invertebrates are broadly defined as

**Table 1** Estuary and bay names and codes for bay trawl sampling stations by the Texas Parks and Wildlife Department fishery-independent monitoring program

| Names | | Codes | | |
| --- | --- | --- | --- | --- |
| Estuary | Bay | Estuary | Bay | Bay Type |
| Sabine-Neches Estuary | Sabine Lake | SN | SN | 1° and 2° |
| Trinity-San Jacinto Estuary | Galveston Bay | TJ | GB | 1° |
| Trinity-San Jacinto Estuary | Trinity Bay | TJ | TB | 2° |
| Lavaca-Colorado Estuary | Lavaca Bay | LC | LB | 2° |
| Lavaca-Colorado Estuary | Matagorda Bay | LC | MB | 1° |
| Guadalupe Estuary | Lower San Antonio Bay | GE | LS | 1° |
| Guadalupe Estuary | Upper San Antonio Bay | GE | US | 2° |
| Mission-Aransas Estuary | Aransas Bay | MA | AB | 1° |
| Mission-Aransas Estuary | Copano Bay | MA | CB | 2° |
| Nueces Estuary | Corpus Christi Bay | NC | CC | 1° |
| Nueces Estuary | Nueces Bay | NC | NB | 2° |
| Laguna Madre | Baffin Bay | LM | BB | 2° |
| Laguna Madre | Upper Laguna Madre | LM | UL | 1° |
| Laguna Madre | Lower Laguna Madre | LM | LL | 1° |

1° = primary bay, 2° = secondary bay

nekton in this study as organisms that are free-swimming at some stage in their life cycle, which is inclusive of mobile epifauna and species with sessile adult stages (e.g., gulf wedge clams, *Rangia cuneata*) captured in otter trawls (Minello et al. 2003; Patrick et al. 2020). However, some highly mobile nekton may actively avoid capture by swimming above or below the approaching otter trawl resulting in inherent biases in gear selectivity and efficiency (Allen et al. 1992; Rozas and Minello 1997; Baker and Minello 2011). Captured fish and invertebrates greater than 5 mm total length were identified to the lowest taxonomic level (typically species) and counted. Individuals who could not be identified to species were grouped into the lowest possible taxon (e.g., Ctenophora). Depth and hydrological data, including salinity (psu), water temperature (°C), dissolved oxygen (mg/L), and turbidity (NTU), were recorded from 0.3 m off the bottom at each sampling site before trawling began.

Diversity indices are univariate metrics used to summarize multivariate community characteristics in a single number and identify the number of species relative to one another. Nekton diversity was calculated using Hill's diversity number one (N1; Hill 1973), which measures the effective number of species in a sample and indicates the number of abundant species. It is calculated as the exponentiated form of the Shannon diversity index (*H′*; Shannon and Weaver 1949):

$$N1 = e^{H'} \tag{1}$$

As diversity decreases, N1 will tend toward 1. *H′* is the average uncertainty per species in an infinite community of species with known proportional abundances and is calculated by:

$$H' = -\sum_{i=1}^{S} p_i \ln p_i \tag{2}$$

where $p_i$ is the proportion of individuals belonging to the *i*th species in the sample and *S* is the total number of species in the sample, otherwise known as species richness. *S* is the simplest measure of diversity and does not consider differences in relative species abundance. Species evenness (*J′*) or the similarity (i.e., uniformity) in species' relative abundance in a community captures another aspect of diversity by determining diversity as a standardized index of relative species abundance (Krebs 1999). Here we calculate Pielou's *J′* (Pielou 1975), which expresses the observed *H′* relative to the maximum value of *H′* (i.e., the total number of species or richness, *S*):

$$J' = \frac{H'}{\ln S} \tag{3}$$

The value of *J′* ranges from 0 to 1, with larger values representing more even distributions in abundance among species.

## Nekton Diversity Among Bays and Estuaries

Univariate two-way analysis of variance (ANOVA) was used to compare nekton diversity metrics (abundance, richness, N1 diversity, evenness) and environmental variables (salinity, water temperature, dissolved oxygen, turbidity, and depth) among estuaries, bays, and sampling months using SAS/ETS 14.3 (SAS Institute Inc. 2017). Nekton abundance was square root transformed to ensure homogeneity of variance and normality of residuals. Estuary and sampling month were modeled as the main fixed effects, with sampling stations representing replicates within bays. A partially hierarchical design was used because bays are unique to estuaries (i.e., bays are nested within estuaries), and each sampling station is unique to each bay within an estuary. Thus, estuaries, bays, and sampling stations are completely nested random effects that can be modeled by the following formula:

$$Y_{ijklm} = \mu + \alpha_j + \beta_k + \beta\gamma_{k(l)} + \varepsilon_{(i)jklm} \tag{4}$$

where $Y_{ijklm}$ is the dependent response variable, $\mu$ is the overall sample mean, $\alpha_j$ is the main fixed effect for the *j*th sampling month from a total of 466 (monthly from 1982–2020; no data were collected in April–May 2020 due to the COVID-19 pandemic), $\beta_k$ is the main fixed effect for the *k*th estuary from a total of seven, $\beta\gamma_{k(l)}$ is the main effect for bays that are nested (or unique) within each estuary and are thus a random effect as denoted by the parentheses around the subscript *l* that represents the 14 bays (Table 1), and $\varepsilon_{(i)jklm}$ is the random error term for each of the *i*th replicate measurements. The expected mean squares for the F-tests were calculated for each source of variation using type III sum of squares ($\alpha = 0.05$). A Tukey's studentized range test was used for post hoc pairwise comparisons of groups for significant main effects.

Within estuaries, the mean and variability in environmental variables shape nekton diversity metrics. The main effect of estuary was significant for nekton abundance, evenness, salinity, temperature, and dissolved oxygen and was marginally non-significant for richness and N1 diversity (Table 2). Sampling month and bay (nested within estuary) had a significant effect on all nekton diversity metrics and environmental variables (except for bay versus water temperature). These results indicate the presence of latitudinal, within-estuary, and seasonal (monthly) variability in nekton diversity and environmental conditions. Overall, the Nueces Estuary had the highest nekton N1 diversity and evenness (tied with the lower Laguna Madre) and second highest richness following the Mission-Aransas Estuary (Table 3). Salinity exhibited the most pronounced difference of any environmental variable across each estuary, ranging from an overall mean of 6.8 psu in the Sabine-Neches Estuary (northeast) to 39.5 psu in the upper Laguna Madre (southwest; Table 3).

**Table 2** Results of two-way, partially hierarchical analysis of variance examining the main effects of sampling month and estuary and bays nested within estuaries on nekton diversity metrics and environmental variables

| Nekton diversity metric | Mean | CV | $R^2$ | P-values | | |
|---|---|---|---|---|---|---|
| | | | | Month | Estuary | Bay (estuary) |
| Abundance ($\sqrt{\text{N}}$/trawl) | 10.02 | 49.38 | 40% | <0.0001 | 0.0198 | <0.0001 |
| Richness (S/trawl) | 7.53 | 28.32 | 53% | <0.0001 | 0.0608 | <0.0001 |
| Diversity (N1/trawl) | 3.71 | 31.40 | 30% | <0.0001 | 0.0575 | <0.0001 |
| Evenness (J'/trawl) | 0.33 | 37.71 | 21% | <0.0001 | 0.0351 | <0.0001 |
| Environmental variable | | | | | | |
| Salinity (psu) | 20.00 | 27.31 | 82% | <0.0001 | 0.0336 | <0.0001 |
| Temperature (°C) | 22.57 | 7.46 | 94% | <0.0001 | 0.0004 | 0.3122 |
| Dissolved oxygen (mg/L) | 7.65 | 14.52 | 61% | <0.0001 | 0.0230 | <0.0001 |
| Turbidity (NTU) | 27.27 | 98.77 | 34% | <0.0001 | 0.4360 | <0.0001 |
| Depth (m) | 2.02 | 15.77 | 62% | <0.0001 | 0.1569 | <0.0001 |

Nekton abundance (N) was square root transformed prior to analysis. *CV* coefficient of variation, $R^2$ percent of model variance explained by data

**Table 3** Tukey's studentized range test results for differences among estuary means

| Abundance ($\sqrt{\text{N}}$/trawl) | LL 7.06 | SN 7.13 | UL 7.41 | LC 8.70 | NC 9.12 | TJ 9.76 | GE 12.17 | MA 13.77 |
|---|---|---|---|---|---|---|---|---|

| Richness (S/trawl) | UL 6.03 | SN 6.28 | LC 6.82 | LL 7.02 | TJ 7.04 | GE 7.25 | NC 8.73 | MA 9.83 |
|---|---|---|---|---|---|---|---|---|
| Diversity (N1/trawl) | GE 3.36 | TJ 3.51 | SN 3.52 | LC 3.52 | UL 3.56 | LL 3.89 | MA 4.06 | NC 4.60 |

| Evenness (J'/trawl) | GE 0.29 | MA 0.29 | TJ 0.32 | LC 0.33 | UL 0.36 | SN 0.37 | LL 0.38 | NC 0.38 |
|---|---|---|---|---|---|---|---|---|

| Temperature (°C) | SN 21.61 | TJ 21.80 | LC 22.45 | GE 22.55 | MA 22.57 | NC 23.18 | UL 23.50 | LL 23.74 |
|---|---|---|---|---|---|---|---|---|

| DO (mg/L) | UL 6.82 | NC 7.22 | LL 7.27 | LC 7.65 | MA 7.66 | SN 7.82 | TJ 7.94 | GE 8.12 |
|---|---|---|---|---|---|---|---|---|

| Salinity (psu) | SN 6.77 | GE 12.37 | TJ 13.25 | LC 20.10 | MA 20.38 | NC 27.01 | LL 33.41 | UL 39.49 |
|---|---|---|---|---|---|---|---|---|

| Turbidity (NTU) | SN 18.46 | LL 20.70 | MA 22.14 | LC 26.23 | NC 28.51 | TJ 29.75 | GE 33.24 | UL 36.56 |
|---|---|---|---|---|---|---|---|---|

| Depth (m) | NC 1.55 | GE 1.61 | LL 1.73 | SN 1.98 | LC 2.02 | UL 2.10 | MA 2.34 | TJ 2.42 |
|---|---|---|---|---|---|---|---|---|

Underlined values are not statistically different ($\alpha = 0.05$). Estuaries in a geographic order from northeast to southwest: *SN* Sabine Lake, *TJ* Trinity-San Jacinto Estuary (Galveston Bay System), *LC* Lavaca-Colorado Estuary (Matagorda Bay System), *GE* Guadalupe Estuary (San Antonio Bay System), *MA* Mission-Aransas Estuary (Aransas Bay System), *NC* Nueces Estuary (Corpus Christi Bay System), *UL* upper Laguna Madre and Baffin Bay, and *LL* lower Laguna Madre

## The Relationship Between Salinity and Nekton Diversity Across Texas Estuaries

Nekton diversity may serve as ecological indicators to create broad whole estuary health indices and assess the impact of freshwater inflow changes. It is necessary to determine how these estuarine organisms respond to salinity to identify the optimal salinity preferences and freshwater inflow requirements that maximize diversity. The analyses below are based on a previous unpublished study (Montagna et al. 2021a).

The functional relationship between benthic communities to salinity has been explained using a nonlinear model in Texas (Montagna et al. 2002) and Florida estuaries (Montagna et al. 2008). The assumption behind the model is that there is an ideal range for ecological drivers and biological diversity responses decline prior to and after reaching a maximum threshold driver value. The relationship resembles a right-skewed bell-shaped curve, such as a log-normal distribution. The nonlinear log-normal equation is fit to the dataset using the Max Bin regression method (Turner and Montagna 2016) and can be used to characterize the nonlinear relationship between nekton diversity metrics and salinity, among other environmental parameters. The shape of this log-normal curve can be predicted with a three-parameter model:

$$Y = a \times \exp\left(-0.5 \times \left(\ln\frac{\left(\frac{X}{c}\right)}{b}\right)^2\right) \quad (5)$$

The three parameters characterize different attributes of the curve, where $a$ is the peak of the dependent value ($Y$), $b$ is the skewness or rate of change of the response as a function of the independent variable ($X$), and $c$ is the optimal value of the independent variable.

All nekton diversity metrics increased rapidly with increasing salinity to a maximal salinity between 10.1 and 16.7 psu and then gradually decreased (Fig. 2, Table 4). The lowest nekton diversity metrics were generally observed at salinities exceeding 40 psu that typically only occur in the hypersaline upper Laguna Madre. While variability within and between estuaries certainly exists, the remarkable consistency in optimal salinity ranges for a variety of nekton diversity metrics is noteworthy and valuable for ecosystem management purposes where the maximum response of a biological indicator to an environmental driver such as salinity is desired (Turner and Montagna 2016).

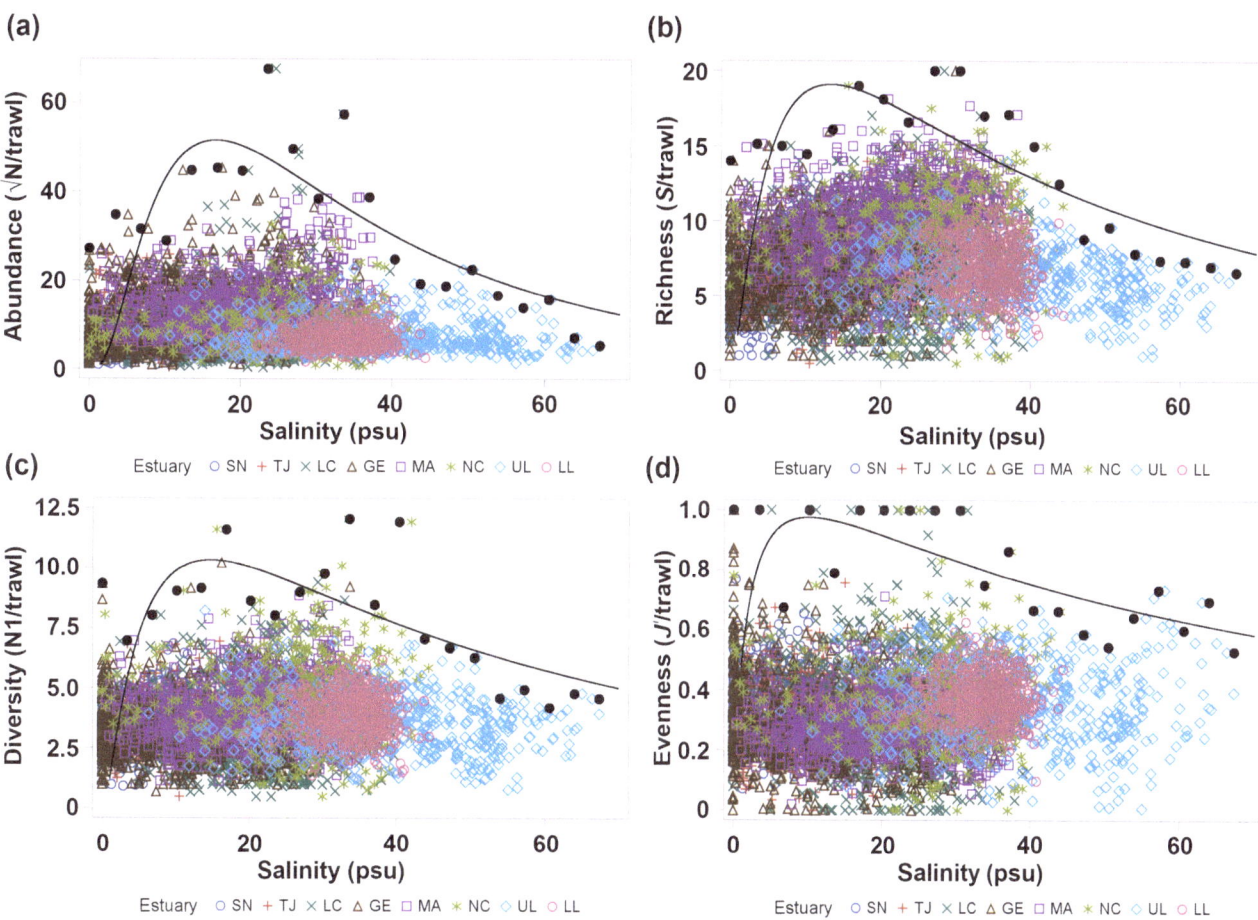

**Fig. 2** Nonlinear Max Bin regression (black line) of nekton diversity metrics to salinity. (**a**) Nekton abundance (N), (**b**) richness (*S*), (**c**) N1 diversity, and (**d**) evenness (*J'*). Abundance was square root transformed prior to analysis. Estuaries (symbols) in a geographic order from northeast to southwest: *SN* Sabine Lake, *TJ* Trinity-San Jacinto Estuary (Galveston Bay System), *LC* Lavaca-Colorado Estuary (Matagorda Bay System), *GE* Guadalupe Estuary (San Antonio Bay System), *MA* Mission-Aransas Estuary (Aransas Bay System), *NC* Nueces Estuary (Corpus Christi Bay System), *UL* upper Laguna Madre and Baffin Bay, and *LL* lower Laguna Madre. Black dots are the maximum value in a salinity bin

**Table 4** Parameter estimates ($a$ = maximum metric value, $b$ = rate of change, and $c$ = salinity yielding maximum value) and standard error of the estimate in parentheses from nonlinear Max Bin regression of nekton diversity metrics to salinity

| Nekton diversity metric | Parameters | | | |
|---|---|---|---|---|
| | $a$ | $b$ | $c$ | $R^2$ |
| Abundance ($\sqrt{N}$/trawl) | 51.67 (5.09) | 0.851 (0.12) | 16.66 (1.93) | 91% |
| Richness ($S$/trawl) | 19.09 (1.24) | 1.255 (0.15) | 13.20 (1.67) | 97% |
| Diversity ($N1$/trawl) | 10.35 (0.73) | 1.298 (0.18) | 14.75 (2.04) | 96% |
| Evenness ($J'$/trawl) | 0.98 (0.06) | 1.921 (0.37) | 10.07 (2.66) | 98% |

Nekton abundance (N) was square root transformed prior to analysis. For all models $P < 0.0001$. $R^2$, percent of model variance explained by data

## Nekton Community Structure Among Bays and Estuaries

While univariate diversity metrics have long been important ecological tools to describe the stability and long-term sustainability of an ecosystem, they contain no information about indicator species or any species of interest (Gonzalez et al. 2021). Moreover, in an estuary, many key species may occur transiently in large numbers, thereby depressing diversity metrics. Thus, multivariate analyses were used to compare nekton community structure among bays and estuaries using PRIMER v7 (Clarke and Gorley 2015). Species abundances were square root transformed to reduce the importance of more abundant species and to allow for changes in rarer species to be statistically discernable. Nekton community structure was analyzed with non-metric multidimensional scaling (nMDS) using a Bray-Curtis similarity matrix among sampling station and month combinations. The resulting two-dimensional nMDS plot spatially represents the nekton community relationship among sampling station-month combinations so that the distances among stations are directly related to the similarities in nekton species compositions among those same stations (Clarke et al. 2014). Relationships within each nMDS were highlighted with a cluster analysis using the group average method based on Bray-Curtis similarity matrices. Cluster analysis was displayed as similarity contours on the nMDS plots using percentage similarity based on species abundances. A similarity percentage (SIMPER) analysis was conducted to determine which species abundances were the most dissimilar among estuaries.

Although 512 species (or the lowest identified taxon) were found, only the 21 most abundant species had at least 1% of the total abundance (Table 5). The most abundant species were the warty comb jelly (or sea walnut, *Mnemiopsis leidyi*) and Atlantic croaker, which accounted for 10.9% and 10.2% of total species abundance, respectively. Warty comb jellies were most abundant in the lower San Antonio Bay of the Guadalupe Estuary and least abundant in Trinity Bay of the Trinity-San Jacinto Estuary. One caveat is that jellyfish are not target species using bottom trawls and may be undercounted due to their fragile nature. Atlantic croakers were the top 5 most abundant species in every bay except for the upper Laguna Madre.

Across the estuaries, patterns of community structure roughly matched the study by Tolan (2013) using TPWD bag seine data and the northeast-southwest salinity gradient present on the Texas coast, with the freshest estuaries on the upper coast having significantly different communities than the more saline estuaries on the lower coast (Fig. 3). The central coast estuaries (Lavaca-Colorado, Guadalupe, Mission-Aransas, and Nueces) showed the most significant degree of overlap in their community structure at the 60% similarity level. The Sabine-Neches Estuary and Baffin Bay of the upper Laguna Madre only overlapped with estuaries at the 50% similarity level, exclusive of the upper and lower Laguna Madre. Differences in community structure across the salinity gradient of estuaries appear to be primarily driven by the relative proportion of only a few discriminating taxa: sea grapes (*Molgula manhattensis*), comb jellies (sea walnuts and unidentified Ctenophora), pinfish, bryozoans (*Bugula neritina*, *Amathia verticillata*, unidentified Bryozoan), spot (*Leiostomus xanthurus*), and gulf wedge clams (Table 6).

An nMDS was also performed on environmental variables, using a normalization transformation and Euclidean distances to create the resemblance matrix. A principal component analysis examined the relationship between bays and environmental variables. Prior to analysis, data were normalized to a mean of 0 and a standard deviation of 1. Principal components with eigenvalues greater than 1 explaining more than 10% of the variance in the data were included. The first principal component (PC1) explained 56.6% of the variance and was represented by high water temperatures and salinities correlated to low dissolved oxygen (Fig. 4). The second principal component (PC2) explained 29.0% of the variance and is represented by low turbidity correlated to deeper water. The largest difference in the spatial patterns was observed for bays within the Nueces Estuary (Corpus Christi and Nueces Bays) and Laguna Madre (Baffin Bay, upper and lower Laguna Madre), which were separated based on high water temperatures and salinities relative to other bays and estuaries (Fig. 4). The upper San Antonio Bay was also separated from the lower San Antonio Bay of the Guadalupe Estuary and other bays based on high turbidity.

Nekton community structure was linked with environmental variables using non-metric multivariate procedures in PRIMER (Clarke and Gorley 2015) to identify environmental drivers of variation in the community assemblages among bays and estuaries. Environmental variables were normalized and used to build a Euclidean distance-based resemblance matrix. A non-parametric form of a Mantel test (RELATE) assessed agreement in the multivariate pattern between the

**Table 5** Dominant species (or lowest identified taxon) captured in bay trawls between 1982 and 2020 by the Texas Parks and Wildlife Department fishery-independent monitoring program

| Rank | Taxa name | SN | GB | TB | LB | MB | LS | US | AB | CB | CC | NB | BB | UL | LL | Mean | % | Cum.% |
|---|---|---|---|---|---|---|---|---|---|---|---|---|---|---|---|---|---|---|
| 1 | Mnemiopsis leidyi | 9.2 | – | <0.1 | 6.9 | 9.2 | 74.7 | 32.1 | 22.7 | 31.0 | 7.1 | 16.2 | 38.7 | 14.1 | 0.5 | 18.7 | 10.9% | 10.9% |
| 2 | Micropogonias undulatus | 17.4 | 22.0 | 27.0 | 11.7 | 25.4 | 23.5 | 23.0 | 25.4 | 28.4 | 19.9 | 12.8 | 3.2 | 2.0 | 5.6 | 17.6 | 10.2% | 21.1% |
| 3 | Leiostomus xanthurus | 7.5 | 5.4 | 3.1 | 17.8 | 17.7 | 13.9 | 2.4 | 52.0 | 29.4 | 42.1 | 8.7 | 5.1 | 3.3 | 3.4 | 15.1 | 8.8% | 29.8% |
| 4 | Lagodon rhomboides | 1.2 | 0.6 | 0.1 | 1.6 | 3.1 | 3.9 | 0.3 | 37.3 | 4.6 | 76.7 | 4.6 | 2.8 | 27.3 | 28.4 | 13.7 | 8.0% | 37.8% |
| 5 | Molgula manhattensis | <0.1 | 2.7 | 1.0 | 47.8 | 26.7 | 32.7 | 12.8 | 2.3 | 4.1 | <0.1 | <0.1 | – | <0.1 | 0.1 | 9.3 | 5.4% | 43.2% |
| 6 | Ctenophora | 0.1 | 40.4 | 35.7 | 11.2 | 10.5 | 16.7 | 7.8 | – | 0.2 | <0.1 | – | <0.1 | <0.1 | 4.1 | 9.0 | 5.2% | 48.4% |
| 7 | Penaeus aztecus | 3.6 | 5.0 | 3.5 | 4.6 | 6.8 | 26.1 | 19.2 | 13.2 | 16.1 | 6.3 | 8.6 | 3.0 | 2.7 | 2.4 | 8.6 | 5.0% | 53.5% |
| 8 | Penaeus setiferus | 9.3 | 15.9 | 14.5 | 7.8 | 4.2 | 7.5 | 16.3 | 3.3 | 12.7 | 1.3 | 16.2 | 2.2 | 0.3 | 0.3 | 8.0 | 4.6% | 58.1% |
| 9 | Amathia verticillata | <0.1 | 4.6 | 0.1 | 7.2 | 27.7 | 12.6 | 1.6 | 20.7 | 11.5 | 9.4 | 4.6 | 2.4 | 0.1 | <0.1 | 7.3 | 4.2% | 62.3% |
| 10 | Bryozoa | <0.1 | 6.2 | 1.1 | – | 0.3 | 0.5 | – | 24.3 | 6.2 | 20.6 | 9.8 | 0.4 | 1.3 | 0.1 | 5.1 | 2.9% | 65.2% |
| 11 | Bugula neritina | – | – | – | – | 2.5 | 16.3 | 5.9 | 23.4 | 15.2 | <0.1 | 1.2 | – | – | – | 4.6 | 2.7% | 67.9% |
| 12 | Anchoa mitchilli | 7.3 | 2.4 | 2.2 | 2.1 | 5.4 | 3.7 | 1.7 | 7.2 | 5.7 | 10.2 | 2.7 | 2.4 | 8.3 | 2.2 | 4.5 | 2.6% | 70.6% |
| 13 | Beroe ovata | 1.3 | 8.3 | 5.3 | 8.0 | 2.7 | 4.1 | 1.4 | 5.4 | 8.9 | 2.3 | 1.4 | 3.2 | 2.9 | 0.0 | 3.9 | 2.3% | 72.8% |
| 14 | Brevoortia patronus | 3.4 | 6.0 | 6.5 | 3.7 | 3.8 | 8.1 | 8.5 | 2.9 | 4.5 | 1.5 | 4.0 | 0.3 | 0.1 | 0.2 | 3.8 | 2.2% | 75.1% |
| 15 | Aurelia aurita | 0.0 | 0.5 | 0.5 | 2.7 | 3.0 | 7.9 | 3.9 | 4.1 | 15.9 | 2.5 | 5.8 | 5.1 | 0.3 | 0.1 | 3.7 | 2.2% | 77.2% |
| 16 | Stomolophus meleagris | 0.0 | 0.2 | 0.1 | 14.6 | 5.1 | 3.8 | 12.5 | 1.8 | 11.6 | 0.5 | 0.7 | 0.2 | 0.1 | 0.1 | 3.7 | 2.1% | 79.3% |
| 17 | Chrysaora quinquecirrha | 0.1 | 14.4 | 6.0 | 1.0 | 2.3 | 4.9 | 2.2 | 2.9 | 7.7 | 1.3 | 1.0 | 4.9 | 0.2 | 0.1 | 3.5 | 2.0% | 81.4% |
| 18 | Bairdiella chrysoura | 0.1 | 0.5 | 0.1 | 1.1 | 0.9 | 2.4 | 2.4 | 6.3 | 5.7 | 2.4 | 6.2 | 1.2 | 2.0 | 5.4 | 2.6 | 1.5% | 82.9% |
| 19 | Callinectes sapidus | 0.6 | 2.2 | 2.5 | 1.9 | 0.9 | 6.0 | 5.7 | 3.1 | 3.2 | 0.7 | 1.6 | 0.7 | 1.7 | 2.6 | 2.4 | 1.4% | 84.3% |
| 20 | Lolliguncula brevis | 0.1 | 1.5 | 0.4 | 2.1 | 8.3 | 0.8 | <0.1 | 4.7 | 1.0 | 8.0 | 1.1 | 0.2 | 0.4 | 0.7 | 2.1 | 1.2% | 85.5% |
| 21 | Rangia cuneata | 7.9 | 0.1 | 10.8 | <0.1 | <0.1 | 0.1 | 9.5 | – | <0.1 | <0.1 | <0.1 | – | <0.1 | – | 2.0 | 1.2% | 86.7% |
| 21 | Subtotal dominant species | | | | | | | | | | | | | | | 163.9 | | 86.7% |
| 491 | Subtotal other species | | | | | | | | | | | | | | | 8.7 | | 13.3% |
| 512 | Total species | | | | | | | | | | | | | | | 172.6 | | 100.0% |

Mean abundances and percent contribution are calculated across all estuaries. Cum%, cumulative percent. Bay codes are listed in a geographic order from northeast to southwest (see Table 1 for estuary and bay code definitions)

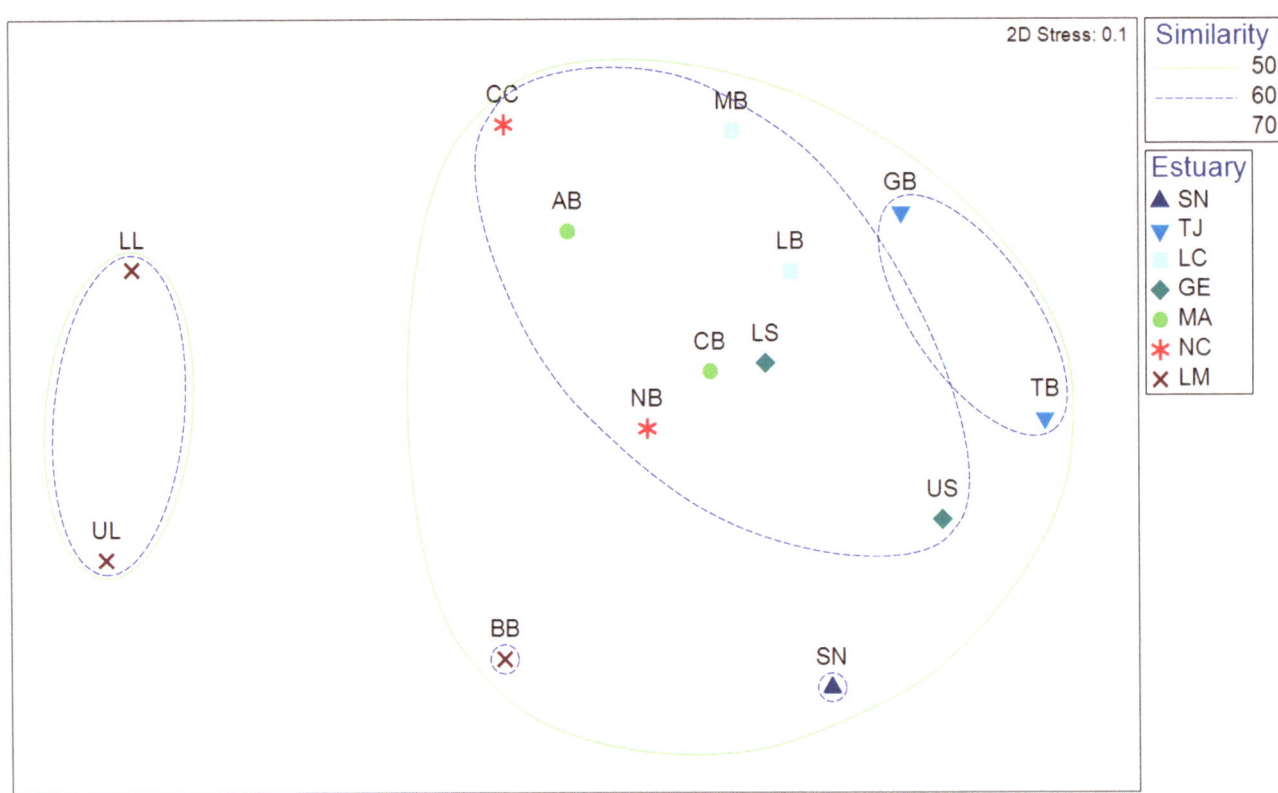

**Fig. 3** Non-metric multidimensional scaling ordination of nekton community structure by estuary (symbols) and bay (labels). Estuary codes are listed in the legend in a geographic order from northeast to southwest (see Table 1 for estuary and bay code definitions). Bray-Curtis cluster analysis similarity contours (lines) represent the percentage similarity at 50, 60, and 70% based on species abundances

**Table 6** Similarity percentage (SIMPER) analysis results using Bray-Curtis dissimilarities among estuaries

| Estuary code | Mean % | Highest contribution | | Second highest contribution | |
| | | Taxa name | % | Taxa name | % |
| --- | --- | --- | --- | --- | --- |
| SN vs. TJ | 40.1% | Ctenophora | 13.2% | *Mnemiopsis leidyi* | 6.8% |
| SN vs. LC | 47.4% | *Molgula manhattensis* | 9.9% | *Amathia verticillata* | 6.3% |
| SN vs. GE | 46.6% | *Molgula manhattensis* | 7.4% | *Mnemiopsis leidyi* | 6.8% |
| SN vs. MA | 50.0% | *Bugula neritina* | 6.2% | *Amathia verticillata* | 5.5% |
| SN vs. NC | 45.5% | *Lagodon rhomboides* | 7.3% | Bryozoa | 6.6% |
| SN vs. LM | 56.2% | *Lagodon rhomboides* | 5.0% | *Rangia cuneata* | 4.9% |
| TJ vs. LC | 40.4% | *Molgula manhattensis* | 8.2% | Ctenophora | 4.9% |
| TJ vs. GE | 40.5% | *Mnemiopsis leidyi* | 11.9% | *Molgula manhattensis* | 5.5% |
| TJ vs. MA | 44.9% | Ctenophora | 8.3% | *Mnemiopsis leidyi* | 7.1% |
| TJ vs. NC | 46.6% | Ctenophora | 9.3% | *Lagodon rhomboides* | 7.1% |
| TJ vs. LM | 58.1% | Ctenophora | 7.9% | *Mnemiopsis leidyi* | 5.2% |
| LC vs. GE | 35.9% | *Mnemiopsis leidyi* | 7.3% | *Bugula neritina* | 4.2% |
| LC vs. MA | 36.2% | *Molgula manhattensis* | 6.7% | *Bugula neritina* | 5.7% |
| LC vs. NC | 37.8% | *Molgula manhattensis* | 10.1% | *Lagodon rhomboides* | 6.2% |
| LC vs. LM | 55.0% | *Molgula manhattensis* | 8.0% | *Amathia verticillata* | 4.2% |
| GE vs. MA | 34.2% | *Leiostomus xanthurus* | 6.1% | Bryozoa | 5.5% |
| GE vs. NC | 44.8% | *Molgula manhattensis* | 6.4% | *Lagodon rhomboides* | 5.6% |
| GE vs. LM | 56.5% | *Molgula manhattensis* | 5.8% | *Mnemiopsis leidyi* | 4.7% |
| MA vs. NC | 32.3% | *Bugula neritina* | 6.8% | *Lagodon rhomboides* | 5.8% |
| MA vs. LM | 51.4% | *Bugula neritina* | 5.6% | *Leiostomus xanthurus* | 5.5% |
| NC vs. LM | 48.6% | *Lagodon rhomboides* | 5.2% | Bryozoa | 4.8% |

For each estuary comparison, the top 2 discriminating species (or lowest identified taxon) contributing most to the dissimilarity are listed and their percent contribution

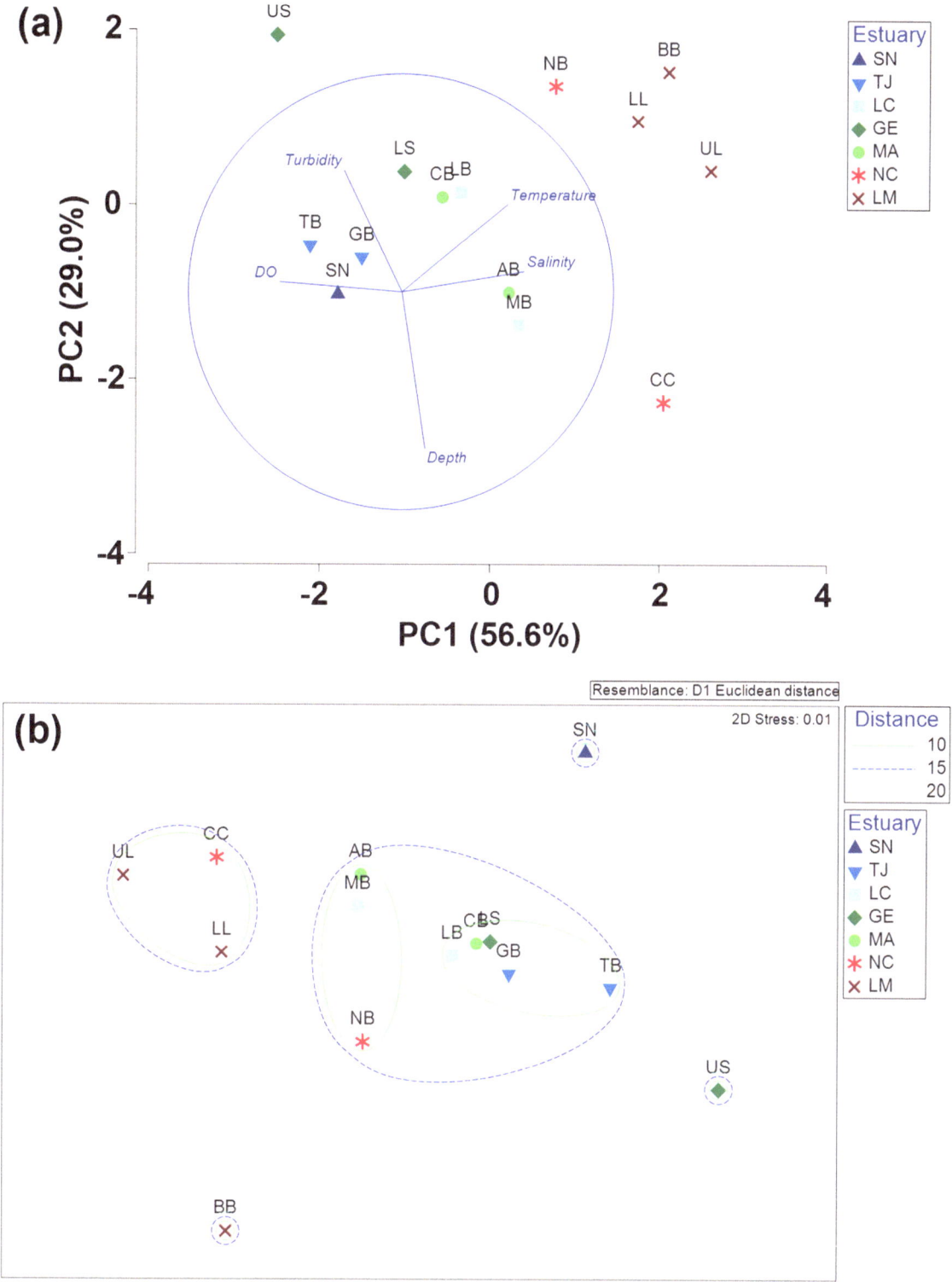

**Fig. 4** (a) Principal component analysis and (b) non-metric multidimensional scaling ordination of the environmental variables by estuary (colored symbols) and bay (labels). Estuary codes are listed in the legend in a geographic order from northeast to southwest (see Table 1 for end in a geographic order from northeast to southwest (see Table 1 for estuary and bay code definitions). Cluster analysis Euclidean distance contours (lines) represent the pairwise distances of 10, 15, and 20 units based on environmental variables. Points within smaller distance contours are considered more similar

**Table 7** Results from BIO-ENV analysis examining the best subset of environmental variables explaining nekton assemblage composition

| No. vars. | $\rho$ | Variables |
|---|---|---|
| 3 | 0.736 | Salinity, water temperature, turbidity |
| 2 | 0.735 | Salinity, water temperature |
| 4 | 0.701 | Salinity, water temperature, dissolved oxygen, turbidity |
| 2 | 0.699 | Water temperature, turbidity |
| 3 | 0.689 | Salinity, water temperature, dissolved oxygen |
| 4 | 0.678 | Salinity, turbidity, depth |
| 2 | 0.671 | Water temperature, dissolved oxygen |
| 3 | 0.669 | Water temperature, dissolved oxygen, turbidity |
| 1 | 0.669 | Salinity |
| 1 | 0.660 | Water temperature |

*No. vars.* number of environmental variables, $\rho$ Spearman rank correlation coefficient

nekton community and environmental resemblance matrices using a suite of random permutations. Following the RELATE procedure, a BIO-ENV analysis was used to identify the best subset of environmental variables so that the Euclidean distances of scaled environmental variables have the maximum Spearman rank correlation with community dissimilarities. The best subset of environmental variables with the maximum Spearman rank correlation with community dissimilarities was salinity, water temperature, and turbidity between bays ($\rho = 0.736$), and the maximum Spearman rank correlation with a single variable was salinity ($\rho = 0.669$; Table 7).

## A Case Study of the Nueces Estuary

There can be environmental heterogeneity within a single estuary, including areas with varying salinity gradients, hydrodynamics, and habitat types. Investigating the influence of freshwater inflow within an estuary allows us to identify specific areas more influenced by freshwater inputs and changes in salinity. This provides a finer-scale understanding of community-level and species-specific responses to localized variations in freshwater inflow and salinity and can guide the development of targeted management strategies to maintain and enhance populations and overall estuarine health.

During the last three decades, the Nueces Estuary, located along the south-central Texas coast, has been the subject of much scientific study regarding freshwater inflow and salinity. Thus, it has some of the best science available to make predictions concerning freshwater inflow to the estuary relating to the estuarine nekton response and serves as a valuable case study. In particular, there is a current concern over rising salinities in this region due to several long-term changes, such as a warmer and dryer climate in the southwest (Seager et al. 2007), which leads to less streamflow (Miller et al. 2021), reduced freshwater inflow from diversions (Asquith et al. 1997), increased evaporation, and industrial brine discharges

from oil and gas production, desalination, or other industrial practices. These potential changes can reduce freshwater inflow and decrease seawater dilution or increase salinity by concentrating the existing salt in the bay. Rising seawater temperatures and salinities have already been documented in the Texas Coastal Bend (Montagna et al. 2011a, b) and are raising a concern about the effects of decreasing freshwater inflow and increased salinity on ecologically and economically important living marine resources of the region.

The Nueces Estuary comprises three major systems: Nueces River Delta, Nueces Bay, and Corpus Christi Bay (Fig. 5). The Nueces River and associated catchment (greater Nueces River Basin) drains 4.3 million hectares (10.6 million acres) of a semi-arid, subtropical region of south Texas into the Nueces Estuary. The Nueces River flows along the southern edge of the Nueces River Delta complex and empties away from the delta directly into Nueces Bay, a secondary bay in the estuary. Nueces Bay is a shallow, well-mixed, wind-driven bay located in a semi-arid zone. Mean annual salinity is approximately 25 and may vary from near fresh (<2) during heavy flood events to hypersaline (>45) during prolonged periods of low inflow.

The Nueces River Delta is approximately 20,000 acres in size and composed of a complex array of channels, pools, marshes, and tidal flats and is one of the most extensive marsh ecosystems on the Texas Gulf Coast. A typical estuary is regularly inundated with salt water from tides and fresh water from river inflows. These freshwater inflows are essential in maintaining the critical balance of salinity needed in the estuary for a healthy environment. However, due to increasing freshwater demands, the lower reaches of the Nueces River system and delta have been subjected to extensive management and alternations (e.g., Montagna et al. 2009; NBBEST 2011), which have dramatically altered the flow regime, ecology, and physical characteristics of the region. Most notably, the combined effects of the enlargement of Lake Corpus Christi in 1958, the construction of the Choke Canyon Reservoir in 1982, and with each drought cycle producing lower inflow, freshwater inflows to the Nueces Delta have significantly been reduced. This activity has reduced the number and volume of flood events that reach the Nueces Estuary and provided this critical habitat with fresh water.

The delta provides a critical transitional environment used by marine and estuarine species at important life cycle stages (Longley 1994). For example, TPWD recognized the ecological value of the Nueces Delta and Bay by closing it to shrimping harvest—the largest commercial fishery in Texas (Montagna et al. 1998)—and designated it as a nursery area for postlarval and juvenile shrimp. As the freshwater inflows diminish, the estuary's function as a healthy, productive biological ecosystem has been significantly reduced. Consequently, this has led to a hypersaline or "reverse estuary" condition where salinity in the Nueces Delta and Bay is generally higher than that of the adjacent bays or Gulf of Mexico (Montagna et al. 2009). Many species can tolerate these harsher environmental conditions, but the

**Fig. 5** The Nueces Estuary comprises the Nueces River Delta, Nueces Bay, and Corpus Christi Bay

prolonged periods of salinity-induced stress lead to lower biological productivity and species diversity. For example, salinity increased to hypersaline conditions (>36 psu) during the drought period of 1988–1990 and consequently reduced shrimp and oyster populations in Nueces Bay. Historically, Nueces Bay supported abundant shrimp and oyster populations, which generally require salinities of 10–20 psu (Montagna et al. 1998). Shell dredging and substrate removal coupled with restriction in freshwater inflow ended the oyster and shell fishery in Nueces Bay, and by 1967, Nueces Bay was considered "fished-out" (Ward 1997). There is no commercial oyster fishery currently in Nueces Bay, nor has there been for several decades, and the frequent high salinities inhibit sustainable oyster production at levels supporting commercial harvest (NBBEST 2011).

Nueces Bay and Delta are linked to the Gulf of Mexico by Corpus Christi Bay, the primary bay within the Nueces Estuary, designated as an estuary of national significance by the US Environmental Protection Agency in 1992. While salinity conditions in Nueces Bay are primarily driven by freshwater inflow from the Nueces River, circulation in Corpus Christi Bay is driven by freshwater inflow, tidal exchange, wind, frontal passages, and Coriolis forces (Ward 1997). The bay has a total open water surface area of 167 square miles and a direct connection to the Gulf of Mexico via the Aransas Pass and Packery Channel inlets for tidal exchange. A dominant feature affecting the salinity regime

and effects of freshwater inflow to the bay is a deep ship channel that runs the entire length of the bay. The Corpus Christi Ship Channel stretches from the Port Aransas Jetties to the Inner Harbor of Corpus Christi adjacent to Nueces Bay and was completed in 1926 to allow ships and their cargo access to the mainland Port of Corpus Christi. The ship channel is 45 feet deep and has changed the water movement and natural circulation patterns within the ~10.5 feet shallow bay (Ward and Armstrong 1997). Freshwater typically moves along the southern part of Nueces Bay and travels along the south of Corpus Christi Bay. Saltwater from the Gulf of Mexico primarily enters through the Aransas Pass inlet, moves along the northwest via the Corpus Christi Ship Channel, and travels along the northern part of Corpus Christi Bay. The movement of lower salinity water along the southern subregion and higher salinity water along the northern subregion of Corpus Christi Bay sets up a gyre that moves counter-clockwise. Thus, the Corpus Christi Ship Channel facilitates the exchange of Gulf water, creating marine conditions in the bay and restricting any significant impact of freshwater inflow from the Nueces River in reducing salinities in Corpus Christi Bay, even during high flow events (i.e., large-scale floods) to the estuary. However, circulation patterns and salinity are greatly altered because of man-made structures and practices, including freshwater diversions, seawater withdrawal for industrial cooling systems, the John F. Kennedy Memorial Causeway crossing the upper Laguna Madre, and the Corpus Christi Ship Channel (Matsumoto et al. 1997).

Diversions exceed return flows in the watersheds of the Nueces Estuary, which have the most significant net difference between return and diversion flows. Water consumption by municipal and industrial users in the greater Corpus Christi area and agricultural users in rural areas reduces freshwater inflow to the Nueces Estuary by about 14% from what it would be without return and diversion flows. Consequently, Nueces Bay often has high salinities, and the Nueces Delta has become a reverse estuary where low salinity water enters the delta from the bay as opposed to freshwater entering the delta from a river source. Therefore, the Nueces Estuary is a prominent example of an estuary at risk because of how changing inflow can severely damage an estuary due to freshwater impoundments and diversions.

## Characterizing Salinity and Freshwater Inflow Regimes

Establishing relationships between salinity and freshwater inflow is essential to determine the inflow volume that will generate the desired salinities (i.e., base flow) for a sound ecological environment based on species diversity and selected indicator species. There is probably nowhere else on the Texas coast as an extensive real-time salinity monitoring program as in Nueces Bay. Several salinity measuring stations ("SALT" stations) have collected salinity daily since the early 1990s. The average monthly salinity from the Texas Coastal Ocean Observation Network SALT03 station (#074) from Nueces Bay and cumulative monthly inflow from the US Geological Survey stream gage on the Nueces River at the saltwater barrier dam at Calallen (#08211500) was used to calculate how much inflow would produce the required salinities empirically. The SALT03 salinity station was used because it was the most appropriate gage based on its historical use for numerous studies and had the best relationship with Nueces River inflow based on its proximity to the Nueces River (Fig. 6). A simple linear regression using the $\log_{10}$ of total monthly inflow and monthly mean salinity showed a strong negative relationship between salinity and inflow (Fig. 6). Outputs from this model can be used to help calculate the inflow needed to achieve the desired salinity for species diversity and selected indicator species in the Nueces Estuary.

**Fig. 6** Linear regression of cumulative monthly inflow from the US Geological Survey (USGS) stream gage on the Nueces River at the saltwater barrier dam at Calallen (#08211500) and average monthly salinity from the Texas Coastal Ocean Observation Network SALT03 station (#074) from Nueces Bay. Salinity = 66.183 – (11.690 × $\log_{10}$(inflow)); $R^2 = 0.58$. Inset: Study map of Nueces Estuary with selected water quality stations (yellow circles)

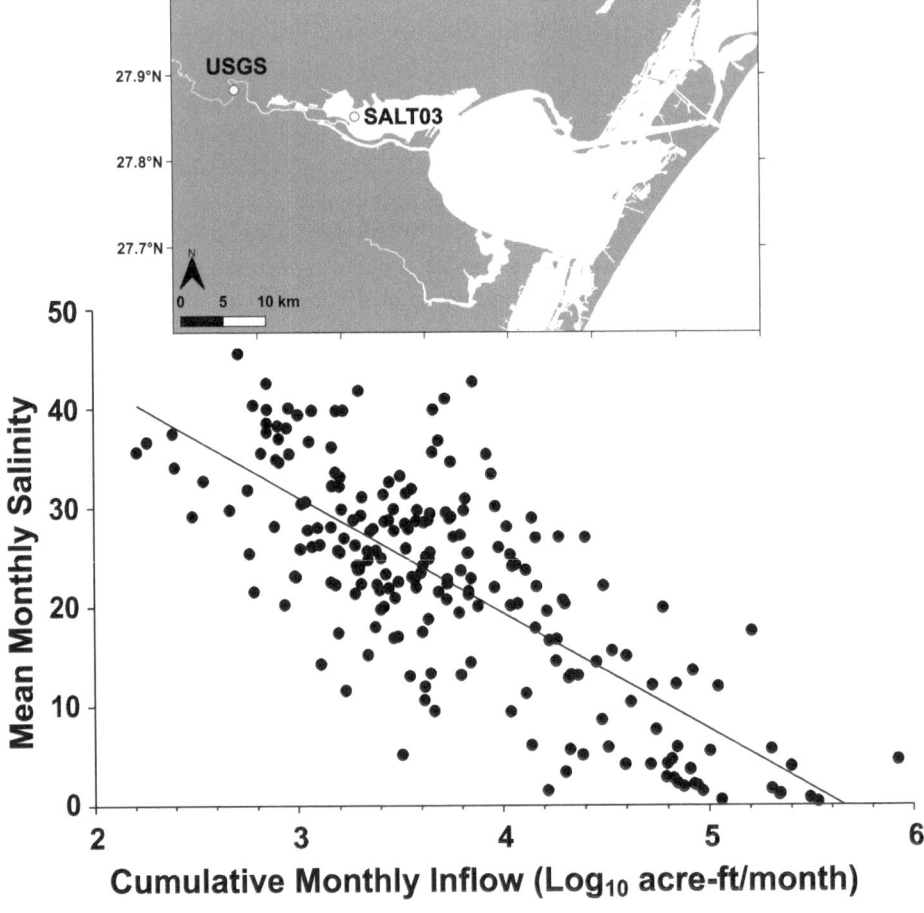

Salinity in Texas estuaries is known to vary spatially and cyclically across years, given interannual patterns in climatic conditions and freshwater flow regimes. TxBLEND, a two-dimensional, depth-averaged hydrodynamic and salinity transport model, is the primary hydrodynamic model used by the Texas Water Development Board (TWDB) to simulate water circulation and salinity condition within the bays and has been used to develop freshwater inflow recommendations for Texas estuaries (see chapter "Hydrology, Circulation, and Salinity" for additional details). Because TxBLEND produces high-resolution, dynamic simulations of horizontal salinity zonation patterns throughout estuaries over long-term periods, the model was used to simulate monthly salinity conditions over a 30-year (1987–2016) period for the Nueces Estuary. The TxBLEND computational grid for the Nueces Estuary contains 11,009 nodes and 19,035 linear triangular elements. In addition to the bays of the Nueces Estuary, the model grid includes Copano and Aransas Bays to the northeast and the upper Laguna Madre and Baffin Bay to the southwest to better simulate water circulation and salinity transport within the estuary.

TxBLEND-modeled salinity outputs were averaged across all nodes within the Nueces Estuary according to month and year to classify climatic periods based on quartile analysis. Average annual salinities below the first quartile (Q1; <25.47) were classified as wet years, salinities above the third quartile in the fourth quartile (Q4; >32.04) were considered dry years, and thus salinities within the interquartile range (Q1–Q3; 25.47–32.04) were classified as average (moderate) years. TxBLEND-modeled salinity outputs were used to classify climatic periods for the entire Nueces Estuary since other salinity measuring stations and TPWD hydrological data focus on more limited and specific areas.

## The Relationship Between Salinity and Nekton Diversity Within the Nueces Estuary

To model the potential impacts of these differing salinity regimes (dry, average, wet) on estuarine nekton diversity and the spatial extent and distribution of suitable nursery habitats for the selected indicator species, fish and invertebrate data were collected using bag seines as part of the TPWD fishery-independent monitoring program (Martinez-Andrade 2015). Bag seines are designed to sample juvenile estuarine fish and invertebrate nekton in shallow habitats (≤2 m depth) along the shoreline that are inherently more sensitive to changes in freshwater inflow and salinity due to their lower overall volumes compared to deeper areas sampled by bay trawls. Moreover, shallow estuarine areas typically exhibit high habitat diversity, including submerged vegetation and tidal flats that provide refuge, foraging opportunities, and nursery grounds for a wide range of nekton. While the bay trawls

sampled across multiple estuaries provided valuable information on broader spatial patterns and regional comparisons, sampling in shallow habitats within a single estuary captures fine-scale variations in nekton responses to specific environmental changes across different regions of the estuary.

Twenty bag seine samples are collected monthly within each major bay system of the Texas coast based on a stratified, random sampling design; however, the monthly sample size for each bay system ranged between 10 and 16 samples prior to 1992. Bag seine samples ($n = 8021$) collected from 1987 to 2016 from the Nueces Estuary that overlapped with the TxBLEND-modeled salinity outputs were analyzed. Bag seines (18.3 × 1.8 m, with 19-mm-stretched nylon mesh in the wings and 13-mm-stretched mesh in the bag) were deployed in shallow habitats (≤2 m depth) and pulled parallel to the shore for 15.2 m (further details of sampling protocols are described in Martinez-Andrade 2015). Captured fish and invertebrates greater than 5 mm total length were identified to the lowest taxonomic level and counted. Similar to bay trawls, hydrological data were recorded from surface waters (0–15 cm) at each sampling site. The minimum and maximum depth (m) of each bag seine haul was also recorded, and for this study, depth is defined as the mean depth of each seine haul.

The TPWD uses a stratified, random sampling design, which means bag seine sampling sites are not sampled consecutively. However, salinity gradients exist in relatively fixed locations, so it is necessary to aggregate bag seine sampling sites to transform the data into a quasi-synoptic sampling design to allow comparison of the change in a given area over time. For analyses, bag seine sampling sites were assigned to four subregions of the Nueces Estuary, namely, Redfish and South Aransas Bays (RB/SAB), North Corpus Christi Bay (NCCB), South Corpus Christi Bay (SCCB), and Nueces Bay (NB) (Fig. 7), which are differentially impacted by changes in freshwater inflow and thus have different long-term annual salinities (Table 8).

The response of nekton N1 diversity (Eqs. 1–2) to spatial and temporal variables was investigated using a generalized additive model (GAM) framework. GAMs were selected due to their ability to account for nonlinear relationships between the response and multiple explanatory variables. Mean annual nekton N1 diversity was highest in Redfish and South Aransas Bays and lowest in Nueces Bay (Table 9); thus, separate models were constructed for each subregion to account for differences in nekton diversity in response to freshwater inflow and salinity. Candidate predictor variables included year, month, depth, and hydrological data recorded at each bag seine sampling site. Collinearity between candidate predictor variables was assessed with Pearson correlation coefficients and variance inflation factors (VIF) using the "corvif" function (Zuur et al. 2009) in R (R Core Team 2020). Thin plate regression splines were estimated for each environmen-

**Fig. 7** The Nueces Estuary and four subregions: Redfish and South Aransas Bays (cyan), North Corpus Christi Bay (red), South Corpus Christi Bay (green), and Nueces Bay (blue). Shaded portions denote waters ≤2 m depth

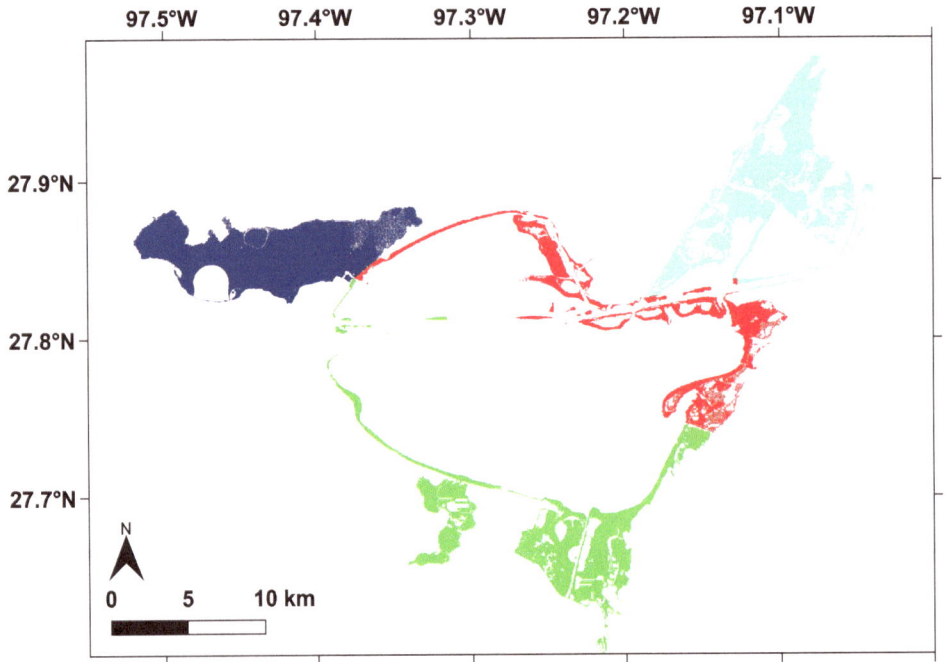

**Table 8** Mean (± standard deviation among years), minimum, and maximum annual salinities recorded at bag seine sampling sites within subregions of the Nueces Estuary between 1987 and 2016 by the Texas Parks and Wildlife Department fishery-independent monitoring program

| Subregion | Code | No. of samples | Annual salinity (psu) |
|---|---|---|---|
| Redfish and South Aransas Bays | RB/SAB | 2110 | 28.11 ± 3.58 (20.20–33.48) |
| North Corpus Christi Bay | NCCB | 2605 | 30.97 ± 3.18 (24.14–35.70) |
| South Corpus Christi Bay | SCCB | 1716 | 32.50 ± 2.90 (27.33–36.99) |
| Nueces Bay | NB | 1590 | 26.75 ± 7.46 (14.18–38.09) |

**Table 9** Mean (± standard deviation among years), minimum, and maximum annual nekton N1 diversity recorded at bag seine sampling sites within subregions of the Nueces Estuary between 1987 and 2016 by the Texas Parks and Wildlife Department fishery-independent monitoring program

| Subregion | Code | No. of samples | Annual Nekton N1 diversity |
|---|---|---|---|
| Redfish and South Aransas Bays | RB/SAB | 2110 | 3.79 ± 0.33 (2.95–4.42) |
| North Corpus Christi Bay | NCCB | 2605 | 3.27 ± 0.29 (2.73–3.88) |
| South Corpus Christi Bay | SCCB | 1716 | 3.21 ± 0.26 (2.63–3.64) |
| Nueces Bay | NB | 1590 | 3.18 ± 0.30 (2.59–3.80) |

tal variable, whereas month was modeled using a cyclic cubic regression spline, which constrains the start and end points of the smooth term to be the same (Wood 2017). To prevent overfitting (i.e., unrealistic ecological responses), the gamma parameter was set to 1.4 in all models (Wood 2017), and each regression spline was automatically penalized from specified maximum degrees of freedom (df = 5, excepting month df = 12) and the degree of smoothing selected by minimizing the restricted maximum likelihood (REML) score (Wood 2011). GAMs were constructed using a gamma distributed response with a log link function in R using the "mgcv" package (Wood 2017), and model selection was based upon an information-theoretic approach through minimization of the second-order Akaike Information Criterion (AIC$_c$; Burnham and Anderson 2002) using the "MuMIn" package (Bartoń 2020). Models with substantial support were selected based on a $\Delta$AIC$_c$ < 2 from the model with the lowest AIC$_c$ and included in model averaging based on Akaike weights ($w$; Burnham and Anderson 2002). Significant predictor variables ($p < 0.05$) with high relative importance (sum of model weights over all models including each explanatory variable) were retained in the final model used for the graphical representation of terms, calculation of deviance explained, and adherence to statistical assumptions of residuals.

Hydrological and depth data were absent for a minor portion of bag seine samples ($n = 56$); thus, a total of 7965 bag seine samples from the Nueces Estuary were analyzed using GAMs. Absolute Pearson correlation coefficients were < 0.21, and variance inflation factors (VIFs) were < 1.70, indicating low collinearity between candidate predictor variables. Spatial, interannual, and intra-annual (seasonal) variability in nekton N1 diversity was detected across Nueces Estuary subregions. Month was retained as a predictor variable in all final models, except for the Redfish Bay/South Aransas Bay subregion, the only subregion where year was retained in the final model (Fig. 8). Salinity was retained only in the models for the Nueces Bay and North Corpus Christi Bay subregions, whereas temperature was the only candidate predictor variable retained in all subregion models (Fig. 9). The GAMs explained between 6.72% and 13.53% of residual deviance, indicating additional factors beyond our models explain a significant portion of each subregion's nekton diversity.

The Nueces Bay and North Corpus Christi Bay subregion models exhibited similar negative linear responses for nekton N1 diversity with increasing salinity. Nekton N1 diversity was higher at salinities below ~27 psu for both subregions. The Nueces River and Corpus Christi Ship Channel in northern Corpus Christi Bay have the greatest effects on water movement and salinity within the Nueces

Estuary, resulting in reduced circulation in the southeast corner of Corpus Christi bay and the formation of two loops on either side of the ship channel. Collectively, these circulation patterns and man-made alterations have created different salinity conditions in northern and southern Corpus Christi Bay, resulting in an increased sensitivity for nekton diversity in North Corpus Christi Bay to changes in salinity compared to the other subregions of the Nueces Estuary (i.e., Redfish and South Aransas Bays and South Corpus Christi Bay).

## Focal Species and Indicators of Estuarine Health

In addition to establishing the relationship between estuarine nekton diversity and salinity, the selection of focal indicator species that show clear and well-documented responses to

**Fig. 8** (**a**) Estimated partial coefficients for the parametric term year on nekton N1 diversity from the final generalized additive models for each subregion of the Nueces Estuary: Nueces Bay (NB, blue), North Corpus Christi Bay (NCCB, red), Redfish and South Aransas Bays (RB/SAB, cyan), and South Corpus Christi Bay (SCCB, green). The base level of 1987 for year is centered, and dashed lines represent 95% confidence limits. (**b**) Estimated response curves (solid black line) of component smooth functions for month on nekton N1 diversity from the final models. Subplots for predictor variables are consistently positioned for each estuary subregion, with missing subplots indicating candidate predictors that were not retained in the final best-fit model. Shaded areas represent 95% confidence limits of uncertainty in the centered smooth. Vertical axes are partial responses (estimated, centered smooth functions) on the scale of the linear predictor. Ticks on the *x*-axis denote values for which there are data. Positive values on the *y*-axis (above the dashed red line) indicate an increase in nekton N1 diversity for each subregion

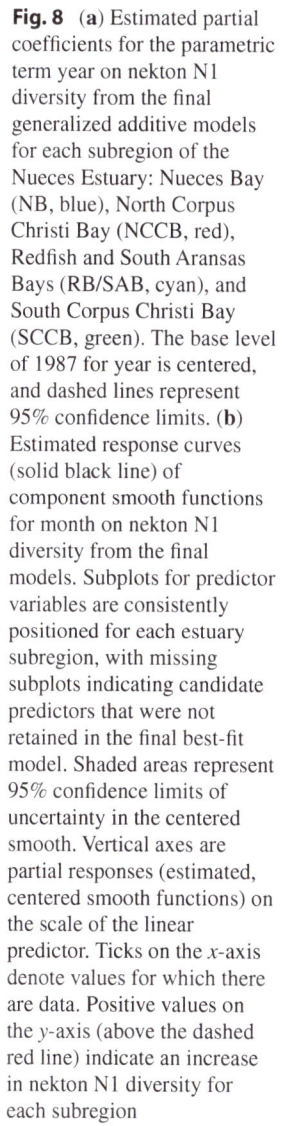
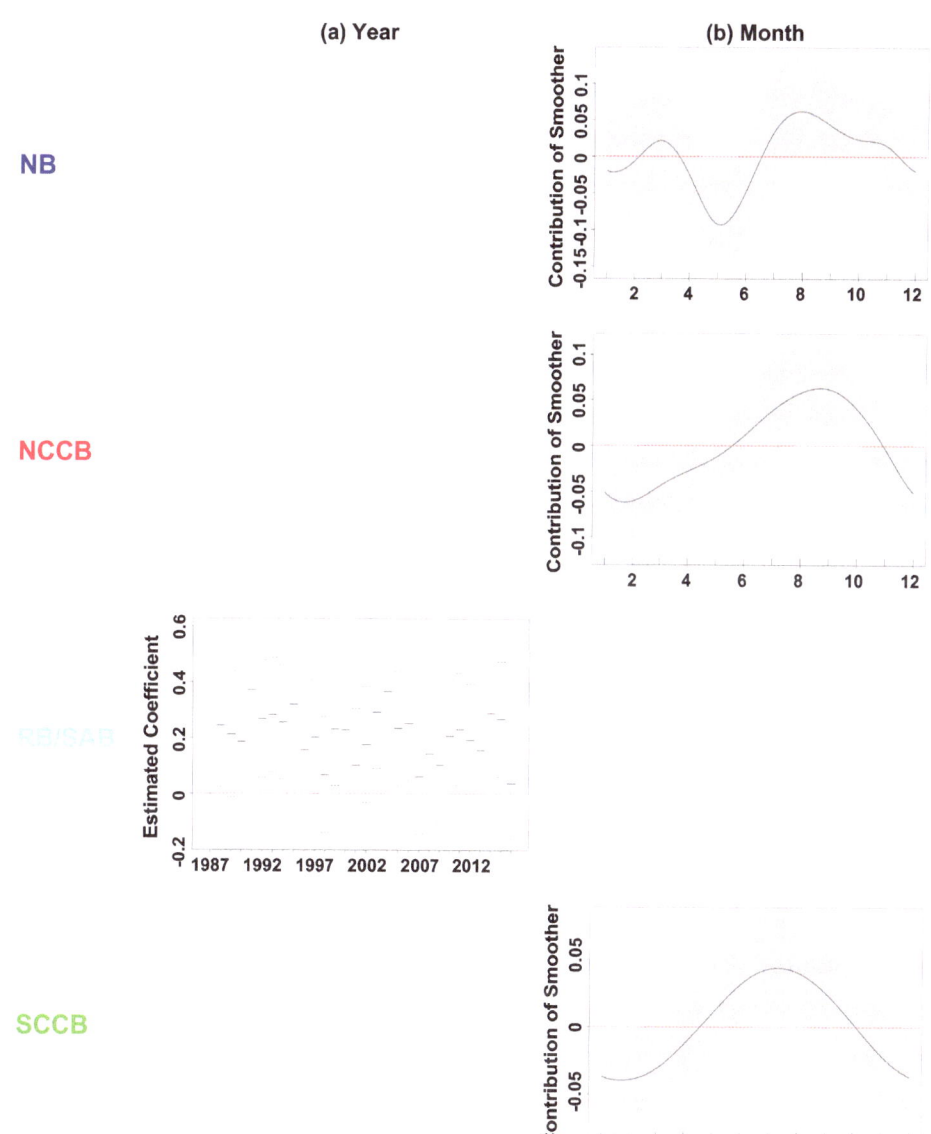

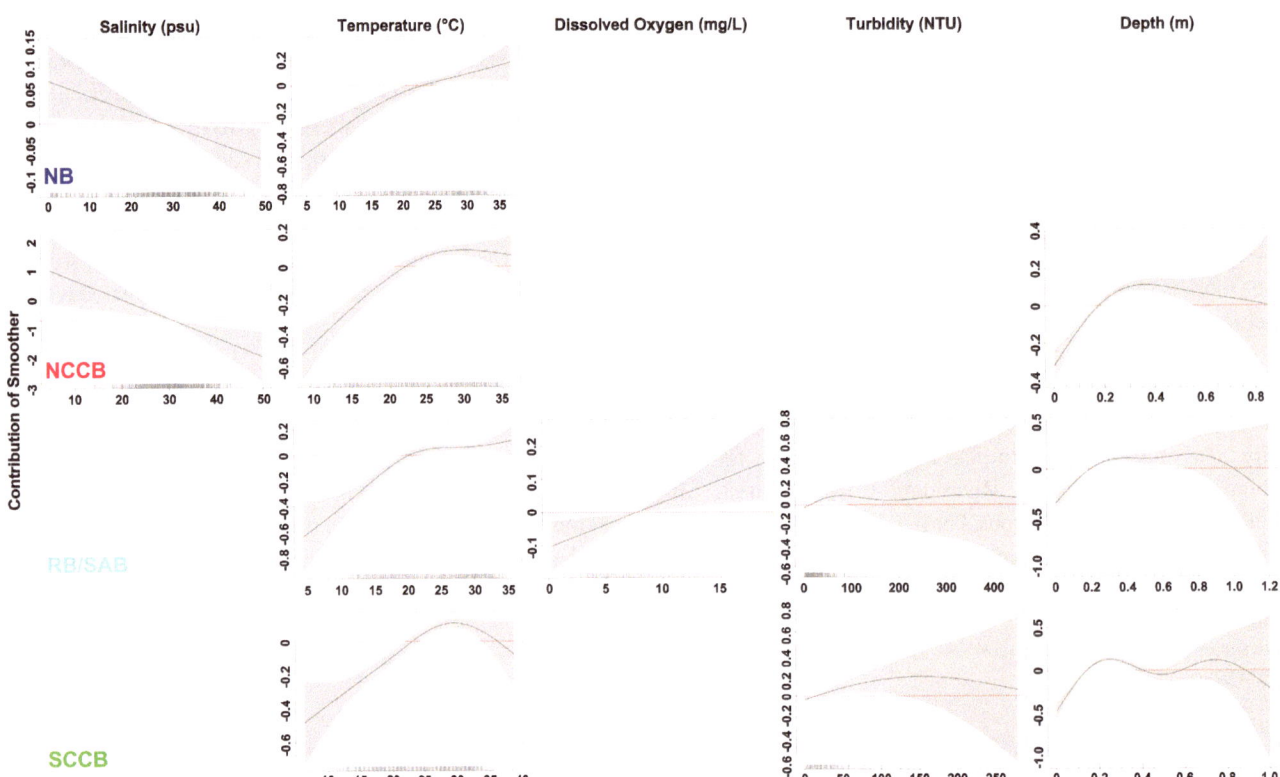

**Fig. 9** Estimated response curves (solid black line) of component smooth functions on nekton N1 diversity from the final generalized additive models for each subregion of the Nueces Estuary: Nueces Bay (NB, blue), North Corpus Christi Bay (NCCB, red), Redfish and South Aransas Bays (RB/SAB, cyan), and South Corpus Christi Bay (SCCB, green). Subplots for predictor variables are consistently positioned for each estuary subregion, with missing subplots indicating candidate pre- dictors that were not retained in the final best-fit model. Shaded areas represent 95% confidence limits of uncertainty in the centered smooth. Vertical axes are partial responses (estimated, centered smooth func- tions) on the scale of the linear predictor. Ticks on the *x*-axis denote values for which there are data. Positive values on the *y*-axis (above the dashed red line) indicate an increase in nekton N1 diversity for each subregion

salinity, and thus freshwater inflow, would enable the cre- ation of base inflow regime recommendations that protect a sound ecological environment to maintain the productivity, extent, and persistence of critical aquatic habitats and spe- cies in the bays and estuaries (Pulich et al. 2002). When selecting indicator species of varying environmental condi- tions, it is important to consider organisms with a broad dis- tribution (Steichen and Quigg 2018). Moreover, the identification of common and abundant indicator species could be used to assess populations of species that are rare or difficult to sample.

Blue crab and Atlantic croaker were selected as freshwa- ter inflow indicator species for habitat suitability analyses based on ecological and economic importance, varying salinity tolerances (Pulich et al. 2002; Tolan 2013), and suit- able sample sizes. Blue crab is an estuarine-dependent, eury- haline crustacean with a complex life history, disperses widely, and spawns year-round (Pile et al. 1996; Blackmon and Eggleston 2001; Reese et al. 2008). Blue crabs support

an economically important commercial fishery in Texas and play a critical role in estuarine food webs by transferring car- bon from the benthos to the nekton (Hines et al. 2003). Atlantic croaker is an estuarine-dependent, euryhaline fish and supports an economically important commercial and recreational fishery in Texas. Adults spawn in the Gulf of Mexico, with peak juvenile recruitment into Texas estuaries in winter (Rooker et al. 1998; Searcy et al. 2007). Atlantic croakers are also a key predator of benthic invertebrates (Pattillo et al. 1997). Although crustaceans and fish are mobile and can move in response to changes in the physical environment, blue crabs and Atlantic croaker have distinct salinity preferences and are found in high abundance (Table 5), making them valuable indicators of freshwater inflow into the Nueces Estuary (Pattillo et al. 1997; Pulich et al. 2002).

A GAM framework was used to investigate the influence of spatiotemporal variables on the probability of capture or occurrence (presence/absence) for each species using a binomially distributed response with a logit link function. Single models were constructed for each indicator species across all subregions to identify areas of highly suitable habitat and regions most vulnerable to changes in salinity and freshwater inflow across the Nueces Estuary. Candidate predictor variables included subregion, year, month, depth, and hydrological data recorded at each bag seine sampling site. Data exploration, GAM construction, and model selection were performed as previously described. The final model was used to evaluate model predictive performance, graphical representation of terms, and habitat suitability mapping.

The predictive performance of GAMs was assessed using cross-validation in two configurations. First, models were refit on a randomly selected 75% training subset of observations, and the resulting model was used to predict the probability of occurrence on the withheld 25% test subset. This process was then repeated ten times. Second, each year was omitted from the training dataset when refitting the model to examine interannual predictive performance. The cross-validation process generated two mean (± standard error) estimates of the area under the receiver operating characteristic curve (AUC) for each final model using the ROCR package (Sing et al. 2005) in R. AUC is a threshold-independent statistic that represents the relationship between the false-positive ratio (1, specificity) and the true-positive ratio (sensitivity) and ranges from 0 (no predictive capability) to 1 (perfect predictive capability). A value of 0.5 indicates that model predictive performance is no better than random, and generally, values from 0.7 to 0.8 are considered acceptable, 0.8 to 0.9 are good, and >0.9 represents excellent model performance (Hosmer and Lemeshow 2000).

Spatially explicit descriptions of nekton communities and distributions of juveniles of many fishery species can help develop conservation and management plans targeting habitats used by fishery species, as well as ecosystem processes and estuarine biodiversity. Areas of high probability of occurrence of juvenile fishery species identified in this study may help determine conservation areas and direct management actions that would benefit these fishery species within this area. In addition, high productivity is a defining characteristic of estuaries, and describing areas of relatively high nekton productivity can help incorporate these areas into efforts to conserve ecosystem processes. Although conservation goals targeting biodiversity should ideally include the entire suite of estuarine biodiversity, the spatial distribution of the entire suite of biodiversity is rarely known across an estuary or embayment.

Following Furey and Rooker (2013), spatial grids of 493 m$^2$ resolution were generated for all areas of the Corpus Christi Bay region ≤2 m depth (total area = 221.94 km$^2$) using bathymetry data from NOAA's National Geophysical Data Center (NGDC 2007). Hydrological data recorded at bag seine sampling sites were interpolated across each grid cell throughout the Nueces Estuary using ordinary kriging from the "automap" package (Hiemstra et al. 2009) in R. Spatial grids were generated for each month during 2007 (wet year), 2008 (average year), and 2009 (dry year) to compare strong interannual contrasts in salinity (annual means of 22.14, 29.27, and 32.70 psu, respectively; Fig. 10).

The final GAMs for blue crab and Atlantic croaker were used to predict the probability of occurrence during months with a high abundance of juvenile recruitment (based on catch per unit effort [CPUE]; individuals per bag seine haul,

**Fig. 10** Mean annual salinities recorded at bag seine sampling sites within the Nueces Estuary between 1987 and 2016 by the Texas Parks and Wildlife Department fishery-independent monitoring program. The dashed red lines indicate quartile values to classify climatic periods for wet (<25.47 psu), average (25.47–32.04 psu), and dry (>32.04 psu) salinity regimes based on TxBLEND-modeled salinity outputs

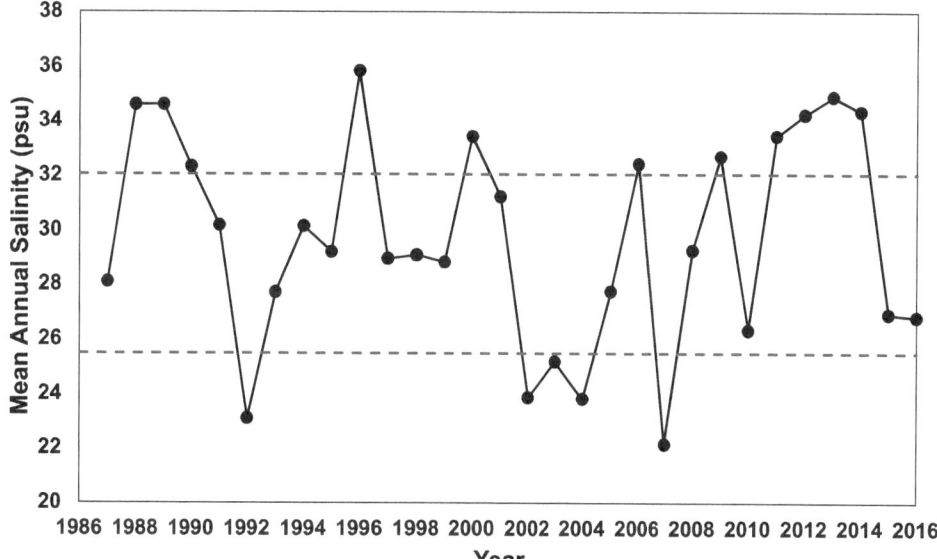

~300 m$^2$ or 0.03 ha) during each year under average environmental conditions and across an increasing salinity gradient or decreasing freshwater inflow (i.e., 1–10 increase in salinity from baseline). Predictions from the fitted GAM models were averaged across high recruitment months each year to examine interannual changes in habitat suitability. The percent change in habitat suitability relative to 2008 (average year) was calculated to identify spatially explicit changes in predicted occurrence throughout the study area. The highest 10% of predicted values for each species in 2008 were used to designate areas of highly suitable habitat each year. Areas of highly suitable habitat within each subregion were then divided by the total amount of available habitat (i.e., ≤2 m depth) within the same subregion (Redfish and South Aransas

Bays, 60.94 km$^2$; North Corpus Christi Bay, 36.05 km$^2$; South Corpus Christi Bay, 56.98 km$^2$; Nueces Bay, 67.97 km$^2$) to provide the proportion of habitat designated as highly suitable habitat in each year and examine predicted changes in habitat along an increasing salinity gradient from baseline conditions.

Spatial, interannual, and intra-annual (seasonal) variability in blue crab and Atlantic croaker occurrence was detected as the subregion, year, and month candidate predictor variables were retained in the final models for both species (Fig. 11). Depth and all hydrological candidate predictor variables were retained within the final models for both species, except for dissolved oxygen in the model for Atlantic croaker (Fig. 12). Blue crab and Atlantic croaker exhibited

**Fig. 11** (**a–b**) Estimated partial coefficients for the parametric terms subregion and year on occurrence (presence/absence) from the final generalized additive models for blue crab and Atlantic croaker. The base levels of Nueces Bay (NB) and 1987 for subregion and year, respectively, are centered, and dashed lines represent 95% confidence limits. (**c**) Estimated response curves (solid black line) of component smooth functions for month on species' occurrence from the final models. Shaded areas represent 95% confidence limits of uncertainty in the centered smooth. Vertical axes are partial responses (estimated, centered smooth functions) on the scale of the linear predictor. Ticks on the *x*-axis denote values for which there are data. Positive values on the *y*-axis (above the dashed red line) indicate an increased probability of occurrence for each species

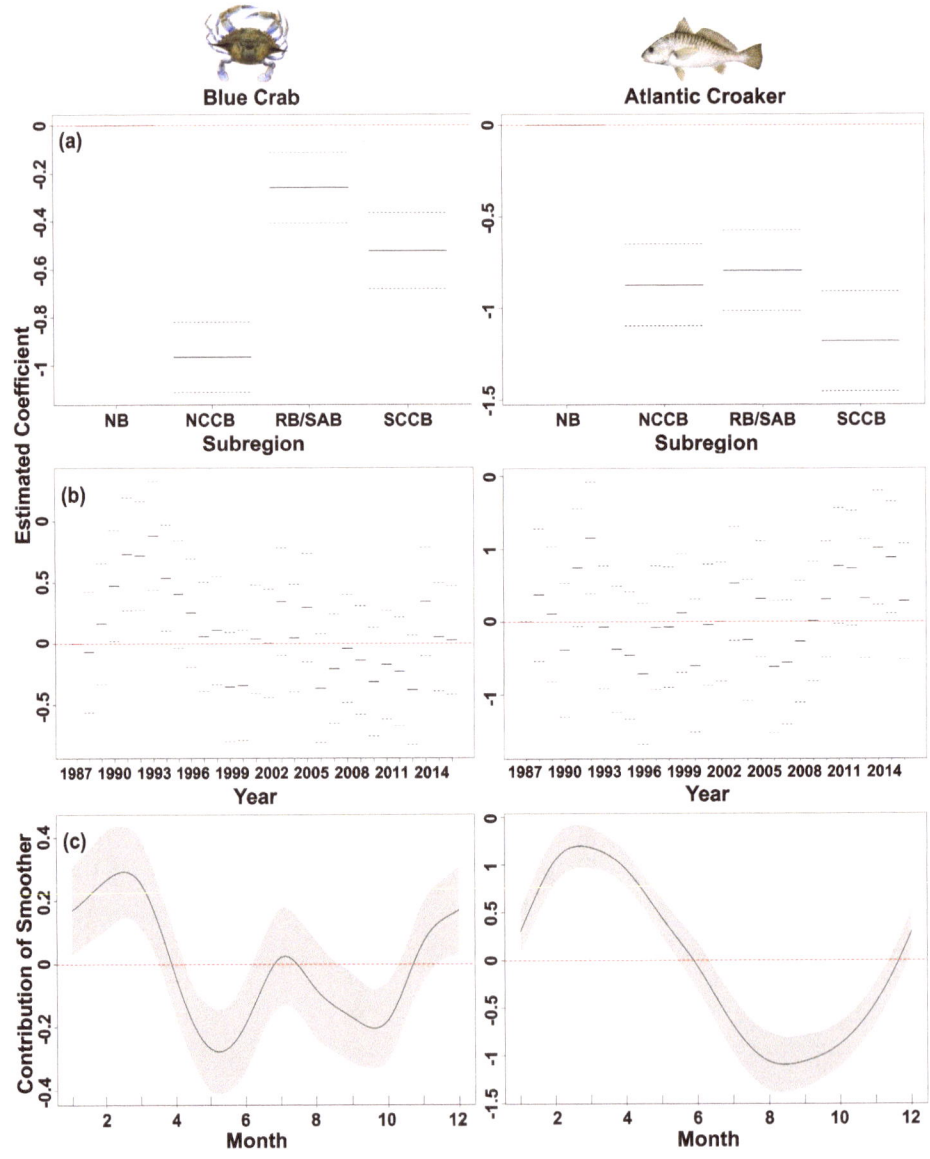

**Fig. 12** Estimated response curves (solid black line) of component smooth functions on occurrence (presence/absence) from the final generalized additive models for blue crab and Atlantic croaker. Subplots for predictor variables are consistently positioned for each species, with missing subplots indicating candidate predictors that were not retained in the final best-fit model. Shaded areas represent 95% confidence limits of uncertainty in the centered smooth. Vertical axes are partial responses (estimated, centered smooth functions) on the scale of the linear predictor. Ticks on the *x*-axis denote values for which there are data. Positive values on the *y*-axis (above the dashed red line) indicate an increased probability of occurrence for each species

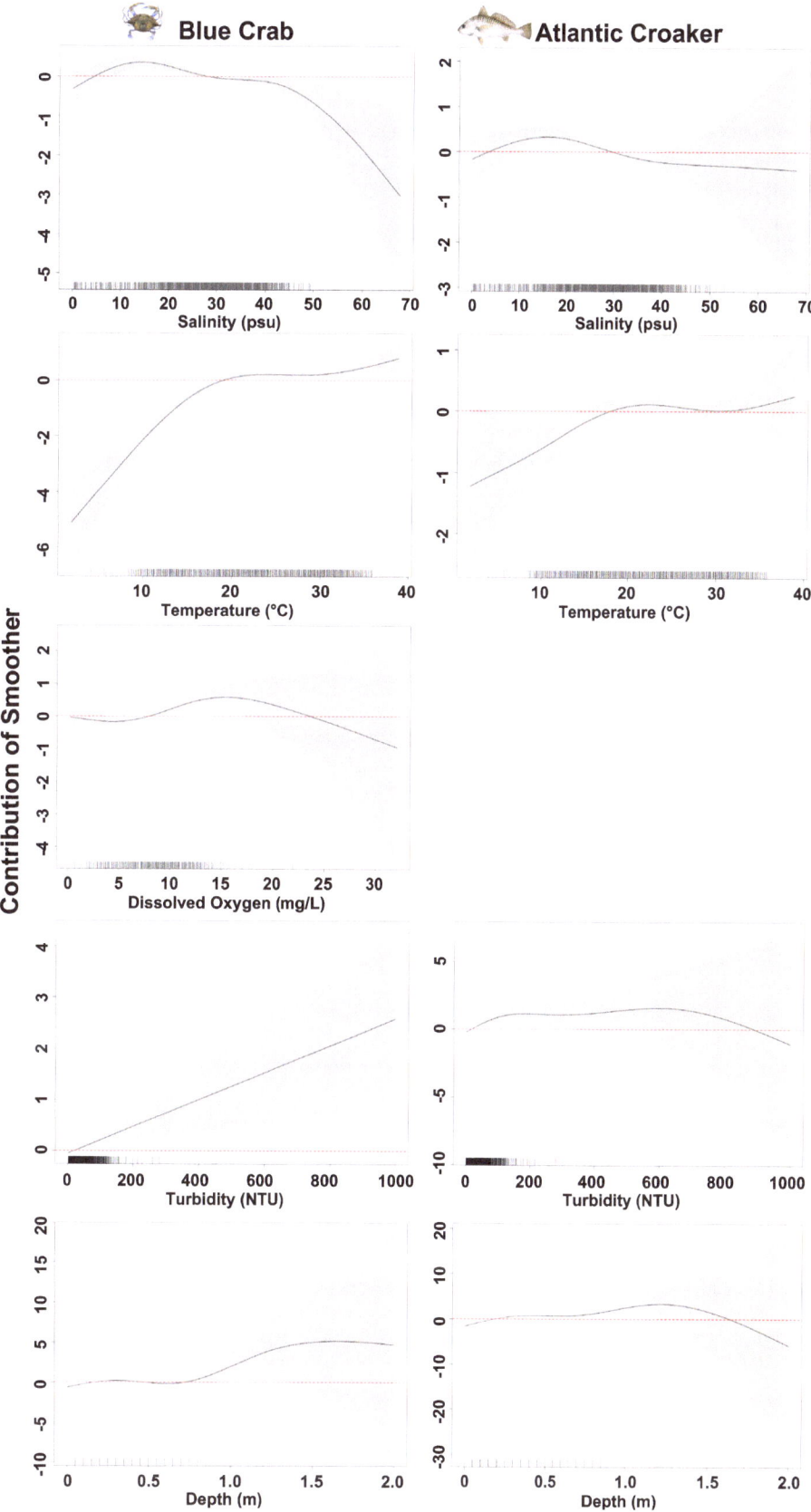

similar response curves for their occurrence across the range of recorded salinity values (Fig. 12). The preferred salinity range for blue crab was approximately 10–20 psu, and occurrence decreased dramatically at higher salinities exceeding 30–40 psu. Similarly, the preferred salinity range for Atlantic croaker was approximately 8–22 psu; however, the response to salinities exceeding 40 psu is highly variable due to sparse numbers of Atlantic croaker captured in hyperhaline conditions.

The final GAMs explained 8.34% and 17.62% of residual deviance for blue crab and Atlantic croaker occurrence, respectively, indicating additional factors beyond our models explain a significant portion of each species' occurrence. Model fit to 75% training–25% test validation data, measured by AUC, was acceptable for Atlantic croaker ($0.775 \pm 0.004$) and marginal for blue crab ($0.678 \pm 0.002$). Cross-validation by year resulted in mean AUC values of $0.659 \pm 0.008$ and $0.784 \pm 0.008$ for blue crab and Atlantic croaker, respectively, closely resembling those generated

during the first validation configuration indicating marginal to acceptable interannual predictive performance.

Predicted maps of habitat suitability for blue crab and Atlantic croaker probability of occurrence under varying salinity regimes were produced to visually demonstrate their distinct freshwater inflow and salinity response. Predicted maps of habitat suitability revealed increasing (2009, dry year) or decreasing (2007, wet year) salinity regimes from the average year (2008) resulted in a reduced probability of occurrence for blue crabs (Fig. 13). In contrast, habitat suitability maps for the dry year had higher probabilities of occurrence for Atlantic croaker than the average or wet year, except for areas of Nueces Bay closest to the Nueces River and therefore most influenced by decreased freshwater inflow and resulting high salinities under dry climatic conditions (Fig. 13). The spatial distribution of the predicted highly suitable habitat was exclusively in Nueces Bay for both blue crab and Atlantic croaker (Fig. 14). The proportion of highly suitable habitats decreased with increasing salinity

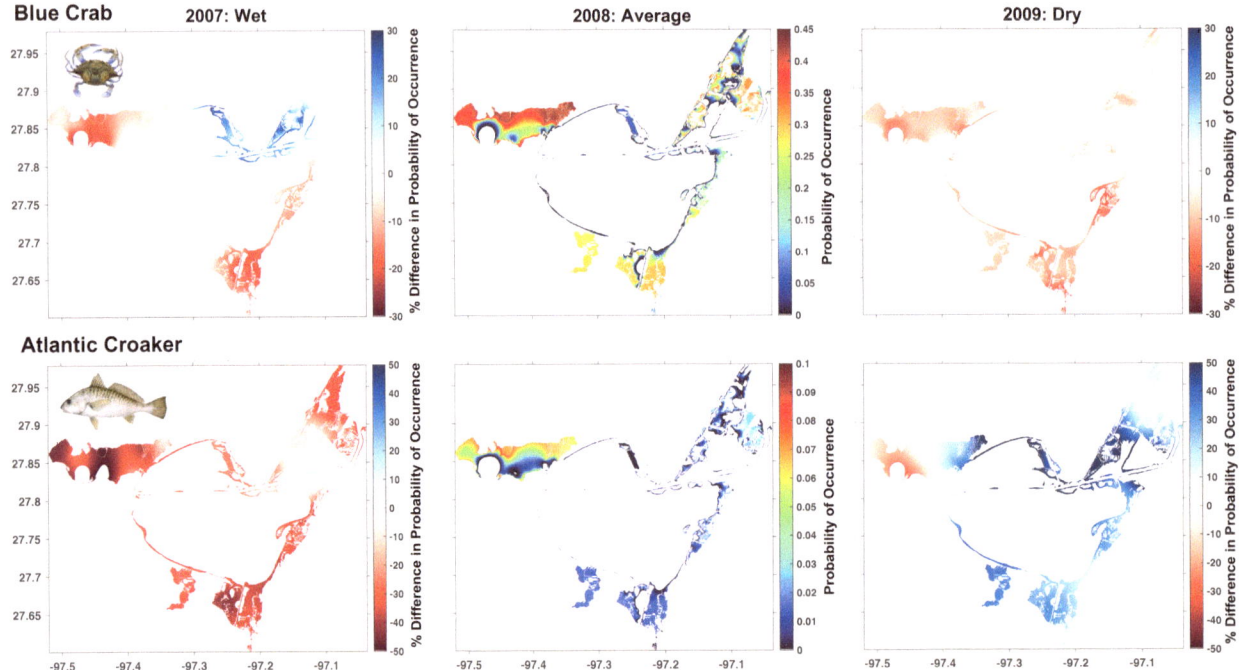

**Fig. 13** Spatial distribution of habitat suitability, defined as the probability of occurrence predicted from the final GAM models, for blue crab and Atlantic croaker across the Nueces Estuary during wet (2007), average (2008), and dry (2009) years. Color scales for wet and dry years represent the percent change in habitat suitability values from the average year. Percent change values beyond ±50% were observed for Atlantic croaker and truncated for clearer illustration

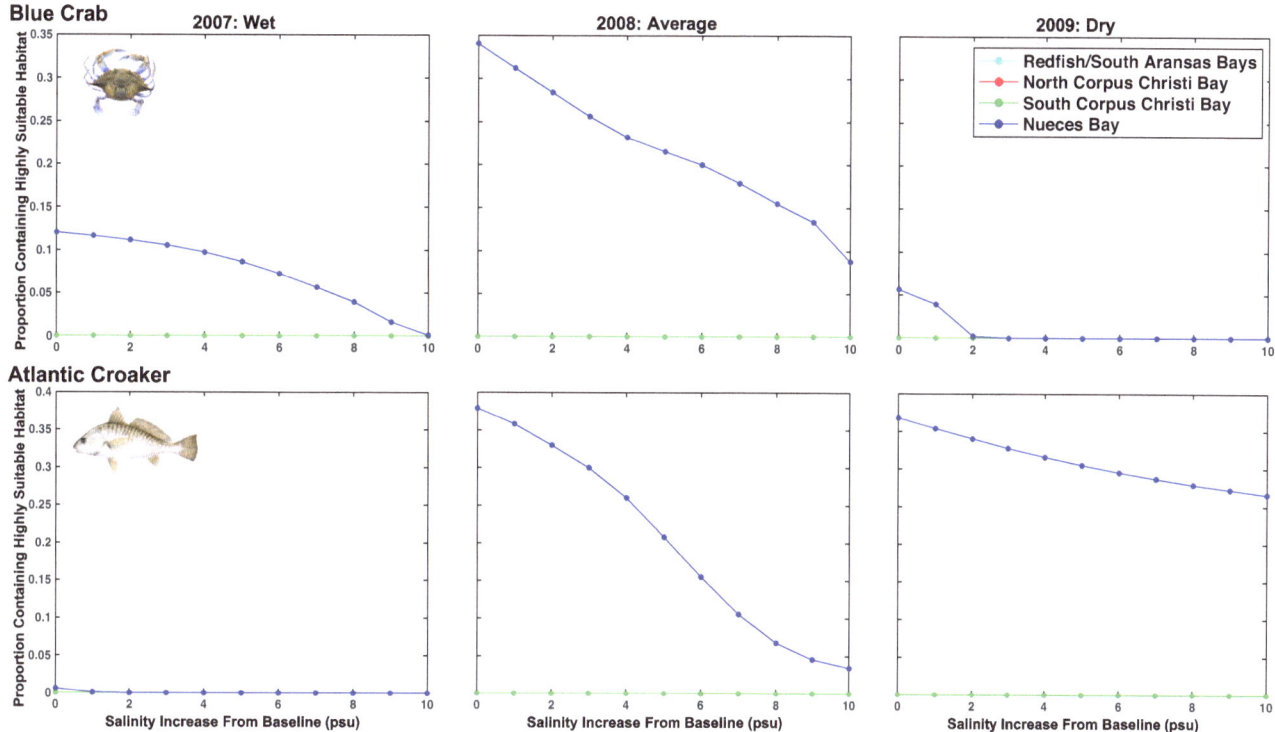

**Fig. 14** The proportion of each subregion within the Nueces Estuary that contained highly suitable habitat (highest 10% of predicted values for blue crab and Atlantic croaker in 2008) along an increasing salinity gradient from baseline conditions during wet (2007), average (2008), and dry (2009) years

from baseline conditions each year for blue crab and Atlantic croaker, demonstrating that freshwater inflow and salinity highly influence these species' occurrence in Nueces Bay.

There are clear and strikingly independent convergences on the ideal salinity and inflow regime needed for highly suitable habitat conditions for a suite of indicator species ranging from marsh plants to vertebrate fish species (Fig. 15). In addition to the nekton indicator species described here, a wealth of scientific studies have identified smooth cordgrass (*Spartina alterniflora*), benthic macroinfauna, and eastern oyster (*Crassostrea virginica*) as primary indicators for establishing a freshwater inflow regime for the Nueces Estuary (see chapters "Submerged Aquatic Vegetation, Marshes, and Mangroves" to "Effects of Climate-Driven Salinity Regimes on Disease Dynamics of the Eastern Oyster, a Key Estuarine Resource and Bioindicator"; NBBEST 2011). These similarities provide a robust suite of indicator species to base recommendations for freshwater inflow and target salinities to produce a sound ecological environment.

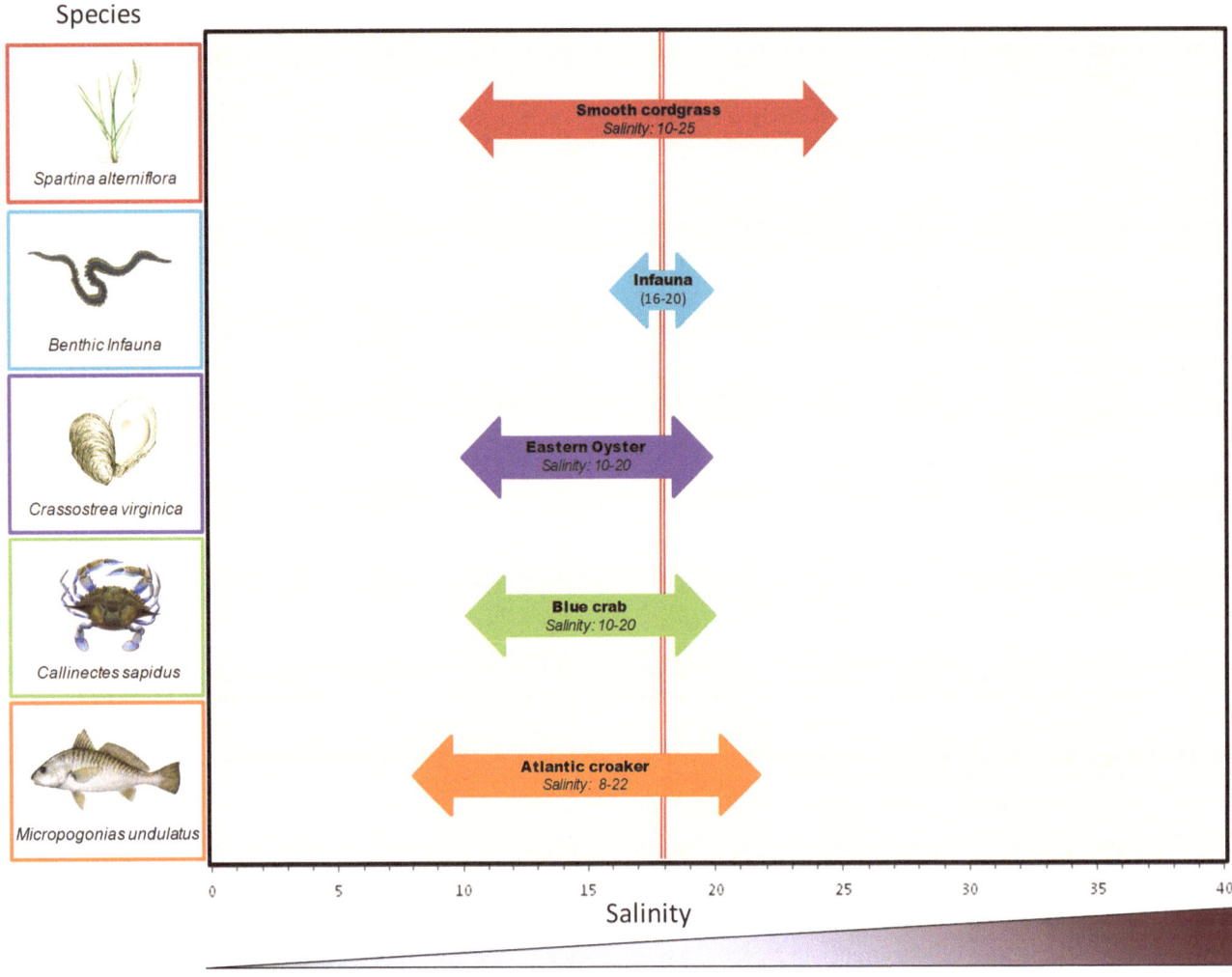

**Fig. 15** Indicator species profile showing salinity preferences in the Nueces Estuary

## Discussion and Recommendations

Anthropogenic pressures and shifts in climate patterns may impact the Texas estuarine fish and invertebrate communities heavily linked to freshwater inflow and salinity levels across a latitudinal gradient (Seager et al. 2007; Tolan 2013). This study used broad-scale diversity and community structure metrics combined with a suite of focal indicator species to determine how changes in multiple abiotic environmental variables, freshwater inflows, and drought and flood events influence Texas estuarine communities across spatial and temporal scales. These patterns indicate that managers may be well-served to incorporate community dynamics, freshwater inflow, and salinity regimes when developing management plans for important fisheries species. With the growing prominence of ecosystem-based fisheries management, it is essential to describe how key factors like salinity affect not just important fisheries species but the entire estuarine communities. Our results highlight how critical long-term fishery-independent datasets are to understanding individual species and community-level responses to environmental variables like salinity. As climate variability increases, resource managers will likely have to develop solutions to meet the specific needs of communities inhabiting them. Resources are often too limited to conserve or restore all estuarine habitats, and being able to focus on the most vulnerable areas and species is critical (Levin and Stunz 2005).

Texas estuaries vary considerably in size, shape, diversity, and water quality. By examining multiple estuaries using a long-term dataset, these results capture the latitudinal, seasonal, and interannual range of environmental conditions, provide insights into the general patterns and trends across different ecosystems, and demonstrate how latitudinal freshwater inflow regimes affect estuarine nekton communities. Across estuaries, salinity was a primary driver of estuarine diversity and community structure; however, nekton diversity and community structure can be highly dynamic. In general, species diversity metrics decreased with increasing salinity; however, community composition shifted across latitudinal salinity regimes so that a suite of species could inhabit each estuary in response to divergent patterns in freshwater inflows and environmental variables. Thus, while salinity is a primary environmental driver, other environmental (e.g., water temperature, habitat type) and ecological (e.g., recruitment) variables are involved.

Within the Nueces Estuary, salinity was also a key driver of nekton diversity for the Nueces Bay and North Corpus Christi Bay subregions, as diversity declined linearly when salinities exceeded 27 psu for both subregions. The spatial distribution of highly suitable habitat was exclusively in Nueces Bay for juvenile blue crab and Atlantic croaker designating critical nursery habitat for these ecologically and economically important indicator species. Moreover, the proportion of highly suitable habitats decreased with increasing salinity from baseline conditions for both indicator species, demonstrating that freshwater inflow and salinity highly influence these species' occurrence in Nueces Bay, particularly near the Nueces River and Delta. These shallow nursery habitats (≤2 m depth) located near the Nueces River mouth contain lower overall volumes and thus require less fresh water to dilute salt water. Therefore, a small amount of fresh water delivered to such vulnerable areas of the Nueces Estuary during droughts may produce significant ecosystem and economic benefits (Montagna et al. 2021b). While we focused on the Nueces Estuary, these indicator species may be beneficial to be used in other Texas estuarine systems and elsewhere with similar environmental conditions, including salinity and freshwater inflow gradients and inter- and intra-annual fluctuations. In addition, the dynamic spatial heterogeneity in the Nueces Estuary helps provide context for some of the unique challenges facing estuaries threatened by decreasing freshwater inflow and increasing temperatures.

As climate variability increases, resource managers will likely have to develop solutions explicit to these estuarine systems to meet the specific needs of the communities inhabiting them. Combining broad-scale diversity and community metrics with fine-scale analyses of key indicator species fosters an ecosystem-based management approach to elucidate the estuarine nekton response to altering freshwater inflow volumes across time and space. This approach also provides opportunities for Gulf-wide assessments to examine the influence of salinity and other abiotic factors, environmental change, and resulting departures of estuarine diversity and community composition from reference conditions.

**Acknowledgments** This publication was funded in part through Contract No. 21-155-007-C879 from the Texas General Land Office (GLO) with Gulf of Mexico Energy Security Act of 2006 funding made available to the State of Texas and awarded under the Texas Coastal Management Program. The views contained herein are those of the authors and should not be interpreted as representing the views of the GLO or the State of Texas. Partial funding was also provided through Contract No. 2120 from the Coastal Bend Bays and Estuaries Program. We thank Dr. Mark Fisher of the Texas Parks and Wildlife Department for contributing bay trawl and bag seine data and Mr. Joe Trungale of Trungale Science and Engineering for contributing TxBLEND-modeled salinity outputs within the Nueces Estuary.

**Data Availability**
The Texas Parks and Wildlife Department fishery-independent monitoring program bay trawl and bag seine survey data used in this chapter is available from the Gulf of Mexico Research Initiative Information and Data Cooperative (https://doi.org/10.7266/m663qn60).

# References

Allen DM, Service SK, Ogburn-Matthews MV (1992) Factors influencing the collection efficiency of estuarine fishes. Trans Am Fish Soc 121(2):234–244

Asquith WH, Mosier JG, Bush PW (1997) Status, trends and changes in freshwater inflows to bay systems in the Corpus Christi Bay National Estuary Program Area. CCBNEP-17. Texas Natural Resource Conservation Commission, Austin, Texas, 47 pp

Baker R, Minello TJ (2011) Trade-offs between gear selectivity and logistics when sampling nekton from shallow open water habitats: a gear comparison study. Gulf Caribb Res 23(1):37–48

Bartoń, K (2020) MuMIn: Multi-Model Inference v 1.43.17

Blackmon DC, Eggleston DB (2001) Factors influencing planktonic, post-settlement dispersal of early juvenile blue crab (*Callinectes sapidus* Rathbun). J Exp Mar Biol Ecol 257:183–203

Breaux N, Lebreton B, Palmer TA, Guillou G, Pollack JB (2019) Ecosystem resilience following salinity change in a hypersaline estuary. Estuar Coast Shelf Sci 225:f106258

Burnham KP, Anderson DR (2002) Model selection and multimodel inference: a practical information-theoretic approach, 2nd edn. Springer, New York, p 496

Chin T, Beecraft L, Wetz MS (2022) Phytoplankton biomass and community composition in three Texas estuaries differing in freshwater inflow regime. Estuar Coast Shelf Sci 277:108059

Clarke KR (1993) Non-parametric multivariate analyses of changes in community structure. Aust J Ecol 18(1):117–143

Clarke KR, Gorley RN (2015) PRIMER v7: User Manual/Tutorial. PRIMER-E Ltd, Plymouth, United Kingdom, 296 pp

Clarke KR, Tweedley JR, Valesini FJ (2014) Simple shade plots aid better long-term choices of data pre-treatment in multivariate assemblage studies. J Mar Biolog Assoc 94(1):1–6

Dance MA, Rooker JR (2016) Stage-specific variability in habitat associations of juvenile red drum across a latitudinal gradient. Mar Ecol Prog Ser 557:221–235

Deegan LA, Finn JT, Ayvazian SG, Ryder-Kieffer CA, Buonaccorsi J (1997) Development and validation of an estuarine biotic integrity index. Estuaries 20:601–617

Dray S, Pélissier R, Couteron P, Fortin MJ, Legendre P, Peres-Neto PR, Bellier E, Bivand R, Blanchet FG, De Cáceres M, Dufour AB (2012) Community ecology in the age of multivariate multiscale spatial analysis. Ecol Monogr 82(3):257–275

Fausch KD, Lyons JO, Karr JR, Angermeier PL (1990) Fish communities as indicators of environmental degradation. Am Fish Soc Symp 8:123–144

Fujiwara M, Zhou C, Acres C, Martinez-Andrade F (2016) Interaction between penaeid shrimp and fish populations in the Gulf of Mexico: importance of shrimp as forage species. PLoS One 11(11):e0166479

Furey NB, Rooker JR (2013) Spatial and temporal shifts in suitable habitat of juvenile southern flounder (*Paralichthys lethostigma*). J Sea Res 76:161–169

Gonzalez LA, Quigg A, Steichen JL, Gelwick FP, Lester LJ (2021) A new approach to functionally assess estuarine fish communities in response to hydrologic change. Estuar Coasts 44:1118–1131

Harris EK, Montagna PA, Douglas AR, Vitale L, Buzan D (2023) Influence of an industrial discharge on long-term dynamics of abiotic and biotic resources in Lavaca Bay, Texas, USA. Environ Monit Assess 195:40

Heim-Ballew H, Olsen Z (2019) Salinity and temperature influence on Scyphozoan jellyfish abundance in the Western Gulf of Mexico. Hydrobiologia 827:247–262

Hiemstra PH, Pebesma EJ, Twenhöfel CJ, Heuvelink GB (2009) Real-time automatic interpolation of ambient gamma dose rates from the Dutch radioactivity monitoring network. Comput Geosci 35:1711–1721

Hill MO (1973) Diversity and evenness: a unifying notation and its consequences. Ecology 54:427–432

Hines AH, Jivoff PR, Bushmann PJ, van Montfrans J, Reed SA, Wolcott DL, Wolcott TG (2003) Evidence for sperm limitation in the blue crab, *Callinectes sapidus*. Bull Mar Sci 72:287–310

Hosmer DW, Lemeshow S (2000) Applied logistic regression, 2nd edn. Wiley, New York, p 392

Imperial MT, Hennessey TM (1996) An ecosystem-based approach to managing estuaries: an assessment of the National Estuary program. Coast Manage 24(2):115–139

Krebs CJ (1999) Ecological methodology, 2nd edn. Benjamin Cummings, Menlo Park, California, p 624

Lellis-Dibble KA, McGlynn KE, Bigford TE (2008) Estuarine fish and shellfish species in U.S. commercial and recreational fisheries: economic value as an incentive to protect and restore estuarine habitat. NOAA Tech. Memo. NMFS-F/SPO-90. U.S. Department of Commerce, National Marine Fisheries Service, Silver Spring, Maryland, 94 pp

Levin PS, Stunz GW (2005) Habitat triage for exploited fishes: can we identify essential "Essential Fish Habitat?". Estuar Coast Shelf Sci 64:70–78

Livernois MC, Fujiwara M, Fisher M, Wells RD (2021) Seasonal patterns of habitat suitability and spatiotemporal overlap within an assemblage of estuarine predators and prey. Mar Ecol Prog Ser 668:39–55

Longley WL (ed) (1994) Freshwater inflows to Texas Bays and Estuaries: ecological relationships and methods for determination of needs. Texas Water Development Board and Texas Parks and Wildlife Department, Austin, Texas, p 386

Mahoney PC, Bishop MJ (2017) Assessing risk of estuarine ecosystem collapse. Ocean Coast Manag 140:46–58

Marshall DA, La Peyre MK, Palmer TA, Guillou G, Sterba-Boatwright BD, Beseres Pollack J, Lebreton B (2021) Freshwater inflow and responses from estuaries across a climatic gradient: an assessment of northwestern Gulf of Mexico estuaries based on stable isotopes. Limnol Oceanogr 66(9):3568–3581

Martin CW, López-Duarte PC, Olin JA, Roberts BJ (2023) Gulf of Mexico estuaries: ecology of the nearshore and coastal ecosystems impacted by the Deepwater Horizon oil spill. Front Environ Sci 11:1203443

Martinez-Andrade F (2015) Marine resource monitoring operations manual. Texas Parks and Wildlife Department, Coastal Fisheries Division, Austin, Texas, 127 pp

Martinez-Andrade F, Campbell FP, Fuls B (2005) Trends in relative abundance and size of selected finfishes and shellfishes along the Texas Coast: November 1975–December 2003. Management Data Series No. 232. Texas Parks and Wildlife Department, Coastal Fisheries Division, Austin, Texas, 140 pp

Matsumoto J, Powell GL, Longley WL, Brock DA (1997) Effects of structure and practices on the circulation and salinity patterns of the Corpus Christi Bay National Estuary Program Area, Texas.

CCBNEP-19. Texas Natural Resource Conservation Commission, Austin, Texas, 169 pp

Miller OL, Putman AL, Alder J, Miller M, Jones DK, Wise DR (2021) Changing climate drives future streamflow declines and challenges in meeting water demand across the southwestern United States. J Hydrol X 11:100074

Minello TJ, Able KW, Weinstein MP, Hays CG (2003) Salt marshes as nurseries for nekton: testing hypotheses on density, growth and survival through meta-analysis. Mar Ecol Prog Ser 246:39–59

Montagna PA, Holt S, Ritter C, Binney K, Herzka S, Dunton K (1998) Characterization of anthropogenic and natural disturbance on vegetated and unvegetated bay bottom habitats in the Corpus Christi Bay National Estuary Program study area. CCBNEP-25. Texas Natural Resource Conservation Commission, Austin, Texas, 130 pp

Montagna PA, Kalke RD, Ritter C (2002) Effect of restored freshwater inflow on macrofauna and meiofauna in upper Rincon Bayou, Texas, USA. Estuaries 25(6B):1436–1447

Montagna PA, Estevez ED, Palmer TA, Flannery MS (2008) Meta-analysis of the relationship between salinity and molluscs in tidal river estuaries of southwest Florida, USA. Am Malacol Bull 24(1):101–115

Montagna PA, Hill EM, Moulton B (2009) Role of science-based and adaptive management in allocating environmental flows to the Nueces Estuary, Texas, USA. In: Brebbia CA, Tiezzi E (eds) Ecosystems and sustainable development VII. WIT Press, Southampton, pp 559–570

Montagna PA, Brenner J, Gibeaut J, Morehead S (2011a) Coastal Impacts. In: Schmidt J, North GR, Clarkson J (eds) The impact of global warming on Texas, 2nd edn. University of Texas Press, Austin, Texas, pp 96–123

Montagna P, Vaughan B, Ward G (2011b) The importance of freshwater inflows to Texas estuaries. In: Griffin RC (ed) Water policy in texas: responding to the rise of scarcity. The RFF Press, Washington, DC, pp 107–127

Montagna PA, Palmer TA, Beseres Pollack J (2013) Hydrological changes and estuarine dynamics. Springer, New York, p 94

Montagna PA, Coffey DM, Jose RH, Stunz G (2021a) Vulnerability assessment of Coastal Bend Bays. Final Report 2120 to the Coastal Bend Bays & Estuaries Program, contract # 2120. Available via https://www.cbbep.org/manager/wp-content/uploads/2120-Final-Report_FINAL.pdf

Montagna PA, McKinney L, Yoskowitz D (2021b) Focused flows to maintain natural nursery habitats. Tex Water J 12:129–139

National Geophysical Data Center (NGDC) (2007) Corpus Christi, Texas 1/3 arc-second MHW Coastal Digital Elevation Model. National Oceanic and Atmospheric Administration National Centers for Environmental Information

Nueces River and Corpus Christi and Baffin Bays Basin and Bay Expert Science Team (NBBEST) (2011) Environmental flows recommendation report. Final Submission to the Environmental Flows Advisory Group, Nueces River and Corpus Christi and Baffin Bays Basin and Bay Area Stakeholders Committee, and Texas Commission on Environmental Quality, 285 pp

Patrick CJ, Yeager L, Armitage AR, Carvallo F, Congdon VM, Dunton KH, Fisher M, Hardison AK, Hogan JD, Hosen J, Hu X (2020) A system level analysis of coastal ecosystem responses to hurricane impacts. Estuar Coasts 43:943–959

Pattillo ME, Czpla TE, Nelson DM, Monaco ME (1997) Distribution and abundance of fishes and invertebrates in Gulf of Mexico Estuaries. Volume II: Species life history summaries. ELMR Report no. 11. NOAA/NOA Strategic Environmental Assessments Division, Rockville, Maryland, 377 pp

Pielou EC (1975) Ecological diversity. Wiley, New York, p 165

Pile AJ, Lipcius RN, Van Montfrans J, Orth RJ (1996) Density-dependent settler-recruit-juvenile relationships in blue crab. Ecol Monogr 66:277–300

Powell GL, Matsumoto J, Brock DA (2002) Methods for determining minimum freshwater inflow needs of Texas bays and estuaries. Estuaries 25:1262–1274

Pulich WM, Tolan JM, Lee WY, Alvis W (2002) Freshwater inflow recommendation for the Nueces Estuary. Texas Parks and Wildlife Department, Resource Protection Division, Austin, Texas, 99 pp

R Core Team (2020) R: a language and environment for statistical computing. R Foundation for Statistical Computing, Vienna, Austria

Reese MM, Stunz GW, Bushon AM (2008) Recruitment of estuarine-dependent nekton through a new tidal inlet: the opening of Packery Channel in Corpus Christi, TX, USA. Estuar Coasts 31:1143–1157

Rooker JR, Holt SA, Soto MA, Holt GJ (1998) Post-settlement patterns of habitat use by sciaenid fishes in subtropical seagrass meadows. Estuaries 21:318–327

Rozas LP, Minello TJ (1997) Estimating densities of small fishes and decapod crustaceans in shallow estuarine habitats: a review of sampling design with focus on gear selection. Estuaries 20:199–213

Seager R, Ting M, Held I, Kushnir Y, Lu J, Vecchi G, Huang HP, Harnik N, Leetmaa A, Lau NC, Li C (2007) Model projections of an imminent transition to a more arid climate in southwestern North America. Science 316:1181–1184

Searcy SP, Eggleston DB, Hare JA (2007) Environmental influences on the relationship between juvenile and larval growth of Atlantic croaker (*Micropogonias undulatus*). Mar Ecol Prog Ser 349:81–88

Shannon CE, Weaver W (1949) The mathematical theory of communication. University of Illinois Press, Urbana, IL, p 144

Sheridan P (2004) Recovery of floral and faunal communities after placement of dredged material on seagrasses in Laguna Madre, Texas. Estuar Coast Shelf Sci 59(3):441–458

Sing T, Sander O, Beerenwinkel N, Lengauer T (2005) ROCR: visualizing classifier performance in R. Bioinformatics 21:3940–3941

Sklar FH, Browder JA (1998) Coastal environmental impacts brought about by alterations to freshwater flow in the Gulf of Mexico. Environ Manag 22:547–562

Statistical Analysis System (SAS) Institute Inc. (2017) SAS/ETS® 14.3 User's Guide. SAS Institute Inc., Cary, North Carolina, 4272 pp

Steichen JL, Quigg A (2018) Fish species as indicators of freshwater inflow within a subtropical estuary in the Gulf of Mexico. Ecol Indic 85:180–189

Thronson A, Quigg A (2008) Fifty-five years of fish kills in coastal Texas. Estuar Coasts 31:802–813

Tolan JM (2013) Estuarine fisheries community level response to freshwater inflows. In: Wurbs R (ed) Water resources planning, development, and management. IntechOpen, London, pp 73–94

Turner EL, Montagna PA (2016) The max bin regression method to identify maximum bioindicator responses to ecological drivers. Ecol Inform 36:118–125

Ward GH (1997) Process and trends of circulation within the Corpus Christi Bay National Estuary Program Study Area. CCBNEP-21. Texas Natural Resource Conservation Commission, Austin, Texas, 380 pp

Ward GH, Armstrong NE (1997) Current Status and Historical Trends of Ambient Water, Sediment, Fish and Shellfish Tissue Quality in the Corpus Christi Bay National Estuary Program Study Area. CCBNEP-13. Texas Natural Resource Conservation Commission, Austin, Texas, 270 pp

Wetz MS, Yoskowitz DW (2013) An 'extreme' future for estuaries? Effects of extreme climatic events on estuarine water quality and ecology. Mar Pollut Bull 69(1–2):7–18

Whaley SD, Burd JJ Jr, Robertson BA (2007) Using estuarine landscape structure to model distribution patterns in nekton communities and in juveniles of fishery species. Mar Ecol Prog Ser 330:83–99

Whaley SD, Shea CP, Burd JJ Jr, Harmak CW (2023) Trophodynamics of nekton assemblages and relationships with estuarine habitat structure across a subtropical estuary. Estuar Coasts 46(2):580–593

Whitfield AK, Elliott M (2002) Fishes as indicators of environmental and ecological changes within estuaries: a review of progress and some suggestions for the future. J Fish Biol 61:229–250

Withers K, Chapman BR, Tunnell JW, Judd FW (2023) The Laguna Madre of Texas and Tamaulipas, Revised ed. Texas A&M University Press, College Station, Texas, 544 pp

Wood SN (2011) Fast stable restricted maximum likelihood and marginal likelihood estimation of semiparametric generalized linear models. J R Stat Soc Ser B Methodol 73:3–36

Wood SN (2017) Generalized additive models: an introduction with R, 2nd edn. CRC Press, Boca Raton, FL, p 496

Zuur AF, Ieno EN, Walker NJ, Saveliev AA, Smith GM (2009) In: Gail M, Krickeberg K, Samet JM, Tsiatis A, Wong W (eds) Mixed effects models and extensions in ecology with R. Springer, New York, p 574

# Nitrogen and Phosphorus Budgets for Texas Estuaries

Danielle A. Marshall ⓘ and Paul A. Montagna ⓘ

## Abstract

Productivity in estuaries is dependent upon nutrient influx and cycling. However, increasing urbanization and socioeconomic activities continue to influence nutrient dynamics and thus estuarine organisms. The western Gulf of Mexico coast lies in a climatic gradient that drives differences in hydrology among Texas estuaries providing a natural experiment to determine how hydrology affects nutrient biogeochemical budgets. One modeling approach is the land-ocean interaction in the coastal zone (LOICZ) guideline. The LOICZ budgets were calculated for all major estuaries in Texas to compare long-term water, salt, and nutrient budgets. The budgets were calculated based on daily averages using various datasets from multiple Texas agencies. Separate nutrient and water budgets were calculated for the Lower Laguna Madre Estuary and the Upper Laguna Madre Estuary. The climatic gradient along the coasts drives the water budgets because precipitation difference is the main factor influencing inflow balance. Inflow balance is the main factor influencing differences in salinity and nutrient concentrations among the estuaries. However, elevated nutrient levels in two estuaries are caused by urbanization, return flows, and agricultural runoff. Draining the adjacent large city of Houston, Texas, USA, the Trinity-San Jacinto Estuary had higher daily average dissolved inorganic phosphorus (DIP) and total phosphorus (TP) levels compared to the other estuaries. These high levels are likely caused by high use of detergents and wastewater return flows that include fecal and urine waste. At the opposite end of the coast, the Upper Laguna Madre Estuary has higher total Kjeldahl nitrogen (TKN) compared to the other estuaries due to its unique features including hypersalinity and low inflow dominated by sewage and agricultural runoff. The Texas budgets are comparable to budgets in estuaries in Asia, South America, and Africa with similar hydrology. This study demonstrates that hydrology controls nutrient transport and altered hydrology can alter the nutrient dynamics. Altered nutrient dynamics can affect biological productivity and water quality in estuaries.

## Keywords

Biogeochemistry · Budgets · LOICZ · Nitrogen · Phosphorus · Nutrients

## Abbreviations

| | |
|---|---|
| DIC | Dissolved inorganic carbon |
| DIN | Dissolved inorganic nitrogen ($NO_3 + NO_2 + NH_4$) |
| DIP | Dissolved inorganic phosphorus |
| GE | Guadalupe Estuary |
| LC | Lavaca-Colorado Estuary |
| LLM | Lower Laguna Madre Estuary |
| LOICZ | Land-Ocean Interactions in the Coastal Zone |
| MA | Mission-Aransas Estuary |
| NC | Nueces Estuary |
| NEM | Net ecosystem metabolism |
| $NH_4$ | Ammonium |
| $NO_2$ | Nitrite |

**Supplementary Information** The online version contains supplementary material available at https://doi.org/10.1007/978-3-031-70882-4_15.

D. A. Marshall · P. A. Montagna (✉)
Texas A&M University-Corpus Christi, Harte Research Institute for Gulf of Mexico Studies, Corpus Christi, TX, USA
e-mail: Paul.Montagna@tamucc.edu

© The Author(s) 2025
P. A. Montagna, A. R. Douglas (eds.), *Freshwater Inflows to Texas Bays and Estuaries*, Estuaries of the World,
https://doi.org/10.1007/978-3-031-70882-4_15

| NO₃ | Nitrate |
| SN | Sabine-Neches Estuary |
| TCEQ | Texas Commission of Environmental Quality |
| TKN | Total Kjeldahl nitrogen |
| TON | Total organic nitrogen (TKN-NH₄) |
| TP | Total phosphorus |
| TPWD | Texas Parks and Wildlife Department |
| TS | Trinity-San Jacinto Estuary |
| TWDB | Texas Water Development Board |
| ULM | Upper Laguna Madre Estuary |
| USGS | United States Geological Survey |
| Y | Nonconservative material of interest |

## Introduction

Surface water management is critical to maintaining healthy ecosystems, protecting human health, keeping up with urban expansion, and increasing economic development. Rapid population growth and urbanization have increased environmental stresses on surface water availability, making water resource management decisions and tools essential to maintaining freshwater inflow needs. As water demand increases, the diversion and capturing of freshwater will decrease environmental flow into the coastal zone and estuaries. Estuarine systems are a critical component to maintaining a healthy coastal ecosystem as they provide ecosystem services including natural habitats for vegetation and wildlife, erosion control, freshwater for human consumption, recreational activities, and act as "buffer zones" to stabilize shorelines from rain and flooding events.

Freshwater inflow in particular plays a vital role in estuarine ecosystems including reducing salinity levels by diluting seawater, providing a flow of land-derived nutrients, and providing an influx of sediment from the watershed to coastal wetlands (Brandes et al. 2009). Watershed inputs from runoff, groundwater, and sewage return flows also contribute to freshwater inflow in estuaries transporting nutrients and sediment from upstream processes down to the coastline. Other sources of nutrients come from tidal exchange, which are also crucial contributors to overall nutrient loads. In recent decades, human activities have greatly increased the nutrient inputs over naturally occurring levels such as urban runoff, burning of fossil fuels, and the increased use of agricultural practices.

The concentration of nutrients in an estuarine process is ever-changing with constant inputs and outputs from river flows and oceanic exchange as well as regeneration and biological uptake (Montagna et al. 2013). Changes in nutrient concentrations can lead to changes of biological processes such as primary and secondary production. Net primary productivity in an aquatic ecosystem is often characterized by the "Redfield ratio" where many essential processes in estuarine environments are limited by the availability of either nitrogen or phosphorus or both simultaneously. The ratio of carbon to nitrogen to phosphorus in both phytoplankton biomass and in dissolved nutrient pools in the ocean is nearly constant at 106:16:1 where nitrogen and phosphorus are limiting factors (Redfield 1958).

Nitrogen and phosphorus are important nutrients that can be found in estuarine waters in various forms. Nitrogen can be found in the form of dissolved organic nitrogen (DON), dissolved inorganic nitrogen (DIN), and total organic nitrogen (TON) including the dissolved and particulate forms of organic nitrogen. Nitrogen undergoes several processes in the nitrogen cycle including nitrogen fixation, nitrification, assimilation, ammonification, denitrification, anammox (anaerobic ammonium oxidation), and nitrogen loss and can be distributed through the uptake of plant biomass, storage in salt marsh sediments, and microbial denitrification (Bowen et al. 2020). Phosphorus can be found in the form of dissolved inorganic phosphorus (DIP), dissolved organic phosphorus (DOP), and total phosphorus (TP) including both DIP and DOP. Phosphorus is important for plant and animal energy transfer processes and is introduced into waterways by weathering, plant uptake, and decomposition by microorganisms. However, nutrient cycling pathways are not exclusively influenced by environmental factors, and human sources also play a significant role in introducing nutrients into a watershed system. Several human activities contribute to nutrient inputs, including agriculture, wastewater treatment, urbanization, and industrial activities. Excess nutrient levels can have detrimental effects on estuarine systems including algae blooms, hypoxia, fish kills, and excessive turbidity. Conversely, nutrient deficiencies can also stress an estuarine system by inhibiting vegetation and algae growth. Therefore, maintaining a balance of nutrients is critical to sustaining an estuary ecosystem and maintaining sustainable water quality parameters.

The distinction between estuaries can often be defined by their salinity and climate gradients that are vital in regulating ecological and biological processes in estuarine systems (Montagna et al. 2013). Texas contains seven major estuaries along the Gulf Coast including the Sabine-Neches Estuary, Trinity-San Jacinto Estuary, Lavaca-Colorado Estuary, Guadalupe Estuary, Mission-Aransas Estuary, Nueces Estuary, and Laguna Madre Estuary. All seven of these estuaries have similar geographical features but differ in hydrology due to a climate gradient (Montagna et al. 2013).

The climatic gradient along the Texas coast provides a unique opportunity to examine hydrological control of nutrient budgets (Montagna et al. 2018). Creating a nutrient budget for nitrogen and phosphorus can also provide an understanding of nutrient dynamics in each of the seven watersheds along the Texas coast. Understanding the interactions between estuaries and coastal inlets and how this ecosystem responds to global climate changes cannot be understood with observational studies alone. The Land-Ocean Interactions in the Coastal Zone (LOICZ) model is a biogeochemical budget calculator on mass balance of specific variables for a defined geographic area and time period (Gordon et al. 1996). This model can be used to identify and quantify the important fluxes in and out of the coastal zone (Gordon et al. 1996). Moreover, knowledge of these nutrient budgets will allow key stakeholders to make more informed decisions on environmental flow standards in Texas.

A recent application of the LOICZ modeling approach was completed in Shenzhen Bay, China, specifically looking at the internal nutrient loading in sediment as it can both release and receive nutrients produced during biogeochemical reactions in water columns across the water-sediment interface (Yan et al. 2021). These researchers hypothesized that internal nutrient loading is one of the important contributing factors to the eutrophication of Shenzhen Bay (Yan et al. 2021). Budget models were developed for salinity, inflow, and the nutrient budget of dissolved inorganic nitrogen (DIN) and phosphorus, considering both wet and dry seasons. This approach allowed the researchers to show quantitatively that internal nutrient loading is the main contributing factor to the total input fluxes of nitrogen and phosphorus in Shenzhen Bay and classify runoff patterns of the bay to provide possible management solutions. Their results demonstrated that internal nutrient loading supplies 65–69% of the total input fluxes of DIN and phosphorus.

Another study was performed in Oso Bay, the northwestern extension of Corpus Christi Bay, using the LOICZ modeling approach to assess the impact of storms on nutrient loadings, specifically before and after Category 2 Hurricane Dolly (Arismendez 2010). Daily salinity and water budgets were created before, during, and after the storm where Oso Bay is the "box" with inputs and outputs from various sources. Then estimates of net ecosystem metabolism and nutrient budgets were calculated under the LOICZ guidelines. The nutrient budgets revealed that there is a change in ecosystem response from autotrophic to heterotrophic then back to autotrophic 2 days after the storm, and there was a spike in nutrient concentration on the day of the hurricane (Arismendez 2010). Moreover, the nutrient budget model provided information that total nutrient loads of 52,701 kg DIN and 81,581 kg DIP were delivered to Oso Bay because of this hurricane (Arismendez 2010).

The objective of the current study is to determine the effect of hydrology on the nitrogen and phosphorus nutrient budgets in estuaries. The study contrasts budgets of the seven major estuaries along the Texas coast that lay in a climatic gradient. The Laguna Madre Estuary was further separated into two systems: the Lower Laguna Madre Estuary and the Upper Laguna Madre Estuary. The budgets were assembled by applying the LOICZ modeling approach to increase the understanding of nitrogen and phosphorus nutrient loading patterns with ever-changing environmental conditions. Results from this study of nitrogen and phosphorus nutrient loading patterns can inform the evaluation and development of freshwater inflow standards and can be useful for planning restoration and management practices to maintain ecosystem health of these estuarine systems.

## Methods

The Texas coastline stretches roughly 590 km (367 miles) in which 21 river basins bring freshwater from rivers, streams, and point and nonpoint sources into seven individual bay systems (TWDB 2022a). In latitudinal order from the northeast near the Louisiana border to the southwest portion near the Mexico border, seven major estuaries were studied. These estuaries include the Sabine-Neches (SN), Trinity-San Jacinto (TS), Lavaca-Colorado (LC), Guadalupe (GE), Mission-Aransas (MA), Nueces (NC), and the Laguna Madre Estuary (LM). The Laguna-Madre Estuary study site was further divided into the Upper Laguna Madre Estuary (ULM) and the Lower Laguna Madre Estuary (LLM) due to the physical separation of the two sections by the 3-km-(20-mile)-long Saltillo Flats. The physical and hydrological characteristics of each estuary including area and volume can be found in Table 1. Maps of each estuary with their corresponding stations that were used in this analysis can be found in the supplemental material (Figs. S1–S8).

## LOICZ Modeling

The general methodology for building LOICZ models includes four main components: a water budget, a salt budget, a budget of nonconservative materials, and the stoichiometric relationships among nonconservative budgets that do not adhere to physical mixing alone (Gordon et al. 1996). The current study focused on creating a series of long-term average daily budgets for conservative and nonconservative

**Table 1** Physical and hydrological characteristics of Texas estuaries

| Estuary | Area (km²) | Volume (10⁹ m³) | Tidal prism (10⁹ m³) |
|---|---|---|---|
| Sabine-Neches | 265 | 0.66 | 0.13 |
| Trinity-San Jacinto | 1456 | 2.71 | 0.45 |
| Lavaca-Colorado | 1115 | 2.43 | 0.22 |
| Guadalupe | 587 | 0.73 | 0.11 |
| Mission-Aransas | 524 | 0.85 | 0.12 |
| Nueces | 571 | 1.28 | 0.23 |
| Upper Laguna Madre | 591 | 0.45 | 0.09 |
| Lower Laguna Madre | 1308 | 1.00 | 0.51 |

Tidal prism is defined as the volume of water between average high and low tides in the estuary (Engle et al. 2007)

materials that are stoichiometrically linked. The modeling approach in this study includes a simple box model where each estuary is represented as the "box" or "system" and components from all inputs (e.g., inflow from gages, stream-flow, return flows, and precipitation) and outputs (e.g., diversions, evaporation, and residual flow) with steady-state assumptions. The data used in this study were compiled from multiple agencies (e.g., TCEQ, USGS, TWDB, and TPWD) to complete the LOICZ nutrient budget model for both nitrogen and phosphorus in all eight major estuaries in Texas, counting the division of the Laguna Madre Estuary.

The water budget (Fig. 1) of freshwater inflows was developed. Tidal or wind-induced inflows and outflows tend to balance out over time and therefore are not important to include in the water budget (Gordon et al. 1996). The salt budget (Fig. 2) was created as a budget of conservative materials due to salt not being either consumed or produced in the system. Consequently, the salt entering the system must equal the outputs of that system. Water budgets were calculated using the following equations (Gordon et al. 1996):

$$\frac{dV_1}{dt} = V_G + V_M + V_F + V_P + V_R - V_E - V_D \qquad (1)$$

Where:

$\frac{dV_1}{dt} =$ water storage represented by the change in the system of interest with time

$V_G =$ gaged flow

$V_M =$ modeled flow

$V_F =$ return flow

$V_P =$ precipitation

$V_R =$ residual flow

$V_E =$ evaporation

$V_D =$ diversion flow

It is important to note that $\frac{dV_1}{dt} = 0$ with the assumption of the steady-state conditions. The concept behind the water budget is that there is conservation of water mass. Therefore, it is assumed that the change of water volume through time is known or the water volume remains constant. For that reason, the net water outflow from the system can be estimated by a difference between inflows and outflows known as the "residual flow." The residual flow was calculated using the following equation (Gordon et al. 1996):

$$V_R = -V_G - V_M - V_F - V_P + V_E + V_D \qquad (2)$$

The freshwater inflow balance was calculated using the following equation:

$$V_B = V_G + V_M + V_F + V_P - V_E - V_D \qquad (3)$$

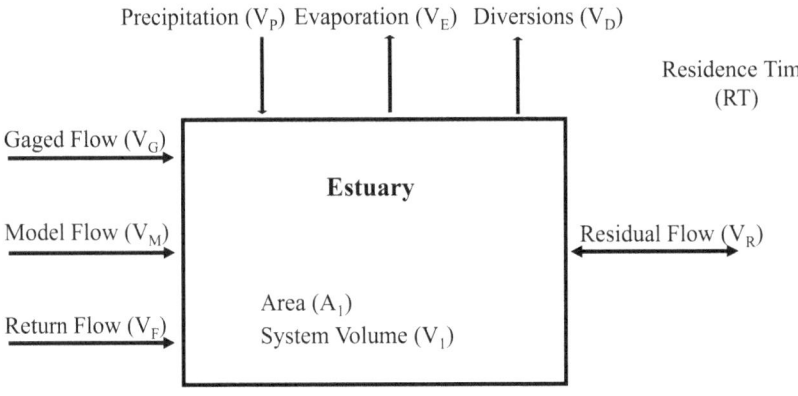

**Fig. 1** Schematic of the water budget using the LOICZ modeling guidelines. The arrows represent the direction and net flow of each component

Precipitation ($V_P$)  Evaporation ($V_E$)  Diversions ($V_D$)

Residence Time (RT)

Gaged Flow ($V_G$)

Model Flow ($V_M$)

Return Flow ($V_F$)

**Estuary**

Area ($A_1$)
System Volume ($V_1$)

Residual Flow ($V_R$)

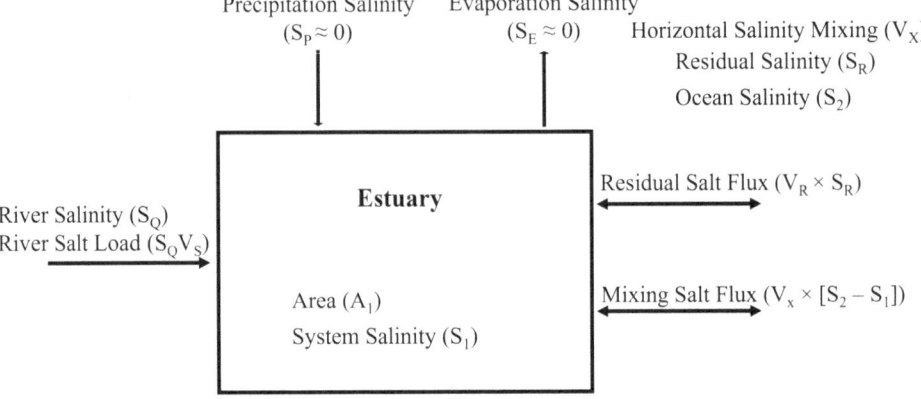

**Fig. 2** Schematic of the salt budget using the LOICZ modeling guidelines. The arrows represent the direction and net flow of each component

Precipitation Salinity ($S_P \approx 0$)  Evaporation Salinity ($S_E \approx 0$)  Horizontal Salinity Mixing ($V_X$)
Residual Salinity ($S_R$)
Ocean Salinity ($S_2$)

River Salinity ($S_Q$)
River Salt Load ($S_Q V_S$)

**Estuary**

Area ($A_1$)
System Salinity ($S_1$)

Residual Salt Flux ($V_R \times S_R$)

Mixing Salt Flux ($V_X \times [S_2 - S_1]$)

Residence time, also referred to as the freshwater flushing time, is defined as the average time it takes to replace the equivalent freshwater in the estuary by freshwater inputs (Solis and Powell 1999). There are many different methods to calculate residence time, and they were all calculated and compared in this paper. Residence times were calculated using all three methods in the following Eqs. (4, 5, 6):

$$RT = \frac{V}{Q} \tag{4}$$

Where:

$V$ = volume of the estuary
$Q$ = freshwater inflow rate into the estuary

$$RT = \frac{V}{V_B} \tag{5}$$

Where:

$V$ = volume of the estuary
$V_B$ = freshwater inflow balance

$$f = \frac{\sigma - s}{s}, RT = \frac{f \times V}{Q} \tag{6}$$

Where:

$f$ = freshwater fraction
$\sigma$ = salinity of receiving waters
$s$ = average salinity of the estuary
$V$ = volume of the estuary
$Q$ = freshwater inflow rate into the estuary

For the salt budgets, salinity data were collected with sondes as practical salinity units (a calculation based on measuring the conductivity of seawater); therefore the units were reported as S (Millero et al. 2008; Lewis 1980). On the other hand, ocean salinity reported as $S_P$ (Intergovernmental Oceanographic Commission 2015) and psu (practical salinity units) are still in common use for these types of measurements. The salt flux is a rate and is reported as g/m³/d. Grams

of salt are estimated from the measurements of practical salinity ($S$). The mixing terms ($V_x$) for the salt budgets that occurs between the bay and the ocean were calculated using the following equation:

$$V_x = (V_R \times S_R) \div (S_1 - S_2) \tag{7}$$

Where:

$V_x$ = horizontal salinity mixing
$S_R$ = residual salinity
$S_1$ = system salinity
$S_2$ = Gulf/ocean salinity

Precipitation and evaporation were not used to calculate the horizontal salinity mixing term because precipitation is freshwater and when evaporation occurs only water evaporates, not salt (Arismendez 2010). When creating the salinity input from rivers, river streamflow data from TCEQ were used to calculate the salt input ($S_Q V_S$) rather than using the inflow data from TWDB. This is because using streamflow data specific to each river segment gives a more accurate calculation of the salt load entering the estuary. However, the streamflow data were not used in the water budget to stay consistent with the TWDB data and avoid calculating freshwater inflow rates twice.

Several offshore salinity station data from TPWD were used in more than one estuary due to some estuaries not having a direct connection to the Gulf, and the stations between the estuaries overlap. For example, the offshore salinity from the Lavaca-Colorado Estuary stations spanned across the Guadalupe Estuary; therefore data from those stations were used for both individual estuary LOICZ budgets in Figs. S3 and S4.

Budgets of nonconservative materials were calculated for nitrogen and phosphorus (Fig. 3). A univariate analysis to investigate the normality and distribution of the data was performed. The univariate test results showed that the nutrient data available from TCEQ contained some outliers that could potentially skew the data. The natural log of the values

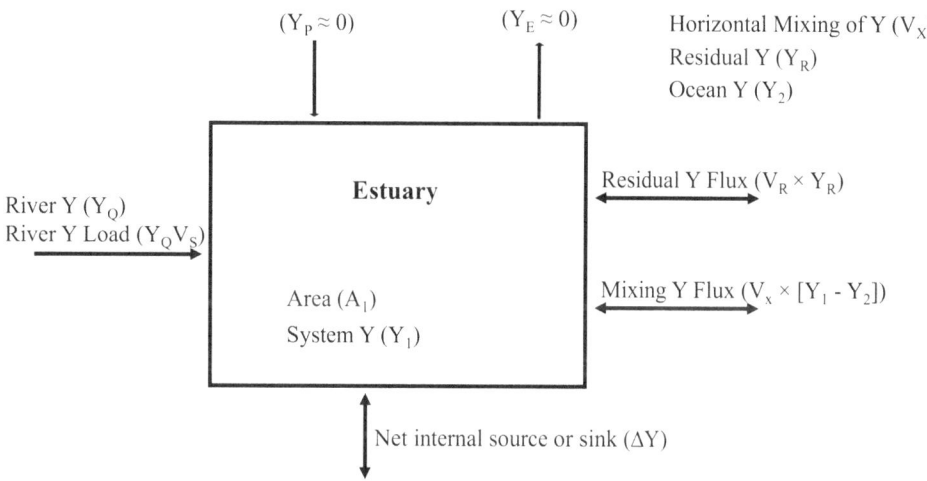

**Fig. 3** Schematic of the nutrient budget using the LOICZ modeling guidelines. $Y$ represents the nonconservative material of interest (nutrients). The arrows represent the direction and net flow of each component

was taken to transform the data and achieve a more normally distributed dataset. The univariate test was then performed again to check the normality and distribution values and compared against the initial univariate test (Figs. S9–S14). The overall average of the naturally logged values of nitrogen and phosphorus was then computed and de-transformed. Lastly, the de-transformed nutrient data then underwent a conversion of their various forms from mg/L to mmol/m³ for the ease of comparisons.

The following equation was used to convert the nutrient values from its initial units of $\frac{mg}{L}$ (as N, P, or PO₄) to $\frac{mmol}{m^3}$:

$$1\frac{mg}{L} \times \frac{1\frac{g}{L}}{1000\frac{mg}{L}} \times \frac{1\frac{mol}{L}}{molar\,mass\left(\frac{g}{L}\right)} \times \frac{1,000,000\frac{mmol}{m^3}}{1\frac{mol}{L}} \quad (8)$$

The following equation was used to calculate DIN:

$$DIN = NO_2 + NO_3 + NH_4 \quad (9)$$

The following equation was used to calculate TON:

$$TON = TKN - NH_4 \quad (10)$$

Like the salt budgets, the streamflow ($V_S$) data from TCEQ rather than the freshwater inflow data from TWDB were used for calculating the river nutrient loading ($Y_Q V_S$) to have a more accurate calculation of nutrients entering the system. Again, this streamflow data was not included in the water budgets to avoid freshwater inflow rates being calculated twice.

## Stoichiometry

Once the water, salt, and nutrient budgets were developed, net ecosystem metabolism (NEM) estimates were calculated using LOICZ guideline assumptions. The following equation was used:

$$[p-r] = -\Delta DIP \times (C:P)_{part} \quad (11)$$

Where:

$[p-r]$= net ecosystem metabolism (NEM) or production minus respiration

$\Delta DIP$= change in DIP (inputs-outputs)

C : P= Redfield ratio (106:1)

To calculate [nfix − denit], the difference between the measured dissolved nitrogen flux and that expected from decomposition and production of organic matter, the estimation of DIN changes expected associated with the ecosystem metabolism ($\Delta DIN_{exp}$) had to be calculated first. The following equation was used to calculate the expected $\Delta DIN$:

$$\Delta DIN_{exp} = \Delta DIP(N:P) \quad (12)$$

Where:

$\Delta DIN_{exp}$= expected $\Delta DIN$

$\Delta DIP$= change in DIP (inputs-outputs)

N : P= Redfield ratio (16:1)

The following equation was used to calculate [nfix − denit]:

$$[nfix - denit] = \Delta N - \Delta P \times (N:P)_{part} = \Delta DIN - \Delta DIN_{exp} \quad (13)$$

Where:

nfix= nitrogen fixation

denit= denitrification$\Delta DIP$= change in DIP (inputs-outputs)

N : P= Redfield ratio (16:1)

## Data Sources

A daily time series of historical freshwater inflow into all Texas bays has been assembled for the period from January 1, 1940, to December 31, 2020. The primary source for this data is the input file for the TWDB TxBLEND salinity circulation model. The data used to develop these historical inflow sets are available on TWDB's Water Data for Texas webpage (https://waterdatafortexas.org/coastal/hydrology/) (TWDB 2022b) and include gage flows, measured at USGS streamflow gage locations; ungaged flows, estimated from a rainfall runoff model (TxRR); and diversion and return flows downstream of USGS gauge locations. Historical freshwater inflows were calculated as described in Guthrie and Lu (2010), Guthrie et al. (2010a, b), Guthrie et al. (2011), Matsumoto et al. (2014), and Schoenbaechler et al. (2011a, b, c).

The evaporation data have been recorded from 1954 to 2020 and were obtained from TWDB's Water Data for Texas webpage (https://waterdatafortexas.org/lake-evaporation-rainfall). The evaporation data were measured by a gridded 1° × 1° quadrangle (Fig. S15).

The precipitation data have been recorded from 1940 to 2020 and were obtained from the Global Historical Climatology Network (GHCN) database found on the NOAA National Centers for Environmental Information (NCEI) website (https://www.ncei.noaa.gov/cdo-web/). This database is a composite of climate records from numerous sources that were subjected to a suite of quality assurance reviews, and the methods are described in Menne et al. (2012). Each estuary has one weather station in which the precipitation data was obtained from Fig. S16.

Salinity data are from two agencies. The primary salinity data for the bay and offshore stations are from field measurements made by TPWD, Coastal Fisheries Program. TPWD collects biological samples monthly (see below) and measures salinity, temperature, dissolved oxygen, and turbidity at all sampling locations. The data series began in 1977. The methods are described in Martinez-Andrade et al. (2005). Additionally, salinity is predicted in the TxBLEND models described above. The remaining salinity values from rivers and streams were obtained by TCEQ's Surface Water Quality Monitoring Information System (SWQMIS) through the TCEQ's Surface Water Quality Web Reporting Tool (https://www80.tceq.texas.gov/SwqmisPublic/index.htm).

Nutrient data were obtained from the TCEQ's Surface Water Quality Monitoring Information System (SWQMIS) through the TCEQ Surface Water Quality Web Reporting Tool. Estuary physical and hydrological attributes such as estuary area, volume, and tidal prism were obtained from Engle et al. (2007). The TCEQ nutrient data consists of routine water quality monitoring, as well as special studies. The dataset contains mostly physical-chemical measurements such as salinity, temperature, nutrients, and some contaminants. While these data are from multiple sources and were collected by multiple programs, all the data would have been associated with a Quality Assurance Project Plan (QAPP).

All data collected, aggregated, and used in this study is available for download (Marshall and Montagna 2023). The overall water-salt-nutrient budgets were calculated with data recorded on various dates in the LOICZ diagrams. Supplements SA–SH list the dates for which data were recorded that went into calculating the overall daily averages for each parameter. It should be cautioned that when the flux estimates were calculated, the margins of error became quite large due to large uncertainties in the input datasets (Kempthorne and Allmaras 1965). In this case, multiplying two averages together such as the loading estimates from rivers or the residual flux will yield an exponentially large variance and in turn large confidence limits that could not have been avoided. In addition, when interpreting the river loading estimates in the bays for nitrogen, phosphorus, and salinity, it is important to consider that there is tidal sloshing near the tidal segments over time. However, these temporal dynamics of tidal sloshing were not captured in these budgets.

## Missing Data

Only a limited amount of salinity, nutrient, and water data were available for this study. All nutrient data along the Gulf stations from TCEQ, including $NO_2$, $NO_3$, $NH_4$, DIP, and TP, were missing from the Lavaca-Colorado Estuary, Guadalupe Estuary, Mission-Aransas Estuary, and Upper Laguna Madre Estuary. In addition, DIP for the Trinity-San Jacinto Estuary in the Gulf stations was missing too. To complete the LOICZ budgets for these estuaries, the missing data had to be supplemented. The supplemented data were obtained from the de-transformed log values of all the existing remaining Gulf stations from all the remaining estuaries' nutrient data ($n = 633$) and averaged. These results were then used to fill in the missing data for budget calculations and analysis (Table 2).

## Results

The full LOICZ budget models for each estuary can be found in the supplemental material (Figs. S17–S40). Presented below is a summary of the LOICZ budget models separated into three sections: water budgets, salt budgets, and nutrient budgets.

## Water Budgets

The water budgets illustrated on average per day the residual flow ($V_R$) of each system was negative for all estuaries except the Upper Laguna Madre Estuary and Lower Laguna Madre (Table 3). Negative results for residual flow indicate residual outflow where water is leaving the system results in a net loss. However, the Upper and Lower Laguna Madre Estuary's average daily residual flow demonstrates seawater inflow into the bay (Figs. S35 and S38) known as residual inflow or net gain. The Sabine-Neches Estuary had the largest daily average residual outflow with a rate of $-46.45 \times 10^6$ m³/d, while the second largest daily average residual outflow is the Trinity-San Jacinto Estuary with a rate of $-40.32 \times 10^6$ m³/d.

**Table 2** Complete table of overall nutrient averages for the Gulf stations including the supplemented missing values in *italics* (abbreviations found in table of abbreviations)

| Estuary | NH₄ | NO₂ | NO₃ | TP | DIP | TKN | DIN | TON |
|---------|------|-------|-------|------|------|-------|-------|-------|
| SN | 5.04 | 12.82 | 8.60 | 3.24 | 0.66 | 52.59 | 26.46 | 47.55 |
| TS | 4.29 | 35.68 | 35.68 | 6.29 | *0.52* | 53.22 | 75.65 | 48.92 |
| LC | *4.55* | *7.66* | *5.54* | *2.69* | *0.52* | *40.18* | *17.75* | *35.63* |
| GE | *4.55* | *7.66* | *5.54* | *2.69* | *0.52* | *40.18* | *17.75* | *35.63* |
| MA | *4.55* | *7.66* | *5.54* | *2.69* | *0.52* | *40.18* | *17.75* | *35.63* |
| NC | 4.28 | 3.86 | 4.02 | 1.48 | 0.49 | 26.43 | 12.16 | 22.15 |
| ULM | *4.55* | *7.66* | *5.54* | *2.69* | *0.52* | *40.18* | *17.75* | *35.63* |
| LLM | 4.06 | 6.11 | 2.92 | 1.61 | 0.43 | 20.56 | 13.09 | 16.50 |

Units: mmol/m³

**Table 3** Summary of the LOICZ water budget for each estuary in Texas

| Estuary | $N$ | $V_G$ | $V_M$ | $V_D$ | $V_F$ | $V_E$ | $V_P$ | $V_I$ | $V_R$ | $V_i$ |
|---------|-----|-------|-------|-------|-------|-------|-------|-------|-------|-------|
| SN | 972 | 38.91 | 9.02 | 3.73 | 2.02 | 0.84 | 1.07 | 660 | −46.46 | 51.03 |
| TS | 972 | 27.64 | 11.50 | 2.15 | 2.74 | 4.65 | 5.25 | 2707 | −40.32 | 47.13 |
| LC | 972 | 8.45 | 3.72 | 0.33 | 0.28 | 3.89 | 3.31 | 2430 | −11.54 | 15.76 |
| GE | 972 | 6.89 | 1.25 | 0.30 | 0.26 | 2.05 | 1.70 | 728 | −7.75 | 10.10 |
| MA | 972 | 0.46 | 1.46 | 0.003 | 0.02 | 1.92 | 1.32 | 847 | −1.34 | 3.26 |
| NC | 972 | 1.64 | 0.39 | 0.23 | 0.40 | 2.34 | 1.20 | 1280 | −1.05 | 3.62 |
| ULM | 972 | 0.04 | 0.81 | 0.001 | 0.37 | 2.42 | 1.07 | 447 | 0.13 | 2.29 |
| LLM | 972 | 1.06 | 0.65 | 0.03 | 0.11 | 4.95 | 2.76 | 997 | 0.41 | 4.57 |

$N$ number of samples, $V_G$ gaged flow, $V_M$ modeled flow, $V_D$ diversions, $V_F$ return flows, $V_E$ evaporation, $V_P$ precipitation, $V_I$ volume, $V_R$ residual flow, $V_i$ freshwater inflow balance. Units: $10^6$ m$^3$/d

The Sabine-Neches Estuary has the highest average gaged flow ($V_G$) of 38.91 m$^3$/d, along with the highest average diversion flow ($V_D$) of 3.73 m$^3$/d. The rates then gradually decrease in a southeast to northwest climate gradient. However, the modeled flow ($V_M$) rate of 11.50 m$^3$/d and return flow ($V_F$) rate of 2.74 m$^3$/d were found to be highest in the Trinity-San Jacinto Estuary. Nevertheless, similar to the other flows, these rates also decreased gradually along the climate gradient. Furthermore, evaporation rates were found to be highest in the Lower Laguna Madre Estuary at 49.5 m$^3$/d, and the highest precipitation average rate was observed in the Trinity San Jacinto Estuary at 5.25 m$^3$/d (Table 3).

Comparing the freshwater residence times (RT) between estuaries, also known as the flushing time, Nueces Estuary has the longest residence time with an average of 353 days (Eq. 4). However, if the Laguna Madre Estuary is not further divided into the Upper Laguna Madre Estuary and Lower Laguna Madre Estuary, this estuary has the longest residence time of 413 days (Eq. 4). The Sabine-Neches Estuary has the shortest residence time at an average of 12.9 days (Eq. 4). Residence times are calculated in various ways in different literature. This research calculated the residence time of each estuary using three different methods (Eqs. 4, 5, 6) to illustrate that each method can yield significantly different results.

In Table 4, the residence times using the freshwater fraction in the third column are significantly shorter compared to the other two methods. Moreover, negative values for residence time are observed in the freshwater fraction method for both the Upper Laguna Madre Estuary and the Laguna Madre Estuary as a whole. These discrepancies arise due to limitations in the model, which relies on detectable differences in salinity among water masses and is only applicable primarily to estuaries with substantial freshwater inflow (Sheldon and Alber 2006). Therefore, it is advised to exercise caution and refrain from using this method in estuaries with low freshwater inflow and minor salinity variations. Conversely, the method based on inflow balance in the second column yields considerably longer residence times.

**Table 4** Residence time calculations from Eqs. 4, 5, 6

| Estuary | Volume/ inflow rate | Volume/ inflow balance | (Volume*freshwater fraction)/inflow rate |
|---------|---------------------|------------------------|-------------------------------------------|
| SN | 12.93 | 14.21 | 8.96 |
| TS | 57.44 | 67.13 | 25.81 |
| LC | 154.20 | 210.53 | 51.92 |
| GE | 72.08 | 93.93 | 29.36 |
| MA | 259.61 | 631.22 | 95.16 |
| NC | 353.38 | 1221.32 | 17.40 |
| ULM | 194.99 | 3380.65 | −37.73 |
| LLM | 218.01 | 2424.66 | 8.34 |
| LM (ULM + LLM) | 412.99 | 5805.31 | −29.39 |

Units: days

**Table 5** Average annual precipitation in Texas estuaries

| Estuary | N | Precipitation (in/yr) | Precipitation ($10^6$ m/s) |
|---------|---|------------------------|-----------------------------|
| SN | 74 | 57.48 | 0.046 |
| TS | 81 | 49.97 | 0.040 |
| LC | 78 | 42.50 | 0.034 |
| GE | 53 | 33.23 | 0.027 |
| MA | 58 | 35.37 | 0.028 |
| NC | 75 | 30.27 | 0.024 |
| ULM | 70 | 24.64 | 0.020 |
| LLM | 71 | 25.14 | 0.020 |

Data obtained for the period 1940–2020 from the NOAA Global Historical Climatology Network (GHCN) database (https://www.ncei.noaa.gov/cdo-web/)

Residence times can help understand the flow dynamics of a system including the flushing time of nutrients and the effects of mixing (Montagna et al. 2013). Shorter residence or flushing times can be beneficial to systems as they are often prone to better water quality due to the high hydrodynamics and exportation (Cabral and Fonseca 2019).

Table 5 showcases the average annual precipitation across various estuaries in Texas. The Sabine-Neches Estuary has the highest average annual precipitation of 57.48 in/yr. (1.460 m/yr), while the Upper Laguna Madre Estuary has the lowest average annual precipitation of 24.64 in/yr. (0.6259 m/y). It is worth noting that these values follow the climate gradient that exists along the Texas coastline, with a gradual decrease in precipitation rates from the northeast to the southwest.

The results of the spearman correlation analysis (Table 6) showed that all estuaries' freshwater inflow balance and precipitation had evidence of a relatively strong monotonic relationship ($p < 0.0001$). These monotonic relationships were positive, indicating that as the precipitation increases the freshwater inflow balance increases. The estuary with the strongest correlation was the Mission-Aransas Estuary ($r = 0.77$, $N = 81$), and the estuary with the weakest correlation was the Sabine-Neches Estuary ($r = 0.42$, $N = 792$).

**Table 6** Spearman correlation analysis results comparing precipitation and the freshwater inflow balance for each individual estuary in Texas

| Estuary | N | r | p |
|---------|-----|------|---------|
| SN | 792 | 0.42 | <0.0001 |
| TS | 745 | 0.63 | <0.0001 |
| LC | 494 | 0.74 | <0.0001 |
| GE | 125 | 0.64 | <0.0001 |
| MA | 81 | 0.77 | <0.0001 |
| NC | 619 | 0.71 | <0.0001 |
| ULM | 272 | 0.71 | <0.0001 |
| LLM | 258 | 0.75 | <0.0001 |

## Salt Budgets

Table 7 provides an overview of the average salinity levels within each estuary, classified by river segment, across the eight bay systems. Notably, in the Nueces Estuary, segment 2484 situated at the Corpus Christi Inner Harbor recorded the highest average salinity of 31.53 S. Conversely, segment 2453E at Arenosa Creek in the Lavaca-Colorado Estuary exhibited the lowest average salinity, reaching 0.00 S (Table 7). Additionally, examining the river salt loading ($S_Q V_S$), which signifies the amount of salt transported into the estuaries by streams or rivers, the Sabine-Neches Estuary segment 601 recorded the highest value of $129.79 \times 10^6$ g/m$^3$/d. In contrast, the Trinity-San Jacinto Estuary demonstrated the smallest salt loading estimate with a negative value of $-4.22 \times 10^6$ g/m$^3$/d. The negative value arises due to the streamflow value being negative, indicating salt outflow from the estuary rather than its inflow via rivers and streams.

Within the bay system, the Upper Laguna Madre exhibited the highest average daily salinity, recording a value of 37.90 S (Table 8). Moving toward the Gulf, the section adjacent to the Lower Laguna Madre Estuary displayed the highest salinity averages outside the estuaries, reaching 33.38 S. Regarding the residual salt flux ($V_R S_R$), all estuaries, except for the Upper Laguna Madre Estuary and Lower Laguna Madre Estuary, demonstrated negative values, indicating the export of salt from the bay system into the Gulf of Mexico. Conversely, the positive values observed in the

**Table 7** An overview of the salt budgets for the rivers within each bay system, as depicted in the LOICZ models

| Estuary | Segment ID | N | Streamflow ($V_S$) | Streamflow Std | River salinity | River salinity Std | River salt loading ($S_Q V_S$) |
|---------|-----------|------|---------|---------|-------|-------|--------|
| SN | 501 | 4559 | 24.05 | 32.59 | 5.25 | 5.40 | 126.30 |
| SN | 601 | 6348 | 18.32 | 18.18 | 7.08 | 6.33 | 129.79 |
| TS | 0801E | 49 | 0.04 | 0.057 | 1.17 | 0.73 | 0.05 |
| TS | 1001 | 2798 | 0.006 | . | 4.51 | 4.44 | 0.03 |
| TS | 1005 | 3461 | 0.01 | 0.0051 | 12.52 | 5.97 | 0.13 |
| TS | 1006 | 4478 | 2.29 | 0.48 | 8.07 | 5.40 | 18.44 |
| TS | 1103 | 1427 | 0.40 | 0.48 | 6.08 | 6.13 | 2.44 |
| TS | 2422B | 234 | 0.17 | 0.54 | 6.70 | 6.00 | 1.15 |
| TS | 2422D | 112 | 0.25 | 0.65 | 2.19 | 2.96 | 0.55 |
| TS | 2424A | 894 | −0.35 | 0.65 | 12.05 | 8.61 | −4.22 |
| LC | 1501 | 203 | 0.56 | 0.86 | 4.79 | 4.77 | 2.70 |
| LC | 2453E | 1 | 0.09 | 0.35 | 0.00 | . | 0.00 |
| LC | 2456A | 12 | 0.04 | 0.072 | 0.58 | 0.57 | 0.02 |
| GE | 1801 | 230 | 5.36 | 2.02 | 1.09 | 0.81 | 5.86 |
| MA | 2003 | 104 | 0.02 | . | 3.71 | 5.29 | 0.08 |
| NC | 2484 | 1916 | 0.03 | . | 31.53 | 4.48 | 0.93 |
| NC | 2485A | 414 | 0.24 | 1.16 | 11.25 | 16.15 | 2.67 |
| ULM | 2204 | 178 | 0.08 | 0.40 | 8.28 | 6.40 | 0.64 |
| ULM | 2204B | 17 | 0.04 | 0.068 | 7.29 | 7.48 | 0.26 |
| ULM | 2492A | 71 | 0.03 | 0.13 | 2.20 | 0.64 | 0.06 |
| LLM | 2201 | 1769 | 0.25 | 0.40 | 15.33 | 10.60 | 3.81 |
| LLM | 2301 | 45 | 4.15 | 10.52 | 6.42 | 5.79 | 26.65 |

N number of samples, *streamflow Std* standard deviation, *river salinity Std* standard deviation. Units: streamflow ($V_S$) = $10^6$ m$^3$/d; streamflow Std = $10^6$; river salinity = S; river salt loadings ($S_Q V_S$) = $10^6$ g/m$^3$/d

**Table 8** A summary outlining the salt budgets within the bay and gulf systems, as documented in the LOICZ models

| | Bay | | | Gulf | | | Flows/fluxes | | | |
|---|---|---|---|---|---|---|---|---|---|---|
| Estuary | $N$ | Sal | Std | $N$ | Sal | Std | $S_R$ | $V_R S_R$ | $V_X$ | $V_X(S_2–S_1)$ |
| SN | 18,098 | 8.19 | 6.76 | 6608 | 26.65 | 4.55 | 17.42 | −809.27 | 43.84 | 809.27 |
| TS | 37,481 | 16.19 | 8.23 | 6730 | 29.40 | 4.54 | 22.79 | −919.00 | 69.56 | 919.00 |
| LC | 30,525 | 21.00 | 8.67 | 6743 | 31.66 | 3.30 | 26.33 | −303.87 | 28.51 | 303.87 |
| GE | 30,602 | 18.76 | 10.52 | 6743 | 31.66 | 3.30 | 25.21 | −195.38 | 15.15 | 195.38 |
| MA | 31,974 | 20.11 | 9.12 | 6789 | 31.75 | 3.45 | 25.93 | −34.80 | 2.99 | 34.80 |
| NC | 23,480 | 30.19 | 6.54 | 6789 | 31.75 | 3.45 | 30.97 | −32.46 | 20.76 | 32.46 |
| ULM | 17,731 | 37.90 | 10.66 | 6789 | 31.75 | 3.45 | 34.83 | 4.60 | 0.75 | −4.60 |
| LLM | 17,731 | 32.10 | 7.36 | 6717 | 33.38 | 2.74 | 32.74 | 13.46 | −10.54 | −13.46 |

$N$ number of samples; *Std* standard deviation. Units: salinity (Sal) = $S$; salinity of the residual flow ($S_R$) = $S$; residual salt flux ($V_R S_R$) = $10^6$ g/m³/d; horizontal salinity mixing ($V_X$) = $10^6$ g/m³/d; mixing salt flux ($V_X(S_2–S_1)$) = $10^6$ g/m³/d

**Table 9** Spearman correlation analysis showcasing the relationship between monthly freshwater inflow balance and average salinity levels

| Estuary | $N$ | $r$ | $p$ |
|---|---|---|---|
| SN | 419 | −0.79 | <0.0001 |
| TS | 526 | −0.71 | <0.0001 |
| LC | 490 | −0.60 | <0.0001 |
| GE | 124 | −0.47 | <0.0001 |
| MA | 81 | −0.49 | <0.0001 |
| NC | 501 | −0.36 | <0.00001 |
| ULM | 269 | −0.14 | 0.0250 |
| LLM | 255 | −0.25 | <0.00001 |

Upper Laguna Madre Estuary and Lower Laguna Madre Estuary suggest that, on average per day, seawater inflow contributes to salt delivery within the bay. In terms of horizontal salt mixing or exchange between estuary water and ocean water ($V_X$), the Lower Laguna Madre Estuary exhibited the highest rate of excess salt removal, with a rate of −10.55 × $10^6$ g/m³/d. On the other hand, the Trinity-San Jacinto Estuary had the lowest salt exchange rate, adding salt to the system, at a rate of approximately 69.56 × $10^6$ g/m³/d (Table 8).

A one-way ANOVA was conducted to compare the average salinity values among the bay systems in each estuary, and the results indicate that there is a significant difference in salinity averages among all estuaries ($p < 0.001$, $N = 139,560$) (Table 9). A Tukey's HSD post hoc test shows a negative monotonic relationship across all Texas estuaries. The Sabine-Neches Estuary has the strongest correlation ($r = −0.79$) between freshwater inflow balance and salinity, while the Upper Laguna Madre Estuary has the weakest correlation ($r = −0.14$) according to Table 9. This indicates that as the freshwater inflow balance decreases, salinity levels tend to increase in all estuaries. The correlation coefficients gradually weaken from northeast to southwest along the Texas coastline. A graphical representation of the average salinity per year displays the clear difference in the salinity averages by estuary and shows great fluctuation from year to year (Fig. 4). Figure 4 also confirms that the Upper Laguna Madre Estuary has and maintains the highest average salinity, while the Sabine-Neches Estuary has the lowest average salinity.

## Nutrient Budgets

When making a comparison of the nutrient loads found in the rivers and stream segments in each estuary, the Trinity-San Jacinto Estuary was the only estuary that had a river segment with negative streamflow causing nutrients to travel up the estuary rather than importing into the bay (Table 10). This river segment is located at the Highland Bayou tidal stream (segment 2424A). A possible reason for this is that there has been a construction to the main channel of the Highland Bayou over time like the construction of the Diversion Canal that has dramatically altered the flow (Texas Community Watershed Partners 2020). However, the estuary with the highest amount of nutrient river loadings is the Sabine-Neches Estuary where the TON loading estimate is 1098.75 × $10^6$ mmol/m³/d (Table 10).

Upon analyzing the nutrient levels in the bay system within each estuary, the Trinity-San Jacinto Estuary had the highest average TP concentration with a recorded value of 5.41 mmol/m³. Similarly, the highest DIP concentration was also found in the Trinity-San Jacinto Estuary with a recorded value of 9.06 mmol/m³ (Table 11). What sets these two nutrient values apart from the rest is that they are significantly higher compared to the other estuaries, while the remaining nutrients have very similar mean values. The Upper Laguna Madre Estuary is also experiencing a similar occurrence, where it is exhibiting a very high average TON concentration with a recorded value of 84.93 mmol/m³, compared to its counterparts, which have values ranging only from 36 to 48 mmol/m³ (as shown in Table 11). Lastly, the average DIN concentrations among the estuaries in Texas are relatively similar; the Trinity-San Jacinto Estuary has the largest mean concentration of DIN with a recorded

**Fig. 4** Average daily salinity values in the bay system of each estuary by year. Abbreviations can be found in the table of abbreviations

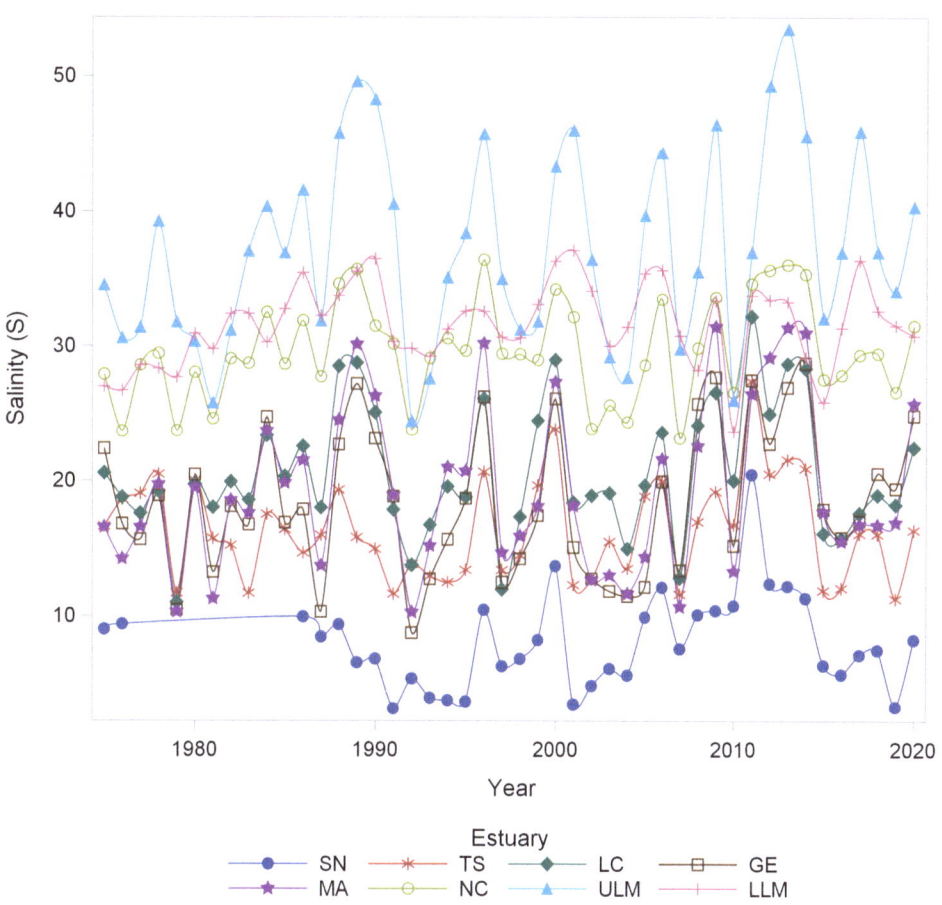

**Table 10** Summary of the LOICZ nutrient budgets in the river system of each estuary

| Estuary | Segment ID | NH$_4$ | NO$_2$ | NO$_3$ | TP | DIP | TKN | DIN | TON | DIN*V$_S$ | TON*V$_S$ | TP*V$_S$ | DIP*V$_S$ |
|---------|-----------|--------|--------|--------|------|------|--------|--------|--------|---------|---------|--------|--------|
| SN | 501 | 4.95 | 2.07 | 4.04 | 2.40 | 0.58 | 50.63 | 11.06 | 45.68 | 266.13 | 1098.75 | 57.62 | 13.89 |
| SN | 601 | 5.36 | 1.99 | 7.02 | 2.26 | 0.58 | 48.20 | 14.37 | 42.84 | 263.26 | 784.86 | 41.40 | 10.60 |
| TS | 1001 | 7.55 | 6.91 | 10.56 | 8.18 | 10.14 | 80.57 | 25.02 | 73.02 | 0.16 | 0.46 | 0.05 | 0.06 |
| TS | 1005 | 13.84 | 8.15 | 14.70 | 13.88 | 18.56 | 85.02 | 36.69 | 71.18 | 0.39 | 0.75 | 0.15 | 0.20 |
| TS | 1006 | 24.11 | 14.17 | 54.96 | 24.78 | 40.48 | 105.69 | 93.23 | 81.58 | 213.16 | 186.52 | 56.65 | 92.54 |
| TS | 1103 | 9.99 | 2.66 | 5.59 | 5.70 | 8.12 | 70.54 | 18.25 | 60.56 | 7.32 | 24.28 | 2.29 | 3.26 |
| TS | 2422B | 6.08 | 3.60 | 21.08 | 6.68 | 4.57 | 84.74 | 30.76 | 78.65 | 5.30 | 13.55 | 1.15 | 0.79 |
| TS | 2422D | 5.72 | 2.73 | 6.32 | 3.51 | 2.68 | 69.52 | 14.77 | 63.80 | 3.70 | 15.99 | 0.88 | 0.67 |
| TS | 2424A | 8.92 | 4.59 | 10.29 | 5.84 | 6.00 | 94.01 | 23.81 | 85.08 | −8.34 | −29.79 | −2.04 | −2.10 |
| LC | 1501 | 6.20 | 4.19 | 18.79 | 7.38 | 3.04 | 82.04 | 29.19 | 75.84 | 16.47 | 42.80 | 4.17 | 1.72 |
| GE | 1801 | 3.92 | 2.38 | 97.96 | 15.60 | 13.68 | 54.11 | 104.26 | 50.19 | 558.34 | 268.80 | 83.56 | 73.26 |
| MA | 2003 | 4.06 | 2.02 | 7.29 | 16.22 | 15.83 | 72.08 | 13.37 | 68.02 | 0.30 | 1.51 | 0.36 | 0.35 |
| NC | 2484 | 9.71 | 4.14 | 10.19 | 3.54 | 1.61 | 53.62 | 24.05 | 43.91 | 0.71 | 1.29 | 0.10 | 0.05 |
| NC | 2485A | 13.90 | 6.84 | 38.50 | 36.90 | 9.79 | 130.19 | 59.25 | 116.28 | 14.06 | 27.59 | 8.76 | 2.32 |
| ULM | 2204 | 3.99 | 2.46 | 8.08 | 6.87 | 2.66 | 98.17 | 14.54 | 94.18 | 1.12 | 7.24 | 0.53 | 0.20 |
| ULM | 2492A | 10.11 | 8.75 | 76.86 | 70.54 | 69.61 | 125.58 | 95.72 | 115.47 | 2.67 | 3.22 | 1.96 | 0.19 |
| LLM | 2201 | 14.68 | 12.28 | 62.62 | 14.41 | 6.31 | 85.72 | 89.58 | 71.05 | 22.26 | 17.66 | 3.58 | 1.57 |
| LLM | 2301 | 6.69 | 3.11 | 9.27 | 7.41 | 1.48 | 99.87 | 19.07 | 93.18 | 79.09 | 386.51 | 30.72 | 6.13 |

Units: $Y$ mmol/m$^3$, $Y$ river loading ($Y$*$V_S$) = $10^6$ mmol/m$^3$/d. $Y$ represents the nonconservative material of interest (nutrient)

value of 22.65 mmol/m$^3$, and the estuary with the lowest recorded average is the Mission-Aransas Estuary with a recorded value of 9.40 mmol/m$^3$.

The stoichiometric calculations reveal that all eight estuaries, including the Laguna Madre Estuary, have positive values for net ecosystem metabolism ($p$–$r$), indicating that

**Table 11** Summary of the nutrient concentrations in bay systems and their fluxes

| Estuary | $TP_1$ | $DIP_1$ | $DIN_1$ | $TON_1$ | $TP_2$ | $DIP_2$ | $DIN_2$ | $TON_2$ | $DIN_R$ | $TON_R$ | $TP_R$ | $DIP_R$ | $DIN*V_R$ | $TON*V_R$ | $TP*V_R$ | $DIP*V_R$ | $V_X[DIN_1-DIN_2]$ | $V_X[TON_1-TON_2]$ | $V_X[DIP_1-DIP_2]$ | $V_X[TP_1-TP_2]$ |
|---|---|---|---|---|---|---|---|---|---|---|---|---|---|---|---|---|---|---|---|---|
| SN | 2.05 | 0.64 | 15.17 | 36.47 | 3.24 | 0.66 | 26.46 | 47.55 | 20.82 | 42.01 | 2.64 | 0.65 | −967.04 | −1951.56 | −122.83 | −30.25 | 437.65 | 67.95 | 5.77 | 23.81 |
| TS | 5.41 | 9.06 | 22.65 | 47.71 | 6.29 | 0.52 | 75.65 | 48.92 | 49.15 | 48.32 | 5.85 | 4.79 | −1981.95 | −1948.24 | −235.87 | −193.19 | 1760.27 | 1736.48 | 97.77 | 176.75 |
| LC | 2.67 | 0.79 | 11.01 | 46.35 | 2.69 | 0.52 | 17.75 | 35.63 | 14.38 | 40.99 | 2.68 | 0.66 | −165.96 | −473.11 | −30.93 | −7.57 | 149.49 | 430.31 | 5.86 | 26.76 |
| GE | 3.09 | 1.29 | 11.66 | 41.15 | 2.69 | 0.52 | 17.75 | 35.63 | 14.71 | 38.39 | 2.89 | 0.91 | −113.98 | −297.58 | −22.39 | −7.02 | −444.36 | 28.78 | −66.23 | −61.16 |
| MA | 2.78 | 1.22 | 9.40 | 41.21 | 2.69 | 0.52 | 17.75 | 35.63 | 13.57 | 38.42 | 2.73 | 0.87 | −18.21 | −51.55 | −3.67 | −1.17 | 17.91 | 50.04 | 0.82 | 3.31 |
| NC | 2.46 | 0.88 | 10.49 | 39.90 | 1.48 | 0.49 | 12.16 | 22.15 | 11.32 | 31.02 | 1.97 | 0.68 | −11.87 | −32.52 | −2.06 | −0.72 | −2.90 | 3.63 | −1.65 | −6.80 |
| ULM | 2.15 | 0.58 | 10.14 | 84.93 | 2.69 | 0.52 | 17.75 | 35.63 | 13.94 | 60.28 | 2.42 | 0.55 | 1.84 | 7.97 | 0.32 | 0.07 | −5.63 | −18.43 | −2.22 | −2.81 |
| LLM | 2.11 | 0.67 | 13.81 | 34.56 | 1.61 | 0.43 | 13.09 | 16.50 | 13.45 | 25.53 | 1.86 | 0.55 | 5.53 | 10.50 | 0.76 | 0.23 | −106.89 | −414.66 | −7.93 | −35.06 |

Units: nutrient concentration $(Y_1)$ = mmol/m³; nutrient concentration $(Y_2)$ = mmol/m³; Gulf nutrient concentration $(Y_R)$ = mmol/m³; residual nutrient concentration $(Y_R)$ = mmol/m³; residual $Y$ flux $(Y*V_R)$ = 10⁶ mmol/m³/d; mixing $Y$ flux $(V_X[Y_1-Y_2])$ = 10⁶ mmol/m³/d. $Y$ represents the nonconservative material (i.e., nutrient) of interest

**Table 12** Summary of the stoichiometry in the LOICZ budgets

| Estuary | ΔDIN | ΔTON | ΔDIP | ΔTP | [p–r] | ΔDIN Expected | nfix–denit |
|---------|------|------|------|-----|-------|---------------|-----------|
| SN | −1496.43 | −3835.18 | −54.74 | −221.85 | 5802.42 | −875.84 | −620.59 |
| TS | −2203.64 | −2160.00 | −288.61 | −295.00 | 30592.47 | −4617.73 | 2414.09 |
| LC | −182.44 | −515.91 | −9.29 | −35.09 | 984.50 | −148.60 | −33.83 |
| GE | −672.32 | −566.37 | −80.28 | −105.95 | 8509.93 | −1284.52 | 612.20 |
| MA | −18.51 | −53.07 | −1.52 | −4.03 | 161.16 | −24.33 | 5.82 |
| NC | −26.63 | −61.40 | −3.09 | −10.92 | 327.22 | −49.39 | 22.76 |
| ULM | −1.94 | −2.48 | −2.07 | −2.17 | 219.43 | −33.12 | 31.18 |
| LLM | −95.82 | −393.67 | −7.48 | −33.53 | 792.39 | −119.61 | 23.78 |

Units: $\Delta Y$ = mmol/m$^3$; net ecosystem metabolism ([$p$–$r$]) = $10^6$ mmol/m$^3$/d; ΔDIN expected = $10^6$ mmol/m$^3$/d; nitrogen fixation minus denitrification (nfix-denit) = $10^6$ mmol/m$^3$/d. $Y$ represents the nonconservative material of interest (nutrient)

they are net autotrophic and consume inorganic material. Additionally, the calculations show that, on average, nitrogen fixation surpasses denitrification in the Trinity-San Jacinto, Guadalupe, Mission-Aransas, Nueces, Upper Laguna Madre Estuary, and Lower Laguna Madre Estuary, as indicated by the negative [nfix–denit] value in the models. However, in the Sabine-Neches and Lavaca-Colorado Estuaries, denitrification exceeds nitrogen fixation, as indicated by a positive value in the models (Table 12). The highest expected ΔDIN values are in the Trinity-San Jacinto Estuary, and the lowest expected ΔDIN values are in the Mission-Aransas Estuary.

The results of the analysis of nonconservative processes suggest that all eight estuaries in Texas function as daily nutrient sinks, demonstrated by the negative $\Delta Y$ values (ΔDIN, ΔTON, ΔTP, ΔDIP). These findings indicate that these estuaries tend to retain nutrients. Additionally, Fig. 5 shows an inverse relationship between the daily average DIP and DIN loadings and their internal fluxes (ΔDIN and ΔDIP), further supporting the notion that these estuaries function as sinks for nitrogen and phosphorus on average. It is important to note that these calculations were for a long-term average; therefore the temporal element or seasonal variation is not captured in these stoichiometric calculations. Nutrient cycling and estuarine processes are dynamic and ever-changing, and because of this, the results could be different for different days, months, or years.

A principal component analysis (PCA) was used to assess the relationship between the daily average nutrients values with inflow balance and salinity (Table 13). The first and second principal component values (PC1 and PC2) explained 59% and 24% of the variability in the data for a total of 83% (Fig. 6). The variable loading plot (Fig. 6a) illustrates that the inflow balance and DIN as well as TP are positively correlated with one another on PC1. In other words, as inflow balance increases, the nutrient values for DIN and TP increase. On the other hand, inflow balance is inversely correlated with salinity. This can be interpreted as when inflow balance increases the salinity decreases which was also confirmed in the post hoc testing of the one-way ANOVA

(Table 9). However, TON is orthogonal to inflow balance. The estuary sample score results (Fig. 6b) follow the climate gradient pattern where the northeastern estuaries appear on the top of the PC1 axis, and the southwestern estuaries appear near the bottom on the PC1 axis. A possible explanation for the Trinity-San Jacinto Estuary (TS symbol) on the far right on the PC2 axis is the very high DIP and TP concentrations the Trinity-San Jacinto Estuary versus the other estuaries (Fig. S41), and being the only estuary along the Texas coast with a river segment exhibiting a negative streamflow rate (Table 10). Nonetheless, this analysis indicates that hydrology is a strong driver of nutrient levels in the Texas estuaries.

Another example where there is a strong connection between hydrology and nutrient concentrations in each estuary across Texas can be found in Figs. 7 and 8. Besides the Sabine-Neches Estuary, nitrite + nitrate (NO$_x$) and salinity have a negative linear relationship across the remaining estuaries along the Texas coast (Fig. 7). Salinity can be a good indicator of hydrological conditions as it has a very close linear relationship with inflow balance. Salinity often reflects the amount to which seawater has been diluted by freshwater inflow (Montagna et al. 2013). Similar results are shown in Fig. 8, but in addition to the Sabine-Neches Estuary, having a positive correlation with total phosphorus (TP) the Upper Laguna Madre has one too unlike the rest of the estuaries displaying a negative correlation with salinity.

## Discussion

By comparing the results of the LOICZ budgets among estuaries, the budgets indicate that the climate gradient plays an important role in the water, salt, and nutrient budgets of Texas estuaries, demonstrating that there is a direct relationship between the water budgets of Texas estuaries and the climate gradient. For example, the Sabine-Neches Estuary had the highest average precipitation, highest residual outflow, and lowest average evaporation rates because it is located at the most northern point on the climate gradient,

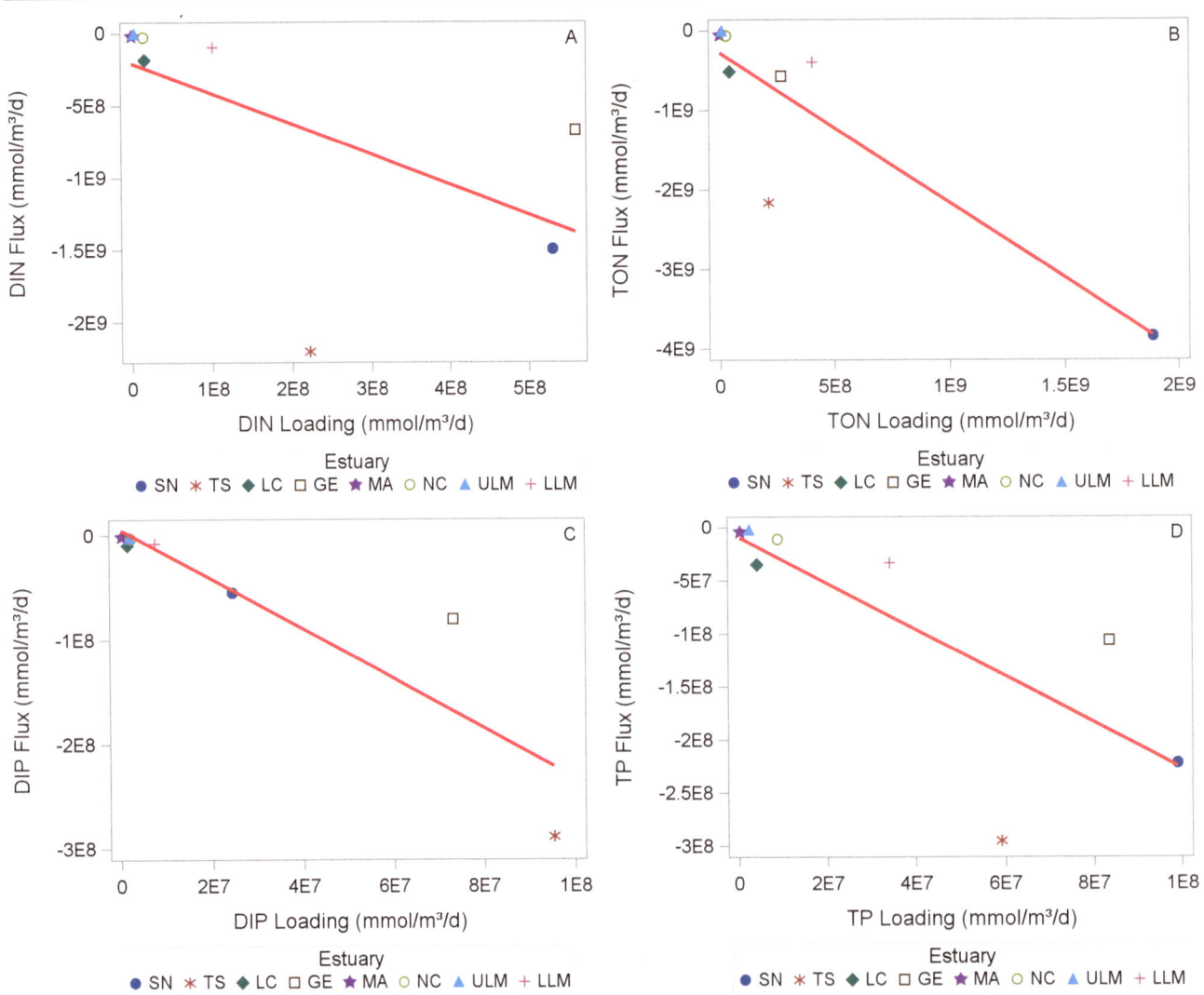

**Fig. 5** Relationship between nutrient loadings and internal fluxes at each estuary along the Texas coast. (**a**) DIN, (**b**) TON, (**c**) DIP, and (**d**) TP. Abbreviations: see table of abbreviations

**Table 13** Summary of key findings from the LOICZ budgets' nutrient concentrations in the bay systems

| Estuary | Inflow balance | RT | Sal | NH$_4$ | NO$_2$ | NO$_3$ | TP | DIP | TKN | DIN | TON |
|---------|----------------|------|-------|--------|--------|--------|------|------|-------|-------|-------|
| SN | $46.46 \times 10^6$ | 12.93 | 8.19 | 4.97 | 2.57 | 7.63 | 2.05 | 0.64 | 41.44 | 15.17 | 36.47 |
| TS | $40.32 \times 10^6$ | 57.44 | 16.19 | 4.57 | 9.39 | 8.69 | 5.41 | 9.06 | 52.28 | 22.65 | 47.71 |
| LC | $11.54 \times 10^6$ | 154.20 | 21.00 | 3.90 | 3.37 | 3.74 | 2.67 | 0.79 | 50.25 | 11.01 | 46.35 |
| GE | $7.75 \times 10^6$ | 72.08 | 18.76 | 3.92 | 3.49 | 4.26 | 3.09 | 1.29 | 45.07 | 11.66 | 41.15 |
| MA | $1.34 \times 10^6$ | 259.61 | 20.11 | 3.58 | 2.64 | 3.18 | 2.78 | 1.22 | 44.78 | 9.40 | 41.21 |
| NC | $1.05 \times 10^6$ | 353.38 | 30.19 | 3.98 | 3.12 | 3.38 | 2.46 | 0.88 | 43.88 | 10.49 | 39.90 |
| ULM | $-0.13 \times 10^6$ | 194.99 | 37.90 | 3.90 | 3.02 | 3.22 | 2.15 | 0.58 | 88.84 | 10.14 | 84.93 |
| LLM | $-0.41 \times 10^6$ | 218.01 | 32.10 | 4.07 | 5.72 | 4.03 | 2.11 | 0.67 | 38.63 | 13.81 | 34.56 |

Units: inflow balance = m$^3$/d; residence time (RT) = d; salinity (Sal) = S; remaining columns without units = mmol/m$^3$. Abbreviations can be found in the table of abbreviations

while the Lower Laguna Madre Estuary had one of the lowest average precipitation rates, had residual inflow from the ocean rather than residual outflow, and highest evaporation rates and is located at the most southern region in Texas. The dominant winds from the southeast also play a role in the

residual inflow from the ocean through wind water transport in the Lower Laguna Madre Estuary, and it is not solely due to low freshwater inflow rates (Opdyke et al. 2024). The salt budgets for each estuary in Texas also had a direct relationship with hydrology. The one-way ANOVA and post hoc

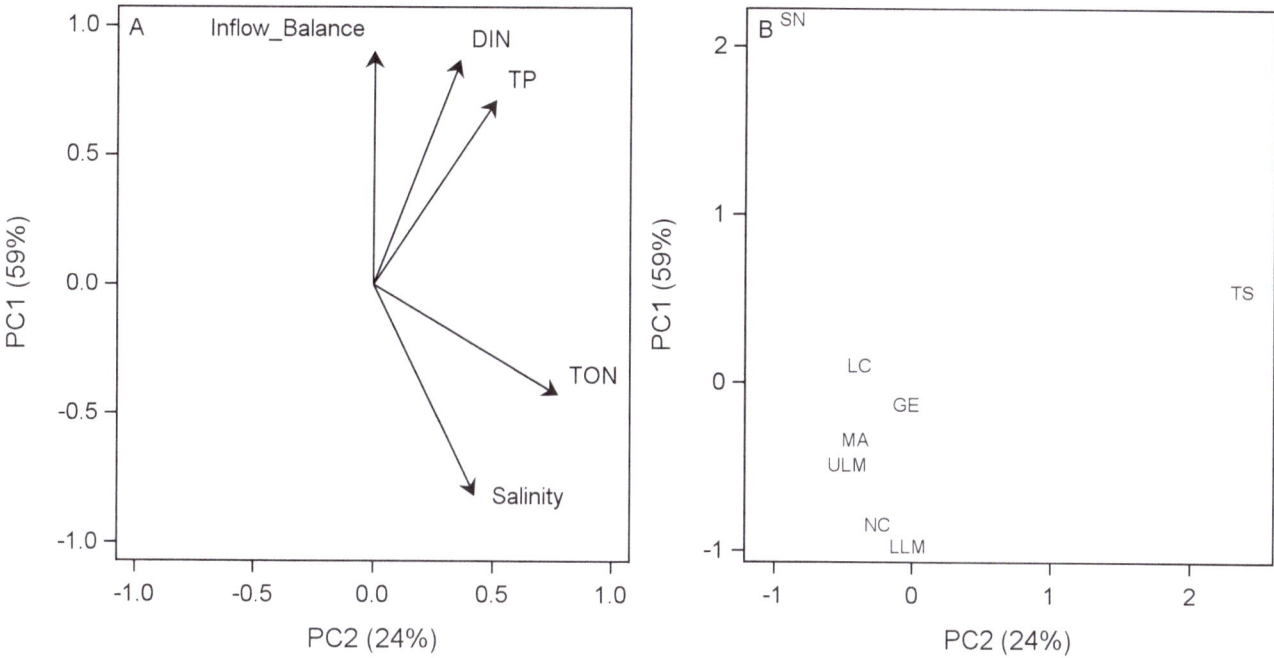

**Fig. 6** Principal component analysis of the nutrient content, salinity, and residual flow in each estuary. (**a**) Variable loads and (**b**) estuary sample scores (see abbreviations in the table of abbreviations)

**Fig. 7** Salinity values versus nitrite (NO₂) + nitrate (NO₃) concentrations in each bay system in estuaries across the Texas coast. Units: salinity = S; NO₂ + NO₃ = mmol/m³

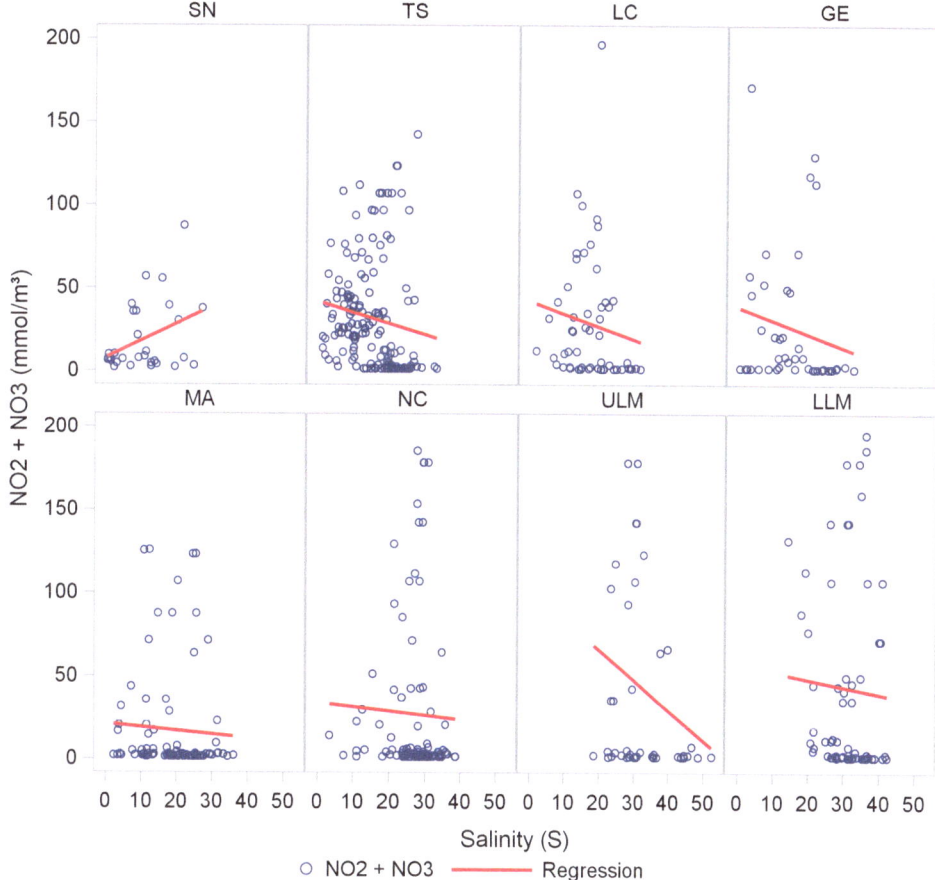

**Fig. 8** Salinity values versus TP values for each estuary across the Texas coast. Units: salinity = S; NO$_2$ + NO$_3$ = mmol/m$^3$

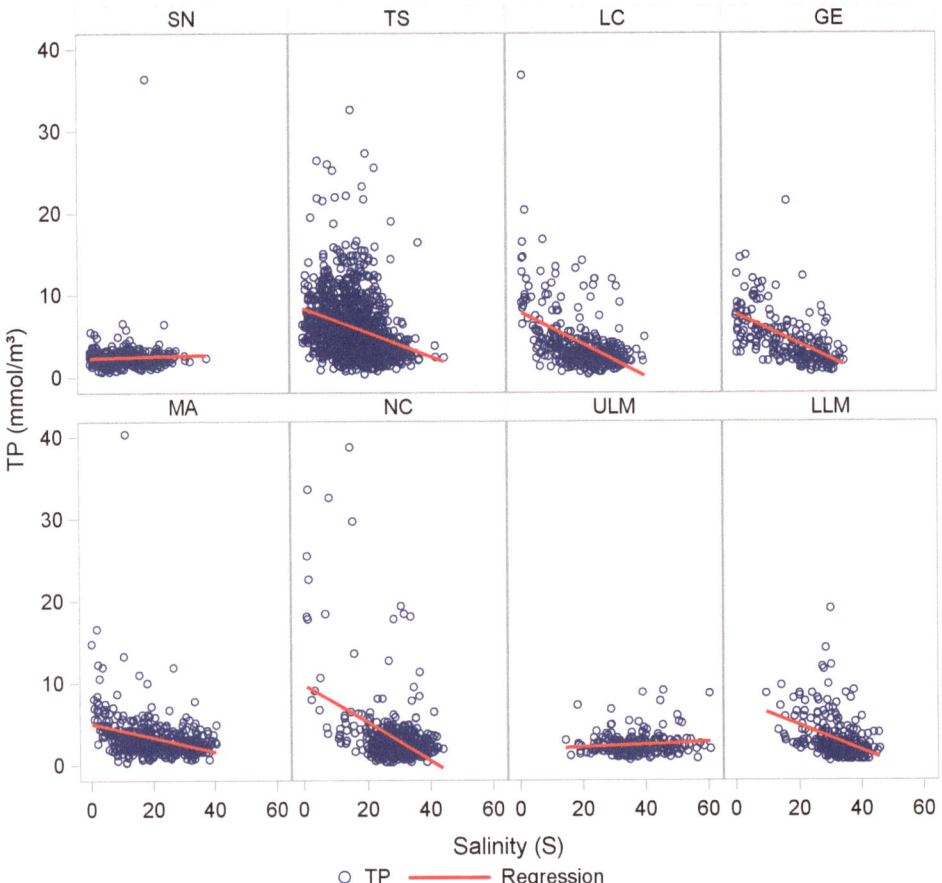

testing confirmed that the average salinity in each bay system shows a pattern that aligns with the climate gradient where the Sabine-Neches Estuary has the lowest salinity values, and the salinity values steadily increase with decreasing rainfall patterns down the Texas coastline to the south. Moreover, the nutrient budgets and the PCA analysis indicated that hydrology was a strong driver of nutrients along the Texas coast specifically for DIN and TP.

The Texas coast is unique because the estuaries lie in a subtropical zone where the latitude is around 30° and contains a long stretch of barrier islands that limit the tidal influences in their bays. In addition, these estuaries are microtidal meaning that their tidal ranges reach less than 2 m, but often it is closer to 0.5 m. Different tidal ranges such as microtidal (0–2 m), mesotidal (2–4 m), and macrotidal (>4 m) have major influences on the type of intertidal zone and biota of these ecosystems (Whitfield and Elliott 2011). This influences these estuaries to rely heavily on freshwater inflow to deliver the essential nutrients for a sustainable ecosystem. Similar research on Texas estuaries' nutrient content and their drivers has found that freshwater inflows contribute most nutrients delivered into the bays, while other factors such as marshes do not make up a significant portion of the nutrient budgets in these estuaries although they still

contribute substantial amounts (Armstrong 1982). Marshes add only a small fraction of the nutrients that reach the estuaries compared to the freshwater inflows that account for over 80% of nutrient transportation. Marsh contributions come from tidal exchange or through flood inundation and dewatering (Armstrong 1982). A possible explanation for the small overall contributions of nutrients from Texas marshes could be due to the small marsh areas available for tidal inundation and the relative infrequency of marsh inundation (Armstrong 1982).

Unlike the unique geomorphological features of the Texas coast, other estuaries' nutrient content is driven by a larger number of factors such as larger tidal regimes, which Texas does not experience. For example, a study in the Hugli (also referred to as "Hooghly") Estuary located near the northern Bay of Bengal near Bangladesh with no barrier islands and dissimilar geomorphology than the Texas coast showed there was evidence that nutrients were driven by tidal influences. The results indicated that DIN and DIP concentrations significantly varied between spring and neap phases as well as high and low tides (Das et al. 2017). The nutrient concentrations were higher during the spring phase and lower during the neap phase. Another study off the coast of California had similar findings that nutrient fluxes were influenced by tidal components, unlike the Texas coast. The Elkhorn Slough Estuary located in Central California, in the temperate region, with a latitude of approximately 36°, has mesotidal ranges and no barrier islands. A study found that 39% of the variation in nutrient concentrations can be explained by tidal components, while only 15% of the variation can be explained by diurnal processes such as nutrient uptake by primary producers (Caffrey et al. 2007). These two separate studies in Bangladesh and California demonstrate that large tidal ranges can play an important role in nutrient dynamics importing nutrients to estuary systems.

However, the climate gradient does not drive nutrient levels alone in Texas, specifically for the Trinity-San Jacinto Estuary and the Upper Laguna Madre Estuary. The Trinity-San Jacinto Estuary had much higher phosphorus levels compared to the other estuaries in Texas (Fig. S41). It is possible that this is due to high population density and urbanization in Harris, Chambers, Galveston, and Brazoria County that surrounds the Trinity-San Jacinto Estuary. According to the 2020 census in Texas, Harris County is ranked #1 with a total population of 4,731,145 individuals (about twice the population of Kansas, USA) and a population density of 2771.7 people per square mile (US Census Bureau 2021). Brazoria County follows close behind ranking #14 in Texas with a total population of 372,031 and a population density of 272.9 people per square mile (US Census Bureau 2021). Having higher population densities in the watershed than the other estuaries in Texas, the Trinity-San Jacinto Estuary is expected to experience more phosphorus inputs into the bay through municipal and industrial sources, which is reflected in the daily average phosphorus levels in its bay system. When there are many wastewater treatment plants concentrated in a certain area, the amount of phosphorus being released into these estuaries can be quite large and significantly affect estuarine processes.

Phosphorus being released from wastewater treatment plants is often derived from detergents. Sodium tripolyphosphate (STPP), a form of phosphate, is a common ingredient in many household items such as laundry and dishwasher detergents and was used more heavily in the past than today. Shortly after World War II, the soap industry began to develop synthetic detergents containing phosphorus to deal with competing resources of natural fats and oils and the poor cleaning performance (Knud-Hansen 1994). By 1959, almost all laundry detergents contained 7–12% phosphorus, and by 1983, over two million tons of phosphorus was used annually in detergents alone (Knud-Hansen 1994). In addition, older wastewater treatment plant systems did not remove phosphorus as effectively as newer tertiary wastewater treatment systems that use a combination of removal methods such as physical and chemical processes.

Phosphorus from human feces and urine could also be another potential source of high phosphorus levels in the Trinity-San Jacinto Estuary. A study quantifying the mass of phosphorus available from human feces and urine illustrates that developed countries generate about 417,000 metric tons of total phosphorus per year (Mihelcic et al. 2011). Only about 0.3 to 1.5 million metric tons of phosphorus are estimated to be recovered from sewer water by wastewater reuse and reclamation annually (Mihelcic et al. 2011; Liu et al. 2008; Cordell et al. 2009). Today most of the phosphorus from human excreta ends up in waterways via wastewater from municipal discharges and/or illicit or malfunctioning septic systems or as sludge in landfills compared to the past (Cordell et al. 2009). Although phosphorus derived from feces and urine in Texas has not been quantified, this could be a plausible explanation for the sources of phosphorus entering the Trinity-San Jacinto Estuary.

The daily average DIP levels across all estuaries in Fig. S41 were calculated with data from 1973 to 1985, and the TP levels were calculated with data from 1969 to 2021 due to data availability (Table SB). This is important because over time there has been changes to ingredients in detergents containing lower or no phosphates, more advanced wastewater treatment technologies, and updated environmental law and policies regarding reducing phosphate use and increasing removal that could not have been captured in these overall daily average phosphorus calculations. To capture these changes over time, the DIP and TP concentrations in the Trinity-San Jacinto Estuary by year were illustrated (Fig. 9). In the late 1970s, there is a spike in both DIP and TP concentrations. Then in the early 1980s through 2021, there is a steady decrease in TP concentrations over time. However, the Trinity-San Jacinto Estuary still exhibited the highest daily average TP levels compared to the other estuaries regardless of inflow rates that could be a potential concern in the future due to population growth, land subsidence, sea-level rise, leaching septic systems, and tropical storm surge.

The Houston metropolitan area has been severely impacted by land subsidence and sea-level rise, potentially surpassing other urban areas in the United States. The coastline bordering Houston has undergone significant shifts, with subsidence of up to 10 feet altering the distribution of wetlands and aquatic vegetation (Coplin and Galloway 1999). Land subsidence can occur due to various factors, including groundwater extraction, oil and gas extraction, and sediment compaction. These processes contribute to rising sea levels, increase the susceptibility of areas to flooding, cause extensive damage to infrastructure, and disrupt nutrient dynamics. Consequently, natural flow patterns are disturbed, leading to changes in sediment transport and nutrient cycles and loss of wetlands. The lower reaches of the San Jacinto River near its confluence with Buffalo Bayou have the most extensive changes in wetlands in the Houston region and have subsided by 3 feet or more by 1978 (Coplin and Galloway 1999). A submerged wetland undergoes changes that affect its floral

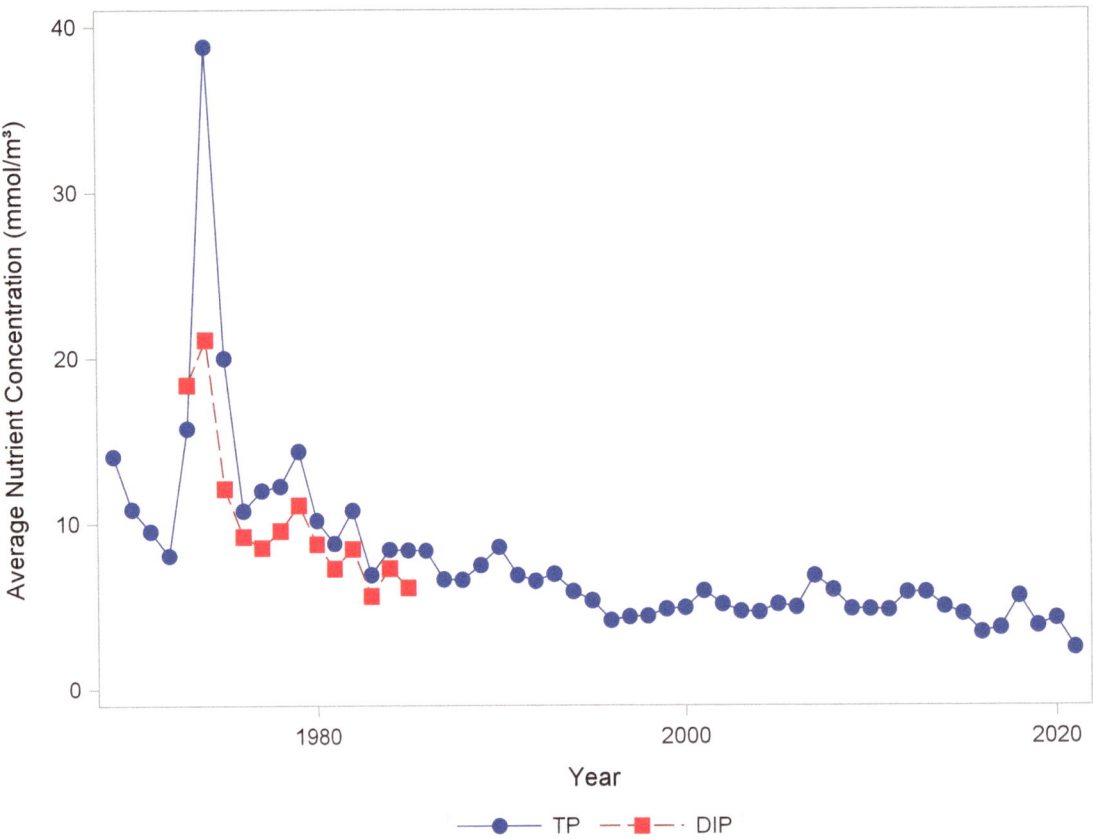

**Fig. 9** Average phosphorus concentrations in the Trinity-San Jacinto Estuary by year. Units: mmol/m³. Abbreviations can be found in the table of abbreviations

community and sediment-trapping capability. These changes in topography and hydrology are a concern and may affect nutrient budgets in the future.

Similar to the Trinity-San Jacinto Estuary, nutrient levels in the Upper Laguna Madre Estuary are also not driven by hydrology alone. The Upper Laguna Madre Estuary, which includes Baffin Bay, exhibits higher TKN concentrations than any other estuary in Texas. This is primarily due to this estuary's unique geographic features and habitats that create an ideal environment for organic nitrogen to proliferate. This estuary system is comprised mainly of tidal mudflats and seagrass beds with inflow dominated by sewage discharge and agricultural runoff. Blue-green algae are often found on the mudflats in the Upper Laguna Madre Estuary, conducting nitrogen fixation and resulting in nitrogen byproducts such as ammonia and organic nitrogen. A short-term assay study of algal mat samples from Aransas Bay, TX, illustrated the average annual input of nitrogen from these algal mats represented an ample contribution of nitrogen to shallow coastal environments (40.6 kg N ha⁻¹) (Gotto et al. 1981). Blue-green algae production on Texas flats, especially for the Upper Laguna Madre region, is significant because these "mats" cover thousands of hectares, and most other plants and algae are unable to grow there due to the harsh hypersa-

line conditions, high temperatures, and alternating cycles of inundation (Pulich and Rabalais 1986).

Baffin Bay is often inundated with reoccurring "brown tide blooms" made up of the phytoplankton species *Aureoumbra lagunensis* (Wetz et al. 2017). The presence of brown tide or *A. lagunensis* in Baffin Bay is a good indicator of high ammonium and organic nitrogen concentrations due to the fact that this phytoplankton species primarily uses ammonium and organic nitrogen as an available nutrient source (Muhlstein and Villareal 2007). When these brown tide blooms occur, much of the seagrass die off (Onuf 1996). The decomposition of seagrass can add an additional source of organic matter into the ecosystem leading to more reduced forms of nitrogen. From 2003 to 2013, the TKN concentrations were very high and exceed TKN concentrations in the other Texas estuaries by two- to fivefold (Wetz et al. 2017). These TKN concentrations have been positively correlated with cyanobacteria bloom biomass (Hicks et al. 2022) and the cell concentration of the brown tide organism *A. lagunensis* (Wetz et al. 2017).

Most of the freshwater inflow into the Upper Laguna Madre system is dominated by wastewater outfalls and agricultural runoff that contains high amounts of organic nitrogen. The results of an isotope mixing model study imply that sewage effluents are comprised mainly of DON and are the primary source of organic nitrogen into the bay (Felix and Campbell 2019). However, seasonal variations in isotopic composition of DON from sewage effluents are lower in the spring, suggesting that there are occasional increases in contributions of DON from agricultural sources (Felix and Campbell 2019). The ratio of DON to DIN entering the bay through inflow such as return flows and livestock waste can be quite high and significant to the estuarine system (Felix and Campbell 2019).

Not only does the wastewater return flows contribute significant amounts of organic nitrogen into the Upper Laguna Madre Estuary, but groundwater can also elevate DON levels by increasing the discharge of septic contaminated groundwater to the bay (Felix and Campbell 2019). A large portion of the Upper Laguna Madre watershed is a rural area containing many septic systems as a cost-effective method of wastewater treatment and storage. These septic systems undergo anaerobic processes resulting in most of the nitrogen to exist as $NH_4$ and DON (Felix and Campbell 2019). Many of these septic systems are situated on porous and permeable soils that are not adequate for proper drain fields, leading to the effluents directly infiltrating the groundwater (Felix et al. 2021). Furthermore, despite the addition of organic nitrogen from multiple sources such as sewage return flows, blue-green algae, seagrass decomposition, and septic-contaminated groundwater, TKN values in the Upper Laguna Madre Estuary can also increase because of the long residence time of the system. This enables organic nitrogen to be retained and nutrient cycling to persist over a long period of time before being flushed out of the system.

The high TKN concentrations in Baffin Bay can be a potential concern for watershed management because *A. lagunensis* relies heavily on reduced nitrogen including DON to proliferate, and they cannot utilize $NO_3$ (DeYoe and Suttle 1994; Wetz et al. 2017). Moreover, most other ecologically "healthy" phytoplankton fail to survive using DON as a nutrient source, allowing these "brown tides" to persist (Felix and Campbell 2019). These conditions can cause more frequent harmful algae blooms in the Upper Laguna Madre Estuary in the future leading to eutrophication, hypoxia, and water quality degradation in this estuarine system.

## Estuary Comparisons

LOICZ budgets for the Texas bay systems from the present study were compared to other estuary bay systems from around the world (Table 14). There is a pattern of lower daily average nitrate concentrations in Texas estuaries of this study compared to the nitrate concentrations reported from other estuaries in Asia, South America, and Africa (Fig. 10). The highest nitrate concentration was recorded in the Yalujiang

Estuary, located in the Northern Yellow Sea, Liaoning Province, China (Liu et al. 2009), while the lowest nitrate concentration was recorded in the Mission-Aransas Estuary, Texas, USA. However, salinity values between Texas estuaries and the other estuary systems vary greatly. These salinity values varied from 6.51 to 37.9 S. The lowest salinity value was in the Yalujiang Estuary with a value of 6.51 S, and the highest salinity recorded was in the Upper Laguna Madre Estuary. The area and volume among all these estuaries also varied greatly where the area ranged 2–3094 $km^2$ and volume ranged 1–23,000 $10^6$ $m^3$. The residual flow values ($V_R$) were similar among Texas estuaries and the Huanghe Estuary (Laizhou Bay, South Bohai, China) and Itaipu Lagoon (Niterói City, Rio de Janeiro Brazil). However, the Changjiang Estuary (East China Sea, China), Cameroon Estuary (Douala, Cameroon, Africa), and Rio-del-Rey Estuary (Ndian département, Cameroon, Central Africa) had significantly higher residual outflow rates. The relationship between nitrate and salinity among all the estuaries in this comparison illustrates a negative exponential decline where nitrate values are negatively correlated with water column salinity values. The nitrate values rapidly declined at first with increasing salinity and then slowly leveled out (Fig. 10).

In this study, nonconservative processes revealed that all eight estuaries in Texas act as a sink for both nitrogen and phosphorus nutrients. However, compared to other estuaries around the world, the results differed, and each system's nutrients can have different source-sink patterns. There is not a single common factor or circumstance to determine the fate of nutrients across all estuarine systems. Determining the main driver of nutrient source-sink behaviors in estuaries is a challenging task. This is due to the complexity of estuarine systems and exposure to many external factors, including the combinations and interactions between them. However, many of these factors can still be identified.

Unlike the unique geomorphological features of the Texas coast, the fate of nutrients in other estuaries is driven by many factors including salt marshes. For instance, marshes in other estuaries could serve as a significant role in nutrient source-sink behavior in a coastal zone contrary to the Texas coast. A study completed on the Mullica River-Great Bay (MRGB) Estuary located in southern New Jersey discovered that the nutrient cycling trends follow the "marsh-estuarine continuum theory" where net nutrient fluxes are a function of marsh-estuarine system's geological age and developmental status. Upper regions of estuaries are assumed to be at earlier stages of development, functioning as net nutrient importers, and lower regions of estuaries are more mature, functioning as net nutrient exporters (Dame 1994; McGuirk Flynn 2008). Previous research conducted in the MRGB Estuary has provided support for this theory by suggesting that the marsh ecosystem exports dissolved organic nitrogen (DON) and dissolved inorganic phosphorus (DIP) while simultaneously

**Table 14** LOICZ budget comparisons between the current study and studies around the world

| Estuary | Area | Volume | Location | Coordinates | Dates | $V_R$ | Sal | $NO_3$ | $NH_4$ | Source |
|---|---|---|---|---|---|---|---|---|---|---|
| Yalujiang Estuary | 170 | 1020 | The Northern Yellow Sea, Liaoning Province, China. Lies between China and North Korea (50–60% in China) | Approximately 39°40′N to 40°10′N and 120°10′E to 120°40′E | 1992, 1994, 1996 | −88.70 | 6.51 | 0.1588 | 0.00313 | Liu et al. (2009) |
| Huanghe Estuary | 3.6 | 7.2 | Laizhou Bay, South Bohai. Originates on the eastern Qinghai-Tibet Plateau flowing through northwestern China | Approximately 37°45′N to 37°55′N and 119°10′E to 119°20′E | 2001–2002 | −8.45 | 26.93 | 0.0665 | 0.00544 | Liu et al. (2009) |
| Changjiang Estuary | 3094 | 15,470 | Located in the Qinghai-Tibet Plateau flowing to East China Sea on the west boundary of the East China Sea shelf | Approximately 27.5°N to 33.7°N and 117.5°E to 125°E | 1997–2001 | −2498.10 | 10.82 | 0.05975 | 0.00566 | Liu et al. (2009) |
| Itaipu Lagoon | 2 | 1 | Located off the southeastern coast of Brazil, in Niterói City, Rio de Janeiro Brazil | 22°55′ to 22°58′S and 43°07′ to 43°03′W | 2005–2006, 2009–2010 | −0.0086 | 30.85 | 0.00642 | 0.029589 | Cerda et al. (2017) |
| Cameroon Estuary | 1500 | 23,000 | Located in Douala, Cameroon, Central Africa, at the northeastern end of the Gulf of Guinea of sub-Saharan Africa | Longitudes 9.25° to 10.00° E, latitudes 3.83° to 4.1° N | Unknown | −553.42 | 12.25 | 0.0597 | Unknown | Gabche and Smith (2002) |
| Rio-del-Rey Estuary | 1350 | 20,000 | Located in the Ndian département, Cameroon, Central Africa, at the northeastern end of the Gulf of Guinea of sub-Saharan Africa | Longitude 8.3° E, latitude 4.8° N | Unknown | −473.97 | 14.55 | 0.0508 | Unknown | Gabche and Smith (2002) |
| SN | 265 | 660 | Texas coast | Approximately 29.9° N, 93.8° W | Table SA–H | −46.41 | 8.19 | 0.00763 | 0.00497 | This paper |
| TS | 1456 | 2707 | Texas coast | Approximately 29.5° N, 94.9° W | Table SA–H | −40.08 | 16.19 | 0.00869 | 0.00457 | This paper |
| LC | 1115 | 2430 | Texas coast | Approximately 28.5° N, 96.4° W | Table SA–H | −11.31 | 21.00 | 0.00374 | 0.0039 | This paper |
| GE | 587 | 728 | Texas coast | Approximately 28.3° N, 96.7° W | Table SA–H | −7.67 | 18.76 | 0.00426 | 0.00392 | This paper |
| MA | 524 | 847 | Texas coast | −Approximately 28.1° N, 97.0° W | Table SA–H | −1.10 | 20.11 | 0.00318 | 0.00358 | This paper |
| NC | 571 | 1280 | Texas coast | Approximately 27.8° N, 97.3° W | Table SA–H | −1.01 | 30.19 | 0.00338 | 0.00398 | This paper |
| ULM | 591 | 447 | Texas coast | Approximately 27.3° N, 97.4° W | Table SA–H | −0.0037 | 37.90 | 0.00322 | 0.0039 | This paper |
| LLM | 1308 | 997 | Texas coast | Approximately 26.3° N, 97.3° W | Table SA–H | 0.022 | 32.10 | 0.00403 | 0.00407 | This paper |

Units: area = $km^2$; volume = $10^6$ $m^3$; $V_R$ (residual flow) = $10^6$ $m^3/d$; Sal = S; $NO_3$ = $mol/m^3$; $NH_4$ = $mol/m^3$

**Fig. 10** NO$_3$ and salinity budget comparisons between the estuaries in this study and other estuaries around the world. Other studies are represented by filled symbols, and this study is represented by outlined symbols. Abbreviations can be found in the table of abbreviations

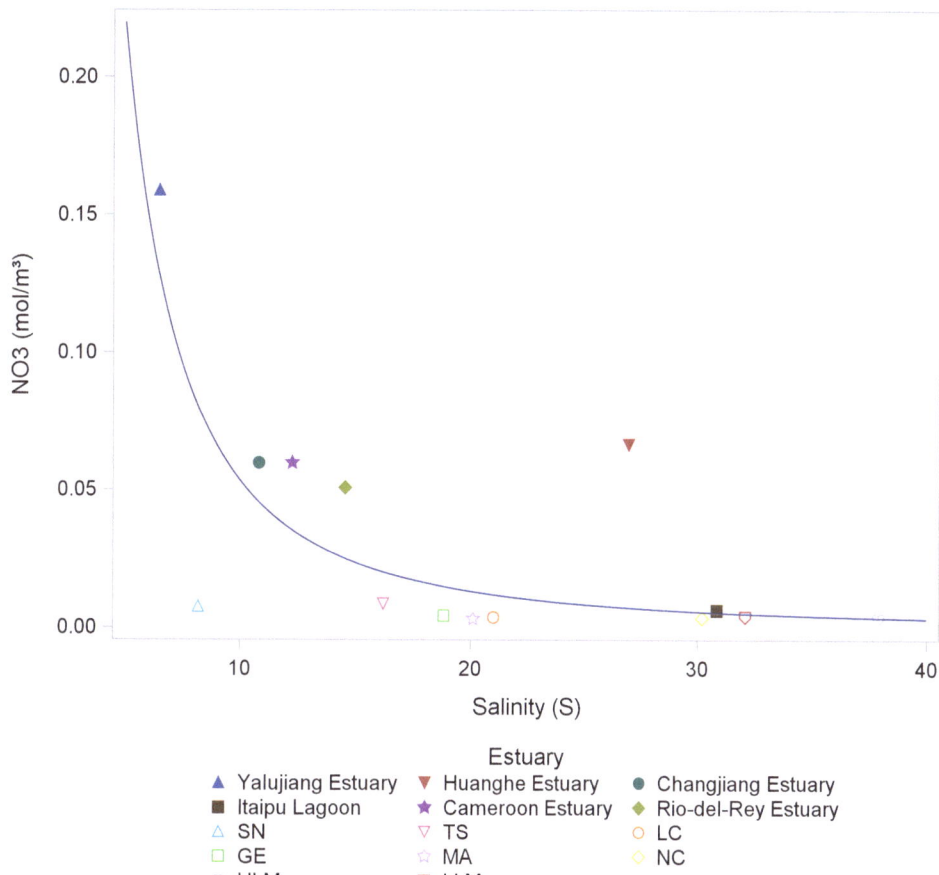

importing nitrate (NO$_3$) and ammonium (NH$_4$). These findings led to the conclusion that marshes in this estuary may experience limitations in terms of nitrogen availability (Durand 1988; McGuirk Flynn 2008). Annually this estuary is a net source of DON, DIP, and a net sink for DIN. However, the estuary also exhibited temporal source-sink patterns. Seasonally, the upper region of the estuary was a source of DON in the fall and spring, but it was a sink in the winter and summer. Moreover, DOP was a net sink in the upper and lower regions of the estuary in the winter and spring and was alternating to a net source of DOP during the fall and summer (McGuirk Flynn 2008).

Marshes and mangroves do not control the biogeochemical processes of nutrient source-sink behavior alone in an estuarine system, and large tidal regimes can play a role unlike the microtidal ranges in Texas with the exception of tropical storm tidal surges. Nutrient exchange between the mangrove and estuary interfaces in the Zhangjiang Estuary of East China is affected by tidal cycling and mixing. Nutrient and water fluxes were positively correlated across the mangrove region during a flood tide event, suggesting that nutrient export was driven mainly by tidal exchange (Wang et al. 2021). The results of the nutrient fluxes suggest that man-

groves were always a source of NH$_4$ and dissolved reactive phosphorus (DRP) but a strong nitrate sink (Wang et al. 2021). Although tidal regimes are a strong influence on nutrients in the Zhangjiang Estuary, there is an alternating source-sink dynamic between seasons (Wang et al. 2021). This is likely due to biological processes including microbial activity and primary production, aside from the strong tidal influences exhibited. In the summer months (June and July), the net import of NO$_3$ was higher than in the spring (April) due to higher temperatures often stimulating NO$_3$ removal by microbial denitrification (Wang et al. 2021).

Additional instants of a source-sink patterns are illustrated in a partition coefficient ($K_d$), the final distribution of phosphorus between the dissolved and particulate phase ($K_d = \dfrac{P_{es}}{C_{es}}$), where a source-sink behavior can happen due to driving factors such as riverine inputs. The results of a $K_d$ fit model study comparing several different estuaries' phosphorus concentrations and patterns suggest that riverine DIP levels are important regulators of DIP behavior. When riverine DIP concentrations increase, estuaries gradually move from being a source for DIP to a sink for DIP (Prastka et al. 1998). Therefore, when there are low riverine DIP concentrations, desorption from river-borne particulate material is predicted to occur, whereas at high DIP concentrations, removal of DIP to particles is predicted to occur (Prastka et al. 1998). Another source-sink alternative pattern can be observed in the San Francisco Estuary, located in California, where temporal patterns of multiple factors influence nutrient dynamics. Nearly two decades of seasonal nutrient distributions illustrate an alternating sink-source dynamic between competing factors, such as river flow being a nutrient source and phytoplankton productivity being a nutrient sink. In the dry, warmer months, the phytoplankton (nutrient sink) activities dominate, and in wetter months, the river flow (nutrient source) dominates, creating a nutrient balance (Peterson et al. 1985).

Residence time, or the time it takes for the estuary to replace the freshwater volume of the estuary, is also a significant determinant of the amount of nutrients that get exported to the coast. In Fourleague Bay, Louisiana, the total nitrogen export was less than 3% at high residence times to greater than 80% at low residence times (Perez et al. 2010). A possible explanation is that as the residence time increases, it slows the flushing rate. When the residence time is long, there is more time for transformation processes to occur, whereas with short residence times and faster flushing, there is limited time for biochemical reactions to take place (Perez et al. 2010). Systems with better water quality often experience shorter residence times that are beneficial to the system due to high hydrodynamics and exportation (Cabral and Fonseca 2019).

Wind-driven sediment resuspension events in the water column have also been observed to alter chemical and biological processes in estuaries. The largest removal episode where nutrients were released to the water column occurred due to a strong wind event on the Neuse River Estuary (NRE), North Carolina. Ammonia's advective flux was two to six times greater through wind events than simple diffusion (Corbett 2010). Another example where wind-induced sediment resuspension altered nutrient dynamics was in Lake Arreso, Denmark. The internal phosphorus loading was 20 to 30 times greater than the release from an undisturbed water column (Søndergaard et al. 1991). In some cases, internal nutrient loading from suspended materials can be the main contributor of nutrients to an estuary system like Shenzhen Bay, where DIN and P from internal nutrients supplied 65 to 69% of the total input fluxes to the system (Yan et al. 2021). Internal nutrient loading from sediments can worsen water quality conditions and contribute to eutrophication, resulting in recovery setbacks. However, managing internal nutrient loads can be a restoration method for eutrophic systems. Nutrient reduction methods include phosphate inactivation agents and have been used for restoration efforts in the Netherlands and New Zealand (Zamparas and Zacharias 2014).

It is known that many naturally occurring environmental factors can influence if an estuary is a source or sink for nutrients; however anthropogenic changes such as construction and damming profoundly affect coastal ecosystems' water quality too. In the São Francisco Estuary, located in East Brazil, there was a decrease in DIN outputs by 94% from 1985 to 2001 that was attributed to the construction of the Xingó Dam in 1995 (Medeiros et al. 2011). The estuary had become oligotrophic at the time of this study in 2011, where most of the nutrients were retained by the dam reservoirs. On the other hand, an analysis of the cumulative load difference before and after dam construction indicated that the water quality in the Geum River Estuary Dam System, Korea, deteriorated considerably since the dam construction. The estuary acted as a sink for $NH_4$ and P due to excessive phytoplankton growth in the reservoir (Jeong et al. 2014). Lastly, a long-term water quality study on the construction of the Wyaralong Dam near the Logan-Albert Estuary in Australia demonstrated that the dam improved the water quality in the upper region of the estuary by lowering the nutrient concentrations and turbidity levels (Eccles et al. 2020). These comparisons indicate that damming can have mixed results on different estuarine systems, which can be detrimental or beneficial to these systems' biological and chemical processes. The effects of marine construction and dams on Texas estuaries have been documented by TWDB's Pollution Response Inventory and Species Mortality (PRISM) database. From 1958 to 1997, one of the leading causes of fish kills in Texas estuaries was anthropogenic factors such as construction of dead-end canals in residential or industrial areas (Thronson and Quigg 2008). Fish mortalities can be indicators of poor water quality due to often being unable to survive the adverse effects of changing nutrient concentrations such as hypoxia and algal blooms and poor water circulation.

Other anthropogenic factors affecting the fate of nutrients in an estuary are urban effluents from wastewater treatment plant outfalls. In the upper region of the Can Gio Estuary in Vietnam, wastewater effluents were the main input of nutrients. However, the ammonia and nitrate concentrations peaked downstream due to export by mangroves (Taillardat et al. 2020). Wastewater treatment plants in the United States alone process about 34 billion gallons (128,702,826 $m^3$) of wastewater daily, including human waste, food, and detergents (US EPA 2022). Depending on the equipment, some treatment facilities can remove more N and P than others. Nonetheless, wastewater is not the only contributor to nutrient loads, and this estuary is another example where there are multiple factors that influence nutrient dynamics in a coastal zone. In some situations, the magnitude at which external factors contribute to nutrients in estuarine systems can change over time from contributing significantly to nutrient loading to contributing very little. Since 1990, estuaries of Waquoit Bay in Massachusetts had an increase of N loadings from wastewater by 80%; however, loads from atmospheric deposition decreased by about 41%. The net result of these changes over time has not significantly changed the total N loads on a decadal scale (Valiela et al. 2016).

## Conclusion

In conclusion, this study has demonstrated important insights between hydrology and the water, salt, and nutrient dynamics in each estuary along the Texas coast. This research shows that precipitation is a strong driver of inflow, which is then a strong driver of salinity and nutrient budgets. However, nutrient dynamics are not always driven by hydrology alone, and this can be seen in the Trinity-San Jacinto Estuary where high P is caused by phosphorus derived from historical detergent use and possible human execrate in a high population density area. It can also be seen in the Upper Laguna Madre Estuary where high TKN is caused by large populations of blue-green algae, decomposition of seagrass, long residence times, and sewage discharge dominated by organic nitrogen. Comparing nutrient budgets in different estuaries around the world, other estuaries are exposed to many different components or combinations of factors that dictate what nutrients will be sources or sinks in an estuary. Among the literature, the factors that impact the fate of nutrients in estuaries are not equal and can have a stronger or weaker influence depending on the estuary, and there can be a combination of different factors. Physical, chemical, and biological processes such as river discharge, tidal exchange, phytoplankton activity, wind-driven resuspension of sediment containing nutrients, and strong seasonal variations in temperature all play a role in the fate of nutrients in an estuarine environment. Moreover, there even appears to be a strong alternating system between the source and sink behaviors of nutrients in some estuaries. In addition, through construction such as damming and wastewater treatment plants, humans can strongly influence nutrient dynamics in a watershed and alter the hydrology and flow of downstream nutrients.

Understanding how nutrients fluctuate and what drives these changes is important for adaptative management practices to promote and sustain healthy ecological conditions. Over time, population densities and total population size will increase, which can be a cause of concern for the phosphorus levels in estuaries such as the Trinity-San Jacinto Estuary. The projected population growth between 2010 and 2050 in Texas is expected to reach 40,502,749 individuals by the end of 2050 (Potter and Hoque 2014). That is almost a 40% increase in the number of people than the current 2020 census recorded with 29,145,505 individuals (US Census Bureau 2021). Land subsidence and sea-level rise are also two interconnected physical factors that have the potential to significantly impact nutrient budgets in estuaries in the future. Understanding the effects of hydrology, human water consumption, and consequently the return flows from wastewater outfalls on water quality in estuaries is essential to watershed management.

## Recommendations

This research provided many insights on biogeochemical drivers in Texas estuaries, but a recommendation can be made based on available data. The process of building these LOICZ models identified critical information gaps. It is recommended to maintain consistent monitoring protocols to create consistent long-term databases of water quality parameters, specifically nutrient concentrations. This allows for more accurate estimates of change trajectories in ecosystem functioning. Due to the lack of consistency, these datasets limited the statistical analyses that could be performed. For example, LOICZ budgets can identify daily and monthly accumulations of conservative and nonconservative materials in each estuary, but this study could not complete it because samples were only recorded less than half the time in a month and sometimes even only once in a month. In addition, LOICZ budgets can also potentially reveal seasonal and spatial variations of nutrients, salinity, and inflow, but could not be completed as the available data samples were inconsistently recorded, with entries only made 1 or 2 months per year and, in some cases, no entries were made in consecutive years. Lastly, these budgets can identify when estuaries switch from sources to sinks depending on interannual variability of nutrients and their transformation processes but were not completed because there was not enough data from year to year for all variables including nutrients, salinity, and river streamflow. It is extremely important to have consistent water quality monitoring in place to be able to identify how these ever-changing environmental conditions are affecting these ecosystems. Enhancing the establishment of consistent data collection protocols moving forward will greatly benefit scientists in conducting further research on nutrient dynamics in coastal zones. This advancement will not only enable regulators and policymakers to make better-informed decisions but also contribute to future efforts aimed at improving water quality in estuaries.

**Acknowledgments** This chapter is based on a thesis (Marshall 2023). Support was provided by the Texas General Land Office via the Gulf of Mexico Energy Security Act of 2006 funding made available to the State of Texas and awarded under the Texas Coastal Management Program number 21-155-007-C879. Additional support was provided by the Jacob & Terese Hershey Foundation and the Crutchfield Fellowship.

# References

Arismendez S (2010) Land-water nutrient coupling processes in central Texas estuaries. Dissertation, Texas A&M University-Corpus Christi

Armstrong NE (1982) Responses of Texas estuaries to freshwater inflows. In: Kennedy VS (ed) Estuarine comparisons. Academic Press, New York, pp 103–120. https://doi.org/10.1016/b978-0-12-404070-0.50013-2

Bowen JL, Giblin AE, Murphy AE, Bulseco AN, Deegan LA, Johnson DS, Nelson JA, Mozdzer TJ, Sullivan HL (2020) Not all nitrogen is created equal: differential effects of nitrate and ammonium enrichment in coastal wetlands. Bioscience 70(12):1108–1119. https://doi-org.manowar.tamucc.edu/10.1093/biosci/biaa140

Brandes RJ, Heitmuller F, Huston R, Jensen P, Kelley M, Manhart F, Montagna PA, Ward G, Wiersema J (2009) Methodologies for establishing a freshwater inflow regime for Texas estuaries within the context of the senate bill 3 environmental flows process (Report No. SAC-2009-03). https://hdl.handle.net/1969.6/94344 accessed 20 Jan 2023

Cabral A, Fonseca A (2019) Coupled effects of anthropogenic nutrient sources and meteo-oceanographic events in the trophic state of a subtropical estuarine system. Estuar Coast Shelf Sci 225:106228. https://doi.org/10.1016/j.ecss.2019.05.010

Caffrey JM, Chapin TP, Jannasch HW, Haskins JC (2007) High nutrient pulses, tidal mixing and biological response in a small California estuary: variability in nutrient concentrations from decadal to hourly time scales. Estuar Coast Shelf Sci 71(3–4):368–380. https://doi.org/10.1016/j.ecss.2006.08.015

Cerda M, Nunes-Barboza CD, Scali-Carvalho CN, De Andrade-Jandre K, Marques AN Jr (2017) Nutrient budgets in the Piratininga-Itaipu lagoon system (southeastern Brazil): effects of sea-exchange management. Latin Am J Aquat Res 41(2):226–238. https://doi.org/10.3856/vol41-issue2-fulltext-3

Coplin LS, Galloway D (1999) Houston-Galveston, Texas. Land subsidence in the United States: US Geological Survey Circular 1182:35–48

Corbett DR (2010) Resuspension and estuarine nutrient cycling: insights from the Neuse River Estuary. Biogeosciences 7(10):3289–3300. https://doi.org/10.5194/bg-7-3289-2010

Cordell D, Drangert JO, White S (2009) The story of phosphorus: global food security and food for thought. Glob Environ Chang 19(2):292–305. https://doi.org/10.1016/j.gloenvcha.2008.10.009

Dame RF (1994) The net flux of material between marsh-estuarine systems and the sea: the Atlantic coast of the United States. In: Mitsch WJ (ed) Global Wetlands Old World and New. Elsevier Science, New York, pp 295–302

Das S, Giri S, Das I, Chanda A, Ghosh A, Mukhopadhyay A, Akhand A, Choudhury SB, Dadhwal VK, Maity S, Kumar TS (2017) Nutrient dynamics of northern Bay of Bengal (nBoB)—Emphasizing the role of tides. Reg Studies Mar Sci 10:116–134. https://doi.org/10.1016/j.rsma.2017.01.006

DeYoe HR, Suttle CA (1994) The inability of the Texas "brown tide" alga to use nitrate and the role of nitrogen in the initiation of a persistent bloom of this organism. J Phycol 30(5):800–806. https://doi.org/10.1111/j.0022-3646.1994.00800.x

Durand JB (1988) Field studies in the Mullica River-Great Bay Estuarine System. Center for Coastal and Environmental Studies, Rutgers University, New Brunswick, New Jersey, pp 1–54

Eccles R, Zhang H, Hamilton D, Maxwell P (2020) Trends in water quality in a subtropical Australian river-estuary system: responses to damming, climate variability and wastewater discharges. J Environ Manag 269:110796. https://doi.org/10.1016/j.jenvman.2020.110796

Engle VD, Kurtz JC, Smith LM, Chancy C, Bourgeois PA (2007) A classification of US estuaries based on physical and hydrologic attributes. Environ Monit Assess 129:397–412. https://doi.org/10.1007/s10661-006-9372-9

Felix JD, Campbell J (2019) Investigating reactive nitrogen sources that stimulate algal blooms in Baffin Bay. Final Report to the Coastal Bend Bays & Estuaries Program, Project # 1818, Publication CBBEP–129. https://www.cbbep.org/manager/wp-content/uploads/1818-Final-Report.pdf accessed 13 Oct 2023

Felix JD, Qiu Y, Cox A, Murgulet D (2021) Quantifying septic effluent nitrogen loading and processing in the Baffin Bay Watershed. Final Report to the Coastal Bend Bays & Estuaries Program, Project # 2005, Publication CBBEP – 144. https://www.cbbep.org/manager/wp-content/uploads/2005-Quantifying-Septic-Effluent-Nitrogen-Loading-and-Processing-in-the-Baffin-Bay-Watershed-04012021-144.pdf accessed 13 Oct 2023

Gabche CE, Smith VS (2002) Water, salt and nutrients budgets of two estuaries in the coastal zone of Cameroon. West African J Appl Ecol 3(1). https://doi.org/10.4314/wajae.v3i1.45581

Gordon DC, Boudreau PR, Mann KH, Ong JE, Silvert WL, Smith SV, Wattayakorn G, Wulff F, Yanagi T (1996) LOICZ biogeochemical modelling guidelines, Vol. 5, Yerseke: LOICZ Core Project, Netherlands Institute for Sea Research. https://www.ferrybox.eu/imperia/md/content/loicz/print/rsreports/report5.pdf

Gotto JW, Tabita FR, Van Baalen C (1981) Nitrogen fixation in intertidal environments of the Texas gulf coast. Estuar Coast Shelf Sci 12(2):231–235. https://doi.org/10.1016/S0302-3524(81)80099-0

Guthrie CG, Lu Q (2010) Coastal hydrology for the Guadalupe Estuary: updated hydrology with emphasis on diversion and return flow data for 2000–2009. https://www.twdb.texas.gov/surfacewater/bays/major_estuaries/guadalupe/doc/TWDB_Hydrology_GuadalupeEstuary_201011.pdf

Guthrie CG, Matsumoto J, Lu Q (2010a) TxBLEND model calibration and validation for the Guadalupe and Mission-Aransas Estuaries. Texas Water Development Board, 46, 1–46. https://www.twdb.texas.gov/surfacewater/bays/major_estuaries/guadalupe/doc/TWDB_TxBLEND_GuadalupeAransas_July2010.pdf

Guthrie CG, Matsumoto J, Lu Q (2010b) TxBLEND model validation for the upper Guadalupe Estuary using recently updated inflow data. Austin, Texas: Texas Water Development Board, 25pp. https://www.twdb.texas.gov/surfacewater/bays/major_estuaries/guadalupe/doc/TWDB_TxBLEND_Guadalupe_Validation_November2010.pdf

Guthrie CG, Schoenbaechler C, Matsumoto J, Lu Q (2011) Comparison of two hydrology datasets, as applied to the TxBLEND model, on salinity condition in Nueces Bay. https://www.tceq.texas.gov/assets/public/permitting/watersupply/water_rights/eflows/20110729nuecesbbest_comparison.pdf

Hicks TD, Kuns CM, Raman C, Bates ZT, Nagarajan S (2022) Simplified method for the determination of total Kjeldahl Nitrogen in wastewater. Environments 9(5):55. https://doi.org/10.3390/environments9050055

Intergovernmental Oceanographic Commission (2015) The International thermodynamic equation of seawater–2010: calculation and use of thermodynamic properties. [Includes corrections up to 31st October 2015]. https://www.teos-10.org/pubs/TEOS-10_Manual.pdf

Jeong YH, Yang JS, Park K (2014) Changes in water quality after the construction of an estuary dam in the Geum River Estuary Dam system. Korea J Coast Res 30(6):1278–1286. https://doi.org/10.2112/JCOASTRES-D-13-00081.1

Kempthorne O, Allmaras RR (1965) Errors of observation. In: Black CA (ed) Methods of soil analysis, Part 1. Physical and mineralogical properties, including statistics of measurement and sampling. American Society of Agronomy, Madison, pp 1–23

Knud-Hansen C (1994) Historical perspective of the phosphate detergent conflict. Conflict Research Consortium. In: Natural Resources and Environmental Policy Seminar of the University of Colorado. University of Colorado, Boulder

Lewis E (1980) The practical salinity scale 1978 and its antecedents. IEEE J Ocean Eng 5(1):3–8

Liu Y, Villalba G, Ayres RU, Schroder H (2008) Global phosphorus flows and environmental impacts from a consumption perspective. J Ind Ecol 12(2):229–247

Liu SM, Hong GH, Zhang J, Ye XW, Jiang XL (2009) Nutrient budgets for large Chinese estuaries. Biogeosciences 6(10):2245–2263. https://doi.org/10.5194/bg-6-2245-2009

Marshall DA (2023) Nitrogen and phosphorus budgets in estuaries of the Texas Coast. Thesis, Texas A&M University-Corpus Christi

Marshall DA, Montagna PA (2023) Nitrogen and phosphorous land-ocean interaction in the coastal zone (LOICZ) budgets for Texas estuaries. Distributed by: Gulf of Mexico Research Initiative Information and Data Cooperative (GRIIDC), Harte Research Institute, Texas A&M University–Corpus Christi. https://doi.org/10.7266/7gx3tv5c

Martinez-Andrade F, Campbell RP, Fuls BE (2005) Trends in relative abundance and size of selected finfishes and shellfishes along the Texas Coast: November 1975-December 2003. Management Data Series No. 232, Texas Parks and Wildlife Department, Coastal Fisheries Division, Austin, Texas. 140 pp

Matsumoto J, Powell GL, Brock DA, Paternostro C (2005) Effects of structures and practices on the circulation and salinity patterns of Galveston Bay, Texas. Texas Water Development Board, Austin, Texas. https://www.twdb.texas.gov/surfacewater/bays/models/doc/GalvesModeling.pdf

Matsumoto J, Guthrie PCG, Crockett D (2014) TxBLEND model extension and salinity validation for the Sabine-Neches estuary: extending simulations through 2013. https://www.twdb.texas.gov/surfacewater/bays/models/doc/TWDB_TxBLEND_SabineTechnicalMemo.pdf

McGuirk Flynn A (2008) Organic matter and nutrient cycling in a coastal plain estuary: carbon, nitrogen, and phosphorus distributions, budgets, and fluxes. J Coast Res 10055:76–94. https://doi.org/10.2112/si55-010.1

Medeiros PRP, Knoppers BA, Cavalcante GH, Souza WFLD (2011) Changes in nutrient loads (N, P and Si) in the São Francisco estuary after the construction of dams. Brazilian Arch Biol Technol 54(2):387–397. https://doi.org/10.1590/s1516-89132011000200022

Menne MJ, Durre I, Vose RS, Gleason BE, Houston TG (2012) An overview of the global historical climatology network-daily database. J Atmos Ocean Technol 29:897–910. https://doi.org/10.1175/JTECH-D-11-00103.1

Mihelcic JR, Fry LM, Shaw R (2011) Global potential of phosphorus recovery from human urine and feces. Chemosphere 84(6):832–839. https://doi.org/10.1016/j.chemosphere.2011.02.046

Millero FJ, Feistel R, Wright DG, McDougall TJ (2008) The composition of Standard Seawater and the definition of the reference-composition salinity scale. Deep Sea Res Part I: Oceanogr Res Papers 55(1):50–72

Montagna PA, Palmer TA, Beseres PJ (2013) Hydrological changes and estuarine dynamics. Springer, New York. https://doi.org/10.1007/978-1-4614-5833-3

Montagna PA, Hu X, Palmer TA, Wetz M (2018) Effect of hydrological variability on the biogeochemistry of estuaries across a regional climatic gradient. Limnol Oceanogr 63:2465–2478. https://doi.org/10.1002/lno.10953

Muhlstein HI, Villareal TA (2007) Organic and inorganic nutrient effects on growth rate–irradiance relationships in the Texas brown-tide alga Aureoumbra lagunensis (Pelagophyceae). J Phycol 43(6):1223–1226

Onuf CP (1996) Seagrass responses to long-term light reduction by brown tide in upper Laguna Madre, Texas: distribution and biomass patterns. Mar Ecol Prog Ser 138:219–231

Opdyke D, Hoffman J, Montagna PA, Trungale JF (2024) Chapter 4, Hydrology, circulation, and salinity. In: Montagna PA, Douglas AR (eds) Freshwater inflow to Texas Bays and Estuaries: a regional-scale review, synthesis, and recommendations. Springer Nature, Cham

Perez BC, Day JW, Justic D, Lane RR, Twilley RR (2010) Nutrient stoichiometry, freshwater residence time, and nutrient retention in a river-dominated estuary in the Mississippi Delta. Hydrobiologia 658(1):41–54. https://doi.org/10.1007/s10750-010-0472-8

Peterson DH, Smith RE, Hager SW, Harmon DD, Herndon RE, Schemel LE (1985) Interannual variability in dissolved inorganic nutrients in Northern San Francisco Bay Estuary. In: Cloern JE, Nichols FH (eds) Temporal dynamics of an estuary: San Francisco Bay. Developments in hydrobiology, vol 30. Springer, Dordrecht. https://doi.org/10.1007/978-94-009-5528-8_3

Potter LB, Hoque N (2014) Texas population projections, 2010 to 2050. Office of the State Demographer. https://demographics.texas.gov/Resources/Publications/2014/2014-11_ProjectionBrief.pdf

Prastka K, Sanders R, Jickells T (1998) Has the role of estuaries as sources or sinks of dissolved inorganic phosphorus changed over time? Results of a Kd study. Mar Pollut Bull 36(9):718–728. https://doi.org/10.1016/s0025-326x(98)00052-6

Pulich W, Rabalais S (1986) Primary production potential of blue-green algal mats on Southern Texas tidal flats. Southwest Nat 31(1):39–47. https://doi.org/10.2307/3670958

Redfield AC (1958) The biological control of chemical factors in the environment. Am Sci 46:205–221

Schoenbaechler C, Guthrie CG, Matsumoto J, Lu Q (2011a) TxBLEND Model calibration and validation for the Laguna Madre Estuary. Tex Water Dev Board, 60pp. https://www.twdb.texas.gov/surfacewater/bays/major_estuaries/laguna_madre/doc/TWDB_TxBLEND_LagunaMadre_20111027.pdf

Schoenbaechler C, Guthrie CG, Matsumoto J, Lu Q, Negusse S (2011b) TxBLEND model calibration and validation for the Lavaca-Colorado estuary and East Matagorda Bay. Austin, Texas: Texas Water Development Board, 72pp. https://www.twdb.texas.gov/surfacewater/bays/major_estuaries/colorado_lavaca/doc/20110214clbbest_twdb_txblend.pdf

Schoenbaechler C, Guthrie CG, Matsumoto J, Lu Q, Negusse S (2011c) TxBLEND model calibration and validation for the Nueces estuary. Austin, Texas: Texas Water Development Board, 72pp. https://www.tceq.texas.gov/assets/public/permitting/watersupply/water_rights/eflows/20110729nuecesbbest_txblend.pdf

Sheldon JE, Alber M (2006) The calculation of estuarine turnover times using freshwater fraction and tidal prism models: a critical evaluation. Estuar Coasts 29:133–146. https://doi.org/10.1007/BF02784705

Solis RS, Powell GL (1999) Hydrography, mixing characteristics, and residence times of Gulf of Mexico Estuaries. In: Bianchi T, Pennock J, Twilley RR (eds) Biogeochemistry of Gulf of Mexico Estuaries. Willey, New York, NY, pp 29–61

Søndergaard M, Kristensen P, Jeppesen E (1991) Phosphorus release from resuspended sediment in the shallow and wind-exposed Lake Arresø, Denmark. Hydrobiologia 228(1):91–99. https://doi.org/10.1007/bf00006480

Taillardat P, Marchand C, Friess DA, Widory D, David F, Ohte N, Nakamura T, Van Vinh T, Thanh-Nho N, Ziegler AD (2020) Respective contribution of urban wastewater and mangroves on nutrient dynamics in a tropical estuary during the monsoon season. Mar Pollut Bull 160:111652. https://doi.org/10.1016/j.marpolbul.2020.111652

Texas Community Watershed Partners (2020) Watershed protection plan for Highland Bayou, Highland Bayou Diversion Canal, Marchand Bayou, Moses Bayou, and unnamed tributary of Moses Lake. Texas A&M Agrilife. https://agrilife.org/highlandbayou/files/2020/11/Highland-Bayou-Coastal-Basin-WPP-draft-for-review-November-2020.pdf Accessed 8 Sep 2022

Texas Water Development Board (2022a) Bays & estuaries. https://www.twdb.texas.gov/surfacewater/bays/index.asp Accessed 11 Jul 2022

Texas Water Development Board (2022b). Guadalupe Estuary (San Antonio Bay). https://www.twdb.texas.gov/surfacewater/bays/major_estuaries/guadalupe/index.asp Accessed 13 Jul 2022

Thronson A, Quigg A (2008) Fifty-five years of fish kills in coastal Texas. Estuar Coasts 31:802–813. https://doi.org/10.1007/s12237-008-9056-5

U.S. Census Bureau (2021) Texas added almost 4 million people in last decade. Census.gov. https://www.census.gov/library/stories/state-by-state/texas-population-change-between-census-decade.html. accessed 6 Oct 2022

U.S. EPA (2022) The sources and solutions: wastewater. https://www.epa.gov/nutrientpollution/sources-and-solutions-wastewater

Valiela I, Owens C, Elmstrom E, Lloret J (2016) Eutrophication of Cape Cod estuaries: effect of decadal changes in global-driven atmospheric and local-scale wastewater nutrient loads. Mar Pollut Bull 110(1):309–315. https://doi.org/10.1016/j.marpolbul.2016.06.047

Wang F, Cheng P, Chen N, Kuo YM (2021) Tidal driven nutrient exchange between mangroves and estuary reveals a dynamic source-sink pattern. Chemosphere 270:128665. https://doi.org/10.1016/j.chemosphere.2020.128665

Wetz MS, Cira EK, Sterba-Boatwright B, Montagna PA, Palmer TA, Hayes KC (2017) Exceptionally high organic nitrogen concentrations in a semi-arid South Texas estuary susceptible to brown tide blooms. Estuar Coast Shelf Sci 188:27–37

Whitfield A, Elliott M (2011) Ecosystem and biotic classifications of estuaries and coasts. Treatise Estuar Coast Sci 1:99–124. https://doi.org/10.1016/b978-0-12-374711-2.00108-x

Yan Q, Cheng T, Song J, Zhou J, Hung CC, Cai Z (2021) Internal nutrient loading is a potential source of eutrophication in Shenzhen Bay, China. Ecol Ind 127:107736. https://doi.org/10.1016/j.ecolind.2021.107736

Zamparas M, Zacharias I (2014) Restoration of eutrophic freshwater by managing internal nutrient loads: a review. Sci Total Environ 496:551–562. https://doi.org/10.1016/j.scitotenv.2014.07.076

# Social and Economic Values of Environmental Flows to the Coast

David W. Yoskowitz ⓘ

## Abstract

Environmental flow (EF) has been shown to have a significant economic value and impact. These values are often in the form of ecosystem services (ES). In Texas, science and policy on EF are well advanced, but operationalizing the protection of it is still lacking. Institutional alignment at a state level is part way there, but more work can be done to facilitate transactions for all water, including EF. Socio-behavioral-economic tools can be used for valuing ES provided by EF and determining the true value of water beyond the cost of water infrastructure.

## Keywords

Ecosystem services · Environmental flows · Water markets

## Abbreviations

EF      Environmental flow
ES      Ecosystem services
NGO     Non-governmental organizations
TEFI    Texas Environmental Flows Initiative
TWDB    Texas Water Development Board
TWT     Texas Water Trade

The views of the author do not necessarily reflect those of Texas Parks and Wildlife Department and its Commission.

D. W. Yoskowitz (✉)
Texas Parks and Wildlife Department, Austin, TX, USA

## Driving Issues

Environmental water, or freshwater inflows, to estuaries has a significant social and economic impact as it does ecological. Water has value in its many different uses, such as drinking, irrigation, cooling, etc. But what is the value of water when it is not diverted for human use, but remains in its body or course for what is referred to as "environmental flow"? Reducing freshwater flow in rivers and therefore reducing inflow into estuaries can lead to a loss of biodiversity, critical habitat, and important commercial and recreational fisheries. While individuals rarely use freshwater flow directly, they benefit from the impact that inflow has on the eventual generation of ecosystem services (Box 1).

### Box 1 What Are Ecosystem Services?

Ecosystem services are the benefits nature provides that sustains mankind (Daily 1997). These benefits are critical for human survival, e.g., clean air to breathe, clean water to drink, the organic (plant and animal) sustenance, and materials needed to build shelter. The list of ecosystem services is much larger. Ecosystem services are created by complex natural systems and processes, e.g., biogeochemical cycles. Harvest, manufacture, and trade of these services are the bases of economies. Thus, valuation of ecosystem services provides a tool for measuring the value of the benefits the natural environment provides.

What to do about environmental flows (instream flows within rivers and freshwater inflow from rivers to bays and estuaries) has developed into an important water policy issue in Texas. In 2007, the state passed legislation (SB 3 2007) that creates a process for considering environmental flow needs in each of the river basins. For several years, the conventional wisdom was that any water that made it to the coast

P. A. Montagna, A. R. Douglas (eds.), *Freshwater Inflows to Texas Bays and Estuaries*, Estuaries of the World,
https://doi.org/10.1007/978-3-031-70882-4_16

was wasted water. In fact, we know that not to be the case (Montagna et al. 2002). Freshwater inflow affects the many different ecological assets that produce essential ecosystem services in estuaries. Some of these services, such as recreational activities, have direct market processes that value can be extracted from. However, many services do not, such as erosion control, nutrient regulation, and aesthetics as examples. It has been a challenge for the water resource management community to elucidate these values to help in decision-making.

Early valuation studies focused on valuing water quality improvement that environmental flow can bring, and this ranged from maintaining and improving water quality in the United States and England (Harris 1984; Desvousges et al. 1987; Green and Tunstall 1991) to protecting instream flow for various recreational and environmental activities including recreational fishing (Daubert and Young 2018) and the protection of critical ecosystems, such as Mono Lake, California (Loomis 1987). Studies have also dealt with protecting instream flow as an alternative to dam building (Gonzalez-Caban and Loomis 1997) and the impact on ecosystem services writ large because of increasing instream flow and water quality (Loomis et al. 2000).

Focus on the value of flow into bays and estuaries, which can be thought of as an extension of the "instream" or "environmental flow" literature, came about a little later. Some of the earliest important work focused on using economic non-market valuation methods, like those applied to water quality studies. Examination of the Keurbooms Estuary near Plettenberg Bay, South Africa, focused on recreational users and what they were willing to pay to ensure water would flow into the estuary. This value was less than what farmers were willing to pay for the water as an input into agriculture (Hosking and du Preez 2004).

By contrast, estimates of environmental services provided by restored instream flows in the Yaqui River Delta in Mexico found that households of the region would be willing to tax themselves by paying an additional 73 pesos ($7 at the time of the study) per month on their water bill to purchase water for environmental flows. At the time of the study, this was approximately 1% of the mean monthly household income ($6100 M or $610 US) of those surveyed (Ojeda et al. 2008).

Many of the challenges that are experienced worldwide are also challenges in Texas: competing demands for water, diminishing per capita water supply, and aging infrastructure. Additionally, Texans want a thriving economy and a healthy water environment, as evidenced by the action taken by the legislature and the activities of non-governmental organizations, businesses, and academic institutions. A critical part of developing and implementing effective decisions for environmental flow in a twenty-first-century Texas is incorporating the "value" of that flow to ecosystems, people, and economics.

## Ecosystem Services and Environmental Flow

Many of the benefits that humans receive from our natural environment, referred to as ecosystem services, are a result of water on the landscape. Rainfall impacts the extent and health of flora and fauna and the benefits that we enjoy. That rainfall moves off the land and, depending on the extent to which it is diverted or impounded, makes its way into aquifers, streams, rivers, and where there is enough sustained flow into the estuaries. The flow of water itself is not an ecosystem service but, through the impact that flow has on structure, function, and processes of riverine and estuarine systems, creates the supply of potential ecosystem services, which are the benefits that humans receive from our natural system (Gopal 2016; Jorda-Capdevila et al. 2016; Pinto et al. 2013; Yoskowitz and Montagna 2009).

Connections between aquatic conditions and ecosystem services—the benefits that humans receive from a well-functioning system—are well understood (Grizzetti et al. 2019). Freshwater inflow into estuaries, the terminus of EF, is critical for ecological health and by extension beneficial for humans (see Chapters "Nutrient-Phytoplankton Dynamics in Texas Estuaries", "Physical and Biogeochemical Conditions and Trends in Texas Estuaries", "Effects of Climate-Driven Salinity Regimes on Disease Dynamics of the Eastern Oyster, a Key Estuarine Resource and Bioindicator", and "Freshwater Inflow and Salinity Shape Nekton Diversity and Community Structure Within Texas Estuaries"). While the quantity of water reaching an estuary to keep it ecologically productive is important, of growing importance is the quality of the water flowing into coastal systems. Upstream human activities of diversions and the quality of return flows, along with changing climate conditions, can lead to additional stress on estuarine systems and the eventual provision of ecosystem services (Booi et al. 2022; Wetz and Yoskowitz 2013; Davis and Kidd 2012).

## Economic Tools to Enhance Environmental Flow in Texas

Texas has benefitted from having addressed issues of water, water rights, and environmental flow for decades (see Chapters "Introduction: History of Inflow Studies in Texas" and "Historical Perspective and Context of Freshwater Inflow Policy and Law in Texas"). However, there is room to advance the recognition of the importance of environmental flow and the application of new policies and tools. Utilizing socio-behavioral-economic tools as part of a multi-pronged approach to address environmental flow can potentially lead to faster action than solely relying on traditional policy development. That is not to say that policy fixes wouldn't be needed in order to provide institutional clarity and support.

One of the most recognizable and understood tools are traditional market transactions where there is a seller of water and a buyer. Why? Because we use this every day in our own lives. We have a demand for a certain product, we search out a seller, and then we purchase it depending on if the price and quality meet our requirements. However, many of the goods and services that we purchase daily have several providers that are easily found. Water for purchase or lease in Texas, especially for environmental flow, does not operate in quite the same way.

Texas has a strong history in water marketing in certain parts of the state where institutional structure and policy have allowed water to flow from willing sellers to willing buyers, including water authorities and state agencies. The best example of water exchange is along the Rio Grande where, through institutional leadership and policy, the Rio Grande Watermaster Program has facilitated the development of this market for permanent, leased, and term water (Leidner et al. 2011; Yoskowitz 1999; Schoolmaster 1991).

While the markets are robust in the Rio Grande with many buyers and sellers of water, the institutional structure and support that exist there are not replicated on other rivers in Texas. For protection of environmental flows, or purchases of water for any purpose, the lack of a defined market can lead to increased search costs which diminishes activity that could benefit environmental flow if water is made available for purchases or lease. The Nature Conservancy of Texas has published an analysis of water markets for environmental flow that could reduce some of the challenges previously encountered in these types of transactions.[1]

An important complement to the purchase or long-term lease of water for environmental purposes is the Texas Water Trust that was established in 1997 as part of Senate Bill 1 of the 75th Texas Legislature (Tex. Water Code § 15.7031). Its primary purpose is to hold water rights placed in the Trust for the purpose of protecting environmental flow to enhance aquatic life and habitat. Although this tool has been underutilized, there is a renewed opportunity to see it live up to its potential with the passage of HB 2225 of the 87th Texas Legislative session. That legislation directs the Texas Parks and Wildlife Department "...to encourage and facilitate the dedication of water rights in the Texas Water Trust through lease, donation, purchase, or other means of voluntary transfer for environmental needs..." (Texas Parks and Wildlife Code § 12.028).

As available water for permanent purchase or long-term lease becomes more difficult, especially with increased demands from growing municipalities and industry, new approaches to securing environmental flow during low-flow events are beginning to evolve. One such approach is the use of option contracts to secure water when it is most needed. A

water option is a financial product that allows interested parties (buyers and sellers) to create a contract that formalizes the terms of the possible future delivery of water. The buyer of the option contract is securing the right – not the obligation – to buy a specific amount of water for a defined price. She can then "exercise" the option when she deems it most appropriate, for example, in times of extremely low flow (McColly et al. 2021).

The "option contract" approach is supported by biophysical evidence that shows small-scale hydrological restoration work can have measurable impacts, especially in times of low flow, such as a drought. Actions that can sustain important nurseries with "focused flows" during times of system stress can help estuaries recover quicker when normal flows return (Montagna et al. 2021). Option contracts are ideally aligned with a focused flow regime because *they allow water to be repurposed for environmental flow protection during periods of drought, which may occur with* increased frequency in Texas occurring in *various regions of the state, while continuing to support other uses during nondrought periods.*

## The Future of Socio-Behavioral-Economic Approaches Addressing Environmental Flow

Socio-behavioral-economic tools can be brought forward to address the lack of surface water that is set aside for environmental flow. But significant challenges exist, not the least of which is the projected increase of population of Texas in 2050 to 42 million (TWDB 2021). With that population growth, per-capita water supplies will shrink particularly without significant increases in new and sustainable sources and conservation measures, which may cause water for the environment to become more marginalized as demand from other sectors increases.

What opportunities are available to advance environmental flow? Conservation efforts are more effective when they are driven by a community of stakeholders. In Texas, the "water" community, ranging from NGOs to academic institutions and government agencies, has made a significant contribution to legislation, professional development, education, and advocacy work. These long-standing efforts resulted in a series of legislative actions from the late 1990s through the early 2000s. This included Senate Bill 1 (75th Legislature) which focused primarily on establishing the regional water planning process that the state uses today and included some limited advancements to provide water for environmental needs (Tate 2002). Senate Bill 2 (77th Legislature) is primarily recognized as the "groundwater" bill that shored up the right of groundwater districts to be established and manage groundwater resources, but it also directs state agencies to establish and maintain monitoring programs to assess levels

---

[1] texas-water-markets-review.pdf (nature.org)

of instream flow that lead to a healthy ecological environment (Ellis and Houston 2002). Then with Senate Bill 3 (80th Legislature), a process of accounting for environmental flow in bay-basin systems through significant stakeholder and science input was established (Rubinstein et al. 2022). Senate Bills 1, 2, and 3 benefited from significant stakeholder involvement, but a critical entity that brought together the science and policy expertise was the Texas Living Waters initiative, a collaboration of conservation groups established in 2001.[2]

The water stakeholder community continues to expand its activity in pressing for better water management including an increased focus on environmental flow that includes the Texas Environmental Flows Initiative (TEFI) which was created in 2014. The 3-year pilot project included environmental NGOs and academic partners to develop the legal and scientific frameworks to implement water transactions for the benefit of priority bays in Texas.[3] A spinoff of the TEFI work was the establishment of Texas Water Trade (TWT) whose goal is to operationalize the best practices in securing water for the environment. Utilizing the "focused flows" approach described above, they have secured over 20,000 acre-feet of interruptible water to be delivered at critical times.[4]

Delivering water for environmental flow will continue to be a challenge as competing demands grow along with population in Texas. It is critical that its value be elucidated to make better decisions about water in all its uses and incentivize the development of new sustainable sources of water along with strong conservation efforts so we can have a Texas of tomorrow that is economically and ecologically robust.

# References

Booi S, Mishi S, Andersen O (2022) Ecosystem services: a systematic review of provisioning and cultural ecosystem services in estuaries. Sustain For 14(12):7252. https://doi.org/10.3390/su14127252

Daily GC (1997) Introduction: what are ecosystem services. Nature's services: Societal dependence on natural ecosystems 1(1)

Daubert JT, Young RA (2018) Recreational demands for maintaining instream flows: a contingent valuation approach. In: Economics of water resources. Routledge, pp 65–75

Davis J, Kidd IM (2012) Identifying major stressors: the essential precursor to restoring cultural ecosystem services in a degraded estuary. Estuar Coasts 35:1007–1017. https://doi.org/10.1007/s12237-012-9498-7

Desvousges WH, Smith VK, Fisher A (1987) Option price estimates for water quality improvements: a contingent valuation study for the Monongahela River. J Environ Econ Manag 14:248–267

Ellis GM, Houston JA (2002) Senate Bill 2: step two towards effective water resource management and development for Texas. State Bar Texas Environ Law J 32:53–72

Gonzalez-Caban A, Loomis J (1997) economic benefits of maintaining ecological integrity of Rio Mameyes in Puerto Rico. Ecol Econ 21:63–75

Gopal B (2016) A conceptual framework for environmental flows assessment based on ecosystem services and their economic valuation. Ecosyst Serv 21(Part A):53–58. https://doi.org/10.1016/j.ecoser.2016.07.013

Green CH, Tunstall SM (1991) The evaluation of river water quality improvements by the contingent valuation method. Appl Econ 23:1135–1146

Grizzetti B, Liquete C, Pistocchi A, Vigiak O, Zulian G, Bouraoui F, De Roo A, Cardoso AC (2019) Relationship between ecological condition and ecosystem services in European rivers, lakes and coastal waters. Sci Total Environ 671:452–465. https://doi.org/10.1016/j.scitotenv.2019.03.155

Harris BS (1984) Contingent valuation of water pollution control. J Environ Econ Manag 19:199–208

Hosking SG, du Preez M (2004) A recreational valuation of the freshwater inflows into the Keurbooms Estuary by means of a contingent valuation study. S Afr J Econ Manag Sci 7:280–298

Jorda-Capdevila D, Rodríguez-Labajos B, Bardina M (2016) An integrative modelling approach for linking environmental flow management, ecosystem service provision and inter-stakeholder conflict. Environ Model Softw 79:22–34. https://doi.org/10.1016/j.envsoft.2016.01.007

Leidner AJ, Rister ME, Lacewell RD, Sturdivant AW (2011) The water market for the middle and lower portions of the Texas Rio Grande Basin. J Am Water Res Assoc 47:597–610. https://doi.org/10.1111/j.1752-1688.2011.00527.x

Loomis JB (1987) Balancing public trust resources of Mono Lake and Los Angeles water rights: an economic approach. Water Resour Res 23:1449–1456

Loomis J, Kent P, Strange L, Fausch K, Covich A (2000) Measuring the total economic value of restoring ecosystem services in an impaired river basin: results from a contingent valuation survey. Ecol Econ 33:103–117

McColly Q, Mace R, Tissot P, Yoskowitz D (2021) Pricing options on water in Texas. Texas Water J 12:91–108. https://doi.org/10.21423/twj.v12i1.7121

Montagna PA, Kalke RD, Ritter C (2002) Effect of restored freshwater inflow on macrofauna and meiofauna in Upper Rincon Bayou, Texas, USA. Estuaries 25:1436–1447

Montagna PA, McKinney L, Yoskowitz D (2021) Focused flows to maintain natural nursery habitats. Texas Water J 12:29–39. https://doi.org/10.21423/twj.v12i1.7123

Ojeda MI, Mayer AS, Solomon BD (2008) Economic valuation of environmental services sustained by water flows in the Yaqui River Delta. Ecol Econ 65:155–166. https://doi.org/10.1016/j.ecolecon.2007.06.006

Pinto R, de Jonge VN, Neto JM, Domingos T, Marques JC, Patrício J (2013) Towards a DPSIR driven integration of ecological value, water uses and ecosystem services for estuarine systems. Ocean Coast Manag 72:64–79. https://doi.org/10.1016/j.ocecoaman.2011.06.016

Rubinstein C, Seaton C, Mace RE (2022) Beyond Senate Bill 3: how to achieve environmental flows in Texas under prior appropriation. Texas Water J 13:13–26. https://doi.org/10.21423/twj.v13i1.7115

S.B. 3, 2007 Biennium, 2007 Reg. Sess (TX 2007) https://capitol.texas.gov/tlodocs/80R/billtext/html/SB00003F.HTM

Schoolmaster A (1991) Water marketing and water rights transfers in the Lower Rio Grande Valley, Texas. Prof Geogr 43:292–296

Tate D (2002) Creating environmental consensus: comparison of Senate Bill 1 water planning to a consensus building process. State Bar Texas Environ Law J 32:85–99

---

[2] See: https://texaslivingwaters.org/projects/

[3] See: https://digital.library.txst.edu/items/4fd4b890-c3f2-44c0-9c3d-19a302ac1601

[4] See: https://texaswatertrade.org/connected-coast/

Texas Water Development Board (TWDB) (2021) Regional water plan – population projections for 2020–2070. Retrieved on September 4, 2023. https://www3.twdb.texas.gov/apps/reports/Projections/2022%20Reports/pop_region

Wetz M, Yoskowitz D (2013) An 'extreme' future for estuaries? Effects of extreme climatic events on estuarine water quality and ecology. Mar Pollut Bull 69:7–18. https://doi.org/10.1016/j.marpolbul.2013.01.020

Yoskowitz D (1999) Spot market for water along the Rio Grande: opportunities for water management. Nat Res J 39(2):345–355

Yoskowitz DW, Montagna PA (2009) Socio-economic factors that impact the desire to protect freshwater flow in the Rio Grande, USA. WIT Trans Ecol Environ 122:547–558. https://doi.org/10.2495/ECO090501

# Summary of Recommendations for the Future

Paul A. Montagna ⓘ and Audrey R. Douglas

## Abstract

While freshwater inflow (FWI) needs for the maintenance of estuary health have long been acknowledged, environmental flow standards or, specifically, FWI standards are still uncommon. Where they exist, they are rapidly evolving over time. Texas, USA, has been working on this problem since the 1960s. Through several iterations, the legal and regulatory framework evolved from a species management approach to an ecosystem-based management one. More recently, based on extensive scientific research, Texas adopted FWI standards for all the major bay systems between 2011 and 2014. After a decade however, new technical questions and problems have arisen. There are at least five major needs for the future: (1) While there is a lot of data, little of it is focused sufficiently to define estuary responses to inflow in a way that there are clear connections between physical characteristics and biological responses, so a state-wide monitoring approach is needed. (2) The physics-based models of circulation that are typically applied to Texas bays and estuaries are not up to the current scientific methods, so 3-D models are needed as well as updated bathymetry, shoreline locations, and salinity monitoring to calibrate and validate the models. (3) More mechanistic studies are needed that specifically link the biological response to physical dynamics, and this will be easier if the prior recommendations are enacted. (4) Because of the semi-arid climate, there will never be enough water to dilute salinity in all bay systems, especially in central and south Texas, so a focused approach to protect key nursery habitats during droughts is needed. (5) Finally, the FWI standards are complex, 3-D, hydrology tables that are difficult to apply, so a simpler approach is needed that specifically is linked to biological outcomes. This last recommendation may be impossible to implement if the first four are not implemented. While enormous progress has been made in Texas, there is still more to do, but the history of these activities can serve as a guide to any organization interested in the conservation, restoration, enhancement, or protection of estuaries anywhere in the world.

## Keywords

Environmental flow standards · Texas · Bays · Estuaries · Freshwater inflow needs

## Abbreviations

| | |
|---|---|
| FWI | Freshwater inflow |
| GPT | Generative pretrained transformer |
| TPWD | Texas Parks and Wildlife Department |
| 2-D | Two-dimensional |
| 3-D | Three-dimensional |

## Introduction and Purpose

There are detailed recommendations for future studies and research needs specific to the topic at the end of each chapter in this book. The purpose of this chapter is to crosswalk those recommendations, weave together major themes, and synthesize multiple recommendations into singular or summary recommendations (Sects. 3–7).

First, a brief review of the specific chapter recommendations is presented. ChatGPT was used to help create a simple summary of the recommendations in Sect. 2. The recommendations were extracted from the chapters and submitted to the software. The resulting file was edited for accuracy and completeness.

P. A. Montagna (✉) · A. R. Douglas
Texas A&M University-Corpus Christi, Corpus Christi, TX, USA
e-mail: Paul.Montagna@tamucc.edu

© The Author(s) 2025
P. A. Montagna, A. R. Douglas (eds.), *Freshwater Inflows to Texas Bays and Estuaries*, Estuaries of the World,
https://doi.org/10.1007/978-3-031-70882-4_17

## Summary of Chapter Recommendations

Texas estuaries, like all estuaries worldwide, are where freshwater and saltwater meet and mix, which creates vibrant and delicate ecosystems. Estuaries are not merely geographical features but life-support systems teeming with biodiversity, providing sustenance for countless species and livelihoods for coastal communities. They are the lifeblood of our coastal regions, and their health is paramount for the well-being of both nature and society.

Hydrology is the main driver of these dynamic environments. There are variable annual and seasonal cycles based on complex interactions between precipitation, freshwater inflow, and the ebb and flow of tides that determine salinity. These natural rhythms that define the character of estuaries are susceptible to external forces caused by human activities such as natural resource extractions. The human influence on estuaries should not be underestimated. Dam construction and wastewater treatment have fundamentally altered the flow of nutrients within watersheds.

Nutrient dynamics are key because nutrients are delivered to the coast by creeks, streams, and rivers and drive primary production at the base of the food web. However, there are both natural and anthropogenic contributors to materials flowing to the coast. The consequences of high human population density are evident in Texas estuaries by the occurrence of high salinity, nutrients, and harmful algal blooms. Understanding the sources and transformations of these nutrients is vital to the management of these ecosystems.

There are looming challenges of land subsidence and sea-level rise, which have the potential to reshape estuarine landscapes and nutrient budgets. The rise of sea levels and the sinking of the land create a complex interplay that demands proactive management strategies to preserve marshes and wetlands, because marshes and wetlands are the nurseries of the estuaries. Seagrasses have been lost in the upper coast because of this combination of land subsidence and sea-level rise.

Consistent and concurrent collection of data is necessary to enhance long-term databases of water quality parameters, particularly salinity, and nutrient and chlorophyll concentrations. Consistent, long-term data is also necessary for biological responses to inflow. The data repositories will serve as invaluable tools for assessing ecosystem changes and guiding informed decision-making.

We need to address information gaps to delve deeper into the dynamics of estuarine ecosystems. Statistical analyses, seasonal variation assessments, and tracking interannual fluctuations are within our reach with comprehensive data collection.

Embracing an ecosystem-based management approach is recommended. We have seen how community dynamics, freshwater inflow, and salinity regimes play integral roles in estuarine health. For those entrusted with the stewardship of these ecosystems, this holistic perspective is necessary.

Adaptive management practices are mandated and necessary. As climate variability intensifies and human populations surge, strategies must evolve to meet the specific needs of our estuarine communities. Focusing resources on the most vulnerable areas and species is necessary where intervention can provide the greatest benefit.

The history of Texas estuaries is one of resilience, adaptation, and interconnectedness. These coastal ecosystems have withstood both natural and man-made disturbances and continue to thrive. It is important to be stewards of our estuaries because they are the lifeblood that sustains our coastal communities, so conservation is important for generations to come. In this shared endeavor, we find hope for a brighter, more sustainable future along the Texas coast.

## Need for Systematic Monitoring

As exemplified by the technical data in all chapters from "Climate Effects on Inflows" to "Nitrogen and Phosphorus Budgets for Texas Estuaries", there is a lot of data available for Texas on all topics of estuarine ecology. Two important findings from all the chapters are that (1) every abiotic and biotic variable is highly variable spatially and temporally and (2) nearly all variables have a changing trajectory over time. Only long-term research can provide answers to fundamental questions on environmental processes, population dynamics, and community structure (Hampton et al. 2019). Yet, most of the data presented is a result of short-term, focused studies, in specific places, which likely provides skewed results. This means there are many data gaps. Specific data gaps are mentioned in all chapters or are evident by their absence in chapters. Often, old data was found but not in a form that was digitally accessible, or the data could not be analyzed because the metadata was incomplete.

### Spatial and Temporal Concerns

Long-term data is required to understand estuary dynamics. Estuaries are dynamic because hydrology is driven by climate cycles, and tides are driven by astronomical cycles. Understanding these rhythms is essential to understanding estuary variability.

To be useful for freshwater inflow analytics, data must also range across the entire estuary gradient from the river mouths to the ocean passes. Fixed sites within estuaries are necessary. One common mistake in experimental design is "tiered sampling" where many sites are sampled, but only a few are followed over time. This approach is useless for estuary studies because spatial dynamics are not static. Spatial variation within estuaries is driven by mixing caused by hydrology, tides, circulation, wind, and weather. While hydrology, tides, and circulation can be governed by physics, they are also driven by wind and weather events. The overlay of wind and weather drives stochasticity (i.e., randomness) of the hydrology and circulation patterns. This randomness is why interpolations between locations are not simple and the entire network should be sampled completely over time.

The long-term wet and dry cycles are driving biological dynamics over the long term, and they have slightly different effects in the different parts of Texas estuaries (Fig. 1). During droughts, the system is essentially oligotrophic because of the lack of nutrient transport. This leads to low primary production, which cannot support a robust food web. The upper reaches of the estuaries, particularly the marshes and secondary bays, are affected the most and can be severely damaged by droughts (Montagna et al. 2009, 2017). When the grazing food chain is enhanced by nutrient flux, there are enhanced levels of productivity in the upper trophic levels as well.

## Experimental Design Concerns

In most places of the world, these two features, sampling over time and along spatial within-estuary gradients, are sufficient for experimental designs of estuary studies. However, because of its large size, Texas has an additional variable, regional-scale climate variability. Thus, studies in Texas must also add among-estuary sampling to the experimental design for estuary dynamic studies. These three elements of the study design (time, within-estuary, and among-estuaries) are available for only two datasets: weather (Chapter "Climate Effects on Inflows") and hydrology (Chapter "Hydrology, Circulation, and Salinity"). The Texas Parks and Wildlife Department (TPWD), Coastal Fisheries Division, dataset is excluded because it is missing one of the elements. The TPWD data is long-term and regional in scale, but each month stations are chosen randomly within estuaries using a probabilistic design; thus, the within-estuary element is lacking. Fixed-point designs are more effective than probabilistic designs for detecting spatial change at smaller

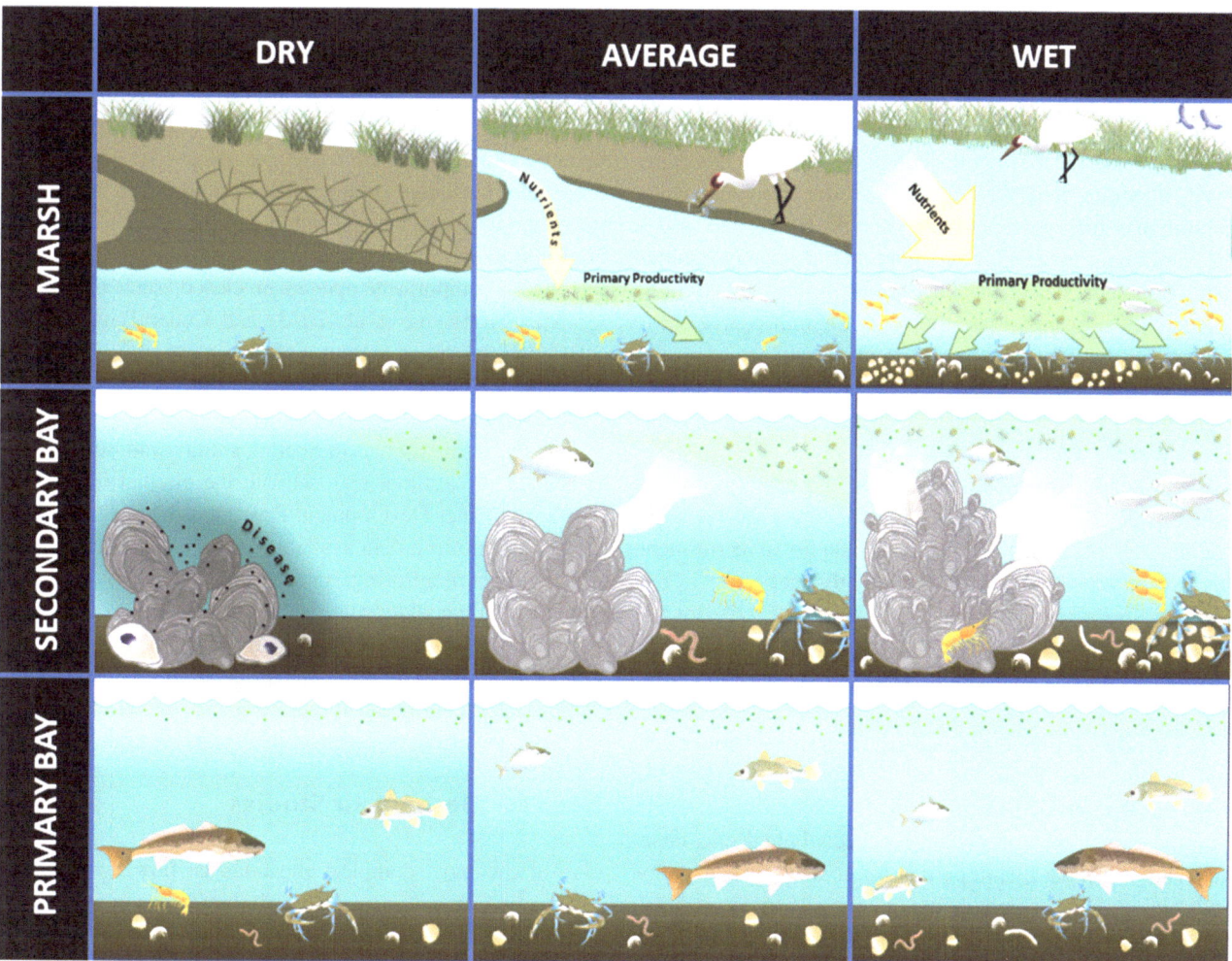

**Fig. 1** Relationship between long-term climatic cycles and biological responses in marshes, secondary bays, and primary bays

scales (Morehead et al. 2008). Thus, the TPWD data is useful for regional-scale dynamics, i.e., among estuaries, but not for within-estuary dynamics. One solution to this problem is to create segments (or area bins) by aggregating stations within areas of similar salinity conditions within each bay (Kurr 2019). This technique effectively turns a random design into a fixed-point design.

Despite the existing coastwide estuary data, long-term, fixed-point, and regional-scale monitoring of a few key variables is still lacking. The most important is salinity, which is easy to monitor with in situ sonde deployments. Next would be the main constituents of freshwater inflow, which are the nutrients, both inorganic and organic nitrogen and phosphorus. Continuous in situ monitoring of nutrients is harder because measurement methods often must be adjusted for salinity range, so most in situ probe methods are only reliable in low salinity waters. Nutrient and salinity budgets are also further complicated by submarine groundwater discharges, which may range from fresh to hypersaline, are highly variable in space and time, and may transport proportionally more nutrients than the rivers (Chapter "Groundwater-Surface Water Interactions in the Coastal Zone"). Chlorophyll is also useful because it is an indicator of the first trophic step of the food chain, which is photosynthesis by algae (Chapters "Nutrient-Phytoplankton Dynamics in Texas Estuaries" and "Plankton Dynamics in Texas Estuaries"). Benthos are important because they are fixed in place, are relatively long-lived, integrate ephemeral changes in the overlying water column over long periods of time, and are the second step in the food chain by grazing on algae that ultimately leads to fish (Chapter "Effect of Freshwater Inflow on Benthic Infauna"). Zooplankton are the water column equivalent to benthos in that they also graze on algae, but they are also ephemeral and are not fixed in place (Chapter "Plankton Dynamics in Texas Estuaries"). More importantly, many zooplankton are meroplankton, which are the larval forms of fish and benthic invertebrates that use the nursery areas of estuaries that depend on lower salinity habitats. Accordingly, zooplankton studies should be focused on larval recruitment, which may be seasonally and spatially distinct.

## Need for New Physical Models

All of the data reported in Chapter "Hydrology, Circulation, and Salinity" are from the old, 2-D (vertically averaged), TxBLEND model. Modern physical models use 3-D grids. Stratification of the water column, especially in deeper channels and near passes, is important for controlling circulation and mixing. For example, hypoxia is known to occur due to stratification in Corpus Christi Bay (Ritter and Montagna 1999) because of complex, thin-layer dynamics (Hodges et al. 2011) of hypersaline water entering Corpus Christi Bay

(Islam et al. 2008). On the other hand, hypoxia has not been observed in Galveston Bay. It is also common in estuaries that less dense freshwater is at the surface, while more dense saltwater is below the surface, forming a saltwater wedge near the mouth of the bay or in the tidal river. Thus, stratification is one process that requires 3-D modeling. Another reason 3-D modeling is necessary is the increasing potential for lower dissolved oxygen in bottom waters with increasing bay temperatures due to climate change (Montagna et al. 2011, Chapter "Effect of Freshwater Inflow on Benthic Infauna"). Furthermore, 3-D modeling will be able to capture the estuary-specific hydrology and hence the probability of hypoxia and other phenomenon occurring on a range of spatial and temporal scales which would not be possible with traditional sampling approaches. The greatest utility of 3-D models will be for modeling flows in passes and navigation channels deeper than the surrounding bathymetry and the effect of those channels on adjacent shallow areas.

New 3-D models will need more data as well. High-quality flow data is needed to calibrate the models, and this kind of data is very expensive to collect. More detailed bathymetry is available and is constantly being updated (Chapter "Hydrology, Circulation, and Salinity"), and it is important that the more detailed bathymetry data be incorporated into modeling. Because of channel dredging, erosion, deposition, subsidence, and sea-level change, new and updated shoreline locations are necessary. In many cases, more tidal exchange is happening in dredged channels than the original natural passes, and the location for the exchange is changing over time. Many central and south Texas estuaries can be temporarily opened or closed, such as the Rio Grande (Montagna et al. 2023) and Cedar Bayou (Ward 2010), but the dynamic nature of passes opening and closing is also not captured in current models and should be included in future 3-D modeling.

Finally, as mentioned in Sect. 3, salinity monitoring data is needed to calibrate and validate the circulation models. The Texas Water Development Board has the most extensive salinity monitoring data available,[1] but there are few permanent salinity monitoring stations throughout the bays and estuaries. Some older stations are inactive, and data gaps exist. New stations should be placed along the estuary salinity gradients from fresh to salt water. Historical or existing stations need continuous monitoring.

## Need for Mechanistic Studies

One of the most complex questions is: how do estuarine organisms respond to freshwater inflow? Certain aspects are well known, such as the importance of turbidity and nutrients

---

[1] https://www.twdb.texas.gov/surfacewater/bays/monitoring/index.asp

in controlling primary production (Chapters "Nutrient-Phytoplankton Dynamics in Texas Estuaries" and "Plankton Dynamics in Texas Estuaries"). The mechanics of photosynthesis are well known and have been modeled extensively. However, physiological stress or adaptation to constantly changing salinities is less well known. Often what is known is not from direct measurements in experiments but inferred from distributions of organisms and communities. One problem with indirect distribution data is that organisms may respond differently to salinity and have different salinity preferences and tolerances, over their life cycle. This is true for the estuarine-dependent species, which are those that spawn offshore, but the juveniles migrate into estuaries seeking nursery habitat. More mechanistic studies are needed that specifically link biological response to salinity and physical dynamics.

Understanding the biophysical processes will lead to two important steps forward. One step is building empirical and predictive relationships for biological response to salinity as has been done with benthic communities (Kim and Montagna 2012). The second step is embedding these biological dynamics within physical models as has been done for oysters (Klinck et al. 2002; Hofmann et al. 1994). Being able to predict biological responses to salinity change will lead to better environmental flow standards because it will enable the standards to reflect biological outcomes. All of these mechanistic modeling suggestions will be easier to achieve if

the prior recommendations to improve monitoring and modeling are enacted first.

## Need for Drought Management

Many of the Texas estuaries are hydrologically neutral or negative (Chapter "Hydrology, Circulation, and Salinity"), especially in the coastal region of central and south Texas, which means there will never be enough water to dilute salinity in the entire bay systems. Thus, drought management will always be a challenge. There is certainly an increased awareness by governments of the extreme hazards that drought pose, as well as an interest in planning (Wilhite 1992). Unlike other major events such as hurricanes or floods, droughts usually evolve slowly, so it is easy to be lulled by a sense of complacency (Fig. 2, Wilhite 1990). Then of course, once the awareness of drought arrives, it can induce panic and poor decision-making. The key is that planning must occur between droughts. Droughts are a natural feature of the climatic region, and most of the current standards include low inflow attainment frequencies for dry periods. In fact, some standards protect no inflows during droughts. The resulting problem is that bay systems can be harmed by droughts and recovery can be slow (Palmer and Montagna 2015; Montagna et al. 2017) or damage can be long-term (Montagna et al. 2009). Thus, protecting key nurs-

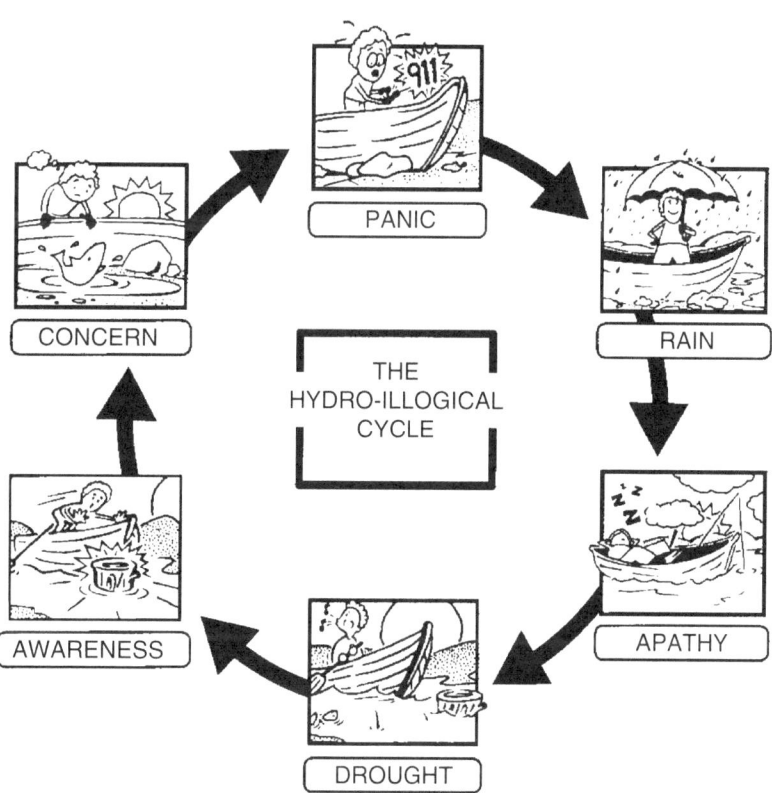

Fig. 2 The hydro-illogical cycle illustrates a common approach to drought management (Wilihite 1990)

ery habitats during droughts is critical. The nurseries typi-cally occur in the upper reaches of estuaries closer to the river mouths, where delivering relatively small volumes of water can measurably dilute salinity. This idea is called the "focused flows" approach to protect nursery habitats during droughts (Montagna et al. 2021). This would protect these habitats from permanent degradation and allow estuaries to recover more quickly when the hydrology returns to average or higher flow conditions. This approach could be applied globally where increasing water infrastructure and deficits are a concern or increasing aridity due to climate change is reducing river flows to coasts. More research on the potential for focused flows and the most critical times and locations for delivery is necessary. That research could inform the implementation of proactive strategies for delivering the needed flows (Chapter "Historical Perspective and Context of Freshwater Inflow Policy and Law in Texas").

The need for biological or ecological mechanistic studies (Sect. 5) and drought management studies stems from the legislative[2] goal that environmental flows should "be ade-quate to support a sound ecological environment and to maintain the productivity, extent, and persistence of key aquatic habitats in and along the affected water bodies." There are well-known, and easy-to-measure, bioindicators for productivity, spatial distribution, and temporal dynamics, and these are the focus of recommendation for consistent long-term monitoring (Sect. 5). But what is a "sound" envi-ronment and how is it related to other terms like ecosystem health? The definition of ecosystem health has a long history for water quality, and there is no reason it cannot be applied to water quantity (Montagna et al. 2013). Ecological health is assessed by determining if indicators of ecological condi-tions are in an acceptable range. Indicators are measures (or metrics) to establish an acceptable range of responses across broad spatial and temporal scales. Condition is the status of ecological function, integrity, and sustainability. Ecological function is judged acceptable when the ecosystem provides important ecological processes. Ecological integrity is acceptable when the ecosystem has a balanced, resilient community of organisms with biological diversity, species composition, structural redundancy, and functional processes comparable to that of natural habitats in the same region. Ecological sustainability is acceptable when an ecosystem maintains a desired state of ecological integrity over time.

## Need for Simpler Standards

The Nueces Estuary was the first to have inflow targets and standards, and they were easy to apply because there was initially only one number for the whole bay each month

(Montagna et al. 2009). Eventually the standards were changed to a "pass through" requirement with different inflow levels required for wet and dry periods and suspen-sion of the pass-through requirement during drought.

The more recently adopted Texas environmental flow standards resulting from 2007 Senate Bill 3 are in Chapter 298[3] of the Texas Commission on Environmental Quality Regulations (Chapter "Introduction: History of Inflow Studies in Texas"). The standards are complex, consisting of multiple tables that describe flow regimes that vary in three dimensions: over time components or climatic periods (such as wet and dry years), over seasons, and spatially within riv-ers, streams, and bay systems. Additionally, terminology in the rules vary. For example, the component climatic periods for inflow regimes are called "wet, average, dry, and subsis-tence" for Lavaca Bay and "level" for Matagorda Bay and Nueces Bay. Seasons are defined by specific months or monthly periods differently in all the bay systems. The attainment frequency aspects of environmental flow stan-dards were based on statistical evaluations of historical occurrence frequencies of different flow-level hydrological categories, which also vary over all systems (Opdyke et al. 2014). The result is a complex set of regulations that are dif-ficult to apply. Also, the concept of attainment frequencies measured over long periods of time means that we never know if the current flow regime is outside the standard or not. In fact, because a large range of flow regimes exist in the standards and they all occur at some point in time, it is easy to imagine that all flow rates can be found in the tables; thus all flow rates can be interpreted as being within the standards.

A simpler and more meaningful approach is needed. It should be simple-to-standardize language relating to wet and dry periods, seasons, and attainment frequencies. However, there are different hydrological responses in different sea-sons in different regions of the state (Chapter "Hydrology, Circulation, and Salinity").

Perhaps a different, more biologically based approach could be adopted. For example, it is possible to calculate how a flow rate number, and its year-to-year variability, would affect salinity and thus biological responses. A good example is the application of the domino theory (Chapter "Introduction: History of Inflow Studies in Texas") to the Caloosahatchee River in Florida (Palmer et al. 2016). Biological resources in estuaries are affected by salinity more than inflow alone, so the links between flow, salinity, and biology will determine the relationship between inflow and living resources. The first step is to identify the resource to be protected. The second step is to identify the salinity range or requirements of the resource in both space and time. The third step is to calculate the flow regime needed to sup-

---

[2]Senate Bill 3, 2007, Texas Water Code 11.002(16).

[3]www.tceq.texas.gov/assets/public/legal/rules/rules/pdflib/298a.pdf

port the required distribution of salinity. More importantly, the current standards should be linked to biological outcomes. This recommendation may be impossible to implement if the first three sections (3–5) are not implemented.

## Summary of the Summary

While enormous progress on environmental flow regulation has been made in Texas dating back to the 1960s (Chapters "Introduction: History of Inflow Studies in Texas" and "Historical Perspective and Context of Freshwater Inflow Policy and Law in Texas"), there is still much more to do. The history of these activities is a guide to any organization (governmental or nongovernmental) interested in the conservation, restoration, enhancement, or protection of estuaries anywhere in the world.

While it is clear that long-term monitoring is necessary to support adaptive management, history tells us that it is very difficult, if not impossible to, to create and sustain these efforts. Long-term monitoring efforts are a hard sell to political appointees and elected officials because they have a stigma as having been started by previous administrations. There is always an interest in something new. This natural resistance to monitoring is a challenge to overcome, or we will never have the data we need to build appropriate inflow standards.

## References

Hampton SE, Scheuerell MD, Church MJ, Melack JM (2019) Long-term perspectives in aquatic research. Limnol Oceanogr 64(S1):S2–S10

Hodges BR, Furnans JE, Kulis PS (2011) Thin-layer gravity current with implications for desalination brine disposal. J Hydraul Eng 137(3):356–371

Hofmann EE, Klinck JM, Powell EN, Boyles S, Ellis M (1994) Modeling oyster populations II. Adult size and reproductive effort. J Shellfish Res 13(1):165–182

Islam MS, Bonner JS, Ojo T, Page C (2008) A mechanistic dissolved oxygen model of Corpus Christi Bay to understand critical processes causing hypoxia. OCEANS 2008, pp 1–9, IEEE

Kim H-C, Montagna PA (2012) Effects of climate-driven freshwater inflow variability on macrobenthic secondary production in Texas lagoonal estuaries: a modeling study. Ecol Model 235–236:67–80

Klinck JM, Hofmann EE, Powell EN, Dekshenieks MM (2002) Impact of channelization on oyster production: a hydrodynamic-oyster population model for Galveston Bay, Texas. Environ Modeling Assess 7:273–289

Kurr E (2019) Focused flows to natural nurseries in Texas estuaries. Masters Thesis, Texas A&M University-Corpus Christi, Corpus Christi, TX. https://www.proquest.com/docview/2240088644/

Montagna PA, Hill EM, Moulton B (2009) Role of science-based and adaptive management in allocating environmental flows to the Nueces Estuary, Texas, USA. In: Brebbia CA, Tiezzi E (eds) Ecosystems and sustainable development VII. WIT Press, Southampton, pp 559–570. https://doi.org/10.2495/ECO090511

Montagna PA, Brenner J, Gibeaut J, Morehead S (2011) Coastal impacts. In: Schmidt J, North GR, Clarkson J (eds) The impact of global warming on Texas, 2nd edn. University of Texas Press, Austin, TX, pp 96–123. http://www.jstor.org/stable/10.7560/723306.8

Montagna PA, Palmer TA, Pollack JB (2013) Hydrological changes and estuarine dynamics. SpringerBriefs in Environmental Sciences, New York, NY

Montagna PA, Sadovski AL, King SA, Nelson KK, Palmer TA, Dunton KH (2017) Modeling the effect of water level on the Nueces Delta marsh community. Wetlands Ecol Manag 25:731–742

Montagna PA, McKinney L, Yoskowitz D (2021) Focused flows to maintain natural nursery habitats. Texas Water J 12(1):129–139

Montagna PA, Palmer TA, Pollack JB (2023) Effect of temporarily opening and closing the marine connection of a river estuary. Estuar Coasts. https://doi.org/10.1007/s12237-022-01159-6

Morehead S, Montagna P, Kennicutt MC II (2008) Comparing fixed-point and probabilistic sampling designs for monitoring the marine ecosystem near McMurdo Station, Ross Sea, Antarctica. Antarctic Sci 20(5):471–484

Opdyke DR, Oborny EL, Vaugh SK, Mayes KB (2014) Texas environmental flow standards and the hydrology-based environmental flow regime methodology. Hydrol Sci J 59:820–830

Palmer TA, Montagna PA (2015) Impacts of droughts and low flows on estuarine water quality and benthic fauna. Hydrobiologia 753:11–129

Palmer TA, Montagna PA, Chamberlain RH, Doering PH, Wan Y, Haunert K, Crean DJ (2016) Determining the effects of freshwater inflow on benthic macrofauna in the Caloosahatchee Estuary, Florida. Integr Environ Assess Manag 12:529–539

Ritter MC, Montagna PA (1999) Seasonal hypoxia and models of benthic response in a Texas bay. Estuaries 22:7–20

Ward GH (2010) A time line of Cedar Bayou. Final report to the Texas Water Development Board, University of Texas at Austin, Austin, TX, USA. https://repositories.lib.utexas.edu/bitstream/handle/2152/69939/0900010973_CedarBayou.pdf accessed 6 September 2022

Wilhite DA (1990) Planning for drought: a process for state government. IDIC Technical Report Series 90-1, University of Nebraska, Lincoln, Nebraska, USA

Wilhite DA (1992) Preparing for drought: a guidebook for developing countries. United Nations Environment Programme, Nairobi, Kenya

Paul A. Montagna ⓘ, William L. Longley, Elizabeth Gomaa ⓘ,
and Jen Corrinne Brown ⓘ

**Correction to:**
**Chapter 1 in: P. A. Montagna, A. R. Douglas (eds.),** *Freshwater Inflows to Texas Bays*
*and Estuaries*, **Estuaries of the World, https://doi.org/10.1007/978-3-031-70882-4_1**

As requested by the editor, Table 7 "Comparison of freshwater inflow requirements to bays and estuaries (million acre-ft/year) in different years" in this chapter has been changed after the initial publication.

Owing to an unfortunate oversight on the part of production, the following reference was missing in the initially published version. The reference list has now been included.

*"Water Resources Engineers (1975) Computer program and documentation for the Dynamic Estuary Model with application to Sabine Lake estuarine system. Final Report to Texas Water Development Board."*

---

The updated version of this chapter can be found at
https://doi.org/10.1007/978-3-031-70882-4_1

© The Author(s) 2025
P. A. Montagna, A. R. Douglas (eds.), *Freshwater Inflows to Texas Bays and Estuaries*, Estuaries of the World,
https://doi.org/10.1007/978-3-031-70882-4_18

# Archived Datasets

| Chapter | Contents | Published | Citation |
|---|---|---|---|
| 3 | Texas climate, climate change, and impacts on inflows to bays and estuaries evaluated through the use of hydrologic model simulations driven by downscaled global climate model simulations and projections | 13-Oct-2023 | John Nielsen-Gammon, Alison A. Tarter. Texas climate, climate change, and impacts on inflows to bays and estuaries evaluated through the use of hydrologic model simulations driven by downscaled global climate model simulations and projections. Distributed by: Gulf of Mexico Research Initiative Information and Data Cooperative (GRIIDC), Harte Research Institute, Texas A&M University–Corpus Christi. https://doi.org/10.7266/qhkhp60n |
| 4 | Freshwater Inflows to Texas Bays and Estuaries: Hydrology, Circulation, and Salinity Data (1977-2018) | 19-Mar-2024 | Trungale, Joe; Hoffmann, Josef; Montagna, Paul A.; Opdyke, Dan 2023. Freshwater Inflows to Texas Bays and Estuaries: Hydrology, Circulation, and Salinity Data (1977-2018). Distributed by: Gulf of Mexico Research Initiative Information and Data Cooperative (GRIIDC), Harte Research Institute, Texas A&M University–Corpus Christi. https://doi.org/10.7266/8KR3TNDT |
| 5 | Submarine groundwater discharge readings in semi-arid estuaries across the Texas coastal bend and associated water quality parameters obtained from 2014-09-29 to 2018-02-04 | 13-Mar-2023 | Murgulet, Dorina, Audrey Douglas, and Cody Lopez. 2023. Submarine groundwater discharge readings in semi-arid estuaries across the Texas coastal bend and associated water quality parameters obtained from 2014-09-29 to 2018-02-04. Distributed by: Gulf of Mexico Research Initiative Information and Data Cooperative (GRIIDC), Harte Research Institute, Texas A&M University–Corpus Christi. https://doi.org/10.7266/SJACB00R |
| 6 | Submerged Lands of Texas | 11-Dec-2023 | W.A. White, T.R. Calnan, R.A. Morton, R.S. Kimble, T.G. Littleton, J.H. McGowen, H.S. Nance, K.E. Schmedes. 2023. Sediment, geochemistry, and benthic macroinvertebrate data from the seven Bureau of Economic Geology (BEG) Submerged Lands of Texas obtained from 1976-03-02 to 1981-05-30. Distributed by: Gulf of Mexico Research Initiative Information and Data Cooperative (GRIIDC), Harte Research Institute, Texas A&M University–Corpus Christi. https://doi.org/10.7266/D1PNX3HD |
| 6, 11 | Formosa Plastics Corporation (FPC) discharge monitoring in Lavaca Bay, Texas, USA from 1993-05-17 to 2020-04-04 | 9-May-2022 | Montagna, Paul, Elizabeth Harris, Audrey Douglas, Lisa Vitale, and David Buzan. 2022. Formosa Plastics Corporation (FPC) discharge monitoring in Lavaca Bay, Texas, USA from 1993-05-17 to 2020-04-04. Distributed by: Gulf of Mexico Research Initiative Information and Data Cooperative (GRIIDC), Harte Research Institute, Texas A&M University–Corpus Christi. https://doi.org/10.7266/DCNHQD59 |
| 7 | Physical, chemical, and biological data for the quantitative description of phytoplankton dynamics in Corpus Christi Bay, Texas from 2016-08-04 through 2018-10-31 | 9-Dec-2021 | Tominack, Sarah A., Kenneth C. Hayes and Michael S. Wetz. 2021. Physical, chemical, and biological data for the quantitative description of phytoplankton dynamics in Corpus Christi Bay, Texas from 2016-08-04 through 2018-10-31. Distributed by: Gulf of Mexico Research Initiative Information and Data Cooperative (GRIIDC), Harte Research Institute, Texas A&M University–Corpus Christi. https://doi.org/10.7266/NCPYG0DH |
| 7 | Monthly discrete hydrographic measurements of Baffin Bay, TX, USA from 2015-01-29 to 2015-12-10 | 4-Oct-2021 | Wetz, Michael. 2021. Monthly discrete hydrographic measurements of Baffin Bay, TX, USA from 2015-01-29 to 2015-12-10. Distributed by: Gulf of Mexico Research Initiative Information and Data Cooperative (GRIIDC), Harte Research Institute, Texas A&M University–Corpus Christi. https://doi.org/10.7266/H3H4EHCJ |
| 7 | Monthly discrete hydrographic measurements of Baffin Bay, TX, USA from 2016-01-21 to 2016-12-14 | 4-Oct-2021 | Wetz, Michael. 2021. Monthly discrete hydrographic measurements of Baffin Bay, TX, USA from 2016-01-21 to 2016-12-14. Distributed by: Gulf of Mexico Research Initiative Information and Data Cooperative (GRIIDC), Harte Research Institute, Texas A&M University–Corpus Christi. https://doi.org/10.7266/VQHNN7HB |

© The Author(s) 2025
P. A. Montagna, A. R. Douglas (eds.), *Freshwater Inflows to Texas Bays and Estuaries*, Estuaries of the World,
https://doi.org/10.1007/978-3-031-70882-4

| Chapter | Contents | Published | Citation |
|---|---|---|---|
| 7 | Monthly discrete hydrographic measurements of Baffin Bay, TX, USA from 2017-02-16 to 2017-12-13 | 6-Oct-2021 | Wetz, Michael. 2021. Monthly discrete hydrographic measurements of Baffin Bay, TX, USA from 2017-02-16 to 2017-12-13. Distributed by: Gulf of Mexico Research Initiative Information and Data Cooperative (GRIIDC), Harte Research Institute, Texas A&M University–Corpus Christi. https://doi.org/10.7266/H65DRM8M |
| 7 | Monthly discrete hydrographic measurements of Baffin Bay, TX, USA from 2018-01-15 to 2018-12-11 | 20-Oct-2021 | Wetz, Michael. 2021. Monthly discrete hydrographic measurements of Baffin Bay, TX, USA from 2018-01-15 to 2018-12-11. Distributed by: Gulf of Mexico Research Initiative Information and Data Cooperative (GRIIDC), Harte Research Institute, Texas A&M University–Corpus Christi. https://doi.org/10.7266/HCK0G337 |
| 7 | Monthly discrete hydrographic measurements of Baffin Bay, TX, USA from 2019-01-16 to 2019-12-05 | 20-Oct-2021 | Wetz, Michael. 2021. Monthly discrete hydrographic measurements of Baffin Bay, TX, USA from 2019-01-16 to 2019-12-05. Distributed by: Gulf of Mexico Research Initiative Information and Data Cooperative (GRIIDC), Harte Research Institute, Texas A&M University–Corpus Christi. https://doi.org/10.7266/H6XCBTQS |
| 7 | Monthly discrete hydrographic measurements of Baffin Bay, TX, USA from 2020-01-14 to 2020-12-01 | 20-Oct-2021 | Wetz, Michael. 2021. Monthly discrete hydrographic measurements of Baffin Bay, TX, USA from 2020-01-14 to 2020-12-01. Distributed by: Gulf of Mexico Research Initiative Information and Data Cooperative (GRIIDC), Harte Research Institute, Texas A&M University–Corpus Christi. https://doi.org/10.7266/DWXDKXJ8 |
| 7 | Monthly discrete water quality measurements of Baffin Bay, Texas between 2013-05-15 and 2013-12-13 | 7-Sep-2021 | Wetz, Michael. 2021. Monthly discrete water quality measurements of Baffin Bay, Texas between 2013-05-15 and 2013-12-13. Distributed by: Gulf of Mexico Research Initiative Information and Data Cooperative (GRIIDC), Harte Research Institute, Texas A&M University–Corpus Christi. https://doi.org/10.7266/JPT4ZJX0 |
| 7 | Monthly discrete water quality measurements of Baffin Bay, Texas, between 2014-01-16 and 2014-12-18 | 7-Sep-2021 | Wetz, Michael. 2021. Monthly discrete water quality measurements of Baffin Bay, Texas, between 2014-01-16 and 2014-12-18. Distributed by: Gulf of Mexico Research Initiative Information and Data Cooperative (GRIIDC), Harte Research Institute, Texas A&M University–Corpus Christi. https://doi.org/10.7266/9P351WAS |
| 7 | Monthly discrete hydrographic measurements of Matagorda Bay, TX, USA between November 2019 and October 2021 | 14-Jun-2023 | Wetz, Michael. Monthly discrete hydrographic measurements of Matagorda Bay, TX, USA between November 2019 and October 2021. Distributed by: Gulf of Mexico Research Initiative Information and Data Cooperative (GRIIDC), Harte Research Institute, Texas A&M University–Corpus Christi. https://doi.org/10.7266/5V78H5A8 |
| 7, 8 | Effect of hydrological variability on the biogeochemistry of estuaries across a regional climatic gradient | 7-Nov-2023 | Montagna, Paul A., Hu, Xinping, Palmer, Terence A., M. Wetz, Michael. 2023. Effect of hydrological variability on the biogeochemistry of estuaries across a regional climatic gradient as seen data obtained from 2013-07-02 to 2016-07-18. Distributed by: Gulf of Mexico Research Initiative Information and Data Cooperative (GRIIDC), Harte Research Institute, Texas A&M University–Corpus Christi. https://doi.org/10.7266/e5zsz0sh |
| 7, 13 | Environmental and phytoplankton data for comparison across a gradient of freshwater inflow within San Antonio Bay, Nueces-Corpus Christi Bay, and Baffin Bay, Texas from 2018-03-13 to 2019-07-11 | 2-May-2022 | Wetz, Michael. 2022. Environmental and phytoplankton data for comparison across a gradient of freshwater inflow within San Antonio Bay, Nueces-Corpus Christi Bay, and Baffin Bay, Texas from 2018-03-13 to 2019-07-11. Distributed by: Gulf of Mexico Research Initiative Information and Data Cooperative (GRIIDC), Harte Research Institute, Texas A&M University–Corpus Christi. https://doi.org/10.7266/Q3Y0RBNM |
| 8 | Water column physical and biogeochemical data in Texas estuaries from 1969-04-17 to 2020-12-10 | 27-Jan-2023 | Hu, Xinping. Water column physical and biogeochemical data in Texas estuaries from 1969-04-17 to 2020-12-10. Distributed by: Gulf of Mexico Research Initiative Information and Data Cooperative (GRIIDC), Harte Research Institute, Texas A&M University–Corpus Christi. https://doi.org/10.7266/6pdpe3ra |

| Chapter | Contents | Published | Citation |
|---|---|---|---|
| 9 | 2-meter Topographic Lidar Digital Elevation Model (DEM) of the Lower Texas Coast | 9-Aug-2021 | Su, L., M. Subedee, M. Dotson, J. Gibeaut, B. Lupher, A. Reisinger, and R. Bezore. 2021. 2-meter Topographic Lidar Digital Elevation Model (DEM) of the Lower Texas Coast. Distributed by: Gulf of Mexico Research Initiative Information and Data Cooperative (GRIIDC), Harte Research Institute, Texas A&M University–Corpus Christi. https://doi.org/10.7266/Z7WG9EGN |
| 10 | Species-Level Seagrass Distribution and Environmental Salinities | 10-May-2021 | Dunton, Ken; Jackson, Kim; Wilson, Sara; Congdon, Victoria; Cuddy, Meaghan; Hall, Wayne; Becker, Madison; Meiman, Joe; Whiteaker, Tim; Bohannon, Patrick; Grubbs, Faye; Hobson, Cindy; The University of Texas at Austin (2018). Seagrass canopy height, water depth, chlorophyll-a concentration and other plant and water quality indicators in Coastal Waters of Texas (NCEI Accession 0181898). [indicate subset used]. NOAA National Centers for Environmental Information. Dataset. https://www.ncei.noaa.gov/archive/accession/0181898 |
| 10 | Freshwater SAV Distribution, *Najas guadalupensis* (Spreng.) Magnus | 5-Sep-2022 | GBIF.org (05 September 2022) GBIF Occurrence Download https://doi.org/10.15468/dl.v4369n |
| 10 | Freshwater SAV Distribution, *Vallisneria americana* Michx. | 5-Sep-2022 | GBIF.org (05 September 2022) GBIF Occurrence Download https://doi.org/10.15468/dl.835ty3 |
| 10 | Freshwater SAV Distribution, *Stuckenia pectinata* (L.) Börner | 12-Aug-2022 | GBIF.org (12 August 2022) GBIF Occurrence Download https://doi.org/10.15468/dl.7ymbxt |
| 10 | Freshwater SAV Distribution, *Heteranthera dubia* (Jacq.) MacMill. | 12-Aug-2022 | GBIF.org (12 August 2022) GBIF Occurrence Download https://doi.org/10.15468/dl.pm7zn6 |
| 10 | Freshwater SAV Distribution, *Myriophyllum spicatum* L. | 12-Aug-2022 | GBIF.org (12 August 2022) GBIF Occurrence Download https://doi.org/10.15468/dl.nd3erg |
| 10 | Freshwater SAV Distribution, *Hydrilla verticillata* (L.f.) Royle | 12-Aug-2022 | GBIF.org (12 August 2022) GBIF Occurrence Download https://doi.org/10.15468/dl.qecnez |
| 10 | Marsh/Seagrass/Mangrove Distribution Data | 21-May-2013 | Office for Coastal Management, 2023: NOAA's Coastal Change Analysis Program (C-CAP) 2016 Regional Land Cover Data - Coastal United States, https://www.fisheries.noaa.gov/inport/item/48336 |
| 11 | Sediment Quality Triad (SQT) Assessment Survey of Lavaca and Matagorda Bays | 1-Sep-2023 | Montagna, P.A., Caillier, J., DeLorenzo, M.E., Pete Key, P. Sediment Quality Triad (SQT) Assessment Survey of Lavaca and Matagorda Bays. Distributed by: Gulf of Mexico Research Initiative Information and Data Cooperative (GRIIDC), Harte Research Institute, Texas A&M University–Corpus Christi. https://doi.org/10.7266/9syzmzrd |
| 11 | Rio Grande water and sediment quality data affecting benthic macrofaunal community obtained from 2000-10-24 to 2005-08-06 | 18-Feb-2023 | Montagna, Paul. 2023. Rio Grande water and sediment quality data affecting benthic macrofaunal community obtained from 2000-10-24 to 2005-08-06. Distributed by: Gulf of Mexico Research Initiative Information and Data Cooperative (GRIIDC), Harte Research Institute, Texas A&M University–Corpus Christi. https://doi.org/10.7266/xzyn67pf |
| 11 | Sampling scales to characterize estuarine macroinfaunal patch dynamics | 26-Sep-2023 | Paul A. Montagna and M. Christine Ritter. Sampling scales to characterize estuarine macroinfaunal patch dynamics, Corpus Christi Bay, Texas obtained from 1996-11-01 to 1997-07-03. Distributed by: Gulf of Mexico Research Initiative Information and Data Cooperative (GRIIDC), Harte Research Institute, Texas A&M University–Corpus Christi. https://doi.org/10.7266/jkp05qj9 |
| 11 | Effects of small freshwater inflow volumes on water and sediment quality | 23-Oct-2023 | Paul A. Montagna, Hannah Ehrmann, Cheyanne M. Olson, Evan L. Turner. Effects of small freshwater inflow volumes on water and sediment quality obtained from bays in Texas from 2015-09-30 to 2016-09-15. Distributed by: Gulf of Mexico Research Initiative Information and Data Cooperative (GRIIDC), Harte Research Institute, Texas A&M University–Corpus Christi. Available from: https://data.gulfresearchinitiative.org/data/HI.x962.000:0018 |

| Chapter | Contents | Published | Citation |
|---|---|---|---|
| 11 | Effect of Freshwater Inflow on Macrobenthos in Minor Bay and River-Dominated Estuaries | 29-Sep-2023 | Paul A. Montagna, Rick Kalke. Effect of freshwater inflow on macrobenthos in minor bay and river-dominated estuaries collected from 2000-10-04 to 2005-08-06, Texas. Distributed by: Gulf of Mexico Research Initiative Information and Data Cooperative (GRIIDC), Harte Research Institute, Texas A&M University–Corpus Christi. https://doi.org/10.7266/btehjm5n |
| 11 | Effects of Hurricane Harvey on benthos of San Antonio Bay | 27-Oct-2023 | Paul A. Montagna. Effects of Hurricane Harvey on benthos of San Antonio Bay, Texas from 2004-01-07 to 2018-10-26. Distributed by: Gulf of Mexico Research Initiative Information and Data Cooperative (GRIIDC), Harte Research Institute, Texas A&M University–Corpus Christi. https://doi.org/10.7266/s1qhhvz2 |
| 11 | Long-term monitoring of water and sediment quality to identify freshwater inflow effects | 23-Oct-2023 | Paul A. Montagna. Long-term monitoring of water and sediment quality to identify freshwater inflow effects from 1984-01-01 to 2019-10-01. Distributed by: Gulf of Mexico Research Initiative Information and Data Cooperative (GRIIDC), Harte Research Institute, Texas A&M University–Corpus Christi. Available from: https://data.gulfresearchinitiative.org/data/HI.x962.000:0019 |
| 12 | Monitoring of *Perkinsus marinus* (Dermo disease) infection intensity in the *Crassostrea virginica*, and water quality in the Mission-Aransas Estuary, 2014-2021 | 16-Feb-2022 | Beseres Pollack, Jennifer, Natasha Breaux, Terence A. Palmer, Kelley B. Savage, Kevin De Santiago, and Danielle Downey. 2022. Monitoring of Perkinsus marinus (Dermo disease) infection intensity in the Crassostrea virginica, and water quality in the Mission-Aransas Estuary, 2014-2021. Distributed by: Gulf of Mexico Research Initiative Information and Data Cooperative (GRIIDC), Harte Research Institute, Texas A&M University–Corpus Christi. https://doi.org/10.7266/JJZYJ9AT |
| 13 | Microbial, physical, and chemical data collected from Galveston Bay research cruises from 2010-01-18 to 2019-11-16 | 1-Jun-2022 | Hillhouse, Jessica, Antonietta Quigg, and Jamie Steichen. 2022. Microbial, physical, and chemical data collected from Galveston Bay research cruises from 2010-01-18 to 2019-11-16. Distributed by: Gulf of Mexico Research Initiative Information and Data Cooperative (GRIIDC), Harte Research Institute, Texas A&M University–Corpus Christi. https://doi.org/10.7266/CTVAS53B |
| 13 | Phytoplankton pigment and secci data collected from Galveston Bay research cruises from 2010-01-18 to 2019-11-16 | 25-Feb-2022 | Hillhouse, Jessica, Antonietta Quigg, and Jamie Steichen. 2022. Phytoplankton pigment and secci data collected from Galveston Bay research cruises from 2010-01-18 to 2019-11-16. Distributed by: Gulf of Mexico Research Initiative Information and Data Cooperative (GRIIDC), Harte Research Institute, Texas A&M University–Corpus Christi. https://doi.org/10.7266/Q0ASPQFS |
| 13 | Harmful algal bloom species determined by Imaging FlowCytobot (IFCB) in Galveston Bay from 2015-04-01 to 2022-09-30 | 27-Jan-2023 | Steichen, Jamie. 2023. Harmful algal bloom species determined by Imaging FlowCytobot (IFCB) in Galveston Bay from 2015-04-01 to 2022-09-30. Distributed by: Gulf of Mexico Research Initiative Information and Data Cooperative (GRIIDC), Harte Research Institute, Texas A&M University–Corpus Christi. https://doi.org/10.7266/emmc0zdt |
| 14 | TPWD fisheries data | 28-Mar-2023 | Coffey, D.M., Stunz, G.W., and P.A. Montagna. 2023. Texas Parks and Wildlife Department fishery-independent monitoring program bay trawl and bag seine surveys, Texas, 1982-2020. Distributed by: Gulf of Mexico Research Initiative Information and Data Cooperative (GRIIDC), Harte Research Institute, Texas A&M University–Corpus Christi. https://doi.org/10.7266/m663qn60 |
| 15 | Nitrogen and Phosphorous Land-Ocean Interaction in the Coastal Zone (LOICZ) Budgets for Texas Estuaries | 16-Feb-2023 | Marshall, Danielle A. and Paul A. Montagna. 2023. Nitrogen and phosphorous land-ocean interaction in the coastal zone (LOICZ) budgets for Texas estuaries. Distributed by: Gulf of Mexico Research Initiative Information and Data Cooperative (GRIIDC), Harte Research Institute, Texas A&M University–Corpus Christi. https://doi.org/10.7266/7gx3tv5c |

## Document Type Abbreviations

| | |
|---|---|
| BBASC | Bay and Basin Stakeholders Committee |
| BBEST | Bay Basin and Bay Expert Science Team |
| SAC | Science Advisory Committee for Texas Environmental Flows |
| TAC | Texas Administrative Code |
| TCEQ | Texas Commission on Environmental Quality |

| Type | URL | Citation |
|---|---|---|
| SAC | https://hdl.handle.net/1969.6/94341 | Science Advisory Committee. 2004. Report on Water for Environmental Flows. Prepared for Senate Bill 1639, 78th Legislature, Study Commission on Water For Environmental Flows, October 26, 2004. |
| SAC | https://hdl.handle.net/1969.6/31367 | Joint Committee on the Study Commission on Water for Environmental Flows. December 2004. Interim Report to 79th Legislature. |
| SAC | https://hdl.handle.net/1969.6/94342 | Science Advisory Committee. 2006. Recommendations of The Science Advisory Committee. Presented to Governor's Environmental Flows Advisory Committee, August 21, 2006. |
| SAC | https://hdl.handle.net/1969.6/94346 | Science Advisory Committee. 2009. Use of Hydrologic Data in the Development of Instream Flow Recommendations for the Environmental Flows Allocation Process and The Hydrology-Based Environmental Flow Regime (HEFR) Methodology. Report # SAC-2009-01-Rev1., April 20, 2009. |
| SAC | https://hdl.handle.net/1969.6/94345 | Science Advisory Committee. 2009. Geographic Scope of Instream Flow Recommendations. Report # SAC-2009-02, April 3, 2009. |
| SAC | https://hdl.handle.net/1969.6/94344 | Science Advisory Committee. 2009. Methodologies for Establishing a Freshwater Inflow Regime for Texas Estuaries Within the Context of the Senate Bill 3 Environmental Flows Process. Report # SAC-2009-03-Rev1., June 5, 2009. |
| SAC | https://hdl.handle.net/1969.6/94355 | Science Advisory Committee. 2009. Fluvial Sediment Transport as an Overlay to Instream Flow Recommendations for the Environmental Flows Allocation Process. Report # SAC-2009-04, May 29, 2009. |
| SAC | https://hdl.handle.net/1969.6/94343 | Science Advisory Committee. 2009. Essential Steps for Biological Overlays in Developing Senate Bill 3 Instream Flow Recommendations. Report # SAC-2009-05, August 31, 2009. |
| SAC | https://hdl.handle.net/1969.6/94347 | Science Advisory Committee. 2009. Nutrient and Water Quality Overlay for Hydrology-Based Instream Flow Recommendations. Report # SAC-2009-06, November 3, 2009. |
| SAC | https://hdl.handle.net/1969.6/94349 | Science Advisory Committee. 2010. Discussion Paper: Moving From Instream Flow Regime Matrix Development to Environmental Flow Standard Recommendations. February 17, 2010. |
| SAC | https://hdl.handle.net/1969.6/94351 | Science Advisory Committee. 2010. Lessons Learned from Initial SB3 BBEST Activities. Report # SAC-2010-01, July 14, 2010. |
| SAC | https://hdl.handle.net/1969.6/94352 | Science Advisory Committee. 2010. Considerations in the Development of an SB3 Work Plan for Adaptive Management. Report # SAC-2010-02, August 20, 2010. |
| SAC | https://hdl.handle.net/1969.6/94350 | Science Advisory Committee. 2010. Considerations of Methods for Evaluating Interrelationships Between Recommended SB-3 Environmental Flow Regimes and Proposed Water Supply Projects. Report # SAC-2010-04, November 12, 2010. |
| SAC | https://hdl.handle.net/1969.6/94353 | Science Advisory Committee. 2011. Discussion Paper: Consideration of Attainment Frequencies and Hydrologic Conditions in Developing and Implementing Instream Environmental Flow Regimes. Report # SAC-2011-01, January 1, 2011. |

© The Author(s) 2025

P. A. Montagna, A. R. Douglas (eds.), *Freshwater Inflows to Texas Bays and Estuaries*, Estuaries of the World, https://doi.org/10.1007/978-3-031-70882-4

| Type | URL | Citation |
| --- | --- | --- |
| SAC | https://hdl.handle.net/1969.6/94354 | Science Advisory Committee. 2011. Use of Hydrologic Data in the Development of Instream Flow Recommendations for the Environmental Flows Allocation Process and the Hydrology-Based Environmental Flow Regime (HEFR) Methodology. Third Edition. Report # SAC-2011-01, March 15, 2011. |
| SAC | https://hdl.handle.net/1969.6/94356 | Science Advisory Committee. 2011. Fluvial Sediment Transport as an Overlay to Instream Flow Recommendations for the Environmental Flows Allocation Process, Addendum. Report # SAC-2011-02, August 31, 2011. |
| BBEST | https://hdl.handle.net/1969.6/94336 | Sabine and Neches Rivers and Sabine Lake Bay Basin and Bay Area Stakeholder Committee and Expert Science Team. 2009. Environmental Flows Recommendation Report. |
| BBEST | https://hdl.handle.net/1969.6/94338 | Trinity and San Jacinto Rivers and Galveston Bay Basin and Bay Expert Science Team. 2009. Environmental Flows Recommendation Report. |
| BBEST | https://hdl.handle.net/1969.6/94332 | Colorado and Lavaca Rivers and Matagorda and Lavaca Bays Basin and Bay Stakeholder Committee and Expert Science Team. 2011. Environmental Flows Recommendation Report. |
| BBEST | https://hdl.handle.net/1969.6/94333 | Guadalupe, San Antonio, Mission, and Aransas Rivers and Mission, Copano, Aransas, and San Antonio Bays Basin and Bay Stakeholder Committee and Expert Science Team. 2011. Environmental Flows Recommendation Report. |
| BBEST | https://hdl.handle.net/1969.6/94335 | Nueces River and Corpus Christi and Baffin Bays Basin and Bay Area Stakeholder Committee and Expert Science Team. 2011. Environmental Flows Recommendation Report. |
| BBEST | https://hdl.handle.net/1969.6/94735 | Brazos River and Associated Bay and Estuary System Stakeholder Committee and Expert Science Team. 2012. Environmental Flow Regime Recommendation Report. |
| BBEST | https://hdl.handle.net/1969.6/94334 | Rio Grande, Rio Grande Estuary, and Lower Laguna Madre Basin and Bay Expert Science Team. 2012. Environmental Flows Recommendations Report. Final Submission to the Environmental Flows Advisory Group, Rio Grande, Rio Grande Estuary, and Lower Laguna Madre Basin and Bay Stakeholders Committee, and Texas Commission on Environmental Quality. |
| BBEST | https://hdl.handle.net/1969.6/94339 | Rio Grande, Rio Grande Estuary, and Lower Laguna Madre Basin and Bay Expert Science Team. 2012. Environmental Flows Recommendations Report. Final Submission to the Environmental Flows Advisory Group, Rio Grande Basin and Bay Area Stakeholders Committee and Texas Commission on Environmental Quality. |
| BBEST | https://hdl.handle.net/1969.6/94340 | Upper Rio Grande Basin and Bay Expert Science Team. 2012. Environmental Flows Recommendations Report. Final Submission to the Environmental Flows Advisory Group, Rio Grande Basin and Bay Area Stakeholders Committee and Texas Commission on Environmental Quality. |
| BBASC | https://hdl.handle.net/1969.6/94330 | Trinity and San Jacinto Rivers and Galveston Bay BBASC. 2010. Report of the Trinity – San Jacinto – Trinity Bay and Basin Stakeholders Committee for Submittal to the Environmental Flows Advisory Group and the Texas Commission on Environmental Quality. |
| BBASC | https://hdl.handle.net/1969.6/94329 | Trinity and San Jacinto Rivers and Galveston Bay BBASC. 2010. Recommended Environmental Flow Standards and Strategies for the Trinity and San Jacinto River Basins and Galveston Bay. |
| BBASC | https://hdl.handle.net/1969.6/94331 | Trinity and San Jacinto and Galveston Bay Basin and Bay Area Stakeholder Committee with support of the Basin and Bay Expert Science Team. 2012. Work Plan Report. |
| BBASC | https://hdl.handle.net/1969.6/94327 | Sabine and Neches Rivers and Sabine Lake Bay Basin and Bay Area Stakeholder Committee. 2010. Recommendations Report. Final Submission to the Environmental Flows Advisory Group and the Texas Commission on Environmental Quality. |
| BBASC | https://hdl.handle.net/1969.6/94328 | Sabine and Neches Rivers and Sabine Lake Bay Basin and Bay Area Stakeholder Committee. 2010. Work Plan. Submission to the Environmental Flows Advisory Group and the Texas Commission on Environmental Quality. |
| BBASC | https://hdl.handle.net/1969.6/94359 | Colorado and Lavaca Basin and Bay Area Stakeholder Committee. 2011. Environmental Flows Recommendation Report. |
| BBASC | https://hdl.handle.net/1969.6/94366 | Colorado and Lavaca Rivers and Matagorda and Lavaca Bays Basin and Bay Area Stakeholder Committee. 2012. Draft Work Plan. |
| BBASC | https://hdl.handle.net/1969.6/94367 | Guadalupe, San Antonio, Mission, and Aransas Rivers and Mission, Copano, Aransas, and San Antonio Bays Basin and Bay Area Stakeholders Committee. 2011. Environmental Flows Standards and Strategies Recommendations Report. |
| BBASC | https://hdl.handle.net/1969.6/94368 | Guadalupe, San Antonio, Mission, and Aransas Rivers and Mission, Copano, Aransas, and San Antonio Bays Basin and Bay Area Stakeholders Committee. 2012. Work Plan for Adaptive Management. |
| BBASC | https://hdl.handle.net/1969.6/94369 | Nueces River and Corpus Christi and Baffin Bay Basin and Bay Area Stakeholder Committee. 2012. Environmental Flow Standards and Strategies Recommendations Report. Final Submission to the Environmental Flows Advisory Group and the Texas Commission on Environmental Quality. |
| BBASC | https://hdl.handle.net/1969.6/94370 | Nueces River and Corpus Christi and Baffin Bay Basin and Bay Area Stakeholder Committee. 2012. Work Plan for Adaptive Management. |

| Type | URL | Citation |
|---|---|---|
| BBASC | https://hdl.handle. net/1969.6/94337 | Brazos River and Associated Bay and Estuary System Basin and Bay Area Stakeholders Committee. 2012. Environmental Flow Standards and Strategies Recommendations Report. |
| BBASC | https://hdl.handle. net/1969.6/94348 | Brazos River and Associated Bay and Estuary System Basin and Bay Area Stakeholders Committee. 2012. Work Plan for Adaptive Management. |
| TCEQ | https://hdl.handle. net/1969.6/94362 | Texas Commission on Environmental Quality. 2010. General Provisions- [Title 30, Texas Administrative Code (TAC), Subsection 298(A)]. |
| TCEQ | https://hdl.handle. net/1969.6/94365 | Texas Commission on Environmental Quality. 2011. Trinity and San Jacinto Rivers, and Galveston Bay - [30 TAC, 298(B)] |
| TCEQ | https://hdl.handle. net/1969.6/94364 | Texas Commission on Environmental Quality. 2012. Sabine and Neches Rivers, and Sabine Lake Bay - [30 TAC, 298(C)] |
| TCEQ | https://hdl.handle. net/1969.6/94358 | Texas Commission on Environmental Quality. 2012. Colorado and Lavaca Rivers, and Matagorda and Lavaca Bays - [30 TAC, 298(D)] |
| TCEQ | https://hdl.handle. net/1969.6/94360 | Texas Commission on Environmental Quality. 2012. Guadalupe, San Antonio, Mission, and Aransas Rivers, and Mission, Copano, Aransas, and San Antonio Bays - [30 TAC, 298 (E)] |
| TCEQ | https://hdl.handle. net/1969.6/94361 | Texas Commission on Environmental Quality. 2014. Nueces River and Corpus Christi and Baffin Bays - [30 TAC, 298(F)] |
| TCEQ | https://hdl.handle. net/1969.6/94357 | Texas Commission on Environmental Quality. 2010. Brazos River and its Associated Bay and Estuary System - [30 TAC, 298(G)] |
| TCEQ | https://hdl.handle. net/1969.6/94363 | Texas Commission on Environmental Quality. 2010. Rio Grande, Rio Grande Estuary, and Lower Laguna Madre - [30 TAC, 298(H)] |

# Archived Oral Histories

| Subject | Date | Interview location | Run length | Citation |
|---|---|---|---|---|
| Ray Allen | 11-May-2022 | Nueces Delta Preserve, Texas | 1:34:20 | Brown, J.C. 2022. Freshwater inflow oral history: Ray Allen. Nueces Delta Preserve, Texas. https://library.tamucc.edu/exhibits/s/thegulf/item/2259 |
| Jim Blackburn | 30-Jun-2022 | Houston, Texas | 0:48:37 | Brown, J.C. 2022. Freshwater inflow oral history: Jim Blackburn. Houston, Texas. https://www.tamucc.edu/library/exhibits/s/thegulf/item/2271 |
| Myron Hess | 1-Oct-2021 | Videoconference | 1:25:12 | Brown, J.C. 2021. Freshwater inflow oral history: Myron Hess. Videoconference. https://library.tamucc.edu/exhibits/s/thegulf/item/2003 |
| Ken Kramer | 14-Dec-2021 | Chappell Hill, Texas | 1:53:31 | Brown, J.C. 2021. Freshwater inflow oral history: Ken Kramer. Chappell Hill, Texas. https://library.tamucc.edu/exhibits/s/thegulf/item/2127 |
| Cindy Loeffler | 27-Mar-2022 | San Marcos, Texas | 1:55:58 | Brown, J.C. 2022. Freshwater inflow oral history: Cindy Loeffler. San Marcos, Texas. https://library.tamucc.edu/exhibits/s/thegulf/item/2237 |
| Larry McKinney | 17-Dec-2020 | Videoconference | 1:47:54 | Brown, J.C. 2020. Freshwater inflow oral history: Larry McKinney. Videoconference. https://www.tamucc.edu/library/exhibits/s/thegulf/item/1525 |
| Paul Montagna-1 | 23-Feb-2017 | Corpus Christi, Texas | 1:39:13 | Brown, J.C. 2017. Freshwater inflow oral history: Paul Montagna-1. Corpus Christi, Texas. https://www.tamucc.edu/library/exhibits/s/thegulf/item/275 |
| Paul Montagna-3 | 15-Apr-2022 | Corpus Christi, Texas | 1:18:27 | Brown, J.C. 2022. Freshwater inflow oral history: Paul Montagna-3. Corpus Christi, Texas. https://library.tamucc.edu/exhibits/s/thegulf/item/2120 |
| John Nielsen-Gammon | 11-Feb-2022 | Videoconference | 0:59:23 | Brown, J.C. 2022. Freshwater inflow oral history: John Nielsen-Gammon. Videoconference. https://library.tamucc.edu/exhibits/s/thegulf/item/2255 |
| Andrew Sansom | 12-Sep-2022 | San Marcos, Texas | 1:20:02 | Brown, J.C. 2022. Freshwater inflow oral history: Andrew Sansom. San Marcos, Texas. https://library.tamucc.edu/exhibits/s/thegulf/item/2252 |
| Joe Trungale | 5-Nov-2021 | Austin, Texas | 1:14:07 | Brown, J.C. 2021. Freshwater inflow oral history: Joe Trungale. Austin, Texas. https://library.tamucc.edu/exhibits/s/thegulf/item/2010 |

© The Author(s) 2025

P. A. Montagna, A. R. Douglas (eds.), *Freshwater Inflows to Texas Bays and Estuaries*, Estuaries of the World, https://doi.org/10.1007/978-3-031-70882-4